AF786829

Asphalt Pavement Materials and Surface

Asphalt Pavement Materials and Surface

Editors

Leilei Chen
Ming Liang
Dongyu Niu
Shisong Ren
Ruxin Jing

MDPI

Basel • Beijing • Wuhan • Barcelona • Belgrade • Novi Sad • Cluj • Manchester

Editors
Leilei Chen
Southeast University
Nanjing
China

Ming Liang
Shandong University
Jinan
China

Dongyu Niu
Chang'an University
Xi'an
China

Shisong Ren
Delft University of
Technology
Delft
Netherlands

Ruxin Jing
Delft University of
Technology
Delft
Netherlands

Editorial Office
MDPI
St. Alban-Anlage 66
4052 Basel, Switzerland

This is a reprint of articles from the Special Issue published online in the open access journal *Coatings* (ISSN 2079-6412) (available at: https://www.mdpi.com/journal/coatings/special_issues/asphalt_pavement_materials_surface).

For citation purposes, cite each article independently as indicated on the article page online and as indicated below:

Lastname, A.A.; Lastname, B.B. Article Title. *Journal Name* **Year**, *Volume Number*, Page Range.

ISBN 978-3-7258-0749-9 (Hbk)
ISBN 978-3-7258-0750-5 (PDF)
doi.org/10.3390/books978-3-7258-0750-5

© 2024 by the authors. Articles in this book are Open Access and distributed under the Creative Commons Attribution (CC BY) license. The book as a whole is distributed by MDPI under the terms and conditions of the Creative Commons Attribution-NonCommercial-NoDerivs (CC BY-NC-ND) license.

Contents

coatings

MDPI

Article

Research on the Fracture Characteristics of Asphalt Mixtures in High Altitude and Cold Regions with Large Temperature Differences

Shuangjie Wang [1,2], Haibo Cao [1,2,*], Tuanjie Chen [2], Wenhao Ke [2] and Wu Bo [3]

1 School of Civil Engineering, Chongqing Jiaotong University, Chongqing 400741, China
2 FHCC State Key Laboratory of Road Engineering Safety and Health in Cold and High-Altitude Regions, Xi'an 710075, China
3 Intelligent Transport System Research Center, Southeast University, Nanjing 211189, China
* Correspondence: bohaichb@163.com

Abstract: Due to the harsh climatic conditions in high altitude and cold regions with large temperature differences, asphalt pavement is generally prone to cracking, and the cracks propagate rapidly, which reduces the service life and service level of the road. The factors influencing the fracture characteristics of asphalt mixtures were analyzed in this paper, and the mixtures with different aggregate gradations from various types of asphalt were prepared. The fracture characteristics were explored using the thermal stress restrained specimen test (TSRST) and low-temperature bending test, and the good consistency of the low-temperature fracture performance was identified according to the results of frost-break temperature, flexural strength, and fracture toughness. The frost-break temperature was confirmed as the best indicator of the material crack resistance and could be used as the index to evaluate the performance of asphalt mixtures at low temperatures. The frost-break temperature of matrix asphalt mixture is 8–10 °C higher than that of modified asphalt mixture, and AC asphalt mixture is 2–4 °C higher than that of SMA asphalt mixture. The excellent asphalt performance has a more important influence on the fracture characteristics of asphalt mixture. The asphalt mixture of the same type had similar fracture toughness at varying notch depths, the most deviation is 3.78% which shows that the initial crack depth has little effect on the fracture toughness of asphalt mixture at low temperature. The results of the study can provide a basis for the selection of asphalt pavement surface materials and the optimization of mixtures in high altitude and cold regions with large temperature differences.

Keywords: asphalt mixture; high altitude and cold regions; large temperature differences; fracture; TSRST; flexural strength; fracture toughness

Citation: Wang, S.; Cao, H.; Chen, T.; Ke, W.; Bo, W. Research on the Fracture Characteristics of Asphalt Mixtures in High Altitude and Cold Regions with Large Temperature Differences. *Coatings* **2023**, *13*, 618. https://doi.org/10.3390/coatings13030618

Academic Editor: Valeria Vignali

Received: 12 January 2023
Revised: 3 March 2023
Accepted: 10 March 2023
Published: 14 March 2023

Copyright: © 2023 by the authors. Licensee MDPI, Basel, Switzerland. This article is an open access article distributed under the terms and conditions of the Creative Commons Attribution (CC BY) license (https://creativecommons.org/licenses/by/4.0/).

1. Introduction

With a climate featuring cold weather, long duration of low temperature, and large temperature differences, the asphalt concrete pavement in the high altitude and cold regions with large temperature differences is generally vulnerable to severe cracks and other distress. Moreover, natural factors such as the strong ultraviolet light, lead to premature aging of the surface layer of the asphalt pavement. This thus greatly shortens the service life of the pavement, increases the maintenance cost of the road, and restricts the highway development in the Qinghai Tibet Plateau and other high altitude and cold regions with large temperature differences.

The factors affecting the low-temperature fracture characteristics of asphalt mixtures can be divided into two categories: internal and external factors. The internal factors refer to the material and structural characteristics of asphalt mixture components, including the aggregate type, asphalt type, asphalt content, gradation type, admixture material, etc. External factors include the traffic load, environmental factors, etc. Vinson et al. claimed that low-temperature cracking and rutting were two temperature-related asphalt concrete pavement damage forms in cold regions. They proposed a design method based on the asphalt binder and aggregate test results, and this method could reduce the occurrence of low-temperature cracks and ruts in asphalt pavement structures [1]. Guy Dore believed that the asphalt concrete pavement in cold regions mainly suffered from low-temperature and frost heaving cracking [2].

In high altitude and cold regions with large temperature differences, owing to the long duration of low temperature, the temperature stress accumulates, and the mixture is in a tensile state for a long period of time. Consequently, the mixture strength decreases continuously, eventually leading to damage. On account of large temperature differences and repeated heating and cooling cycles in this area, the failure strain of the material is reduced despite the small temperature stress. In addition, the aging effect of the material further gives rise to fatigue cracking. The faster the cooling rate, the greater the shrinkage strain rate of the mixture, the more difficult it is to exert the stress relaxation of the material, and the higher the accumulated temperature stress, the more likely it is to be destroyed. Different types of asphalt have varied viscosities and toughness. The asphalt with a greater viscosity exhibits better toughness under low-temperature conditions, and it can bond more tightly to aggregates. Besides, asphalt mixtures made from high-viscosity asphalt have better low-temperature performance. Aggregate is the skeleton and main body of asphalt mixtures. Asphalt mixtures prepared from mineral aggregates with different gradation designs are composed of particles of various sizes, and they have different asphalt contents. In addition, the embedded extrusion force and internal friction resistance between mineral aggregates of different size also vary. Therefore, the gradation design directly affects the performance of asphalt mixtures [3].

Research on temperature fatigue of asphalt mixtures began in the 1970s in the United States. Shahin et al. established the temperature fatigue damage model of asphalt pavement by using the phenomenon method [4]. Mahboub et al. studied the formation process of temperature fatigue cracks in asphalt pavement, as well as the factors contributing to fissure generation. Their research results showed that the crack development process notably influenced the prediction of pavement fatigue life [5]. Lytton et al. analyzed the fatigue mechanism of pavements with increasing number of temperature fatigue effects and established the corresponding temperature fatigue life equation. According to their study result, the temperature fatigue was ascribed to both thermal fatigue and low-temperature cracking [6,7]. Eshan V. Dave held that the one-dimensional stress evaluation model of temperature cracks in the AASHTO Mechanical Experience Pavement Design Guide was oversimplified and unable to characterize the nonlinear fracture behavior of asphalt concrete, so it was not sound. Based on the nonlinear finite element method, he proposed an interactive temperature cracking prediction model. This model included a time and temperature-dependent viscoelastic material model, as well as a viscous material fracture model that could accurately simulate the initiation and propagation of temperature load-induced cracks [8].

Wei Youpo et al. simulated the high and cold conditions in Qinghai-Tibet region with large temperature differences and conducted indoor flexural strength tests of asphalt mixtures. They also studied how the flexural strength of asphalt mixtures changed with different gradations, oil-stone ratios, temperatures, etc. Their research results help perfect the current asphalt pavement design method, improve the performance of asphalt pavement in permafrost areas, and prolong the service life of asphalt pavement. Moreover, their findings take on great application value to construction projects, and can guide the design of asphalt pavement structures in the cold highland [9,10]. Guo Bo and Li Dongqing et al. investigated the influence of different material compositions on the high- and low-temperature performance and water stability of asphalt mixtures in regions with large temperature differences. They evaluated the high temperature stability, low-temperature crack resistance, and water stability of asphalt mixtures using the dynamic stability, low-temperature failure strain, and freeze-thaw splitting strength indexes, respectively. Through range analysis and experimental verification, the order of sensitivity of factors affecting the high- and low-temperature performance of asphalt mixtures was determined. Finally, the optimal mix proportion was obtained, which provided a basis for the raw material selection and composition design of asphalt mixtures in areas with large temperature differences [11,12]. In the research of Hao Peiwen, the crack resistance of asphalt mixtures was assessed by the low-temperature crack resistance coefficient. A larger crack resistance coefficient suggested better low-temperature crack resistance [13]. Lu Songtao et al. explored the temperature fatigue life of asphalt concrete through low circle temperature fatigue testing, and made an analysis using the temperature fatigue damage model based on dissipated energy. The results revealed a linear relationship between the dissipated energy and the plastic strain of asphalt mixtures [14]. Guo Weiwei et al. examined the influence of different aging degrees on the low-temperature performance of asphalt mixtures through the splitting test, beam bending test and low-temperature shrinkage test. They also characterized the low-temperature crack resistance of asphalt mixtures under different aging conditions [15].

Zhang Yi studied the asphalt mixture after short-term aging and long-term aging with 0 °C bending creep test and −10 °C bending test, and analyzed the low-temperature crack resistance of asphalt mixture under different aging conditions. Based on the experimental results, the key properties of low-temperature cracking resistance, aging resistance and frost resistance in permafrost areas were combined to propose the design method of asphalt mixture ratio for permafrost areas [16]. Fu Qiang et al. constructed a numerical simulation finite element model based on fracture mechanics to analyze the propagation process of surface cracks and base reflection cracks. The results suggested a nonlinear relationship between surface cracks and the stress intensity factor in the propagation process when the reference temperature was lower than 5 °C. The comprehensive research results above show that the asphalt mixture will produce temperature fatigue when the temperature cycle changes, and the effect of temperature fatigue is greater when the reference temperature is lower, so the role of temperature in the analysis of the effect of freeze-thaw cycles on asphalt mixtures cannot be ignored [17].

In high altitude and cold regions with large temperature differences, asphalt mixtures are apt to crack, so the low-temperature crack resistance of mixtures has vital influence on asphalt pavement performance in such areas. Existing domestic and abroad methods for assessing the low-temperature performance of asphalt mixtures mainly include the low-temperature bending test, notch low-temperature bending test, asphalt mixture splitting test, shrinkage coefficient test, relaxation test, and the thermal stress restrained specimen test (TSRST). The United States Highway Strategic Research Program (SHRP) examined the performance of the above-mentioned approaches in simulating field conditions, the applicability of their test results to the mechanical model, their practicality in aging and wet conditions, their operation, and the instrument cost. The results show that the TSRST can well simulate the field conditions and accurately evaluate the low-temperature cracking performance of the asphalt mixture. Meanwhile, the TSRST can directly measure the temperature-stress curve of the asphalt mixture during the cooling process, and obtain the values of such indices as frost-break temperature and frost-break stress [18–20].

Previous research results have shown that severe cold and large temperature difference conditions have a notable effect on the low-temperature performance of asphalt mixtures. However, further research on the fracture characteristics of asphalt mixtures under said conditions is still needed. In this paper, we simulated the conditions in the high altitude and cold regions with large temperature differences, and prepared a variety of asphalt mixtures with different aggregate gradations from various types of asphalt. Then the TSRST and low-temperature bending test were carried out to study how the fracture characteristics of asphalt mixtures varied with the asphalt type, mixture type, and cooling rate. This study provides a foundation for the optimal design of asphalt pavement surface mixtures in high altitude and cold regions with large temperature differences.

2. Asphalt Mixture Composition Design

In this study, the effect of cold and large temperature differences on the fracture performance of fine-graded asphalt mixture (AC-13), medium-graded asphalt mixture (AC-16), and stone matrix asphalt (SMA-13) was investigated. These three materials are widely applied to the asphalt pavement upper layer. Matrix asphalt and SBS-modified asphalt were selected as binders.

2.1. *Raw Material Technical Standard*

The raw materials used in this research mainly included matrix and modified asphalt, coarse and fine aggregates.

No. 90 matrix asphalt (unmodified asphalt) and SBS-modified asphalt widely used in high altitude and cold regions in China were compared in this study. The specific technical indicators of the two types of asphalt are shown in Table 1.

Table 1. Technical index value of asphalt.

Test Content		Unit	Asphalt Type	
			No. 90	SBS Modified
Penetration (25 °C)		0.1 mm	89.3	73.3
Softening point (Ring-ball method)		°C	46.0	64.0
Ductility (10 °C)		cm	92.0	79.0
Density (25 °C)/		g·cm^{-3}	0.998	1.101
Solubility (trichloroethylene)		%	99.8	99.21
RTFOT (163 °C)	Quality loss	%	0.23	0.06
	Penetration ratio	%	77	73
	Ductility (10 °C)	cm	32	26
	Softening point	°C	50.3	68.7

The basalt stone and 0–3 mm machine-made sand were taken as coarse and fine aggregates, respectively. The performance of the coarse and fine aggregates was measured. Their main technical indexes are listed in Table 2, which meet the technical requirement.

Table 2. Technical index value of aggregates.

Test Content	Technical Requirement	Aggregate Specifications/mm				
		13.2–16	9.5–13.2	4.75–9.5	2.36–4.75	0–3.00
Apparent relative density	≮2.6	2.954	2.956	2.948	2.932	2.726
Needle-like content/%	≯15	5	6	3	/	/
Soft stone content/%	≯3	1.8	0.8	/	/	/
Crushing value of stone/%	≯26	14	13	/	/	/
Sturdiness/%	≯12	3	3	3	3	14
Water absorption/%	≯2	0.82	0.79	0.91	0.76	/
Sand equivalent/%	≮60	/	/	/	/	87

2.2. *Mix Design*

The mix ratio design method for asphalt mixtures in "Technical Specification for Construction of Highway Asphalt Pavement" (JTG F40-2004) [21] was adopted in this study. According to the screening results of raw materials and the grading range and mineral aggregate grading selection method stipulated in the current specification, the gradations of the three types of asphalt mixtures were determined. The design gradations of asphalt mixture are shown in Table 3.

Table 3. Gradation range.

Mixture Type	Mass Percentage (%) Passing through Each Hole Sieve (mm)										
	19.0	**16.0**	**13.2**	**9.5**	**4.75**	**2.36**	**1.18**	**0.60**	**0.30**	**0.15**	**0.075**
AC-13	100	100	94.2	76.4	44.2	31	24	17	12.3	9.5	7.4
SMA-13	100	100	90.3	61.3	29.7	21.7	18.6	15.7	13.8	11.9	10.3
AC-16	100	96.9	87.6	73.7	52.9	29.2	19.7	14.8	10.1	8.3	7.2

The Marshall test was carried out, and the asphalt content in each mixture was obtained. The results are shown in Table 4.

Table 4. Marshall test results.

No.	Asphalt	Asphalt-Stone Ratio (%)	VV (%)	VMA (%)	VFA (%)	MS (kN)
AC-13	Matrix	5.1	4.0	15.2	73.7	8.95
	SBS	5.2	4.2	16.1	73.9	12.8
SMA-13	Matrix	6.1	3.0	17.5	82.9	8.7
	SBS	6.4	3.5	18.2	80.8	11.8
AC-16	Matrix	4.7	3.8	14.9	74.5	8.74
	SBS	5.1	4.0	15.4	74.0	11.8

From the test results, it can be seen that under the same mix type, the asphalt dosage, stability and asphalt saturation of the modified asphalt mix were higher than those of the matrix asphalt mix, and the asphalt dosage of SMA-13 was higher than those of AC-13 and AC-16. The Marshall test results met the requirements of "Technical Specification for Construction of Highway Asphalt Pavement" (JTG F40-2004). The asphalt mixture specimens were prepared according to the determined mineral aggregate gradation and asphalt content. Then their fracture characteristics under large temperature differences were investigated.

3. Fracture Characteristics of Asphalt Mixtures

3.1. Test Scheme

(1) Thermal stress restrained specimen test (TSRST) for asphalt mixtures

The TSRST device consists of a refrigeration system, a load/displacement system, a data acquisition and control system, and a high-precision displacement sensor, with test accuracy higher than 0.0025 mm. This TSRST instrument could simulate the actual temperature change and the forces acting on the mixture, and well reflect the low-temperature cracking performance of the asphalt mixture.

According to the mix proportion determined in Section 2.2, the beam specimens of AC-13, SBS-modified AC-13, AC-16, SBS-modified AC-16, SMA-13, and SBS-modified SMA-13 asphalt mixtures were prepared. All samples had a size of 40 mm × 40 mm × 200 mm. Before the test, the beam specimen was suspended inside the TSRST device. The temperature of the specimen was measured in real time by four temperature sensors attached to the four surfaces of the specimen. A temperature sensor was clamped by a fixed clip and hung inside the TSRST device to measure the internal ambient temperature. Displacement sensors were installed on both sides of the specimen. During the test, the liquid nitrogen tank kept introducing liquid nitrogen to the interior of the TSRST instrument, so as to constantly lower the internal temperature. As the temperature declined, the beam specimen continued to shrink. Once the temperature contraction stress of the specimen was greater than the ultimate tensile strength of the beam, the specimen cracked.

During the experiment, the specimen was cooled at different cooling rates (5 °C/h, 10 °C/h, and 15 °C/h). The initial temperature of the test was set at 10 °C and the test was conducted after the temperature stabilized. In the cooling process, the lowest temperature reached −50 °C. The test ceased if the specimen fractured. If the specimen did not brake, it was left in the −50 °C environment for an hour, and then the test automatically stopped.

(2) Low-temperature bending test

Under the action of environmental factors in high altitude and cold regions with large temperature differences, small cracks may occur after the asphalt mixture pavement is laid. These small cracks result in stress concentration at the bottom of the mixture, and then expand due to the influence of prolonged low temperature and large temperature differences. In order to simulate this phenomenon, the low-temperature bending testing of pre-cracked beams was conducted to evaluate the low-temperature crack resistance of the mixture in the paper. The test process is detailed in Standard Test Methods of Bitumen and Bituminous Mixtures for Highway Engineering (JTG E20-2011) [22].

In accordance with the mix proportion determined in Section 2.2, the beam specimens of SBS-modified AC-13, SBS-modified AC-16, and SBS-modified SMA-13 asphalt mixtures were manufactured. All samples had a size of 250 mm × 30 mm × 35 mm. In addition, the beam specimens were subject to pre-cracked treatment according to the test requirements. There were five notch depths: 0, 4, 8, 12, and 16mm. The test temperature were 20, 10, 0, −10, and −20 °C. The test loading rate was 50 mm/min.

3.2. Thermal Stress Restrained Specimen Test (TSRST) for Asphalt Mixture Specimens

3.2.1. TSRST Results and Evaluation Indicators

Four evaluation indexes, namely, frost-break temperature, frost-break stress, transformation point temperature, and temperature curve slope dS/dT, were obtained by TSRST. The frost-break temperature is the temperature when the asphalt mixture breaks, and it is the minimum temperature that the mixture may withstand. The lower the fracture temperature, the better the low-temperature crack resistance of the mixture. The frost-break stress is the maximum stress when the asphalt mixture breaks during the cooling process, which reflects the mixture strength in low-temperature shrinking. The transition point temperature can reflect the rheological properties of asphalt mixtures during cooling. Asphalt mixtures with a lower transition point temperature have a stronger stress relaxation ability and better low-temperature performance. The temperature-stress curve slope dS/dT refers to the slope of the linear part, which represents the temperature stress growth rate. The test results are summarized in Table 5.

Table 5. TSRST results of asphalt mixture specimens.

Cooling Rate	Mixture Type	Frost-Break Temperature (°C)	Frost-Break Stress (MPa)	Transition Point Temperature (°C)	dS/dT
5 °C/h	AC-13	−25.80	3.62	−19.20	0.24
	SBS-AC-13	−36.10	3.52	−20.70	0.23
	AC-16	−23.54	3.23	−15.20	0.21
	SBS-AC-16	−32.41	4.14	−20.30	0.28
	SMA-13	−29.01	3.53	−19.80	0.23
	SBS-SMA-13	−38.35	4.95	−21.20	0.25
10 °C/h	AC-13	−24.50	3.56	−18.63	0.31
	SBS-AC-13	−31.62	4.91	−20.47	0.26
	AC-16	−22.89	4.07	−18.90	0.28
	SBS-AC-16	−30.67	4.53	−19.60	0.23
	SMA-13	−26.24	3.14	−17.70	0.24
	SBS-SMA-13	−35.58	4.24	−20.20	0.25
15 °C/h	AC-13	−23.83	3.35	−15.20	0.23
	SBS-AC-13	−30.06	4.45	−19.30	0.21
	AC-16	−22.30	3.82	−19.70	0.30
	SBS-AC-16	−28.70	4.11	−20.80	0.26
	SMA-13	−26.01	2.82	−15.70	0.36
	SBS-SMA-13	−33.15	4.97	−18.70	0.27

Each of the four evaluation indexes obtained has its own significance. To simplify the analysis, the most representative index was used to evaluate the low-temperature performance of the mixture. While choosing the most representative analysis index, we tried to minimize the loss of information contained in the original index. Therefore, based on the test results, the frost-break temperature, frost-break stress, transition point temperature, and temperature curve slope were analyzed using the principal component analysis method. The analysis results are shown in Tables 6 and 7.

Table 6. Correlation coefficient matrix.

Index	Frost-Break Temperature	Frost-Break Stress	Transition Point Temperature	dS/dT
Frost-break temperature	1.000	−0.535	0.608	−0.019
Frost-break stress	−0.535	1.000	−0.689	0.121
Transition point temperature	0.608	−0.689	1.000	−0.177
dS/dT	−0.019	0.121	−0.177	1.000

Table 7. Variance decomposition principal component extraction analysis results.

Principal Component	Initial Eigenvalue			Extract Square and Load		
	Value	Percentage of Variance	Summation Percentage	Total	Percentage of Variance	Summation Percentage
Frost-break temperature	2.252	56.290	56.290	2.252	56.290	56.290
Frost-break stress	1.002	24.804	81.094	0.992	24.804	81.094
Transition point temperature	0.462	11.562	92.656	-	-	-
dS/dT	0.294	7.344	100.000	-	-	-

In some cases, eigenvalues can be regarded as an indicator to measure the influence of principal components. Therefore, the principle of extracting the number of principal components is that the corresponding eigenvalue of the principal components is greater than 1 and the cumulative percentage is greater than 80%. According to the analysis results, the first principal components (Frost-break temperature) reflected 56.290% of the information, and the first two principal components reflected 81.094% of the information, and their eigenvalues were both greater than 1. Among the four indexes, the frost-break temperature had the largest weight coefficient. This suggests that the frost-break temperature is the best indicator of low-temperature performance and can be used to evaluate the mixture specimens in TSRST. Meanwhile, considering that the frost-break temperature is directly related to the low-temperature cracking of asphalt pavement and has a clear physical significance, it is recommended to use the frost-break temperature as the evaluation index of the low-temperature performance of asphalt mixtures.

3.2.2. Effect of the Asphalt Type

The influence of the asphalt type and cooling rate on the frost-break temperature of mixtures was studied. The covariance analysis was performed by taking the frost-break temperature as the dependent variable. The analysis results are shown in Table 8.

Table 8. Covariance analysis of the influence of the asphalt type (AC-13).

Source of Variance	Quadratic Sum	Degree of Freedom	Mean Square	F Value	*p* Value
Asphalt type	33.704	2	16.852	17.915	0.021
Asphalt type × Cooling rate	4.799	2	2.400	2.551	0.225

$\alpha = 0.05$ was considered statistically significant. According to the above table, the p value for the interaction between the asphalt type and cooling rate is $p = 0.225 > 0.05$, so the covariance analysis was made. In the meantime, the asphalt type has a $p = 0.021 < 0.05$, indicating that the asphalt type greatly affects the frost-break temperature.

Based on the test and analysis results, the asphalt type notably affected the frost-break temperature of mixtures. The matrix asphalt had a higher frost-break temperature than the modified asphalt. Compared with the matrix asphalt, the modified asphalt had a higher viscosity, as well as better toughness at low temperatures. In addition, the mixture prepared from modified asphalt had relatively larger viscous resistance. The modified asphalt could improve the adhesion between asphalt and aggregates, so the mixture prepared from modified asphalt had better low-temperature performance than that made from matrix asphalt.

3.2.3. Effect of the Mixture Type

The impact of the mixture type and cooling rate on the frost-break temperature of mixtures composed of the same asphalt type was examined. The frost-break temperature was taken as the dependent variable for variance analysis. The analysis results are described in Table 9.

Table 9. Covariance analysis of the effect of the mixture type (matrix asphalt).

Source of Variance	Quadratic Sum	Degree of Freedom	Mean Square	F Value	*p* Value
Mixture type	7.607	2	3.804	9.992	0.047
Mixture type × Cooling rate	0.782	2	0.391	1.027	0.457

A significance level of $\alpha = 0.05$ was used. In the above table, the interaction between the mixture type and the cooling rate has a $p = 0.457 > 0.05$, so the covariance analysis of the data was adopted. Meanwhile, the mixture type has a $p = 0.047 < 0.05$, suggesting that the mixture type significantly influences the frost-break temperature.

According to the test and analysis results, the mixture type had a marked effect on the frost-break temperature. The frost-break temperature of the SMA asphalt mixture was lower than that of the AC asphalt mixture, and the AC-13 mixture had a lower frost-break temperature than that of the AC-16 asphalt mixture. The reason is that the SMA asphalt mixture has a dense skeleton structure, and it is formed by a large number of coarse aggregates, filled by sufficient fine aggregates. Due to the high content of coarse aggregates, mineral powder, and asphalt, the SMA asphalt mixture has better low-temperature performance than the AC asphalt mixture with a suspended-dense structure. The gradation of AC-16 mixtures is thicker than that of AC-13 mixtures. Previous studies have confirmed that asphalt mixtures with a smaller nominal maximum particle size and a finer aggregate gradation have better low-temperature crack resistance. Therefore, the AC-13 asphalt mixture has a lower frost-break temperature than the AC-16 asphalt mixture.

3.2.4. Effect of the Cooling Rate

The effect of the mixture type and cooling rate on the frost-break temperature of mixtures made of the same asphalt type was analyzed. The oblique variance analysis was performed by taking the frost-break temperature as the dependent variable. The analysis results are summarized in Table 10.

Table 10. Covariance analysis of the effect of the cooling rate (SBS-modified asphalt).

Source of Variance	Quadratic Sum	Degree of Freedom	Mean Square	F Value	*p* Value
Cooling rate	37.250	1	37.250	77.115	0.003
Mixture type × Cooling rate	1.392	2	0.696	1.441	0.364

$\alpha = 0.05$ was taken as the significance level in this study. It can be seen from the above table that the *p* value of the interaction between the mixture type and the cooling rate is $p = 0.364 > 0.05$, so the covariance analysis was made. At the same time, the cooling rate has a $p = 0.003 < 0.05$, indicating that the cooling rate has marked influence on the frost-break temperature. As the cooling rate increases, the frost-break temperature of the mixture shows an ascending trend, because the stress accumulates faster in the mixture. The temperature stress generated in the mixture cannot be released in time through deformation. As a result, the mixture fractures earlier.

3.3. Low-Temperature Bending Test for Pre-Cracked Asphalt Mixture Beams

3.3.1. Load Deflection Curve Analysis

(1) Test results at different temperatures

SBS-modified AC-13, SBS-modified AC-16, and SBS-modified SMA-13 asphalt mixtures were tested for the notched beam bending analysis. The beam notch was 4 mm deep and the test temperatures were 20, 10, 0, −10 and −20 °C. The load deflection curve is drawn in Figures 1–3.

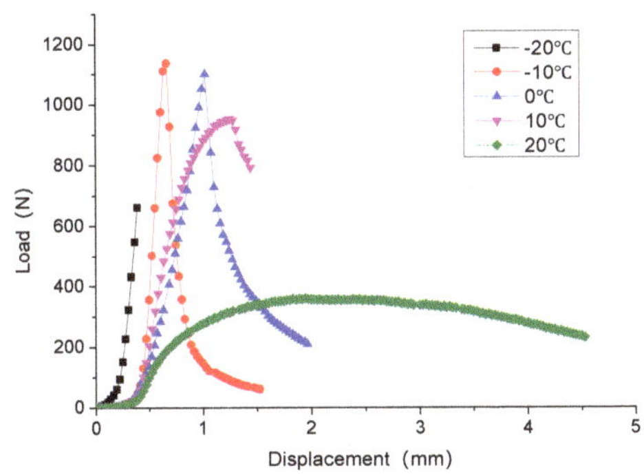

Figure 1. Load-displacement diagram of AC-13 at different temperatures.

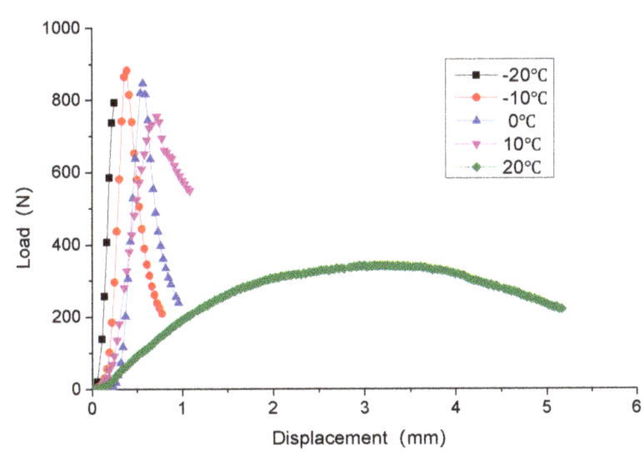

Figure 2. Load-displacement diagram of AC-16 at different temperatures.

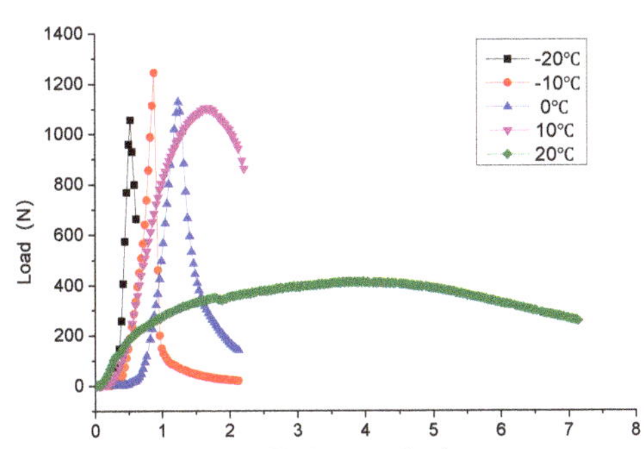

Figure 3. Load-displacement diagram of SMA-13 at different temperatures.

As shown in the above figures, the change rate of the load-displacement curves increases with the declining temperature, and the mixture transforms from the viscoelastic state to the elastic state. At −20 °C, both AC-13 and AC-16 break. As the temperature continues to rise, the three curves tend to coincide before the specimen failure. This demonstrates a large difference in the deformation ability and crack resistance among the three types of asphalt mixtures at low temperatures. At high temperatures, their deformation ability is similar. According to the change trend of load-displacement curves at different temperatures, and the maximum load and displacement at failure, it is found that the SMA-13 mixture has the best low-temperature performance (Compared with AC-13, the performance is improved about 6.48%), followed by AC-13 and AC-16 successively.

(2) Test results at different notch depths

The pre-cracked beam bending test of SBS-modified AC-13, SBS-modified AC-16, and SBS-modified SMA-13 asphalt mixtures was conducted at the temperature of −10 °C. The depths of the beam notch were 0, 4, 8, and 12 mm. The load deflection curve is shown in Figures 4–7.

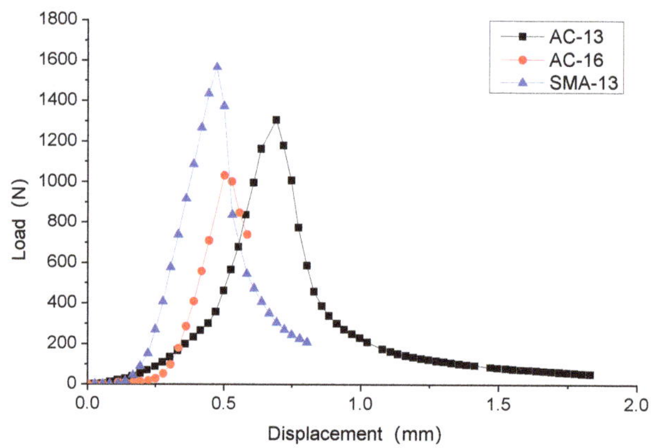

Figure 4. Load-displacement diagram of asphalt mixture beams with a notch depth of 0 mm.

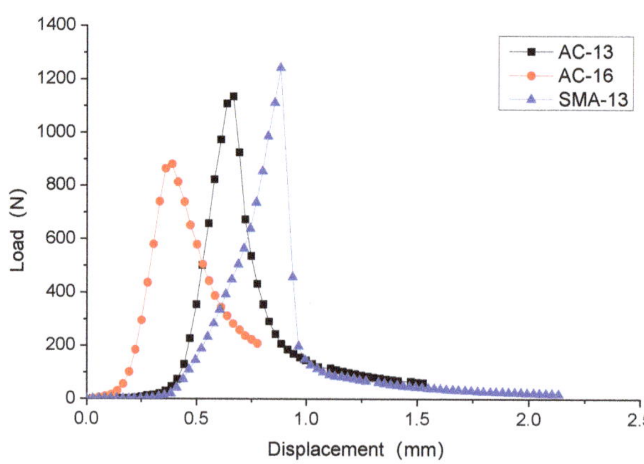

Figure 5. Load-displacement diagram of asphalt mixture beams with a notch depth of 4 mm.

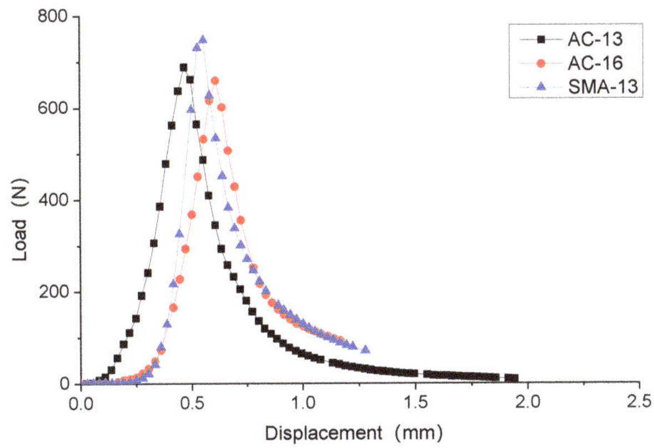

Figure 6. Load-displacement diagram of asphalt mixture beams with a notch depth of 8 mm.

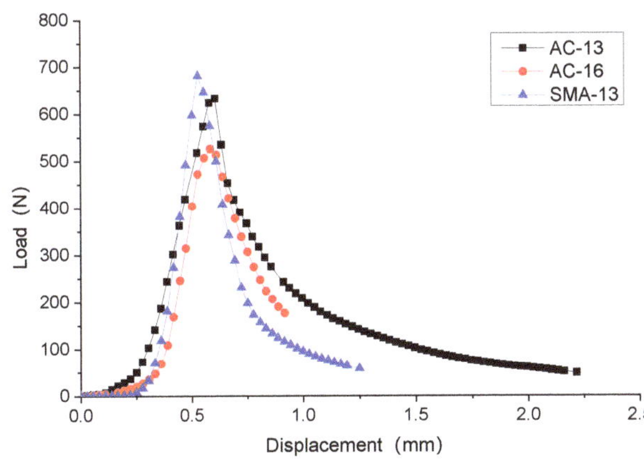

Figure 7. Load-displacement diagram of asphalt mixture beams with a notch depth of 12 mm.

In the above figures, no evident change is observed in the variation rate of the three load-displacement curves at different notch depths. This shows that the deformation resistance and low-temperature crack resistance of mixtures are not correlated with the notch depth. When the un-notched asphalt mixture specimens and the specimens with shallow notches break, their displacement is discrete. However, as the notch depth increases, the displacement tends to concentrate. The load values at the failure of the three asphalt mixtures are close at different notch depths. Specifically, SMA-13 has the largest load at failure, followed by AC-13 and AC-16 successively.

3.3.2. Flexural Strength Analysis

(1) Test results at different temperatures

The flexural strength of the three asphalt mixtures with a notch depth of 4 mm at different test temperatures is illustrated in Figure 8.

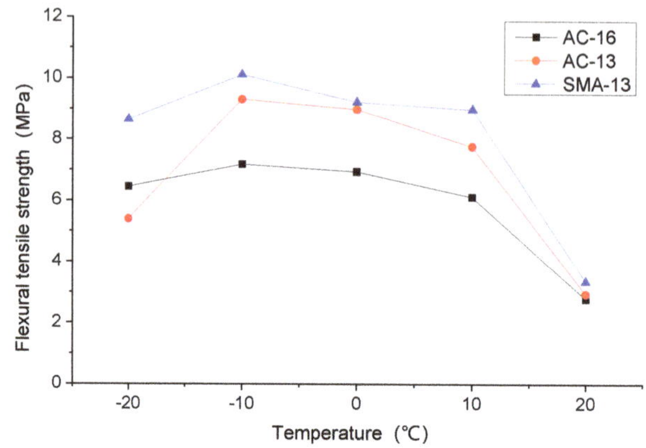

Figure 8. Flexural strength of asphalt mixtures at different test temperatures.

According to the above figure, the flexural strength of AC-13, AC-16, and SMA-13 asphalt mixtures has a similar change trend with temperature. Specifically, their flexural strength increases first and then decreases with the reducing temperature. Their flexural strength is ranked as AC-16 < AC-13 < SMA-13. In the range of −20 °C~ −10 °C, the flexural strength increases with the rising temperature. As the temperature climbs from −10 to 10 °C, the flexural strength decreases. When the temperature rises from 10 to 20 °C, the flexural strength decreases at a faster speed. Therefore, it can be considered that 10~20 °C is a sensitive temperature range for flexural tensile strength. At −10 °C, the flexural tensile strength of asphalt mixture reaches the maximum, which can be called embrittlement temperature. When the temperature is higher than the embrittlement temperature, the material shows notable yield failure characteristics. In the crack propagation process, when the flexural strength increases with the dropping temperature, there is a relatively slow subcritical expansion stage. When the temperature is reduced below the embrittlement temperature, the material has remarkable brittle failure features. The subcritical expansion stage lasts for a short time, and the flexural strength decreases as the temperature declines. At a temperature above 10 °C, the asphalt mixture is viscoelastic, with strong rheological properties. The decrease rate of the flexural strength with the increasing temperature above 10 °C is different from that below 10 °C.

(2) Test results at different notch depths

The flexural strength of three asphalt mixtures with different notch depths at the test temperature of −10 °C is shown in Figure 9.

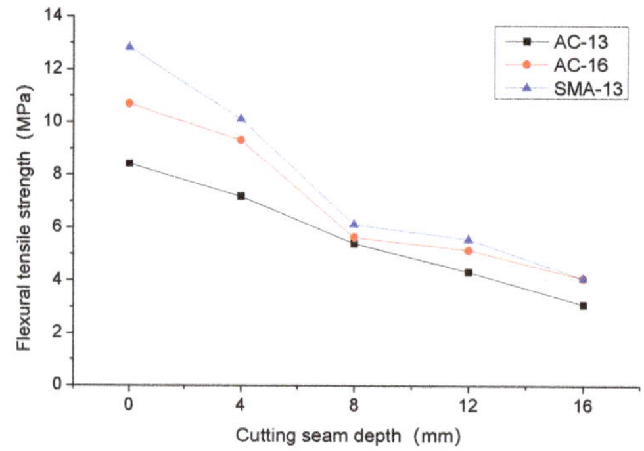

Figure 9. Flexural strength of asphalt mixtures at different notch depths.

It can be seen from the above figure that the flexural strength of the mixtures decreases with the increasing notch depth, and the difference in flexural strength among the three asphalt mixtures gradually narrows.

3.3.3. Fracture Toughness Analysis

Fracture toughness is defined as the toughness of the material to resist crack propagation fracture. The bending test of the pre-cracked beam specimens was carried out, and the fracture toughness was calculated.

(1) Asphalt mixture fracture toughness at different temperatures

The fracture toughness of three asphalt mixtures with a notch depth of 4 mm at different temperatures is demonstrated in Table 11.

Table 11. Fracture toughness of asphalt mixtures at different temperatures (MPa$\sqrt{mm}$).

Temperature (°C)	AC-13	AC-16	SMA-13
20	10.67	10.08	12.16
10	28.17	22.24	32.61
0	32.61	25.20	33.50
−10	33.80	26.09	36.76
−20	19.57	23.42	31.43

In the above table, SMA-13 has the largest fracture toughness at different test temperatures, followed by AC-13 and AC-16 successively. This shows that SMA-13 has the best crack resistance, followed by AC-13. AC-16 has the worst crack resistance. Based on the variation in fracture toughness with temperature, the three types of asphalt mixtures have the maximal fracture toughness at the test temperature of −10 °C.

(2) Asphalt mixture fracture toughness at different notch depths

The fracture toughness of three asphalt mixtures with different notch depths at the test temperature of −10 °C is described in Table 12.

Table 12. Fracture toughness of asphalt mixtures at different notch depths (MPa$\sqrt{mm}$).

Notch Depth (mm)	AC-13	AC-16	SMA-13
4	33.80	26.09	36.76
8	29.19	27.92	31.73
12	35.35	29.74	38.15
16	37.84	28.76	37.84

According to the above table, for AC-13 asphalt mixtures, except for the fracture toughness with 8mm notch depth, the fracture toughness with other notch depths is close. The deviation between the maximum and minimum values is 11.95%. For AC-16 asphalt mixtures, the fracture toughness values at different notch depths are close, and the deviation between the maximum and the minimum is 14.00%. For SMA-13 asphalt mixtures, the fracture toughness values at notch depths of 0, 4, and 12 mm are similar, but differ from that at the notch depth of 8 mm. The deviation between the maximum and the minimum is 3.78%. Excluding the influence of test errors, the same type of asphalt mixture has similar fracture toughness at different notch depths. The results show that the initial crack depth has little effect on the fracture toughness of asphalt mixture at low temperature.

4. Conclusions

In this paper, we analyzed the factors influencing the facture characteristics of asphalt mixtures in high altitude and cold regions with large temperature differences, and simulated the conditions in these regions. A variety of asphalt mixtures with different aggregate gradations were manufactured from various asphalt types. Besides, the TSRST and low-temperature bending test were conducted to study the fracture characteristics of asphalt mixtures in high altitude and cold regions with large temperature differences. The main conclusions are as follows:

(1) The low-temperature performance of asphalt mixture evaluated according to the frost-break temperature, flexural strength, and fracture toughness is in good consistency. The low temperature, large temperature differences, cooling rate and mixture raw materials have an important influence on the fracture characteristics of asphalt mixtures.
(2) The weight coefficient of the frost-break temperature obtained by the TSRST test is 56.290%, which can best reflect the crack resistance of the material. Moreover, the frost-break temperature is directly related to the low-temperature cracking of asphalt pavement, and it has a specific physical significance. Therefore, the frost-break temperature is recommended to assess the low-temperature performance of asphalt mixtures.
(3) The asphalt type and mixture type have a significant effect on the fracture characteristics of asphalt mixture. The frost-break temperature of matrix asphalt mixture is 8–10 °C higher than that of modified asphalt mixture, and AC asphalt mixture is 2–4 °C higher than that of SMA asphalt mixture. The excellent asphalt performance has a more important influence on the fracture characteristics of asphalt mixture.
(4) In the low-temperature bending test of notched beams, the flexural strength decreases with the increasing notch depth. SMA-13 has the best low-temperature performance, followed by AC-13 and AC-16 successively. The fracture toughness values of the same asphalt mixture at different notch depths are close, and the deviation between the maximum and the minimum is 3.78%, so the initial crack depth has little effect on the fracture toughness of asphalt mixture at low temperature.

Author Contributions: S.W.: Conceptualization, Methodology, Validation, Investigation, Writing—original draft. H.C.: Conceptualization, Writing—review and editing, Supervision, Project administration, Resources, Funding acquisition. T.C.: Methodology, Writing—review and editing. W.K.: Methodology, Writing—review and editing. W.B.: Writing—review and editing, Supervision. All authors have read and agreed to the published version of the manuscript.

Funding: This research was funded by National Key R&D Program of China (Grant No. 2021YFB2601200), National Natural Science Foundation of China (No. 52178419), Transportation Industry Key Science and Technology Projects of China (No. 2020-MS1-059) and Youth Science and Technology Innovation Fund project of FHCC (No. YGY2019QC-02).

Institutional Review Board Statement: Not applicable.

Informed Consent Statement: Not applicable.

Data Availability Statement: Data will be made available on reasonable request.

Acknowledgments: Thanks is also to Biao Ding who proposed the idea of the performance test in the current study.

Conflicts of Interest: The authors declare no conflict of interest.

References

1. Vinson, T.S.; Hicks, R.G.; Janoo, V.C. Low temperature cracking and rutting in asphalt concrete pavements. In *Roads and Airfields in Cold Regions*; Technical Council on Cold Regions Engineering Monograph; American Society of Civil Engineers: Reston, VA, USA, 1996; pp. 203–248.
2. Dore, G. Cold Region Pavement. *J. Glaciol. Geocryol.* **2002**, *24*, 593–600.
3. Cao, H.; Chen, T.; Zhu, H.; Ren, H. Influence of Frequent Freeze-Thaw Cycles on Performance of Asphalt Pavement in High-Cold and High-Altitude Areas. *Coatings* **2022**, *12*, 752. [CrossRef]
4. Shahin, M.Y.; McCullough, B.F. *Damage Model for Predicting Temperature Cracking in Flexible Pavements*; University of Texas: Austin, TX, USA, 1974.
5. Little, D.N.; Mahboub, K. Engineering properties of first generation plasticized sulfur binders and low temperature fracture evaluation of plasticized sulfur paving mixtures. *Transp. Res. Rec.* **1985**, *1034*, 103–111.
6. Luo, X.; Luo, R.; Lytton, R.L. Characterization of fatigue damage in asphalt mixtures using pseudostrain energy. *J. Mater. Civ. Eng.* **2012**, *25*, 208–218. [CrossRef]

Xiao, F.; Zhao, W.; Amirkhanian, S.N. Laboratory investigation of fatigue characteristics of rubberized asphalt mixtures containing warm asphalt additives at a low temperature. *J. Test. Eval.* **2011**, *39*, 1.

Eshan, V.D.; Leon, S.; Park, K. Thermal Cracking Prediction Model and Software for Asphalt Pavements. In Proceedings of the Congress of Transportation and Development Institute of the American Society of Civil Engineers 2011, Chicago, IL, USA, 13–16 March 2011; pp. 667–676.

Wei, Y.; Hu, H.; Chu, K. Asphalt mixture's flexural-tensile strength's influencing factor in plateau area. *West. China Commun. Sci. Technol.* **2009**, *5*, 5–7.

0. Kliewer, J.E.; Zeng, H.; Vinson, T.S. Aging and low-temperature cracking of asphalt concrete mixture. *J. Cold Reg. Eng.* **1996**, *10*, 134–148. [CrossRef]

1. Guo, B.; Gao, N. Effect of component factors of asphalt mixture on pavement performance in areas with big temperature difference. *Road Mach. Constr. Mech.* **2017**, *34*, 81–85.

2. Li, D.; Meng, Q.; Fang, J. Test and Research on performance of Frost Thawing Resistance of Asphalt Mixture. *Highway* **2007**, *12*, 145–147.

3. Hao, P.; Zhang, D. Research on Evaluation Target for the Cracking Resistance Property of Bituminous Mixture in Low Temperature. *Highway* **2000**, *45*, 63–67.

4. Lu, S.; Tian, X.; Zheng, J. Simulation Research on Thermal Fatigue Damage to Bituminous Mixture. *Highway* **2005**, *2*, 116–118.

5. Guo, W.; Zhang, Z. Influence of ray aging on low temperature performance of asphalt mixture in plateau-cold region. *J. Chongqing Jiaotong Univ. (Nat. Sci.)* **2012**, *31*, 51–53.

6. Zhang, Y. *Study on Low Temperature Crack Resistance and Mix Design Method of Asphalt Mixture in Permafrost Regions*; Chang'an University: Xi'an, China, 2004.

7. Fu, Q. Analysis of temperature fatigue expansion life of asphalt pavement cracks. In Proceedings of the 10th China Association for Science and Technology Annual Conference of China Association for Science and Technology and Henan Provincial People's Government, Zhengzhou, China, 1 September 2008; pp. 9–11.

8. Krcmarik, M.; Varma, S.; Kutay, M.E.; Jamrah, A. Development of Predictive Models for Low-Temperature Indirect Tensile Strength of Asphalt Mixtures. *J. Mater. Civ. Eng.* **2016**, *28*, 04016139. [CrossRef]

9. Choudhary, J.; Kumar, B.; Gupta, A. Evaluation of engineering, economic and environmental suitability of waste filler incorporated asphalt mixes and pavements. *Road Mater. Pavement Des.* **2021**, *22*, S624–S640. [CrossRef]

0. Wen, H. Use of fracture work density obtained from indirect tensile testing for the mix design and development of a fatigue model. *Int. J. Pavement Eng.* **2013**, *14*, 561–568. [CrossRef]

1. *JTG F40-2004*; Technical Specifications for Construction of Highway Asphalt Pavements. Ministry of Communications: Beijing, China, 2004.

2. *JTG E20-2011*; Standard Test Methods of Bitumen and Bituminous Mixtures for Highway Engineering. Ministry of Transportation: Beijing, China, 2011.

Disclaimer/Publisher's Note: The statements, opinions and data contained in all publications are solely those of the individual author(s) and contributor(s) and not of MDPI and/or the editor(s). MDPI and/or the editor(s) disclaim responsibility for any injury to people or property resulting from any ideas, methods, instructions or products referred to in the content.

coatings

MDPI

Article

Encapsulation for Sensing Element and Its Application in Asphalt Road Monitoring

Chuanyi Ma [1], Xue Xin [2,*], Ning Zhang [1], Jianjiang Wang [2], Chuan Wang [1], Ming Liang [2,*], Yunfeng Zhang [2] and Zhanyong Yao [2]

1 Shandong Hi-Speed Group Co., Ltd., Jinan 250014, China
2 School of Qilu Transportation, Shandong University, Jinan 250002, China
* Correspondence: xinxue@mail.sdu.edu.cn (X.X.); ming.liang@sdu.edu.cn (M.L.); Tel.: +86-1866-1628-291 (X.X. & M.L.)

Abstract: The internal pavement structure is a "black box"; an accurate strain response for th pavement interlayer structure under vehicle load is hard to obtain by conventional road surfac detection methods. This is due to the true strain field of the pavement structure, which means tha the service state of the pavement cannot be accurately evaluated. This paper proposes an innovativ strain sensor based on a carbon nanotube and epoxy (CNT/EP) composite to solve the current strai monitoring problem in asphalt pavement health monitoring. The CNT/EP composite encapsulatio method was proposed, and the I-shaped strain sensor for asphalt pavement structure was develope The strain–resistance response characteristics of the self-developed sensor were further investigate using a universal testing machine. The encapsulated sensor was used to monitor the strain of th asphalt mixture by means of a laboratory asphalt concrete beam and a practical pavement fiel The results showed that the encapsulation method proposed in the study is suitable for CNT/E material, which could guarantee the survivability and monitoring effectiveness of the self-develope sensor under the harsh environment of high temperature and pressure of asphalt mixture pavin The resistance of encapsulated sensor presents a linear relationship with strain. The laboratory an practical paving verified the feasibility of the self-sensor for strain monitoring of asphalt pavemen Compared to other post-excavating buried sensors, the self-developed sensor can be embedded i the pavement interlayer as the asphalt mixtures paving process, which can obtain the real strai response of pavement structure and reduce the perturbation of the sensor to the dynamic response o the pavement.

Keywords: pavement health monitoring; CNT/EP composites; embedded strain sensor; encapsulatio

Citation: Ma, C.; Xin, X.; Zhang, N.; Wang, J.; Wang, C.; Liang, M.; Zhang, Y.; Yao, Z. Encapsulation for Sensing Element and Its Application in Asphalt Road Monitoring. *Coatings* **2023**, *13*, 390. https://doi.org/10.3390/coatings13020390

Academic Editor: Andrea Simone

Received: 23 December 2022
Revised: 2 February 2023
Accepted: 5 February 2023
Published: 8 February 2023

Copyright: © 2023 by the authors. Licensee MDPI, Basel, Switzerland. This article is an open access article distributed under the terms and conditions of the Creative Commons Attribution (CC BY) license (https://creativecommons.org/licenses/by/4.0/).

1. Introduction

By the end of 2021, the total mileage of China's road network will have reache 5.28 million kilometers and road maintenance expenditure will be approximately RM 7.4 trillion [1]. Over the past few decades, approximately 90% of the work has involve the maintenance and upgrading of existing roads. It is expected that more and more road will need to be maintained and upgraded in the future [2]. Thus, it is necessary to obtai the road structure conditions to judge whether the road needs to be maintained or no Timely maintenance can help to reduce the maintenance and upgrading costs and exten the life of roads from the view of the full life cycle [3,4]. A prerequisite for obtaining th structure conditions is pavement monitoring. The growth of the traffic volume has force the world to focus on road surface monitoring and providing a sustainable transportatio network [5]. Generally, macroscopic surface investigation and external nondestructiv testing techniques are used to evaluate the performance of asphalt pavements. For exampl digital image processing techniques are used to assess pavement damage and surfac texture [6,7]. A falling weight deflectometer (FWD) is applied to detect road surfac bending and further calculate pavement modulus [8–10]. In addition, ground penetrating radar is used to detect the thickness and density of the pavement. However, pavement structure is a "black box"; the traditional conventional external inspection methods cannot obtain the real strain/stress condition of interlayer structure [11,12], which lacks the actual pavement internal/interlayer structure stress–strain field. As a result, the traditional detection methods from the road surface are not enough to accurately assess the pavement conditions [13,14].

To obtain the interlayer strain/stress of road structures, many researchers proposed embedding sensors in road structures [15–19]. Liu et al. [20] developed a new encapsulated rubber strain sensor for asphalt pavement vertical strain monitoring. Their testing results show that the sensor has good sensing performance, but the sensitivity of the sensor is poor. Wu et al. [21] used a fiber-encapsulated long-range strain sensor with a larger contact surface of the sensing part subjected to temperature, making it more sensitive to temperature and more accurate in monitoring results. Liang et al. [22] used carbon fiber composites to encapsulate the fiber optic sensor, and the results showed that the sensor has good linearity and repeatability. Sohel et al. [23] developed carbon fiber composites for the health monitoring of civil engineering structures, which were found to be highly ductile and greatly prevented the sudden collapse of the structure. However, there is a lack of sufficient information in the literature on the effect of environmental and service conditions on the self-induced properties of the developed composite. Jiang Wei et al. [24] made use of lead zirconate titanate (PZT) powder and PVDF to prepare composite sheet material and then encapsulated the composite sheet material with 3D printed cylindrical packaging structure to prepare a piezoelectric sensing device, which was buried in asphalt mixture and tested the output effectiveness of its sensing signal by simulating vehicle load. These reports indicated the good feasibility of applying sensors for civil structures, and the ideal monitoring effect of the structure service state was obtained. However, the research on asphalt pavement structure monitoring sensors was insufficient. Few studies concentrate on preliminary embedding sensor techniques during the asphalt mixture paving process. The embedded sensor in the asphalt pavement interlayer struggles to survive since the temperature and pressure are high under the mixing, paving, and compaction of the asphalt mixture.

In recent years, the development of composite materials with self-sensing properties has provided a new idea to address the above-mentioned limitations of traditional strain sensors. Smart composite material is a new type of functional material that has self-perception of external factors such as force, deformation, temperature, humidity, etc. Composite materials could obtain information about the strain, stress, and other mechanical parameters of the measured structure by means of establishing a correlation between external factors and the electrical signal output of the composite smart material [25]. However, the composite material would be limited according to its application environment. For example, the harsh construction environment makes the survival rate of ordinary composite smart-material-based sensors greatly reduced in the field of civil engineering monitoring. In addition, the temperature and humidity of the environment in which the sensor's work environment varies greatly, which also has a greater impact on the accuracy of sensor signal monitoring [26,27].

For the harsh monitoring environment of asphalt pavement, the authors have carried out a series of related research in the development of new sensors for pavement [28–31]. Smart-material-based strain sensing elements have been developed, and the high accuracy, as well as viability of the sensing elements, were verified. On the basis of our previous research work, the encapsulation technique of smart composite materials and the shape design of the resulting sensor were emphasized in this paper. Afterward, the response of strain with resistance was measured by a universal testing machine. The encapsulated smart composite materials-based sensor was used on a road construction site in order to verify the viability and stability of the developed sensors. It is expected that this work will provide a new approach to asphalt pavement service state monitoring. The flowchart about this paper's methodology and the expected results, please see as Figure 1.

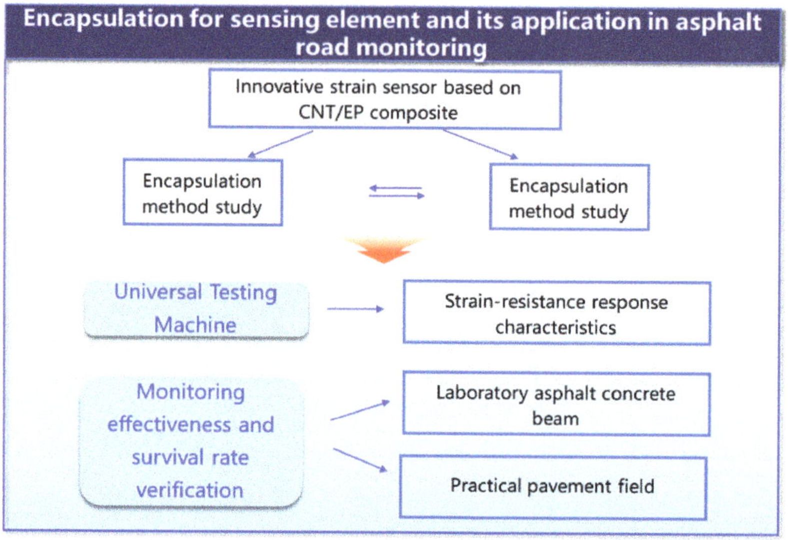

Figure 1. Flowchart about this paper's methodology and the expected results.

2. Preparation and Encapsulation of Smart Material-Based Strain Sensors

2.1. Preparation of Carbon Nanotube/Epoxy Resin (CNT/EP) Composites

Epoxy resin/multi-walled carbon nanotube (MWCNT) composites were prepared in the study as the sensing materials. MWCNT is obtained from high-temperature cracking of acetylene catalyzed by nickel-based catalysts. The selected MWCNT is of 95 wt% purity, with a specific surface area greater than 165 m^2/g. It is noteworthy that the MWCNT shows excellent electrical conductivity of 200 s/cm, which is expected to be beneficial to the conductivity sensitivity. The epoxy resin is bisphenol A type epoxy resin with an epoxy value of 0.48–0.54 eq/100 g and a viscosity of 12,000 mPa.s.

In order to form an effective conductive network, the MWCNT should be uniformly dispersed in the epoxy polymer matrix. The authors' studies have shown that the dispersion degree of CNT plays a key role in the response of strain–resistance [22]. Therefore, in order to overcome the non-uniform dispersion phenomenon of MWCNT caused by the van der Waals forces due to their huge specific surface area and nanoparticle size, the MWCNT were firstly ground in an agate mortar for 30 min in this study, followed by thorough mixing with the dispersant of N,N-Dimethylformamide(DMF). The MWCNT/DMF mixtures were mixed with mechanical stirring for 20 min, and then sonicated for 1 h with an ultrasonic disperser. After that, the mixed MWCNT/DMF suspension was added to the epoxy resin and continued to be sonicated for 1 h. The MWCNT/DMF/EP mixture was put in a vacuum drying oven for 1 h at 80 °C for DMF evaporation. Finally, the curing agent was added and stirred for 10 min to produce MWCNT/EP composites. The strain–resistance response of the composites was tested on equipment shown in Figure 2 (equipment parts: a—strain console, b—developed MWCNT/EP composite rod, c—strain gauge, d—resistance measuring device, e—electrical resistance recorder, and f—strain gauge recorder).

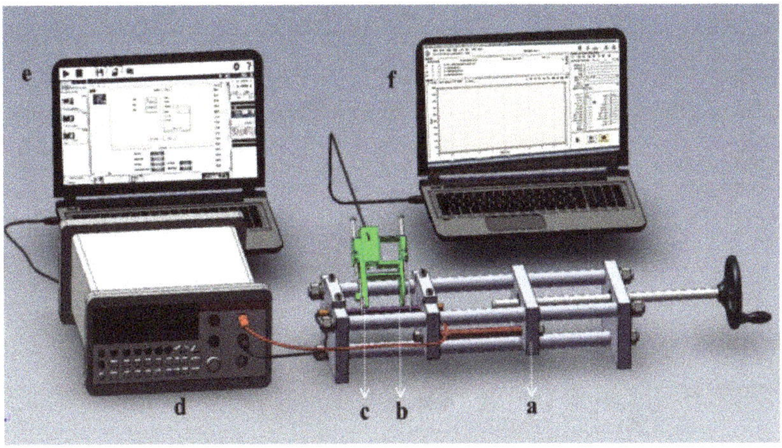

Figure 2. Testing equipment for strain–resistance response of CNT/EP composites.

2.2. Encapsulating MWCNT/EP Composite Rod by Epoxy for Manufacturing Strain Sensing Element

The MWCNT/EP suspensions showed the viscous flow state before curing, and they could be shaped with certain shape characteristics by casting and molding in the mold. The molding and encapsulating procedure of the MWCNT/EP composite rod are shown in Figure 3. The most common forming methods of conductive composite materials include surface conductive film forming method, conductive filler dispersion coating method, conductive material layer method, etc. However, when these methods are used for performance testing after sample preparation, conductive glue such as silver adhesive electrode, aluminum foil adhesive electrode, or the tin welding method is generally used to connect the wires. The material samples prepared by this method may have some shortcomings, such as uneven surface integrity, a long curing period of the conductive adhesive, poor adhesion effect, unstable contact electrical signal, and so on. Moreover, the production process is complex, and the production efficiency is low. So, in the preparation device, conducting wires were embedded in the mold through cap plugs at both ends of the mold. Both sides of the cover plug were provided with a feeding port and a discharge port for the composite self-sensing material to enter and exit. The shape of the cured composite was maintained by means of a fixed tube whose inner diameter was close to the outer diameter of the formed tube.

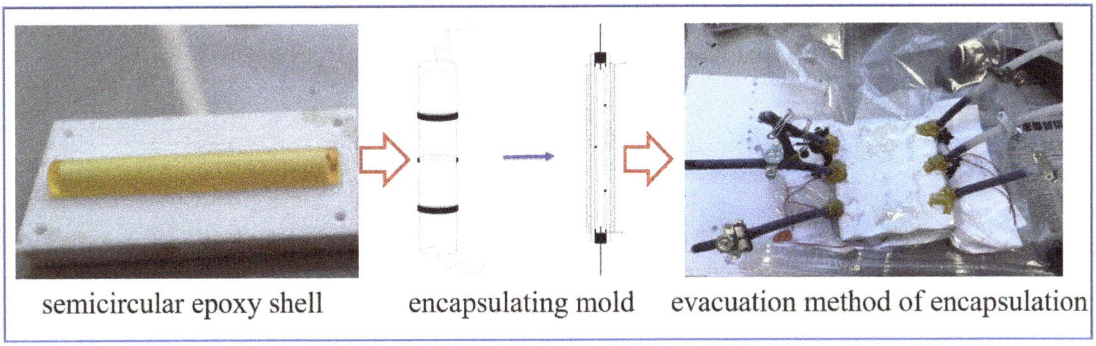

Figure 3. The molding and encapsulating procedure of MWCNT/EP composite rod.

To ensure the co-axiality of the MWCNT/EP composite rod and its covered epoxy shell, the semicircular epoxy was first pre-casted in the PTFE mold. The semicircular mold is designed to ensure that the semicircular epoxy has the same internal diamete as the semicircular epoxy and the same thickness as the packaging layer to ensure tha the composite rod is at the center of the packaged sensing element. Three semicircula epoxy parts were used to support the MWCNT/EP composite rod along the axial directio The two ends of the black rod prepared by MWCNT/EP were set with a cap plug with hole in the middle and, thus, the viscous epoxy could inject into one hole and outflow i another hole through the cap plug with holes. The cap plug can be connected to the mol to ensure the sealing effect. After the forming inner tube was installed, the lead outle sample injection hole, sample outlet hole, cap plug, and the forming inner tube were seale by a vacuum sealant so as to prevent the packaging material from flowing out of the ga resulting in an uneven surface of the sample. Lastly, the encapsulating epoxy material were injected into the hole and cured into shape to complete the sensor encapsulation. Afte the sensor was assembled with the molding mold, it was put into the sealed vacuum ba and connected to the vacuum pumping device. Then the vacuum pump was opened, an the packaging material flowed through the injection hole on the cap plug structure. Unde the vacuum negative pressure condition, the gas inside the mold cavity was removed, an the packaging material was driven to flow, infiltrate, and finally fill the entire cavity. I the process of injection of packaging materials, the whole sensor and part of the electrod wires at both ends were sealed in the air-tight packaging material through the moldin mold and the vacuum device so as to ensure that the sensor itself is completely isolate from the influence of the external environment.

The encapsulation process, with the help of mold, makes the integrated sensin element with the electrode wires at both ends within the encapsulation material. Thi encapsulated structure could ensure the sensing element is completely insulated fror the influence of moisture from the external environment. The excellent waterproofnes and airtightness of the encapsulated materials can help to prevent the sensor from bein eroded by the harsh external environment during service, thus solving the problems of lov sensor survival rate, short survival period, the drift of the received electrical signal, an undesirable measurement accuracy.

After curing at room temperature for 24 h, the encapsulated MWCNT/EP composit sensor in the mold was placed in an oven and cured for 2 h at 80 °C. The strain sensor ca be obtained after removing the cap plugs and protective sleeves at both ends. In additior the thickness of the covered epoxy film can be adjusted, and the multilayer of encapsulate materials can be used based on the same method, which improves the high-temperatur resistance characteristics and reduces the humidity sensitivity of the sensor. When th sensing element was manufactured, the I-shaped aluminum accessories were installed o the bottom of the encapsulated cylindrical sensing element, and thus the I-shaped senso was obtained (shown in Figure 4).

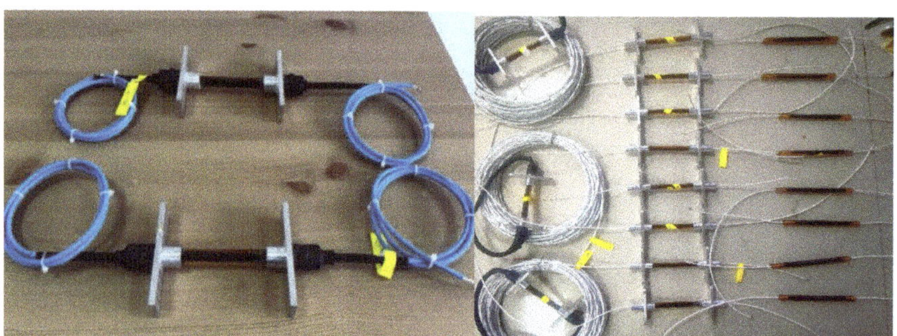

Figure 4. The self-developed I-shaped strain sensor made by encapsulated cylindrical MWCNT/EP composite.

3. Strain–Resistance Response Characteristics of the Self-Developed I-Shaped Sensor

3.1. Embedding Process of Sensor in Asphalt Concrete Slab

In order to figure out the adaptability and survivability of the self-developed strain sensor (Figure 4) in asphalt pavement and its resistance to high temperature and pressure during the asphalt mixture paving process, the self-developed strain sensor was embedded in the bottom of the double-layer asphalt concrete slab (size of 300 mm × 300 mm × 100 mm) in a laboratory. Two holes were set on the two sides of the mold to prepare the double-layer asphalt concrete slab, and to ensure the wire of the sensor could go through. The sensors were placed on the bottom of the asphalt mixture mold, and they needed to be firstly covered by part of the asphalt mixture that was further compacted. Afterward, the asphalt mixtures that had been calculated in volume and quantity were added into the mold, which was compacted for 50 rolls in each direction by a wheel mill to ensure that the asphalt mixture reached the specified compaction level. Subsequently, the mold was removed, and the asphalt concrete slab was cut into 10 cm lengths per beam according to the area where the sensor was located. The embedding process of the sensor in asphalt concrete slab is shown in Figure 5. All the embedded strain sensors have been proven to survive effectively in a laboratory. It is concluded that the sensor can withstand high temperatures and pressure during the asphalt mixture compaction. Thus, it is reasonable to deduce that the self-developed I-shaped strain sensor made by encapsulated MWCNT/EP composite is of good adaptability and survivability, which can adapt to the harsh environment of asphalt pavement construction. The accuracy and durability of the sensor need to be further studied.

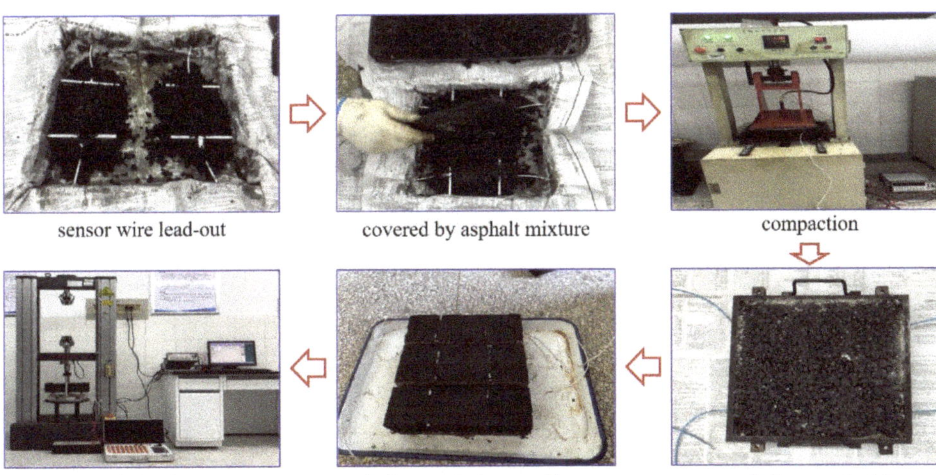

Figure 5. Embedding process of sensor in asphalt concrete slab.

3.2. Strain Monitoring Effectiveness of the Sensor in Asphalt Concrete

The monitoring effectiveness of deformation, the viability of monitoring elements, and so on are the key indicators of monitoring sensors, which could provide guidance on the embedding design and process of the sensor in asphalt pavement. When the asphalt concrete slab embedded with a self-developed sensor was prepared, the sensor reliability and monitoring effectiveness of deformation were analyzed by a universal testing machine. The commercial metal foil strain gauge was pasted on the bottom of the asphalt concrete beam with an embedded strain sensor to compare the difference in deformation strain at the same position of the beam. The asphalt beam was placed on the universal testing machine to conduct the three-point bending test. The resistance of the sensor was acquired synchronously by connecting a data acquisition instrument during strain loading (shown in Figure 6).

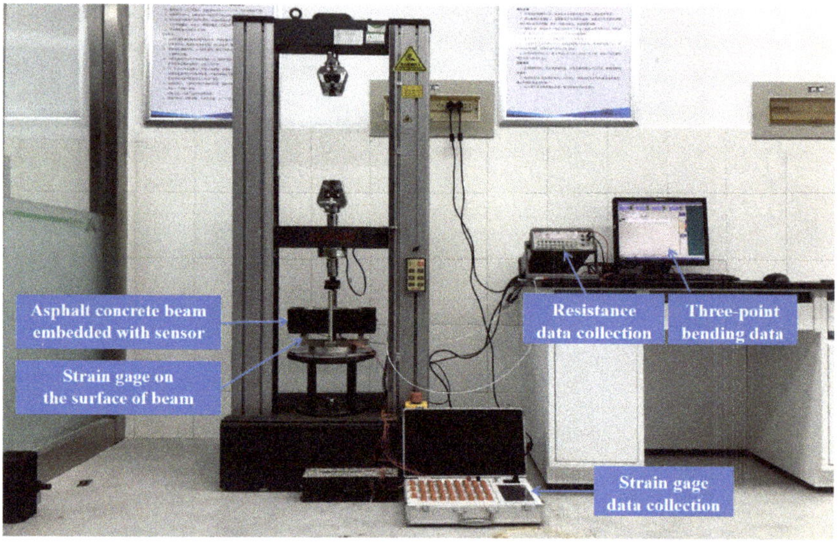

Figure 6. Strain monitoring calibration test of sensor on universal testing machine.

The strain control mode on the universal testing machine takes the program displacement control and the loading speed is 0.02 mm/min. When the displacement reaches 0.2 mm, it is maintained for 10 s. Then, the loading continues to 0.4 mm and the displacement is maintained for 10 s. Repeat the step until the displacement reaches 3.0 mm. The loading curve is shown in Figure 7.

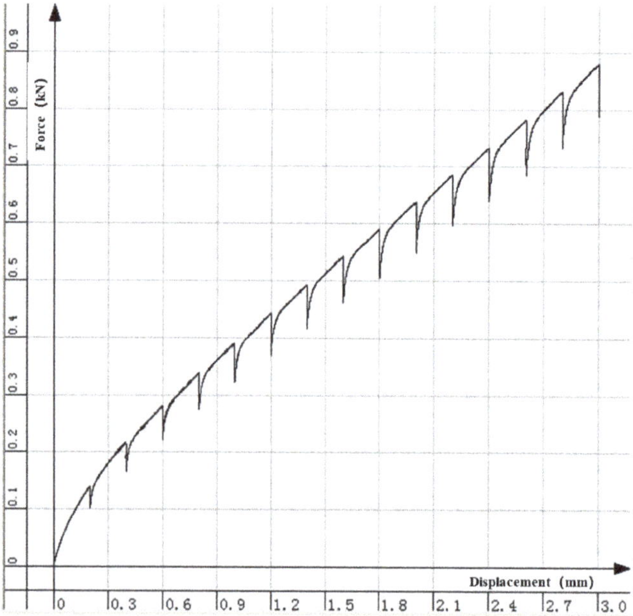

Figure 7. Strain control loading method.

The relationship between strain and resistance changing rate collected simultaneously under the step-by-step loading is shown in Figure 8. It can be seen that the resistance changing rate $\Delta R/R_0$ shows an increasing trend accordingly with the gradual increase in the deformation of the asphalt mixture beam. Furthermore, the strain–resistance changing response curve is plotted using the strain gauge data from the bottom of the asphalt mixture beam as the calibration strain of the embedded sensor. It can be seen that the resistance changing rate $\Delta R/R_0$ of the strain sensor shows a good linear relationship with the strain, and the fitting linear equation can be obtained.

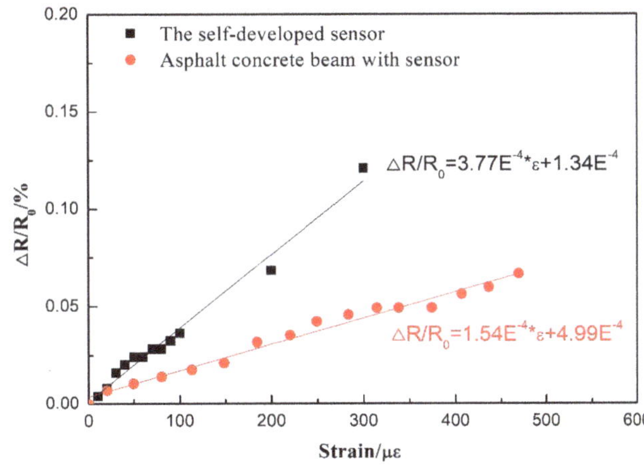

Figure 8. Force–electric response during step-by-step loading.

The strain sensor has good effectiveness in monitoring the deformation of the asphalt mixture beam. By comparing the fitted curves of sensor response before and after embedding, it can be seen that the value of $\Delta R/R_0$ for the embedded sensor is about 0.65 times that before embedding at the experimentally selected depth and loading rate. Therefore, the embedded strain sensors in asphalt pavement can be calibrated by multiplying the correction factor on the sensor room calibration results for practical engineering applications. The strain calibration test of asphalt concrete beam embedded with a sensor also verified that the self-developed I-shaped strain sensor made by encapsulated MWCNT/EP composite has stable sensing characteristics when employed for pavement strain monitoring, although it suffered the high temperature and pressure during asphalt mixtures compaction.

3.3. Application of Self-Developed Sensor in Practical Pavement Interlayer Strain Monitoring

On the basis of laboratory strain monitoring effectiveness of the sensor in asphalt concrete, the self-developed sensor was applied in practical pavement construction, aiming to verify its engineering performance [32,33]. The engineering field application of the self-developed sensor is shown in Figure 9. The position of the sensor in the lane should be accurately located in practical pavement engineering in order to make it coincide with the road lane wheel tracks. Then, the distance between every sensor needs to be determined, and the location of each sensor is coded. The sensors were pre-installed and positioned according to the location marker, and the wire was led through the curbstone gap. The hot asphalt mixtures were taken out from the truck and used to cover the installed sensor. The covered asphalt mixtures were further compacted using a rubber hammer. Eventually, as the paving truck moved, the sensor was buried in the interlayer of the asphalt pavement structure. After compaction, the monitoring results verified the good survivability and monitoring effectiveness of the self-developed sensor. Compared to other post-excavating buried sensors, the self-developed sensor can be embedded in the pavement interlayer as the asphalt mixtures paving process, which can obtain the real strain response of pavement structure and reduce the perturbation of the sensor to the dynamic response of the pavement. It also provides the possibility of significantly simplifying the strain sensor embedding process.

Figure 9. Engineering field application validation of strain sensors.

4. Conclusions

In order to solve the problem of direct monitoring of the internal strain response of the asphalt pavement, this study innovatively developed a strain sensor based on MWCNT/EP composite materials. The encapsulation techniques of MWCNT/EP composite rod by epoxy, the embedding process of the sensor in asphalt concrete, the strain monitoring effectiveness of the sensor in asphalt concrete, and the application of a self-developed sensor in the practical pavement were investigated systematically. The conclusions are summarized as follows:

- The self-developed embedded strain sensor can detect the strain changes in the asphalt pavement structure. When it is subjected to an external load, the conductive path is changed accordingly in conductive MWCNT/EP composite material, causing changes in the resistance of the sensor.
- The strain monitoring effectiveness test showed that the sensor has a high sensitivity to strain, and the resistance value of the sensor is linearly related to the changes in the external strain.
- By comparing with the commercial strain sensor, the self-developed embedded strain sensor has the advantage of deformation coordination and consistency with the road structure because of the encapsulating materials and packaging techniques studied in this paper, which can help to improve the monitoring accuracy.
- The embedding of the sensor in asphalt concrete slab and its application in practical pavement both showed that the sensor could effectively survive after being subjected to high temperature and heavy pressure during asphalt pavement compaction, which verified the feasibility of a self-developed sensor in asphalt pavement strain monitoring.

Author Contributions: C.M., X.X. and M.L.: Data curation, Writing—original draft; Writing—review & editing. N.Z. and J.W.: Data curation; C.W.: Investigation. Y.Z. and Z.Y.: Investigation. We confirm that the order of authors listed in the manuscript has been approved by all named authors. All authors have read and agreed to the published version of the manuscript.

Funding: The authors acknowledge the financial support of the Qilu Young Scholars Program of Shandong University, Natural Science Foundation of Shandong Province (CN) (No. ZR2020ME244), Jinan Research Leader Studio (202228101) and National Key R&D Program "Transportation Infrastructure" "Reveal the list and take command" project (2022YFB2603300).

Institutional Review Board Statement: Not applicable.

Informed Consent Statement: Not applicable.

Data Availability Statement: Not applicable.

Conflicts of Interest: The authors declare no conflict of interest.

References

1. China Highway Network Website. Available online: http://www.chinahighway.com/ (accessed on 12 November 2022).
2. Di Graziano, A.; Marchetta, V.; Cafiso, S. Structural health monitoring of asphalt pavements using smart sensor networks: A comprehensive review. *J. Traffic Transp. Eng.* **2020**, *7*, 639–651. [CrossRef]
3. Corticelli, R.; Pazzini, M.; Mazzoli, C.; Lantieri, C.; Ferrante, A.; Vignali, V. Urban Regeneration and Soft Mobility: The Case Study of the Rimini Canal Port in Italy. *Sustainability* **2022**, *14*, 14529. [CrossRef]
4. Tiboni, M.; Rossetti, S.; Vetturi, D.; Torrisi, V.; Botticini, F.; Schaefer, M.D. Urban policies and planning approaches for a safer and climate friendlier mobility in cities: Strategies, initiatives and some analysis. *Sustainability* **2021**, *13*, 1778. [CrossRef]
5. Shtayat, A.; Moridpour, S.; Best, B.; Shroff, A.; Raol, D. A review of monitoring systems of pavement condition in paved and unpaved roads. *J. Traffic Transp. Eng.* **2020**, *7*, 629–638. [CrossRef]
6. Du, Z.; Yuan, J.; Xiao, F.; Hettiarachchi, C. Application of image technology on pavement distress detection: A review. *Measurement* **2021**, *184*, 109900. [CrossRef]
7. Mataei, B.; Nejad, F.M.; Zakeri, H. An improved multiresolution technique for pavement texture image evaluating. *Multimed. Tools Appl.* **2023**, *82*, 3007–3031. [CrossRef]
8. Salour, F.; Erlingsson, S. Investigation of a pavement structural behaviour during spring thaw using falling weight deflectometer. *Road Mater. Pavement Des.* **2013**, *14*, 141–158. [CrossRef]
9. Jung, F.W.; Stolle, D.F. Nondestructive Testing with Falling Weight Deflectometer on Whole and Broken Asphalt Concrete Pavements. *Transp. Res. Rec.* **1992**, *1377*, 183–192.
10. Salour, F.; Erlingsson, S. *Impact of Groundwater Level on the Mechanical Response of a Flexible Pavement Structure: A Case Study at the Torpsbruk Test Section along County Road 126 Using Falling Weight Deflectometer*; VTI: Sydney, Australia, 2014.
11. Li, W. A theoretical analysis of the evolution of traffic and transportation structure development. *Water Transp. Manag.* **2000**, *7*, 2–7.
12. Fang, L.; Guo, B.; Zheng, Y.; Zhang, X. Research on intelligent monitoring technology for asphalt pavement construction. *China Meas. Test.* **2022**, *47*, 122–125.
13. Yang, Y.; Wang, L.; Wei, J.; Ma, S. Data collection and analysis of dynamic response of typical pavement structures under heavy load. *Road Traffic Technol.* **2010**, *27*, 11–16.
14. Zeng, J.; Xiao, C.; Li, B.; Xu, T. Study on the effect of temperature on the dynamic response of asphalt pavement under FWD loading. *Highw. Eng.* **2015**, *40*, 107–111, 161.
15. Mustafa, S.; Sekiya, H.; Maeda, I.; Takaba, S.; Hamajima, A. Identification of external load information using distributed optical fiber sensors embedded in an existing road pavement. *Opt. Fiber Technol.* **2021**, *67*, 102705. [CrossRef]
16. Mustafa, S.; Sekiya, H.; Morichika, S.; Maeda, I.; Takaba, S.; Hamajima, A. Monitoring internal strains in asphalt pavements under static loads using embedded distributed optical fibers. *Opt. Fiber Technol.* **2022**, *68*, 102829. [CrossRef]
17. Xiang, P.; Wang, H. Optical fibre-based sensors for distributed strain monitoring of asphalt pavements. *Int. J. Pavement Eng.* **2016**, *19*, 842–850. [CrossRef]
18. Li, Q.; Cary, C.; Combs, S.; Garg, N. Evaluation of asphalt concrete layer response using asphalt strain gauges and fiber optic strain gauges. In Proceedings of the International Conference on Transportation and Development, Houston, TX, USA, 26–29 June 2016; pp. 42–53.
19. Han, M.; Muhammad, Y.; Wei, Y.; Zhu, Z.; Huang, J.; Li, J. A review on the development and application of graphene based materials for the fabrication of modified asphalt and cement. *Constr. Build. Mater.* **2021**, *285*, 122885. [CrossRef]
20. Liu, Y. *Development and Research of Rubber Encapsulated FBG Asphalt Pavement Vertical Strain Sensor*; Harbin Institute of Technology: Harbin, China, 2008.
21. Wu, B.; Lin, Z.; Zeng, Y.; Lin, Z.; Lu, H. Experimental study on the temperature sensitivity coefficient of fiber-encapsulated long-scale strain sensors. *Adv. Laser Optoelectron.* **2022**, *59*, 142–149.
22. Liang, L.; Liu, D.; Li, P. Research on carbon fiber composite encapsulated fiber optic sensors. *J. Wuhan Univ. Technol.* **2008**, *30*, 55–57.
23. Rana, S.; Subramani, P.; Fangueiro, R.; Correia, A.G. A review on smart self-sensing composite materials for civil engineering applications. *AIMS Mater. Sci.* **2016**, *3*, 357–379. [CrossRef]

24. Li, P.; Jiang, W.; Lu, R.; Yuan, D.; Shan, J.; Xiao, J. Design and durability of PZT/PVDF composites based on pavement perception. *Constr. Build. Mater.* **2022**, *323*, 126621. [CrossRef]
25. Hou, S.; Lei, J.; Ou, J. Vehicle load monitoring for asphalt concrete pavement based on smart aggregates. *J. Vib. Shock* **2014**, *33*, 42–47.
26. Zhao, Q.; Wang, L.; Zhao, K.; Yang, H. Development of a Novel Piezoelectric Sensing System for Pavement Dynamic Load Identification. *Sensors* **2019**, *19*, 4668. [CrossRef] [PubMed]
27. Wang, Q. *Research on the Electrical Conductivity and Mechanosensitivity of Carbon Fiber Epoxy Resin Composites*; Chongqing University: Chongqing, China, 2006.
28. Xin, X.; Liang, M.; Yao, Z.; Su, L.; Zhang, J.; Li, P.; Sun, C.; Jiang, H. Self-sensing behavior and mechanical properties of carbon nanotubes/epoxy resin composite for asphalt pavement strain monitoring. *Constr. Build. Mater.* **2020**, *257*, 119404. [CrossRef]
29. Xin, X.; Yu, R.; Su, L.; Qiu, Z.; Yang, C.; Liang, M.; Ji, P.; Zhong, G.; Yao, Z. Dynamic Mechanical and Chemorheology Analysis for the Blended Epoxy System with Polyurethane Modified Resin. *J. Renew. Mater.* **2022**, *10*, 1081. [CrossRef]
30. Xin, X.; Qiu, Z.; Luan, X.; Ding, X.; Liang, M.; Yao, Z.; Li, Y. Novel conductive polymer composites for asphalt pavement structure in situ strain monitoring: Influence of CB/CNT and GNP/CNT nano/micro hybrid fillers on strain sensing behavior. *IEEE Sens. J.* **2022**, *10*, 3945–3956. [CrossRef]
31. Xin, X.; Luan, X.; Su, L.; Ma, C.; Liang, M.; Ding, X.; Yao, Z. The innovative self-sensing strain sensor for asphalt pavement structure: Substitutability and synergy effects of graphene platelets with carbon nanotubes in epoxy composites. *Front. Mater.* **2022**, *9*, 824364. [CrossRef]
32. Guo, L.; Chen, X.; Yu, J.; Tang, Y.; Liu, R.; Rogers, R.; Leidy, J.; Claros, G. Pavement deflection vehicle weighing method with embedded piezoelectric sensor. *SPIE Proc.* **2005**, *5758*, 471–478.
33. Tan, Z.; Li, H.; Xu, X.; Wu, C.; Liu, F. Study on the arrangement scheme of asphalt pavement mechanical response monitoring system. *J. Wuhan Univ. Technol.* **2017**, *41*, 528–532.

Disclaimer/Publisher's Note: The statements, opinions and data contained in all publications are solely those of the individual author(s) and contributor(s) and not of MDPI and/or the editor(s). MDPI and/or the editor(s) disclaim responsibility for any injury to people or property resulting from any ideas, methods, instructions or products referred to in the content.

coatings

MDPI

Article

Preparation and Performance Test of UV Resistant Composite-Modified Asphalt

Quanjun Shen [1], Shijie Ma [2], Yaohui Yang [1], Liang Fan [2], Yongzhen Li [2,*] and Pinhui Zhao [3]

1 Innovation Research Institute of Shandong Expressway Group Co., Ltd., Jinan 250098, China
2 Shandong Transportation Institute, Jinan 250102, China
3 School of Transportation Engineering, Shandong Jianzhu University, Jinan 250101, China
* Correspondence: lyztgzy1@126.com

Abstract: Ultraviolet radiation is the main cause of degradation in asphalt pavement. To improve the performance of the pavement used in the strong ultraviolet (UV) region of the western plateau, China, this study explores the effects of adding nano-montmorillonite and carbon black to SBS-modified asphalt. Through conventional index detection, dynamic shear rheological tests, low-temperature bending creep tests, and UV aging tests, the high- and low-temperature performance, fatigue performance, UV aging resistance, and other aspects of the asphalt were studied. Various performance and price factors were considered in the optimization of various UV resistant composite-modified asphalt formulas. Increasing the contents of nano-montmorillonite and carbon black increases the high-temperature performance and the UV aging resistance but reduces the low-temperature and fatigue performance of asphalt; hence, their total content should be limited to <4%. Nano-montmorillonite has a better high-temperature performance and UV aging resistance than carbon black and is also less favorable to low-temperature and fatigue performance. Hence, it is recommended that its content not exceed 3%. UV resistant composite-modified asphalt has obvious advantages in high-temperature performance and UV aging resistance compared with conventional SBS-modified asphalt, and its low-temperature performance meets the use requirements of the strong-UV areas in the western plateau.

Keywords: nano-montmorillonite; carbon black; SBS-modified asphalt; ultraviolet aging

Citation: Shen, Q.; Ma, S.; Yang, Y.; Fan, L.; Li, Y.; Zhao, P. Preparation and Performance Test of UV Resistant Composite-Modified Asphalt. *Coatings* **2023**, *13*, 239. https://doi.org/10.3390/coatings13020239

Academic Editor: Claudio Lantieri

Received: 29 November 2022
Revised: 10 January 2023
Accepted: 16 January 2023
Published: 19 January 2023

Copyright: © 2023 by the authors. Licensee MDPI, Basel, Switzerland. This article is an open access article distributed under the terms and conditions of the Creative Commons Attribution (CC BY) license (https://creativecommons.org/licenses/by/4.0/).

1. Introduction

The diseases afflicting asphalt pavement in China's western plateau are mostly related to intense ultraviolet radiation. Hence, improving the anti-ultraviolet aging ability of asphalt pavement is one way to alleviate pavement diseases. At present, the most commonly used types of road asphalt in China are mainly road petroleum asphalt and polymer-modified asphalt. Most of the polymers used are thermoplastic elastomers, which have a low UV resistance.

Research on the aging of polymer-modified asphalt has always been a hot topic. M. Giunta [1] proposed a new viscoelastic plastic constitutive model for asphalt concrete that was verified by experiments and met the general thermodynamic requirements. This model can characterize the nonlinear time-varying characteristics of asphalt materials. J. Zhu [2] described the history of asphalt polymer modifications in chronological order, discussed the advantages and disadvantages of the plastic and thermoplastic elastomers commonly used in asphalt modification, and proposed that functionalization is a promising method for improving the performance of the polymers currently used and developing new polymer modifiers. C. Celauro [3] studied the short-term aging effects when using different types of modifiers (such as polymers recovered from waste). The results showed that polymer modification can reduce the aging effect, that is, the increase of the elastic modulus experienced by the original adhesive. G. Sun [4] has aged styrene butadiene styrene (SBS) modified asphalt, recycled polyethylene (RPE) modified asphalt, and its base asphalt through thermal oxidation aging and all-weather aging tests. The results showed that all-weather aging runs through almost the entire depth of the base asphalt. On the other hand, the SBS's space network structure can significantly reduce the oxidation depth of SBS asphalt, while the mitigation effect of the RPE asphalt on all-weather aging was weakened. I. Joohari [5] analyzed whether the aging of mixed a PMB of styrene butadiene styrene (SBS) and linear low density polyethylene (RLLDPE) recovered from waste plastics was caused by asphalt aging, polymer degradation, or both. The results showed that asphalt aging caused by thermal oxidation was dominant, and polymer degradation had the least impact on asphalt hardening. N.I.M. Yusoff [6] investigated the effect of mixing different percentages of nano silica (NS) with PMB under the conditions of no aging and aging. The results show that adding NS to PMB could improve its viscoelastic and anti-aging properties. I. Binti Joohari [7] evaluated the effects of styrene butadiene styrene (SBS), rubber crumbs (CR), ethylene vinyl acetate (EVA), and various polyethylene modifiers on the rheological properties of mixed polymer-modified asphalt (PMB) using the multiple stress creep and recovery (MSCR) and linear amplitude scanning (LAS) tests. G. Sun [8] conducted a fatigue healing fatigue (FHF) test to study the effects of temperature and thermal oxidation aging on the self-healing ability of styrene butadiene styrene (SBS) modified asphalt (SBS-MB) and polyethylene (PE) modified concrete (PE-MB). The results showed that the self-healing ability of PMB gradually weakened with the progression of aging. It is worth noting that thermal oxidative aging has a completely different effect on the self-healing ability of SBS-MB and PE-MB at lower temperatures. Y. Tian [9] selected four types of adhesives for aging tests over different periods of time. The results showed that polymer degradation mainly occurred at an early age, and that the oxidation of asphalt phase was the main process of long-term aging. R. Kleizienė [10] extended the regeneration of standard polymer-modified asphalt (PMB) by introducing nano materials such as nano SiO_2 and nano TiO_2 into traditional regeneration agents. The most promising results were obtained after the aging PMB was regenerated with a modified soft asphalt containing 6% nano TiO_2. It can be seen from the above studies that the aging of asphalt is still dominated by thermal oxygen aging, but that the aging effects of ultraviolet light cannot be ignored.

There has been much research on improving the anti-ultraviolet ability of asphalt pavement. The most common method is to modify asphalt by adding materials with anti-ultraviolet properties. The majority of Chinese research has focused on improving the ultraviolet resistance of asphalt by using the strong ultraviolet-absorption properties of carbon black and the thermal resistance of nanomaterial intercalation structures. Some scholars have studied the dry preparation of a carbon black asphalt mixture and successfully paved a surface test road with a UV resistance function [11–13]. The research of Li et al. [14] showed that carbon black improves the high-temperature and anti-aging performance of asphalt but reduces its low-temperature performance. The research of Zhang et al. [15] revealed how carbon black can improve the high-temperature and anti-aging performance and storage stability of styrene butadiene styrene block copolymer (SBS) modified asphalt. Fa et al. [16] modified SBS with nanographene oxide, which effectively improved the UV resistance performance of SBS-modified asphalt. Lu et al. [17] proved the feasibility of a technical scheme for a hydrotalcite/waste rubber powder composite-modified asphalt. Their results proved that hydrotalcite improves the UV resistance of the composite-modified asphalt. Sui et al. used different combinations of nanomaterials, such as SiO_2, ZnO, TiO_2, organic vermiculite, and organic montmorillonite, to develop a nanomaterial/SBS composite-modified asphalt, which demonstrated excellent high- and low-temperature performance [18–22].

Following from the above research, this study aimed to improve the performance of the asphalt pavement used in the strong ultraviolet region of the western plateau, China. Nano-montmorillonite and carbon black were used to amend SBS-modified asphalt to prepare a UV resistant composite-modified asphalt. Through conventional index detection, dynamic shear rheological tests, low-temperature bending creep tests, and UV aging tests, the high-

and low-temperature performance, fatigue performance, and UV resistance were studied and analyzed. The aim was to develop an asphalt with excellent high- and low-temperature performance and outstanding UV resistance for use in the western plateau area.

2. Materials and Methods

2.1. Materials

2.1.1. SBS-Modified Asphalt

The physical properties of the SBS-modified asphalt are shown in Table 1; all met the relevant specifications in [23].

Table 1. Basic physical properties of the SBS-modified asphalt.

Property	Specification	Measured Value
Penetration (25 °C) (0.1 mm)	40–60	54
Penetration index	≥0	0.6
Softening point (°C)	≥60	75.2
Ductility (5 °C) (cm)	≥20	33
Apparent viscosity (135 °C) (Pa·s)	≤3	1.975
Recovery elastic (RE) (%)	≥75	90
Flash point (°C)	≥230	280
Storage stability (segregation, 48 h softening point difference) (°C)	≤2.5	0.3
Mass loss (%)	−1.0–1.0	−0.070
Penetration ratio (%)	≥65	86
Ductility (5 °C) (cm)	≥15	21
Performance grade	/	PG76-22

2.1.2. Carbon Black

Carbon black was prepared by pyrolysis of waste tires at a production site in Shanghai, China; its technical indicators are shown in Table 2. Its price is 4500 yuan (645.3 U.S. dollars) per ton.

Table 2. Carbon black test results.

Property	Specification	Property
Ash content (%)	≤18.5	13.2
Oil absorption value (mL/100 g)	≥7.0	11.5
Sieve residue (%)	/	0.008
Fineness (μm)	/	10
Moisture (%)	≤3.0	0.5
Specific surface area (m^2/g)	/	120
pH	≥6.0	7

2.1.3. Nano-Montmorillonite

The physical and chemical properties of nano-montmorillonite are shown in Table 3. Its price is 28,000 yuan (4015.2 U.S. dollars) per ton.

Table 3. Physical and chemical properties of nano-montmorillonite.

Property	Measured Value
Appearance (color)	White powder
Particle size (200 mesh pass rate) (%)	97.5
Density (g/cm^3)	1.8
Montmorillonite content (%)	≥95
pH	8.0

2.2. Sample Preparation

Firstly, the SBS-modified asphalt was heated to a molten state in an oven at 163 °C. Then, carbon black was added slowly and uniformly while stirring at 1000 r/min. After 30 min of stirring, the stirring speed was increased to 2000 r/min, and then nano-montmorillonite was added slowly. After all the carbon black was added, it was uniformly stirred for 1 h, and then the mold was poured for inspection.

2.3. Orthogonal Experimental Design

The formulation of UV resistant composite-modified asphalt with different combinations of raw materials was conducted via an orthogonal test design. The relevant indexes were then tested. The specific test scheme is shown in Table 4.

Table 4. The specific test schemes.

Recipe	SBS-Modified Asphalt (%)	Carbon Black (%)	Nano-Montmorillonite (%)
A	100	0	0
B	99	0	1
C	98	0	2
D	97	0	3
E	96	1	3
F	97	1	2
G	98	1	1
H	99	1	0
I	95	2	3
J	96	2	2
K	97	2	1
L	98	2	0
M	94	3	3
N	95	3	2
O	96	3	1
P	97	3	0

2.4. Test Methods

(1) Routine index detection

According to the relevant provisions and requirements of the specifications in [24], the softening point, apparent viscosity (135 °C), and ductility (5 °C) index of the UV resistant composite-modified asphalt were determined.

(2) Dynamic shear rheological tests

A dynamic shear rheometer was used (TR2000, TA Company, Boston, MA, USA) to carry out temperature scanning, frequency scanning, and time scanning tests.

Temperature sweep tests: During this test, the control strain was 1%, the diameter of the parallel plate was 25 mm, the distance between the upper and lower parallel plates was 1 mm, and the frequency was fixed at 10 rad/s. The test was carried out in a temperature range of 28–82 °C with a step size of 6 °C. The complex shear modulus G^*, phase angle δ, and rutting factor $G^*/\sin\delta$ were collected.

Frequency sweep tests: The strain was controlled at 5%, the temperature was set to 40 °C, 50 °C, 60 °C, 70 °C, and 80 °C, and the frequency scanning was carried out within the frequency range of 0.01–10 Hz. The complex shear modulus G^* and phase angle δ were collected.

Time sweeps tests: The temperature was set at 25 °C, the frequency was 10 rad/s, the strain was controlled at 10%, and the complex shear modulus G^* under different loading times was measured.

(3) Low-temperature bending creep tests: A low-temperature bending beam rheometer was used (TE-BBR-T, Cannon, Cranberry Township, PA, USA). The test temperature was −12 °C, and the creep stiffness modulus s and creep rate m were collected.

(4) UV aging tests: A melted asphalt sample was evenly distributed on a flat-bottom disc and then placed in a self-developed ultraviolet aging oven (Figure 1). The power of the ultraviolet lamp was 400 W, and the axis of the lamp was 40 cm away from the sample surface. The manufacturer of ultraviolet lamp is the China Guangzhou Langpu

Optoelectronic Technology Co., Ltd., Guangzhou, China. The power supply voltage is 220 V, the maximum working temperature is 45 °C, the maximum service life is 8000 h, the length is 330 mm, and the pipe diameter is 28 mm. The oven was continuously ventilated to prevent thermal aging due to excessive heat. The temperature was kept at about 35 °C for 96 h. The aged samples were used in 60 °C temperature scanning tests, low-temperature bending creep tests, and 135 °C viscosity tests. The relevant test parameter settings were consistent with the other test methods described in this section.

(a) (b) (c)

Figure 1. Ultraviolet aging oven: (**a**) design schematics, (**b**) actual equipment, (**c**) test photos.

3. Results

3.1. Optimization of UV Resistant Composite-Modified Asphalt Formula

The influences of nano-montmorillonite and carbon black on the performance of SBS-modified asphalt are shown in Figures 2 and 3. Figure 2 shows the sample with only nano-montmorillonite added, corresponding to Schemes A, B, C, and D. Figure 3 shows the sample with only carbon black added, corresponding to Schemes A, H, L, and P.

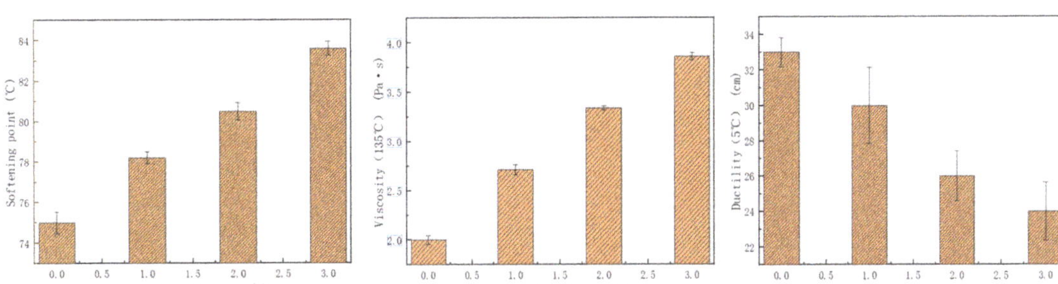

(a) (b) (c)

Figure 2. Effects of nano-montmorillonite on the (**a**) softening point and (**b**) viscosity and (**c**) ductility of SBS-modified asphalt.

It can be seen from Figures 2 and 3 that mixing with nano-montmorillonite and carbon black had the same effect on the performance of SBS-modified asphalt. With increases in the mixing amount, the softening point and 135 °C viscosity of the SBS-modified asphalt gradually increased, and the increased apparent viscosity presented a greater resistance to flow deformation. Under a given load, relatively weak deformation occurred, and the elastic part could recover readily, which is conducive to a good high-temperature anti-rutting performance. However, the ductility at 5 °C decreased gradually, the ability to resist deformation at low temperatures decreased, and the low-temperature crack resistance decreased. This shows that both nano-montmorillonite and carbon black can improve the high-temperature performance of SBS-modified asphalt, but not its low-temperature performance. According to the literature [25–27], nano-montmorillonite has a lamellar structure with upper and lower spacings that are uniformly dispersed in asphalt by stirring and shearing. The lamellar structure is interlocked with SBS molecules, resulting in a change in its microstructure, which is conducive to its storage stability. The intercalation effects of montmorillonite and the light components in asphalt and the barrier to oxygen diffusion improve the anti-aging performance of UV resistant composite-modified asphalt. Carbon black contains active agent components that can easily promote the cross-linking reaction of the modifier in SBS-modified asphalt and improve the high-temperature and water-loss resistance of SBS-modified asphalt [15].

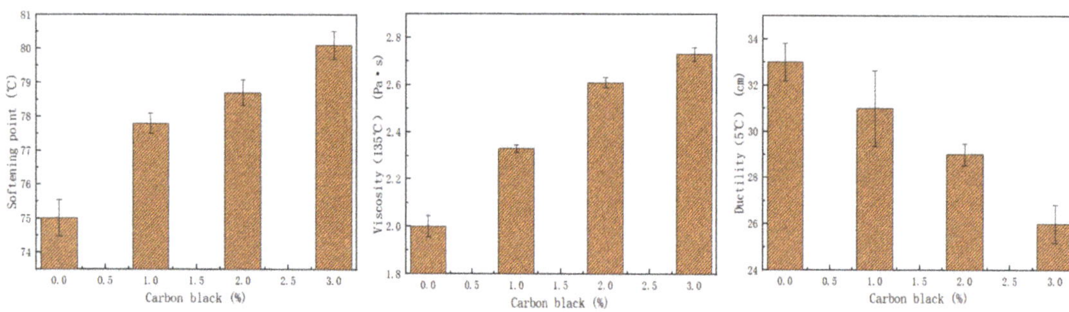

(a) (b) (c)

Figure 3. Effects of carbon black on the (**a**) softening point and (**b**) viscosity and (**c**) ductility of SBS-modified asphalt.

It can be seen from Table 5 that, at the same dosage, the influence of nano-montmorillonite on the performance index of the SBS-modified asphalt is greater than that of carbon black, which is due to the larger specific surface area of nano-montmorillonite and its intercalation effect. The climate of the western plateau is not only exposed to strong ultraviolet radiation but also large temperature differences. This requires the composite-modified asphalt to have an excellent high-temperature performance and a good low-temperature performance and UV aging resistance.

Table 5. Physical properties of composite-modified asphalt.

Recipe	Softening Point (°C)	Apparent Viscosity (135 °C) (Pa·s)	Ductility (5 °C) (cm)
E	86.4	3.94	22
F	84.2	3.37	26
G	81.3	2.79	27
I	87.3	4.03	21
J	86.1	3.91	23
K	82.2	3.17	25
M	89.4	4.23	18
N	86.8	3.95	19
O	82.9	3.62	22

The price of nano-montmorillonite is 28,000 yuan (4015.2 U.S. dollars) per ton, while the price of carbon black is 4500 yuan (645.3 U.S. dollars) per ton. The price of nano montmorillonite is 6.2 times that of carbon black at the same dosage. Therefore, the formula of the composite-modified asphalt was preliminarily determined as Schemes E, F, G, J, K and O based on conventional indexes and cost.

3.2. Performance of UV Resistant Composite-Modified Asphalt

3.2.1. High-Temperature Performance

The temperature sweep test results for the samples of the preferred formulas are shown in Figure 4. The rutting factor G */sinδ of the sample of the optimized formula is better than that of Scheme A (SBS-modified asphalt) without nano-montmorillonite and carbon black. The rutting factor G */sinδ increases with the total amount of nano-montmorillonite and carbon black, and the larger the rutting factor, the stronger the resistance to the flow dynamic deformation under load. The rutting factor G */sinδ of Scheme E (3% nano-montmorillonite + 1% carbon black) is greater than that of Scheme O (1% nano-montmorillonite + 3% carbon black). When the total content is 4%, the rutting factor of Scheme E (3% nano-montmorillonite + 1% carbon black) is greater than that of Scheme J (1% nano-montmorillonite + 3% carbon black content). This indicates that, at a given content of nano-montmorillonite and carbon black, the contribution of nano-montmorillonite to the high-temperature performance of the composite-modified asphalt is greater than that of carbon black. The reason is that nano-montmorillonite is an inorganic material that is uniformly dispersed in SBS-modified asphalt after shearing and forms intercalated nano-microstructures with the SBS, which restricts the movement of the SBS molecular chain and prevents the deformation of the SBS-modified asphalt under high-temperature stress. Carbon black adsorbs light components in the asphalt via its strong adsorption function so that the asphalt has a strong deformation resistance. Under high-temperature stress, the light components adsorbed by carbon black will precipitate, altering the high-temperature performance.

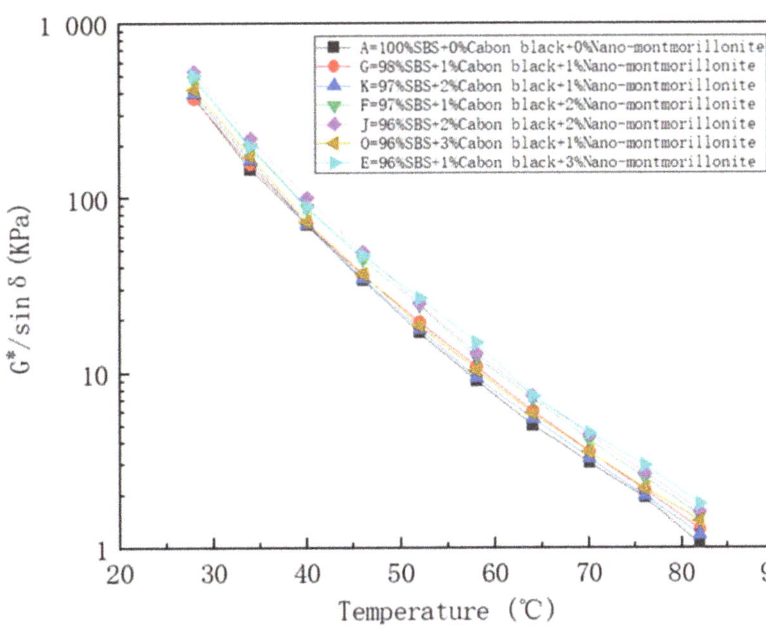

Figure 4. Relationships between complex modulus and temperature in DSR temperature scanning tests of various samples.

Frequency sweep tests were carried out on samples of the preferred formula. The test data at different temperatures and frequencies were converted using the principle of time-temperature equivalence, allowing a complex modulus master curve at 60 °C to be established. The results are shown in Figure 5, which shows that, in general, the complex modulus of the asphalt mixed with nano-montmorillonite and carbon black in the low-frequency region is better than that of the SBS-modified asphalt, indicating that the high-temperature rutting resistance of the preferred samples is higher than that of the SBS-modified asphalt. However, the complex shear modulus of the samples with high nano-montmorillonite contents in the high-frequency region is less than that of the SBS-modified asphalt, which indicates that those samples have excellent high-temperature rutting resistance but insufficient shear deformation resistance at low temperatures, so the content of nano-montmorillonite should be limited.

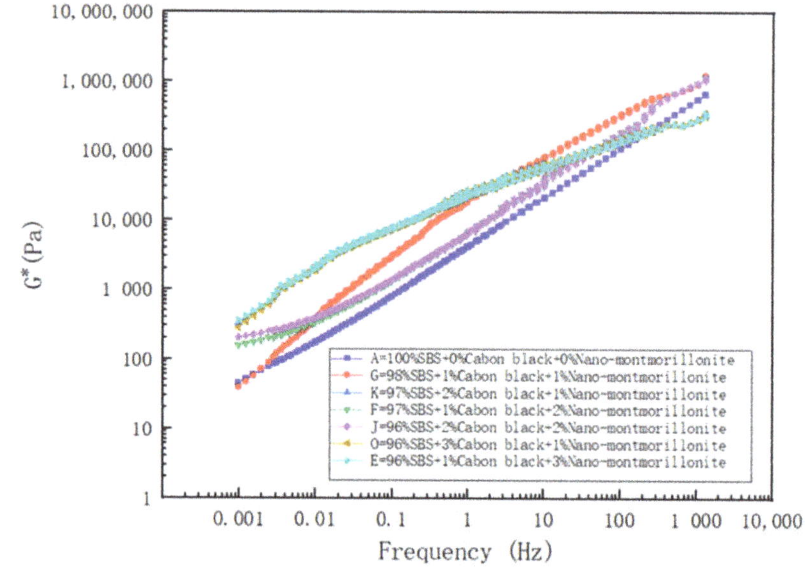

Figure 5. Main curves of the complex shear modulus of asphalt of different schemes.

3.2.2. Low-Temperature Performance

The low-temperature performance results for the samples of the preferred formulas are shown in Figures 6 and 7. They show that, after adding nano-montmorillonite and carbon black, the creep rate m of the SBS-modified asphalt decreases, and the creep stiffness s increases. The carbon black content of the samples in Schemes E, F, and G is 1%, while the contents of nano-montmorillonite are 3%, 2%, and 1%, respectively. Through comparative analysis of the creep rate m and creep stiffness s of Schemes E, F and G, it can be seen that, with increases in the nano-montmorillonite content, the creep rate m gradually decreases and the creep stiffness s gradually increases, indicating that the SBS-modified asphalt becomes brittle after adding nano-montmorillonite. The amount of nano-montmorillonite in Schemes G, K, and O is 1%, and the amounts of carbon black are 1%, 2%, and 3%, respectively. By comparing the creep rate m and creep stiffness s of Schemes G, K, and O, it can be seen that, with increases in the amount of carbon black, the creep rate m gradually decreases and the creep stiffness s gradually increases, indicating that the addition of carbon black also increases the brittleness of the SBS-modified asphalt. Comparing Schemes F and K, it can be seen that the low-temperature flexibility reduction of the SBS-modified asphalt caused by adding nano-montmorillonite is greater than that caused by adding carbon black at the same dosage. According to the conventional indexes and rheological tests, the addition of nano-montmorillonite and carbon black can improve the high-temperature performance of asphalt and reduce its low-temperature flexibility. The reason is that the main components of nano-montmorillonite are kaolinite minerals, while those of carbon black are simple carbon particles. Both exhibit characteristics of good weather resistance and poor plasticity similar to those of inorganic materials, which are favorable to the high-temperature performance and unfavorable to the low-temperature performance of the SBS-modified asphalt. Therefore, the total content of nano-montmorillonite and carbon black should be controlled to be within 4%.

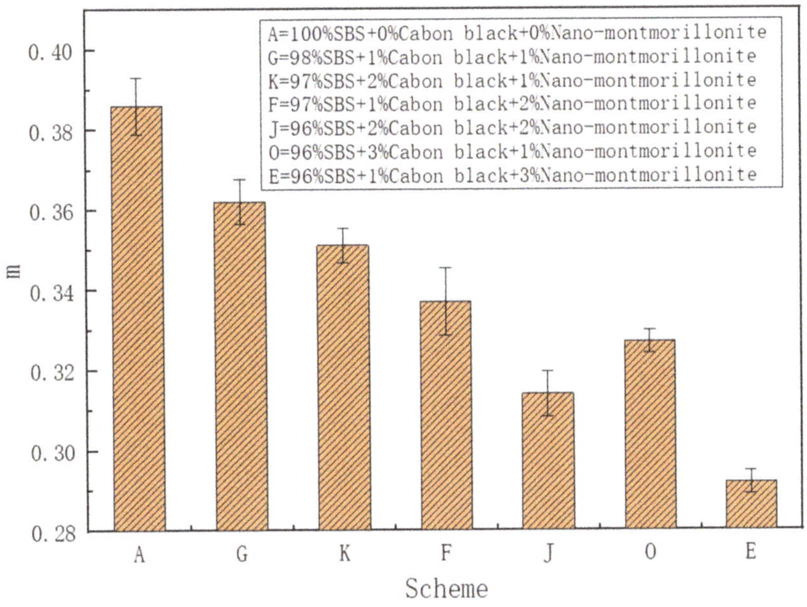

Figure 6. Low-temperature creep rates m of asphalt from different schemes.

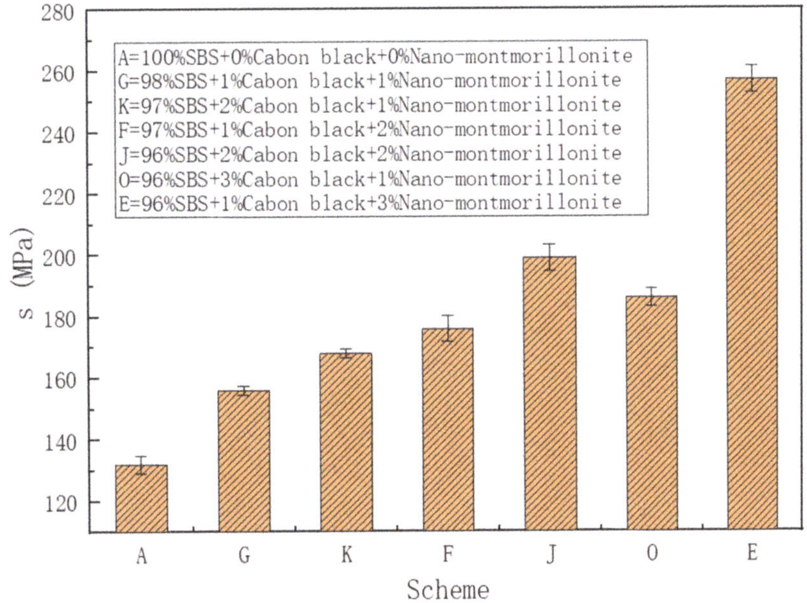

Figure 7. Low-temperature creep stiffness modulus s of asphalt from different schemes.

3.2.3. Fatigue Performance

At present, time scanning tests are mainly used to evaluate the fatigue performance of asphalt. There are two data processing methods, using (1) the complex shear modulus and (2) the cumulative dissipation energy. The fatigue life of asphalt is taken as the corresponding load action times when the two decrease to 50% of their initial values [28–30]. The first method was used in this test to evaluate the fatigue performance of the UV resistant composite-modified asphalt. The test results are shown in Figures 8 and 9.

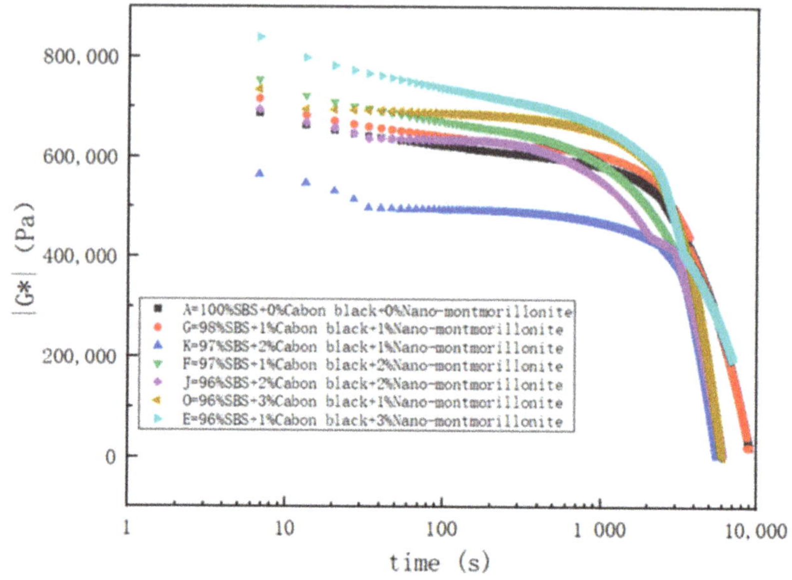

Figure 8. Fatigue curves as a function of time.

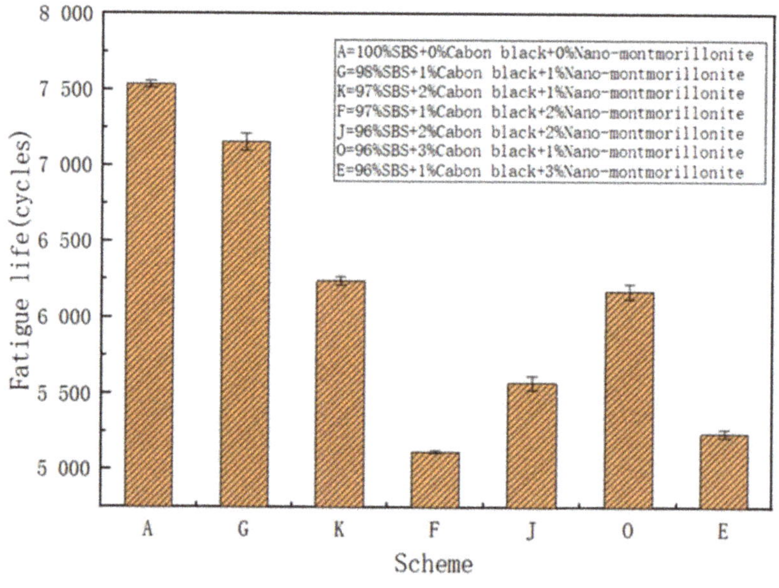

Figure 9. Fatigue life of asphalt from different schemes.

It can be seen from Figure 8 that the fatigue performance of Scheme A was greater than those of other schemes. By comparing the fatigue performance of Schemes E, F, G, K, and O, it can be seen that the fatigue life of the composite-modified asphalt gradually decreases with the content of carbon black or nano-montmorillonite. This is related to nano-montmorillonite and carbon black improving the high-temperature performance and the disadvantage of the low softness of the SBS-modified asphalt. That is, the asphalt becomes hard and brittle, the number of load-bearing actions decreases, and the durability becomes poor. At the same dosage, the specific surface area of nano-montmorillonite is larger, and it has an intercalation structure, which creates a greater binding force with the SBS modifier and reduces the fatigue life.

3.2.4. UV Aging Resistance

Based on the UV aging tests, the complex shear modulus G *, creep stiffness s, creep rate m, and apparent viscosity (135 °C) η before and after aging were measured. The ultraviolet aging resistance of the preferred asphalt was evaluated using four indexes: the complex shear modulus rate of increase (MIR), the creep stiffness increasing ratio (SIR), the creep rate decreasing ratio (MRR), and the viscosity aging index (VAI). The calculation method is shown in Formulas (1)–(4), and the test results are shown in Figure 10.

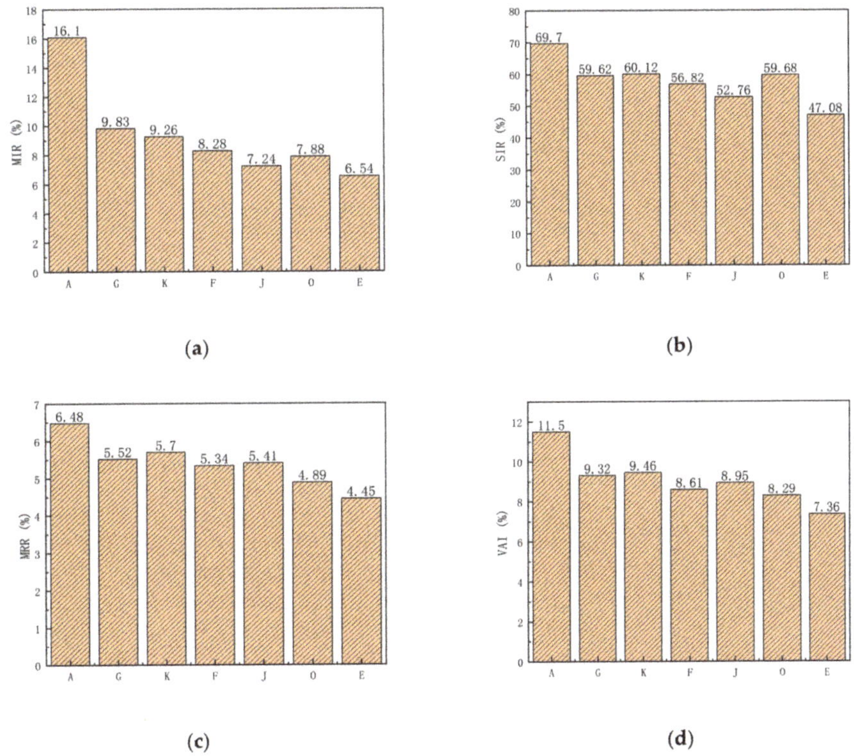

Figure 10. Ultraviolet aging resistance indexes of asphalt of different schemes: (**a**) complex shear modulus increasing ratio (MIR); (**b**) creep stiffness increase ratio (SIR); (**c**) creep rate reduction ratio (MRR); (**d**) viscosity aging index (VAI).

$$MIR = \left(\frac{G_a^* - G_f^*}{G_f^*}\right) \times 100, \quad (1)$$

where

- G_a^* is the complex shear modulus of the original asphalt;
- G_f^* is the complex shear modulus of the aged asphalt.

$$SIR = \left(\frac{s_a - s_f}{s_f}\right) \times 100, \quad (2)$$

where

- s_f is the creep stiffness of the original asphalt;

- s_a is the creep stiffness of the aged asphalt.

$$MRR = \left(\frac{m_f - m_a}{m_f}\right) \times 100, \tag{3}$$

where

- m_f is the creep rate of the aged asphalt;
- m_a is the creep rate of the original asphalt.

$$VAI = \left(\frac{\eta_a - \eta_f}{\eta_f}\right) \times 100, \tag{4}$$

where

- η_a is the apparent viscosity (135 °C) of the aged asphalt;
- η_f is the apparent viscosity (135 °C) of the original asphalt.

It can be seen from Figure 10 that, after ultraviolet aging, the complex modulus, creep stiffness, and 135 °C viscosity η of the different asphalt samples are greater than those of fresh asphalt, with decreased creep rates. The MIR, SIR, MRR, and VAI values of the asphalt samples containing nano-montmorillonite and carbon black are lower than those of the SBS-modified asphalt. This indicates that, although the materials are mixed with an anti-ultraviolet material (nano-montmorillonite and carbon black), ultraviolet aging cannot be prevented; however, that of the SBS-modified asphalt can be alleviated. Comparing the UV aging evaluation indexes of Schemes E, F, G, K, and O, it can be seen that, with increases in the nano-montmorillonite or carbon black content, the UV aging resistance of the composite-modified asphalt is enhanced. At the same dosage, the contribution of nano-montmorillonite to the anti-UV aging ability is stronger than that of carbon black. The greater the total amount of nano-montmorillonite and carbon black, the stronger the anti-ultraviolet aging ability. This is because the nano-montmorillonite lamellar structure has the function of preventing oxygen and heat from diffusing into the asphalt. Combined with the ultraviolet-absorption ability of carbon black, the two synergistically improve the anti-ultraviolet aging performance of the composite-modified asphalt.

In summary, the general rule is that nano-montmorillonite and carbon black are favorable to the high-temperature and UV aging performance of UV resistant composite-modified asphalt and unfavorable to its low-temperature and fatigue performance. The greater the content of nano-montmorillonite, the more obvious the decline in the fatigue performance; its price is also higher than that of carbon black. Considering the performance and price factors, Schemes G, J, K, and O were finally selected.

4. Conclusions

(1). Nano-montmorillonite and carbon black are beneficial to the high-temperature performance and UV aging resistance of the composite-modified asphalt, but unfavorable to its low-temperature performance and fatigue performance. The total amount should be controlled to be within 4%. Nano-montmorillonite has a better ability to improve the high-temperature and anti-ultraviolet aging performance than carbon black and is also less favorable to the low-temperature and fatigue performance. It is recommended that its amount should not exceed 3%.

(2). Based on factors such as the performance and prices of nano-montmorillonite, carbon black, and SBS composite-modified asphalt, Schemes G, J, K, and O provide the best performance.

(3). Compared with SBS-modified asphalt, UV resistant composite-modified asphalt has obvious advantages in high-temperature performance and UV aging resistance, and its low-temperature performance meets the use requirements of the western plateau and other strong-UV areas.

This research is only limited to the initial formulation of UV resistant composite-modified asphalt determined in the laboratory, and has not been applied in actual projects. Research on the storage stability of the UV resistant composite-modified asphalt is still lacking, the evaluation methods for the samples are still limited, and the relevant properties have not been characterized by microscopic means such as infrared spectroscopy, scanning electron microscopy, etc. The actual factory production of composite-modified asphalt has not been studied.

Author Contributions: Conceptualization, Q.S. and S.M.; methodology, L.F.; validation, Y.Y. and Y.L.; investigation, Y.L.; resources, Y.Y.; data curation, S.M.; writing—original draft preparation, Y.L.; writing—review and editing, Q.S., S.M., L.F. and P.Z.; project administration, Q.S.; funding acquisition, Y.Y. All authors have read and agreed to the published version of the manuscript.

Funding: This research was funded by General Program of Shandong Natural Science Foundation, grant number (ZR202190324) and National Natural Science Foundation of China, grant number (51608511).

Institutional Review Board Statement: Not applicable.

Informed Consent Statement: Not applicable.

Data Availability Statement: All of the data, models, and codes that support the findings of this study are available from the corresponding author upon reasonable request.

Acknowledgments: The authors thank Jin-Cheng Wei of Shandong Transportation Institute for his technical support.

Conflicts of Interest: The authors declare no conflict of interest.

References

1. Giunta, M.; Pisano, A. One-Dimensional Visco-Elastoplastic Constitutive Model for Asphalt Concrete. *Multidiscip. Model. Mater. Struct.* **2006**, *2*, 247–264. [CrossRef]
2. Zhu, J.; Birgisson, B.; Kringos, N. Polymer modification of bitumen: Advances and challenges. *Eur. Polym. J.* **2014**, *54*, 18–38. [CrossRef]
3. Celauro, C.; Saroufim, E.; Mistretta, M.C.; La Mantia, F.P. Influence of Short-Term Aging on Mechanical Properties and Morphology of Polymer-Modified Bitumen with Recycled Plastics from Waste Materials. *Polymers* **2020**, *12*, 1985. [CrossRef]
4. Sun, G.; Li, B.; Sun, D.; Yu, F.; Hu, M. Chemo-rheological and morphology evolution of polymer modified bitumens under thermal oxidative and all-weather aging. *Fuel* **2021**, *285*, 118989. [CrossRef]
5. Joohari, I.; Maniam, S.; Giustozzi, F. Effect of Long-Term Aging on Polymer Degradation and Fatigue Resistance of Hybrid Polymer-Modified Bitumen. *J. Mater. Civ. Eng.* **2022**, *34*, 04022286. [CrossRef]
6. Yusoff, N.; Alhamali, D.I.; Ibrahim, A.; Rosyidi, S.; Hassan, N.A. Engineering characteristics of nanosilica/polymer-modified bitumen and predicting their rheological properties using multilayer perceptron neural network model. *Constr. Build. Mater.* **2019**, *204*, 781–799. [CrossRef]
7. Joohari, I.B.; Giustozzi, F. Oscillatory shear rheometry of hybrid polymer-modified bitumen using multiple stress creep and recovery and linear amplitude sweep tests. *Constr. Build. Mater.* **2022**, *315*, 125791. [CrossRef]
8. Sun, G.; Ma, J.; Sun, D.; Li, B.; Ling, S.; Lu, T. Influence of thermal oxidative aging on temperature induced self-healing transition of polymer modified bitumens. *Mater. Des.* **2020**, *192*, 108717. [CrossRef]
9. Tian, Y.; Li, H.; Sun, L.; Zhang, H.; Harvey, J.; Yang, J.; Yang, B.; Zuo, X. Laboratory investigation on rheological, chemical and morphological evolution of high content polymer modified bitumen under long-term thermal oxidative aging. *Constr. Build. Mater.* **2021**, *303*, 124565. [CrossRef]
10. Kleizienė, R.; Panasenkienė, M.; Zofka, A.; Vaitkus, A. Nanobased rejuvenators for polymer-modified bitumen under long-term ageing conditions. *Constr. Build. Mater.* **2022**, *341*, 127474. [CrossRef]
11. Cong, P.; Xu, P.; Chen, S. Effects of carbon black on the anti aging, rheological and conductive properties of SBS/asphalt/carbon black composites. *Constr. Build. Mater.* **2014**, *52*, 306–313. [CrossRef]
12. Yu, H.; Meng, X.; Gan, X.; Li, C.; Gan, Y. Experimental study on the first engineering process of pyrolysis carbon black modified asphalt surface layer of waste tire by dry casting method. *Highway* **2022**, *67*, 7–15.
13. Meng, X.; Gan, X.; Gan, Y. Bloomberg Indoor preparation process parameters and road performance test and evaluation of asphalt mixture modified by pyrolysis carbon black of dry waste tire. *Highway* **2022**, *67*, 62–70.
14. Li, Y.; Chen, H.; Li, C.; Song, W. Indoor test study on the performance of waste tire pyrolysis carbon black modified asphalt. *Sino Foreign Highw.* **2021**, *41*, 296–301. [CrossRef]

15. Zhang, W.; Yuan, Z. Influence of white carbon black on SBS-modified asphalt and asphalt mixture performance. *J. Shandong Univ. Technol. Nat. Sci. Ed.* **2019**, *33*, 1–4. [CrossRef]
16. Fa, C. *Study on the Aging Resistance of Graphene Oxide/SBS Composite and Its Modified Asphalt*; Chang'an University: Xi'an, China, 2021. [CrossRef]
17. Lu, Y.; Yu, L.; Yang, J.; Zeng, W.; Lu, J. Preparation and performance of Hydrotalcite/waste rubber powder composite modified asphalt. *Pet. Refin. Chem. Ind.* **2020**, *51*, 81–86.
18. Sui, N.; Lang, J. Effect of nanocomposites on thermal and UV aging properties of SBS-modified asphalt. *J. Wuhan Univ. Technol. Traffic Sci. Eng. Ed.* **2022**, *46*, 319–322, 329.
19. Liao, M.; Liu, Z.; Gao, Y.; Liu, L.; Xiang, S. Study on UV aging resistance of nano-TiO_2/montmorillonite/styrene-butadiene rubber composite modified asphalt based on rheological and microscopic properties. *Constr. Build. Mater.* **2021**, *301*, 124108. [CrossRef]
20. Ameli, A.; Babagoli, R.; Khabooshani, M.; AliAsgari, R.; Jalali, F. Permanent deformation performance of binders and stone mastic asphalt mixtures modified by SBS/montmorillonite nanocomposite. *Constr. Build. Mater.* **2020**, *239*, 117700. [CrossRef]
21. Yang, X.; Shen, A.; Jiang, Y.; Meng, Y.; Wu, H. Properties and mechanism of flame retardance and smoke suppression in asphalt binder containing organic montmorillonite. *Constr. Build. Mater.* **2021**, *302*, 124148. [CrossRef]
22. Ye, F.; Yin, W.; Lu, H.; Dong, Y. Property improvement of Nano-Montmorillonite/SBS modified asphalt binder by naphthenic oil. *Constr. Build. Mater.* **2020**, *243*, 118200. [CrossRef]
23. *JTG F40-2004*; Technical Specification for Construction of Highway Asphalt Pavement. People's Communications Press: Beijing, China, 2004.
24. *JTG E20-2011*; Test Code for Asphalt and Asphalt Mixture of Highway Engineering. People's Communications Press: Beijing, China, 2011.
25. Ren, Z.; Li, Y. Study on the influence of montmorillonite on the aging resistance of asphalt. *Highway* **2022**, *67*, 24–31.
26. Wang, X.; Zhang, J.; Xiao, W. Preparation and performance of sodium montmorillonite modified asphalt. *Highway* **2012**, *10*, 130–133.
27. Zhu, S.; Liu, D.; Xue, Y. Study on low temperature cracking resistance of montmorillonite modified asphalt mixture. *Highway* **2007**, *9*, 150–152.
28. Yang, G.; Wang, X.; Shi, J. Evaluation of fatigue resistance of hard asphalt mixtures based on dissipative energy method. *J. Build. Mater.* **2018**, *21*, 913–919, 925.
29. Meng, Y.; Zhang, X. Fatigue performance of modified asphalt based on cumulative dissipated energy ratio. *J. South China Univ. Technol. Nat. Sci. Ed.* **2012**, *40*, 99–103.
30. Zhang, H.; Xu, G.; Chen, X.; Wang, R.; Zhou, W. Study on fatigue performance of aging asphalt with different test methods. *J. Build. Mater.* **2020**, *23*, 168–175.

Disclaimer/Publisher's Note: The statements, opinions and data contained in all publications are solely those of the individual author(s) and contributor(s) and not of MDPI and/or the editor(s). MDPI and/or the editor(s) disclaim responsibility for any injury to people or property resulting from any ideas, methods, instructions or products referred to in the content.

coatings

Article

Performance and Verification of High-Modulus Asphalt Modified by Styrene-Butadiene-Styrene Block Copolymer (SBS) and Rock Asphalt

Yuxin Li [1], Xiangpeng Yan [2,*], Jianmin Guo [1], Wenjuan Wu [2], Wencheng Shi [1], Qinsheng Xu [2] and Zhengjun Ji [2,*]

1 Shandong Hi-Speed Company Limited, Jinan 250014, China
2 Shandong Transportation Institute, Jinan 250102, China
* Correspondence: yanxiangpeng@sdjtky.cn (X.Y.); ssacool@126.com (Z.J.)

Abstract: High asphalt grade and poor high-temperature performance are the primary reasons for the permanent rutting deformation of asphalt pavement. However, the low grade of asphalt and the poor low-temperature performance and fatigue life of the mixture can easily lead to the low-temperature cracking of asphalt pavement. With the rapid increase in road traffic, volume, and traffic load, the performance requirements of road asphalt materials are becoming higher and higher. High-modulus asphalt has excellent temperature stability and good fatigue resistance. However, high-modulus asphalt is expensive, so its use can greatly increase the pavement cost, restricting its wide application in road engineering. It is necessary to find an economical way to produce modified asphalt to meet the current road requirements. The aim of this study is to investigate the effects of styrene-butadiene-styrene block copolymer (SBS) and rock asphalt on the road performance of modified high-modulus asphalt, in which the replacement level of SBS and rock asphalt below 8 wt.% are compared. Apart from the conventional performance measurements, such as softening point, penetration, ductility and viscosity, thermal storage stability and rheological properties are also measured. The test results show that the composite modification of SBS and North American rock asphalt can effectively improve the high-temperature resistance and reduce the temperature sensitivity of 50# matrix asphalt, but it has no obvious improvement on its low-temperature performance. The preferred ternary blending system containing 4~6 wt.% SBS and 6~8 wt.% rock asphalt was obtained by performance analysis. It was verified that the performances of high-modulus asphalt mixture with the ternary blending asphalt above all meet the requirements of high-modulus asphalt mixture performance index.

Keywords: high-modulus asphalt; SBS; rock asphalt; high- and low-temperature performance; creep performance; performance verification

Citation: Li, Y.; Yan, X.; Guo, J.; Wu, W.; Shi, W.; Xu, Q.; Ji, Z. Performance and Verification of High-Modulus Asphalt Modified by Styrene-Butadiene-Styrene Block Copolymer (SBS) and Rock Asphalt. *Coatings* **2023**, *13*, 38. https://doi.org/10.3390/coatings13010038

Academic Editor: Valeria Vignali

Received: 12 August 2022
Revised: 21 October 2022
Accepted: 8 November 2022
Published: 26 December 2022

Copyright: © 2022 by the authors. Licensee MDPI, Basel, Switzerland. This article is an open access article distributed under the terms and conditions of the Creative Commons Attribution (CC BY) license (https://creativecommons.org/licenses/by/4.0/).

1. Introduction

The rapid growth of highway traffic volume, the continuous increase in super-heavy loads, and the complex and bad traffic environment lead to higher requirements for asphalt concrete pavement. According to statistics, more than 80% of the maintenance and repair of asphalt pavement is caused by the permanent deformation of pavement rutting [1]. Compared with other pavement performance issues, the rutting disease of asphalt pavement is the most harmful [2]. In order to improve the rutting resistance of asphalt concrete pavement, in the selection of raw materials, in addition to selecting the ore aggregate with hard stone, wear resistance, no impurities, no weathering and good particle shape, we should also select a high-performance asphalt with a high softening point, high consistency and good temperature stability, which can maintain sufficient viscosity under high-temperature conditions and ensure that the mixture has good low-temperature cracking resistance in a low-temperature environment.

In recent years, the application research direction of improving the high-temperature performance of asphalt mixture is mainly reflected in the selection of asphalt, such as styrene-butadiene-styrene block copolymer (SBS) modified asphalt, natural rock asphalt-modified asphalt [3], low-grade hard asphalt [4–6], plus anti-rutting agent or high-modulus agent [7–11]. However, in engineering application, it is found that a single-modified asphalt, such as SBS-modified asphalt, cannot meet the increasing high-temperature performance requirements [12]. The asphalt modified with pure natural rock asphalt significantly improves the high-temperature performance, but reduces the low-temperature performance to almost the same extent [13]. Low-grade hard asphalt improves the high-temperature performance of the mixture, but also significantly reduces the low-temperature performance, and the low-temperature cracking of the pavement is serious [14,15]. The cost of adding high-modulus agent or anti-rutting agent is high, which increases the difficulty of mixing and the probability of inaccurate measurement. Therefore, considering both performance and cost factors, natural rock asphalt and SBS modifier were used to modify the matrix asphalt, so as to obtain a composite-modified asphalt with low price and high road performance, and apply it to the construction of expressway.

Rock asphalt is a kind of natural asphalt. It has the advantages of high nitrogen content, good durability and good compatibility with base asphalt. The preparation process and application process are simple. It is commonly used in the modification of asphalt and asphalt mixture. It can significantly improve the high-temperature stability and shear strength of asphalt pavement, reduce rutting deformation and prolong the service life of pavement. SBS polymer modifier can significantly improve the temperature sensitivity, stability, durability, adhesion and aging resistance of asphalt. It has been found [16–20] that the composite modification of asphalt with rock asphalt and SBS modifier can significantly improve the viscosity and softening point of asphalt binder, and improve the high-temperature stability and water stability of asphalt mixture. For example, Bulgis et al. [21] studied the influence of Buton rock asphalt particles on the compressive stress-strain behavior of asphalt mixture. The results showed that the compressive strength of the mixture mixed with butun rock asphalt is higher than that of the mixture without modifier, and the number of cracks is fewer. Suaryana [22] studied the performance of Buton rock asphalt in Stone Mastic Asphalt (SMA) pavement and the results showed that the addition of Buton rock asphalt can improve the road performance of SMA mixture. Li et al. [23] used dynamic shear rheometer (DSR) and bending beam rheometer (BBR) tests to study the rheological properties of asphalt modified by rock asphalt and found that the addition of rock asphalt can increase the strength of asphalt materials and reduce the low-temperature relaxation potential of asphalt mixture. Compared with the performance before and after long-term aging, rock asphalt can slow down the aging rate of asphalt mixture. Huang et al. [24] have systematically studied the high-temperature performance, water stability and fatigue resistance of rock asphalt-modified asphalt mixture by using sk70#, Zhonghai 70# asphalt as matrix asphalt and Qingchuan rock asphalt, North American rock asphalt and star rock asphalt as modifiers. The research shows that the high-temperature resistance, water stability and fatigue resistance of rock asphalt-modified asphalt mixtures have been improved to varying degrees. Li et al. [25] have compared and analyzed the road performance of matrix asphalt mixture, BRA rock asphalt mixture and SBS-modified asphalt mixture. The research shows that BRA rock asphalt can significantly improve the deformation resistance, water loss resistance, low-temperature crack resistance and fatigue resistance of asphalt mixtures. According to the above research and the application of rock asphalt, so far, compared with Qingchuan rock asphalt and Buton rock asphalt, the composite modification technology of North American rock asphalt and SBS has not been reported and has not been widely used. The research on North American rock asphalt is mostly in the indoor test stage. Therefore, according to the current situation, this paper studies and evaluates the road performance of rock asphalt and SBS composite modified asphalt. It provides a theoretical basis for the large-scale application of North American rock asphalt in highway construction.

2. Materials and Experimental

2.1. Raw Materials

The 50 mesh North American rock asphalt powder and SBS modifier are used to modify 50# matrix asphalt for the preparation of high-modulus asphalt in this paper. The selected raw materials are tested according to Chinese standards JTG E20-2011 [26] and SH/T 1610-2011 [27]. The test results meet the basic road requirements of The Technical Specifications for Construction of Highway Asphalt Pavements (JTG F40-2004) [28] as shown in Tables 1–3.

Table 1. The basic index of 50# matrix asphalt.

Test Project	Test Values	Threshold Values
Softening point/°C	49.6	45~60
Penetration/mm	45.3	40~60
Ductility (10 °C)/cm	15.2	≥15
135 Brookfield viscosity/Pa.s	0.5	<3.0
RTFOT ductility (15 °C)/cm	10.1	≥10

Table 2. The basic index of SBS modifier.

Test Project	Structure	Volatile/%	Styrene Content/%	Molecular Weight/10,000	Melt Mass Flow Rate (MFR)/(g/10 min)	Tensile Strength/MPa	300% Constant Elongation Stress/MPa	Elong-ation at Break/%	Yellow Index	Else
Test values	Linear type	0.4	40	15	0.09	21.4	3.1	736	0.6	Unsaturated

Table 3. The basic index of North American rock asphalt powder.

Test Project	Color	Density/g.cm^{-3}	Softening Temperature/°C	Asphalt Content (Combustion Method)/%	Water Content/%	Ash Content/%
Test values	Black-brown	1.18	175	86.6	0.8	8

2.2. Preparation of High-Modulus Asphalt

The preparation of high-modulus modified asphalt is mainly divided into two steps. The first step is the preparation of SBS-modified asphalt. 2%, 4% and 6% of SBS modifier are added to 1000 g of 50# matrix asphalt by external mixing methods to complete the preliminary modification of base asphalt. The second step is the preparation of composite modified high-modulus asphalt. 4%, 6% and 8% dosages of 50-mesh North American rock asphalt powder were added to the SBS-modified asphalt by external admixture methods for compound modification to complete the preparation of high-modulus asphalt. The modifier blending scheme was given in Table 4. Here, the reference group MA0 denotes the 50# matrix asphalt. S02, S04 and S06 denote the SBS-modified asphalt using 2%, 4% and 6% of SBS modifier, respectively. S2R4, S2R6 and S2R8 denote the composite-modified high-modulus asphalt using 4%, 6% and 8% of 50-mesh North American rock asphalt on the basis of 2% SBS-modified asphalt. S4R4, S4R6 and S4R8 denote the composite-modified high-modulus asphalt using 4%, 6% and 8% of 50-mesh North American rock asphalt on the basis of 4% SBS-modified asphalt. S6R4, S6R6 and S6R8 denote the composite-modified high-modulus asphalt using 4%, 6% and 8% of 50-mesh North American rock asphalt on the basis of 6% SBS-modified asphalt.

Table 4. The modifier blending scheme.

Sample	SBS Modifier/%	Rock Asphalt Powder/%
MA0	0	0
S02	2	0
S04	4	0
S06	6	0
S2R4	2	4
S2R6	2	6
S2R8	2	8
S4R4	4	4
S4R6	4	6
S4R8	4	8
S6R4	6	4
S6R6	6	6
S6R8	6	8

2.3. Mixture Design

EME-20 is selected for asphalt mixture gradation. According to the aggregate screening results, the composite gradation of the mixture is designed as shown in Table 5, and the grading curve is shown in Figure 1. The optimum asphalt aggregate ratio of the designed mixture is 4.8%.

Table 5. Synthetic Gradation Design of High-Modulus Asphalt Mixture.

Seive Size (mm)	31.5	26.5	19	16	13.2	9.5	4.75	2.36	1.18	0.6	0.3	0.15	0.075
Upper limit	100	100	100	-	-	82.0	64.0	43.0	-	-	-	-	8.0
Lower limit	100	100	90.0	-	-	66.0	41.0	28.0	-	-	-	-	6.0
Design median	100	100	95.0	-	-	74.0	52.5	35.5	-	-	-	-	7.0
Synthetic grading	100	100	93.8	88.5	82.1	71.4	46.6	32.4	26.2	17.6	13.0	8.4	6.3

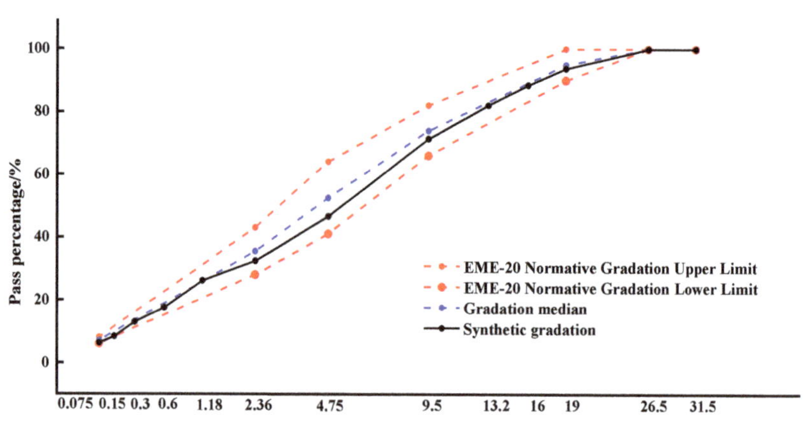

Figure 1. Synthetic gradation curves of high-modulus asphalt mixtures.

2.4. Test Methods

2.4.1. Softening Point, Penetration, Ductility and Viscosity Tests

Softening point refers to the temperature measured when the asphalt specimen is softened and sagged by heat, which represents the temperature stability of asphalt. The test was carried out according to the test specification (JTG E20-2011) [26]. The asphalt sample in the metal ring of prescribed size was placed in 5 °C water or 32.5 °C glycerol, and heated at a rate of 5 ± 0.5 °C/min until the steel ball sank to the prescribed distance (25.4 mm). The temperature was recorded and expressed in °C.

Penetration is one of the main quality control indexes of asphalt. The relative hardness and consistency of asphalt and its ability to resist shear failure can be characterized. The test was carried out according to the test specification (JTG E20-2011) [26]. A 100 g standard cone was sunk vertically into a bitumen sample insulated at 25 °C for 5 s, and the depth was recorded in 1/10 mm units.

Ductility is an important index to evaluate the plasticity of asphalt. The test was carried out according to the test specification (JTG E20-2011) [26]. The asphalt was poured into a standard specimen, and the specimen was drawn to fracture at a speed of 50 mm/min under the corresponding test temperature (5 °C, 10 °C, 15 °C or 25 °C). The length at this time was recorded and expressed as cm.

Viscosity is an index to characterize the viscosity of asphalt, the main basis for the classification of modern asphalt, and a main control index for the production and construction of asphalt mixture. There are many viscosity indicators, and 135 °C kinematic viscosity is selected to study in this paper. The kinematic viscosity at 135 °C is often measured by a capillary viscosity meter according to the test specification (JTG E20-2011) [26].

2.4.2. Segregation Softening Point Test

Segregation softening point test is usually used to evaluate the thermal storage stability of SBS-modified asphalt according to Chinese standard JTG E20-2011 [26]. The separation tube containing about 50 g of asphalt was sealed and put in an oven at 163 ± 5 °C for 48 ± 1 h, and then was moved to the freezer for at least 4 h. After the asphalt is solidified, it is taken out and divided into three equal parts, and the upper and lower softening points are measured according to the softening point test method (T0606), and the difference is taken. When the value is less than 2.5 °C, it is considered that the SBS-modified asphalt segregation is qualified.

2.4.3. Dynamic Shear Rheometer (DSR)

The dynamic shear rheometer (DSR) is usually used to evaluate the high-temperature stability of asphalt binder. According to Chinese standard JTG E20-2011 [26], a fully automatic dynamic shear rheometer was used to determine the dynamic shear modulus and phase angle of asphalt. The instrument applies periodic sinusoidal shear stress to asphalt samples through the axis of rotation. The asphalt sample is placed between two 25 mm plates, one plate is fixed, the other plate is rotated around the central axis at a speed of 10 rad/s. It goes through the A-B-A-C-A cycle. When the stress (strain) is applied to the sample, DSR can calculate the strain (stress) response generated by the sample, and the complex shear modulus can be calculated from the stress and the corresponding strain.

2.4.4. Bending Beam Rheometer (BBR)

Bending Beam Rheometer (BBR) is usually used to evaluate the low-temperature performance of asphalt materials. According to Chinese standard JTG E20-2011 [26], a fully automatic bending beam rheological tester was used to measure the flexural creep stiffness modulus S and creep rate m. The bending beam rheometer applies loads to asphalt samples through load sensors and records deflection changes over time. The instrument can automatically collect the creep stiffness modulus S of sample at 8 s, 15 s, 30 s, 60 s, 120 s and 240 s, and calculate the value of creep rate m when the test is carried out at the test temperature. The asphalt should be kept warm in absolute ethanol at a constant temperature before the start of the test. Taking into account the low-temperature climate in cold regions, the test temperature is selected as −12 °C and −18 °C.The S and m values with the load acting time of 60 s are taken in the test.

2.4.5. Pneumatic Rheological Rebound Test

The maximum creep deformation when loaded and the ability to recover from deformation when unloaded are unique properties of each thermoplastic material. The LTI-210 asphalt quality rapid testing equipment can quickly evaluate the mechanical response and road performance of asphalt materials at a certain temperature. It mainly uses nitrogen-loading technology to load asphalt specimens at 25 °C in a circular area for a period of time. The laser measurement system of the equipment measures and records the deformation (displacement) of the loading center. After the loading is finished, the deformation of asphalt materials begins to recover. The creep and creep recovery ability of asphalt under single stress or multiple stress conditions can be measured.

3. Results and Discussions

3.1. Penetration, Softening Point, Ductility and Viscosity Index

The basic performance indicators of modified asphalt with different contents are analyzed. Figure 2 exhibits the effect of SBS and rock asphalt on the softening point, penetration, ductility, and viscosity of high-modulus asphalt. The additions of 2%, 4% and 6% SBS modifier cause obvious increases in softening point of approximately 9.07%~34.5% higher than that of MA0, and achieve penetrations 13.7%~25.9% lower than that of MA0. It indicates that the high-temperature performance is significantly improved by the addition of SBS modifier. As for ductility and viscosity, continuous increases occur with the increase in SBS modifier content. However, when the SBS content reaches 6%, the viscosity index exceeds the requirement of index ≤3 Pa·s in technical code for construction of asphalt pavement JTG F40-2004 [28]. If the viscosity of modified asphalt is too large, it may increase the difficulty of the process of transportation and mixing, which is not conducive to the transportation and construction of asphalt mixture.

The addition of rock asphalt improves the softening point and viscosity and decreases the penetration and ductility. In comparing the results of the binary blend containing SBS modifier, the ternary blends containing both SBS and rock asphalt show better performance. Considering S2R4 as an example, increases of approximately 10% and 22.2% in softening point and viscosity and decreases of 4.9% and 14.2% in penetration and ductility are achieved compared to those of S02. It indicates that the addition of rock asphalt further improves the high- and low-temperature performance of the composite-modified asphalt. Moreover, the higher the dosage of rock asphalt, the better the improvement effect.

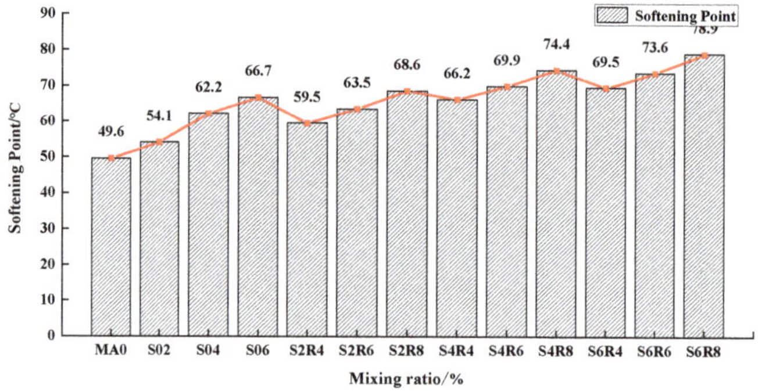

(**a**) Softening point

Figure 2. *Cont.*

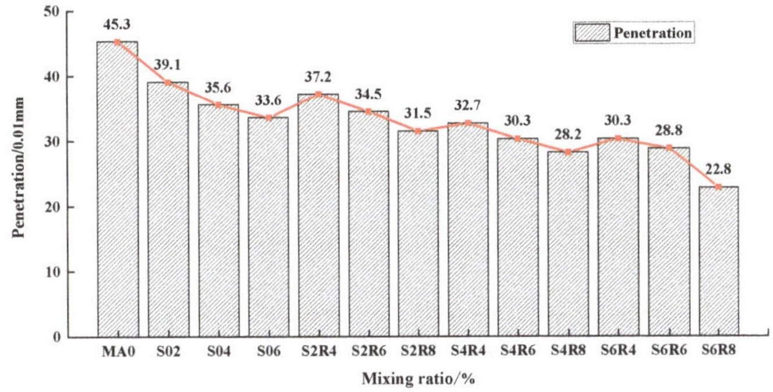

(**b**) Penetration

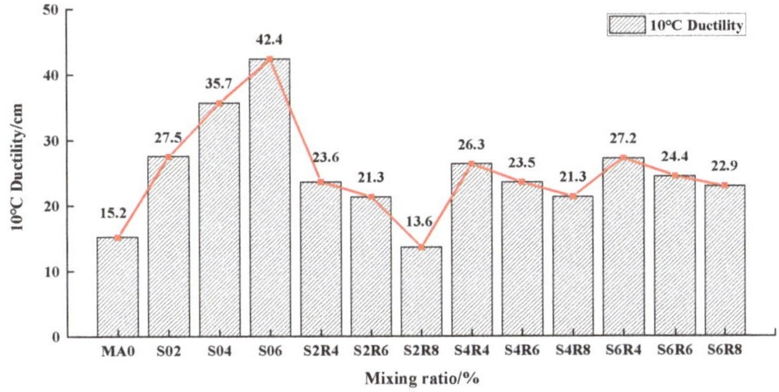

(**c**) Ductility

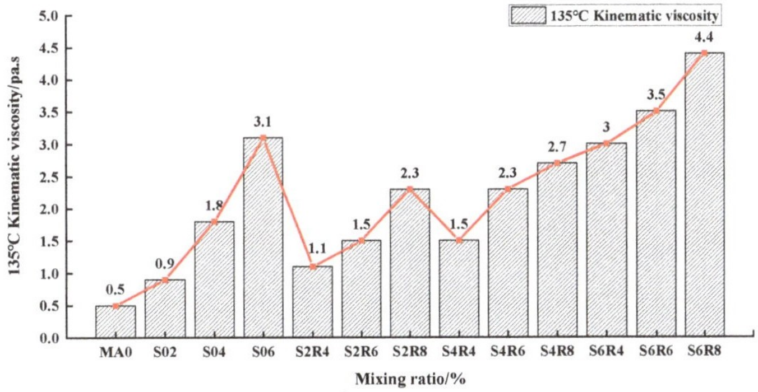

(**d**) Viscosity

Figure 2. The change law of soft point, penetration, ductility and viscosity for modified asphalt.

3.2. Thermal Storage Stability of Asphalt

Thermal storage stability is an important indicator to ensure that modified asphalt does not segregate or lose its excellent performance during storage and transportation. The thermal stability of the prepared composite-modified asphalt was analyzed as shown in Figure 3. Results show that the softening point difference of 50# matrix asphalt between top and bottom is 0 °C, and there is no segregation. The inclusion of SBS increases the difference of the softening point; the higher the content of SBS, the greater the difference in softening point. When the SBS modifier content is 4% and 6%, the softening point difference exceeds the specified value ≤2.5 °C index in the construction technical specification JTG F40-2004 [28], resulting in segregation.

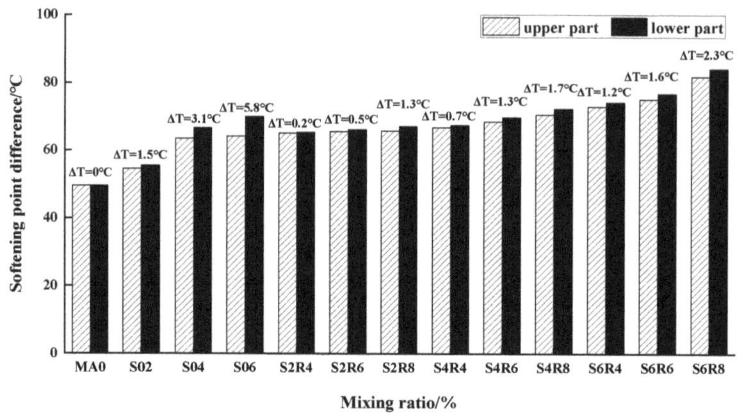

Figure 3. The difference in change of softening point for modified asphalt.

However, when rock asphalt is used, the asphalt segregation condition is improved and the softening point difference is gradually reduced compared to the SBS series. Taking the rock asphalt content of 8 wt.% as an example, the S- and R-containing ternary composite-modified asphalt achieve differences of softening point 13.3%~60.3% lower than those of the SBS series, which indicates that the ternary blending system containing SBS and rock asphalt behaves well in improving the thermal storage stability, better than the effect of SBS modifier alone. When the content of SBS modifier is fixed, the softening point difference of composite-modified asphalt increases with the increase in rock asphalt content.

3.3. *Analysis on Rheological Properties of Asphalt*

3.3.1. High-Temperature Rheological Properties

The research shows that the dynamic shear rheometer (DSR) can be used to characterize the viscoelastic characteristics of asphalt under medium- and high-temperature conditions [29], and the rutting factor G*/sinδ was taken as an index for evaluating and controlling the high-temperature rutting resistance of asphalt materials. The research shows that the dynamic shear rheometer (DSR) can be used to evaluate the viscoelastic characteristics of asphalt under medium- and high- temperature conditions [29], and the rutting factor G*/sinδ is taken as an index for evaluating and controlling the high-temperature rutting resistance of asphalt materials. The effects of SBS and rock asphalt on the high-temperature rheological properties of modified asphalt are exhibited in Figure 4. It can be seen in Figure 4 that the rutting factor G*/sinδ of all series of modified asphalt decreases with the increase in temperature. The effects of SBS and rock asphalt on the rutting factor of modified asphalt are exhibited in Figure 4. The effects of SBS and rock asphalt on the high-temperature rheological properties of all series of modified asphalt decreases are exhibited in Figure 4. The rutting factor G*/sinδ rheological properties of all series of modified asphalt are exhibited in Figure 4. It can be seen in Figure 4 that the rutting factors G*/sinδ of all series of modified asphalt decrease with the increase in temperature. The smaller the asphalt rutting factor, the greater the loss of shear flexibility of asphalt [17], the fewer elastic components contained in asphalt, and the worse the rutting resistance of asphalt. With the incorporation of SBS modifier, the rutting resistance is improved, with larger rutting factor values than that of MA0. The combination of SBS and rock asphalt produces better effects than that of SBS alone in the rutting resistance improvement. When 4%~6% of SBS is added, the high-temperature grade of modified asphalt can reach 75~85 °C, which is about 10~20 °C higher than that of MA0. There is no obvious change in the high-temperature grade of ternary blending modified asphalt containing SBS and rock asphalt, but significant increases by approximately 1~3 times are achieved in rutting factor values compared to those of asphalt containing SBS alone.

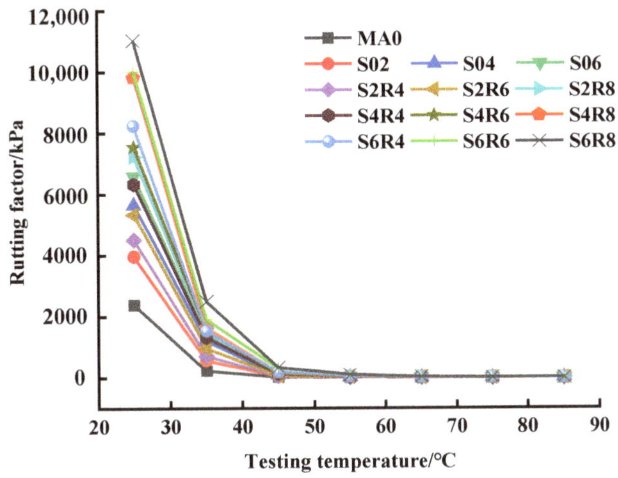

Figure 4. The change of rutting factor with temperature.

3.3.2. Low-Temperature Rheological Properties

The bending creep stiffness modulus S and creep rate m are usually used to evaluate the low-temperature crack resistance of asphalt. The larger the stiffness modulus S value of asphalt material, the more obvious the brittleness of asphalt, the easier the pavement is to crack and damage. The smaller the m value of the asphalt creep rate, the greater the stiffness of the material when the temperature decreases, the greater the tensile force in the material, and the greater the possibility of low-temperature cracking.

The effects of SBS and rock asphalt on the low-temperature rheological properties of modified asphalt are exhibited in Figures 5 and 6. It can be seen that with the inclusion of SBS, the stiffness modulus of modified asphalt gradually decreases and the creep rate continuously increases at −12 °C and −18 °C. When SBS is added in 50# matrix asphalt, decreases of 8.2%~27.6% and 5.0%~13.9% in stiffness modulus at −12 °C and −18 °C, respectively, are achieved, and increases of 0.9%~15.9% and 9.58%~32.5% in creep rate at −12 °C and −18 °C, respectively, are obtained. This indicates that the addition of SBS modifier reduces the tensile force of 50# matrix asphalt at low-temperature conditions and effectively improves the low-temperature cracking of asphalt. However, for the ternary blending asphalt system containing SBS and rock asphalt, the values of stiffness modulus S increase obviously and the values of creep rate m decrease compared to the binary blend containing SBS alone. The addition of rock asphalt increases the brittleness of ternary blending composite-modified asphalt, reduces the relaxation stress, and increases the possibility of low-temperature cracking of asphalt.

According to the Superpave specification of SHRP [30], the creep stiffness modulus S should be less than 300 MPa, and the creep rate m should be greater than 0.300. It can be seen in Figure 5 that at −12 °C, the S of MA0 is less than 300 MPa and the m is less than 0.300. When the SBS is added, the S and m of SBS- all meet the requirements at −12 °C. When the dosage of SBS reaches 4% and 6%, the S and m of SBS meet the requirements at −18 °C. when rock asphalt is added to the modified asphalt by SBS, the S and m of all the ternary blending series do not meet the requirements at −18 °C. At −12 °C, the S and m of the ternary blending system containing 4~6 wt.% SBS and 6~8 wt.% rock asphalt can meet the requirements. This shows that although the addition of rock asphalt weakens the low-temperature cracking resistance of composite modified asphalt, it still has advantages compared to 50# matrix asphalt.

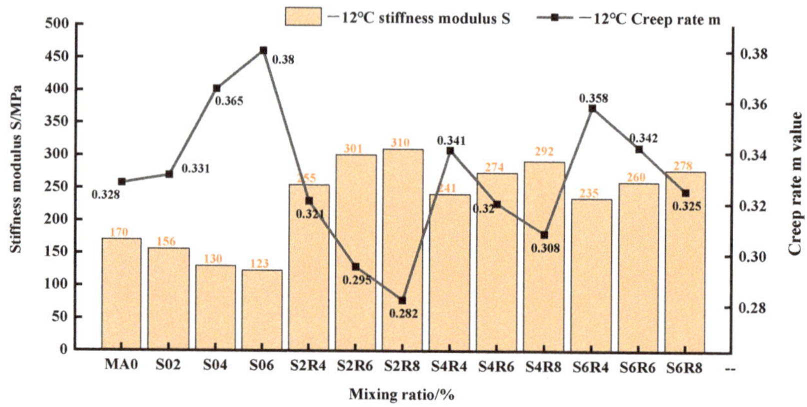

Figure 5. The results of stiffness modulus and creep rate for modified asphalts under −12 °C.

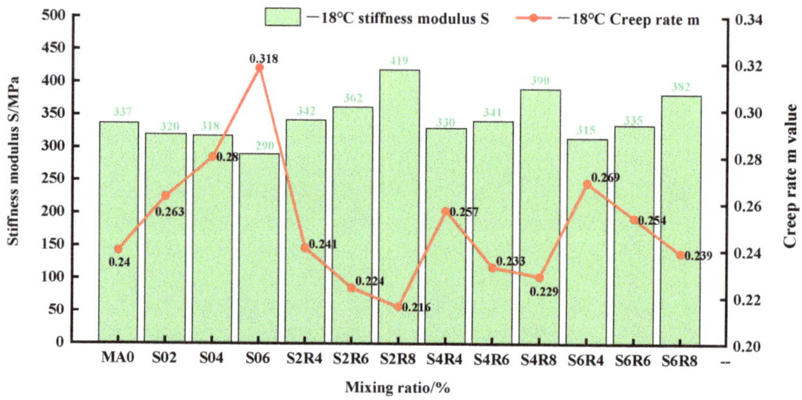

Figure 6. The results of stiffness modulus and creep rate for modified asphalts under −18 °C.

3.3.3. Pneumatic Rheological Rebound Performance

The pneumatic asphalt rheological rebound test is carried out using the detection equipment of lti-210 asphalt quality control system to quickly measure the creep and creep recovery capacity of asphalt under the same temperature (25 °C) and stress (9.75 Pa·S). The PG classification, maximum creep deformation and elastic recovery rate of asphalt binder are exhibited in Table 4, Figures 6 and 7. It can be seen from the results that 43.6%~57.9% decreases in the maximum creep deformation and 55.5%~198.5% increases in the creep recovery rate are achieved when 2%~6% SBS modifier is added to 50# matrix asphalt. This shows that the stiffness and modulus of the SBS-modified asphalt are larger, the asphalt is harder, and the viscoelastic performance of the asphalt is better compared to those of MA0.

As shown in Table 4 and Figure 6, the lowest maximum creep deformation and the highest deformation recovery rate of modifier asphalt are obtained with SBS and rock asphalt additions. The addition of SBS at 6% and rock asphalt at 8% leads to a reduction of 87.5% in maximum deformation and an increase of 32.5% in deformation recovery rate, respectively, compared to those of S06. As more rock asphalt is added, the maximum creep deformation is decreased and the deformation recovery rate is increased. The incorporation of rock asphalt can sufficiently improve the pneumatic rheological rebound performance of modified asphalt containing SBS.

In comparing the results of PG grading analysis in Table 6 by S04 and S4R6, Rock asphalt has a better effect in improving the high- and low-temperature grade of modified asphalt than SBS does. The high- and low-temperature grades of S04 are 76 °C and −28 °C, respectively, 1 level higher than those of MA0. In addition, with the incorporation of rock asphalt at 6%, the high-temperature classification of asphalt is improved by 2 grades, while the low-temperature grade has no change compared to those of MA0. It is concluded that the presence of rock asphalt improves the high-temperature stability and viscoelastic properties of composite-modified asphalt, but has no obvious improvement on its low-temperature performance. Although the low-temperature performance is attenuated, it can still maintain the original level with addition of rock asphalt.

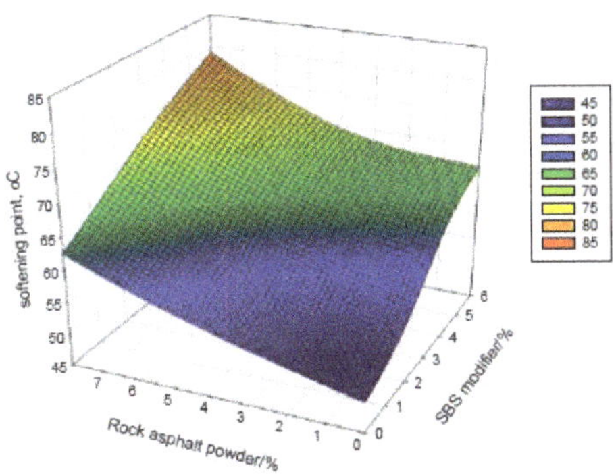

(**a**) Softening point

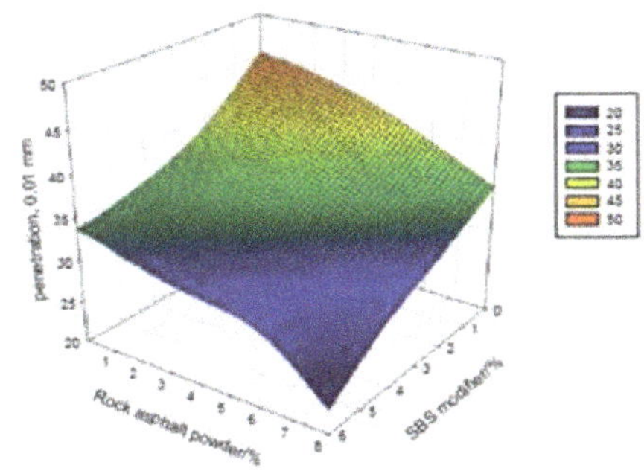

(**b**) Penetration

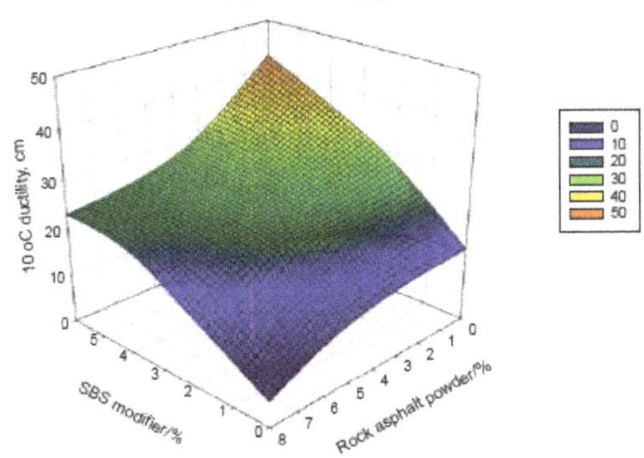

(**c**) Ductility

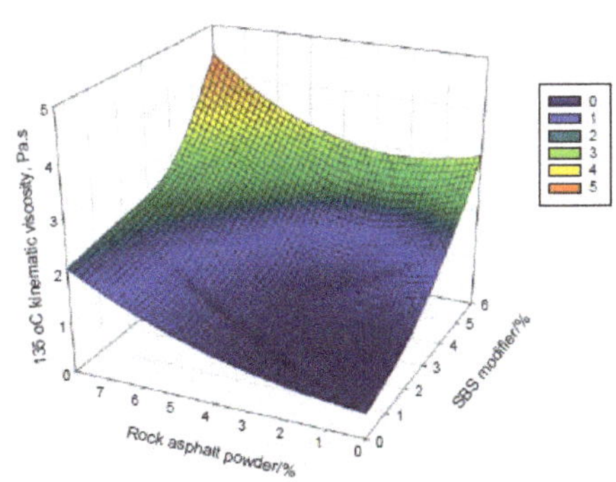

(**d**) Viscosity

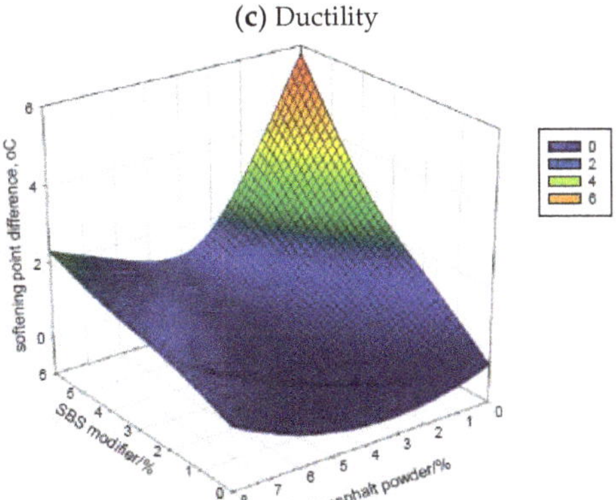

(**e**) Softening point difference

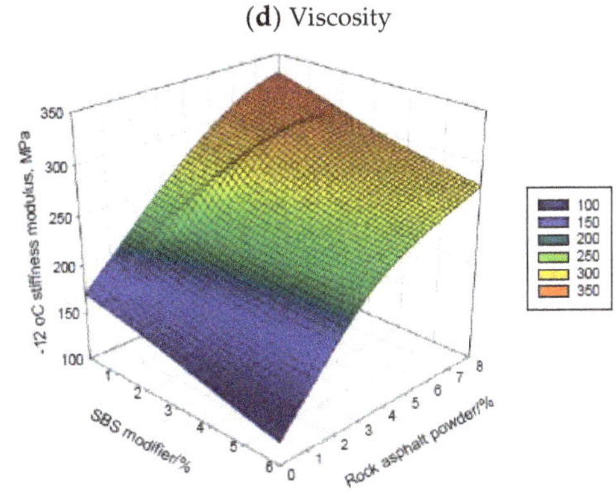

(**f**) −12 °C stiffness modulus

Figure 7. *Cont.*

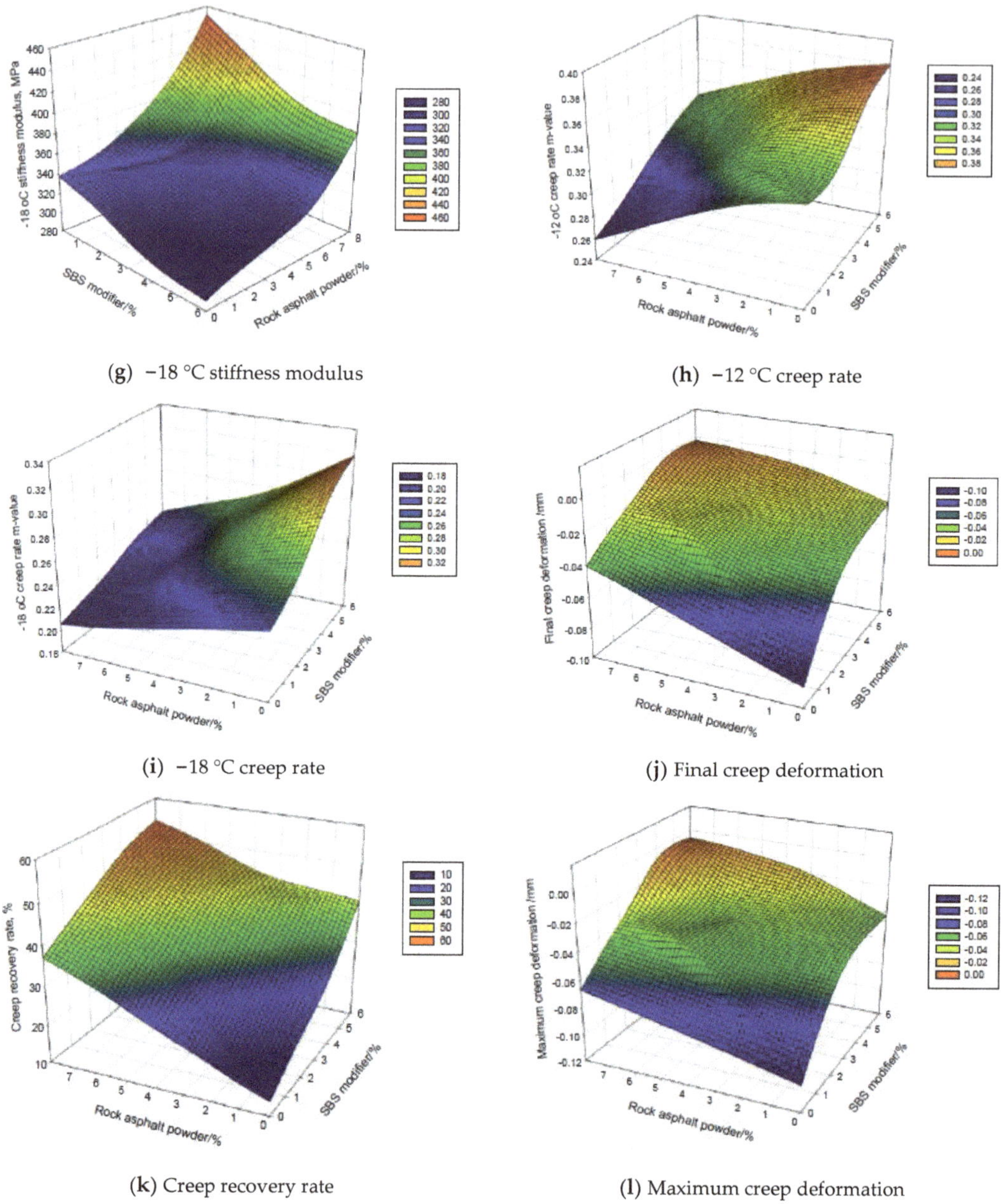

(**g**) −18 °C stiffness modulus (**h**) −12 °C creep rate

(**i**) −18 °C creep rate (**j**) Final creep deformation

(**k**) Creep recovery rate (**l**) Maximum creep deformation

Figure 7. The multivariate analysis results.

3.3.4. Multivariate Analysis

In order to better analyze the effect of the content of rock asphalt and SBS on high-modulus modified asphalt, multivariate analysis was carried out. The analysis results are shown in Figure 7. It gives SBS modifier and Rock asphalt powder component effects. It can be seen that the effect of composite modification of SBS and North American rock asphalt is better than SBS or rock asphalt alone in the performance improvement on the matrix asphalt. According to the performance characteristic of the modified asphalt by multivariate analysis, the ternary blending system containing 4~6 wt.% SBS and 6~8 wt.%

rock asphalt is preferred. When the content of SBS modifier is 4%~6% and the content of rock asphalt is not less than 6%, the performance indicators of modified asphalt meet the requirements of DB 37/T 3564-2019 and EN14023 specifications for high-modulus asphalt.

Table 6. The creep deformation and recovery rate of different asphalt samples.

Samples	Final Creep Deformation/mm	Creep Recovery Rate/%	Maximum Creep Deformation/mm	PG
MA0	−0.0868	13.7	−0.1006	PG 70-22
S02	−0.0446	21.3	−0.0567	PG 70-22
S04	−0.0304	28.3	−0.0424	PG 76-28
S06	−0.0252	40.9	−0.0426	PG 76-28
S2R4	−0.0319	32.4	−0.0472	PG 76-22
S2R6	−0.0237	39.8	−0.0394	PG 76-16
S2R8	−0.0216	44.2	−0.0387	PG 76-16
S4R4	−0.0221	41.9	−0.0380	PG 76-22
S4R6	−0.0193	46.6	−0.0361	PG 82-22
S4R8	−0.0043	50.5	−0.0087	PG 82-22
S6R4	−0.0087	44.8	−0.0158	PG 82-22
S6R6	−0.0049	49.9	−0.0098	PG 82-22
S6R8	−0.0027	54.2	−0.0059	PG 82-22

3.4. *Performance Verification of Composite Modified Asphalt Mixture*

To further verify the performance of the preferred modified high-modulus asphalt, the performances of the high-modulus asphalt mixture with asphalt S4R6, S4R8, S6R6 and S6R8 are evaluated. According to JTG E20-2011 [26] and GB/T 36143-2018 [31], the high-temperature, low-temperature, water stability, modulus and fatigue performance indicators of composite-modified asphalt mixture are verified. The results indicate that the performance indicators of the mixture meet the requirements of GB/T 36143-2018 [31]. The test results are shown in Table 7.

Table 7. Test results of compound-modified asphalt mixtures.

Technical Index	Requirements	Test Results				
		MA0	S4R6	S4R8	S6R6	S6R8
Void ratio/%	≤4	3.46	3.52	3.50	3.55	3.45
Freeze-thaw splitting tensile strength/%	≥80	79.8	86.5	88.5	88.4	87.6
60 °C dynamic stability times/mm	≥4000	3150	4117	4345	4200	4375
Dynamic modulus (45 ± 0.5 °C, 10 ± 0.1 Hz)/MPa	≥4000	3250	4010	4080	4040	4150
Fatigue life (15 ± 0.5 °C, 10 ± 0.1 Hz, 230 με)/cycles	$\geq 10^6$	865,245	2,424,220	2,362,530	2,260,360	2,116,500
Low-temperature bending test failure strain (−10 ± 0.5 °C)/με	≥2000	1864.56	2510.60	2493.98	2385.16	2149.32

4. Conclusions

Based on the limited testing results, the following conclusion can be drawn.

(1) The composite modification of SBS and North American rock asphalt can effectively improve the high-temperature resistance and reduce the temperature sensitivity of 50# matrix asphalt, but it has no obvious improvement on the low-temperature performance. It is found that the reasonable content of SBS and rock asphalt can better reflect the thermal storage stability of composite-modified asphalt. The preferred ternary blending system containing 4~6 wt.% SBS and 6~8 wt.% rock asphalt is obtained by performance analysis.

(2) When the content of SBS modifier is 4%~6% and the content of rock asphalt is not less than 6%, the performance indicators of modified asphalt meet the requirements of DB

37/T 3564-2019 and EN14023 specifications for high-modulus asphalt. The test results meet the requirements of GB/T 36143-2018 specification for high-modulus asphalt mixtures.

(3) This study shows that the 50# matrix asphalt can be enhanced in its traditional performance, thermal storage stability and rheological properties using common modifiers. Moreover, the ternary blending system containing SBS and rock asphalt can achieve good performance, indicating their potential to reduce construction costs in road engineering.

Author Contributions: Conceptualization, Y.L., X.Y. and W.S.; methodology, Y.L.; software, X.Y. and Z.J.; validation, Y.L. and X.Y.; formal analysis, X.Y.; investigation, J.G. and Z.J.; resources, Q.X.; data curation, X.Y., J.G. and Z.J.; writing—original draft preparation, W.W. and Q.X.; writing—review and editing, W.W.; visualization, Q.X.; supervision, Y.L.; project administration, W.S.; funding acquisition, X.Y. All authors have read and agreed to the published version of the manuscript.

Funding: This research was funded by the National Key R & D Program of China, grant No. 2018YFB1600103, Shandong Provincial Natural Science and Foundation, grant No. ZR2020QE271 and Shandong Provincial Key Research and Development Program, grant No. 2019GSF109020.

Institutional Review Board Statement: Not applicable.

Informed Consent Statement: Not applicable.

Data Availability Statement: Data sharing is not applicable to this article.

Conflicts of Interest: The authors declare no conflict of interest.

References

1. Shen, J.; Li, F.; Chen, J. *Early Damage and Preventive Measures of Highway Asphalt Pavement*; People's Communications Press: Beijing, China, 2004.
2. Wang, Z. *Research on New Preventive Maintenance Technology of Asphalt Pavement Based on Lateral Redistribution of Traffic Load*; Chang'an University: Xi'an, China, 2012.
3. Song, X.; Fan, L.; Shen, Q. Research on road performance of domestic natural rock asphalt modified asphalt mixture. *Highway* **2009**, *12*, 135–139.
4. Bai, M.; Jiang, H.; Ma, Q. Application of hard asphalt mixture in overhaul of national and provincial trunk highways. *Theor. Res. Urban Constr.* **2019**, *13*, 125–126. [CrossRef]
5. Yu, P. *Development and Application of Special Hard Asphalt for High Modulus Asphalt Mixtures*; Chang'an University: Xi'an, China, 2014.
6. Gu, S.; Sun, X.; Han, P. Application of low-grade hard asphalt particles in high-modulus asphalt mixture. *Pet. Asph.* **2014**, *28*, 42–45.
7. Shi, X.; Yang, J.; Chen, X. Study on high and low temperature performance of high modulus asphalt mixture. *China Foreign Highw.* **2013**, *33*, 309–312.
8. Huang, H. Influence of direct injection high modulus modifier dosage on road performance of asphalt mixture. *Transp. World.* **2021**, *79*, 4–6.
9. Guan, Z.; Liu, X.; Li, S.; Wu, C.; Xu, X.; Cheng, S. Study on high temperature properties of high modulus modified asphalt binder and mixture. *Guangdong Highw. Transp.* **2020**, *46*, 1–6.
10. Yang, P. *Research on the Characteristics of High Modulus Asphalt and Its Mixture*; South China University of Technology: Guangzhou, China, 2012.
11. Wei, O. *Research on Anti-Rutting Performance of High Modulus Asphalt Concrete*; Northeastern University: Shenyang, China, 2010.
12. Zhang, Q.; Fan, W.; Wang, T.; Nan, G. Research on the properties and microstructure of SBS modified AH-70 asphalt before and after emulsification. *J. China Univ. Pet.* **2011**, *35*, 146–151.
13. Liu, S. *Theoretical Research on Buton Rock Asphalt Modified Asphalt Mixture and Asphalt Mixture Proportion Design*; Tongji University: Shanghai, China, 2006.
14. Zou, G.; Li, W.; Xu, J. Research on low temperature rheological behavior of low-grade asphalt mortar. *Zhong Wai Highw.* **2014**, *34*, 210–214. [CrossRef]
15. Yu, S.J. Production and Application of No. 30 Hard Asphalt. *Pet. Asph.* **2013**, *27*, 45–49.
16. Du, S.W. Properties and mechanism of rock asphalt SBS composite modified asphalt mixture. *J. Build. Mater.* **2012**, *15*, 871–874.
17. Lu, M.; Jiang, Y.; Chen, J. Research on high-performance asphalt mixture modified by SBS + rock asphalt. *Urban Roads Bridges Flood Control* **2011**, *12*, 135–138+5. [CrossRef]
18. Wang, S. *Research on Road Performance of Qingchuanyan Asphalt/SBR Composite Modified Asphalt and Its Mixture*; Chongqing Jiaotong University: Chongqing, China, 2019.
19. Yang, Y. *Research on the Technical Performance of Qingchuan Rock Asphalt Modified Asphalt and Its Mixture*; Zhengzhou University: Zhengzhou, China, 2017.
20. Tao, Z. *Research on the Application Technology of North American Rock Asphalt/SBS Composite Modified Asphalt*; South China University of Technology: Guangzhou, China, 2015.

21. Bulgis; Tjaronge, M.W.; Adisasmita, S.A.; Hustim, M. Effect of Buton granular asphalt on compressive stress-strain behavior of asphalt emulsion mixture. *IOP Conf. Ser. Mater. Sci. Eng.* **2017**, *271*, 012069. [CrossRef]
22. Suaryana, N. Performance evaluation of stone matrix asphalt using Indonesian natural rock asphalt as stabilizer. *Int. J. Pavement Res. Technol.* **2016**, *9*, 387–392. [CrossRef]
23. Li, R.X.; Karki, P.; Hao, P.; Bhasin, A. Rheological and low temperature properties of asphalt composites containing rock asphalts. *Constr. Build. Mater.* **2015**, *96*, 47–54. [CrossRef]
24. Huang, G.; He, Z.; Zhang, Z.; Yang, Y. Research on the performance of asphalt modified mixture against rutting rock. *China Foreign Highw.* **2009**, *29*, 187–192.
25. Li, R.X.; Hao, P.W.; Wang, C. Performance Evaluation of BRA Modified Asphalt Based on Analysis of Rheological Property. *Adv. Mater. Res.* **2012**, *374–377*, 1385–1390.
26. *JTG E20-2011*; Highway Engineering Asphalt and Asphalt Mixture Test Regulations. People's Communications Press: Beijing, China, 2011.
27. *SH/T 1610-2011*; Thermoplastic Elastomer Styrene-Butadiene Block Copolymer (SBS). People's Communications Press: Beijing, China, 2013.
28. *JTG F40-2004*; Technical Specification for Highway Asphalt Pavement Construction. People's Communications Press: Beijing, China, 2004.
29. Cai, H.; Xu, J.; Liu, Y.; Mei, S. Study on the correlation between high temperature index of asphalt and its high temperature road performance. *New Build. Mater.* **2011**, *38*, 65–68.
30. Asphalt Institute. *Superpave Mix Design, Superpave Series No.2 (SP-2)*; Federal Highway Administration: Frankfort, KY, USA, 2003.
31. *GB/T 36143-2018*; Durable High Modulus Asphalt Mixture for Road Engineering. People's Communications Press: Beijing, China, 2013.

Disclaimer/Publisher's Note: The statements, opinions and data contained in all publications are solely those of the individual author(s) and contributor(s) and not of MDPI and/or the editor(s). MDPI and/or the editor(s) disclaim responsibility for any injury to people or property resulting from any ideas, methods, instructions or products referred to in the content.

coatings

Article

Mechanical Response Analysis of Asphalt Pavement Structure with Embedded Sensor

Pengcheng Wang [1], Guoqiang Zhong [2], Xue Xin [3,*], Fei Xiao [1], Ming Liang [3,*], Chao Wang [2], Yuepeng Jiao [3], Yanli Zhu [3], Shang Liu [2] and Hao Wang [4]

1 Shandong High-Speed Infrastructure Construction Co., Ltd., Jinan 250014, China
2 Shandong Provincial Communications Planning and Design Institute Group Co., Ltd., Jinan 250101, China
3 School of Qilu Transportation, Shandong University, Jinan 250002, China
4 Shandong Hi-Speed Jiwei Expressway Co., Ltd., Jinan 250200, China
* Correspondence: xinxue@mail.sdu.edu.cn (X.X.); ming.liang@sdu.edu.cn (M.L.); Tel.: +86-1866-162-8291 (M.L.)

Abstract: Long-term and real-time monitoring of asphalt pavement can be carried out by using embedded sensors to perceive and predict structural damage during pavement operation period, so as to avoid sustained development of damage. However, the influence of embedded sensors on the mechanical properties of asphalt pavement structure and the structural optimization of sensing elements needs to be further studied. Based on the finite element numerical simulation method, static load model and three-point bending test mode were conducted with three "pavement-sensor" coupling model without sensor, with embedded I-shape sensor, with embedded corrugated-shape sensor. Three simulated conditions were studied comparatively of the sensing element embedding effect on the mechanical response of asphalt pavement structure. Results show that the sensing elements embedded with the two structures have a certain influence on the stress and strain field of asphalt concrete. Within the range of 60–100 mm the asphalt mixture is in a state of tension; the stress values increase with depth and show a maximum tensile stress state at the bottom of the beam. In the compression zone, the strain of the I-shape sensing element embedded is closer to that of the strain without the sensing element embedded. Along the axis of the two sensing elements, the axial strain of the I-shape sensing element is smoother and uniform, which ensures the deformation coordination in the road state. The optimal length L of the sensing element is 14 cm, the diameter φ of the sensor is 10 mm, and the I-beam length GL is 10 cm.

Keywords: embedded; sensing elements; asphalt pavement; mechanical response

Citation: Wang, P.; Zhong, G.; Xin, X.; Xiao, F.; Liang, M.; Wang, C.; Jiao, Y.; Zhu, Y.; Liu, S.; Wang, H. Mechanical Response Analysis of Asphalt Pavement Structure with Embedded Sensor. *Coatings* **2022**, *12*, 1728. https://doi.org/10.3390/coatings12111728

Academic Editor: Claudio Lantieri

Received: 15 October 2022
Accepted: 9 November 2022
Published: 12 November 2022

Publisher's Note: MDPI stays neutral with regard to jurisdictional claims in published maps and institutional affiliations.

Copyright: © 2022 by the authors. Licensee MDPI, Basel, Switzerland. This article is an open access article distributed under the terms and conditions of the Creative Commons Attribution (CC BY) license (https://creativecommons.org/licenses/by/4.0/).

1. Introduction

In recent years, to improve the performance of pavement and extend the service life of pavements, researchers have carried out a lot of studies on the structural mechanics, environmental and traffic monitoring for the structural health monitoring (SHM) of pavements [1–3]. However, traditional external inspection methods cannot obtain mechanical information inside of the pavement structure directly and effectively [4,5]. In order to accurately obtain and assess the damage state and the health performance in the pavement structure, sensors need to be embedded in the pavement structure in order to directly obtain mechanical information such as stress–strain of the pavement [6–8]. The long-term and real-time monitoring of embedded sensing elements and the mechanical–structural response relationships can be used to perceive and predict the structural damage that may be incurred during the operation of the pavement so as to avoid the continuous development of damage to the whole pavement structure [9–11].

To monitor the strain conditions of the pavement structure and pavement materials under actual loading, sensors and sensing devices are very important. Several special peculiarities are necessary for pavement sensors to deal with the harsh condition of pavement such as high-temperature resistance, compaction resistance and corrosion resistance, etc. Literature reported the sensors using for pavement structure monitoring, including fiber Bragg grating sensor (FBG) [12], strain gauge sensor [13], temperature sensor [14], pressure sensor [15], humidity sensor and so on.

As an embedded element, the sensor should be deformation coincident with the pavement, which means when the pavement deforms under the traffic load or long period fatigue, the embedded sensor should follow synchronous deformation to obtain the value of the mechanical response. Otherwise, when the modulus of the sensor is rather higher than the pavement, the deformation value of sensor will be lower than the pavement or even there is no deformation; when the modulus of sensor is rather lower than the pavement, the deformation value of sensor will be higher than the pavement. Some researchers [16,17] have focused the importance of deformation coincident between the embedded sensor and matrix. In Zhen Liu et al.'s study [18], the encapsulating material with an elastic modulus of 40 GPa for the FBG sensor and a metallic tube with an elastic modulus of 40 GPa for R sensor were established. One study about the effect of the modulus ratio of the FBG sensor to that of the asphalt mixture on the stability of the regression model suggested that the stability worsens as the modulus ratio increases at the same temperature. Zejiao Dong et al. [19] constructed a certain amount of 4-point bending beams filled with random aggregates and asphalt mortar and embedded with the FBG sensor utilizing the FEM. The simulation results illustrated the diverse effects of the different geometries and moduli of embedded sensors on the stress and strain states of the asphalt mixture. However, in these studies, the modulus of sensor is much higher than the asphalt beam according to the encapsulating material which should be improved with the deformation coincident.

The authors' team have carried out a series of related studies in the previous period [20,21]. The strain sensor based on the polymer sensing materials have been developed and verified with the high accuracy and shown the viability for the SHM of asphalt pavement under severe construction and complex operating environments [11,22]. How does the embedding of the sensor element affect structural mechanics of the asphalt pavement, and which kind of sensor element structure can reduce the impact on the pavement structure to a minimum while ensuring the effects of monitoring? To solve these problems, this paper will use ABAQUS software to conduct the statically step-by-step loading method to investigate the impact of embedded sensing elements on the structure of the asphalt pavement and the mechanical changes of the sensing elements themselves by constructing with two different models of sensing elements and asphalt concrete beams. Furthermore, the structural optimization analysis of the sensing element was also carried out.

2. Experimental Design or Methodology

2.1. Modelling Dimensions and Meshing

The simulation parameters were optimized by two steps. The first step is: three simulated conditions (without sensor, with embedded I-shape sensor, with embedded corrugated-shape sensor) were studied comparatively of the sensing element embedding effect on the mechanical response of asphalt pavement structure. After the analysis conclusion of the preferable I-shaped sensor, the paper continues to analyze the structural optimization design of the I-shaped sensing element, in order to compare and study the influence of the sensing element structure on the asphalt mixture matrix, mainly including sensing element length L, sensing element diameter φ, and I-shaped length GL.

Three simulation conditions were built according to the asphalt concrete three-point bending test. Condition 1 was a model of asphalt concrete without sensing element embedded, condition 2 was a model of asphalt concrete with an I-shaped sensing element embedded, condition 3 was a model of asphalt concrete with a corrugated-shape sensing element embedded, in which the sensing element was placed at the bottom centre of the asphalt concrete. The dimension of the asphalt concrete beam has a length of L = 300 mm, width of W = 100 mm, and height of H = 100 mm. The I-beam-shaped sensing element is endowed with the diameter of φ = 10 mm, length of L = 140 mm and the aluminium I-beam size of 70 mm × 20 mm × 5 mm. The corrugated-shape sensing element is endowed with the diameter of φ = 10 mm, outer diameter of φ = 16 mm, length of L = 140 mm and evenly distributed with 7 bumps with the bump breadth of 5 mm. The asphalt concrete beams and sensor elements are solid homogeneous bodies with C3D8R units (8-node hexahedral linear reduced integration units). The approximate mesh size for the asphalt concrete beam is 5 mm × 5 mm × 5 mm with a total of 24,000 meshes, and the approximate mesh size for the two shaped sensing elements is 1 mm with a total of approximately 26,000 meshes. The model diagrams of the asphalt concrete, I-beam sensing element and corrugated-shape sensing element are shown in Figure 1.

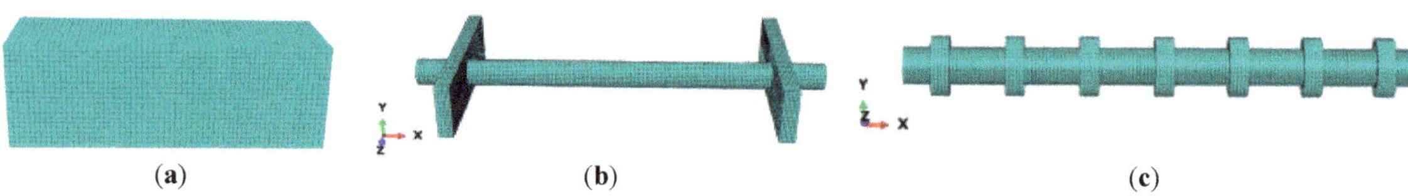

Figure 1. Component modeling and meshing, (**a**) asphalt concrete, (**b**) I-shaped sensor, (**c**) corrugated-shape sensor.

2.2. Material Parameters

2.2.1. Material Parameters of Asphalt Concrete [23,24]

In this paper, asphalt concrete is assumed to be an isotropic material with Young's modulus of asphalt concrete as 1500 MPa, Poisson's ratio of 0.3, power law multiplier of 6.536×10^{-11}, equivalent force order of 0.937 and time order of -0.592 in creep parameters.

2.2.2. Embedded Sensing Element

Based on the self-developed mechanically sensitive sensor element [25], an isotropic homogeneous elastomer with density of 1500 kg/m^3, modulus of elasticity of 1400 MPa and Poisson's ratio of 0.3 is assumed. The I-beam aluminium alloy is also assumed to be an isotropic homogeneous elastomer with the density of 800 kg/m^3, modulus of elasticity of 72,000 MPa and Poisson's ratio of 0.3 [26–29].

2.3. Boundary Conditions and Load Application

The boundary conditions of the asphalt concrete were imposed according to a three-point bending test. In the model, the asphalt beam was at a free state for the surrounding four sides and fixed constraint at the 30 mm from the edge in the bottom. During the loading, the vertical downward displacement constraint at the top surface were conducted with a statically step-by-step loading mode of loading rate of 0.1 mm/min. The loading target displacement is 1 mm with a total time of 600 s and fixed incremental steps of 12 s.

The stress S11 and strain E11 at the same displacement constraint were obtained along the transverse direction of the asphalt concrete. Six feature points were chosen at 20, 40, 60, 80, 90, and 100 mm from the top of the asphalt concrete in the centre of the asphalt concrete profile, as shown in Figure 2.

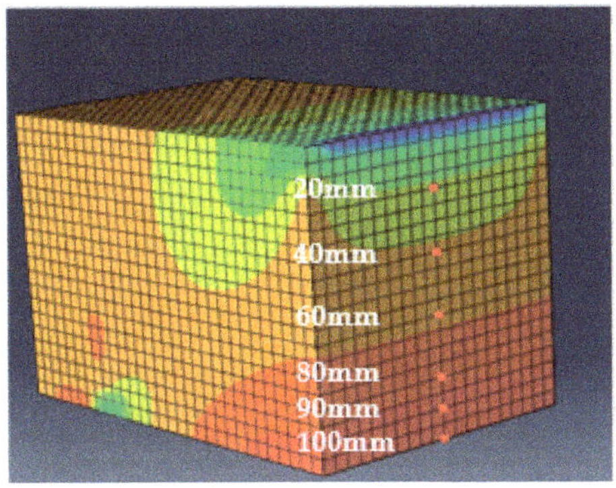

Figure 2. The position of the point in the model.

3. Results and Discussion

3.1. Effect of Embedded Sensing Element on Stress–Strain Field

The schematic diagram of horizontal tensile stress S11, horizontal tensile strain E11 for the exterior of asphalt mixture beam and the central section of the asphalt mixture beam without embedded sensor elements and with embedded sensor elements (I-shaped and corrugated-shape sensor) after the static step-by-step loading simulation are shown in Figure 3.

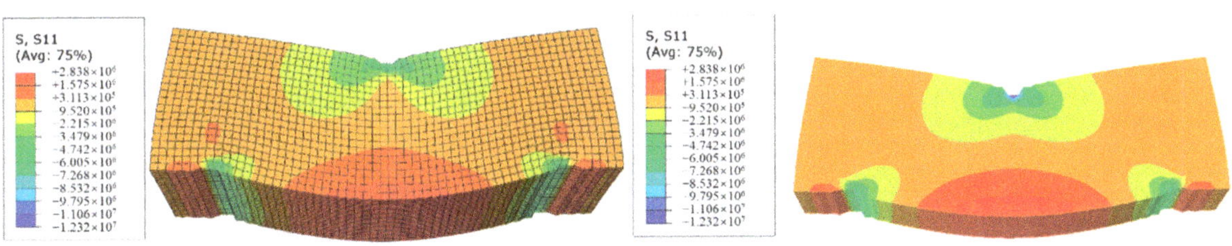

(**a**)

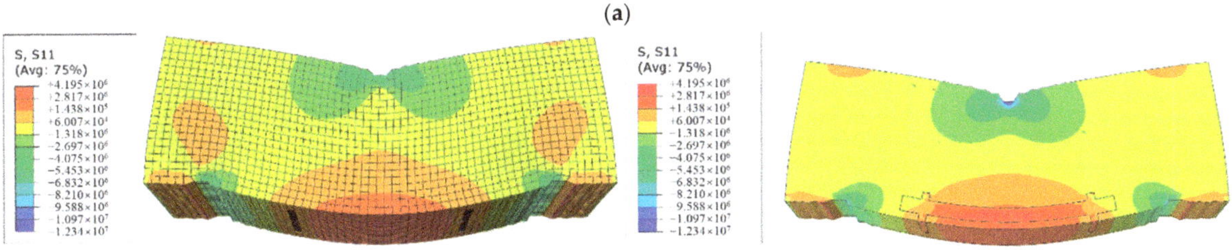

(**b**)

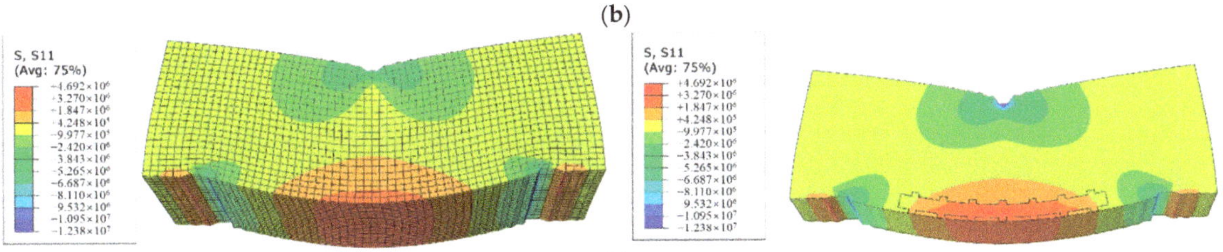

(**c**)

Figure 3. *Cont.*

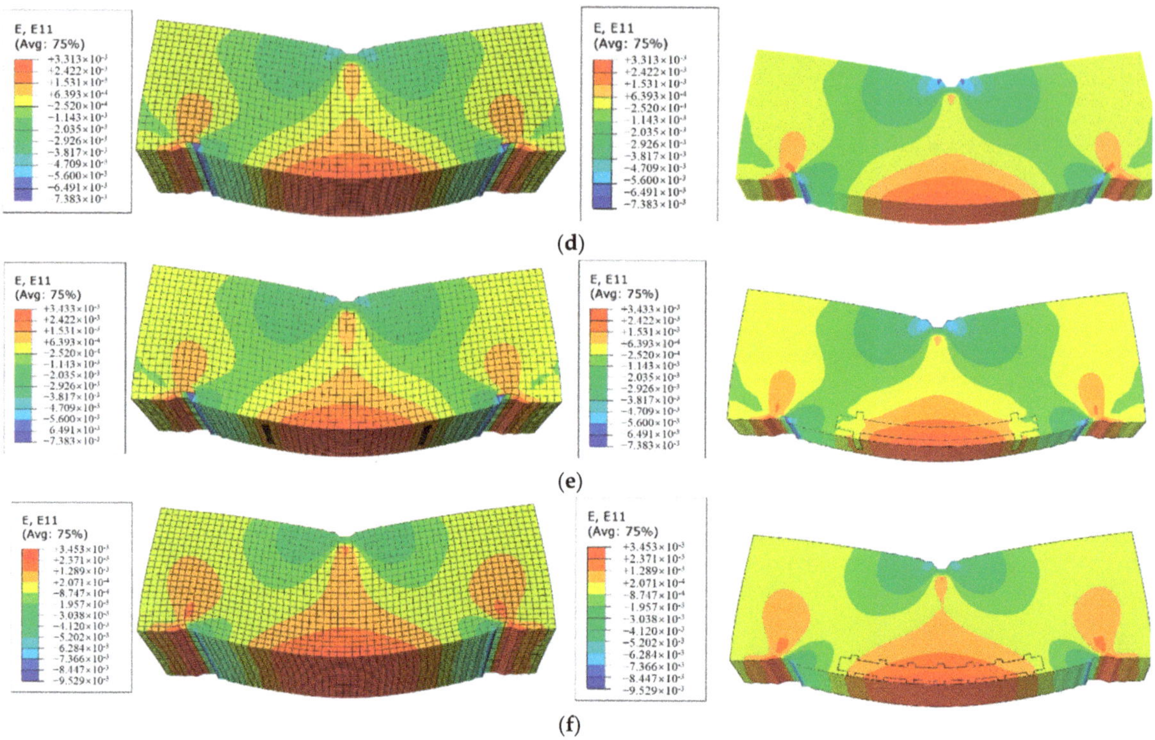

Figure 3. S11 and E11 of asphalt beam under static progressive loading condition, (**a**) S11 without embedded sensors, (**b**) S11 with I-shaped sensor, (**c**) S11 with corrugated-shape sensor, (**d**) E11 without embedded sensors, (**e**) E11 with I-shaped sensor and (**f**) E11 with corrugated-shape sensor.

It can be seen from the figure that the asphalt beams under three conditions show different degrees of loading stresses and strains. The maximum tensile stresses and maximum tensile strains of the asphalt mixture beams with embedded sensors are greater than that in the un-embedded condition, especially for the stress response. The sensor was located at the bottom of the asphalt mixture beam, which just at the location of the maximum tensile stress and maximum tensile strain [30,31].

To describe the tensile stresses and tensile strains at different locations within the asphalt mixture beam and explore the effect of the embedding of sensor elements on the stresses and strains within the structure of the mixture beam more clearly, the variation of S11 and E11 for the selected feature points with vertical loading displacement, are shown in Figure 4.

It can be seen from Figure 4 that in the static step-by-step loading mode, the stress and strain of the asphalt mixture at different locations increase gradually with the increasement of the applied displacement. Within the range of 0–40 mm, the asphalt mixture beam is in a compressed state with negative stress rates. Around the middle of the asphalt mixture beam, the tensile stresses tend towards zero showing the state of neither tensile nor compressive. Within the range of 60–100 mm, the asphalt mixture is in a state of tension the stress values increase with depth and show a maximum tensile stress state at the bottom of the beam, which is in accordance with the stress conditions during the operation of the asphalt pavement. By comparing Figure 4a–d, the embedment of the sensor element did not have a significant effect on the tensile stresses inside the asphalt mix beams [22].

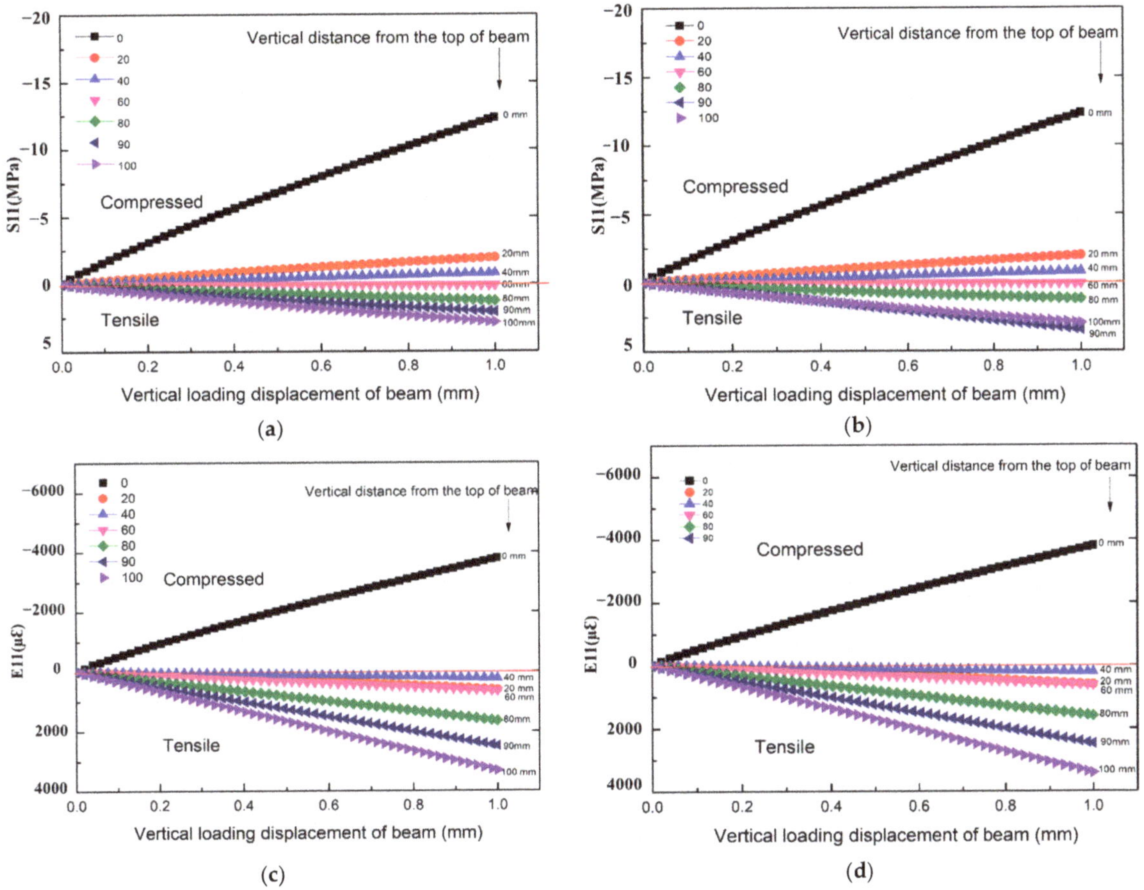

Figure 4. Change of S11 and E11 with the vertical loading displacement, (**a**) S11 of beam without sensor, (**b**) S11 of beam with I-shaped sensor, (**c**) E11 of beam without sensor, (**d**) E11 of beam with I-shaped sensor.

In order to quantitatively compare the effects of the two sensors embedded in different structures on the stress and strain fields in the internal region of the asphalt mixture trabecular, the changes in S11 and E11 with the vertical distance from the top of the trabecular are shown in Figure 5.

It can be seen from Figure 5 that the embedded sensors with two different structures have no great influence on the stress and strain of the asphalt mixture except for the embedded sensing element area. However, compared with the stress and strain curves in Figure 5, the tensile stresses of working condition 2 (embedded I-shape sensor) are much closer to those of working condition 1 (without sensor) than those of working condition 3 (embedded corrugated-shape sensor), indicating that the stress fields of embedded I-shaped sensing element are closer to those of un-embedded sensing element. As for the strain at the embedded area, the differences among the three conditions are almost negligible. Figure 6 shows the tensile strain curves at 90 mm from the top of asphalt mixture (the center of sensor element) under three couple models. It can be seen from the diagram that with the increase in the applied displacement, the tensile strain changes are almost the same under three conditions.

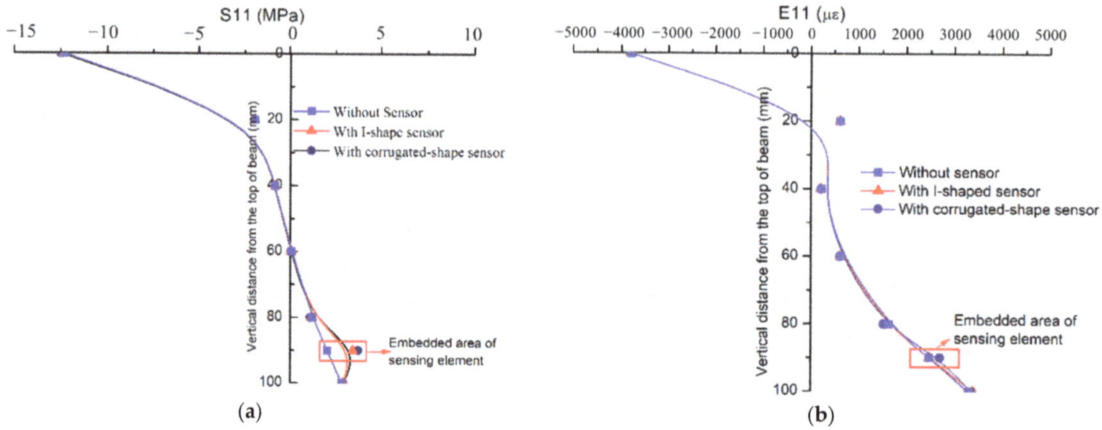

Figure 5. Change in S11 and E11 with the vertical distance from the top of beam, (**a**) S11, (**b**) E11.

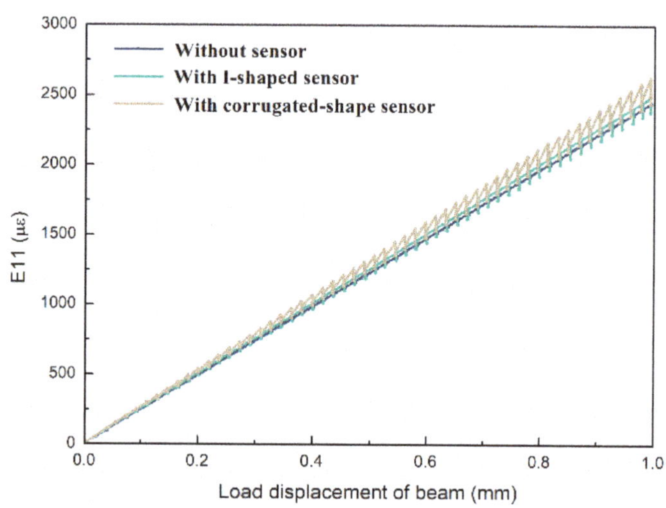

Figure 6. E11 comparison at 90 cm under the three coupling models.

The tensile strain without sensing element of asphalt mixture increases linearly, while the one with sensing element increases in a fluctuating form. Moreover, the tensile strain of the embedded I-shaped sensor is closer to that of the without sensor than that of the embedded corrugated-shape sensor, indicating that the embedded I-shaped sensor has less influence on the asphalt mixture. By comparing, the influence of two embedded sensors with different shapes on the structure of asphalt mixture trabecular is relatively small, but the tensile stress and strain are closer to the asphalt mixture when the I-shaped sensor is embedded. In addition, sensors usually work in different asphalt layers with different gradations and structures, the shape and material of the I-shaped sensor can better ensure the effectiveness, stability and cooperative deformability with the asphalt mixture. Under the actual stress state of the pavement, the pavement structure information obtained by the sensor can better reflect the actual situation of the asphalt pavement [31].

3.2. Horizontal Tensile Strain E11 along the Central Axis of the Sensing Element

To study the stress state of the internal structure of the sensor, the points on the central axis of the sensing element are taken, respectively, as shown in Figure 7.

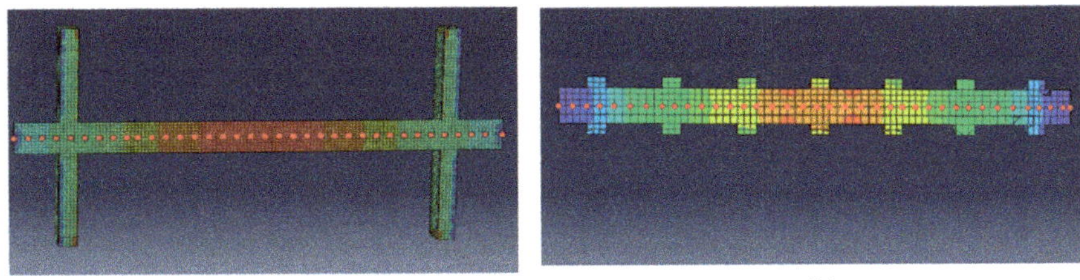

(**a**) (**b**)

Figure 7. Axial point of sensing element, (**a**) points of the I-shaped sensor, (**b**) points of the corrugated-shape sensor.

Figure 8a shows the variation of the horizontal tensile strain at each point on the axis of the sensing element.

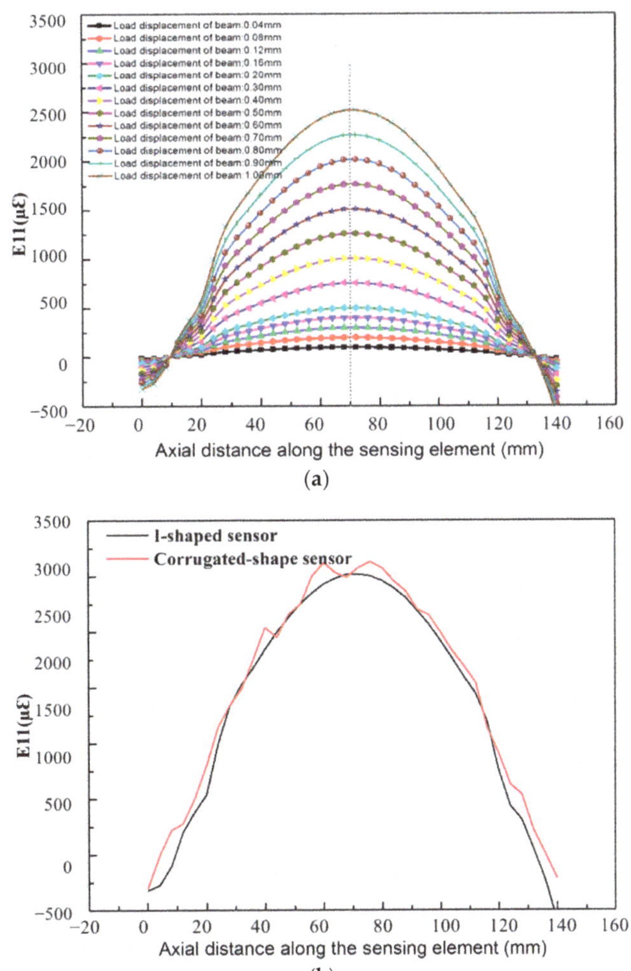

Figure 8. The change of horizontal tension strain E11 along the axial distance of the sensor, (**a**) I-shaped sensing element, (**b**) comparison of two structural sensors.

The horizontal tensile strain E11 on the sensing element gradually increases as the load displacement at the top of the trabecular increases step by step. The horizontal tensile strain shows a trend of increase and then decrease along the axial distance of the sensing element at the central axis of the sensing element, and the maximum horizontal tensile strain occurs at the axial center of the sensing element. This variation law of tensile strain is consistent with the asphalt mixture trabecular itself, which fully verifies the deformation compatibility of the sensing element and the asphalt mixture matrix.

By comparing the embedded sensing elements with two different structures, the largest strain of the sensing element is at the center of the sensing element, and there is a transverse compression area at the end. However, the transition of strain curve is smoother when the I-shaped sensor is embedded. When the corrugated-shape sensing element is embedded, there are many inflection points in the strain curve. If the corrugated-shape sensing element is embedded in the road, it may lead to the internal deformation of the sensing element that cannot be completely consistent when it is subjected to the road load, thereby affecting the monitoring accuracy.

3.3. Structure Optimization Design of Embedded Sensor

The research on the influence of two kinds of sensing elements on the internal mechanical characteristics of asphalt mixture and the deformation characteristics of the sensing element structure show that the I-shaped sensing element is superior to the corrugated-shaped sensing element. Therefore, this paper will continue to analyze the structural optimization design of I-shaped sensing element, in order to compare and study the influence of the sensing element structure on asphalt mixture matrix, mainly including sensing element length L, sensing element diameter φ, and I-shaped length GL. Combined with the actual engineering application, the working condition parameters are selected as shown in Table 1, and the structural optimization modeling examples and simulation results are shown in Figures 9 and 10, respectively.

Table 1. I-shape sensing element structure optimization design parameters.

Structure Optimization	Parameters			
Length of sensor L/cm	6	10	14	18
Diameter of sensor φ/mm	6	10	14	/
Length of I-beam GL/cm	3	7	10	/

Figure 9. Schematic diagram of structural optimization modeling for the I-sensor.

Shear stress describes the internal stress concentration of asphalt mixture caused by the embedding of sensing elements. Stress concentration will lead to the slip of the interface between asphalt pavement and sensing element, resulting in loose and falling off of asphalt pavement structure, and then causing pavement diseases. It can be seen from Figure 10a that, with the increase in the sensor length L, the maximum horizontal tensile stress S11 gradually decreases, and the maximum shear stress S12 gradually increases.

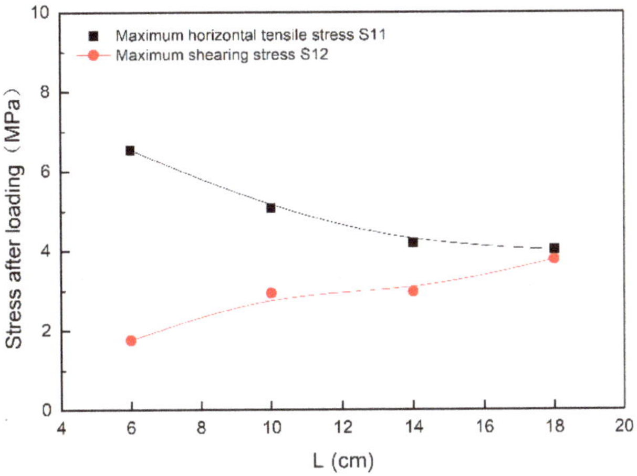

(a)

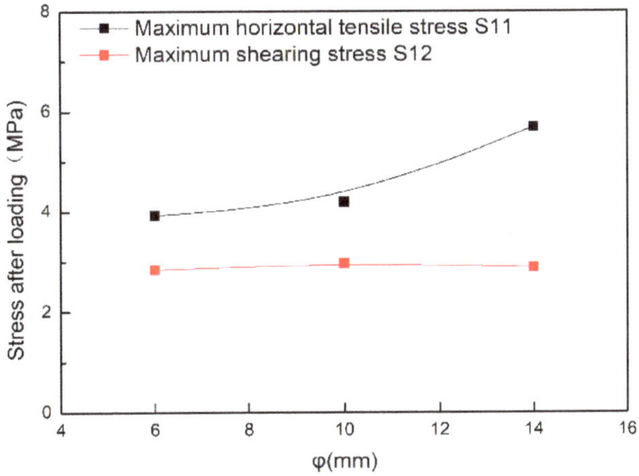

(b)

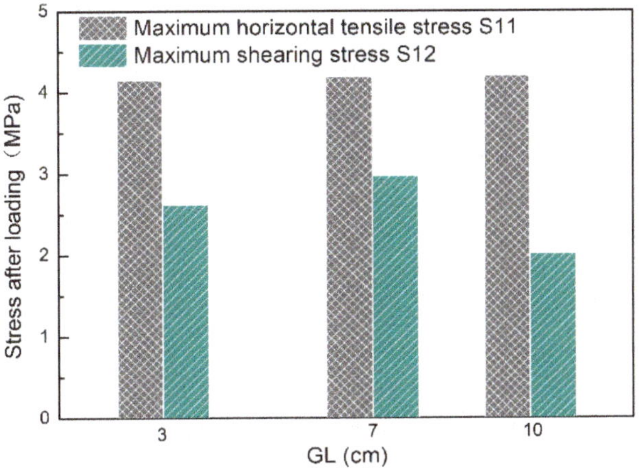

(c)

Figure 10. Optimization of I-sensor structure, (**a**) sensing element length optimization, (**b**) sensing element diameter optimization, (**c**) I-beam length optimization.

However, both S11 and S12 have inflection points when L = 14 cm. When L increases from 14 to 16 cm, the decrease in S11 slows down, and the increase in S12 grows significantly. In a comprehensive comparison, the optimal length L of the sensing element is 14 cm. Similarly, the optimal diameter φ of the sensor is 10 mm, and the optimal I-beam length GL is 10 cm.

4. Conclusions

1. In this work, static load model and three-point bending test mode were conducted with three "pavement sensor" coupling model of without sensor, embedded I-shape sensor, embedded corrugated-shape sensor based on the finite element numerical simulation. Three simulated conditions were studied comparatively of the sensors embedding effect on the mechanical response of the asphalt pavement structure.
2. The sensors embedded with the two structures have little influence on the stress and strain field of asphalt concrete. Within the range of 60–100 mm, the asphalt mixture is in a state of tension; the stress values increase with depth and show a maximum tensile stress state at the bottom of the beam. In the compression zone, the strain information is consistent for both the sensors modeled but more consistent for the I-shaped sensor with the actual pavement information.
3. Along the axis of the two sensing elements, the axial strain of the I-shaped sensor is smoother and uniform, which ensures the deformation coordination in the road state.
4. The stress concentration phenomenon occurred on the surface of both sensing elements, but considering that, under actual action, the sensing element embedded in the pavement is subjected to repeated vehicle loads from the pavement, the use of I-shaped sensors can meet the requirements of both durability and structure working.
5. The optimal length L of the sensing element is 14 cm, diameter φ of the sensor is 10 mm, I-beam length GL is 10 cm.

The encapsulating material with an elastic modulus of the sensor have significant impact on the deformation coincident with the pavement, we should choose the encapsulating material with modulus of as close as the asphalt concrete as well as the proper geometries of sensors. Even though the study of embedded sensors has obtained some development, the authors will continue to carry out some simulation work mainly focusing on two areas in the future. One is the asphalt beam with the embedded sensor at the loading stage of dynamic cyclic loading. The other one is to establish the hierarchical model of road structure and insert the sensor elements to simulate the response.

Author Contributions: P.W., X.X. and M.L.: Data curation, Writing—original draft, Writing—review and editing. G.Z. and F.X.: Data curation. C.W. and Y.J.: Investigation. Y.Z.: Investigation. S.L.: Methodology. H.W.: Funding acquisition. All authors have read and agreed to the published version of the manuscript.

Funding: This research was funded by [Natural Science Foundation of Shandong Province (CN)] grant number [ZR2020ME244], [Transportation Technology Project of Shandong Province], [Qilu Young Scholars Program of Shandong University] grant number [20209900060], [Fundamental Research Funds of Shandong University] grant number [No. 2020GN059].

Institutional Review Board Statement: Not applicable.

Informed Consent Statement: Not applicable.

Conflicts of Interest: The authors declared that they have no conflict of interest to this work. We declare that we do not have any commercial or associative interest that represents a conflict of interest in connection with the work submitted.

References

1. Brownjohn, J.M.W. Structural health monitoring of civil infrastructure. Philosophical Transactions of the Royal Society A: Mathematical. *Phys. Eng. Sci.* **2007**, *365*, 589–622.
2. Meng, L.; Wang, L.; Hou, Y.; Yan, G. A Research on Low Modulus Distributed Fiber Optical Sensor for Pavement Material Strain Monitoring. *Sensors* **2017**, *17*, 2386. [CrossRef] [PubMed]
3. Timm, D.H.; Priest, A.L.; Mcewen, T.V. *Design and Instrumentation of the Structural Pavement Experiment at the NCAT Test Track*; America National Center for Asphalt Technology (US): Auburn, AL, USA, 2004.
4. Barriera, M.; Pouget, S.; Lebental, B.; Van Rompu, J. In Situ Pavement Monitoring: A Review. *Infrastructures* **2020**, *5*, 18. [CrossRef]
5. Sony, S.; Laventure, S.; Sadhu, A. A literature review of next-generation smart sensing technology in structural health monitoring. *Struct. Control Health Monit.* **2019**, *26*, e2321. [CrossRef]
6. Zappino, E.; Carrera, E. Advanced modeling of embedded piezo-electric transducers for the health-monitoring of layered structures. *Int. J. Smart Nano Mater.* **2020**, *11*, 325–342. [CrossRef]
7. Yan, J.; Luo, D.; Li, Y.; Cafiso, A. Influence of Sensing Element on Mechanical Properties of Asphalt Pavement Substrate. *J. Transp. Res.* **2021**, *7*, 115–123.
8. Mustafa, S.; Sekiya, H.; Maeda, I.; Takaba, S.; Hamajima, A. Identification of external load information using distributed optical fiber sensors embedded in an existing road pavement. *Opt. Fiber Technol.* **2021**, *67*, 102705. [CrossRef]
9. Ko, J.M.; Ni, Y.-Q. Technology developments in structural health monitoring of large-scale bridges. *Eng. Struct.* **2005**, *27*, 1715–1725. [CrossRef]
10. Di Graziano, A.; Marchetta, V.; Cafiso, S. Structural health monitoring of asphalt pavements using smart sensor networks: A comprehensive review. *J. Traffic Transp. Eng. Engl. Ed.* **2020**, *7*, 639–651. [CrossRef]
11. Alavi, A.H.; Hasni, H.; Lajnef, N.; Chatti, K. Continuous health monitoring of pavement systems using smart sensing technology. *Constr. Build. Mater.* **2016**, *114*, 719–736. [CrossRef]
12. Liu, H.; Ge, W.; Pan, Q.; Hu, R.; Lv, S.; Huang, T. Characteristics and analysis of dynamic strain response on typical asphalt pavement using Fiber Bragg Grating sensing technology. *Constr. Build. Mater.* **2021**, *310*, 125242. [CrossRef]
13. Bruno, S.; Del Serrone, G.; Di Mascio, P.; Loprencipe, G.; Ricci, E.; Moretti, L. Technical Proposal for Monitoring Thermal and Mechanical Stresses of a Runway Pavement. *Sensors* **2021**, *21*, 6797. [CrossRef] [PubMed]
14. Lu, R.; Jiang, W.; Xiao, J.; Xing, C.; Ruan, C.; Li, Y.; Wu, W. Temperature characteristics of permeable asphalt pavement: Field research. *Constr. Build. Mater.* **2022**, *332*, 127379. [CrossRef]
15. Raskar, D.; Gadkari, R.; Israji, A.; Mahadik, S. Pavement Pressure Sensor Based Battery Charging and Street Lightning System. *Int. Res. J. Innov. Eng. Technol.* **2022**, *6*, 163.
16. Ye, Z.; Cai, Y.; Liu, C.; Lu, K.; Ildefonzo, D.G.; Wang, L. Optimization of Embedded Sensor Packaging Used in Rollpave Pavement Based on Test and Simulation. *Materials* **2022**, *15*, 2283. [CrossRef] [PubMed]
17. Otto, G.G.; Simonin, J.M.; Piau, J.M.; MattarValente, A. Weigh-in-motion (WIM) sensor response model using pavement stress and deflection. *Constr. Build. Mater.* **2017**, *156*, 83–90. [CrossRef]
18. Liu, Z.; Gu, X.; Wu, C.; Ren, H.; Zhou, Z.; Tang, S. Studies on the validity of strain sensors for pavement monitoring: A case study for a fiber Bragg grating sensor and resistive sensor. *Constr. Build. Mater.* **2022**, *321*, 126085. [CrossRef]
19. Dong, Z.; Ma, X.; Gong, X.; Oeser, M. Theoretical evaluation of the measurement accuracy of fiber Bragg grating strain sensors within randomly filled asphalt mixtures based on finite element simulation. *Struct. Control Health Monit.* **2018**, *25*, e2057. [CrossRef]
20. Xin, X.; Qiu, Z.; Luan, X.; Ding, X.; Liang, M.; Yao, Z.; Li, Y. Novel Conductive Polymer Composites for Asphalt Pavement Structure in Situ Strain Monitoring: Influence of CB/CNT and GNP/CNT Nano/Micro Hybrid Fillers on Strain Sensing Behavior. *IEEE Sens. J.* **2022**, *22*, 3945–3956. [CrossRef]
21. Su, L.; Luan, X.; Qiu, Z.; Liang, M.; Rong, Y.; Xin, X.; Yao, Z.; Ma, C. Sensing Performance and Optimizing Encapsulation Materials of a Coordinated Epoxy-Encapsulated Sensor for Strain Monitoring of Asphalt Pavement Layered Structures. *IEEE Sens. J.* **2022**, *22*, 9811–9823. [CrossRef]
22. Tian, G.; Dong, Z.; Hu, Q. Analysis of cooperative deformation between fiber grating sensor and asphalt mixture. *J. Harbin Inst. Technol.* **2009**, *41*, 73–76.
23. Xin, X.; Liang, M.; Yao, Z.; Su, L.; Zhang, J.; Li, P.; Sun, C.; Jiang, H. Self-sensing behavior and mechanical properties of carbon nanotubes/epoxy resin composite for asphalt pavement strain monitoring. *Constr. Build. Mater.* **2020**, *257*, 119404. [CrossRef]
24. Xin, X.; Rong, Y.; Su, L.; Yao, Z. Dynamic Mechanical and Chemorheology Analysis for the Blended Epoxy System with Polyurethane Modified Resin. *J. Renew. Mater.* **2022**, *10*, 1081. [CrossRef]
25. Xin, X.; Luan, X.; Su, L.; Ma, C.; Liang, M.; Ding, X.; Yao, Z. The Innovative Self-Sensing Strain Sensor for Asphalt Pavement Structure: Substitutability and Synergy Effects of Graphene Platelets with Carbon Nanotubes in Epoxy Composites. *Front. Mater.* **2022**, *9*, 824364. [CrossRef]
26. Awed, A.M.; Tarbay, E.W.; El-Badawy, S.M.; Azam, A.M. Performance characteristics of asphalt mixtures with industrial waste/by-product materials as mineral fillers under static and cyclic loading. *Road Mater. Pavement Des.* **2022**, *23*, 335–357. [CrossRef]
27. Varma, S.; Kutay, M.E. Viscoelastic Nonlinear Multilayered Model for Asphalt Pavements. *J. Eng. Mech.* **2016**, *142*, 04016044. [CrossRef]

28. Li, Q.; Wang, L.; Li, Z. Finite element analysis of asphalt mixture rutting based on dynamic modulus. *Shanxi Archit.* **2019**, *45*, 154–156.
29. Mishra, M.; Lourenço, P.B.; Ramana, G.V. Structural health monitoring of civil engineering structures by using the internet of things: A review. *J. Build. Eng.* **2022**, *48*, 103954. [CrossRef]
30. Liu, W.; Wang, H.; Zhou, Z.; Xing, X.; Cao, D.; Jiang, Z. Optical fiber-based sensors with flexible encapsulation for pavement behavior monitoring. *Struct. Control Health Monit.* **2015**, *22*, 301–313. [CrossRef]
31. Yao, X. Development and Application of Monitoring data Analysis System for Asphalt Pavement Structure Information. Ph.D. Thesis, Harbin Institute of Technology, Harbin, China, 2018; pp. 5–16.

coatings

Article

Aggregate Gradation Variation on the Properties of Asphalt Mixtures

Yanchao Yue [1], Moustafa Abdelsalam [2,*] and M. S. Eisa [2,*]

1 School of Human Settlements and Civil Engineering, Xi'an Jiaotong University, Xi'an 710054, China
2 Civil Engineering Department, Benha Faculty of Engineering, Benha University, Benha 13518, Egypt
* Correspondence: eng_mostafaabdelsalam@bhit.bu.edu.eg (M.A.); mohamedeisa524@bhit.bu.edu.eg (M.S.E.)

Abstract: National highway projects present a crucial role in economic growth, as they have a great influence on the national income. Therefore, the decision makers plan to construct these projects at a rapid rate. To achieve the just aforesaid, the utilization of asphalt of adequate quality and gradation is essential. The key problem which lies in recent decades is that many types of asphalt mixtures are rejected and reconstructed in the site due to the gradation variation of aggregates in the asphalt mixture which waste raw materials, cost and time. Thus, this research seeks to assess the possibility of accepting asphalt mixes with aggregates gradation variation (within the range from +4% above the upper specification limit to −2% below the lower specification limit). A wearing surface mix with gradation 3B was prepared according to the Egyptian code. The gradation variation was presented as the aggregate gradation is out of the specification limits during Hot Mix Asphalt (HMA) production. The aggregate gradations lie above and below the upper and lower specification limits, respectively, by ±2%, ±4% and ±6%. The design gradation of the control mix was included as a reference case. The different mix properties were measured using the Marshall Mix design method. Then, the performance of HMA mixes was evaluated under the effects of high temperature and water cycles through applying wheel loading tracking and Indirect Tensile Strength (ITS) tests. The results show that the 3B mixes with a gradation within a range of +4% to −2% of the upper and lower specification limits recorded the lowest rutting depth and the highest water damage resistance in hot regions compared to ordinary asphalt mixes. In summary, the new aggregate gradation limits will provide a reference for the design of asphalt mixture in hot climate regions.

Keywords: laboratory evaluation; Hot Mix Asphalt (HMA); aggregate gradation; Marshall test; rutting; water effect

Citation: Yue, Y.; Abdelsalam, M.; S. Eisa, M. Aggregate Gradation Variation on the Properties of Asphalt Mixtures. *Coatings* **2022**, *12*, 1608. https://doi.org/10.3390/coatings12111608

Academic Editors: Leilei Chen and Andrea Simone

Received: 13 September 2022
Accepted: 20 October 2022
Published: 22 October 2022

Publisher's Note: MDPI stays neutral with regard to jurisdictional claims in published maps and institutional affiliations.

Copyright: © 2022 by the authors. Licensee MDPI, Basel, Switzerland. This article is an open access article distributed under the terms and conditions of the Creative Commons Attribution (CC BY) license (https://creativecommons.org/licenses/by/4.0/).

1. Introduction

National highways are the basic axes of economic development and growth and bring important social benefits. They are of vital importance in order to make a nation grow and develop. In addition, providing access to employment, social, health and education services makes a road network crucial to fighting against poverty [1]. Thus, the government has been motivated to finish these projects rapidly to reflect their performance. However, the rapid completion of such projects has a negative effect on the final product due to the aggregate gradation variation that occurs at execution and production processes. On the other hand, the construction process of asphalt pavement typically includes mixing, transportation, paving and compaction stages [2]. There are many details in each process that are difficult to accurately control, which results in the presence of uncertain factors in the construction processes [3,4]. Indeed, it is a well-documented fact that inconsistency in asphalt design and construction parameters has consistently caused premature performance failures [5]. Therefore, exploring the negative implications associated with variability in asphalt mix design and construction parameters may be worthwhile [6].

Asphalt mixtures used in modern pavement construction are mainly comprised of air void, asphalt binder, coarse aggregates, fine aggregates and filler [7–9]. Asphalt binder cements aggregates, and filler to form a multiphase composite with air voids [10]. To ensure a sound pavement performance, the skeletal structure should consist of well-interlocked aggregates [11]. As an essential parameter for asphalt mixture, the gradation is closely related to the quality and performance of the pavement [12]. Gradation variation in the asphalt mixture may result in many distresses of pavement [13].

Many types of asphalt mixtures are rejected and reconstructed in the site due to the gradation variation of aggregates in the asphalt mixture, which waste raw materials, cost and time [14]. Thus, this study seeks to evaluate the possibility of accepting mixes with aggregates gradation variation (within the range from +4% above the upper specification limit to −2% below the lower specification limit).

The pavement design and construction literature shows that many studies on the effect of the gradation variation on HMA properties, or rather the effect of variance of gradation that may occur during production on HMA properties have been undertaken in the past [15–20].

Zhang et al. [15] evaluated the aggregate gradation in asphalt mixtures on the characterize load carrying capacity and rutting resistance. Zhang recorded that the aggregates retaining on sieve sizes of 2.36 and 4.75 mm provide more than 50% contribution to resist load and rutting, and the aggregates retaining on sieve sizes of 1.18, 0.6 and 0.3 mm provide more than 50% contribution to strength the structure. The Influence of Aggregate Gradation on Clogging Characteristics of Porous Asphalt Mixtures was evaluated by martin et al. [16], results revealed that aggregate gradation was found to be strongly correlated to the macrotexture depth of the porous pavement and the permeability of the mixes both before and after clogging. Abo-Qudais et al. [17] study the Effect of aggregate properties on asphalt mixtures stripping and creep behavior and results documented that unconditioned HMA specimens prepared using basalt aggregate resist creep better than those prepared using limestone. However, after conditioning, mixes prepared using basalt were less resistant to creep strain than those prepared using limestone aggregate. Percent absorbed asphalt was found to be directly related to stripping resistant. Also, mixes prepared using aggregate following American Society for Testing and Materials (ASTM) upper limit of dense aggregate gradation presented the highest resistance to stripping. Moghaddam et al. [18] Study the effects of using different types of additives and aggregate gradation on fatigue and rutting resistance of Asphalt Concrete (AC) mixtures and it is concluded that that fatigue and rutting resistance of AC mixture could be enhanced considerably by utilization of different aggregate gradation and types of additives such as fibers. Bazi [19] reported that variability of gradation has a considerable effect on the performance of pavement. To enhance resistance to asphalt pavement rutting, Lv et al. [20] was of the opinion that adjusting the percentage of particles passing the 4.75 mm sieve should be considered. Amir Golalipour et al. [21] examined the effect of aggregate gradation variation on rutting characteristics of asphalt mixture and concluded that the aggregate gradation played a considerable role in resisting the permanent deformation of pavement. Yu et al. [22] investigated the influence of the aggregate gradation during construction on the asphalt pavement performance and reported that the rutting resistance increased first and then decreased with regard to the changing of the gradation from fine to coarse. In research prepared by Moustafa et al. [23], performed a laboratory study to evaluate the performance of asphalt mixtures using a composite mixture of diatomite powder and lignin fiber. The results showed that the addition of diatomite powder and lignin fiber are greatly enhanced the overall performance of asphalt mixes, and the compound mixture was more effective for improving the overall asphalt performance than either lignin fiber or diatomite powder separately. Elliott et al. [24] reported that the mixtures with lower limit gradations almost had the greatest effect on HMA properties. Moreover, they found that asphalt mixtures gradations deflected to lower limit gradation showed the best performance. Awan et al. [25] applied Multi Expression Programming to predict the output parameters Marshall Flow and Marshall Stability of Asphalt Pavements, it is demonstrated that the novelette models have produced results that are consistent with the experimental data and function equally well for unknown data as well as The models developed have successfully incorporated input parameters and have the capability to predict the trends of MS and MF for flexible pavements, as revealed from the parametric study. Rafiq et al. [26] conducted a comparative comparison among Hot Mix Asphalt (HMA) and Reclaimed Asphalt Pavement (RAP) through using Life Cycle Cost (LCC) and the results denoted that the total LCC measurement, a total of 14% cost reduction was reported using RAP as compared to HMA. Moreover, the two materials (HMA and RAP) are manufactured in different types of manufacturing plants. Thus, in analyzing the cost difference between the two chosen manufacturing plants for virgin materials and RAP, a total of 57% cost reduction was observed for a RAP manufacturing plant. Besides this, no cost difference was observed in the rest of the phases, such as manpower, materials transportation, and construction activities, as the same procedures and types of machinery are used.

Accordingly, this study is concerned with the aggregate gradation variation that occurred to HMA. Limited studies have been conducted to determine how far to deviate beyond the upper and lower specification limits of aggregate gradations to achieve an acceptable performance level for HMA mixtures. Therefore, this study investigated the performance of HMA mixtures in the case of the aggregate gradation variation beyond the upper and lower specification limits.

The main objectives of this study are:

- To establish deviations from the upper and lower specification limits of the aggregate gradation curve that are not inimical to the performance of HMA mixtures. Thereafter, the properties of HMA mixtures prepared thereof would also be evaluated.
- Relative to the control mix, the performance of the best HMA mix that lies above the upper specification limit and the best mix beneath the lower specification limit would be compared via wheel tracking and indirect tensile strength tests.
- Determination of the HMA mixture with an optimized skeletal structure capable of providing enhanced resistance to high temperature induced rutting distress and low temperature engendered thermal cracking.

2. Materials and Methods

2.1. Materials

Coarse and Fine aggregates, with particles either retained or passing, respectively, on the 4.75 mm (No. 4) sieve were used. Similarly, aggregate particles finer than the 0.075 mm (No. 200) sieve served as fillers. All the aggregates were uniformly graded, conforming to the accepted gradation limits specified by the Egyptian Code [27]. Two types of coarse dolomite, grade (I) and grade (II) with the physical properties shown in Table 1 were also used as components in the asphalt. Fine siliceous sand with a bulk specific gravity of 2.65 g/cm^3 and limestone dust with a bulk specific gravity 2.85 g/cm^3 (filler) were used as asphalt concrete ingredients. The two types of coarse aggregate were obtained from the "ATAKA" quarry, Suez, Eygt, where the natural siliceous sand obtained from "ELREHAB" quarry, Cairo, Eygt. The binder used is Suez asphalt cement with 60–70 penetration grades, and physical properties shown in Table 2.

Table 1. Properties of coarse aggregate.

Test No.	Test	Designation No.	Values		Specification Limits
			Type 1 (Grade I)	Type 1 (Grade II)	
1	Bulk specific gravity (gm/cm^3)	AASHTO T-85	2.52	2.5	N/A
	Saturated surface dry specific gravity (gm/cm^3)		2.59	2.57	
	Apparent specific gravity (gm/cm^3)		2.7	2.6	
2	absorption %	AASHTO T-85	2.56	2.67	≤5
3	Los angles abrasion test %	AASHTO T-96	22.2	24.5	≤40
4	Stripping Test %	AASHTO T-182	>95	>95	≥95

Table 2. Physical properties of asphalt binder.

Test No.	Test	Designation No.	Values	Specification Limits
1	Penetration test (0.1 mm)	T-49	62	60–70
2	Softening Point (°C)	T-53	52	45–55
3	Flash point (°C)	T-48	+250	+250
4	Kinematics viscosity (cSt)	T-201	395	+320
5	Ductility (cm)	T-51	+100	≥ 95

2.2. Mix Gradation

The asphaltic wearing surface mix (mix 3B) tested in this study consists of 30% grade I coarse aggregate, 20% grade II coarse aggregate, 15% natural sand, 30% crushed sand and 5% limestone dust as a mineral filler. The design gradation of the control asphalt concrete mix (M0) was presented in Table 3. The design gradation of the control asphalt concrete mix (M0) is shown in Table 3, and the design gradation lies within the limits of Egyptian standard specifications for wearing surface (mix 3B) [27]. To highlight the different phases obtainable during asphalt production, aggregate gradation of mixtures was also extended (1) above and (2) below the upper and lower specification limits of the Egyptian standard specifications for wearing surface (mix 3B) [27]. For the three mixes with gradations lying above the upper specification limit, the deviations were +2% (M1), +4% (M2) and +6% (M3) Conversely, the gradations of the three mixes below the lower specification limit by −2%, −4% and −6% were denoted as M4, M5, and M6, respectively.

Table 3. Different gradations of asphalt concrete mixes.

Sieve Size	3B Asphalt Concrete Mixes							
	M0	M1	M2	M3	M4	M5	M6	Specification Limits
1″	100	100	100	100	98	96	94	100–100
3/4″	100	100	100	100	98	96	94	100–100
1/2″	85	100	100	100	74	72	69	75–100
3/8″	73	100	100	100	59	58	55	60–85
No. 4	48	66	77	88	34	33	32	35–55
No. 8	29	42	49	56	20	19	18	20–35
No. 30	18	26	31	35	10	9	9	10–22
No. 50	11	19	22	26	6	5	5	6–16
No. 100	8	14	17	19	4	3	3	4–12
No. 200	5	10	11	13	2	2	2	2–8

3. Experimental Work

The test program involves four stages, as displayed in Figure 1.

- In the first stage, after selection materials, five types of aggregate characterization tests will be applied; Los Angeles abrasion, water absorption, specific gravity, stripping value and selection of design gradation to attain the condition of stone-on-stone contact were undertaken. As well as the chosen asphalt binder will be evaluated through applying penetration, softening point, flash point, viscosity and ductility tests.
- In the second stage, the control asphalt mix (M0) and the other five asphalt mixes (from M2 to M6) were designed and prepared according to the Egyptian code [16].
- In the third stage, the effect of the aggregate gradation variation on HMA properties was examined based on the Marshall Mix Design Method.
- In the final stage, wheel loading tracking and indirect tensile strength tests were conducted on the M0 and samples obtained from the best mixes lying above and below the upper specification limit, and the lower specification limit, respectively.

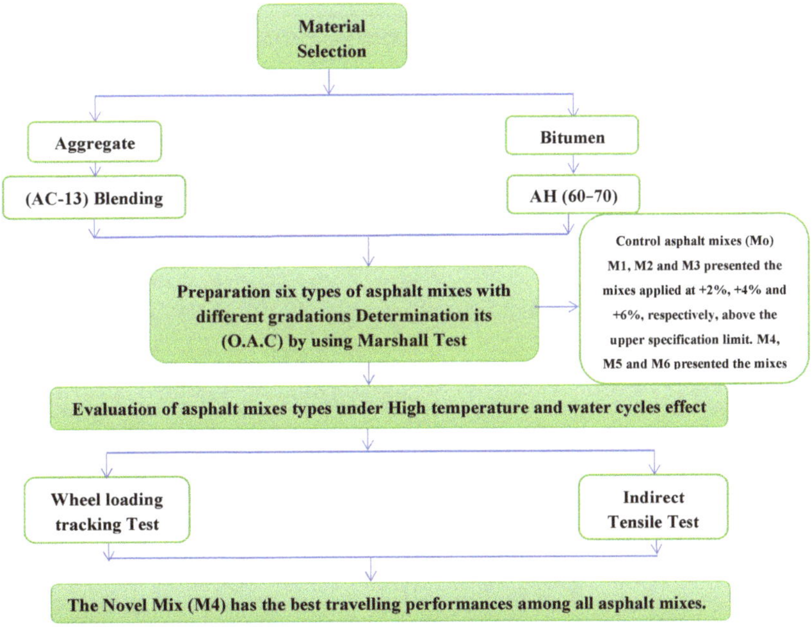

Figure 1. Flow chart of experimental works.

3.1. Marshall Test

The resistance of asphalt mixtures to plastic flow was determined in the current test by measuring the stability (kg) and flow values (mm) for each mix by using the Marshall apparatus, Marshall machine model TO-550-1 which obtained from USA. Marshall Test was performed using three 10 cm × 6 cm cylindrical specimens from each mix. After 24 h of specimen compaction, they were immersed in a water bath at 60 °C for 30 min before the Marshall test commenced. The test criterion used was the AASHTO T-245, 75-blow Marshall Compaction [28], highlighted in Figures 2 and 3. For the stability component of the Marshall test, all specimens were weighed in air and submerged in water, with the maximum load (kg) designated as the stability value. Note that a correction factor was implemented for specimens with thickness differing from 6 cm. On the other hand, the deformation undergone by specimens during loading to the peak value was measured using the flow meter and reported as the flow value.

Figure 2. Adding bitumen to aggregates.

Figure 3. Three Marshall Samples for each mix.

3.2. Wheel Loading Tracking Test

This test investigates the capability of a pavement to withstand rutting, and was performed according to the Egyptian Code specifications. Figure 4 shows the wheel tracking test machine used, which manufactured in Egypt according to BS-EN 12697-22 specifications. One slab measuring 440 mm × 330 mm × 50 mm according to the LTG 2015 [29,30] was prepared for each mixture, and tested at 60 °C under a wheel load of 53.5 kg. The track depth was recorded at regular intervals up to 45 min using a spring-less dial gauge. A description of the test setup is shown in Figure 5.

Figure 4. Wheel loading tracking device.

Figure 5. Slab under wheel load.

3.3. Indirect Tensile Strength

Tensile characteristics of bitumen mixtures were determined according to AASHTO T-283 test method, by loading the Marshall specimen along with its diametric plan with a constant rate, producing uniform stress [31]. In this test, two sets of specimens from each mixture (the control mix, the best mix that lies above the upper specification limit and the best mix that lies below the lower specification limit) were fabricated and evaluated. While one set of specimens was conditioned by soaking in water at 60 °C for 24 h, the other set was left unconditioned. The ratio of the average indirect tensile strength of the conditioned samples to the average indirect tensile strength of the unconditioned specimens was recorded as the Tensile Strength Ratio (TSR).

4. Results

4.1. Marshall Test Results

Marshall Test was conducted to determine the OAC for each mix, as reported in Table 4. Then evaluate the properties of the control mix and other mixes at different gradations. Different gradations showed different views of the aggregate gradation variation during asphalt production. The mixes of M0 through M6 for 3B mix showed this difference of gradations as shown in Figure 6. Where M0 presented the mix applied at the design gradation curve (control mix). M1, M2 and M3 presented the mixes applied at +2%, +4% and +6%, respectively, above the upper specification limit. M4, M5 and M6 presented the mixes applied at −2%, −4% and −6%, respectively, below the lower specification limit. The previous mixes of M0 through M6 differed at the OAC. The Marshall properties (stability, flow, bulk specific gravity, air voids, voids in mineral aggregate and voids filled with asphalt) were measured for all the mixes studied as shown in Table 4. Then the data of the results would be collected and analyzed.

Table 4. Effect of aggregate gradation variation on the investigated mixes.

Properties \ Mixes	M0	M1	M2	M3	M4	M5	M6	Specification Limits *
% OAC	4.5	4.3	4.0	3.5	5.0	5.2	5.5	3%–6%
Stability (Kg)	1203	1182	1175	985	1306	1318	1334	900 kg (min)
Flow (mm)	3.2	3.6	4.0	5.2	2.8	1.9	1.6	2–4 mm
Stiffness (kg/mm)	376	328	294	189	466	694	834	300–500 kg/mm
Bulk specific gravity (G_{mb}) (gm/cm^3)	2.345	2.358	2.361	2.350	2.305	2.263	2.245	–
% Air voids in total mix (V_a)	3.54	4.35	4.22	6.0	4.44	5.92	7.66	3%–5%
% Air voids in mineral aggregate (VMA)	16.44	15.8	15.43	15.39	18.3	19.96	20.84	–
% Air voids filled with asphalt (VFA)	87.5	72.5	72.7	61.4	75.7	65.3	63.2	–

* Egyptian code of practice (ECP 104) limits [27].

According to Table 4, the aggregate gradation variation above the upper specification limit had a negative effect on the mix stability. The stability value for the control mix (M0) was 1203 Kg. The following two mixes (M1 and M2), in which the gradation variation was increased to +2% and +4% above the upper specification limit respectively, the mix stability almost was not affected as it decreased by 1.7% and 2.3%. For the following mix (M3), as the gradation was increased to +6% above the upper specification limit, the mix stability decreased by 18%, reaching its lowest value of 985 kg but remained higher than the minimum stability value (900 kg).

Concerning values of flow for 3B mixes that lie above the upper specification limit the flow value for M0 was 3.2 mm. For the following two mixes (M1 and M2), as the gradation variation was increased above the upper specification limit by 2% and 4% respectively, the flow increased by 12.5% and 20% but still comply with the specification (2 mm ≤ Flow ≤ 4 mm). Mix (M3) lies beyond the specification requirement for flow, as its flow was (5.2 mm).

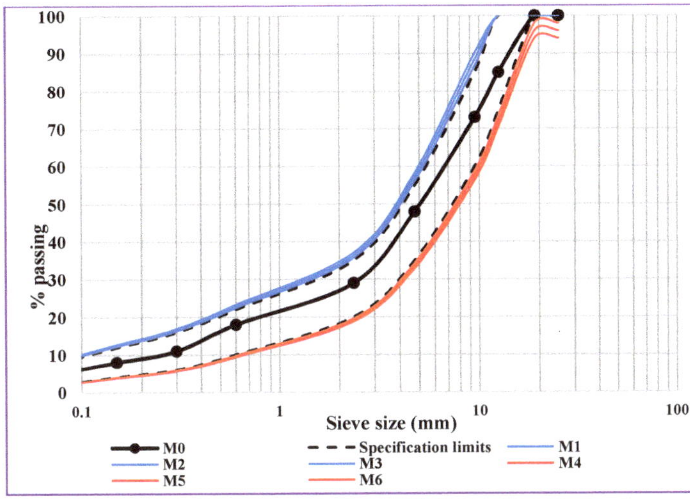

Figure 6. Different gradations for 3B mix.

Values of mix stability for 3B mixes that lie below the lower specification limit are displayed in Table 4. The mix stability for the following three mixes (M4, M5, and M6) increased by 8.5%, 9.6%, and 10% respectively as the gradation variation was below the lower specification limit by −2%, −4% and −6%, achieving stability values of 1306 kg at −2%, 1318 kg at −4%, and 1334 kg at −6% below the lower specification limit.

Based on the results in Table 4 show values of flow for 3B mixes that lie below the lower specification limit. The flow value for M0 was 3.2 mm. For the following mix (M4), in which the gradation was decreased to −2% below the lower specification limit, the flow decreased, reaching 2.8 mm, but remained higher than the minimum flow value (2 mm). This value decreased the flow by 12.5% compared with the control mix (M0). The following two mixes (M5 and M6) gave flow values beyond the specification value for flow (> 4 mm) according to the Egyptian code.

4.2. *Performance Evaluation Tests*

As M2 mix was the highest diffracted mix above the lower specification limit for 3B mixes gradation that comply with the specification requirements and M4 was the highest diffracted mix below the lower specification limit for 3B mixes gradation that comply with the specification requirements, M0, M2, and M4 were chosen to conduct to the performance evaluation tests (wheel load tracking and indirect tensile test).

4.2.1. Wheel Loading Tracking Test Results

Figure 7 presents the rutting depth test results of the three mixes; M0 (Control), M2 (the best mix that lies at +4% above the upper specification limit) and M4 (the best mix that lies at −2% below the lower specification limit). Figure 7 shows a rutting value of 3.92 mm for M0 mix, 4.85 mm for M2 mix, and 4.34 mm for M4 mix. Relative to the M0, the increase in rutting depth was 18% for the M2 and 9.5% for the M4. This implies that compared to the mix M2, the mix M4 had a superior rutting resistance. It is attributed to the increase in stability by 11.1% and the reduction in flow by 30% for M4 when compared with M2.

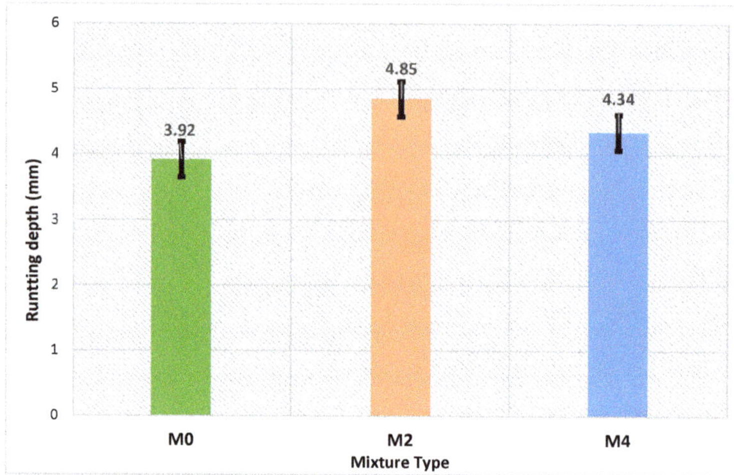

Figure 7. Rutting depth results.

Accordingly, it could be concluded that the 3B mix at −2% below the lower gradation limit showed high resistance to rutting phenomena compared with the 3B mix at +4% above the upper gradation limit.

4.2.2. Indirect Tensile Strength Test

The TSR results of the three mixes are shown in Figure 8. The results shown in Figure 8 indicate that while the TSR for the mix M0 was 83.87%, a value of 81.62% was recorded for the mix M2. Compared to the mix M0, this represents about 2.7% reduction in the TSR. Figure 8 also shows that at a TSR value of 82.84%, the TSR of the mix M4 was slightly better than that of the mix M2. The main inference from these results is that comparing mix M2 and mix M4, the latter possesses a better resistance to moisture induced damage.

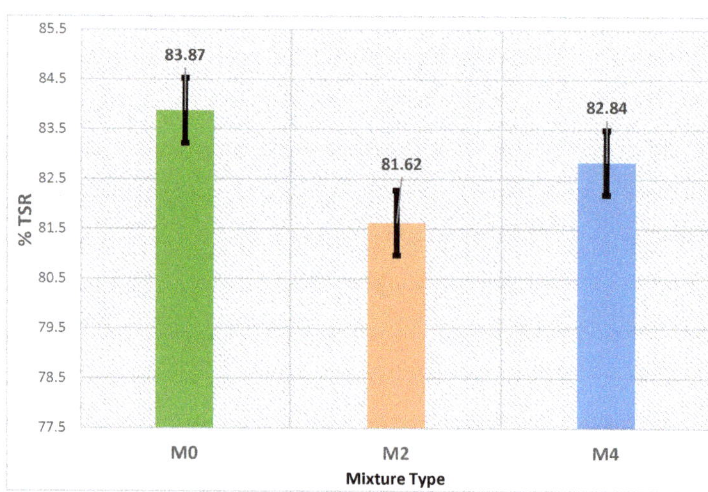

Figure 8. Tensile Strength Ratio (TSR) results.

Accordingly, it could be concluded that the 3B mix at −2% below the lower gradation limit showed good resistance to moisture damage phenomena compared with the 3B mix at +4% above the upper gradation limit.

5. Discussion

This section aims to introduce the analysis of the main results of physical properties of six types of asphalt mixes, as well as the pavement performance under high temperature and water cycles effects.

5.1. Analysis of Marshall Test Results

Based on the results of the Marshall test mentioned in Table 4 and Figure 6, it can be concluded that the aggregate gradation variation over the upper specification limit causes an increased in the mixture stability and decreased the flow. On the contrast, the variation of aggregate gradation lower than specification limit result in decreased the stability and increased the flow of mixtures.

The variation in aggregate gradation above the upper specification limit leads to a damaging influence on the mixture stability and it may be due to the excessive asphalt in the mix due to the 30% reduction in the voids filled with asphalt (VFA) as a result to the fineness of the stone matrix. On contract, the flow of mixtures which prepared above the upper specification limit was increased above the upper specification limit. The increase in flow is attributed to the excessive asphalt in the mix as it is the same reason of stability reduction.

The aggregate gradation variation under the lower specification limit leads to a significant effect on the stability of mixes. This increase in stability is attributed to the adequate asphalt content in the mixes due to the increase in the voids in stone matrix (VMA) so the voids filled with asphalt (VFA) included the entire content of the asphalt. The flow values of mixes that lie under the lower specification limit was decreased, and it is may be because of the absence of excessive asphalt in the mixes and the reduction in fines in the stone matrix.

5.2. Analysis of Pavement Performance

5.2.1. Wheel Loading Tracking Test

The outcomes of wheel loading tracking test revealed that the permanent deformation of M2 and M4 mix is slightly increased comparing with Control asphalt mixture.

The main reason for increasing the rutting depth for M2 and M4 mix is that the total air voids among the skeleton of aggregate is increased which lead to increase the bitumen absorbed by the mixes, thus the rutting depth of M2 and M4 mixes is slightly increased.

5.2.2. Indirect Tensile Test

The indirect tensile strength results from the soaking group in water and unconditional group, and the tensile strength ratio (TSR) are displayed in Figure 8. Results observed that the moisture damage resistance of M2 and M4 was increased as opposed to Control asphalt Mixture.

It may be due to the fineness of its stone matrix, that led to an increase in the cohesion between the stone matrix and the low asphalt content that was affected harmfully by the elevated temperature of the water path. The tensile strength of asphalt mixes was increased, accordingly; as well as, the increase of adhesion force among asphalt mastic and aggregate led to the improvement of the anti-shear strength and rutting of the asphalt mixes.

Generally, based on this discussion, it could be concluded that the 3B mix at −2% below the lower gradation limit presented a good performance against moisture induced damage compared with the 3B mix at +4% above the upper gradation limit.

6. Conclusions

This paper presented a novel aggregate gradation variation with range (2%, 4%, and 6%) from the upper and lower specification limits of aggregate. The HMA mixtures were prepared and performed in laboratory according to the Egyptian Code specifications, afterward Marshall, wheel loading tracking and Indirect Tensile Strength (ITS) tests were applied in order to evaluate the HMA properties, high temperature performance, and water stability of different asphalt mixes, respectively. Based on the research outcomes, the following conclusions can be drawn.

Both of M2 (aggregate gradation within +4% the upper specification limit) and M4 (aggregate gradation within −2% the lower specification limit) significantly enhances the asphalt mixes performance. M2 mix has clearly enhanced the water stability performance, but the enhancement of rutting resistance is limited. M4 mix has a great effect on improving both of high temperature performance and water damage resistance. The rutting resistance of the mix M4 was superior to that of the mix M2. The resistance of mix M4 to moisture damage was higher than that of the mix M2. In summary, compared to the mix M2, the capacity of the mix M4 to enhance the service life and ride quality of pavements is determined to be higher.

7. Recommendation and Future Works

The following main recommendations can be outlined:

- For cold climates, the study should be performed and evaluated.
- Study the influence of OAC variation with the aggregate gradation variation on the performance of asphalt mixes.
- The economic considerations should be studied extensively regarding the possibility of accepting down to −2% variation in 3B mix gradation below the lower specification limits.

Author Contributions: Conceptualization and design of experimental, M.S.E. and M.A.; performed the experimental, validation and formal analysis, M.S.E. and M.A. Project Administration, M.S.E., M.A. and Y.Y.; Writing—Original Draft Preparation, M.S.E. and M.A.; Writing-Review and Editing, M.S.E. and M.A.; Supervision, M.S.E., M.A. and Y.Y. All authors have read and agreed to the published version of the manuscript.

Funding: This research received no external funding.

Institutional Review Board Statement: Not applicable.

Informed Consent Statement: Not applicable.

Data Availability Statement: Not applicable.

Acknowledgments: The authors are thankful to the teamwork of General Authority for Roads, Bridges and Land Transport-GARBLT, Egypt.

Conflicts of Interest: The authors declare no conflict of interest.

References

1. Porter, M. The competitive advantage of the inner city. In *The City Reader*; Routledge: London, UK, 2015; pp. 358–371.
2. Abdullah, B.A.O.; Jakarni, F.M.; Al-Shakhrit, A.K.S.; Masri, K.A. Usage of recycled glass in hot mix asphalt: A review. *Construction* **2021**, *1*, 29–34. [CrossRef]
3. Erdem, C.; Harvey, J.T. Evaluation of laboratory, construction, and performance variability by bootstrapping and monte carlo methods for rutting performance prediction of heavy vehicle simulator test sections. *J. Transp. Eng.* **2011**, *137*, 897.
4. Bazi, G.M.; Sebaaly, P.E.; Weitzel, D.; Elicegui, M. A Laboratory study of construction variability impacts on fatigue and thermal cracking resistance of HMA mixtures. *Int. J. Pavements* **2006**, *5*, 37–49.
5. Elwardany, M.; Planche, J.-P.; King, G. Universal and practical approach to evaluate asphalt binder resistance to thermally-induced surface damage. *Constr. Build. Mater.* **2020**, *255*, 119331. [CrossRef]
6. Gao, Y.; Hou, K.; Jia, Y.; Wei, Z.; Wang, S.; Li, Z.; Ding, F.; Gong, X. Variability evaluation of gradation for asphalt mixture in asphalt pavement construction. *Autom. Constr.* **2021**, *128*, 103742. [CrossRef]
7. Fan, Z.; Du, C.; Liu, P.; Wang, D.; Oeser, M. Study on interfacial debonding between bitumen and aggregate based on micromechanical damage model. *Int. J. Pavement Eng.* **2022**, *23*, 340–348. [CrossRef]
8. Yue, Y.; Abdelsalam, M.; Khater, A.; Ghazy, M. A comparative life cycle assessment of asphalt mixtures modified with a novel composite of diatomite powder and lignin fiber. *Constr. Build. Mater.* **2022**, *323*, 126608. [CrossRef]
9. Khater, A.; Luo, D.; Abdelsalam, M.; Yue, Y.; Hou, Y.; Ghazy, M. Laboratory evaluation of asphalt mixture performance using composite admixtures of lignin and glass fibers. *Appl. Sci.* **2021**, *11*, 364. [CrossRef]

10. Shahnewaz, S.M.; Masri, K.A.; Ghani, N.A.A.A. Porous asphalt modification using different types of additives: A review. *Construction* **2021**, *1*, 44–53. [CrossRef]
11. Wang, D.Y.; Kan, L.; Xu, C. Evaluation of rutting resistance on asphalt mixture based on aggregate contact characteristics. *South China Univ. Technol. Nat. Sci. Ed.* **2012**, *40*, 121–126.
12. Husain, N.M.; Karim, M.R.; Mahmud, H.B.; Koting, S. Effects of aggregate gradation on the physical properties of semiflexible pavement. *Adv. Mater. Sci. Eng.* **2014**, *2014*, 529305. [CrossRef]
13. Li, S.; Tan, H.Y.; Wang, N.; Hong, M.; Li, L.; Cheung, F.; Feng, Y. Research progress and future development for reliability of asphalt pavement structure. *China J. Highw. Transp.* **2016**, *29*, 4305074.
14. Hussein, A.A.; Jaya, R.P.; Hassan, N.A.; Yaacob, H.; Huseien, G.F.; Ibrahim, M.H.W. Performance of nanoceramic powder on the chemical and physical properties of bitumen. *Constr. Build. Mater.* **2017**, *156*, 496–505. [CrossRef]
15. Zhang, Y.; Luo, X.; Onifade, I.; Huang, X.; Lytton, R.L.; Birgisson, B. Mechanical evaluation of aggregate gradation to characterize load carrying capacity and rutting resistance of asphalt mixtures. *Constr. Build. Mater.* **2019**, *205*, 499–510. [CrossRef]
16. Martin, W.D.; Putman, B.J.; Neptune, A.I. Influence of aggregate gradation on clogging characteristics of porous asphalt mixtures. *J. Mater. Civ. Eng.* **2014**, *26*, 04014026. [CrossRef]
17. Abo-Qudais, S.; Al-Shweily, H. Effect of aggregate properties on asphalt mixtures stripping and creep behavior. *Constr. Build. Mater.* **2007**, *21*, 1886–1898. [CrossRef]
18. Moghaddam, T.B.; Mohamed, R.K.; Mahrez, A. A review on fatigue and rutting performance of asphalt mixes. *Sci. Res. Essays* **2011**, *6*, 670–682.
19. Bazi, G.M. Impact of Construction Variability on Pavement Performance. Ph.D. Thesis, University of Nevada, Reno, Nevada, 2006.
20. Lv, Q.; Huang, W.; Zheng, M.; Sadek, H.; Zhang, Y.; Yan, C. Influence of gradation on asphalt mix rutting resistance measured by Hamburg Wheel Tracking test. *Constr. Build. Mater.* **2020**, *238*, 117674. [CrossRef]
21. Golalipour, A.; Jamshidi, E.; Niazi, Y.; Afsharikia, Z.; Khadem, M. Effect of aggregate gradation on rutting of asphalt pavements. *Procedia-Soc. Behav. Sci.* **2012**, *53*, 440–449. [CrossRef]
22. Yu, H.; Yang, M.; Qian, G.; Cai, J.; Zhou, H.; Fu, X. Gradation segregation characteristic and its impact on performance of asphalt mixture. *J. Mater. Civ. Eng.* **2021**, *33*, 04020478. [CrossRef]
23. Abdelsalam, M.; Yue, Y.; Khater, A.; Luo, D.; Musanyufu, J.; Qin, X. Laboratory study on the performance of asphalt mixes modified with a novel composite of diatomite powder and lignin fiber. *Appl. Sci.* **2020**, *10*, 5517. [CrossRef]
24. Elliott, R.P.; Ford, M.C., Jr.; Ghanim, M.; Tu, Y.F. Effect of aggregate gradation diffraction on asphalt concrete mix properties. *Transp. Res. Rec.* **1991**, *1317*, 52–60.
25. Awan, H.H.; Hussain, A.; Javed, M.F.; Qiu, Y.; Alrowais, R.; Mohamed, A.M.; Fathi, D.; Alzahrani, A.M. Predicting marshall flow and marshall stability of asphalt pavements using multi expression programming. *Buildings* **2022**, *12*, 314. [CrossRef]
26. Rafiq, W.; Musarat, M.; Altaf, M.; Napiah, M.; Sutanto, M.; Alaloul, W.; Javed, M.; Mosavi, A. Life cycle cost analysis comparison of hot mix asphalt and reclaimed asphalt pavement: A case study. *Sustainability* **2021**, *13*, 4411. [CrossRef]
27. Egyptian Code of Practice (ECP 104). In *Urban and Rural Roads, Edition 1: Road Materials and Their Tests (Part Four)*; Ministry of Housing, Utilities and Urban Communities: Cairo, Egypt, 2008.
28. American Association of State Highway and Transportation Officials. *Standard Method of Test for Resistance to Plastic Flow of Bituminous Mixtures Using Marshall Apparatus. T-245*; AASHTO: Washington, DC, USA, 2015.
29. *Laboratory Tests Guide, General Authority for Roads, Bridges and Land Transport*; Today's News: Cairo, Egypt, 2015.
30. Eisa, M.S.; Basiouny, M.E.; Youssef, A.M. Effect of using various waste materials as mineral filler on the properties of asphalt mix. *Innov. Infrastruct. Solutions* **2018**, *3*, 27. [CrossRef]
31. American Association of State Highway and Transportation Officials. *Standard Method of Test for Resistance of Compacted Asphalt Mixtures to Moisture-Induced Damage. T-283*; AASHTO: Washington, DC, USA, 2018.

coatings

MDPI

Article

A High Proportion Reuse of RAP in Plant-Mixed Cold Recycling Technology and Its Benefits Analysis

Yanan Li [1], Yuzhen Zhang [2] and Shucai Zhang [1,*]

1 State Key Laboratory of Safety and Control for Chemicals, SINOPEC Research Institute of Safety Engineering Co., Ltd., Qingdao 266100, China
2 College of Chemistry and Chemical Engineering, China University of Petroleum (East China), Qingdao 266580, China
* Correspondence: zhangsc.qday@sinopec.com; Tel.: +86-13854259198

Abstract: The concept of the "no-waste city" has focused increasing attention on the recycling of solid waste. One such waste is reclaimed asphalt pavement (RAP), which is generated during road maintenance. The potential to reuse this resource has attracted extensive attention in recent years. This paper explores this concept via a case study of the reconstruction of two sections of the Beijing-Taipei Expressway (from Bengbu to Hefei, sections K69–K69 + 500 and K69 + 500–K69 + 900). The upper base layer of one section was paved with a novel mixture of emulsified asphalt, mixed with a high proportion of RAP made using plant-mixed cold recycling technology (EAPM-HP_{RAP}). For comparison, the upper base layer of the other section was paved with a conventional large-stone porous asphalt mix (LSPM). The proportions of the components of EAPM-HP_{RAP} were optimized via laboratory-based proportioning design followed by proportioning verification. The results showed that the high-temperature stability, water damage resistance and pavement strength of the EAPM-HP_{RAP} met the specifications of relevant engineering standards. Next, the economic and environmental benefits of this novel approach were estimated. The approach was estimated to save CNY (China Yuan) 1.5–1.8 million in engineering costs per km of road (roadbed width = 27.5 m) and CNY 158–189 million for the whole project (105 km in length). It was also estimated to reduce energy consumption equivalent to 67.41 tons of standard coal per km. Further calculations showed that every km of pavement could reduce CO_2 emissions by 176.6 tons, SO_2 emissions by 0.6 tons, NO_X emissions by 0.5 tons, ash emissions by 17.6 tons and soot emissions by 1.0 tons compared with conventional methods. For the whole road section, this is equivalent to reducing CO_2 emissions by 18,543 tons, SO_2 emissions by 60.2 tons, NO_X emissions by 52.5 tons, ash emissions by 1848 tons, and soot emissions by nearly 105 tons. In summary, it is feasible for EAPM-HP_{RAP} to be used as the upper base layer in highway renovation projects. It reduces the need to mine new ores and allocate land to RAP storage, which is associated with soil and water pollution due to chemical leaching from aged asphalt. This approach provides great economic and environmental benefits compared with the use of conventional pavement technology.

Keywords: solid waste; high proportion reuse; reclaimed asphalt pavement; plant-mixed cold recycling technology; benefits analysis

Citation: Li, Y.; Zhang, Y.; Zhang, S. A High Proportion Reuse of RAP in Plant-Mixed Cold Recycling Technology and Its Benefits Analysis. *Coatings* **2022**, *12*, 1283. https://doi.org/10.3390/coatings12091283

Academic Editors: Valeria Vignali and Giorgos Skordaris

Received: 1 July 2022
Accepted: 27 August 2022
Published: 2 September 2022

Publisher's Note: MDPI stays neutral with regard to jurisdictional claims in published maps and institutional affiliations.

Copyright: © 2022 by the authors. Licensee MDPI, Basel, Switzerland. This article is an open access article distributed under the terms and conditions of the Creative Commons Attribution (CC BY) license (https://creativecommons.org/licenses/by/4.0/).

1. Introduction

Asphalt is one of the most important products of the petroleum refining industry. Road asphalt products are obtained by the distillation, solvent extraction or oxidation of the residual oil obtained after the vacuum distillation of crude oil. According to data from the National Bureau of Statistics, China's petroleum asphalt output in 2020 was more than 60 million tons; however, the domestic output does not meet the domestic demand and China imported nearly five million tons of petroleum asphalt in 2020.

Asphalt pavement is being increasingly used with the rapid development of the national economy because of its many advantages. China's total highway mileage was

5.2 million km at the end of 2021, which included 161,000 km of expressways. Due to load magnitude and load repetition associated with temperature and environmental factors, such as ultraviolet radiation, oxygen and moisture, the asphalt in pavement structures ages slowly and continuously. Its light components, such as saturated and aromatic components, are gradually converted into heavy components, such as resins and asphaltenes, resulting in increasing hardening and brittleness and reduced bonding performance. This makes asphalt pavement prone to rutting, surface aggregate loss, potholes, cracks and other problems, which, in turn, reduce poor driving comfort, safety and road usability. Hence, it is necessary to repair or re-pave degraded asphalt pavement to reduce these impacts. A huge amount of reclaimed asphalt pavement is generated during the maintenance of old asphalt pavement. This contains a large amount of aged asphalt that contains toxic and harmful substances, such as anthracene, naphthalene and pyridine, which can cause serious pollution to soil and water. Moreover, discarding RAP wastes valuable non-renewable resources, such as asphalt and mineral materials. Therefore, asphalt pavement recycling technology (APRT) that can realize the reuse of RAP has attracted increasing research attention [1–3].

APRT refers to the process of excavating, recycling, crushing, and screening old asphalt pavement, and then mixing it with new asphalt, new aggregate (when necessary), and recycling agents (when necessary). There is a complete set of processes for mixing the appropriate proportions of new asphalt mixture and re-paving and forming new pavement layers according to certain performance requirements. APRT not only recycles RAP resources but also reduces the consumption of non-renewable resources, such as asphalt, avoids land use and environmental pollution caused by RAP stacking, and can greatly reduce engineering costs [4–6]. Wang et al. used the PaLATE method (Pavement Life-cycle Assessment Tool for Environmental and Economic Effects) to estimate the energy consumption and carbon dioxide emissions of the RAP-added mixture and the new HMA (Hot Mixture Asphalt) and evaluated the environmental benefits of using RAP. The results showed that producing a mixture containing 30% RAP required only 84% of the energy and produces 80% of carbon dioxide emissions compared to using 100% primary aggregates [7].

The plant-mixed cold recycling technology using emulsified asphalt with a high proportion of RAP (EAPM-HP_{RAP}) is a kind of APRT in which a series of construction operations, such as mixing, paving and rolling, can be carried out under normal temperature conditions. Therefore, in addition to the advantages of APRT described above, it can also reduce energy consumption and pollutant emissions, and protect the health of construction workers [8–10].

This study investigated the feasibility of using EAPM-HP_{RAP} in order to improve RAP recycling efficiency. The determination of the optimal proportions of EAPM-HP_{RAP} was first carried out in the laboratory. Then, the EAPM-HP_{RAP} and large stone porous asphalt mixture (LSPM) were used as an upper pavement base to pave two test sections and conduct performance tests. The results show that the high-temperature stability, water-damage resistance and pavement strength can meet the required specifications when the EAPM-HP_{RAP} is used as a base. Finally, the economic and environmental benefits of RAP-reuse technology were estimated based on the two test sections [11].

2. Materials and Methods

2.1. Reclaimed Asphalt Pavement

The characteristics of RAP have very important impacts on the performance of the final mixture [12,13]. In this study, RAP was collected from materials milled during an overhaul of the Beijing-Taipei Expressway between Bengbu to Hefei. The RAP was sieved according to standard JTG E42 (Test Methods of Aggregate for Highway Engineering) [14] in order to understand the particle size distribution (gradation). The results are listed in Table 1.

Table 1. Particle size distribution of the studied RAP.

Screen Size (mm)	26.5	19	9.5	4.75	2.36	1.18	0.6	0.3	0.15	0.075
Passing rate (%)	100	97.4	80	52.2	29	18.7	11	6.3	4.1	2.6

The parameters of the RAP were tested in accordance with the requirements of *Test Methods of Aggregate for Highway Engineering* and the results are listed in Table 2.

Table 2. Aggregate properties of the studied RAP.

Aggregate Type	Parameter (%)	Result	Standard
Coarse aggregate	Asphalt content	3.8	T 0722-1993
	Needle-like content	6.6	T 0312-2005
	Crushed stone value	16.2	T 0316-2005
Fine aggregate	Asphalt content	4.9	T 0722-1993
	Needle-like content	7.6	T 0722-1993
	Angularity	33.5	T 0345-2005
	Sand equivalent	69.3	T 0334-2D05

2.2. Recycling Asphalt from the RAP

The recycled asphalt was extracted from the RAP using tetrachloroethylene solvent. The penetration of the recycled asphalt was 22 and the ductility was only 36.2 cm, which indicates that the asphalt was seriously aged after more than ten years of use. Despite this, it still had a significant impact on the diffusion process of the newly added asphalt binder [15].

2.3. Mineral Aggregate

In this study, limestone with a particle size of 10–20 mm from Zhangdian Hutian, Jinan, was used as fresh aggregate and limestone mineral powder produced in Pingyin, Jinan, as ore powder. The density, hydrophilicity coefficient (H–C) and methylene blue value (MBV) of the mineral powder were tested (Table 3).

Table 3. Properties of the mineral powder.

Parameter	Result	Standard
Density (g.cm^{-3})	2.673	T 0352-2000
H-C	0.671	T 0353-2000
MBV (g/kg)	1.00	T 0349-2005

2.4. Cement

Cement not only improves the early strength of EAPM but also promotes the demulsification of emulsified asphalt and enhances the post-demulsification interfacial bonding performance between asphalt and fresh aggregate, thereby improving the high-temperature stability of EAPM [16]. Ordinary Portland cement (PO. 32.5; Shanshui Brand, Sunnsy Group, Jinan, China) was used in this study. Its properties are shown in Table 4.

2.5. Emulsified Asphalt

The emulsified asphalt (EA) used for the APRT was produced by a colloid mill in the laboratory. Its properties are shown in Table 5.

Table 4. Properties of the Portland cement used in the EAPM.

Parameter	Requirement	Result
Fineness (80 μm, %)	≤10	0.7
Standard consistency water consumption (%)	≤28	28
Initial setting time (min)	≥90	167
Final setting time (min)	≤600	223
Specific surface area (m^2/kg)	300–450	328
3-day flexural strength (MPa)	>2.5	3.0
3-day flexural strength (MPa)	>5.5	6.4
3-day compressive strength (MPa)	>11.0	15.5
28-day compressive strength (MPa)	>32.5	34.3

Table 5. Properties of emulsified asphalt.

Parameter		Unit	Requirement	Result	Experimental Method
Demulsification speed		—	Slow or moderate	Slow crack	T 0658
Charge of the particle		—	Cationic (+)	Cationic (+)	T 0653
Engla viscosity E25		—	2–30	4	T 0622
Adhesion to coarse aggregates		—	≥2/3	Qualified	T 0654
Adhesion to fine aggregates		—	Mix well	Mix well	T 0659
Residue content by evaporation		%	≥55	62	T 0651
Sieve residue (1.18 mm)		%	≤0.1	0	T 0652
Residual asphalt	Penetration (25 °C)	0.1 mm	45–150	62	T 0604
	Ductility (15 °C)	cm	≥40	88.2	T 0605
Storage stability	1 d, 25 °C	%	≤1	0.8	T 0655
	5 d, 25 °C	%	≤5	3	T 0655

3. Results and Discussion

3.1. Proportioning Design

To determine the optimal material ratio for achieving the best pavement performance, a proportioning design was carried out in the laboratory [17,18].

3.1.1. Experimental Gradation

Based on the sieving results of RAP and mineral aggregates, and referring to the requirements of JTG F41 (*Technical Specification for Highway Asphalt Pavement Recycling*) [19] for EAPM with a medium granular-gradation range, the mixture ratio was determined to be RAP: 10–20 mm limestone: mineral powder: cement = 80:16:2:2. The experimental gradation data and corresponding curve are shown in Table 6 and Figure 1.

Table 6. Data of the experimental gradation.

Screen Size (mm)	0.075	0.15	0.3	0.6	1.18	2.36	4.75	9.5	13.2	16	19	26.5
Upper limit of grading	8	14.5	21	30.6	41.2	50	65	80	86.6	93.2	100	100
Lower limit of grading	2	2.5	3	8.6	14.2	20	35	60	70	80	90	100
Experimental gradation	3	4	6	9	15	27	50	74	85	92	97	100

3.1.2. Mixing Proportion

Referring to the conventional dosages used in actual pavement works, the amount of EA selected was 4.0%, while external water dosages of 1.5%, 2.0%, 2.5%, 3.0% and 3.5% were trialled. Specimen mixtures were prepared using the experimental gradation determined above. Then, compaction tests were performed on samples and their maximum dry density was tested in accordance with JTG E40 (*Test Methods of Soils for Highway Engineering*) [20]. Figure 2 shows the relationship between the maximum dry density and amount of external water.

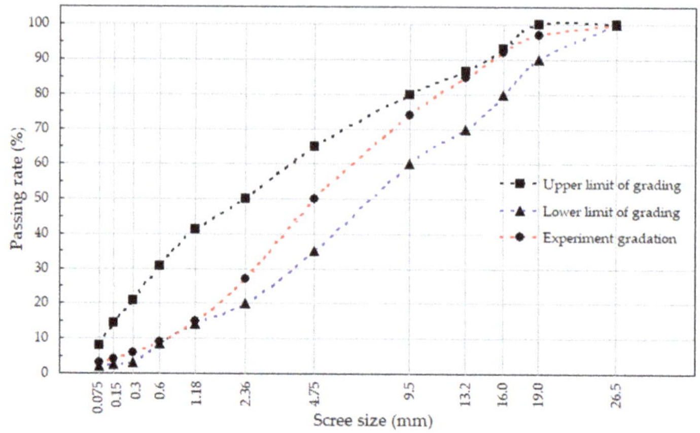

Figure 1. Grade curve of the experimental gradation.

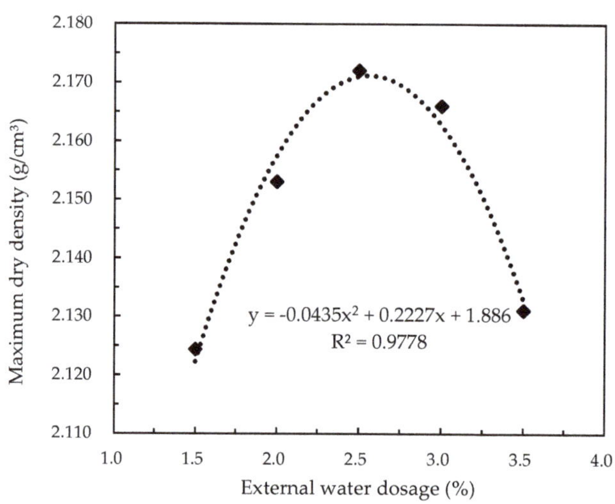

Figure 2. Influence of external water dosage on the maximum dry density of specimens.

The maximum dry density first increased and then decreased with increases in the amount of external water, reaching a peak of 2.172 g/cm^3 at an external water content of 2.5%, representing an increase of 2.2 percentage points compared to the lowest value. Therefore, the optimal external water content was determined to be 2.5% and the optimum total liquid content was 6.5%. At a total liquid content of 6.5% and cement content of 2.0%, specimen mixtures were prepared at EA contents of 3.0%, 3.5%, 4.0%, 4.5% and 5.0%. Then, relative performance tests were carried out, including bulk specific density, Marshall stability and residual Marshall stability (Table 7).

Figure 3 is based on the data in Table 7.

It can be seen that both the Marshall stability and residual Marshall stability peaked at an EA content of 4.0%. Hence, according to the analysis, the optimal content of EA was 4.0% and the optimal external water dosage was 2.5%.

Table 7. Volume Index and Marshall Stability of Specimens.

EA Content (%)	External Water Dosage (%)	Bulk Specific Density	Theoretical Maximum Relative Gravity	Void Ratio (%)	Marshall Stability (kN)	Residual Marshall Stability (kN)
3.0	3.5	2.255	2.512	9.8	12.7	14.1
3.5	3.0	2.258	2.521	10.1	14.5	17.1
4.0	2.5	2.266	2.559	10.6	14.9	18.3
4.5	2.0	2.263	2.507	9.6	14.1	17.5
5.0	1.5	2.261	2.496	9.3	12.5	13.8

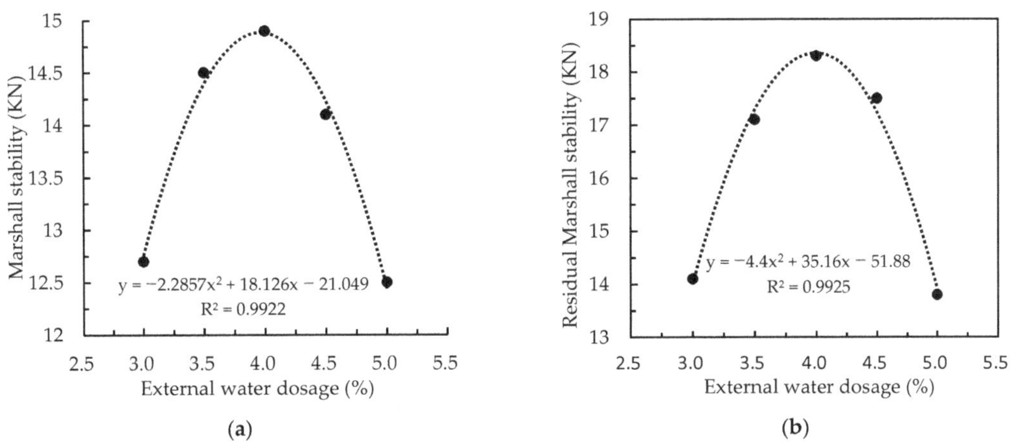

Figure 3. Relationships between external water dosage and the Marshall stability (**a**) and residual Marshall stability (**b**).

3.1.3. High-Temperature Stability

An asphalt mixture is a typical viscoelastic material that is prone to flow deformation under high-temperature conditions [21]. Therefore, the repeated loading of road vehicles, especially heavy-duty overloaded vehicles, can lead to irreversible deformation of the road surface, typically rutting damage [22]. In this paper, the dynamic stability at 60 °C was used as an index to evaluate the high-temperature stability of EAPM. *Dynamic stability* refers to the number of standard axle loads a mixture is subjected to for each 1 mm deformation under high-temperature conditions (generally selected as 60 °C). Rutting specimens (dimensions = 300 × 300 × 80 mm) were prepared in the laboratory by the wheel rolling method according to the experimental gradation and mixing proportion determined above. Rutting tests were carried out in accordance with the requirements of JTG E20 (Standard Test Methods of Bitumen and Bituminous Mixture for Highway Engineering) [23]. The test temperature was 60 °C and the wheel pressure was 0.7 MPa.

Studies have shown that rutting generally occurs on days when the average maximum temperature on the road surface is above 28 °C for seven consecutive days. The test results show that the dynamic stability of the emulsified asphalt mixed with 80% of RAP and made by the cold regeneration technique was 2320 times/mm, which meets the requirements of JTG F41 (*Technical Specification for Highway Asphalt Pavement Recycling*) and shows that the anti-rutting performance is fully qualified.

3.1.4. Water Damage Resistance

Freeze-thaw splitting tests were conducted to measure the effect of freeze-thaw cycling on the asphalt mixtures under specified conditions. The splitting strength ratios of the specimens before and after water damage were determined to evaluate the water damage resistance of the asphalt mixtures. The specimens were divided into two groups for the tests. The first set of samples was used to measure the splitting tensile strength R_{T1} without freeze-thaw cycling, and the second set was used to determine the splitting strength R_{T2} with freeze-thaw cycling. The freeze-thaw cycling was carried out in accordance with the requirements of standard JTG E20 (*Standard Test Methods of Bitumen and Bituminous Mixture for Highway Engineering*). The *TSR* (Tensile strength ratio) can be calculated by Equation (1):

$$TSR = \frac{R_{T2}}{R_{T1}} \times 100 \quad (1)$$

where R_{T1} and R_{T2} are the splitting strengths (MPa) of specimens without and with freeze-thaw treatment, respectively, and TSR is the ratio of R_{T2} to R_{T1} (%).

The test results show that the splitting strength of the specimen without freeze-thaw was 1.17 MPa, while that after freeze-thaw was 0.99 MPa. The TSR was 84.6%, which meets the requirements of standard JTG F41 (*Technical Specification for Highway Asphalt Pavement Recycling*).

3.2. Proportioning Verification

On the basis of the proportioning design, two test sections (from K69 to K69 + 500 and from K69 + 500 to K69 + 900) were paved in combination with a reconstruction project in the Bengbu to Hefei section of the Beijing-Taipei Expressway. Firstly, a 16 cm asphalt surface layer and an 18 cm upper base layer of the old road were milled off. Then, a cold regeneration technique was used to make two emulsified asphalt mixtures mixed with high proportions of RAP (EAPM-HP_{RAP}) or LSPM. The mixtures were used as the upper base layers of the two test sections, upon which the original asphalt surface structure was re-paved. Figure 4 shows the pavement structures of the two test sections.

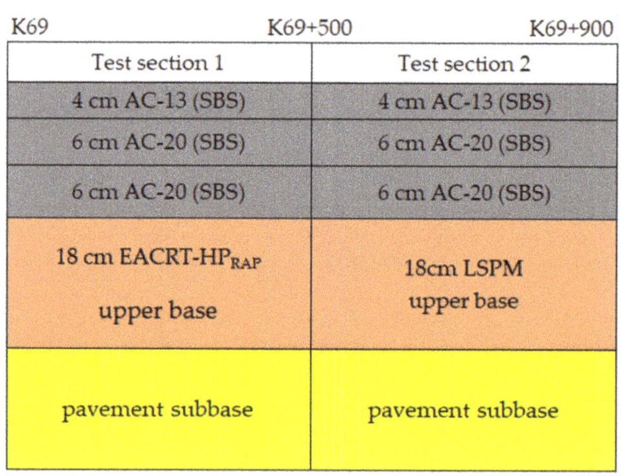

Figure 4. Schematic diagram of the pavement structures of the two test sections.

3.2.1. Compressive Strength

Preliminary specimens of the upper base layer were obtained by a core drilling machine at the four sampling points of K69 + 100 (overtaking lane), K69 + 100 (carriage lane), K69 + 200 (overtaking lane) and K69 + 200 (carriage lane) after the construction of the upper base layer was completed. Both ends of the drilled core samples were cut off to obtain a cylindrical specimen with a size of $\varphi100 \times 100$ mm. Compression tests were carried out according to the requirements of standard JTG E20 (*Standard Test Methods of Bitumen and Bituminous Mixture for Highway Engineering*). The test temperature was 20 °C and the loading rate was 2 mm/min. The compressive strength is calculated by Equation (2):

$$R_c = \frac{4P}{\pi d^2} \quad (2)$$

where R_c is the compressive strength of the specimen (MPa), P is the load (N) at which the specimen fails, and d is the specimen diameter (mm). The results are shown in Table 8.

Table 8. Compressive strength of the test section.

Road Location	Maximum Load (kN)	Compressive Strength (MPa)
K 69 + 100 (overtaking lane)	26.41	3.36
K 69 + 100 (carriage way)	32.21	4.09
K 69 + 200 (overtaking lane)	29.12	3.70
K 69 + 200 (carriage way)	26.14	3.32
Average value	28.49	3.62

The results show that the compressive strength meets the specifications, which shows that it is feasible to use emulsified asphalt cold recycling technology with 80% RAP as the upper base layer in the expressway renovation project.

3.2.2. Deflection Value

The deflection values of the two sections were tested after the construction of the upper base layers was completed (Figure 5).

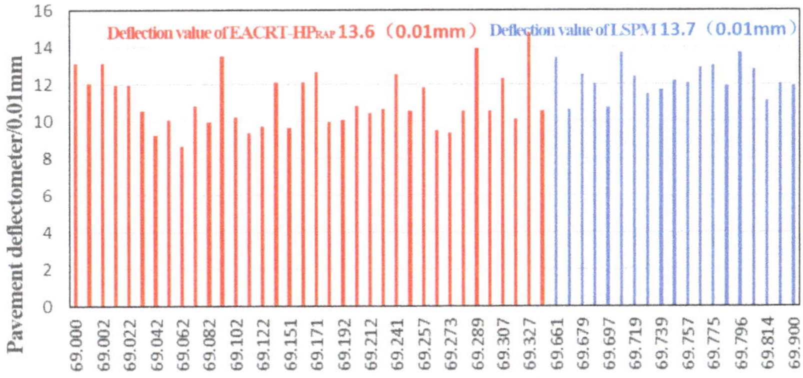

Figure 5. Comparison of the deflection values of the two test sections.

It can be seen that the deflection values of the sections with EAPM-HP_{RAP} and LSPM base layers were low and similar, which indicates that the layers had the same rigidity. Hence, it is feasible to use emulsified asphalt cold recycling technology with 80% RAP as the upper base layer in the expressway renovation project.

The results show that when the EAPM-HP_{RAP} was used as the upper base layer in the expressway renovation project, its high-temperature stability, water damage resistance and pavement strength met the requirements of the relevant specifications.

3.3. Benefits Analysis

Emulsified asphalt cold regeneration technology can be used in multiple construction operations, such as the mixing, paving and rolling of mixtures under normal temperature conditions, without the need to heat the aggregates and asphalt, as is required with traditional pavement construction technology. This not only reduces the amount of construction equipment needed but also greatly reduces the costs of manpower and material resources. More importantly, it can reduce energy consumption and the emissions of pollutants, such as soot, SO_2 and CO_2 [11,24,25].

3.3.1. Resource Savings

The reconstruction project was used as a case study to estimate the economic benefits of the approach. The project had a total length of 105 km and a roadbed width of 27.5 m. According to the experimental gradation and the mixing proportion, we know that the thickness of the EAPM-HP_{RAP} upper base layer was 18 cm, and 80% RAP was used in the gradation during construction, which had a density of 2.2 g/cm^3 (2.2 t/m^3).

According to the above data, it can be inferred that Mz (total mass of mixture required for each 1 km of pavement base) is 27.5 × 1.0 × 0.18 × 2.2 × 1000 = 10,890 t, while the M_R (total mass of RAP reused) is 10,890 × 80% = 8712 t. Hence, using an EAPM-HP_{RAP} upper base layer can save 8712 tons of mineral aggregate per km of road. According to a rough calculation, at an aggregate price of CNY 120–150 /ton, the cost saving is CNY 1.05–1.31 million of aggregate per km; such that the whole road section (105 km) would save CNY 110–138 million.

Moreover, the reuse of RAP reduces the need to mine new ore, avoids the occupation of land by RAP accumulation, and removes the pollution risks to soil and water caused by toxic and harmful substances, such as anthracene, naphthalene, and pyridine, leaching from aged asphalt [26,27].

3.3.2. Energy Savings

As RAP does not need to be heated to mix EAPM-HP_{RAP}, an equivalent LSPM mixture will consume more energy for aggregate and asphalt heating and to evaporate the water in the aggregates. The reconstruction project used mobile plant-mixed cold regeneration equipment (German Wirtgen KMA220). The mechanical energy consumption during the mixing process was not considered. Only the energy consumed by heating the mixture was calculated and analysed because both construction schemes require mixing of the mixture. Therefore, the energy consumption per ton of LSPM is the energy that can be saved per ton of EAPM-HP_{RAP}. The densities of the EAPM-HP_{RAP} and LSPM were both calculated as 2.2 g/cm^3 (2.2 t/m^3).

According to the test section construction data, the oil-to-stone ratio of LSPM was 4.3% and the moisture content of the aggregate was 0.5%. The aggregate was heated from 25 °C to 160 °C and the asphalt was heated from 25 °C to 175 °C. An asphalt mixing plant was used during construction (Marini 4000 MAC320) and its fuel utilization efficiency and heat exchange efficiency were calculated as 95% and 80%, respectively.

Combining the above construction parameters and taking the production of 1 ton of LSPM as the benchmark, the energy consumption was estimated (Table 9).

Table 9. Energy consumption used to heat 1 ton of LSPM.

Parameters	Specific Heat (J/kg·K)	Mass (kg)	Initial Temperature (°C)	Discharge Temperature (°C)	Energy Required (MJ)	Standard Coal Equivalent (kg)
Dry aggregate	920	953.98	25	160	118.484	4.05
Water	2260 kJ/kg	4.79	25	100	10.825	0.37
Asphalt	1340	41.23	25	175	8.287	0.28
Total	—	1000	—	—	137.596	4.70

Note: 6.19 kg of standard coal is needed to produce 1 ton of LSPM after conversion according to the fuel utilization rate and the heat exchange efficiency of the mixing process.

According to the calculations in the previous section, the total mass of mixture required for each 1 km of pavement base is 10,890 tons. The heating energy consumption of LSPM per km is 67.41 tons of standard coal. According to the market prices of other energy sources and their equivalence to standard coal, the energy and cost savings per km of EAPM-HP_{RAP} were obtained (Table 10).

Table 10. Energy consumed per km of LSPM pavement.

Energy Source	Calorific Value (MJ/kg; MJ/m^3)	Standard Coal Equivalent (kgce/kg)	Energy Saving (t)	Price (CNY/t)	Cost Saving (Thousands of CNY)
Standard coal	29.30	1.000	67.41	—	—
Diesel	42.65	1.457	46.27	8946	414
Petroleum	43.07	1.471	45.83	10,247	470
Electric	3600 kJ/(kWh)	0.123	5.48×10^5 KWh	0.7715	423

Note: The electricity price is the average of the peak and segment prices for single-system industrial and commercial electricity ($\leq$1 kV).

From the perspective of reducing energy consumption alone, and ignoring the additional resource savings, some CNY 414,000–470,000 of energy cost can be saved per km of EAPM-HP$_{RAP}$ pavement. The whole reconstruction project (105 km) would save nearly CNY 50 million in energy costs. A comprehensive calculation of the saved ore and energy costs shows that the use of EAPM-HP$_{RAP}$ as the upper base layer saves CNY 1.5–1.8 million per km and CNY 158–189 million for the whole project (105 km). Hence, the use of EAPM-HP$_{RAP}$ can generate considerable economic benefits while also reusing RAP resources.

3.3.3. Emissions Reduction

With the intensification of the global greenhouse effect, carbon emissions have attracted increasing attention from the international community and domestic experts, scholars and governments. On 31 December 2020, the Ministry of Ecology and Environment announced the "Measures for the Administration of Carbon Emissions Trading (Trial)", which came into force on 1 February 2021.

It is reported that the combustion of 1 ton of standard coal will emit 260 kg of ash, 15 kg of soot, 2.62 tons of CO_2, 8.5 kg of SO_2 and 7.4 kg of nitrogen oxides into the atmosphere. The pollutant emissions reduction achieved by using EAPM-HP$_{RAP}$ as an upper base layer instead of conventional asphalt was calculated based on the data from the previous section. The results are shown in Table 11.

Table 11. Emission reductions achieved by using EAPM-HP$_{RAP}$ instead of regular asphalt as an upper base layer.

Emissions	CO_2 (t)	SO_2 (t)	NO_X (t)	Ash (t)	Soot (t)
1 ton of standard coal	2.62	0.0085	0.0074	0.26	0.015
1 km of EAPM-HP$_{RAP}$	176.6	0.6	0.5	17.6	1.0
Whole road section	18,543.0	60.2	52.5	1848.0	105.0

Paving 1 km of road with EAPM-HP$_{RAP}$ as the upper base layer could reduce CO_2 emissions by 176.6 tons, SO_2 emissions by 0.6 tons, NO_X emissions by 0.5 tons, ash emissions by 17.6 tons and soot emissions by 1.0 tons. Using it in the whole road section could reduce CO_2 emissions by 18,543 tons, SO_2 emissions by 60.2 tons, NO_X emissions by 52.5 tons, ash emissions by 1848 tons, and soot emissions by nearly 105 tons.

It is certain that the application and promotion of EAPM-HP$_{RAP}$ will play a very positive role in achieving carbon reduction goals under the new development philosophy. The national carbon emission rights trading market was launched on 16 July 2021, and the closing price of carbon emission allowances (CEA) was CNY 60 per ton on 22 April 2022. The power generation industry became the first industry to be included and it is foreseeable that, with the promotion, implementation and improvement of the carbon emission trading system across the country, industries such as petrochemicals and transportation will also be included in the trading market. The advantages of emissions reduction will generate considerable economic benefits with the implementation of the carbon emissions trading system.

4. Conclusions

1. Using the cold regeneration technique with emulsified asphalt mixed with a high proportion of RAP is a feasible way to pave the upper base layer of expressway renovation projects. With an experimentally optimized gradation and mixing proportion, its high-temperature stability, water damage resistance and pavement strength can meet the requirements of relevant specifications.
2. The EAPM-HP_{RAP} mixture reduces the need to mine new ore resources and allocate land for RAP storage. It also mitigates the soil and water pollution risks caused by the aged asphalt contained in RAP.
3. Compared with conventional pavement technology, EAPM-HP_{RAP} generates considerable economic benefits while reusing RAP resources. This study indicates that engineering costs of CNY 1.5–1.8 million per km of pavement can be saved, while the whole expressway renovation project (105 km) could save CNY 158–189 million by using EAPM-HP_{RAP} as the upper base layer.
4. The reuse of RAP can save energy equivalent to 67.41 tons of standard coal per km when used as the upper base layer in an expressway renovation project. Further calculations show that every km of pavement (at a roadbed width of 27.5 m) could reduce CO_2 emissions by 176.6 tons, SO_2 emissions by 0.6 tons, NO_X emissions by 0.5 tons, ash emissions by 17.6 tons and soot emissions by 1.0 tons. Using it for the whole road section could reduce CO_2 emissions by 18,543 tons, SO_2 emissions by 60.2 tons, NO_X emissions by 52.5 tons, ash emissions by 1848 tons, and soot emissions by nearly 105 tons. These energy and emissions savings are very significant, indicating that the proposed approach warrants further development.

Author Contributions: Methodology, Y.L. and Y.Z.; validation, Y.L. and S.Z.; data curation, Y.L. and S.Z.; writing—original draft, Y.L.; writing—review and editing, Y.Z. and S.Z. All authors have read and agreed to the published version of the manuscript.

Funding: This research received no external funding.

Institutional Review Board Statement: Not applicable.

Informed Consent Statement: Not applicable.

Data Availability Statement: Not applicable.

Conflicts of Interest: The authors declare no conflict of interest.

References

1. Apeagyei, A.K.; Diefenderfer, B.K. Evaluation of cold in-place and cold central-plant recycling methods using laboratory testing of field-cored specimens. *J. Mater. Civ. Eng.* **2013**, *25*, 1712–1720. [CrossRef]
2. Wang, T.; Xiao, F.; Zhu, X.; Huang, B.; Wang, J.; Amirkhanian, S. Energy consumption and environmental impact of rubberized asphalt pavement. *J. Clean. Prod.* **2018**, *180*, 139–158. [CrossRef]
3. Kim, Y.; Lee, H.D.; Heitzman, M. Dynamic modulus and repeated load tests of cold in-place recycling mixtures using foamed asphalt. *J. Mater. Civ. Eng.* **2009**, *21*, 279–285. [CrossRef]
4. Shingles, R.A. Application of reclaimed asphalt pavement and recycled asphalt Shingles in hot-mix asphalt. *Citeseer*. 2014, pp 26–29. Available online: https://citeseerx.ist.psu.edu/viewdoc/download?doi=10.1.1.671.363&rep=rep1&type=pdf (accessed on 25 August 2022).
5. Davidson, J.; Blais, C.; Croteau, J. Review of in-place cold recycling/reclamation in Canada. In Proceedings of the Annual Conference and Exhibition of the Transportation Association of Canada, Transportation Innovation-Accelerating the Pace, Quebec, QC, Canada, 19–22 September 2004; pp. 238–245.
6. Pan, J.S.; Li, Y.J.; Gu, J.F.; Hu, M.J. Research and application prospect of computer simulation on heat treatment process. *J. Shanghai Jiaotong Univ. (Sci.)* **2000**, *5*, 1–13.
7. Wang, T.; Hao, P. Benefits in Energy Saving and CO_2 Reducing by Using Reclaimed Asphalt Pavement. *J. China Foreign Highw.* **2013**, *33*, 325–330.
8. Turk, J.; Pranjić, A.M.; Mladenovič, A.; Cotič, Z.; Jurjavčič, P. Environmental comparison of two alternative road pavement rehabilitation techniques: Cold-in-place-recycling versus traditional reconstruction. *J. Clean. Prod.* **2016**, *121*, 45–55. [CrossRef]
9. Gao, L.; Li, C.; Li, Y. A new upper bound on the infinity norm of the inverse of Nekrasov matrices. *J. Appl. Math.* **2014**, *125*, 89–94. CrossRef]

10. Wang, J.; Yuan, J.; Xiao, F.; Li, Z.; Wang, J.; Xu, Z. Performance investigation and sustainability evaluation of multiple-polymer asphalt mixtures in airfield pavement. *J. Clean. Prod.* **2018**, *189*, 67–77. [CrossRef]
11. Cross, S.A.; Chesner, W.H.; Justus, H.G.; Kearney, E.R. Life-Cycle environmental analysis for evaluation of pavement rehabilitation options. *Transp. Res. Rec.* **2011**, *2227*, 43–52. [CrossRef]
12. Sivilevičius, H.; Bražiūnas, J.; Prentkovskis, O. Technologies and Principles of Hot Recycling and Investigation of Preheated Reclaimed Asphalt Pavement Batching Process in an Asphalt Mixing Plant. *Appl. Sci.* **2017**, *7*, 1104. [CrossRef]
13. Gao, L.; Li, H.; Xie, J.; Yu, Z.; Charmot, S. Evaluation of pavement performance for reclaimed asphalt materials in different layers. *Constr. Build. Mater.* **2018**, *159*, 561–566. [CrossRef]
14. Research Institute of Highway Ministry of Transport. *Test Methods of Aggregate for Highway Engineering*; China Communications Press: Beijing, China, 2005.
15. Karlsson, R.; Isacsson, U. Application of FTIR-ATR to Characterization of Bitumen Rejuvenator Diffusion. 2003, 15, 157–165. *J. Mater. Civ. Eng.* **2003**, *15*, 157–165. [CrossRef]
16. Lin, J.; Huo, L.; Xu, F.; Xiao, Y.; Hong, J. Development of microstructure and early-stage strength for 100% cold recycled asphalt mixture treated with emulsion and cement. *Constr. Build. Mater.* **2018**, *189*, 924–933. [CrossRef]
17. Tarefder, R.A.; Zaman, M.; Sisson, R. Field and laboratory properties of lime-treated cold in-place recycled asphalt pavements. *J. Test. Eval.* **2006**, *34*, 31–41. [CrossRef]
18. Lee Hosin, D.; Kim, Y. Influences of binder and RAP temperatures and foaming water content on cold in-place recycling mix design process using foamed asphalt. *Airfield Highw. Pavement* **2006**, 377–387. [CrossRef]
19. Research Institute of Highway Ministry of Transport. *Technical Specification for Highway Asphalt Pavement Recycling*; China Communications Press: Beijing, China, 2008.
20. Research Institute of Highway Ministry of Transport. *Test Methods of Soils for Highway Engineering*; China Communications Press: Beijing, China, 2007.
21. Lytton, R.L.; Zhang, Y.; Gu, F.; Luo, X. Characteristics of damaged asphalt mixtures in tension and compression. *Int. J. Pavement Eng.* **2018**, *19*, 292–306. [CrossRef]
22. Javilla, B.; Fang, H.; Mo, L.; Shu, B.; Wu, S. Test evaluation of rutting performance indicators of asphalt mixtures. *Constr. Build. Mater.* **2017**, *155*, 1215–1223. [CrossRef]
23. Research Institute of Highway Ministry of Transport. *Standard Test Methods of Bitumen and Bituminous Mixture for Highway Engineering*; China Communications Press: Beijing, China, 2011.
24. Thenoux, G.; Gonzalez, A.; Dowling, R. Energy consumption comparison for different asphalt pavements rehabilitation techniques used in Chile. *Resour. Conserv. Recycl.* **2007**, *49*, 325–339. [CrossRef]
25. Alkins, A.E.; Lane, B.; Kazmierowski, T. Sustainable pavements: Environmental, economic, and social benefits of in situ pavement recycling. *Transp. Res. Rec.* **2008**, *2084*, 100–103. [CrossRef]
26. Li, Y.; Wu, D. Application of Ammonia Desulfurization Technology in Energy Saving and Carbon Emission Reduction of Sulfur Plant. *Saf. Health Environ.* **2021**, *21*, 42–45.
27. Zhu, J.; Liu, L.; Gao, T. Investigation of the Status of Soil and Groundwater in a Gas Station. *Saf. Health Environ.* **2021**, *21*, 34–39.

coatings

MDPI

Article

Decision-Making for Typical Pavement Structure Based on Life-Cycle Economic Evaluation and Key Performance Indicators

Ming Zhang [1,*], Chong Wang [1], Lulu Fan [2] and Junyan Yi [3,*]

[1] School of Engineering, Changchun Normal University, 677#Changji North Road, Changchun 130032, China
[2] Shenzhen Municipal Engineering Corp., Shenzhen 518000, China
[3] School of Transportation Science and Engineering, Harbin Institute of Technology, 73#Huanghe Road, Harbin 150090, China
* Correspondence: zhangming@ccsfu.edu.cn (M.Z.); yijunyan@hit.edu.cn (J.Y.)

Abstract: The economy and rationality of the typical structural scheme for asphalt pavement are directly related to the formation of its service target. Based on the original highway asphalt pavement structure in Jilin Province, the evaluation index system was established by the research of the literature, investigation and Delphi method, and the results of expert consultation were tested by indicators such as enthusiasm, authority and coordination. The Kendall coordination coefficient was 0.803, indicating a strong degree of evaluation consistency. Then, using the finite element method and pavement economic analysis method, the key parameters of pavement structure asphalt, pavement design performance and economy are calculated. Finally, the key indicators in the pavement design process are quantified by using analytic hierarchy process (AHP) and gray relational theory, the weight of each key indicator and the correlation degree of the pavement structure are calculated, and pavement structure 1 is determined to be the best solution. The results show that a decision model for typical structural types of asphalt pavement can be established by using an improved gray relational analytic hierarchy process. In this study, the decision-making model can quickly and conveniently determine the most suitable typical pavement structure, the selected structures are checked, and the calculation results meet the requirements of the specification. However, given the limited reference cases, the proposed pavement structure decision-making method needs to be verified and improved through more practical applications.

Keywords: pavement structure; analytic hierarchy process; gray relational theory; Abaqus; decision

Citation: Zhang, M.; Wang, C.; Fan, L.; Yi, J. Decision-Making for Typical Pavement Structure Based on Life-Cycle Economic Evaluation and Key Performance Indicators. *Coatings* **2022**, *12*, 1124. https://doi.org/10.3390/coatings12081124

Academic Editor: Valeria Vignali

Received: 23 June 2022
Accepted: 3 August 2022
Published: 5 August 2022

Publisher's Note: MDPI stays neutral with regard to jurisdictional claims in published maps and institutional affiliations.

Copyright: © 2022 by the authors. Licensee MDPI, Basel, Switzerland. This article is an open access article distributed under the terms and conditions of the Creative Commons Attribution (CC BY) license (https://creativecommons.org/licenses/by/4.0/).

1. Introduction

In the decision-making process of traditional pavement structures, the method of comparing pavement performance indicators is usually used to determine the appropriate pavement structure, but this evaluation method is relatively single and incomplete. As we all know, in addition to the road service performance index, economic rationality is another important determinant for comparing and selecting various design schemes. Therefore, in the field of road design, technical evaluation and economic evaluation of the design scheme are inseparable when designers are faced with deciding upon the type of pavement structure. However, there are few research methods and models using comprehensive design indicators for analysis, and many studies on pavement structures still rely on a single indicator, such as the thickness of the asphalt layer structure and mechanical indicators. This design method is obviously not comprehensive [1–3]. There is a certain correlation between the economy and the life of the pavement, for example, low initial cost, high post-maintenance cost, high initial cost, long pavement service life and low post-maintenance cost [4,5]. In addition, the quantity of high-quality asphalt produced in China is limited, imported asphalt is expensive, and the pavement structure and cost of asphalt pavement are closely related. In this case, it is necessary to select a combination type of pavement structure with different asphalt layer thicknesses according to technical and economic analyses. In the design and decision-making related to high-grade highway pavement, comprehensively considering various factors and selecting a reasonable type of pavement structure is a problem to be solved.

To seek a pavement structure with long life while taking into account cost, many scholars have conducted in-depth research in recent years. In terms of key indicators of pavement performance, various factors have been studied: Pan et al. [6] proposed that rutting is the most severe pavement deterioration of expressways. Fardin [7] proposed that fatigue cracking is more damaging than rutting on the roller-compacted concrete-base composite pavement. Li et al. [8] proposed that cracking in asphalt concrete layers is among the driving modes of flexible pavement deterioration. Cho et al. [9] proposed that delamination or debonding problems are particularly more severe for asphalt pavements that are subjected to heavy vehicle loads, especially horizontal surface shear forces that are due to braking and turning of vehicle tires. Yang et al. [10] proposed that shear fatigue damage is one of the main asphalt pavement failure modes. Interlayer fatigue resistance is also an important property in ensuring the durability and stability of asphalt pavements. In terms of economic performance indicators, additional work has been performed: Chong et al. [11] concluded that pavements without the need for reconstruction in the analysis period are generally more economical and environmentally friendly than those having to be reconstructed. However, designs that are too conservative lead to higher costs, greater energy consumption and greenhouse-gas emissions. Han et al. [12] established a decision model that fully considers the comprehensive maintenance benefit–cost ratio during the whole life cycle of the road. To overcome problems of experience-led manual decision-making, the decision-making between pavement conditions and maintenance plans was based on a data mining technique. Shon et al. [13] presented a pavement management framework that involves multiscale decisions, including system-level budget allocations, group-level inspections and facility-level maintenance, rehabilitation and reconstruction strategies. Considerable data analyses [14–16] reveal many cases of being focused on choosing typical pavement structures, but there are very few studies that take into account both work performance and economic performance and involve quantitative analysis.

Analytic Hierarchy Process (AHP) is a decision analysis method combining qualitative and quantitative approaches to solve complex multi-objective problems. Gray theory proposes the generalization of gray correlation analysis for each subsystem and intends to seek the numerical relationship between each subsystem (or factor) in the system through a certain method. The analytic hierarchy process and gray relational theory have been widely used in the field of pavement design. Han et al. [17] presented a decision-making method for asphalt pavement maintenance using an improved weight random forest algorithm based on correlation analysis and the AHP. Esmaeeli et al. [18] used the fuzzy real options methodology to help highway management organizations select the optimum alternative for project implementation. Yu et al. [19] presented and demonstrated a methodology for evaluating a microsurfacing treatment of asphalt pavement based on the gray system model and gray relational degree theory. Wang et al. [20] combined the results of gray relational analysis and the gray prediction methodology to establish gray-model-based smoothness predictions by using influencing factors similar to those used in the *Mechanistic-Empirical Pavement Design Guide*. However, the combination of the AHP and gray theory is rarely applied in the field of typical pavement structure decision-making. The judgment matrix in the AHP is completely determined by expert experience, and it is difficult to exclude the influence of subjective factors on the index weight. The gray correlation method is calculated based on the real data of the scheme, and the calculation results are relatively objective. When the traditional AHP calculates the comprehensive weight, it is only a simple synthesis of the index results and does not organically integrate the two methods to obtain the weight. To comprehensively consider these factors and choose a reasonable type of pavement, the new AHP combined with gray relational theory is a good way to solve this problem. At the same time, in the analysis of various factors of road construction investment decisions, the pavement performance index and economic evaluation index can be quantified, but the judgment matrix can only be used for qualitative analysis. Therefore, after the improvement of gray relational theory, the AHP is a new method to effectively address those complex problems that cannot be fully analyzed by using quantitative methods. It is a decision-making method combining quantitative and qualitative analyses. The characteristics of AHP, gray relational theory and improved AHP are summarized in Table 1.

Table 1. Comparison of this study with existing research methods.

Research Methods		Application Areas	Features
Existing research	Analytic hierarchy process (AHP)	Asphalt pavement maintenance decision-making method	The calculation is simple, and the obtained results are simple and clear, which is a concise and practical decision-making method. However, AHP has less quantitative data and more qualitative components, making it difficult to be convincing.
	Gray relational analysis (GRA)	Asphalt pavement micro-surface treatment evaluation method	Calculated according to the real data of the scheme, the calculation results are relatively objective.
		Smoothness prediction based on gray model	
In this study	AHP + GRA	Typical pavement structure decisions	The combination of AHP and GRA is equivalent to effectively combining quantitative calculation and qualitative analysis, which reduces the influence of AHP on decision-making to a certain extent and makes decision-making results more convincing.

In summary, to improve the performance of pavement and reduce the cost of road construction and maintenance, we propose an improved AHP decision-making method to determine the optimal pavement structure.

2. Typical Pavement Structure Decision Process

Design evaluation can accurately reflect the quality of engineering projects. Through reasonable design evaluation methods, quantitative analysis of various indicators of highway construction can be conducted to effectively compare schemes and improve the quality and economic benefits of pavement engineering. The typical structure evaluation process for asphalt pavement in this study is as follows. First, the evaluation index layer and criterion layer of a typical pavement structure are constructed through literature and project historical data reviews. Then, according to the expert's suggestion, the hierarchical structure model of the index system is constructed, and the comprehensive weight of each index is calculated based on the AHP. Next, the finite element (FE) model is used to build the model and calculate the performance and economic indicators. Finally, the gray relational degree is calculated for multiple indicators through the gray relational analysis method, and the one with the highest relational degree is the optimal solution. Figure 1 shows the evaluation process for the typical structure of the optimal pavement.

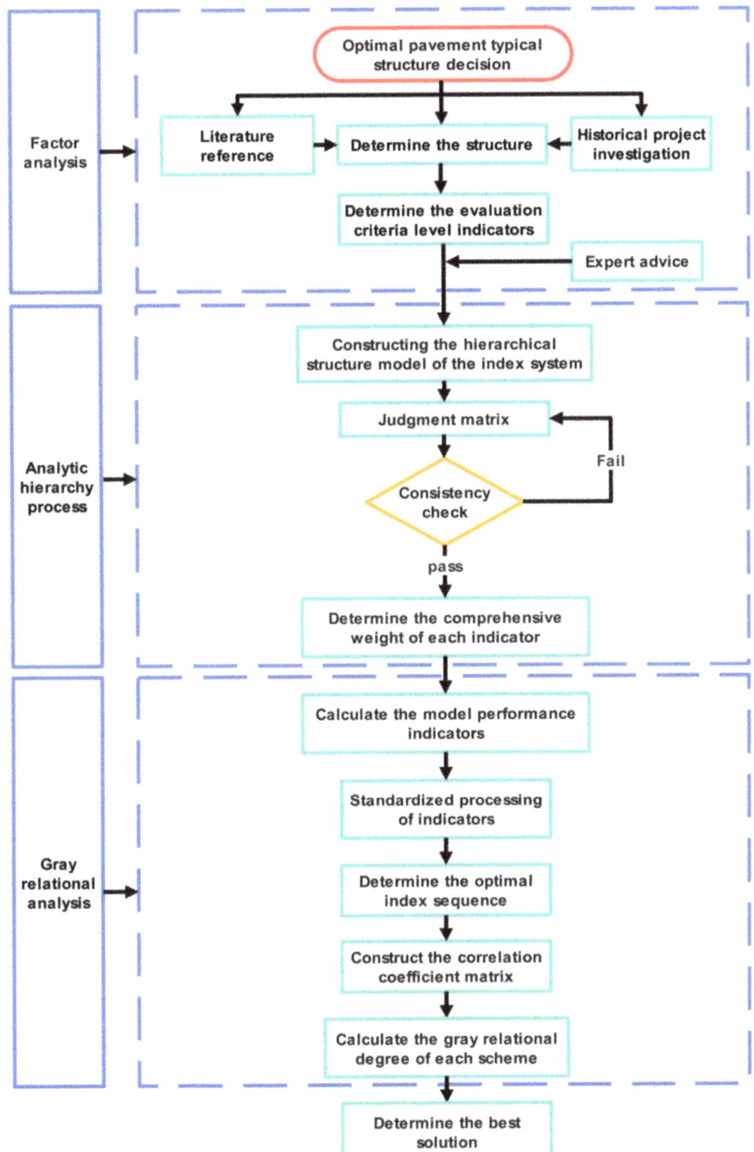

Figure 1. Optimal pavement structure evaluation process.

3. Establishment of Evaluation System for Typical Pavement Structure

3.1. Optimal Decision-Making Process for Typical Pavement Structure

Because there is no clear and easy-to-use decision-making method for pavement structure, the pavement structure scheme of expressways in Jilin Province is not unified. Therefore, this study selects four commonly used highway pavement structure schemes in Jilin Province. The four schemes taken in this study are all of this structure, and the structure is shown in Figure 2.

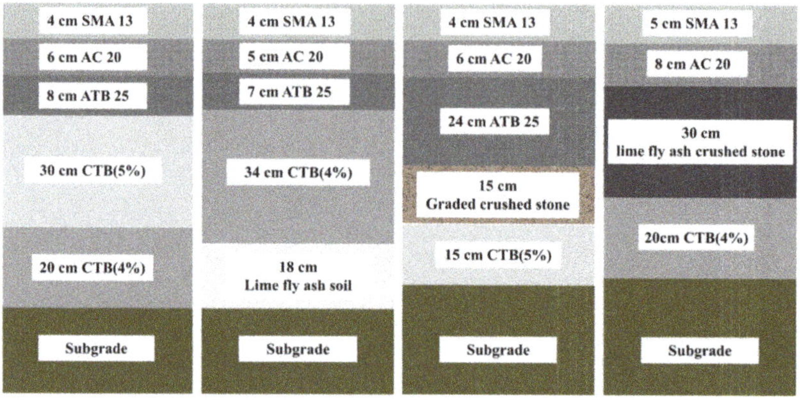

Figure 2. Four typical pavement structure schemes.

3.2. Determination of the Evaluation Indicators

Pavement performance provides a basis for the comparison of the selection of pavement design schemes, but it is not the only reason for deciding on the choice of pavement type. The selection of typical pavement structures should also account for factors such as initial investment and total investment in the whole life cycle. A reasonable pavement design scheme should unify technical rationality and economic rationality.

3.2.1. Technical Indicators

From the perspective of technical rationality, there are multiple factors that need to be considered when selecting the type of pavement, mainly fatigue performance, antirutting performance and lateral deformation performance, such as rutting and sliding damage in high-temperature areas and fatigue cracking in low-temperature and heavy-traffic areas.

Therefore, according to the above performance requirements, six pavement performance parameters are proposed: stress and strain at the bottom of the asphalt layer, permanent deformation of the asphalt layer, the compressive strain on the top surface of the subgrade and horizontal shear stress and strain of the pavement.

3.2.2. Economic Indicators

From the perspective of economic rationality, the economic evaluation of the design scheme is an important basis for the comparison and selection of pavement structures. However, because of the various asphalt pavement structure schemes, the building materials and construction methods used are also different, but for a high-grade highway to be built, they play the same role or yield the same benefits. For the four schemes with the same benefits but different input costs, the present value method of total life cycle cost is most suitable when evaluating their economic advantages and disadvantages. The present value method of the total life cycle cost entails comparing the expenses invested by each program at different times, according to the social discount rate stipulated by the state, to the present value. The scheme with a low present value has an economic advantage. When choosing the type of pavement, the main factors considered are the initial investment and the present value of the total cost of the whole life cycle.

3.3. Factor Analysis

The relative importance of the indicators in the criterion layer and the target layer is scored by searching the literature and research by road design departments and discussing and communicating with relevant road experts.

3.3.1. Expert Scoring System

Two rounds of expert correspondence were conducted in this study, and the first questionnaire was established based on the initially constructed index system. Experts assigned evaluation indices according to their importance, familiarity, judgment basis and academic level. The expert scoring criteria are summarized in Table 2.

Table 2. Expert scoring criteria.

Degree of importance	Not important	Less important	Generally	Important	Very important
Score	Below 60	60–70	70–80	80–90	90–100
Degree of familiarity	Unfamiliar	Less familiar	Generally	Familiar	Very familiar
Score	0.2	0.4	0.6	0.8	1
		Judgment basis			
	Theoretical analysis			**Work experience**	
Large 0.3	Middle 0.2	Little 0.1	Large 0.45	Middle 0.35	Little 0.2
	Peer understanding			**Intuitive sense**	
Large 0.2	Middle 0.15	Little 0.1	Large 0.05	Middle 0.05	Little 0.05
		Academic levels			
Professional title Score	Professor 1	Associate Professor 0.8	Lecturer 0.6	Teaching assistant 0.4	Ordinary worker 0.2

In order to ensure the validity of the results of this study, the research team reviewed a large number of studies and comprehensively screened the evaluation indicators in combination with expert opinions. The analysis results of the first round of surveys are shown in Table 3. The average score of the Antifreeze performance of the pavement indicator is less than 70 points, and the coefficient of variation is greater than 0.25; therefore, the indicator is canceled. Table 4 shows the results of the second round of expert consultation, and all indicators meet the screening requirements.

Table 3. Results of the first round of expert consultation.

Evaluation Index	Sample Size	Minimum	Maximum Value	Average	Standard Deviation	Coefficient of Variation
Asphalt mixture layer bottom strain	20	85	96	89.95	4.045	4.497%
Base layer bottom stress	20	81	93	87.3	4.156	4.761%
Asphalt mixture layer permanent deformation	20	88	99	93.45	3.591	3.842%
Subgrade surface compressive strain	20	76	88	82.45	3.79	4.597%
Pavement surface compressive strain shear stress	20	70	85	76.95	5.689	7.393%
Pavement surface shear strain	20	68	85	77.3	5.555	7.186%
Antifreeze performance of pavement	20	50	94	69.25	18.45	26.643%
Initial investment	20	68	90	80.75	7.055	8.737%
Full life cycle cost present value	20	78	90	83.75	3.726	4.449%

Table 4. Results of the second round of expert consultation.

Evaluation Index	Sample Size	Minimum	Maximum Value	Average	Standard Deviation	Coefficient of Variation
Asphalt mixture layer bottom strain	20	86	96	89.8	3.708	4.129%
Base layer bottom stress	20	80	93	86.1	4.436	5.152%
Asphalt mixture layer permanent deformation	20	88	100	93.6	3.315	3.542%
Subgrade surface compressive strain	20	75	88	81.1	4.291	5.291%
Pavement surface compressive strain shear stress	20	70	84	77.45	4.861	6.276%
Pavement surface shear strain	20	69	83	74	3.756	5.075%
Initial investment	20	70	90	80.6	6.916	8.581%
Full life cycle cost present value	20	78	90	83.8	3.708	4.425%

3.3.2. Consistency Evaluation of Expert Consultation Results

(1) The degree of positivity

The degree of positivity is usually determined by the positivity coefficient, which is equal to the number of questionnaires returned divided by the number of questionnaires published. The positive coefficient of the two rounds is 100%.

(2) Degree of authority

The degree of authority usually depends on the judgment basis of experts, the familiarity with indicators and the academic level. Table 4 lists the scoring criteria for the importance of indicators, professional familiarity and academic level. Authority coefficient = (judgment coefficient + familiarity coefficient + academic level)/3. The authoritative coefficient of this study is 0.918.

(3) Degree of coordination

Using the Kendall coordination coefficient to test the consistency of the research evaluation, it can be seen from Table 5 that the Kendall coordination coefficient test of the two survey results showed significance ($p = 0.000 < 0.05$), which means that the evaluations of the 20 experts are related. In addition, the coefficient of variation of the evaluation results of all indicators is less than 0.09, indicating that the coordination of experts in the two rounds is good. Judging from the results of the second expert consultation after index optimization, the Kendall coordination coefficient is 0.803, which is greater than 0.8, indicating a strong degree of evaluation consistency.

Table 5. Kendall coordination coefficient analysis results.

Number of Investigations	Number of Experts	Number of Evaluation Indicators	Kendall Coordination Coefficient	p
The first time	20	10	0.658	0
The second time	20	9	0.803	0

We use the AHP to calculate the importance weight value of the criterion layer to the target layer and the importance weight value of the index layer to the criterion layer. Based on the above principles and indicators, an evaluation index system for typical optimal pavement structure is established, as shown in Figure 3.

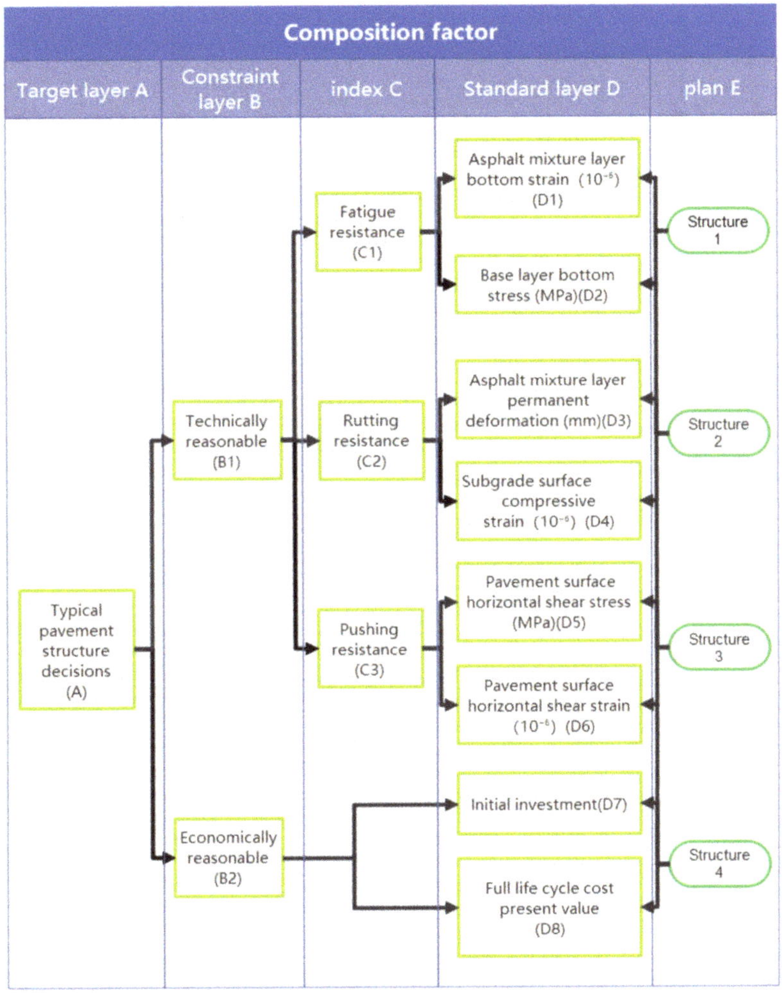

Figure 3. Evaluation index system for typical optimal pavement structure.

3.4. Analytic Hierarchy Process

Before calculating the weights, it is necessary to perform a consistency check on the matrix in advance. First, the matrix needs to be normalized, then the maximum eigenvalue $\lambda_{\max}$ is calculated, and the consistency index C.I. is calculated according to the maximum eigenvalue and the judgment matrix order n. Finally, the average random consistency index R.I. is introduced, and the test coefficient C.R. is calculated. When C.R. < 0.10, the test is passed; otherwise, it is not passed, in which case the comprehensive judgment matrix is adjusted according to the empirical method. The average random consistency index values are listed in Table 6. These were calculated as follows:

$$\lambda_{\max} = \sum_{i=1}^{n} \frac{(AW)_i}{nW_i} \tag{1}$$

$$\text{C.I.} = \frac{\lambda_{\max} - n}{n - 1} \tag{2}$$

$$\text{C.R.} = \frac{\text{C.I.}}{\text{R.I.}} \tag{3}$$

Table 6. Average random consistency index values.

Order	1	2	3	4	5	6	7	8	9
R.I.	0.00	0.00	0.58	0.90	1.12	1.24	1.32	1.41	1.45

After checking the matrix consistency with Formulas (1)–(3), the comprehensive weight of the subindicators to the total target layer is finally established. This is calculated as $W = (w_1, w_2, \ldots, w_n)$, where $w_i \geq 0$ and $\sum_{i=1}^{n} w_i = 1$.

3.5. *Relevance Calculation*

To screen for the optimal scheme, based on the complex pavement structure index and the fuzzy evaluation characteristics, we adopted the gray relational analysis method to calculate the degree of correlation between the design scheme and the ideal design scheme and judge the degree of closeness of the connection according to the correlation coefficient.

If the decision-making system of the pavement structure design scheme consists of m schemes and n indicators, in this study, there are 4 schemes and 8 indicators, then the n index values of the I schemes can form a set $X_i = (x_{i1}, x_{i2}, \ldots, x_{in})$ $(i = 1, 2, \ldots, m)$, according to the n original index values of m schemes, and the following original scheme index matrix can be formed:

$$X = \begin{bmatrix} X_1 \\ X_2 \\ \vdots \\ X_m \end{bmatrix} = \begin{bmatrix} x_{11} & x_{12} & \cdots & x_{1n} \\ x_{21} & x_{22} & \cdots & x_{2n} \\ \vdots & \vdots & \vdots & \vdots \\ x_{m1} & x_{m2} & \cdots & x_{mn} \end{bmatrix} \tag{4}$$

3.5.1. Normalization of Index Values

To eliminate the influence of dimension, the extremum method is used to normalize matrix X. In this study, the smaller the value of the evaluation index, the better the pavement structure scheme, and the quantitative index can be calculated according to

$$X'_{mk} = \frac{M_k - x_{mk}}{M_k - m_k} \tag{5}$$

$$m_k = \min(x_{1k}, x_{2k}, \ldots, x_{mk})(k = 1, 2, \ldots, n) \tag{6}$$

The normalized matrix is given by

$$X' = \begin{bmatrix} X'_1 \\ X'_2 \\ \vdots \\ X'_m \end{bmatrix} = \begin{bmatrix} x'_{11} & x'_{12} & \cdots & x'_{1n} \\ x'_{21} & x'_{22} & \cdots & x'_{2n} \\ \vdots & \vdots & \vdots & \vdots \\ x'_{m1} & x'_{m2} & \cdots & x'_{mn} \end{bmatrix} \tag{7}$$

3.5.2. Determine the Optimal Solution Index Set

The smaller the value of the evaluation index is, the better. Therefore, the optimal value is $x_{0j} = \min(x_{1j}, x_{2j}, \ldots, x_{mj})$; then the optimal index value is extremized to obtain the optimal value set $X'_0 = (x'_{01}, x'_{02}, \ldots, x'_{0n})$.

3.5.3. Construction of the Correlation Coefficient Matrix

The following formula can be used to solve for the correlation coefficient between the original solution and the optimal solution:

$$\alpha_i(j) = \frac{\min_i \min_j \left| x'_0(j) - x'_i(j) \right| + \rho \cdot \max_i \max_j \left| x'_0(j) - x'_i(j) \right|}{\left| x'_0(j) - x'_i(j) \right| + \rho \cdot \max_i \max_j \left| x'_0(j) - x'_i(j) \right|} \quad (8)$$

where $\rho \in [0, 1]$ is the resolution coefficient.

In summary, the correlation coefficient matrix can be obtained by using

$$E = \begin{bmatrix} E_1 \\ E_2 \\ \vdots \\ E_m \end{bmatrix} = \begin{bmatrix} \alpha_{11} & \alpha_{12} & \cdots & \alpha_{1n} \\ \alpha_{21} & \alpha_{22} & \cdots & \alpha_{2n} \\ \vdots & \vdots & \vdots & \vdots \\ \alpha_{m1} & \alpha_{m2} & \cdots & \alpha_{mn} \end{bmatrix} \quad (9)$$

3.5.4. Determination of Weighted Gray Relevance Degree

R_i ($i = 1, 2, \ldots, m$) is the correlation degree of X_i to X_0. According to the row vector E_i of the correlation coefficient and the weight vector W of each subindicator, the correlation degrees of each scheme can be obtained by using

$$R_i = E_i \times W = \begin{bmatrix} \alpha_{i1} & \alpha_{i2} & \cdots & \alpha_{in} \end{bmatrix} \times \begin{bmatrix} w_1 \\ w_2 \\ \vdots \\ w_n \end{bmatrix} \quad (10)$$

The correlation degree R_i reflects the closeness of the relationship between X_i and X_0 and the correlation between a single scheme and the best scheme. Therefore, the pros and cons of each scheme can be sorted according to the correlation degree.

4. Results and Discussion

4.1. Weight Ranking of Typical Pavement Structure Evaluation Index Based on the AHP

According to Tables 7 and 8, the order of importance of key indicators for pavement structure decision-making is as follows: D3 > D1 > D2 > D8 > D4 > D7 > D5 = D6. In the traditional AHP, the order of the weights occupied by the four schemes is structure 1 > structure 4 > structure 2 > structure 3.

Table 7. Evaluation index weight.

Index	Weight	Order
(D1) Asphalt mixture layer bottom strain	0.256	2
(D2) Base layer bottom stress	0.128	3
(D3) Asphalt mixture layer permanent deformation	0.288	1
(D4) Subgrade surface compressive strain	0.096	5
(D5) Pavement surface compressive strain shear stress	0.032	7
(D6) Pavement surface shear strain	0.032	7
(D7) Initial investment	0.042	6
(D8) Full life cycle cost present value	0.125	4

Table 8. The traditional AHP method calculates the weight results.

Type	Structure 1	Structure 2	Structure 3	Structure 4
Weight	0.3373	0.2694	0.1227	0.2707

4.2. Calculation of Relevance Degree of Typical Pavement Structure Based on the Gray Relational Method

4.2.1. Calculation of the Technical Performance Index of Pavement

We used the FE software Abaqus 6.14-4 for numerical simulation calculation. The FE analysis was performed for the four given typical asphalt pavement structures. The key parameters of pavement fatigue performance, rutting performance and pushing resistance were calculated.

Loads and Forms of Action

The pavement model uses a 100-kN single-axis double wheel as the design axle load, accounting for the horizontal load generated by braking. According to the relevant research results, the friction coefficient of the asphalt pavement was taken as 0.4, and the calculated axle load parameters were determined according to Table 9.

Table 9. Design axle load parameters.

Design of Axle Load (kN)	Tire Ground Pressure (MPa)	Equivalent Radius of Single Wheel Grounding (mm)	Two-Wheel Center Distance (mm)	Pavement Horizontal Load (MPa)
100	0.7	213.0	319.5	0.28

Pavement Geometry Model and Boundary Conditions

Figure 4 shows the pavement structure used for the FE calculation. The dimensions of the three-dimensional model were 6 m × 6 m × 3 m. Four typical structures commonly used for expressways in Jilin Province were selected. Table 10 summarizes the layer thicknesses and material properties of the simulated pavement structure. The modulus and Poisson's ratio of each layer material in the pavement structure use the values recommended by the Asphalt Pavement Design Specification [21].

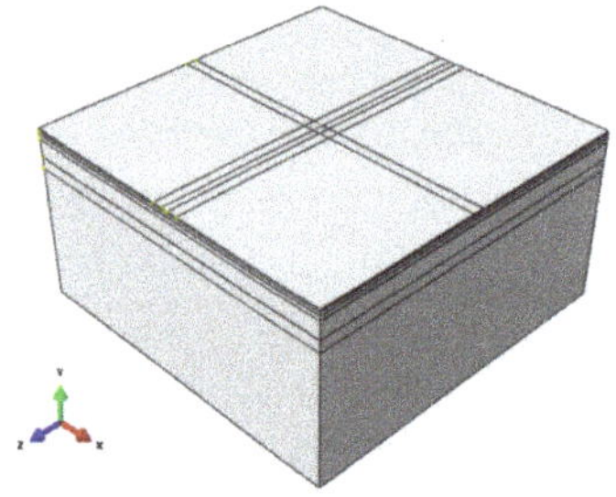
(a)

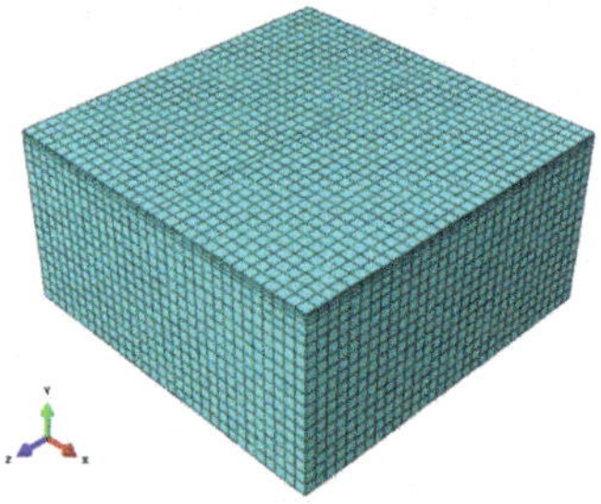
(b)

Figure 4. Asphalt pavement structure model (**a**) and grid division diagram (**b**).

To facilitate the calculation, the FE model has four constraints: There is no constraint on the surface of the model, there is no displacement on the two surfaces along the X-axis and Z-axis directions, and the bottom surface of the model has no displacement along the Y-axis direction. The asphalt material and soil in the pavement structure are nonlinear materials. The viscoelastic material of the asphalt pavement is calculated by using the Prony series [22]. The parameters of the Prony model are listed in Table 11. The elastic–plastic material of the subgrade soil is calculated by using the Drucker–Prager model. According to the Test Methods of Soils for Highway Engineering JTG E40-2007 [23], the cohesion and internal friction angle of the subgrade soil were measured through triaxial experiments. The parameters of the Drucker–Prager model are listed in Table 10.

Table 10. Pavement structure layer thickness and material composition.

Project	Material Type	Modulus (MPa)	Poisson's Ratio	Structure 1	Structure 2	Structure 3	Structure 4
Asphalt surface	Stone matrix asphalt with modified bitumen (SMA-13)	1400	0.35	40 mm	40 mm	40 mm	50 mm
	AC-20	1200	0.3	60 mm	50 mm	60 mm	80 mm
	AC-25	1000	0.3	-	70 mm	-	-
Asphalt base layer	Asphalt treated base (ATB-25)	1200	0.3	80 mm	-	240 mm	-
Inorganic binding materials	Graded crushed stone	500	0.35	-	-	150 mm	-
	Cement-stabilized crushed stone 4%	1500	0.25	500 mm	340 mm	150 mm	-
	Lime fly ash crushed stone	1400	0.25	-	-	-	500 mm
	Lime fly ash soil	750	0.25	-	180 mm	-	
Subgrade	Drucker–Prager	40	0.4	Cohesion	17.1 kPa	Friction angle	11.89°

Table 11. Prony model parameters.

Modulus (Pa)	2.857×10^9	4.086×10^9	4.473×10^9	3.303×10^9	1.578×10^9
Time (s)	2×10^{-5}	2×10^{-4}	2×10^{-3}	2×10^{-2}	2×10^{-1}

Pavement Performance Index Calculation Results

In addition to the permanent deformation index of the asphalt pavement, the calculation results of the FE model are summarized in Table 12.

Table 12. Calculation results using FE pavement performance indicators.

Technically Reasonable Index		Structure			
		1	2	3	4
Fatigue resistance	Asphalt mixture layer bottom strain ($\times 10^{-6}$)	27.074	30.380	46.438	34.000
	Base layer bottom stress (MPa)	0.071	0.060	0.094	0.093
Rutting resistance	Subgrade surface compressive strain ($\times 10^{-6}$)	98.790	136.900	134.600	130.800
Pushing resistance	Pavement surface horizontal shear stress (MPa)	0.139	0.276	0.272	0.117
	Pavement surface horizontal shear strain ($\times 10^{-6}$)	70.620	139.800	138.100	59.280

Similarly, a permanent deformation prediction model [24] for the hot mix asphalt (HMA) layer can be expressed as

$$R_{ai} = 2.31 \times 10^{-8} k_{Ri} T_{pef}^{2.93}\, p_i^{1.80}\, N_{e3}^{0.48}\, (h_i/h_0)\, R_{0i} \tag{11}$$

where h_i is the midpoint depth of each layer, p_i is the layered top compressive stress, R_{0i} is the rut test permanent deformation, k_{ri} is the correction factor, h_a is the asphalt mixture layer thickness (in mm), T_{pef} is the asphalt mixture layer equivalent temperature (in °C), P_i is the vertical compressive stress (in MPa) of the top surface of the ith layer of the asphalt mixture, which is calculated by using BISAR3.0 software, h_0 is the rut test piece thickness (in mm) and N_{e3} represents the equivalent design axle load cumulative design action times on the design lane within the design life. According to the prediction of the Jilin Provincial Highway Survey and Design Institute, N_{e3} is 103,200,000 times. Table 13 lists the calculation results for the permanent deformation of the asphalt mixture layer.

Table 13. Asphalt mixture layer permanent deformation.

Technically Reasonable Index	Structure			
	1	2	3	4
Permanent deformation (mm)	10.840	10.510	12.180	5.920

Figure 5 shows that the tensile strain at the bottom of the asphalt layer of pavement structure 1 is the lowest, while the tensile stress of the base layer of structure 2 is the lowest. By considering the two indicators comprehensively, structure 2 is the best in terms of fatigue resistance of the pavement structure. The main reason for this is that a reasonable pavement structure will effectively reduce the fatigue strength of the asphalt mixture layer and the base and prolong the service life of the pavement structure. The asphalt layer of structure 1 is slightly thicker than that of structure 2, while the thickness of the base layer is thinner than that of structure 2, which will cause the stress of the pavement base to be slightly higher than that of structure 2. The tensile strain at the bottom of the asphalt layer and the tensile stress at the bottom of the base layer of structure 3 is the highest because a thicker asphalt layer is used, and a graded crushed stone flexible layer is set. The minimum thickness of the asphalt layer is structure 4, which leads to a high tensile stress of the subbase, and the pavement structure is very prone to fatigue damage. From the point of view of the fatigue resistance of the pavement structure, structures 2 and 1 are superior.

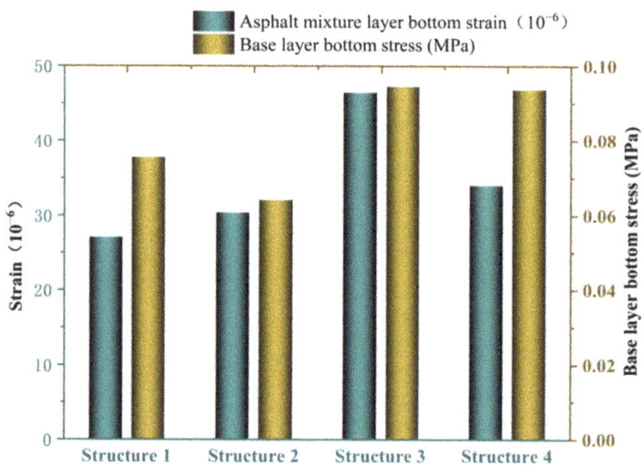

Figure 5. Antifatigue performance parameters for pavement structure.

Figure 6 shows that the permanent deformation value of the asphalt layer of structure 3 is the highest, with the permanent deformation of structures 1 and 2 being somewhat less and the permanent deformation of pavement structure 4 being the least. The main reason for this difference is that the asphalt layer of structure 1 is the thickest, the thicknesses of the asphalt layer of structures 2 and structure 3 are basically the same, and the asphalt layer of structure 4 is the thinnest. This result is consistent with the previous study that indicated that the permanent deformation increases with the thickness of the asphalt layer [25]. In terms of the key parameters of the compressive strain on the top surface of the subgrade, structure 1 has the best performance, and the compressive strain on the top surface of the subgrade of structure 2 is the highest. Designers generally assume that smaller permanent deformations will reduce rutting damage in the pavement while also reducing costs. However, if the thickness of the asphalt layer is too thin, the pavement base is prone to fatigue cracking damage under repeated loads, which has a great impact on the service life of the pavement and subsequent maintenance. In addition, the low compressive strain on the top surface of the subgrade will effectively reduce the structural rutting of the asphalt pavement, which will improve the service life and service level of the pavement. Permanent deformation and vertical compressive strain on the top surface of the subgrade are the main factors affecting rutting. From the aspect of rutting performance, although the deformation performance of structure 4 is more prominent in the asphalt layer, the vertical compressive strain of the subgrade top surface is not ideal, and structure 1 has the most balanced performance.

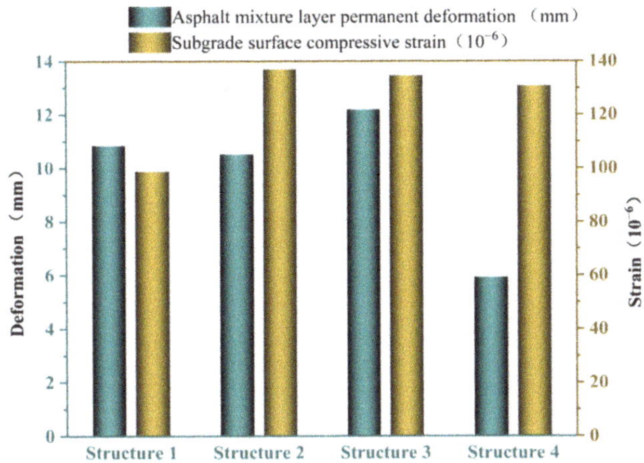

Figure 6. Antirutting performance parameters for pavement structure.

Figure 7 shows that the horizontal shear stress and strain of structure 4 surface are the lowest, while that of structures 1, 2 and 3 are higher. The main reason for this difference is that the pavement of structure 4 is a double-layer asphalt structure, and the thickness of the asphalt layer is only 13 cm. After the asphalt layer and the base layer are bonded, the overall strength is improved, making it difficult to concentrate the horizontal shear stress of the road surface, and the overall stress and strain are low. However, the asphalt layer of the other structures is thicker, and the shear stress and strain of the road surface are higher. However, from the standpoint of antirutting performance of the pavement structure alone, structures 4 and 1 are better.

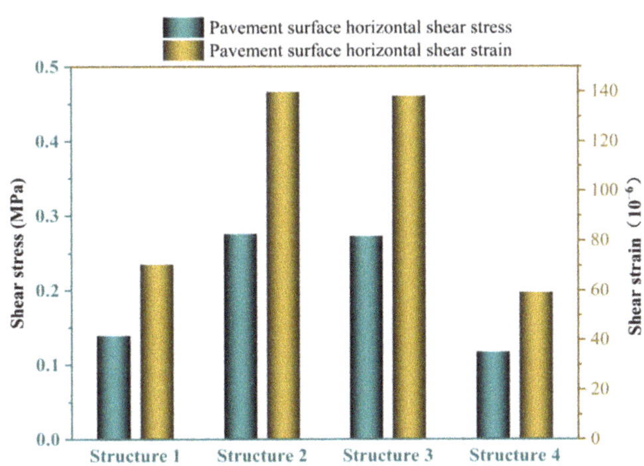

Figure 7. Antipushing performance parameters for pavement structure (in driving direction).

In this study, many factors affecting pavement performance have been considered. In addition to the above-mentioned key parameters of pavement performance, the economic performance of the structural scheme should also be considered, and the optimal structural scheme should be finally determined.

4.2.2. Calculation of Economic Indicators

The expenses incurred at different times during the analysis period are converted into present value according to a predetermined discount rate, and by converting into a single present value, the pros and cons of the schemes in the whole life cycle can be compared on the basis of equivalent value. Because the case is focused on the same expressway and the same traffic volume was used, it was assumed that the annual average daily traffic volume (*AADT*) is 5000 vehicles/day and that the annual average traffic volume growth rate was 10%. For the convenience of analysis, a service life of 15 years was assumed, and the analysis period of each scheme was chosen to be the same. According to data from the United States, the actual cost of government investment is 4–5% [26], so the discount rate in this study was set as 4%.

According to relevant research [27,28], the following formula can be used to calculate the present value of expenses:

$$PWC_{x_1,n} = IC_{x_1} + \sum_{t}^{n} pwf_{i,t}(RC_{x1,t} + MC_{x1,t} + UC_{x1,t}) - pwf_{i,n} \times SV_{x1,n} \quad (12)$$

$PWC_{x1,n}$ is the present value of the total cost of scheme x_1 in n years during the analysis period, which can be obtained by using the present value coefficient of the discount rate i in year t, which can be calculated by using

$$pwf_{i,t} = \frac{1}{(1+i)^t} \quad (13)$$

IC_{x1} is the initial construction cost of solution x_1, $RC_{x1,t}$ is the reconstruction cost of solution x_1 in year t, $MC_{x1,t}$ is the maintenance cost of solution x_1 in year t, $UC_{x1,t}$ is the user fee of solution x_1 in year t and $SV_{x1,t}$ is the residual value of the scheme x_1 at the end of the analysis period.

The pavement residual value [29] at the end of the analysis period can be calculated according to

$$SV = \left(1 - \frac{L_A}{L_B}\right) C_r \quad (14)$$

where the number of years from the construction year of the last L_A remodel to the end of the life cycle, L_B, is the expected service life of the reconstruction measure, and C_r is the construction cost of the reconstruction measure.

The reconstruction period can be calculated according to the pavement condition index (*PCI*). When $PCI < 70$, the original asphalt pavement needs to be reconstructed.

According to the literature [30], we can obtain the maintenance cost MC_i and pavement condition index *PCI*. Substituting

$$\lambda = a_1 h^{b_1} ESAL^{c_1} \quad (15)$$

$$\eta = a_2 h^{b_2} ESAL^{c_2} \quad (16)$$

$$\zeta = a_3 h^{b_3} ESAL^{c_3} \quad (17)$$

gives the life factor β and shape factor α:

$$\beta = K_{\gamma\beta} a_4 h^{b_4} ESAL^{c_4} l_0^d \quad (18)$$

$$\alpha = K_{\gamma\alpha}\lambda\left\{1 - \exp\left[-(\eta/l_0)^{\zeta}\right]\right\} \quad (19)$$

where h is the thickness of the pavement asphalt layer (in cm); *ESAL* is the daily equivalent axle load times (in times/day/lane); l_0 is the initial deflection (0.01 mm); a, b, c and d are the regression coefficients or indices, the values of which can be selected according to Table 14 [27]; the standard axle load is 100 kN; and $K_{\gamma\alpha} = 0.897$ and $K_{\gamma\beta} = 0.904$ are the environmental impact coefficients in the Jilin area [31].

Table 14. *PCI* decay equation regression parameter values.

Formula	*a*	*b*	*c*	*d*
18	15.7238	0.5861	−0.2064	-
19	119.66	−0.1124	−0.1053	-
20	1.5247	−0.1016	−0.2292	-
21	0.6536	0.3449	−0.0255	−0.981

Then, by using these life factors, *PCI* can be expressed as

$$PCI = 100\left\{1 - \exp\left[-(\alpha/y)^{\beta}\right]\right\} \quad (20)$$

Finally, the maintenance cost can be obtained as

$$MC_i = 0.34 + 3.44 \times 10^{-6}(100 - PCI_i) \times AADT_i \quad (21)$$

The user fee consists of three items: fuel consumption, tire consumption and warranty material consumption. According to the literature [32,33], fuel consumption can be expressed by

$$F_L = a_1 + b_1 IRI(1/(1000\ \text{km})) \quad (22)$$

wear cost of vehicle tires can be expressed by

$$T_C = a_2 + b_2 IRI(\text{piece}/(1000\ \text{km})) \quad (23)$$

and the material consumption cost can be expressed by

$$\text{Passenger car}: P_C = C_0 C_{kM}^{k_P} e^{C_q \bullet - IRI}\ (\text{proportion of new car price}/(1000\ \text{km})) \quad (24)$$

$$\text{Truck}: P_C = C_0 C_{kM}^{k_P}(1 + C_q IRI)\ (\text{proportion of new car price}/(1000\ \text{km})) \quad (25)$$

where *IRI* is flatness; a_1, a_2, b_1 and b_2 are regression coefficients; C_0 and C_q are coefficients; K_p is the vehicle age index and C_{km} is the average cumulative mileage (in km) for the vehicle. The regression coefficients are given in Table 15.

Table 15. Parameters for calculating fuel consumption, wheel consumption and material consumption.

Car Class	Vehicle Type	a_1	b_1	a_2	b_2	C_0 ($\times 10^{-6}$)	C_q ($\times 10^{-3}$)	k_p	C_{kM} ($\times 10^4$)
2	Passenger car	9.78	0.182	0.0466	0.0071	12.95	17.81	0.308	15
3	Bus	23.8	0.294	0.0739	0.0016	1.87	4.63	0.483	50
4	Small truck	17.42	0.568	0.0669	0.0107	1.87	327.33	0.371	30
5	Medium truck	23	0.434	0.0653	0.0012	1.87	327.33	0.371	30
6	Heavy truck	19	0.298	0.1556	0.0034	5.52	45.9	0.371	30
7	Container	35.39	0.893	0.2155	0.0053	5.52	20.35	0.371	30

According to the types of vehicles in the Chinese Highway Asphalt Pavement Design Specification [24], the prices of fuels, tires and new cars in the market were investigated. These are summarized in Table 16.

Table 16. Fuel, tire and new car prices.

Car Class	Vehicle Type	Percentage	Fuel Unit Price (RMB/L)	Tire Unit Price (RMB/Article)	New Car Price 10,000 RMB/Vehicle
2	Passenger car	14.4	8 (gasoline)	300	20
3	Bus	12.6	8 (gasoline)	300	30
4	Small truck	2.7	5 (diesel)	1000	10
5	Medium truck	0	5 (diesel)	1000	25
6	Heavy truck	11.7	5 (diesel)	1200	40
7	4-axle semi-trailer truck	4.2	5 (diesel)	1200	40
8	5-axle semi-trailer truck	6.5	5 (diesel)	1200	40
9	6-axle semi-trailer truck	32.8	5 (diesel)	1200	40
10	Double-front-axle semi-trailer	15.1	5 (diesel)	1200	40
11	Full trailer	0	5 (diesel)	1200	40

The design width of the pavement was 15 m, the calculated length was assumed to be 1 km, and the calculated area of the pavement was 15,000 m^2. According to the Jilin provincial engineering cost information network [34], the price of pavement structure materials was investigated. These values are summarized in Table 17. We used the following formula to calculate the initial investment C (in RMB) in pavement:

$$C = A \times (P_i + H_G \times P_G) \tag{26}$$

where A is the calculated area of the pavement (in m^2), P_i is the unit price of each layer material (in RMB/m^2), H_G is the thickness of the graded crushed stone layer (in cm) and P_G is the unit price of the graded crushed stone (in RMB/m^2).

Table 17. Pavement material prices.

Pavement Material	Price (RMB/m^2)	Pavement Material	Price (RMB/m^2)
5 cm fine-grained asphalt concrete	34.36	15 cm cement-stabilized crushed stone 4%	21.98
4 cm fine-grained asphalt concrete	27.49	20 cm cement-stabilized crushed stone 4%	29.31
5 cm mesograin asphalt concrete	28.58	30 cm cement-stabilized crushed stone 4%	44.43
6 cm mesograin asphalt concrete	34.30	34 cm cement-stabilized crushed stone 4%	50.12
7 cm coarse-grained asphalt concrete	37.58	20 cm lime fly ash crushed stone	24.35
8 cm coarse-grained asphalt concrete	42.95	30 cm lime fly ash crushed stone	37.04
24 cm coarse-grained asphalt concrete	128.85	18 cm lime fly ash soil	14.76
15 cm graded crushed stone	12.94	-	-

As can be seen from Figure 8, the initial investment of structure 4 is the least, and the initial investment price of structure 3 is the highest. In the pavement structure, the asphalt concrete material is the most expensive, and the asphalt surface layer used in structure 4 is the thinnest, so the overall construction price is the lowest; the asphalt layer used in structure 3 is the thickest, so the initial investment is the highest. From the perspective of the whole life cycle, the present value of the cost of structure 1 is the lowest. According to the source of funds and the financing situation, by considering the feasibility of short-term investment, it is easy to choose a plan with low initial investment; from a long-term perspective, it is easy to choose a plan with a low present value of the total cost of the life cycle.

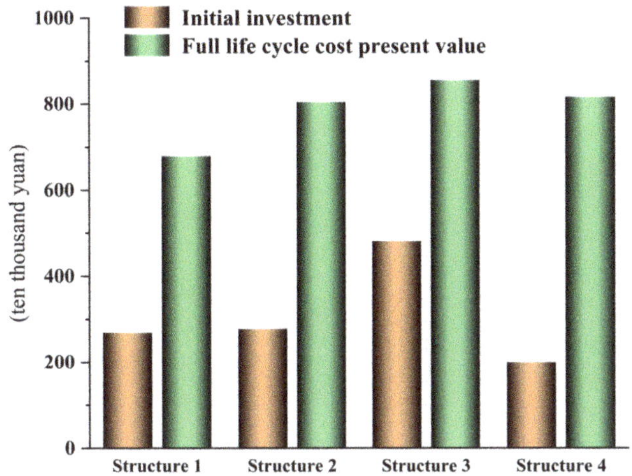

Figure 8. Initial investment and present value of life-cycle cost for four pavement structures.

4.2.3. Calculation of Relevance Degree of Comprehensive Evaluation of Pavement Structure Based on Gray Relevance Theory

According to the performance calculation results for the four types of pavement schemes as the original index values, the matrix can be constructed in Table 18.

Table 18. Initial matrix of key parameters of pavement structure.

Key Indicator		D1	D2	D3	D4	D5	D6	D7	D8
X_i	1	27.074	0.075	10.840	98.790	0.139	70.620	268.700	678.800
	2	30.380	0.064	10.510	136.900	0.276	119.421	237.800	803.930
	3	46.438	0.094	12.180	134.600	0.272	138.100	480.200	855.120
	4	34.000	0.094	5.920	130.800	0.117	59.280	199.400	816.379

The smaller the value of the above eight road performance indicators, the better, and an optimal value was set as the reference sequence $[X_0]$. First, the original data were subjected to dimensionless processing according to Equation (5). The index normalization matrix is shown in Table 19.

Table 19. Normalization matrix.

Matrix		Normalized Results for Each Indicator							
	i	1	2	3	4	5	6	7	8
X_i	0	0.000	0.000	0.000	0.000	0.000	0.000	0.000	0.000
	1	0.000	0.337	0.786	0.000	0.141	0.141	0.247	0.000
	2	0.171	0.000	0.733	1.000	1.000	1.000	0.277	0.710
	3	1.000	1.000	1.000	0.940	0.979	0.979	1.000	1.000
	4	0.358	0.966	0.000	0.840	0.000	0.000	0.000	0.780

Figure 9 shows the correlation degree with different resolution coefficients. From the calculation results, when a = 0.5, the degree of discrimination becomes more obvious.

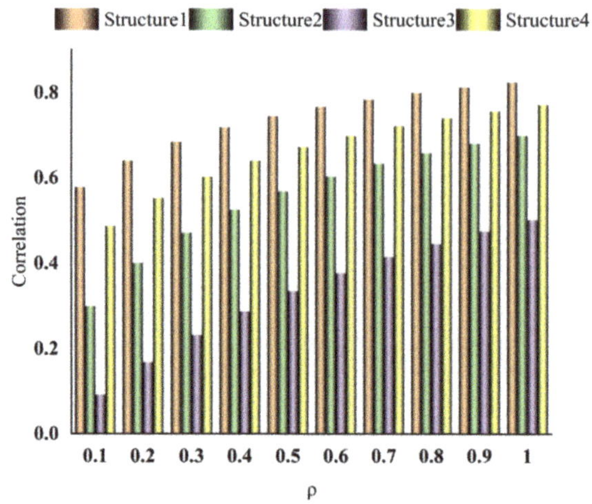

Figure 9. Correlation at different resolution coefficients.

The correlation coefficient matrices are summarized in Table 20. The correlations of the four pavement structures are calculated according to Formula (10), and the final correlations are summarized in Table 21.

Table 20. Correlation coefficient matrix.

Matrix	The Correlation Coefficient of Each Indicator								
	i	1	2	3	4	5	6	7	8
E_i	1	1.000	0.598	0.389	1.000	0.780	0.780	0.670	1.000
	2	0.745	1.000	0.405	0.333	0.333	0.333	0.643	0.413
	3	0.333	0.333	0.333	0.347	0.338	0.338	0.333	0.333
	4	0.583	0.341	1.000	0.373	1.000	1.000	1.000	0.391

Table 21. Correlation of four typical structures calculated by the improved AHP method.

Type	Structure 1	Structure 2	Structure 3	Structure 4
Relevance	0.744	0.574	0.335	0.672

According to the results in Figure 10, $R_1 > R_4 > R_2 > R_3$. Therefore, pavement structure scheme 1 is the closest to the optimal scheme, and it is the best design scheme, and the difference between scheme 3 and the optimal design scheme is the greatest. Although the calculation results using the AHP method were roughly the same as those calculated by using the AHP and gray theory methods, the correlation between the traditional AHP structure 2 and structure 4 is very close and difficult to distinguish. The improved AHP combines the key parameters of the pavement structure, and the correlation degree becomes more obvious.

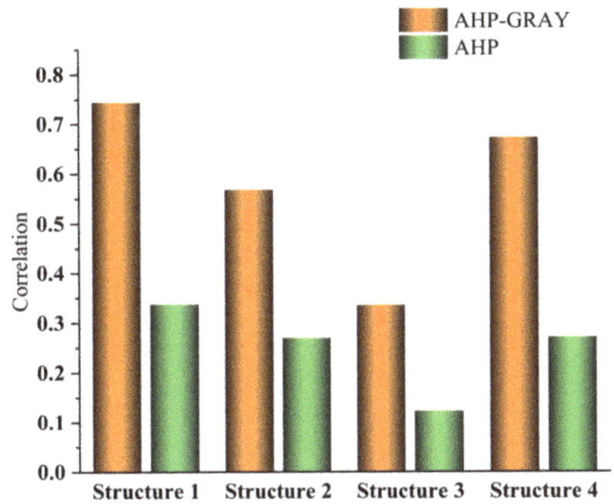

Figure 10. Comparison of correlation results between AHP and improved AHP.

4.3. *Optimal Selection Results and Discussion*

Conducting research on the typical structure of expressways in Jilin Province is needed to improve the local road service level and provide cost savings for infrastructure construction. Finally, through the analysis and evaluation of the pavement structure in Jilin Province and via a comparison of the present value of the cost in the whole life cycle of the structure scheme, the improved AHP can effectively integrate the two most critical indicators involved in road construction and enable quantitative analysis of the calculation results through gray theory. This method of pavement structure decision-making, which integrates pavement performance and economics, can be applied in various regions of the world, facilitating the identification of the most cost-effective pavement structure scheme.

5. Failure Mechanism Analysis

5.1. *Cement-Treated Material Base Layer*

The fatigue cracking life prediction for the cement-treated layer can be calculated according to

$$N_{f2} = k_a k_{T2}^{-1} 10^{a - b\frac{\sigma_t}{R_S} + K_c - 0.57\beta} \quad (27)$$

Table 22 lists the parameters required to calculate the fatigue life model.

Table 22. Fatigue life model parameters of cement-treated layer.

Field Comprehensive Correction Factor	K_c	**−1.43**
Seasonal frozen soil area adjustment factor	K_a	0.65
Temperature adjustment factor	K_{t2}	0.852
Fatigue test regression parameter	a	13.24
Fatigue test regression parameter	b	12.52
Indentation stress of cement-treated layer	σ_t	0.1543
Cement-treated layer bending strength	R_s	1.80
Target reliability indicator	β	1.65
Fatigue cracking life of cement-treated layer (axis)	N_{f2}	4.81×10^9

According to the design of the pavement structure, the fatigue cracking life (axial order) of the inorganic binder stable layer is greater than the design cumulative axis of 103,200,000 times (from the prediction of traffic based on the design report), meeting the specification design requirements.

5.2. HMA Layer Permanent Deformation

Table 23 lists the parameters required to calculate the permanent deformation of the asphalt layer.

Table 23. Permanent deformation prediction model parameters for the HMA layer.

Calculation Parameter	SMA-13			AC-20			ATB-25
i	1	2	3	4	5	6	7
Layer thickness (mm)	15	10	15	20	20	20	80
h_i (mm)	15	20	33	50	70	90	140
p_i (MPa)	0.700	0.699	0.697	0.686	0.663	0.626	0.580
R_{Oi}	1.8	1.8	1.8	2.0	2.0	2.0	3.0
K_{ri}	3.80	5.23	7.11	7.36	6.22	4.74	1.94
R_{ai}	0.81	0.74	1.49	2.23	1.77	1.21	2.60
h_a (mm)	180						
T_{pef} (°C)	17.78						
h_O (mm)	50						
N_{e3}	103,200,000						

The permanent deformation of the HMA layer (in mm) can be calculated according to

$$R_a = \sum_{i}^{n} R_{ai} = 10.84 \text{ mm} < [R_a] = 15 \text{ mm} \tag{28}$$

where R_{ai} is the permanent deformation of each layer (in mm) and $[R_a]$ is the permissible permanent deformation in the design code (in mm).

According to the research in this study, the optimal pavement structure is obtained, and the permanent deformation of the HMA layer and the fatigue of the cement-treated layer can meet the specification requirements.

6. Conclusions

Decision-making for typical pavement structures based on life-cycle economic evaluation and key performance indicators using the AHP and gray relational theory have been investigated. The following conclusions can be made:

(1) In this paper, the Delphi method and analytic hierarchy process are used to construct a decision-making index system for a typical pavement structure. The expert consultation was divided into two rounds. After the first round of the survey, we found that the average score of the antifreeze performance of the pavement index was low, and the coefficient of variation was greater than 0.25. The index was canceled, and the evaluation index system was optimized. After the second round of the survey, according to the survey data, the coordination coefficient increased by 24% compared with the first round, reaching 0.803, and the difference was statistically significant, ensuring the authority, reliability and consistency of the results.

(2) Combined with the four typical structures used by the highway design department of Jilin Province, the finite element model was established by ABAQUS software, and the performance evaluation indexes required by the AHP were calculated. At the same time, according to the material price, design service life, traffic volume, initial cost of road construction and the present value of the whole life cycle cost, the economic evaluation indicators required by the AHP were calculated.

(3) Using the improved gray relational AHP, an evaluation model of the asphalt pavement structure is established. The model organically combines the intermediate process of the two methods of the AHP and gray correlation, which reflects not only the objective situation but also the preference of human subjective factors for evaluation indicators. The evaluation results obtained in this way are more scientific and fairer.

(4) The key technical indicators for the pavement are the permanent deformation of the asphalt layer and the tensile stress and strain at the bottom of the asphalt layer; the key economic indicator for the pavement is the present value of the full life cost. In decision-making for pavement structure, the key technical indicators of pavement performance are more important than the key economic indicator.
(5) This study used the AHP-GRA method to make decisions on typical pavement structures in Jilin Province, and the results show that Pavement Structure 1 is an ideal choice for performance and economy. The fatigue cracking and permanent deformation of structure 1 are checked and calculated, which conforms to the requirements of the Chinese Asphalt Pavement Design Specification.
(6) The improved AHP is used to comprehensively analyze the two key indicators, and the results show that structure 1 is the most suitable typical pavement structure for Jilin Province when the initial budget is relatively sufficient. However, if the initial budget is insufficient, then structure 4 can be considered.

Note that the conclusions obtained in this study are mainly applicable to the four typical structural schemes of the designated Jilin Province. However, when decisions about pavement structure are required in road construction, the analysis scheme presented in this paper can be used to determine a reasonable pavement structure.

Author Contributions: Data curation, M.Z.; Investigation, C.W.; Methodology, L.F.; Project administration, J.Y.; Writing—Original draft, M.Z.; Writing—Review & editing, J.Y. All authors have read and agreed to the published version of the manuscript.

Funding: This research was funded by [Jilin Provincial Department of Education Science and Technology Research Project] grant number [JJKH20220842KJ] and [Science and Technology Program of Shenzhen] grant number [JSGG20201103100601004]. The APC was funded by [Jilin Provincial Department of Education Science and Technology Research Project] grant number [JJKH20220842KJ].

Institutional Review Board Statement: Not applicable.

Informed Consent Statement: Not applicable.

Data Availability Statement: Not applicable.

Conflicts of Interest: The authors declare no conflict of interest.

References

1. Hu, X.; Zhong, S.; Walubita, L.F. Three-dimensional modelling of multilayered asphalt concrete pavement structures: Strain responses and permanent deformation. *Road Mater. Pavement Des.* **2015**, *16*, 727–740. [CrossRef]
2. Erlingsson, S. Rutting development in a flexible pavement structure. *Road Mater. Pavement Des.* **2012**, *13*, 218–234. [CrossRef]
3. Wistuba, M.P.; Walther, A. Consideration of climate change in the mechanistic pavement design. *Road Mater. Pavement Des.* **2013**, *14*, 227–241. [CrossRef]
4. Amini, A.A.; Mashayekhi, M.; Ziari, H.; Nobakht, S. Life cycle cost comparison of highways with perpetual and conventional pavements. *Int. J. Pavement Eng.* **2012**, *13*, 553–568. [CrossRef]
5. Santos, J.; Ferreira, A. Life-cycle cost analysis system for pavement management at project level. *Int. J. Pavement Eng.* **2013**, *14*, 71–84. [CrossRef]
6. Pan, Y.; Shang, Y.; Liu, G.; Xie, Y.; Zhang, C.; Zhao, Y. Cost-effectiveness evaluation of pavement maintenance treatments using multiple regression and life-cycle cost analysis. *Constr. Build. Mater.* **2021**, *292*, 123461. [CrossRef]
7. Fardin, H.E.; dos Santos, A.G. Predicted responses of fatigue cracking and rutting on Roller Compacted Concrete base composite pavements. *Constr. Build. Mater.* **2021**, *272*, 121847. [CrossRef]
8. Ozer, H.; Al-Qadi, I.L.; Singhvi, P.; Bausano, J.; Carvalho, R.; Li, X.; Gibson, N. Prediction of pavement fatigue cracking at an accelerated testing section using asphalt mixture performance tests. *Int. J. Pavement Eng.* **2018**, *19*, 264–278. [CrossRef]
9. Cho, S.-H.; Karshenas, A.; Tayebali, A.A.; Guddati, M.N.; Kim, Y.R. A mechanistic approach to evaluate the potential of the debonding distress in asphalt pavements. *Int. J. Pavement Eng.* **2017**, *18*, 1098–1110. [CrossRef]
10. Yang, K.; Li, R.; Yu, Y.; Pei, J.; Liu, T. Evaluation of interlayer stability in asphalt pavements based on shear fatigue property. *Constr. Build. Mater.* **2020**, *258*, 119628. [CrossRef]
11. Chong, D.; Wang, Y.; Dai, Z.; Chen, X.; Wang, D.; Oeser, M. Multiobjective optimization of asphalt pavement design and maintenance decisions based on sustainability principles and mechanistic-empirical pavement analysis. *Int. J. Sustain. Transp.* **2018**, *12*, 461–472. [CrossRef]

12. Han, C.; Ma, T.; Chen, S. Asphalt pavement maintenance plans intelligent decision model based on reinforcement learning algorithm. *Constr. Build. Mater.* **2021**, *299*, 124278. [CrossRef]
13. Shon, H.; Lee, J. Integrating multi-scale inspection, maintenance, rehabilitation, and reconstruction decisions into system-level pavement management systems. *Transp. Res. Part C Emerg. Technol.* **2021**, *131*, 103328. [CrossRef]
14. Geng, H.; Clopotel, C.S.; Bahia, H.U. Effects of high modulus asphalt binders on performance of typical asphalt pavement structures. *Constr. Build. Mater.* **2013**, *44*, 207–213. [CrossRef]
15. Lv, S.; Yuan, J.; Peng, X.; Zhang, N.; Liu, H.; Luo, X. A structural design for semi-rigid base asphalt pavement based on modulus optimization. *Constr. Build. Mater.* **2021**, *302*, 124216. [CrossRef]
16. Jiang, X.; Zeng, C.; Gao, X.; Liu, Z.; Qiu, Y. 3D FEM analysis of flexible base asphalt pavement structure under non-uniform tyre contact pressure. *Int. J. Pavement Eng.* **2017**, *20*, 999–1011. [CrossRef]
17. Han, C.; Ma, T.; Xu, G.; Chen, S.; Huang, R. Intelligent decision model of road maintenance based on improved weight random forest algorithm. *Int. J. Pavement Eng.* **2022**, *23*, 985–997. [CrossRef]
18. Esmaeeli, A.N.; Heravi, G. A decision support framework for economic evaluation of flexible strategies in pavement construction projects. *Int. J. Pavement Eng.* **2019**, *20*, 1342–1358. [CrossRef]
19. Yu, J.; Zhang, X.; Xiong, C. A methodology for evaluating micro-surfacing treatment on asphalt pavement based on gray system models and gray rational degree theory. *Constr. Build. Mater.* **2017**, *150*, 214–226. [CrossRef]
20. Wang, K.C.; Li, Q.; Hall, K.D.; Elliott, R.P. Experimentation with Gray Theory for Pavement Smoothness Prediction, Transportation Research Record. *J. Transp. Res. Board* **2007**, *1990*, 3–13. [CrossRef]
21. *JTG D50-2006*; Specification for Design of Highway Asphalt Pavement. Ministry of Transport of the People's Republic of China: Beijing, China, 2006.
22. Park, S.W.; Schapery, R. Methods of interconversion between linear viscoelastic material functions. Part I—A numerical method based on Prony series. *Int. J. Solids Struct.* **1999**, *36*, 1653–1675. [CrossRef]
23. *JTG E40-2007*; Test Methods of Soils for Highway Engineering. Ministry of Transport of the People's Republic of China: Beijing, China, 2007.
24. *JTG D50-2017*; Specification for Design of Highway Asphalt Pavement. Ministry of Transport of the People's Republic of China: Beijing, China, 2017.
25. Xu, Q.; Mohammad, L.N. Modeling Asphalt Pavement Rutting under Accelerated Testing. *Road Mater. Pavement Des.* **2008**, *9*, 665–687. [CrossRef]
26. Ferreira, A.; Santos, J. Life-cycle cost analysis system for pavement management at project level: Sensitivity analysis to the discount rate. *Int. J. Pavement Eng.* **2013**, *14*, 655–673. [CrossRef]
27. Li, N.; Huot, M.; Haas, R. Cost-Effectiveness-Based Priority Programming of Standardized Pavement Maintenance. *Transp. Res. Rec. J. Transp. Res. Board* **1997**, *1592*, 8–16. [CrossRef]
28. Smith, K.L.; Titus-Glover, L.; Darter, M.I.; Von Quintus, H.; Stubstad, R.; Scofield, L. Cost–Benefit Analysis of Continuous Pavement Preservation Design Strategies versus Reconstruction, Transportation Research Record. *J. Transp. Res. Board* **2005**, *1933*, 83–93. [CrossRef]
29. Al-Suleiman, T.I.; Shiyab, A.M. Prediction of Pavement Remaining Service Life Using Roughness Data—Case Study in Dubai. *Int. J. Pavement Eng.* **2003**, *4*, 121–129. [CrossRef]
30. Sun, L. *Structural Behavior Study for Asphalt Pavement*, 1st ed.; China Communications Press: Beijing, China, 2005.
31. *JTG 003-86*; Standard for Climatic Zoning of Highways. Ministry of Transport of the People's Republic of China: Beijing, China, 1987.
32. Watanatada, T.; Dhareshwar, A.M.; Lima, P.R. *Vehicle Speeds and Operating Costs: Models for Road Planning and Management*; International Bank for Reconstruction and Development: Washington, DC, USA, 1987.
33. Dickey, J.W.; Miller, L.H. *Road Project Appraisal for Developing Countries*; National Academic: New York, NY, USA, 1984.
34. Jilin Provincial Engineering Cost Information Network. Pavement Material Price. 2022. Available online: https://www.cczjxxw.com/dishi/sz/jgxx/scj.asp (accessed on 1 April 2022).

coatings

MDPI

Article

Performance and Simulation Study of Aged Asphalt Regenerated from Waste Engine Oil

Fan Zhang [1,2], Haibin Li [2,*], Xiaolong Zou [2], Canyang Cui [2], Yaping Shi [1], Hongwei Wang [1] and Fayong Yang [3]

1 School of Urban Construction, Xi'an Kedagaoxin University, Xi'an 710109, China
2 School of Architecture and Civil Engineering, Xi'an University of Science and Technology, Xi'an 710054, China
3 Xi'an Highway Research Institute Co., Ltd., Xi'an 710065, China
* Correspondence: lihaibin1212@126.com

Abstract: In order to reuse waste resources (waste engine oil and waste asphalt mixture), the regeneration process of waste oil on aged asphalt is systematically explained. The BA was treated by aging test, and the basic mechanical properties, molecular dynamics simulation, and infrared spectrum test were carried out on this basis. The results showed that the WEO can restore the physical properties of the aged asphalt, and the recommended amount of WEO is 3%. The density of the asphalt model corresponds to the actual situation of the asphalt. The aged asphalt components are more aggregated. After the WEO was added, the components of asphalt aggregation were reduced and the diffusion ability was improved. Finally, infrared spectroscopy tests were conducted on asphalt specimens, and the results showed that the process of rejuvenated asphalt was dominated by physical reactions. Consequentially, the results of this study build a bridge between the performance and simulation of aged asphalt rejuvenated by WEO. Consequentially, the results of this research promote the recycling of WEO and waste asphalt pavement materials, ultimately advocating the sustainability of pavement construction.

Keywords: aged asphalt; waste engine oil; physical properties; molecular dynamics simulation; molecular model

Citation: Zhang, F.; Li, H.; Zou, X.; Cui, C.; Shi, Y.; Wang, H.; Yang, F. Performance and Simulation Study of Aged Asphalt Regenerated from Waste Engine Oil. *Coatings* **2022**, *12*, 1121. https://doi.org/10.3390/coatings12081121

Academic Editor: Valeria Vignali

Received: 11 July 2022
Accepted: 3 August 2022
Published: 5 August 2022

Publisher's Note: MDPI stays neutral with regard to jurisdictional claims in published maps and institutional affiliations.

Copyright: © 2022 by the authors. Licensee MDPI, Basel, Switzerland. This article is an open access article distributed under the terms and conditions of the Creative Commons Attribution (CC BY) license (https://creativecommons.org/licenses/by/4.0/).

1. Introduction

With the continuous increase in the number of automobiles around world, while satisfying the convenience of transportation, the production of WEO has also increased year by year. According to statistics, at least 1.2 billion liters of WEO can be produced in China every year, but the utilization rate of WEO is less than 20% [1]. The WEO is applied to the field of rejuvenated asphalt [2–4].

On the other hand, with the increase of road service time, a large number of functional and structural diseases appear on the asphalt pavement, and a large amount of rejuvenated asphalt pavement (RAP) is produced in the annual maintenance and rebuilding. At present, the materials used for asphalt regeneration all over the world mainly include: soft asphalt, foam asphalt, asphalt lotion, rejuvenator, etc., while rejuvenator with good performance has shortcomings such as high price, high production cost, and easy to cause environmental pollution [5–7]. With the popularization of the concept of recycling and sustainable utilization of waste materials, asphalt pavement recycling technology has received extensive attention in recent years. How to efficiently utilize road waste materials is a prevalent issue of concern for researchers [8–10]. Relevant scholars have applied WEO to the field of road asphalt recycling and have made great breakthroughs [11].

Al-Mutlaq et al. [12] used WEO as asphalt modifier and adopted chlorination to improve the modification effect of WEO. With the FTIR, DSR, and other test methods, the modification of asphalt with different contents of waste motor oil and chlorine was evaluated. The research results showed that the addition of WEO can improve the viscosity of asphalt, which was beneficial to the high temperature performance of asphalt, and increased the thermal stability of asphalt mixture. Zamhari et al. [13] studied the feasibility of using the used lubricating oil as the rejuvenator of aged asphalt. After adding 5% and 10% WEO to PAV aging asphalt, it was found that adding an appropriate amount of WEO can restore the viscosity, penetration, and rheological properties of aged asphalt. The application of WEO in engineering had relatively significant environmental benefits.

El-Shorbagy et al. [14] used WEO and waste cooking oil (WCO) to rejuvenate the performance of aged asphalt. The study showed that for the same aged asphalt, 3.5%–4% WCO or 5.5%–6% WEO can restore the penetration and softening point of the aged asphalt to the same level, the rejuvenated asphalt from WCO showed better fatigue crack resistance. Mohi Ud Din and Mir [15] studied the effect of WEO on asphalt performance, carried out Marshall stability test, indirect tensile test, and elastic modulus test, and evaluated the performance of the mixture. As the dosage of waste oil increased, the crack resistance of the asphalt mixture decreased. Liu et al. [16] used WEO as modifier to prepare modified asphalt and studied the effect of different content of waste motor oil on the rheological properties of asphalt. WEO had a great influence on the base asphalt (BA). Tabatabaee and Kurth [10] found that the addition of vegetable oil and aromatic oil to the aged asphalt after rejuvenation reduced the intensity of the C=O and S=O peak areas, which means that the waste oil can reduce the asphaltene content. Mamun et al. [17] suggested using different types of waste oil to rejuvenate aged asphalt and studied the rejuvenation effect of WEO and waste cooking oil on aged asphalt. The moisture sensitivity, indirect tensile strength, and resilient modulus of the various asphalt mixtures were analyzed. The results showed that WEO can be used as asphalt rejuvenator.

The continuous change of microstructure reflects the change of macro performance. Therefore, in recent years, researchers have gradually turned to the study of asphalt microstructure. Molecular dynamics simulation technology, as one of the high-precision micro research methods, can predict the macro physical properties of asphalt and provide a theoretical basis for the study of asphalt performance [18–20]. Su et al. [21] used molecular dynamics to calculate the modulus of SBS modifier, nano zinc oxide modifier, and nano zinc oxide/SBS modifier with different sizes and analyzed the influence of different modifiers on the mechanical properties of BA. They found that SBS modifier and nano zinc oxide modifier improved the elastic modulus, shear modulus, and bulk modulus of BA, among which nano zinc oxide/SBS modifier improved the overall performance the most. Zhu et al. [22] studied the strengthening mechanism of asphalt fillers, constructed a composite system of asphalt/silica nanoparticles using molecular dynamics software, verified its density and glass transition temperature, and predicted its mechanical modulus. Fallah et al. [23] constructed different aged asphalt models by adding oxygen atoms to the matrix asphalt model for molecular dynamics simulation to study the effect of aging on the macroscopic mechanical properties of asphalt. Based on the four components of asphalt, Qu et al. [24] established asphalt models with different aging states, simulated the mechanical properties of aged asphalt, and constructed models of aged asphalt and aggregate to study the adhesion of the two during the aging process. It was found that the parameters at the molecular level correlate well with the macroscopic properties of asphalt.

In addition to the single use of waste oil to rejuvenated asphalt, related scholars have begun to study the effect of waste oil and other additives on the performance of asphalt. Seidel and Haddock [25] used WEO and high-density polyethylene (HDPE) as asphalt rejuvenated materials. When the WEO and HDPE dosage ware 7.5% and 4.0%, respectively, the anti-rutting, anti-fatigue, and water stability performance were significantly improved. Li et al. [26] studied the effect of WEO and WCO on rejuvenated and aged asphalt and formulated a new type of rejuvenating agent. The results showed that the rejuvenation effect of the rejuvenator is good after compounding.

To sum up, the use of WEO in road engineering at domestic and foreign level focused on the performance recovery and mechanism research of rejuvenated asphalt. The results have shown that use of WEO improves the performance of aged asphalt. The degree of recovery and improvement of the basic properties of aged asphalt by WEO was analyzed with the help of changes in basic indicators before and after rejuvenation. The rejuvenation mechanism of the WEO on the aged asphalt was revealed through microscopic tests such as SEM and FTIR. It was concluded that WEO can provide light components to the aged asphalt, thus restoring its basic properties. However, there are few studies on the rejuvenation mechanism of aged asphalt from the molecular level. Therefore, the use of molecular dynamics simulation to carry out research on the properties of aged asphalt rejuvenated from WEO is a direction worthy of further development. Accordingly, on the basis of asphalt physical properties tests, this study used molecular dynamics simulation to explain the interaction between WEO and aged asphalt from a molecular point of view through the aggregation and diffusion degree of asphalt components.

2. Materials and Methods

2.1. Materials

2.1.1. Asphalt

The base asphalt (BA) used in this study is 70#A. According to "Standard Test Methods of Bitumen and Bituminous Mixtures for Highway Engineering" (JTG E20-2011), the physical properties of BA were tested. The physical properties test results are listed in Table 1, and all properties meet the requirements of the specification.

Table 1. Performance index of 70# BA.

Items		Units	Measured Value	Technical Specification	Test Method [27]
Penetration (25 °C, 100 g, 5 s)		0.1 mm	65.2	60–80	T 0604-2011
Ductility (10 °C)		cm	102	≥25	T 0605-2011
Softening point		°C	51.6	46.0	T 0606-2011
Viscosity (135 °C)		Pa·s	0.755	-	T 0625-2011
RTFOT (165 °C, 5 h)	Mass Change/%	%	0.316	±0.8	T 0609-2011
	Residual penetration ratio	%	68	≥61.0	
	Residual ductility (15 °C)	cm	10	≥15.0	

2.1.2. Waste Engine Oil

The WEO used in this study was selected from an auto repair and maintenance shop in Xi'an. It was liquid at 25 °C, with a brown-red appearance and a pungent odor. Because the composition of WEO is complex and accompanied by insoluble impurities, the WEO is filtered to remove impurities [28]. The filtration apparatus used in this paper consists of a conical flask with a funnel. The diameter of the filter paper is 11 cm, the pore size is 15–25 μm, and it takes about 12 h to filter once. The filtration process is shown in Figure 1, and the basic composition of WEO is shown in Table 2.

Figure 1. Filter waste engine oil device.

Table 2. Composition of waste engine oil.

Item	Main Ingredient	Additive	Others
WEO	base oil	Anti-aging agent, dispersant, emulsifier, etc.	Metal, dust, moisture, etc.

2.2. Preparation of Aged and Rejuvenated Asphalt

The samples used in this paper are base asphalt, aged asphalt, and rejuvenated asphalt. The preparation process of the samples is as follows:

(1) Adjust the oven temperature control switch to control the temperature at 163 °C ± 1 °C, adjust the film oven to a horizontal state and place it in the sample pan to preheat.
(2) Melt the 70# BA and pour it into the sample dish, the quality of the asphalt in each sample dish is controlled to be 50 ± 0.5 g.
(3) Put the sample pan into the preheated film oven turntable at the predetermined position and conduct an aging test for 5 h, 9 h, and 13 h according to the requirements of the test regulations to obtain an aged asphalt sample.
(4) Add the WEO (1%, 2%, 3%, 4%) to the aged asphalt, stir evenly to obtain the rejuvenated asphalt.

After getting the samples, the optimal addition of WEO is obtained through the basic mechanical property test. With the help of Material studio 2019 software (BIOVIA), the model of base asphalt, aged asphalt, and rejuvenated asphalt can be constructed, and the macro phenomenon is explained through its molecular aggregation and diffusion. Finally, the infrared spectrum test is carried out to systematically analyze the interaction process between WEO and aged asphalt.

2.3. Methods

2.3.1. Physical Performance of Asphalt

In this study, macro (penetration, softening point, ductility, and viscosity) and micro performance tests refer to "Standard Test Methods of Bitumen and Bituminous Mixtures for Highway Engineering" (JTG E20-2011). In this paper, the samples of BA, aged asphalt, and rejuvenated asphalt were tested. The flowchart of this paper is shown in Figure 2.

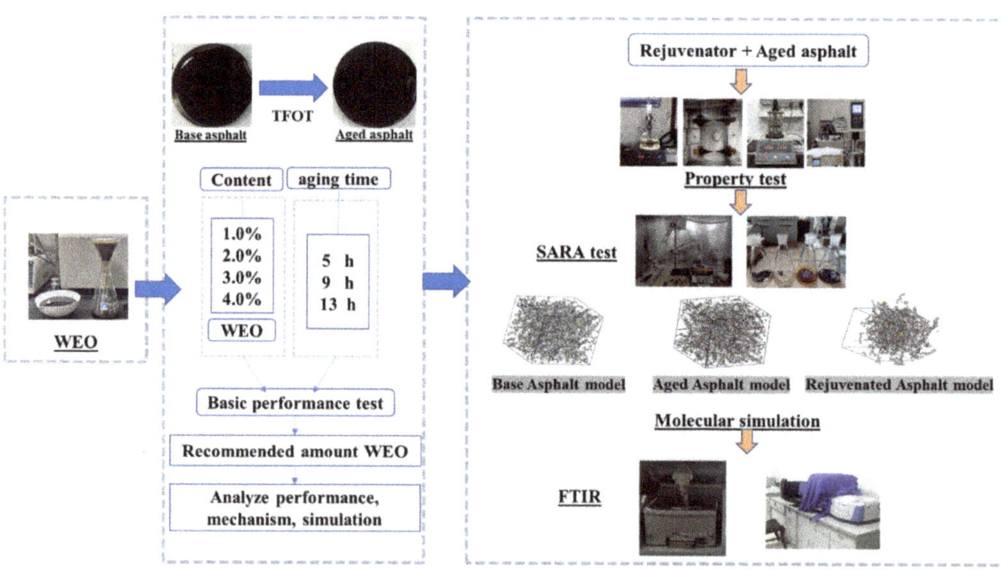

Figure 2. Technical route.

2.3.2. Gas Chromatograph–Mass Spectrometer (GC-MS) Test

A GC-MS (8890B-5977B, Agilent Co., Ltd., Germany) test was carried out on WEO to determine its composition and lay a foundation for the modeling. The gas chromatography–mass spectrometry (GC-MS) test instrument was an Agilent 8890B-5977B gas chromatography–mass spectrometer from Agilent, the scanning mode was full scan mode (SCAN), the mass scanning range was m/z 100–800, and the scanning frequency was 3.2 scan/s.

2.3.3. Fourier Transform Infrared (FTIR) Test

FTIR tests were carried out on BA, aged asphalt, and rejuvenated asphalt. In this experiment, the Varian600-IR series Fourier transform infrared spectrometer (Varian technologies China Co., Ltd., Beijing, China) was used. The test spectral range was 500–4000 cm^{-1}, and the instrument used OPUS software for data acquisition and analysis.

2.3.4. Saturates, Aromatics, Resins, Asphaltenes (SARA) Test

The SARA test was applied to determine the proportion of the components of asphalt, so as to obtain the change of composition before and after asphalt modification, which can also provide a reference basis for modeling. The samples of BA, aged asphalt, and rejuvenated asphalt were tested. The test was carried out in accordance with the method of (JTG E20-2011) t 0618-1993.

2.3.5. Asphalt Aged Test

The aging simulation of BA was carried out. The aging times were 5 h, 9 h, and 13 h. The asphalt aging instrument was 82 asphalt film Oven. The test was carried out according to the requirements of T0609-2011 in "Standard Test Methods of Bitumen and Bituminous Mixtures for Highway Engineering" (JTG E20-2011).

2.4. Molecular Model Determination and Parameter Selection

2.4.1. Asphalt Molecule Selection

Due to the complexity of asphalt composition, it can be divided into four components: asphaltenes, saturates, aromatics, and resins. In this paper, the asphalt binder represented by the 12-component asphalt molecular model proposed by Li and Greenfield [24] was simulated. The molecular formula of the 12 molecules is shown in Figure 3. The model contains three kinds of asphaltenes (AS), two kinds of saturates (SA), two kinds of aromatics (AR), and five kinds of resins (RE). The atomic number of each model before and after asphalt aging is shown in Table 3.

Table 3. Molecular composition of asphalt.

Type		BA	Aged Asphalt
asphaltenes	AS-1	$C_{42}H_{55}O$	$C_{42}H_{46}O_5$
	AS-2	$C_{66}H_{81}N$	$C_{66}H_{67}NO_7$
	AS-3	$C_{51}H_{62}S$	$C_{51}H_{54}O_5S$
saturates	SA-1	$C_{30}H_{62}$	$C_{30}H_{62}$
	SA-2	$C_{35}H_{62}$	$C_{35}H_{62}$
aromatics	AR-1	$C_{35}H_{44}$	$C_{35}H_{36}O_4$
	AR-2	$C_{30}H_{46}$	$C_{30}H_{42}O_2$
resins	RE-1	$C_{40}H_{59}N$	$C_{40}H_{55}NO_2$
	RE-2	$C_{29}H_{50}O$	$C_{18}H_{10}O_2S_2$
	RE-3	$C_{36}H_{57}N$	$C_{36}H_{53}NO_2$
	RE-4	$C_{40}H_{60}S$	$C_{40}H_{56}O_3S$
	RE-5	$C_{18}H_{10}S_2$	$C_{29}H_{49}O_2$

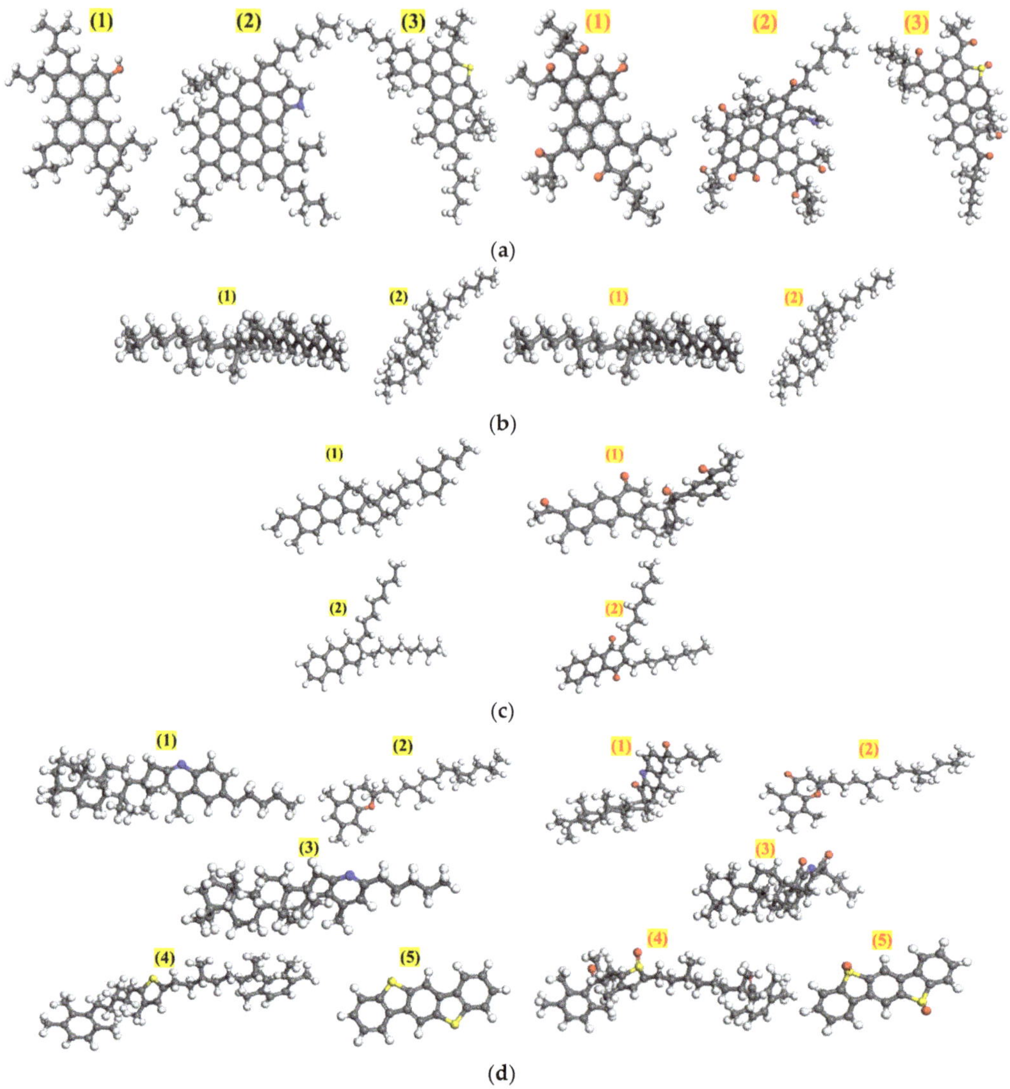

Figure 3. Molecular structures of base and aged asphalt SARA fraction. (**a**) Asphaltenes—As. (**b**) Saturates—Sa. (**c**) Aromatics—Ar. (**d**) Resins—RE. (Gray is carbon atom, white is hydrogen atom, blue is nitrogen atom, red is oxygen atom, and yellow is sulfur atom) (Black numbers represent BA components, and red numbers represent aged asphalt components).

2.4.2. Parameter Selection

The molecular dynamics simulation of the model of asphalt binder and rejuvenator adopted Materials Studio 2019 software to verify the accuracy of the molecular model by density and judged whether the molecular model is stable by energy. By means of the diffusion coefficient and radial distribution function, the diffusion and aggregation degree of the asphalt model were analyzed.

(1) Radial Distribution Function (RDF)

RDF is a measure of the distribution probability of other particles around the reference particle, reflecting the aggregation of the selected molecule with the specified reference molecule. That is to say, the thickness of thin spherical shell interlayer δr tends to 0, the radial distribution function diagram shows the probability of particles appearing on the sphere at the distance from the nucleus r, as shown in Figure 4.

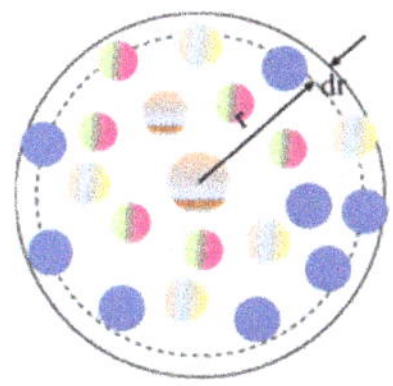

Figure 4. RDF schematic.

Assuming that the number of molecules in the range of $r - r + dr$ around the reference molecule in the system is dN, the radial distribution function is defined as [16]:

$$g(r) = \frac{dN}{\rho 4\pi r^2 dr} \tag{1}$$

In Equation (1), r is the distance between particles, N is the number of particles, and ρ is the average density of the system.

(2) Mean Square Displacement (MSD)

The mean square displacement is used to evaluate the degree of diffusion of the asphalt system and can also be used to calculate the diffusion coefficient of the asphalt system. The stronger the movement ability in the system, the larger the mean square displacement of the particle, as shown in the *MSD* calculation Formula (2):

$$MSD = \lim_{t\to\infty} \frac{d}{dt} \sum_{i=1}^{N\alpha} \left\langle [r_i(t) - r_i(0)]^2 \right\rangle \tag{2}$$

In Equation (2), r_i (t) is the displacement of the particle at a certain time, r_i (0) is the displacement at the beginning of the particle, and N is the number of diffused particles in the system. The *MSD* results of each model can be obtained by analyzing the model with the analysis of force module in the software and calculating the movement of base asphalt, aged asphalt, and rejuvenated asphalt.

The diffusion coefficient of the four components can be calculated through the mean square displacement. The diffusion coefficient represents the diffusion capacity of the material. The reaction degree of the solute in the solvent can be analyzed by using the diffusion coefficient. When the mean square displacement curve shows a linear trend, the equation simplifies to Equation (3):

$$D = \frac{1}{6T} MSD \tag{3}$$

In Equation (3), T is the total time for the movement of atoms or molecules, which is 1/6 of the slope of the simplified linear curve.

3. Results and Discussion

3.1. Evaluation of Physical Properties of Rejuvenated Asphalt

In this section, asphalt before and after aging and rejuvenated asphalt were selected as test materials, and the changes of basic performance indicators were analyzed by different aging time and different WEO content as control factors, and the WEO content with the best rejuvenation effect was recommended. The aging times were 5 h, 9 h, and 13 h, and the content of WEO was 1%, 2%, 3%, and 4%. The basic performance test results are shown in Figure 5. The penetration test results at 25 °C are shown in Figure 5a, softening point test results are shown in Figure 5b, the ductility test results at 10 °C are shown in Figure 5c, and the rotational viscosity test results at 135 °C are shown in Figure 5d.

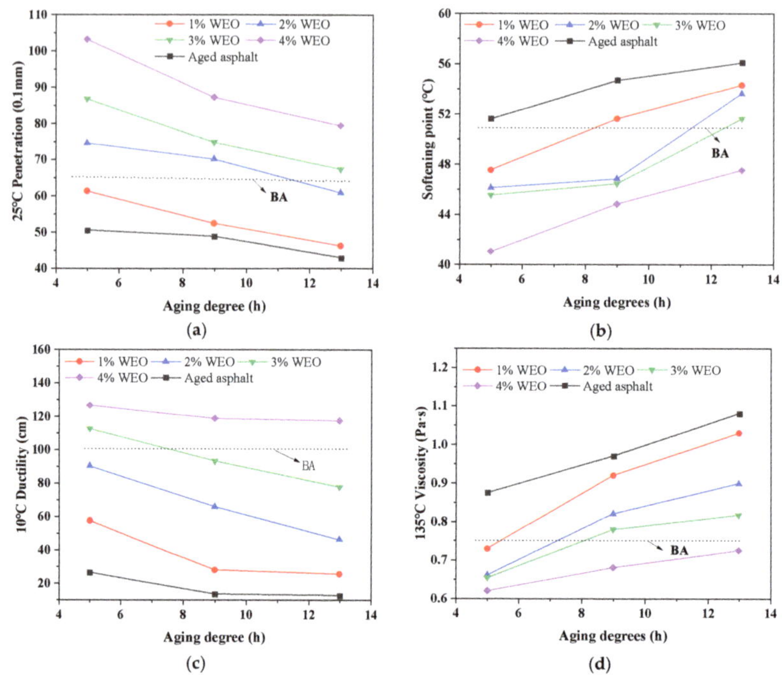

Figure 5. Test results of different indexes of asphalt. (**a**) Penetration. (**b**) Softening point. (**c**) Ductility. (**d**) 135 °C viscosity.

3.1.1. Penetration

As can be seen from Figure 5a, the penetration of the aged asphalt showed a decreasing trend compared with the BA, which was due to the corresponding changes in the asphalt components during the thermal-oxidative aging process, which was macroscopically manifested by the hardening and brittleness of the asphalt. At the same aging time, the penetration of the aging asphalt was restored to different degrees after the addition of WEO and the corresponding needle penetration index increased to a greater extent as the amount of WEO was increased. The penetration showed a consistent decreasing trend with increasing aging process under the same dose, and the recommended dose of WEO is 2% and 3%.

3.1.2. Softening Point

From Figure 5b, it can be seen that the softening point of the aged asphalt showed an increasing trend compared with that of the original asphalt. The reason for this is that the softening point increased due to the reduction of light oil content during the aging process, which led to the softening point of the asphalt not easily flowing at high temperature when measured by the Universal Method. At the same aging time, the softening point index of the aging asphalt was reduced to different degrees after the addition of WEO. The softening point showed a consistent increasing trend with the increasing aging process under the same dose, so the recommended dose of WEO is 2%, 3%, and 4%.

3.1.3. Ductility

As can be seen from Figure 5c, the asphalt ductility showed a decreasing trend compared with the BA. The reason for this was that the asphalt became brittle before and after aging, and it was easy to form stress concentration and fracture during the tensile process, which led to a decrease in ductility. At the same aging time, the softening point of the aging asphalt was restored to varying degrees after the addition of WEO, and the corresponding softening point index increased with the increase of WEO admixture. With the same dose, as the aging process increases, the ductility shows a consistent downward trend; from this index, the recommended WEO dosing is 3% and 4%.

3.1.4. Viscosity

As can be seen from Figure 5d, the rotational viscosity of the aged asphalt at 135 °C showed an increasing trend compared to the original asphalt, indicating that the fluidity of the asphalt became weaker at high temperatures before and after aging, which was due to the increase in the proportion of solid components inside the asphalt, thus hindering the rotation of the asphalt, thus leading to an increase in rotational viscosity. However, the addition of WEO can effectively alleviate the problem of poor flowability of asphalt due to aging, with the increasing amount of WEO, the flowability of asphalt can be further improved.

3.1.5. Optimum Dosage of Waste Engine Oil

In summary, WEO can significantly improve various properties of aged asphalt. In addition, it can be found that the improvement effect of WEO is different under different dosages. From the analysis of basic mechanical properties, it can be seen that under 6 aging times, with the increase of aging time, the performance of asphalt is significantly reduced. After adding WEO, the basic properties of the aged asphalt were recovered. Among them, in terms of penetration, softening point, and viscosity, WEO can improve the workability of aged asphalt, but the excess WEO can improve the workability of aged asphalt. The WEO has a great influence on the softening point, so the amount of WEO should be strictly controlled.

Taking the basic performance index of BA as a reference and taking the regeneration effect under 9 h and 13 h aging time as the judgment basis, comprehensively considering various performances to select the optimal amount of WEO and analyze the results of the penetration, we recommend the best mixture of WEO to be 2% and 3%; according to the analysis of softening point, the recommended optimum dosage of WEO is 2%, 3%, and 4%; and according to the analysis of ductility index, the recommended optimum dosage of WEO is 3% and 4%; from the analysis of the results of the rotational viscosity, it is recommended that the optimal content of WEO is 2% and 3%; and the optimal content of WEO is finally determined to be 3%. The determination of the optimal content of WEO will lay the foundation for the establishment of the later rejuvenated asphalt model.

3.2. Molecular Dynamics Simulation Analysis

3.2.1. WEO Model Construction

In order to analyze the WEO component and molecular formula, gas chromatography–mass spectrometry test, the test results are shown in Figure 6.

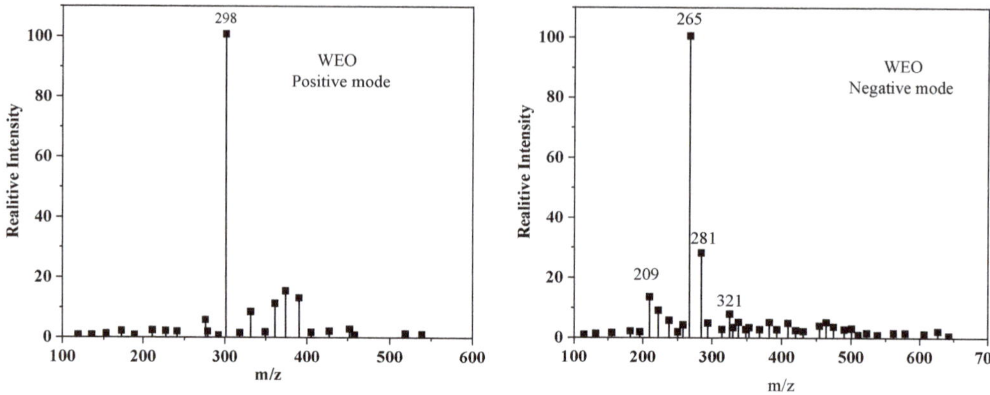

Figure 6. Molecular weight of WEO.

It can be seen from Figure 6 that the WEO has many peaks and irregular distribution, indicating that the waste oil is composed of various compounds. In addition, according to the test results, the molecular weight of WEO is concentrated between 300 and 700 g/mol, indicating that WEO is composed of substances with smaller molecular weights. By consulting related literatures [29,30], the main components of WEO are aromatic hydrocarbons ($C_{n+6}H_{2n+6}$), paraffin oil (C_nH_{2n+2}), and polyolefin n(C_2H_4), and the structure of these compounds is related to the lightweight group of asphalt. Fractions (aromatic hydrocarbons) are similar. WEO is rich in aromatic hydrocarbon molecules, which not only play an important role in the recovery of aged asphalt properties, but also infer its molecular structure, which lays the foundation for the study of molecular dynamics. The composition of WEO is similar to the light components of asphalt, and the molecular formula of WEO is determined to be a benzene ring and a pentyl group [30]. The specific molecular model is shown in Figure 7.

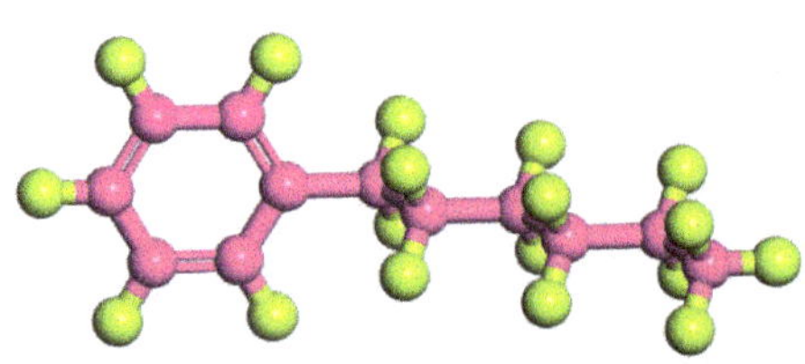

Figure 7. Molecular model of WEO.

3.2.2. Construction of the Asphalt Model

When constructing the asphalt molecular model, the content of the four components in the model must be determined. The test results of the four components of the BA aged asphalt are shown in Figure 8.

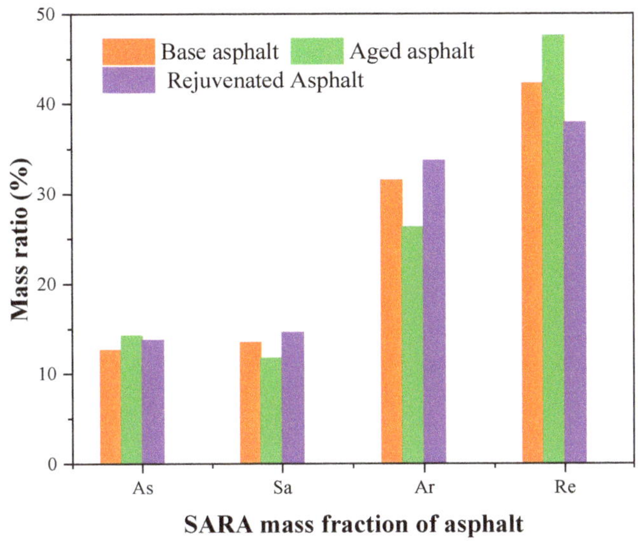

Figure 8. The result of SARA.

It can be seen from Figure 8 that the components of the asphalt change significantly after aging. Compared with the BA, the As and Re of the aged asphalt increased by 12.6% and 12.5%, respectively, and the Sa and Ar decreased by 12.6% and 16.5%, respectively. 28. Compared with the aged asphalt, the Sa and Ar of the asphalt regenerated by adding waste oil increased by 23.7% and 28.1%, respectively. The change of asphalt composition will lead to the change of asphalt performance. After aging, the flow components of asphalt will decrease, resulting in the decrease of penetration and the increase of viscosity and softening point. The addition of waste oil timely replenishes the light components lost due to the aging process, so that the properties of the aged asphalt can be restored.

Through the molecular assembly module in the molecular dynamics software, according to the four component test results, the BA model, the aged asphalt, and the regenerated asphalt molecular model are shown in Figure 9.

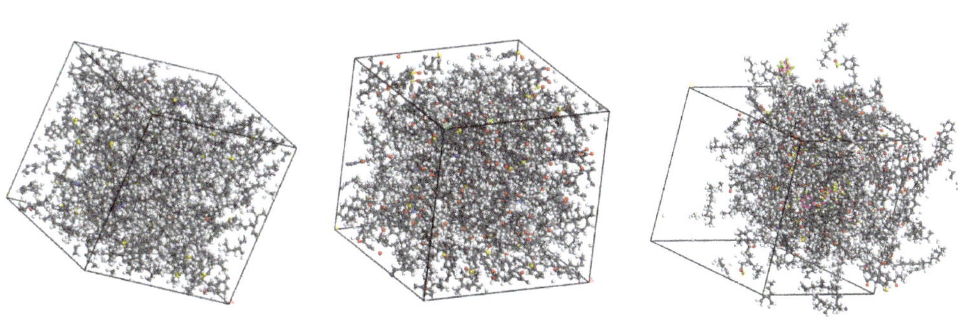

(**a**) (**b**) (**c**)

Figure 9. Asphalt molecular model. (**a**) BA. (**b**) Aged asphalt. (**c**) Rejuvenated asphalt.

3.2.3. Model Accuracy Analysis

As a basic index of asphalt material, density can be used as a direct index to verify the accuracy of simulation methods and force field parameters. Density values of the asphalt at 298 K (25 °C) were calculated using the NPT ensemble. The simulation trend is shown in Figure 10. The results show that the density of BA and aged asphalt increased with the increase of simulation time, and the density gradually stabilized at about 100 ps. With the further increase of simulation time, the density of BA stabilized at 0.92 g/cm^3, the density of aged asphalt was stable at 1.14 g/cm^3, which is in good agreement with the actual situation.

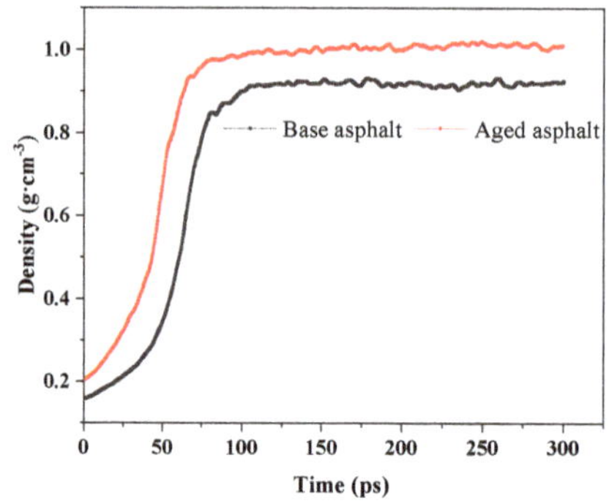

Figure 10. Density changing during simulation.

The total energy of the system includes potential energy, kinetic energy, and bond energy, among which the chemical bond energy is relatively stable from beginning to end, the intramolecular energy does not change significantly during the lifetime, and there is no obvious resistance in the process of molecular dynamics simulation. It is a spontaneous diffusion process. The energy change of the BA system is shown in Figure 11. It can be seen from the figure that the energy of the system is relatively stable.

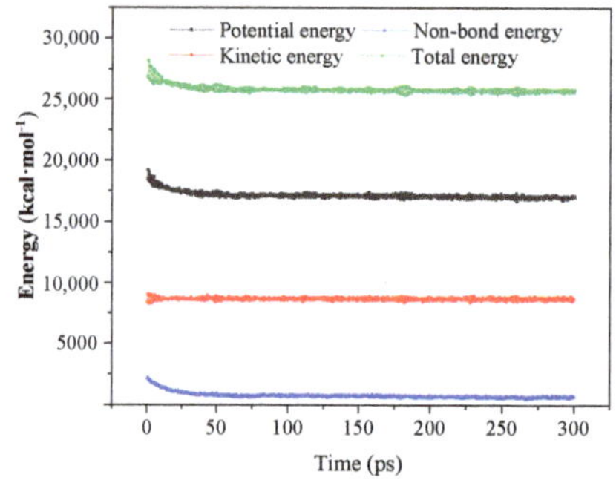

Figure 11. Energy changing during simulation.

3.2.4. Asphalt Agglomeration and Diffusion Capacity Analysis

Figure 11 shows the simulation results of the radial distribution function of each component in different asphalt models. From Figure 12, it can be seen that the aggregation of each component in the asphalt before and after aging changed, which is mainly manifested in the significant increase of g(r) at the same distance; the macroscopic performance shows that the asphalt viscosity increased, which corresponds to the macroscopic mechanical properties test results. Since the addition of the regenerant to the asphalt system reduced the g(r) of each component in the asphalt system, it shows that the addition of the regenerant can effectively alleviate the aggregation of the components of the asphalt. The reason is that the regenerant can effectively improve the proportion of each component of asphalt.

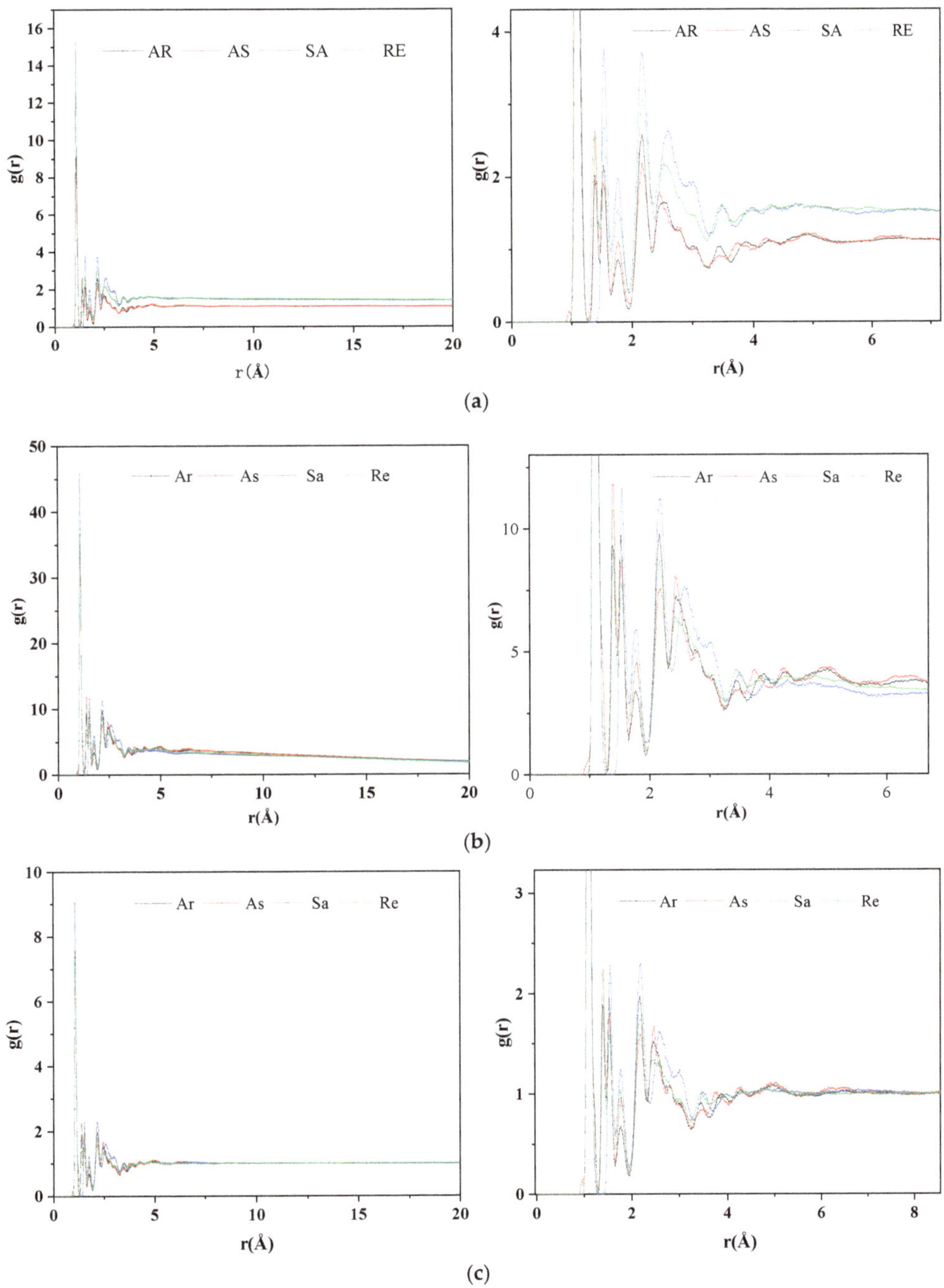

Figure 12. RDF results for asphalt. (**a**) BA. (**b**) Aged asphalt. (**c**) Rejuvenated asphalt.

We used the analysis of the forcite module in the software to analyze the model, calculate the movement of the four components of BA, aged asphalt, and rejuvenated asphalt, and fit the data in the time range of 0 ps to 150 ps with a straight line. The results are shown in Figure 13.

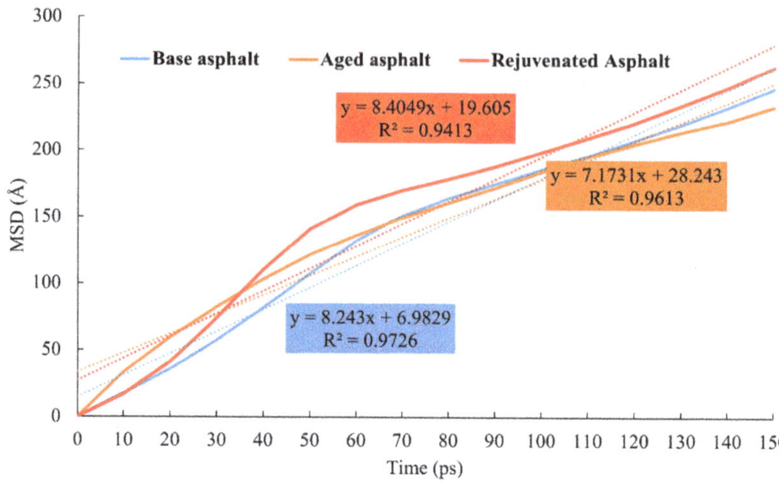

Figure 13. MSD results for asphalt.

As shown in Figure 13, as the simulation time continues to increase, the MSD of each asphalt shows an upward trend, indicating that each molecule in the asphalt system has been in motion. When the time is 70 ps, the mean square displacement curve is linear. The MSD fitting of each asphalt in the range of 0–150 ps shows the diffusion coefficient of each asphalt. The calculation results of the diffusion coefficient are shown in Table 4 below. Compared with aged asphalt, the diffusion coefficient of BA and rejuvenated asphalt increased by 14.9% and 17%, respectively, indicating that the diffusion capacity of rejuvenated asphalt is the best. The addition of waste oil can increase the movement of asphalt components. The macro performance is that the addition of waste oil can effectively alleviate the performance defects of aged asphalt, that is, the recovery of various performance indicators.

Table 4. Diffusion coefficient of asphalt.

Items	$D/10^{-2}$ mm^2/s
BA	1.374
Aged asphalt	1.196
Rejuvenated asphalt	1.401

3.3. Mechanism Analysis

On the basis of basic performance test and molecular simulation, it is also necessary to analyze the mechanism of regeneration of aged asphalt with WEO. In this section, the FTIR test was used to analyze the changes of functional groups before and after the addition of WEO in order to judge the impact of the addition of WEO on its molecular structure, and then through the changes of macro performance indicators and molecular dynamics simulation results, the causes of macro performance changes and the accuracy of molecular dynamics simulation can be determined.

The FTIR of different asphalts are shown in Figure 14. Figure 14a shows the test results of BA, and Figure 14b shows the test results of aged asphalt and rejuvenated asphalt.

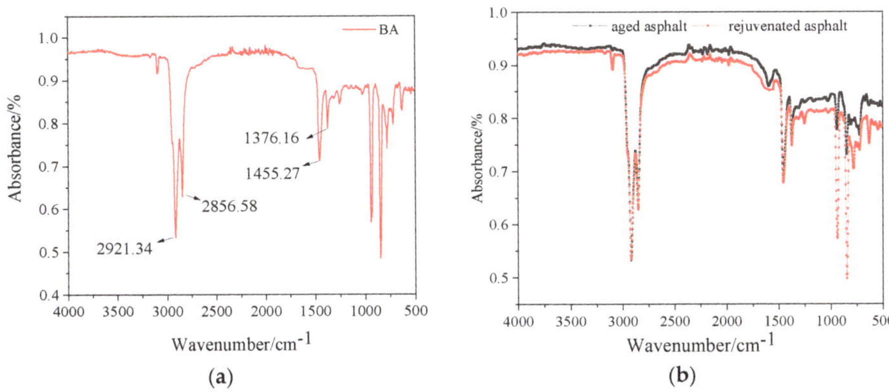

Figure 14. Infrared spectrum test results. (**a**) Base asphalt. (**b**) Aged asphalt and rejuvenated asphalt.

It can be seen from Figure 14a that there are six strong absorption peaks in the functional group region of FTIR of 70# base asphalt. Through the comparison of characteristic absorption peaks, the absorption peak of BA at wave number 3100–3300 cm^{-1} is caused by CH stretching vibration. The strong absorption peaks at 2921.34 cm^{-1} and 2856.58 cm^{-1} are caused by the stretching vibration of the C-H bond in the methylene group. The absorption peak at 1455.27 cm^{-1} and 1376.16 cm^{-1} is the result of bending vibration in C-H plane. The absorption peak at 1630.89 cm^{-1} is the result of the stretching vibration of the C=C double bond. The range 400–1300 cm^{-1} is the infrared fingerprint frequency region, in which small changes in the molecular structure will cause changes in its peaks. It was found that 70# BA has more miscellaneous peak changes in the fingerprint area, indicating that it has more internal components. However, it can be preliminarily judged from the functional group region that the mechanism asphalt is composed of saturated and unsaturated carbon chains [31].

It can be seen from Figure 14b that after the aging of the BA, an obvious characteristic peak appeared at 1690 cm^{-1}. The characteristic peak is carbonyl (C=O) stretching vibration, which is the characteristic peak of asphalt aging. It indicates that the asphalt produces chemical structures such as carbonyl acids or ketones during aging. The existence of this group in asphalt indicates that under the action of high temperature and oxygen in the air, asphalt will be oxidized to produce hydroperoxide, which will decompose into substances with carbonyl functional groups and small molecular free radicals. Asphalt aging is an oxidation process. During the aging process of asphalt, oxygen-containing groups increase, and aromatics and colloids are transformed into asphaltene. It will lead to the increase of viscosity and the decrease of penetration and ductility of asphalt [32].

In addition, by comparing the infrared spectra of the fingerprint area, there is an obvious absorption peak at 500–1000 cm^{-1}. Aromatic hydrocarbons are in this area, indicating that adding WEO can timely supplement the light components, so as to improve the compatibility and anti-aging performance of aged asphalt. The addition of WEO can restore the road performance of aging asphalt, increase the service life of road asphalt, and have good regeneration effect. Meanwhile, by comparing the FTIR of BA and aged asphalt, it was found that the FTIR of rejuvenated asphalt from WEO had no obvious changes. This shows that the regeneration of aged asphalt by WEO is mainly physical. In addition, the FTIR test results match with the molecular simulation results, indicating that molecular dynamics can be used to characterize the aging and regeneration of asphalt [33].

To sum up, it can be concluded that there is no obvious chemical structure change in the regeneration process of WEO. The light component of aged asphalt is supplemented by WEO, and the proportion of light component has a great impact on the macro energy of asphalt. Due to the increase of light component, the viscosity of aged asphalt decreases, the penetration increases, the ductility increases, the softening point decreases, and the diffusion capacity of asphalt increases, which shows that the FTIR results are consistent with the molecular simulation results.

4. Conclusions

Applications of MD simulation in asphalt materials were widely investigated in this study. The development of molecular simulation method provides a reliable way from micro-scale to correlate macro performances and properties with micro mechanisms and molecular structures. The following conclusions can be drawn:

(1) WEO can effectively improve the basic physical properties of aged asphalt. With the increase of WEO content, the low temperature performance of aged asphalt increases, and the high temperature performance is slightly lost. The performance of rejuvenated asphalt can be restored to the level of the original asphalt, and the basic performance is comprehensively considered. For each test index, the final recommended optimal dosage of WEO is 3%.

(2) The asphalt model is established by selecting 12 components. By analyzing the density and capacity of the aging asphalt model and the matrix asphalt model, it can be concluded that with the increase of simulation time, the density of BA and aged asphalt shows an increasing trend, and the density gradually stabilizes at about 100 ps. With the further increase of simulation time, the density of BA is stable at 0.92 g/cm^3, and the density of aged asphalt is stable. At 1.14 g/cm^3, it has a good consistency with the actual situation.

(3) The g(r) of each component in the asphalt system decreased when the WEO was added, indicating that the addition of the rejuvenate can effectively alleviate the aggregation of the components of the asphalt. The reason is that the regenerant can effectively improve the proportion of each component of the asphalt.

(4) The diffusion ability of rejuvenated asphalt is the best, followed by BA, followed by aged asphalt, indicating that the addition of WEO increased the movement of each component of asphalt. The macroscopic performance is that the addition of WEO can effectively alleviate the poor workability of aged asphalt.

(5) The regeneration of aged asphalt by WEO is mainly physical. Compared with BA, the content of carbonyl in aged asphalt increases and the absorption peak of regenerated asphalt decreases at this position. The addition of WEO increases the content of aromatics in aged asphalt, which is consistent with the enhancement of diffusion coefficient of dynamic simulation results. The addition of WEO increases the light components in the aged asphalt, changes the proportion of each component of the asphalt, and the macro performance is the recovery of each performance, which corresponds to the enhancement of the diffusion coefficient of the simulation results.

In this study, the molecular dynamics theory is further combined with the research of pavement materials, and the research ideas of pavement materials are broadened. The models of base asphalt, aged asphalt, and recycled asphalt were successfully established. At the same time, from the perspective of molecular motion, based on dynamics and thermodynamics, the interaction between basic asphalt, aged asphalt, and recycled asphalt is explained. Combined with macro test and infrared spectrum test, the correctness of molecular dynamics simulation results is verified again, and the mechanism of recycled asphalt is further clarified. With the continuous development of modern computer technology, the application of modern science and technology to the research of pavement materials will continue to deepen, which will make it possible to reveal the mystery of the interaction between regeneration and asphalt from a micro perspective and through the intersection of disciplines.

Author Contributions: F.Z.: conceptualization, methodology, writing—original draft. H.L.: supervision, project administration, resources, funding acquisition. X.Z.: conceptualization, writing—review and editing. C.C.: methodology, writing—review and editing. Y.S.: writing—review and editing.

supervision. H.W.: supervision, project administration. F.Y.: validation, investigation. All authors have read and agreed to the published version of the manuscript.

Funding: This research was funded by Key Research and Development Project of Shaanxi Province (2022SF-328), the Science and Technology Project of Henan Department of Transportation (2020J-2-3), Science and Technology Project of Shaanxi Department of Transportation (No. 19-10K, No. 19-28K).

Institutional Review Board Statement: Not applicable.

Informed Consent Statement: Not applicable.

Data Availability Statement: Data will be made available on reasonable request.

Acknowledgments: We express our sincere gratitude to the experts, teachers, and students who have provided help for this article.

Conflicts of Interest: The authors declare no conflict of interest.

References

1. Kuczenski, B.; Geyer, R.; Zink, T.; Henderson, A. Material flow analysis of lubricating oil use in California. *Resour. Conserv. Recycl.* **2014**, *93*, 59–66. [CrossRef]
2. Durga Prashanth, L.; Palankar, N.; Ravi Shankar, A. A study on the effect of rejuvenators in reclaimed asphalt pavement based stone mastic asphalt mixes. *Int. J. Pavement Res. Technol.* **2019**, *12*, 9–16. [CrossRef]
3. Al-Saffar, Z.H.; Yaacob, H.; Mohd Satar, M.K.I.; Saleem, M.K.; Jaya, R.P.; Lai, C.J.; Shaffie, E. Evaluating the chemical and rheological attributes of aged asphalt: Synergistic effects of maltene and waste engine oil rejuvenators. *Arab. J. Sci. Eng.* **2020**, *45*, 8685–8697. [CrossRef]
4. DeDene, C.D.; You, Z. The performance of aged asphalt materials rejuvenated with waste engine oil. *Int. J. Pavement Res. Technol.* **2014**, *7*, 145–152.
5. Qin, Y.; Meng, Y.; Lei, J.; Fu, T.; Xu, R.; Yang, X.; Zhao, Q.; Xi, C. Study on the microscopic characteristics and rheological properties of thermal-oxidative aged and virgin-old recycled asphalts. *Eur. Polym. J.* **2021**, *154*, 110499. [CrossRef]
6. Ma, Y.; Hu, W.; Polaczyk, P.A.; Han, B.; Xiao, R.; Zhang, M.; Huang, B. Rheological and aging characteristics of the recycled asphalt binders with different rejuvenator incorporation methods. *J. Clean. Prod.* **2020**, *262*, 121249. [CrossRef]
7. Fang, Y.; Zhang, Z.; Yang, J.; Li, X. Comprehensive review on the application of bio-rejuvenator in the regeneration of waste asphalt materials. *Constr. Build. Mater.* **2021**, *295*, 123631. [CrossRef]
8. Yu, X.; Li, Y. Optimal percentage of reclaimed asphalt pavement in central plant hot recycling mixture. *J. Wuhan Univ. Technol. Mater. Sci. Ed.* **2010**, *25*, 659–662. [CrossRef]
9. Victory, W. A review on the utilization of waste material in asphalt pavements. *Environ. Sci. Pollut. Res.* **2022**, *29*, 27279–27282. [CrossRef]
10. Tabatabaee, H.A.; Kurth, T.L. Analytical investigation of the impact of a novel bio-based recycling agent on the colloidal stability of aged bitumen. *Road Mater. Pavement Des.* **2017**, *18*, 131–140. [CrossRef]
11. Qurashi, I.A.; Swamy, A.K. Viscoelastic properties of recycled asphalt binder containing waste engine oil. *J. Clean. Prod.* **2018**, *182*, 992–1000. [CrossRef]
12. Al-Mutlaq, S.; Mahal, E.; Yahya, R.; Mahal, A. Effect of chlorination on the assessment of waste engine oil modified asphalt binders. *Pet. Sci. Technol.* **2019**, *37*, 617–628. [CrossRef]
13. Zamhari, K.; Hermadi, M.; Fun, C. Characteristics of rejuvenated bitumen with used lubricating oil as rejuvenating agent. In Proceedings of the International Conference On Sustainable Infrastructure and Built Environment in Developing Countries, Bandung, Indonesia, 12–14 October 2009.
14. El-Shorbagy, A.M.; El-Badawy, S.M.; Gabr, A.R. Investigation of waste oils as rejuvenators of aged bitumen for sustainable pavement. *Constr. Build. Mater.* **2019**, *220*, 228–237. [CrossRef]
15. Mohi Ud Din, I.; Mir, M.S. Laboratory study on the use of reclaimed asphalt pavement and copper slag in warm mix asphalt pavements using waste engine oil as a rejuvenator. *Int. J. Pavement Res. Technol.* **2022**, *15*, 547–559. [CrossRef]
16. Liu, S.; Meng, H.; Xu, Y.; Zhou, S. Evaluation of rheological characteristics of asphalt modified with waste engine oil (WEO). *Pet. Sci. Technol.* **2018**, *36*, 475–480. [CrossRef]
17. Mamun, A.A.; Al-Abdul Wahhab, H.; Dalhat, M. Comparative evaluation of waste cooking oil and waste engine oil rejuvenated asphalt concrete mixtures. *Arab. J. Sci. Eng.* **2020**, *45*, 7987–7997. [CrossRef]
18. Guo, F.; Zhang, J.; Pei, J.; Zhou, B.; Hu, Z. Study on the mechanical properties of rubber asphalt by molecular dynamics simulation. *J. Mol. Model.* **2019**, *25*, 365. [CrossRef]
19. Guo, F.; Zhang, J.; Pei, J.; Ma, W.; Hu, Z.; Guan, Y. Evaluation of the compatibility between rubber and asphalt based on molecular dynamics simulation. *Front. Struct. Civ. Eng.* **2020**, *14*, 435–445. [CrossRef]
20. Chen, Z.; Pei, J.; Li, R.; Xiao, F. Performance characteristics of asphalt materials based on molecular dynamics simulation–A review. *Constr. Build. Mater.* **2018**, *189*, 695–710. [CrossRef]

21. Su, M.; Si, C.; Zhang, Z.; Zhang, H. Molecular dynamics study on influence of Nano-ZnO/SBS on physical properties and molecular structure of asphalt binder. *Fuel* **2020**, *263*, 116777. [CrossRef]
22. Zhu, X.; Du, Z.; Ling, H.; Chen, L.; Wang, Y. Effect of filler on thermodynamic and mechanical behaviour of asphalt mastic: A MD simulation study. *Int. J. Pavement Eng.* **2020**, *21*, 1248–1262. [CrossRef]
23. Fallah, F.; Khabaz, F.; Kim, Y.-R.; Kommidi, S.R.; Haghshenas, H.F. Molecular dynamics modeling and simulation of bituminous binder chemical aging due to variation of oxidation level and saturate-aromatic-resin-asphaltene fraction. *Fuel* **2019**, *237*, 71–80. [CrossRef]
24. Qu, X.; Liu, Q.; Guo, M.; Wang, D.; Oeser, M. Study on the effect of aging on physical properties of asphalt binder from a microscale perspective. *Constr. Build. Mater.* **2018**, *187*, 718–729. [CrossRef]
25. Seidel, J.C.; Haddock, J.E. Rheological characterization of asphalt binders modified with soybean fatty acids. *Constr. Build. Mater.* **2014**, *53*, 324–332. [CrossRef]
26. Li, H.; Dong, B.; Wang, W.; Zhao, G.; Guo, P.; Ma, Q. Effect of waste engine oil and waste cooking oil on performance improvement of aged asphalt. *Appl. Sci.* **2019**, *9*, 1767. [CrossRef]
27. Li, H.; Zhang, F.; Feng, Z.; Li, W.; Zou, X. Study on waste engine oil and waste cooking oil on performance improvement of aged asphalt and application in reclaimed asphalt mixture. *Constr. Build. Mater.* **2021**, *276*, 122138. [CrossRef]
28. Li, D.D.; Greenfield, M.L. Chemical compositions of improved model asphalt systems for molecular simulations. *Fuel* **2014**, *115*, 347–356. [CrossRef]
29. Xu, G.; Wang, H. Molecular dynamics study of oxidative aging effect on asphalt binder properties. *Fuel* **2017**, *188*, 1–10. [CrossRef]
30. Tang, B.; Ding, Y.; Zhu, H.; Cao, X. Study on agglomeration variation pattern of asphalt molecules. *China J. Highw. Transp.* **2013**, *26*, 50.
31. Zhang, M.; Hao, P.; Dong, S.; Li, Y.; Yuan, G. Asphalt binder micro-characterization and testing approaches: A review. *Measurement* **2020**, *151*, 107255. [CrossRef]
32. Zhang, H.; Liu, H.; Zhang, Z. Study on the mechanism of the repeated asphalt ageing and recycling based on the macro-performance. *Road Mater. Pavement Des.* **2016**, *17*, 920–932. [CrossRef]
33. Xiao, M.; Fan, L. Ultraviolet aging mechanism of asphalt molecular based on microscopic simulation. *Constr. Build. Mater.* **2022**, *319*, 126157. [CrossRef]

coatings MDPI

Article

Mechanical and Self-Healing Performance of Yellow River Alluvial Silt Treated with Composite Flexible Curing Agent

Zhiyi Sai [1], Lin Wang [2], Hongchao Han [1], Wenjuan Wu [2,*], Zhaoyun Sun [2], Jincheng Wei [2], Lei Zhang [2], Guiling Hu [3] and Hao Wu [4]

1 Shandong Hi-Speed Company Limited, Jinan 250014, China; jtgsdqgkj@163.com (Z.S.); 18765863568@139.com (H.H.)
2 Shandong Transportation Institute, Jinan 250102, China; wanglin@sdjtky.cn (L.W.); andersontwo@163.com (Z.S.); weijincheng@sdjtky.cn (J.W.); realchungsdu@163.com (L.Z.)
3 School of Transportation Engineering, Shandong Jianzhu University, Jinan 250101, China; huguilingtech@foxmail.com
4 Qingdao Highway & Bridge Group Co., Ltd., Qingdao 266033, China; z5hao@126.com
* Correspondence: wuwenjuan@sdjtky.cn

Abstract: The silt in the Yellow River alluvial plain has low clay content, low cohesion and poor structure. Its stability has always been a difficult problem in the engineering field. In order to improve the engineering properties of the silt in the alluvial plain of the Yellow River, a new type of silt composite flexible curing agent was prepared by using sintered red mud and matrix asphalt as the main materials to comprehensively stabilize the silt. The aim of this study was to investigate the effects of sintered red mud-asphalt composite flexible curing agent on aged mechanical properties of treated silt, in which the replacement levels of the flexible curing agent below 10% by weight are compared. Apart from the compressive strength, the drying shrinkage, low temperature freeze-thaw and high temperature self-healing ability are measured. The test results show that the flexible curing agent has a positive effect on improving the mechanical properties of stabilized silt. The flexible curing agent series exhibit higher compressive strength, better water stability, resistance to freeze-thaw and high temperature self-healing ability, and lower drying shrinkage compared to silt and cement stabilized silt. The preferred dosage 4%~6% of the flexible curing agent is obtained by mechanical property analysis. The SEM images show that the incorporation of the flexible curing agent helps the silt form dense cementation and non-connected microporous structure, that is beneficial to the improvement of water stability and frost resistance. The asphalt component in the flexible curing agent can reorganize and diffuse in the soil, fill the internal pores and micro cracks, and realize the repair of soil damage and structural reinforcement.

Keywords: silt; stabilized silt; sintered red mud-asphalt composite flexible curing agent; mechanical properties; microstructure

Citation: Sai, Z.; Wang, L.; Han, H.; Wu, W.; Sun, Z.; Wei, J.; Zhang, L.; Hu, G.; Wu, H. Mechanical and Self-Healing Performance of Yellow River Alluvial Silt Treated with Composite Flexible Curing Agent. *Coatings* **2022**, *12*, 870. https://doi.org/10.3390/coatings12060870

Academic Editor: Philipp Vladimirovich Kiryukhantsev-Korneev

Received: 8 April 2022
Accepted: 16 June 2022
Published: 20 June 2022

Publisher's Note: MDPI stays neutral with regard to jurisdictional claims in published maps and institutional affiliations.

Copyright: © 2022 by the authors. Licensee MDPI, Basel, Switzerland. This article is an open access article distributed under the terms and conditions of the Creative Commons Attribution (CC BY) license (https://creativecommons.org/licenses/by/4.0/).

1. Introduction

The silt in the alluvial plain of The Yellow River is mainly distributed in the middle and lower reaches of the Yellow River in Shandong, Anhui, Henan, Hebei, etc. These soils have a high silt content (>60%), lack agglomerate structure, and are highly susceptible to loss [1–3]. Historically, because the Yellow River carried a large amount of sediment from the Loess Plateau, and then experienced the advance and retreat of seawater, the oscillating subsidence of the crust and the repeated swings of the river body, shaping into the Yellow River alluvial and sedimentary plain [4,5]. Due to its special origin, The particle distribution and structure of silt in the Yellow River alluvial plain are unique. Its particle gradation is poor, the powder content is too high, and the clay content is low, which makes it difficult to stabilize the binder and compact. Moreover, the capillary water in the silt can rise high and the capillary effect is strong, which is easier to make moisture accumulation in roadbed in seasonally frozen areas. It can result in severe frost heave damage and large post-construction settlement [6–9]. These problems make it difficult for silt to be directly used for roadbed filling. At present, Through technical means such as improved solidification, compaction process optimization and engineering protection measures, the large-scale application of silt to roadbed filling can be realized.

Many scholars have carried out a lot of research on the improvement and stability of silt. The traditional solidifying materials of soil are all solid inorganic binders. Good results have been achieved by using lime, cement and fly ash to improve the soil [10]. It can convert loose soil particles into dense cementitious materials through a series of physical and chemical reactions to improve the strength and durability of silt [11–14]. In the long-term engineering practice, people gradually realize that although these traditional solidified materials can improve the strength of silt, their large dry shrinkage and temperature shrinkage are easy to cause crack, resulting in decrease in compressive strength, impermeability, frost resistance and resistance. It is considered to be the material with large shrinkage and poor water stability among various semi-rigid materials [15–17]. In order to make up for these deficiencies, experts and scholars in related fields are actively committed to the research and development of new soil stabilization technologies.

The researches have been made rapid progress in organic compound curing agents, biological enzyme curing agents and composite curing agents. Organic compound-based soil curing agents are generally liquid and are mainly composed of one or more combinations of water glass, epoxy resin, polymer materials and ionic curing agents [18,19]. This kind of curing agent can promote the exchange of charge in soil particles and soil moisture, and then promote the ionic reaction between the two, and finally play a role in weakening the capillary, pores and surface tension water absorption capacity in the soil. It makes the soil easier to drain and consolidate under the action of external force [20–24]. Ding Rui [25] used X-ray diffraction and X-ray photoelectron spectroscopy to prove that the surface of clay particles would undergo chemical reaction with water glass, and speculated that the chemical reaction may increase the cementation between clay particles. He Jun et al. [26] used water glass-alkali slag-slag to solidify silty clay with high water content, and explored the strength characteristics of the solidified silty clay. Zhao et al. [27] utilized different types of ionic curing agents to cure the expansive soil and evaluated the physical properties such as the expansion rate, liquid-plastic limit and linear shrinkage rate of the cured soil. The change of ion concentration in pore water before and after curing of expansive soil was measured by atomic absorption spectrometry. The expansion potential of expansive soil was obtained by the concentration of cations in pore water. The curing effect of different ionic curing agents was determined. Lynn et al. [28] studied the mechanism of ionic curing agent strengthening montmorillonite by means of chromatography, X-ray diffraction, and titration analysis.

The biological enzyme curing agent is liquid and formed by fermentation of organic matter. Catalyzed by biological enzymes, the adhesion between soil particles will be strengthened when the soil is mechanically compacted, thereby improving the soil engineering properties. The commonly used biological enzyme curing agents mainly include Terrazyme and Permazyme. Greeshma et al. [29] treated high liquid limit montmorillonite from Kerala, India with different concentrations of tyranase and conducted unconfined compressive strength tests on the improved soils with different curing times. The best dosage of the biological enzyme was obtained. Cheng et al. [30] explored the effects of urease concentration, ambient temperature, oil pollution and freeze-thaw cycles on the urease-induced calcite precipitation process through experiments. It proved that this precipitation mode significantly improved the unconfined compressive strength of the soil and its durability under freeze-thaw cycle erosion. Sun [31] used Pyase to solidify Shanghai mixed fill and obtained the change law of the compactness and strength of the soil and gave the optimal dosage of Pyase to solidify the soil. In addition, Peng et al. [32] used the enzyme to solidify coarse sand, fine sand, surface sand and sandy loam, and studied the strength characteristics of the four kinds of enzyme-solidified soils through the unconfined compressive strength test.

Composite soil stabilizers are new types of solidifying materials, generally in solid and liquid forms, and are prepared from two or more compounds in a certain proportion [33–35]. Dong Jinmei et al. [36] used cement-modified polyvinyl alcohol (SH) to solidify light soil and discussed the influence of age, SH content and soil composition on the strength characteristics of solidified light soil. Liu Chengbin [37] used slag composite curing agent to solidify saline soil and evaluated the unconfined compressive strength, water stability, durability and salt swelling of the solidified soil through experiments.

Researchers have carried out a lot of experimental research and practical work on the solidification and stabilization materials of silt, and have achieved good theoretical analysis and experimental results. The solidified materials have gradually developed from traditional inorganic and organic types to inorganic-organic composite and biological improvement cured. However, there are still problems of single improvement effect and high technical difficulty in popularization and application of silt solidification in engineering application practice. And research on self-repair is relatively lacking Therefore, it is very necessary to further study the changes of the road performance of the solidified silt under the unfavorable conditions of actual work, such as water, temperature and natural or load effects, on the basis of the basic performance research on the strength and deformation of the solidified silt.

Sintered red mud is a solid industrial waste residue discharged from the production of alumina by the alkaline process. It has high calcium oxide and silicon oxide content, small particles and a network structure inside, which has strong adsorption capacity and certain hydration activity [38,39]. Asphalt is a complex mixture of hydrocarbons and their derivatives with different molecular weights. It is a temperature-sensitive material with flow self-healing properties [40–42]. Combined with the characteristics of sintered red mud particles and asphalt materials, the two materials were mixed and ground under a certain process to prepare a new type of silt curing agent, and the silt was solidified and stabilized through physical-chemical comprehensive action. The preparation process of the new curing agent can realize the value-added utilization of industrial solid waste—sintered red mud, and at the same time convert asphalt into solid powder form at normal temperature, which is beneficial to the engineering practice and construction quality control of solidified silt. On the basis of evaluating the basic mechanical properties of the composite flexible curing agent stabilized silt, the mechanical properties of stabilized silt under test conditions such as water softening, drying shrinkage, low temperature freeze-thaw, and high temperature self-healing were further studied, and the effect of the composite flexible curing agent content on water stability of stabilized silt, frost resistance and damage repair performance was analyzed. Scanning electron microscope (SEM) was used to observe the microscopic morphology of silt, cement stabilized silt and the composite flexible curing agent stabilized silt, respectively, to analyze the influence of different materials on soil structure and pore characteristics, and to explore the mechanism of solidification and stability.

2. Materials and Experimental

2.1. Materials

The soil used in the study was silt taken from the Yellow River alluvial plain in Dezhou, Shandong Province, China, located approximately 15 km away from the bank of the Yellow River. The particle analysis was conducted as shown in Figure 1. The coefficient of nonuniformity of the soil is 4.8, and its coefficient of curvature is 1.9. The basic physical index properties of the soil were given in Table 1. According to the Test Methods of Soils for Highway Engineering (JTG 3430-2021) [43], the test soil is low-liquid-limit silt containing sand and belongs to the C3 category of fine-grained soil filler.

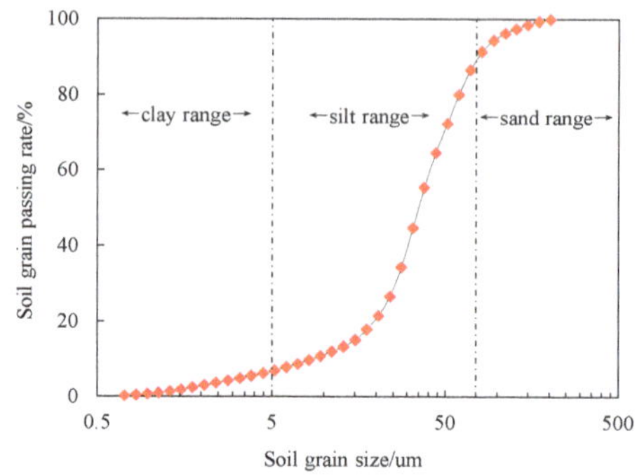

Figure 1. Size distribution of silt.

Table 1. The basic physical index properties of silt soil.

Property	Specific Gravity	Liquid Limit/%	Plastic Limit/%	Plasticity Index	Maximum Dry Density/g·cm^{-3}	Optimum Moisture Content/%
Value	2.7	28.4	19.8	8.6	1.78	15.1

The morphological characteristics of the Yellow River alluvial silt were observed by scanning electron microscope (SEM) in Figure 2. The silt has high particle roundness, uniform particle size, and the particles are connected in an overlapping manner. During rolling, it is difficult to form effective particle embedding between particles, and mutual dislocation between particles is easy to occur under the action of external force.

Figure 2. SEM images of Yellow River alluvial silt particles.

Portland cement PI42.5, Sintered red mud and base asphalt powder were used in the tested mixtures, and their properties are shown in Tables 2–4, respectively. The chemical composition of sintered red mud is mainly CaO and SiO_2, and it has certain hydration activity. The penetration of the base asphalt is 70 (0.1 mm), and the softening point is 46.5 °C.

Table 2. The properties of Portland cement.

Test Project	Specific Gravity	Specific Surface Area m^2/g	SiO_2%	Al_2O_3%	MgO%	Fe_2O_3%	CaO%	Na_2O%	SO_3%
Value	3.1	0.35	23.55	5.64	1.67	2.85	64.17	0.26	0.49

Table 3. The properties of sintered red mud.

Test Project	Water Content/%	Bulk Density g/cm^3	Specific Surface Area m^2/kg	SiO_2%	Al_2O_3%	Fe_2O_3%	CaO%	Na_2O%
Value	<3	0.6-0.8	550–600	18.87	9.6	13.25	41.1	3.63

Table 4. The properties of base asphalt.

Property	Density at 15 °C g/cm^3	Penetration/0.1 mm	Softening Point/°C	Ductility at 15 °C/cm	Dynamic Viscosity at 60 °C/Pa·S
Value	1.033	70	46.5	>100	246

The composite flexible curing agent was developed with sintered red mud and No. 70 base asphalt as the main materials. First, the sintered red mud was dried and ground to less than 120 mesh for use. Then, the base asphalt was heated to a certain temperature and the ground sintered red mud was put into in proportion. After fully stirring for 120 s, it was cooled to room temperature. Finally, an appropriate amount of dispersant was put in, and the mixture was crushed to less than 0.075 mm with a pulverizer to obtain the composite flexible curing agent.

2.2. Sample Preparation

The samples used in the study are compacted by Proctor method according to Chinese standard Test Methods of Soils for Highway Engineering JTG 3430-2021 [43] and Test methods of materials stabilized with Inorganic Binders for Highway Engineering JTG E51-2009 [44] with 96% compaction. The formed samples are placed in a standard curing room for curing. According to engineering experience and previous studies [45], the cement content is limited to 5%. The stabilized soil scheme was given in Table 5. Here, the reference group F-0 denotes the stabilized silt with 5% cement. F-2, F-4, F-6 and F-8 denotes the compound stabilized silt using 2%, 4%, 6%, 8% flexible curing agent on the basis of 5% cement.

Table 5. The stabilized soil scheme.

Sample	Cement/%	Flexible Curing Agent/%
F-0	5	0
F-2	5	2
F-4	5	4
F-6	5	6
F-8	5	8

2.3. Test Methods

2.3.1. Compressive Strength

The compressive strength of soil specimens (Φ39.1 mm × 80 mm) are tested at 3, 7 and 28 standard curing ages using a compression testing machine according to the Chinese standard JTG E51 T0805(2009) [44]. The average value of at least three specimens is reported as the compressive strength test result of the specimen group.

Another set of specimens are prepared to soak in water for one day at the last day of 3, 7 and 28 standard curing ages. The compressive strength after immersion are tested.

2.3.2. Drying Shrinkage

The drying shrinkage are measured in accordance with JTG E51 T0854 (2009) [44]. All drying shrinkage samples (100 mm × 100 mm × 400 mm) are placed in a room with 20 ± 2 °C and 60 ± 5% relative humidity after standard curing for 6 d and soaking in water for 1 d. Measurements are carried out until 30 days drying period.

2.3.3. Low Temperature Freeze-Thaw

The low temperature freeze-thaw of soil specimens (Φ150 mm × 150 mm) are tested at 28 d. according to the Chinese standard JTG E51 T0858 (2009) [44]. The samples are first placed in a room with −18 °C for 16 h, and then melt in 20 °C water tank for 8 h after freezing. The freeze -thaw cycles are performed 5 times.

2.3.4. High Temperature Self-Healing

The high temperature self-healing samples (Φ150 mm × 150 mm) are first loaded to 0.9 times the maximum load, and then heated to 49 ± 1 °C after sealing. Finally, it returned to 20 °C after keeping for 2 h at 49 ± 1 °C. The heating cycles are performed 5 times.

2.3.5. Scanning Electron Microscopy (SEM)

The morphology of soil specimens are analyzed by Field Emission Scanning Electronic Microscopy (Sigma 500, Carl Zeiss AG, Oberkochen, Germany). The microstructural differences of silt, cement stabilized silt, the flexible curing agent stabilized silt and the flexible curing agent stabilized silt after heating are observed.

3. Results and Discussion

3.1. Compressive Strength and Water Stability

Figure 3 shows the effect of the flexible curing agent on the compressive strength of silt at 3, 7 and 28 days. The compressive strengths increase with ages in all series. The compressive strengths at 3 and 7 days are increased rapidly, but tend to be stable at 28 days. The addition of the flexible curing agent has a significant effect on the early strength of stabilized silt. Figure 4 shows the normalized compressive strength of all stabilized soil compared to 0%. The results show that the ternary cementitious system containing the flexible curing agent and cement in silt soil behaves obviously better than the effect of cement alone. The compressive strength of F-0 is 0.80 and 0.91 MPa at the 7 and 28 days, respectively (Table 6). The addition of 2%, 4%, 6% and 8% of the flexible curing agent cause increase respectively about 17.5%, 56.3%, 67.5%, 61.3% at 7 days, and 11.0%, 44.0%, 58.2%, 50.5% at 28 days. As demonstrated in Figure 4, the highest compressive strength rates of stabilized silt are achieved when F-6 blend is added. This incorporation led to approximately 58%–68% of strength increase higher than that of F-0 at 7 and 28 days. The compressive strength of F-8 decreases compared with that of F-6. This phenomenon is attributed to plasticity enhancement with the increase of asphalt content.

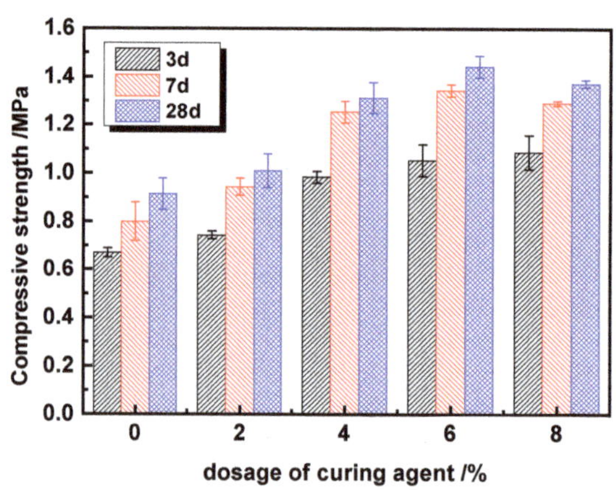

Figure 3. Effect of dosage of the flexible curing agent on compressive strength of stabilized silt with curing ages.

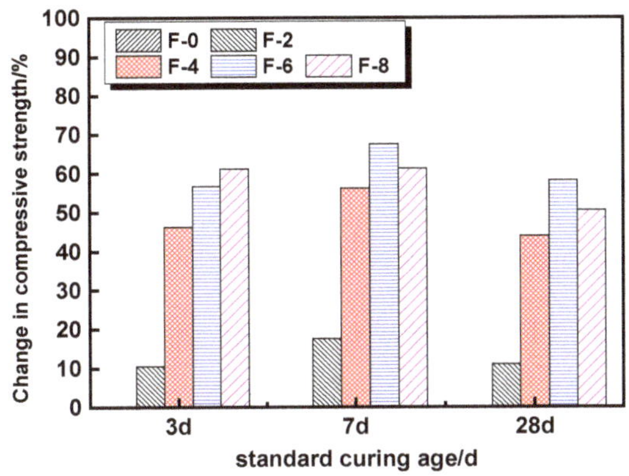

Figure 4. Normalized compressive strength of all stabilized soil related to F-0.

In order to investigate the water stability of the flexible curing agent stabilized silt, the water immersion strength test was carried out. Table 6 shows the strength change before and after immersion for all series at 3, 7 and 28 days. After immersion in water, the compressive strengths of silt are decreased compared with standard curing. This result is consistent with those observed previously by other researcher [46,47]. This is attributed to that the ingress of water damages the internal structure of the solidified soil, the gel material is peeled off from the soil particles, and the number of pores in the sample increases. However, the addition of the flexible curing agent reduces the strength loss rate significantly, as shown in Figure 5. The strength loss rate of F-0 is 61.2%, 51.3% and 50.5% at 3, 7 and 28 days, respectively. However, the addition of 2%, 4%, 6% and 8% of the flexible curing agent cause reduction in strength loss rate respectively about 35.9%, 65.0%, 73.5%, 68.3% at 3 days, about 52.2%, 76.6%, 81.1%, 80.3% at 7 days, and about 60.8%, 85.0%, 86.3%, 81.2% at 28 days. The lowest compressive strength loss rates of stabilized silt are achieved when 6% flexible curing agent blend is added. This incorporation led to approximately 81-86% of strength increase higher than that of F-0 at 7 and 28 days. And the strength loss rate of F-8 increases slightly compared to that of F-6, which is similar to the result in standard compressive strength. As can be seen in Figure 5, after 7 days of curing age, the change range of immersion strength loss rate is small in all series, indicating that prolonging the curing time is conducive to enhancing the water stability of the stabilized silt. Moreover, when the flexible curing agent dosage is higher than 4%, the immersion strength loss rates of stabilized silt gradually tend to be stable.

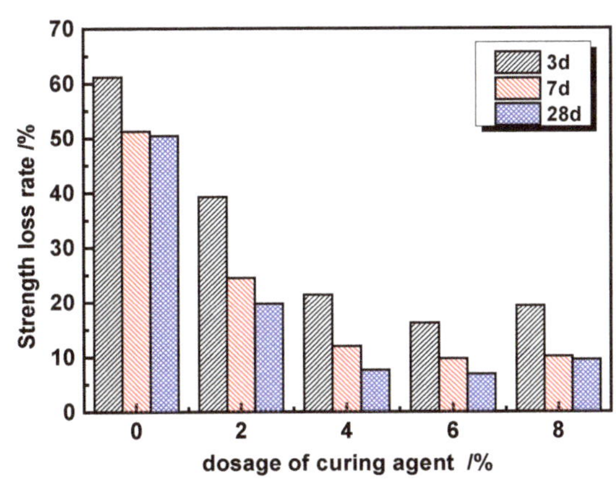

Figure 5. Strength loss rate of stabilized silt after immersion.

Table 6. Standard and immersion compressive strength values of all specimens.

Sample	Compressive Strength/MPa											
	3 d				7 d				28 d			
	Standard Curing	Standard Devia-tion	Immersion	Standard Devia-tion	Standard Curing	Standard Deviation	Immersion	Standard Devia-tion	Standard Curing	Standard Deviation	Immersion	Standard Deviation
F-0	0.67	0.020	0.26	0.022	0.80	0.080	0.39	0.050	0.91	0.065	0.45	0.070
F-2	0.74	0.015	0.45	0.030	0.94	0.035	0.71	0.055	1.01	0.070	0.81	0.045
F-4	0.98	0.025	0.77	0.020	1.25	0.045	1.10	0.025	1.31	0.065	1.21	0.015
F-6	1.05	0.065	0.88	0.070	1.34	0.025	1.21	0.05	1.44	0.045	1.34	0.020
F-8	1.08	0.070	0.87	0.055	1.29	0.01	1.16	0.01	1.37	0.015	1.24	0.060

This phenomenon with higher compressive strengths and water stabilities of the flexible curing agent stabilized silt series is caused for two reasons. One is attributed to the sintered red mud in the flexible curing agent. Sintered red mud is an active component and can stimulate the cement hydration reaction. The other is attributed to the asphalt that works in the soil. The asphalt particles connect the hydration products and the silt particles to form a cohesive skeleton structure under the action of physical compaction.

3.2. Drying Shrinkage

The effects of the flexible curing agent on drying shrinkage of stabilized silt are illustrated in Figures 6–9. Figure 6 shows the variation of loss rate with time. The change of water loss rate with time for silt soil is very different from that of stabilized soil. On the first day of the test, the water loss rate of silt soil reaches 11.65%, accounting for 88% of its final water loss rate at 12 days. As to stabilized silt, the water loss rate changes the most in 7 days. The water loss rate of F-0 is 14.9% at 7 test days, accounting for 89.9% of its final water loss rate. The water loss rate of F-2, F-4 and F-6 is 12.2%, 11.9% and 11.0% respectively at 7 days, and accounting for 87.5%, 86.6% and 85.5% of their final water loss rate at 28 days. The result shows that the addition of cement increases the water loss of silt soil, but the addition of the flexible curing agent decreases the water loss of cement stabilized silt. Compared to the silt soil, all the flexible curing agent series show better water retention in the early stage, which can inhibit the occurrence of shrinkage cracks in the specimen to a certain extent. It is recommended to take maintenance measures within 7 days after the completion of construction to avoid excessive water loss and increase dry shrinkage strain.

Figure 7 exhibits the shrinkage strain variation of silt and stabilized soil with time. The drying shrinkage of silt develops quickly and diverges from the flexible curing agent stabilized silt at early test ages. The dry shrinkage strain of silt soil mainly occurred in the first day of the dry shrinkage test. the shrinkage strain of silt soil reaches 1302.2×10^{-6} $\mu\varepsilon$, accounting for 89.5% of its total strain at 12 days. In Figure 7, we observe that the dry shrinkage strain of stabilized silt mainly occurred in the first 14 days, and the flexible curing agent series demonstrate lower drying shrinkage value than silt and cement stabilized silt at early age. The shrinkage strain of F-0 is 1130.8×10^{-6} $\mu\varepsilon$ at 14 test days, accounting for 89.3% of its total strain. The shrinkage strain of F-2, F-4 and F-6 is 634.3×10^{-6} $\mu\varepsilon$, 535.6×10^{-6} $\mu\varepsilon$ and 443.8×10^{-6} $\mu\varepsilon$ respectively at 14 days, and accounting for 87.3%, 85.4% and 84.4% of their total strain. The incorporation of the flexible curing agent can effectively decrease the shrinkage strain of silt. And the best performance is obtained by the addition 6% curing agent whose drying shrinkage observably decreased and was lowest.

Figure 7 exhibits the shrinkage coefficient variation of silt and stabilized soil with time. With the increase of time, the drying shrinkage coefficient shows a change rule that increases first and then stabilizes for all series. The average shrinkage coefficient of silt, F-0, F-2, F-4 and F-6 is $183.2 \times 10^{-6} \cdot °C^{-1}$, $172.1 \times 10^{-6} \cdot °C^{-1}$, $171.9 \times 10^{-6} \cdot °C^{-1}$, $134.5 \times 10^{-6} \cdot °C^{-1}$ and $122.7 \times 10^{-6} \cdot °C^{-1}$, respectively. The incorporation of 4%–6% curing agent led to approximately 22%–29% of average shrinkage coefficient decrease than that of F-0. It shows that the stabilized silt with the flexible curing agent has better shrinkage resistance.

It is obvious that the drying shrinkage of the material is closely related to the loss of moisture. Figure 9 shows the relationship between water loss rate and dry shrinkage strain of silt and stabilized silt. The water loss rate of stabilized silt in the early stage has little effect on its dry shrinkage strain, but it is larger in the later stage. When the water loss rates of stabilized silt are less than 10%, their dry shrinkage strain hardly increase, but when the water loss rates exceed 10%, the dry shrinkage strains increase rapidly. The loss of moisture in the early stage is mainly the surface moisture of the test specimens and the free water inside the specimens, the loss of that has little effect on the internal pore structure of specimens. In the later stage, the capillary water and adsorbed water inside the specimens are lost, making the water film on the surface of the soil particles thinner, the spacing between the particles smaller, and the molecular force increased. The loss of capillary water and adsorbed water inside the specimens is the main cause of dry shrinkage.

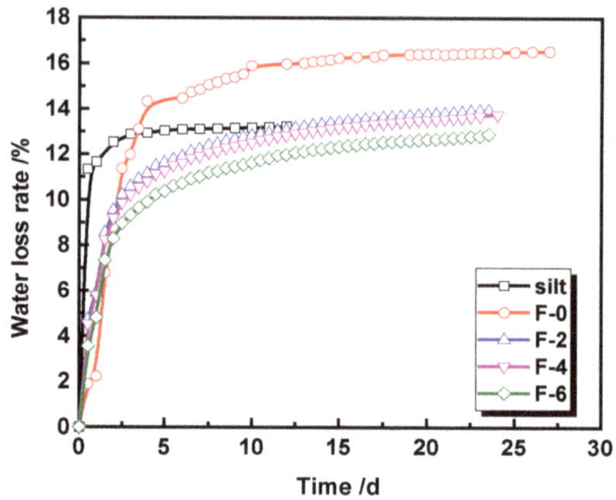

Figure 6. Variation of water loss rate with time.

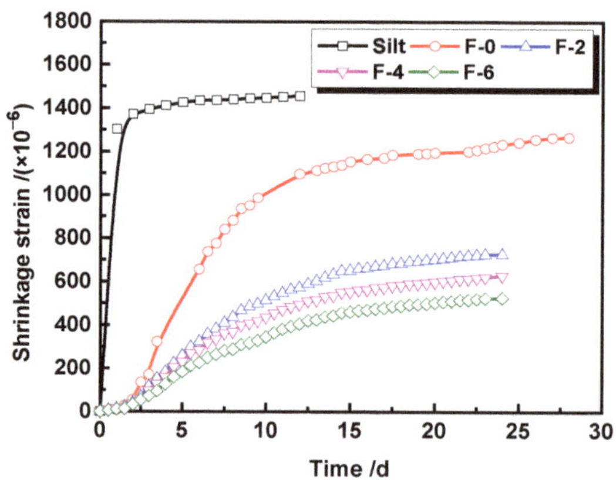

Figure 7. Variation of shrinkage strain with time.

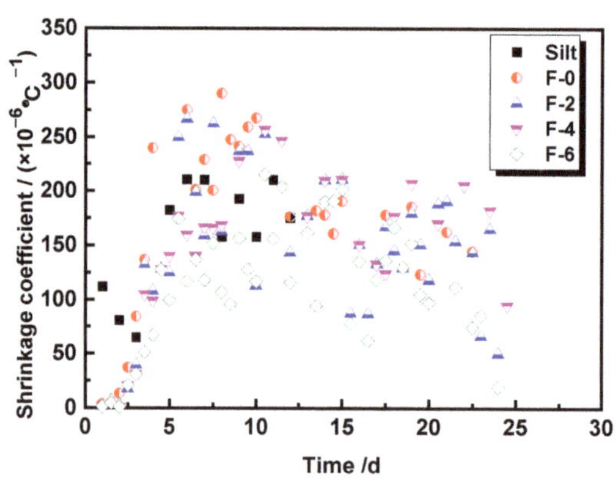

Figure 8. Variation of shrinkage coefficient with time.

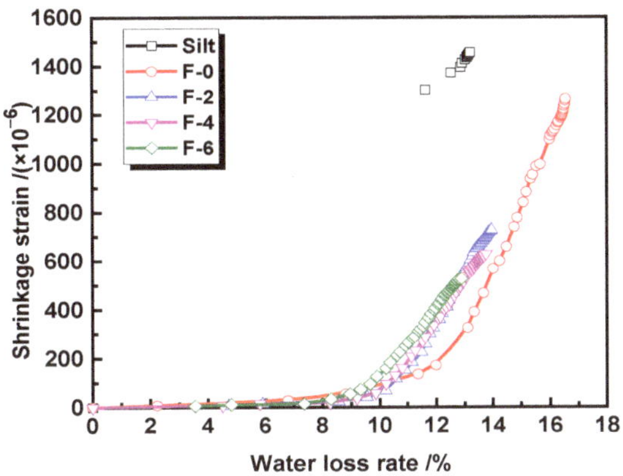

Figure 9. Relationship between water loss rate and dry shrinkage strain.

3.3. *Low Temperature Freeze-Thaw*

The mechanical behavior of the flexible curing agent stabilized silt under low temperature freeze-thaw was analyzed. The crack and damage occurred in F-0 after only 1 freeze- thaw cycle (Figure 10), so 2, 4, 6, 8% curing agent stabilized silt was selected for low-temperature freeze-thaw. The stress-strain curves of curing agent stabilized silt under standard curing and low temperature freeze-thaw are compared in Figure 11. The results indicate that the strength and stiffness of curing agent stabilized silt all decrease after 5 freeze-thaw cycles. The strength change rate of F-2, F-4 and F-6 is −20.5%, −12.9%, −10.1%, and −14.6% respectively as seen in Figure 12. Compared to the cement stabilized silt (F-0), the incorporation of the flexible curing agent improves the freeze resistance ability. The best performance of freeze resistance is obtained by the addition 6% curing agent whose strength loss observably decreased and the strength loss rate was lowest. This is contributed to that the active components of sintered red mud in the flexible curing agent component stimulate the cement hydration reaction, and the asphalt particles in the flexible curing agent blocks the capillary connection in the structure through granulation dispersion and viscous adsorption, reducing the porosity. However, the strength loss increased by the addition of 8% curing agent. This difference is caused by the overdose of the flexible curing agent which makes the form of agglomerate structure and microporosity, increases water absorption and decreases the freeze resistance ability.

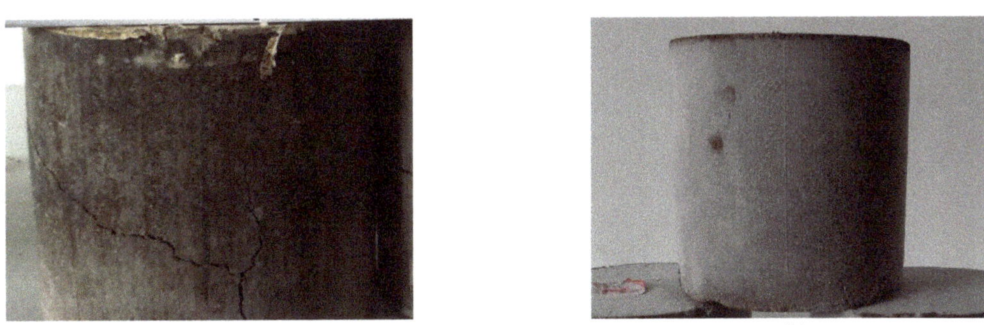

(a) (b)

Figure 10. Soil sample state after low temperature freeze-thaw. (**a**) Cement stabilized; (**b**) flexible curing agent stabilized.

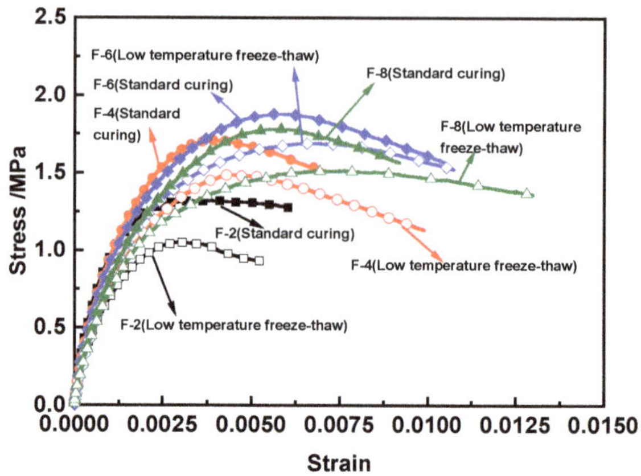

Figure 11. Stress-strain curve of stabilized silt under standard curing and low temperature freeze-thaw.

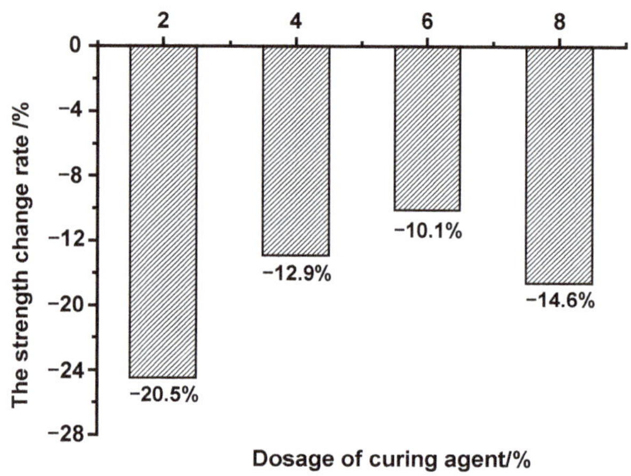

Figure 12. The strength change rate of the flexible curing agent stabilized silt.

3.4. *High Temperature Self-Healing*

The damaged asphalt material can be reorganized and self-healed under high temperature conditions [40,42]. The high temperature self-healing test was conducted to explore the effect of asphalt self-healing on soil remediation. The stress-strain curve of the flexible curing agent stabilized silt under standard curing and high temperature are compared in Figure 13. It can be seen that the slopes of the σ-ε curves of the high-temperature self-healing test are all smaller than those of the standard curing σ-ε curve, and the peak values of the curve gradually increase beyond those of the standard curing curve. The results indicate that the compressive modulus of the stable silt decreases in the high temperature healing test, and the compressive strength is enhanced with the increase of the flexible curing agent content. The strength change rate of F-2, F-4 and F-6 is −8.3%, −2.3%, +8.0%, and +12.9% respectively as seen in Figure 14. The self-healing properties of asphalt components play a role. The stabilized silt with 6%–8% curing agent shows a good self-healing and reinforcing effect. The interface surface energy of asphalt and soil particles is changed significantly under high temperature, where the diffusion and reorganization of asphalt molecules are stimulated. The reconstituted asphalt fill and adhere to the micro-cracks in the soil, which enhance compactness and improve the microstructure in damaged stabilized soil. There is a decrease in modulus with the addition of the flexible curing agent. The modulus of F-2, F-4 and F-6 is 1321 MPa, 1118 MPa, 1013 MPa and 907 MPa respectively.

This result is attributed to that the increase of asphalt content enhances the plasticity of the soil.

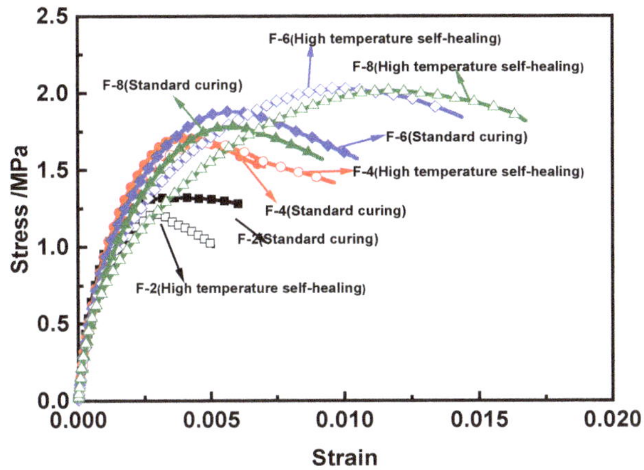

Figure 13. Stress-strain curve of the flexible curing agent stabilized silt under standard curing and high temperature.

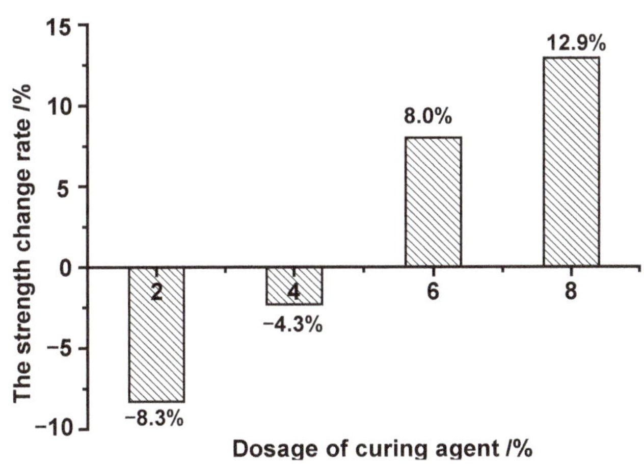

Figure 14. The strength change rate of the flexible curing agent stabilized silt.

3.5. Microstructure

The microstructures of silt, cement stabilized silt, the flexible curing agent stabilized silt, the flexible curing agent stabilized silt after heating at 28 days are shown in Figure 15. In considering the microstructure of silt, high particle roundness and overlapping connected manner are observed between particles. The clay content in the soil is small, there is no obvious bonding between particles, and the structure is loose and granular, and the pores between the particles are large. During rolling, it is difficult to form effective particle embedding between particles, and mutual dislocation between particles is easy to occur under the action of external force. A lot of flocculated calcium silicate hydrate gel and acicular ettringite are found on the surface of cement-stabilized silt particles in Figure 15b, forming a network structure. Despite this, the bonding between soil particles and hydration products is not tight, and there are more connected intergranular pores, resulting in low structural density. In the flexible curing agent stabilized silt, the asphalt component has obvious bonding effect with hydration products and soil particles, as shown in Figure 15c,d. It forms many disconnected pores uniformly distributed non-connected micropore structures, which improve the microstructure development of silt.

Figure 15. SEM morphology photographs of different samples. (**a**) Silt; (**b**) Cement stabilized silt; (**c**) Flexible curing agent stabilized silt; (**d**) Flexible curing agent stabilized silt after heating.

4. Conclusions

Based on the limited testing results, the following conclusion can be drawn.

(1) The inclusion of the flexible curing agent has a significant contribution in improving the compressive strength, water stability and resistance ability to drying shrinkage and freeze-thaw because of cementing effect of the cement hydrates and asphalt. For the ternary cementitious system containing the flexible curing agent and cement in silt soil, the system behaves well in improving the mechanical properties, better than the effect of cement alone. Moreover, it has good high temperature self-healing ability for the addition of asphalt, which will play an important role in the recovery of soil damage.

(2) Asphalt is a viscoelastic plastic material. An appropriate amount of asphalt can make the soil obtain a stable asphalt cementitious structure and improve the performance of the soil together with the cement hydration products. However, the increase of the asphalt content will enhance the plasticity of the soil and reduce the mechanical properties such as strength, water stability and resistance to low temperature crack. The preferred dosage 4%~6% of the flexible curing agent is obtained by mechanical property analysis.

(3) The SEM images show that the incorporation of the flexible curing agent helps the silt form dense cementation and non-connected microporous structure, that is beneficial to the improvement of water stability and frost resistance. The asphalt component can reorganize and diffuse, filling internal pores and micro-cracks, and achieving soil damage repair and structural reinforcement.

Based on the actual application conditions, the mechanical properties evolution test and analysis of the flexible solidified silt under indoor simulation conditions are carried out in this paper. Next, related research work on dynamic loading and fatigue performance under traffic load will be carried out in combination with the application horizon of solidified silt.

Author Contributions: Conceptualization, L.W. and W.W.; methodology, Z.S. (Zhiyi Sai); software, W.W.; validation, Z.S. (Zhiyi Sai), H.H. and G.H.; formal analysis, H.W.; investigation, L.Z.; resources, L.W.; data curation, J.W.; writing—original draft preparation, W.W.; writing—review and editing, W.W.; visualization, J.W.; supervision, J.W.; project administration, Z.S. (Zhaoyun Sun); funding acquisition, G.H. All authors have read and agreed to the published version of the manuscript.

Funding: This research was funded by the National Key R&D Program of China, grant No. 2018YFB1600103, the National Natural Science Foundation of China, grant No. 42107213, Shandong Provincial Natural Science and Foundation, grant No. ZR2020QE271 and Shandong Provincial Key Research and Development Program, grant No. 2019GSF109020.

Institutional Review Board Statement: Not applicable.

Informed Consent Statement: Not applicable.

Data Availability Statement: Not applicable.

Conflicts of Interest: The authors declare no conflict of interest.

References

1. Zhanyong, Y. Engineering Properties on Yellow River Alluvial-Plain Soil. Ph.D. Thesis, Tianjin University, Tianjin, China, 2006.
2. Li, X.; Li, S.; Kai, Y.; Zhu, C. Test study of changing rules of excess pore water pressure during dynamic consolidation at subgrade of expressway in Yellow River flood area. *Rock Soil Mech.* **2011**, *32*, 2815–2820.
3. Jia, Z.; Zhu, H.; Shang, Q.; Fang, J.; Yao, Z. Study on fundamental characteristics and construction techniques of silty soil subgrade in Yellow River flood field. *J. Highw. Transp. Res. Dev.* **2008**, *25*, 52–57.
4. Zhang, H.; Sun, R.; Dong, H. Study on the dynamic characteristics of silt impact rolling in the yellow area. *J. China Foreign Highw.* **2018**, *38*, 39–43.
5. Wang, Q.; Zhang, W.; Xi, W. Study of plate loading test of silt foundation in alluvial plains of Yellow River. *Constr. Technol.* **2011**, *40*, 133–135.
6. Zhu, J.; Li, X.; Wang, S. Increasing and dissipation regularity of the pore water pressure under the dynamic compaction action to the alluvial and diluvium soil of Yellow River. *Chin. J. Rock Mech. Eng.* **2003**, *22*, 500–504.
7. Song, X.; Zhang, H.; Wang, S. Hydrophilic characteristics and strength decay of silt roadbed in Yellow River alluvial plain. *Chin. J. Geotech. Eng.* **2010**, *32*, 1594–1602.
8. Al-Rawas, A.A.; Taha, R.; Nelson, J.D. A comparative evaluation of various additives used in the stabilization of expansive soils. *Geotech. Test. J.* **2002**, *25*, 199–209.
9. Kamon, M.; Gu, H.; Masahiro, I. Improvement of Mechanical Properties of Ferrum Lime Stabilized Soil with the Addition of Aluminum Sludge. *J. Soc. Mater. Sci. Jpn.* **2009**, *50*, 47–53. [CrossRef]
10. Ning, B. Permeability and Mechanical Behaviors of Stabilized Silt by Inorganic Soil Stabilizer. Ph.D. Thesis, Zhengzhou University, Zhengzhou, China, 2020.
11. Petry, T.; Das, B. Evaluation of Chemical Modifiers and Stabilizers for Chemically Active soils-Clays. *Transp. Res. Rec. J. Transp. Res. Board* **2001**, *1757*, 43–49. [CrossRef]
12. Fei, X.; Zekkos, D.; Li, L.; Woods, R.; Sanford, L. Geo-characterisation of lime water treatment sludge. *Environ. Geotech.* **2017**, *4*, 209–219. [CrossRef]
13. Bell, F.G. An assessment of cement-PFA and lime-PFA used to stabilize clay-size materials. *Bull. Eng. Geol. Environ.* **1994**, *49*, 25–32. [CrossRef]
14. Kumar, B.R.P.; Sharma, R.S. Effect of fly ash on engineering properties of expansive soils. *J. Geotech. Geo-Environ. Eng.* **2005**, *130*, 764–767. [CrossRef]
15. Brouwers, H.; Augustijn, D.; Krikke, B.; Honders, A. Use of cement and quicklime to accelerate ripening and immobilize contaminated dredging sludge. *J. Hazard. Mater.* **2007**, *145*, 8–16. [CrossRef] [PubMed]
16. Huang, X.; Zhou, G. Harding Mechanism of cement-stabilized soil. *Chin. J. Geotech. Eng.* **1994**, *4*, 62–68.
17. Phetchuay, C.; Horpibulsuk, S.; Arulrajah, A.; Suksiripattanapong, C.; Udomchai, A. Strength development in soft marine clay stabilized by fly ash and calcium carbide residue based geopolymer. *Appl. Clay Sci.* **2016**, *127*, 134–142. [CrossRef]
18. Xue, L. Test and study of ISS soil stabilizing agent used for the road base. *Northeast. Highw.* **2002**, *25*, 29–30.
19. Wang, J.; Zhang, M.; Yan, Y. Study on ISS of Improving the Capability of Soil Based on Tests in Laboratory. *Buiding Tech. Dev.* **2005**, *32*, 27–28.
20. Wang, Y.; Jia, J.; Zhang, L. Microstructure and strength feature of ISS stabilized soil. *J. South China Univ. Technol.* **2002**, *30*, 96–104. [CrossRef]
21. Rauch, A.F.; Katz, L.E.; Liljestrand, H.M. *Analysis of the Mechanisms and Efficacy of Three Liquid Chemical Soil Stabilizers*; Center for Transportation Research, The University of Texas at Austin: Austin, TX, USA, 2003.
22. Urena, C.; Azanon, J.M.; Corpas, F.; Nieto, F.; Leon, C.; Perez, L. Magnesium hydroxide, seawater and olive mill wastewater to reduce swelling potential and plasticity of bentonite soil. *Constr. Build. Mater.* **2013**, *45*, 289–297. [CrossRef]

23. Lv, Q.F.; Shen, B.; Wang, S.; Meng, H. Strength characteristics and solidification mechanism of sulphate salty soil solidified with sodium silicate. *Rock Soil Mech.* **2016**, *37*, 687–693.
24. Cui, D.; Xiang, W. Pore diameter distribution test of red clay treated with Iss. *Rock Soil Mech.* **2010**, *31*, 3096–3100.
25. Ding, R. Experimental study of the surface reactions between clay minerals and water-glass. *Chin. J. Geotech. Eng.* **1999**, *21*, 78–81.
26. He, J.; Shi, X.; Su, Z. Strength properties of dredged silt at high water content treated with sodium silicate, soda residue and group granulated blastfurnace slag. *J. Eng. Geol.* **2019**, *27*, 729–736.
27. Zhao, H.H.; Ge, L.; Petry, T.M. Effects of chemical stabilizers on an expansive clay. *KSCE J. Civ. Eng.* **2014**, *18*, 1009–1017. [CrossRef]
28. Katz, L.; Rauch, A.; Lilijestrand, H.; Harmon, J. Mechanisms of soil stabilization with liquid ionicstabilizer. *Transp. Res. Rec. J. Transp. Res. Board* **2001**, *1757*, 50–57. [CrossRef]
29. Greeshma, N.E.; Lamanto, T.S.; Chandrakaran, S.; Sankar, N. Enzyme stabilization of high liquid limit clay. *Electron. J. Geotech. Eng.* **2014**, *19*, 6989–6995.
30. Cheng, L.; Shahin, M.A.; Mujah, D. Influence of key environmental conditions on microbially induced cementation for soil stabilization. *J. Geotech. Geo-Environ. Eng.* **2017**, *143*, 1–11. [CrossRef]
31. Sun, Z.; Lv, W.; Sun, G. Investigation on the enhancing and acting mechanism of PERMA-ZYME on densification and strength of soil. *New Build. Mater.* **2010**, *37*, 87–90.
32. Peng, H.; Guo, L.; Peng, Y. Relation of soil granule fineness fractal dimension and unconfined compression strength of soil stabilized by pema-zyme. *China Civ. Eng. J.* **2010**, *43*, 588–592.
33. Xu, R.; Shao, Y. Experimental study on treating the silt foundation of sea dike in Wenzhou peninsula construction project. *J. Zhejiang Univ.* **2005**, *31*, 475–478.
34. Huang, X.; Xu, S.; Ning, J. Experimantal research on stabilized soft soils by alumina bearing modifier. *Chin. J. Rock Mech. Eng.* **2007**, *26*, 156–161.
35. Hou, H.; Zhou, M.; Zhang, D. Characteristics of soil stabilized by HAS stabilizer and ITS engineering use. *Ind. Constr.* **2006**, *36*, 32–34.
36. Dong, J.; Wang, P.; Chai, S. Strength Characteristics of Lightweight Soil Amended by Cement-Modified Poly (vinyl alcohol). *J. Build. Mater.* **2011**, *14*, 576–580.
37. Liu, C. Study on Mechanism and Performance of the Coastal Saline Soil Cured by Slag Composite Curing Agent. Ph.D. Thesis, University of Science and Technology Beijing, Beijing, China, 2015.
38. Ma, S.; Sun, Z.; Wei, J.; Zhang, X.; Zhang, L. Utilization of Modified Red Mud Waste from the Bayer Process as Subgrade and Its Performance Assessment in a Large-Sale Application. *Coatings* **2022**, *12*, 471. [CrossRef]
39. Sun, Z. Research on the Engineering Technology and Environmental Impact of Bayer Red Mud Filling Subgrade. Ph.D. Thesis, Shandong University, Ji'nan, China, 2017.
40. Zhu, Y.; Zhang, H. Review of self-healing capacity of asphalt. *Mater. Rev.* **2015**, *29*, 86–91.
41. Wang, L. Evaluation Methodology of Temperature Susceptibility for Paving Asphalt. Ph.D. Thesis, China University of Petroleum, Dongying, China, 2009.
42. Sun, Y. Research on Self-Healing Properties of Asphalt and Asphalt Mixtures. Ph.D. Thesis, Wuhan University of Technology, Wuhan, China, 2017.
43. *Test Methods of Soils for Highway Engineering*; JTG 3430-2021; China Communication Press Co., Ltd.: Beijing, China, 2021.
44. *Test Methods of Materials Stabilized with Inorganic Binders for Highway Engineering*; JTG E51-2009; China Communication Press Co., Ltd.: Beijing, China, 2009.
45. Song, B.; Lv, G.; Zhu, W. Strength and deformation characteristics of silt modified with slaked lime or cement. *China Harb. Eng.* **2018**, *38*, 42–46.
46. Fan, L.; Mou, S.; Li, Y. Strength and Water Stability of Compound Stabilized Silt with Emulsified Asphalt. *Highw. Eng.* **2019**, *44*, 178–238.
47. Fan, L.; Zhou, S.; Hou, J.; Wang, L. Composite Stabilization of Silty Soil near the Yellow River: Two Methods and Performance Comparison. *J. Yangtze River Sci. Res. Inst.* **2021**, *38*, 118–124.

coatings

MDPI

Article

Influence of Frequent Freeze–Thaw Cycles on Performance of Asphalt Pavement in High-Cold and High-Altitude Areas

Haibo Cao [1,2], Tuanjie Chen [2,3,*], Hongzhou Zhu [1] and Haisheng Ren [3]

1 School of Civil Engineering, Chongqing Jiaotong University, Chongqing 400741, China; bohaichb@163.com (H.C.); zhuhongzhouchina@cqjtu.edu.cn (H.Z.)
2 CCCC First Highway Consultants Co., Ltd., Xi'an 710075, China
3 Intelligent Transport System Research Center, Southeast University, Nanjing 211189, China; ren_hs510@seu.edu.cn
* Correspondence: chentuanj@163.com

Abstract: This study explores the temperature changes and freeze–thaw cycles in certain typical high-altitude areas, finding that these areas encounter more than 120, or even more than 200, freeze–thaw cycles per year. Such frequent freeze–thaw cycles deliver significant impact on the performance of asphalt pavements, with cracks becoming a typical problem in high-altitude areas. Such factors as cold weather, large temperature differences, and frequent freeze–thaw cycles have adverse effects on the stress of asphalt pavement materials, resulting in cracks in pavements. By simulating the conditions of such frequent freeze–thaw cycles, this study explores the law of changes in the performance of roads made from asphalt and asphalt mixtures, as well as the low-temperature crack resistance properties of asphalt and asphalt mixtures in frequent freeze–thaw cycles. It is found that the performance of the three different types of asphalt binders used in the test shows basically no change after 50 freeze–thaw cycles, and the asphalt types have a significant effect on the low-temperature performance of asphalt mixtures. The modified asphalt shows a higher viscosity than the matrix asphalt, with better toughness than that of the matrix asphalt at low temperature. Frequent freeze–thaw cycles significantly influence the low-temperature splitting tensile strength and water stability of asphalt mixtures; with increased freeze–thaw cycles, the splitting strength and freeze–thaw splitting tensile strength ratio will gradually decrease to a significant level. The freeze–thaw conditions are found delivering remarkable influence on the low-temperature splitting tensile strength and water stability of asphalt mixtures. The research results of this study provide a basis for the selection of asphalt pavement materials as well as the optimal design of mixtures in high-altitude area like the Qinghai-Tibet Plateau.

Keywords: high altitude area; low temperature performance; frequent freeze–thaw cycles; crack resistance

Citation: Cao, H.; Chen, T.; Zhu, H.; Ren, H. Influence of Frequent Freeze–Thaw Cycles on Performance of Asphalt Pavement in High-Cold and High-Altitude Areas. *Coatings* **2022**, *12*, 752. https://doi.org/10.3390/coatings12060752

Academic Editors: Paolo Castaldo and Giorgos Skordaris

Received: 28 March 2022
Accepted: 24 May 2022
Published: 31 May 2022

Publisher's Note: MDPI stays neutral with regard to jurisdictional claims in published maps and institutional affiliations.

Copyright: © 2022 by the authors. Licensee MDPI, Basel, Switzerland. This article is an open access article distributed under the terms and conditions of the Creative Commons Attribution (CC BY) license (https://creativecommons.org/licenses/by/4.0/).

1. Introduction

With an average altitude of more than 4000 m, the Qinghai-Tibet Plateau is a typical high-cold and high-altitude area, showcasing such significant plateau climate characteristics as long low-temperature durations, high freeze–thaw cycle frequencies, and strong solar radiation. As a result, asphalt layers in this area are subject to temperature fatigue cracking and extreme temperature freezing, leading to such problems as cracking and asphalt layer cracking due to factors including semi-rigid bases and permafrost subgrades. These problems are widespread and difficult to eradicate; especially, frequent and violent freeze–thaw cycles can greatly accelerate the degradation of pavement performance. With increased service life of the roads, the cost of road maintenance will also go up, thus restricting the development of highways in high-cold and high-altitude areas, including the Qinghai-Tibet Plateau. A survey conducted on pavement usage on the G109 Qinghai-Tibet National Highway found that the typical problems along the highway included transverse cracks, longitudinal cracks, network cracks, potholes, and so on. The cracked asphalt pavements will allow surface water to penetrate into the cracks. Coupled with the action of low temperatures and frequent freezing and thawing, water will be frozen and accumulated in the mixtures, leading to frost heaving and looseness. Continuous development will soften the base layer and cause water damage of the surface layer. Under repeated actions on traffic loads, there will be further serious secondary pavement problems, such as frost boiling, potholes, net cracks, and so on.

The factors affecting the low-temperature cracking resistance of asphalt mixtures can be divided into two types: internal and external factors. The former showcases the material characteristics and structural characteristics of asphalt mixtures, including aggregate type, asphalt type, asphalt content, gradation type, external admixture materials, etc. The latter factors include traffic loads, environmental factors, etc. The low-temperature cracking resistance of asphalt mixtures depends to a large extent on the low-temperature properties of asphalt materials, the bonding strength of asphalt and mineral aggregates, the type of gradation, and the uniformity of asphalt mixtures. In a strategic highway research program conducted in the United States, the restrained sample temperature stress test and the indirect tensile test were made to evaluate the low-temperature crack resistance of asphalt concrete. The indirect tensile test (IDT) is currently most widely used abroad to characterize the low-temperature performance of asphalt mixtures. The cracking test method is necessary in mechanical empirical pavement design and widely used by Strategic Highway Research Program (SHRP), American Association of State Highway and Transportation Officials (AASHTO), and National Cooperative Highway Research Program (NCHRP) to predict the low-temperature tensile strength and creep compliance of mixtures. Other evaluation indicators can also be obtained through an IDT test. For example, fracture energy and fracture work are used as evaluation indicators to evaluate fatigue cracking of asphalt mixtures [1–3]. Hao et al. used the low-temperature crack resistance coefficient to evaluate the crack resistance of asphalt mixtures, finding that the higher the crack resistance coefficient, the better the low-temperature crack resistance [4]. The effects of different aging degrees on the low-temperature properties of asphalt mixtures have also been studied, with the low-temperature crack resistance properties of asphalt mixtures under different aging conditions identified [5,6]. Francesca Russo et al. [7] focused on the investigation of the rheological properties using a dynamic shear rheometer and carrying out a frequency sweep test at temperatures ranging from 0 to 50 °C in increments of 10 °C.

In a related study on the generation process and influencing factors of cracks when asphalt pavements are subjected to temperature fatigue, Mahboub et al. demonstrated that the development process of cracks has a great impact on the fatigue life of pavements [8] by analyzing the calculation method of the loss energy in the temperature fatigue process of asphalt mixtures under different load fatigue forms. Analyses of fatigue mechanisms of road surfaces showcase that with increased temperature fatigue actions, the damage degree will also increase, and the temperature fatigue can be regarded as a thermal fatigue and low-temperature cracking. Based on these results, the corresponding temperature fatigue life equation was established; the temperature fatigue life of asphalt concrete was examined in low-frequency loading frequency temperature fatigue tests; and the temperature fatigue damage model based on dissipated energy was used to analyze the test results. The damage expressed by the sub-meter dissipative energy has a good linear relationship with the plastic strain of asphalt mixtures [9–14]. Using a numerical simulation finite element model based on fracture mechanics, Fu et al. analyzed the propagation process of road surface cracks and base reflective cracks, finding that both the nonlinear relationship of stress intensity factor and the crack propagation speed would mount with the decrease of the reference temperature [15]. Zhan et al. applied thermal-mechanical coupling solution technology to address the temperature stress surface crack problem of asphalt pavements under the action of low temperatures and large temperature differences. Numerical analyses demonstrate that large temperature differences are an important cause of asphalt pavement damage in high-cold areas [16].

Given the special climatic characteristics, such as high cold and large temperature differences, in high-cold and high-altitude areas, the indoor flexural tensile strength test on asphalt mixtures has been carried out to study the law of changes in flexural tensile strength of asphalt mixtures affected by gradation, oil-stone ratio, temperature, and other factors. The test showed that the effects of the materials and the types of the mixtures have an impact on the performance of the mixtures [17,18]. According to the influence of changes in the composition factors of asphalt mixture materials on their high- and low-temperature performance, water stability, and other elements in the areas with large temperature differences, the order of the sensitivity of the factors affecting such performance of the mixtures was determined [19,20]. The effects of asphalt types and gradation types on the water stability and high-temperature performance of asphalt mixtures in the Qinghai-Tibet high-cold areas were explored with the Marshall test, freeze–thaw splitting test, and rutting test [21]. In addition, Tang et al. verified the linear viscoelastic mechanical behavior of asphalt binders by simulating the creep and rheological properties of asphalt binders before and after aging under extreme temperature conditions [22]. Combined with the characteristics of the Tibet area, it is proposed that high-altitude and low-temperature climates are the key factor affecting pavement structures, whereas the adaptability of typical pavement structures in high-cold and high-altitude areas is examined in mineral gradation ranges [23–25].

It can be concluded from the above research that asphalt mixtures will bear temperature damage when the temperature cycle changes; such damage will have a greater impact at low reference temperatures. Therefore, the impact of asphalt mixtures cannot be ignored. At present, the research on the crack resistance of asphalt mixtures in high-cold and high-altitude areas generally adopts a low-temperature bending test, which has a low correlation with regional climate characteristics and a relatively single evaluation index. Few studies have been done on the crack resistance of materials. Based on the climatic and environmental characteristics of high-cold and high-altitude areas as well as cracks and problems in asphalt pavements, this study will, by simulating cold conditions, large temperature differences, frequent freeze–thaw cycles, changing asphalt types, grading, and other factors, implement some asphalt performance tests, thermal stress restrained specimen tests (TSRST), and freeze–thaw splitting tests, so as to explore the change law of road performance of asphalt and asphalt mixtures and examine the low-temperature crack resistance performance of asphalt mixtures under the condition of frequent freeze–thaw cycles. The research results can provide a basis for selecting asphalt pavement surface materials and get the optimal design of mixtures in high-cold and high-altitude areas.

2. Freeze–Thaw Cycles and Pavement Cracking in High-Altitude Areas

2.1. Statistics of Freeze–Thaw Cycle in High-Altitude Areas

The average altitude of the Qinghai-Tibet Plateau is higher than 4000 m, with low annual average temperatures and large temperature differences between day and night. As shown in Figure 1, for the temperature changes on a typical sunny day in Nagqu in January 2021, the lowest temperature appeared at 6:00 to 7:00 in the morning, whereas the highest temperature appeared at approximately 14:00 in the afternoon. Taking the daily maximum temperature higher than 0 °C and the daily minimum temperature lower than 0 °C as the freeze–thaw conditions for pavements, respectively, it can be found that one freeze–thaw cycle occurs per day.

In order to identify the characteristics of freeze–thaw cycles in high-cold and high-altitude areas, the observation data in three temperature observation stations in Wudaoliang (altitude 4680 m), Nagqu (altitude 4350 m), and Lhasa (altitude 3650 m) are used. The period from the daily maximum temperature higher than 0 °C to the minimum temperature lower than 0 °C is taken as one freeze–thaw cycle; and the number of freeze–thaw cycles in the three observation stations spanning 36 years from 1980 to 2015 is counted, with the statistical results shown in Figure 2, which shows that the annual freeze–thaw cycles in this high-cold and high-altitude area mostly exceed 120. In addition, with the increase in altitude, the number of freeze–thaw cycles would show an increasing trend. The melt cycle will inevitably impose an adverse effect on the asphalt pavements in the area.

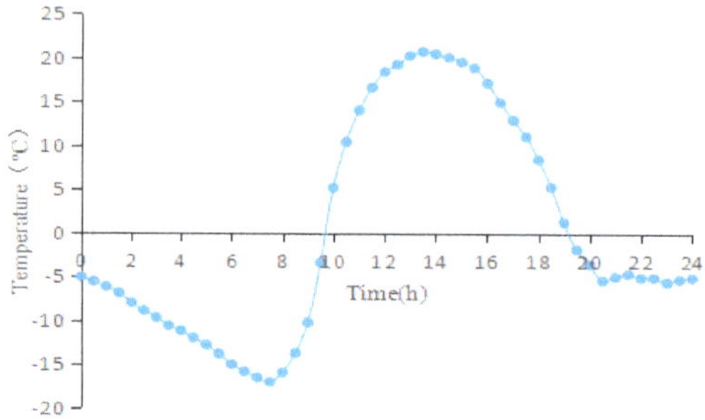

Figure 1. The trend of daily temperature variation in Nagqu, Tibet in January.

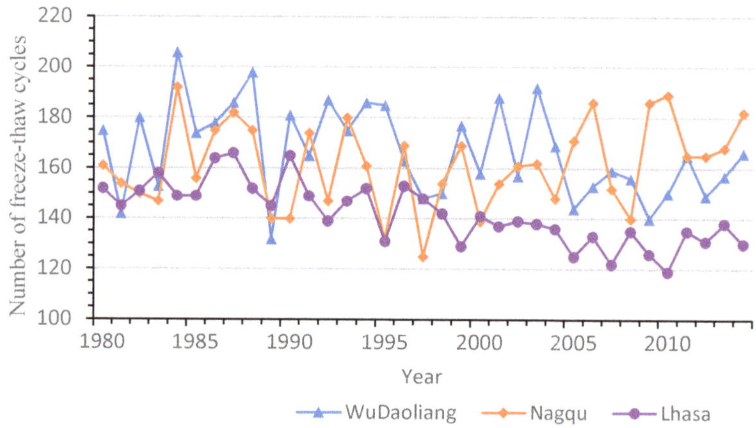

Figure 2. Number of freeze–thaw cycles in typical high-cold and high-altitude areas.

2.2. Investigation of Pavement Cracking in High Altitude Areas

The road conditions of Gongyu Expressway, G214, G109, and Lhasa-Gongga Airport Special Expressway in high-cold and high-altitude areas were investigated, and it was found that the main problem types of asphalt pavements include cracks, subsidence, ruts, potholes, etc., among which cracks are the most prominent problems. The main performance is shown in Figure 3.

Figure 3. Lateral cracking and network cracking problems of asphalt pavement. (**a**) is transverse crack; (**b**) is net crack.

The reasons for the cracks are analyzed. Because the project is located in a high-cold and high-altitude zone, there is a big temperature difference between day and night. Under the conditions of cold, large temperature differences, and frequent freeze–thaw cycles, when the ambient temperature decreases, the asphalt pavement will become warmer inside. When the temperature-shrinkage stress inside the pavement exceeds the ultimate tensile strength of asphalt concrete, the pavement will become cracked. In addition, the cracking of the asphalt surface due to the cracking of the base layer is also a cause of such cracks.

3. Experimental Programs

3.1. Materials

Based on the types of asphalt used in the high-cold and high-altitude areas in China, No. 90 matrix asphalt, No. 110 matrix asphalt, and SBS modified asphalt are adopted for comparative analysis in the study. The specific technical indicators of such asphalt are tested and summarized in Table 1.

Table 1. Technical index value of asphalt.

Test Content		Unit	Asphalt Type		
			No. 90	No. 110	SBS Modified
Penetration (25 °C)		0.1 mm	89.3	100.8	73.3
Softening point (Ring-ball method)		°C	46.0	43.4	64.0
Ductility (10 °C)		cm	92.0	125.0	79.0
Density (25 °C)/		$g{\cdot}cm^{-3}$	0.998	0.977	1.101
Solubility (trichloroethylene)		%	99.8	99.87	99.21
RTFOT (163 °C)	Quality loss	%	0.23	0.12	0.06
	Penetration ratio	%	77	74	73
	Ductility (10 °C)	cm	32	25	26
	Softening point	°C	50.3	52.3	68.7

The coarse aggregate is basalt crushed stone, with its main technical indicators shown in Table 2, which meet the technical requirements of "Technical Specification for Highway Asphalt Pavement Construction" (JTG F40-2004) [26].

Table 2. Technical index value of coarse aggregate.

Test Content	Technical Requirement	Aggregate Specifications/mm			
		13.2–16	9.5–13.2	4.75–9.5	2.36–4.75
Apparent relative density	≥2.6	2.954	2.956	2.948	2.932
Needle-like content/%	≤15	5	6	3	/
Soft stone content/%	≤3	1.8	0.8	/	/
Crushing value of stone/%	≤26	14	13	/	/
Sturdiness/%	≤12	3	3	3	3
Water absorption/%	≤2	0.82	0.79	0.91	0.76

The fine aggregate is made of 0~3 mm machine-made sand, and the filler is limestone ore powder. All the technical indicators of the fine aggregate and the ore powder meet the requirements of "Technical Specification for Highway Asphalt Pavement Construction" (JTG F40-2004).

3.2. Mix Design

Two types of asphalt mixtures, i.e., AC-13 and AC-16, are selected in this study, with the design gradation shown in Table 3.

Table 3. Design gradation of asphalt mixture.

Mix Type	Mass Percentage (%) Passing through Each Hole Sieve (mm)								
	16.0	**13.2**	**9.5**	**4.75**	**2.36**	**1.18**	**0.30**	**0.15**	**0.075**
AC-13	100	94.2	76.4	44.2	31	24	12.3	9.5	7.4
AC-16	96.9	87.6	73.7	52.9	29.2	19.7	10.1	8.3	7.2

No. 90 matrix asphalt and SBS modified asphalt are used to prepare asphalt mixture specimens, and the oil-to-stone ratios of matrix asphalt AC-13 and SBS-modified AC-13 are 5.1% and 5.2%, respectively. The best oil-to-stone ratio of SBS-modified AC-16 is 5.1%.

3.3. *Test Methods*

3.3.1. Freeze–Thaw Cycles Method

According to the temperature distribution (−35 °C to 50 °C) in the high-cold and high-altitude areas, the temperature condition of the freeze–thaw cycle test is set as −35 °C to 50 °C. Different asphalt samples and asphalt mixture samples are placed in an environmental test chamber. When the temperature in the chamber reaches 50 °C, we kept the temperature constant for 1.5 h. We then set the cooling rate (8.0 °C/h) to reduce the temperature in the environmental test chamber, waited for it to drop to −35 °C, and kept it constant for 1.5 h. Next, we set the heating rate (8.0 °C/h) to raise the temperature in the environmental test chamber; when it reaches 50 °C, it was kept constant for 1.5 h. The above freeze–thaw cycle test steps was repeated for 10, 20, 30, and 50 times, separately.

3.3.2. Asphalt Performance Test

According to the "Standard Test Methods of Bitumen and Bituminous Mixtures for Highway Engineering" (JTG E20-2011) [27], tests for the asphalt softening point, penetration, and ductility indexes are performed on the original asphalt and the asphalt samples after freeze–thaw cycles.

The BBR test uses the bending beam rheometer device, and the test is a method to better evaluate the stiffness and creep rate of asphalt at low temperatures, as shown in Figure 4. This test applies the theory of engineering beams to measure the stiffness of the asphalt trabecular specimen under creep load and simulates the creep load with the stress generated in the road surface when the temperature drops; two evaluation parameters, i.e., creep stiffness and m value, are obtained through experiments. Creep stiffness indicates the ability of asphalt to resist permanent deformation, whereas the m value indicates the rate of change of asphalt stiffness under load. The asphalt trabeculae are formed in a rectangular aluminum mold with a size of 125 mm × 12.5 mm × 6.25 mm. Before the test, the asphalt trabeculae are placed in a constant temperature bath for 60 min. The samples are then subjected to BBR test at a temperature of −15 °C.

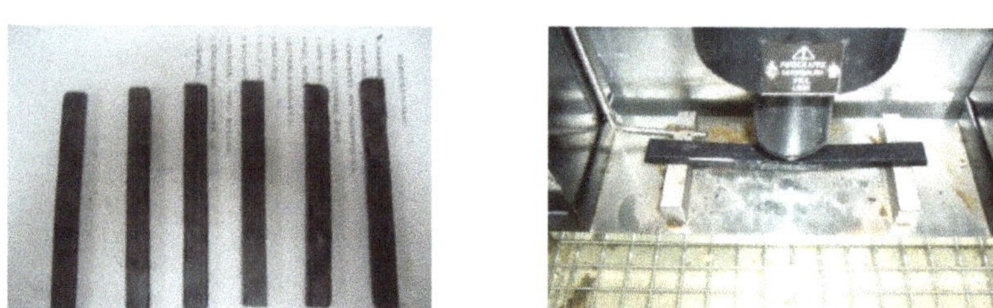

(a) (b)

Figure 4. Low temperature BBR test. (**a**) is BBR test sample; (**b**) is BBR test loading.

The DSR test uses the dynamic shear rheometer device, and the test is implemented to test the viscosity and elasticity of asphalt mortars by measuring the viscous and elastic properties of thin asphalt cement samples sandwiched between a shock plate and a fixed plate. The complex modulus G* and phase angle δ of asphalt samples are measured with a CSA dynamic shear rheometer, as shown in Figure 5, and the rutting factor G*/sinδ is calculated. The DSR test is carried out on the original asphalt of No. 110 and the asphalt samples after freeze–thaw cycles. This test is mainly purposed to measure the temperature sweep. The temperatures used in the temperature sweep test are 46, 52, 58, and 64 °C; the load frequency is ω = 10 rad/s; and the strain control mode is used during the test, with γ = 12%.

Figure 5. DSR test.

3.3.3. Thermal Stress Restrained Specimen Test (TSRST)

The TSRST test can truly simulate the actual temperature changes and the actual stress condition of mixtures, so it can truly reflect the low-temperature cracking performance of asphalt mixtures. The test equipment and the fracture states of the samples are shown in Figure 6. Before the test, the trabecular specimen is suspended inside the restraint instrument; four temperature sensors are attached to the four surfaces of the specimen, so as to observe its temperatures in real time; and one temperature sensor is hung on the fixing clip, so as to measure the environment in the restraint instrument. Temperature displacement sensors are installed on both sides of the specimen, which is pulled tight by manual pulleys. During the test, a liquid nitrogen tank continuously transports liquid nitrogen into the restraint instrument. As the temperature decreases, the trabecular specimen will continue to shrink. When the temperature-shrinkage stress of the specimen gets greater than the ultimate tensile strength of the trabecular, the specimen will become cracked.

(a) (b) (c)

Figure 6. TSRST sample and fracture state. (**a**) is TSRST equipment; (**b**) is TSRST loading; (**c**) is TSRST sample.

Three kinds of asphalt mixtures are adopted, namely matrix asphalt AC-13, SBS modified asphalt AC-13, and SBS modified asphalt AC-16, which are prepared as 40 mm ×40 mm × 200 mm trabecular specimens. During the cutting process, efforts are made to keep the four edges and corners of the specimen vertical. In order to firmly attach the upper and lower planes of the test sample to the chassis of the TSRST experimental instrument, Devcon plastic steel repair agent is utilize to bond the two ends of the test sample to the chassis of the instrument. A self-pressurized liquid nitrogen tank is adopted for cooling at a rate of 10 °C/h. The lowest temperature in the test can reach −50 °C. During the cooling process, the test will stop when the specimen breaks down in the middle of the sample. If the specimen does not break, the environment temperature is kept at −50 °C for one hour, and then the test will automatically stop. The freeze–break temperature and freeze–break stress of the asphalt mixture are tested.

3.3.4. Freeze–Thaw Splitting Test

As one of the most widely used test methods, the splitting test characterizes the low-temperature cracking of asphalt mixtures. The freeze–thaw splitting test is used to perform freeze–thaw cycles on asphalt mixtures under specified conditions, so as to determine their strength ratio before and after water damage. The test is carried out by referencing the Standard Test Methods of Bitumen and Bituminous Mixtures for Highway Engineering (JTG E20-2011).

Taking three kinds of asphalt mixtures, namely matrix asphalt AC-13, SBS modified asphalt AC-13, and SBS modified asphalt AC-16, as the research objects, the Marshall test compactor is used to prepare the test samples. The diameter of the test sample is 101.6 mm, and the height is 63.5 mm. The test samples are first vacuum-saturated and then packed into plastic bags; multiple freeze–thaw cycles are performed according to the set conditions. The split tensile strength and freeze–thaw split tensile strength ratio of the asphalt mixtures are tested at a test temperature of −15 °C. The test process is illustrated in Figure 7.

Figure 7. Low temperature freeze–thaw splitting test.

4. Test Results and Discussion

4.1. Asphalt Performance Test

4.1.1. Penetration, Softening Point, and Ductility Test

The asphalt samples are tested for penetration (P) at 25 °C and ductility (D) and softening point (SP) at 10 °C under different freeze–thaw cycles, with the test results shown in Table 4.

Table 4. Three indexes of asphalt under different high and low temperature cycles.

Freeze–Thaw Cycles	No. 90 Asphalt			No. 110 Asphalt			SBS Modified Asphalt		
	P/0.1 mm	D/cm	SP/°C	P/0.1 mm	D/cm	SP/°C	P/0.1 mm	D/cm	SP/°C
0	89.3	92	46	100.8	125	43.4	73.3	79	64
10	88.7	91.6	46.1	98.5	122.6	43.9	72.6	77.8	63.3
20	89.1	87.5	46.3	97.7	129.2	43.1	71.2	76.5	64.4
30	89	88.5	45.2	98.9	123.5	44.2	71.9	75.5	64.9
50	87.6	88.1	46.8	97.9	121.8	44	70.7	75.9	65.7
Maximum rate of change (%)	−1.90	−4.24	1.74	−2.88	3.36	1.38	−3.55	−4.43	2.66

As seen from the test results, compared with the test data of the three major indexes of asphalt under no freeze–thaw cycles, the test data of the three major indexes for No. 90 matrix asphalt, No. 110 matrix asphalt, and SBS modified asphalt show few changes, with the rate of change lower than 5%. Under different freeze–thaw cycles, the performance of asphalt basically shows no change before and after the test, mainly because the asphalt undergoes only a physical phase change under the freeze–thaw cycle conditions, with transfer or change delivered in its internal chemical composition.

4.1.2. BBR Test

No. 110 asphalt is selected to prepare test samples, and BBR is used to measure the bending stiffness modulus S and the m value of asphalt after freeze–thaw cycles. The test temperature is set to −15 °C. Table 5 shows the test data of stiffness modulus S and m value of No. 110 matrix asphalt before and after 10, 20, 30, and 50 freeze–thaw cycles.

Table 5. BBR test results of No. 110 asphalt binder.

Freeze–Thaw Cycles	0		10		20		30		50	
Test index	S/MPa	m	S/MPa	m	S/MPa	m	S/MPa	m	S/MPa	m
Test results	187	0.361	176	0.370	182	0.362	186	0.359	183	0.356

It can be observed from the test results that after these freeze–thaw cycles, the difference between the stiffness modulus and the m value of the asphalt becomes extremely low. After 10 freeze–thaw cycles, the index difference is the highest, the difference in stiffness modulus is 5.7%, and the difference in m value is 2.6%. Given the possible errors in the test process, it can be concluded that the low-temperature rheological properties of the asphalt have not changed after the freeze–thaw cycles; in other words, the freeze–thaw cycles deliver no effect on the low-temperature properties of the asphalt materials.

4.1.3. DSR Test

No. 110 asphalt is selected to prepare test samples, and DSR is used to measure the complex modulus G^* and phase angle δ of the asphalt after 10, 20, 30, and 50 freeze–thaw cycles; the rutting factor $G^*/\sin\delta$ is then calculated. In the test, the scanning temperature is set to 46, 52, 58, and 64 °C; the loading frequency is $\omega = 10$ rad/s; and the strain control mode is adopted, with $\gamma = 12\%$. The test results are shown in Table 6, and the relationship between the rutting factor and temperatures is shown in Figure 8.

Table 6. DSR test results of No. 110 matrix asphalt.

Test Index	Temperature (°C)	Freeze–Thaw Cycles				
		0	10	20	30	50
G^* (kPa)	46	17.08	17.39	17.25	16.02	16.58
	52	7.18	7.31	7.07	7.36	7.39
	58	3.2	3.16	3.26	3.3	3.34
	64	1.47	1.34	1.51	1.51	1.5
δ(°)	46	82.22	82.92	79.75	84.69	83.25
	52	84.39	83.24	86.92	82.28	84.17
	58	86.02	86.95	88.17	83.87	87.32
	64	87.31	88.25	85.13	87.4	86.53
$G^*/\sin\delta$ (kPa)	46	17.24	17.52	17.53	16.09	16.70
	52	7.22	7.36	7.08	7.43	7.43
	58	3.2	3.16	3.27	3.31	3.34
	64	1.47	1.34	1.52	1.52	1.52

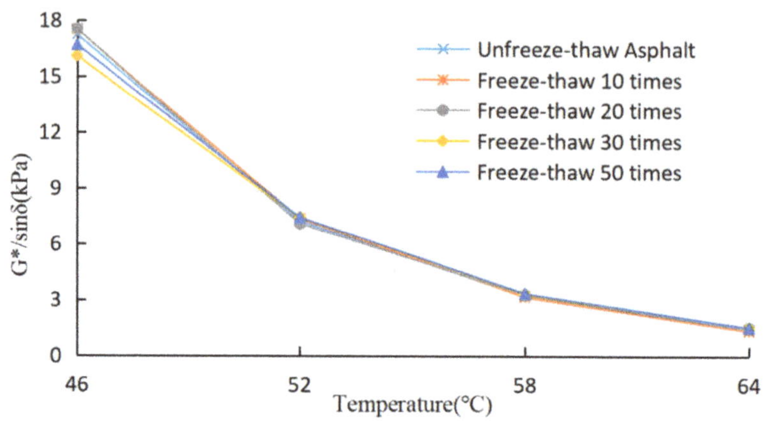

Figure 8. Relationship between rutting factor and temperature under high and low temperature cycle of asphalt.

The test results demonstrate that with the increase in temperature, the complex modulus of the asphalt before and after the freeze–thaw cycles would decrease, the phase angle would increase, and the rutting factor would decrease. This is due to the transformation of asphalt from a highly elastic state at low temperature to a viscous fluid state at high temperature. If the viscous component in the viscoelastic properties of asphalt increases, the elastic component will decrease. During this process, the maximum shear stress would go down, whereas the maximum shear strain will go up. After 10, 20, 30, and 50 freeze–thaw cycles, the rutting factor of the asphalt shows very little change compared with the data of the original asphalt. It can be concluded that the rheological properties of the asphalt itself do not change after the freeze–thaw cycles.

4.2. Asphalt Mixture Performance Test

4.2.1. TSRST Test

Three kinds of asphalt mixtures, namely matrix asphalt AC-13, SBS modified AC-13, and SBS modified AC-16, are used to prepare the samples, and the temperature stress test for the restrained samples is carried out at a cooling rate of 10 °C/h. Two indicators, i.e., freeze–break temperature and freeze–break stress, are obtained from the test. The test results are illustrated in Figures 9 and 10.

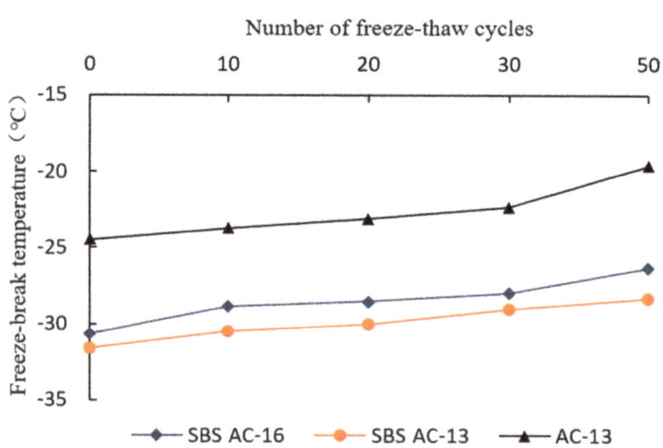

Figure 9. Freeze–break temperature test results.

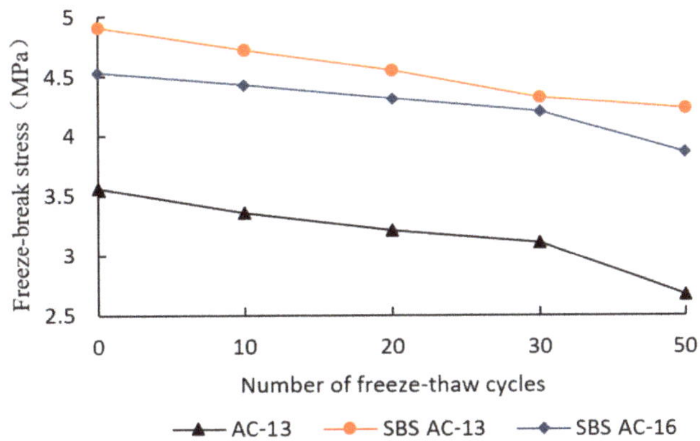

Figure 10. Freeze–break stress test results.

As observed, the overall trends of the two indicators of the asphalt mixtures, i.e., freezing–breaking temperature and freezing–breaking stress, are similar under different conditions. The types of mixtures have a certain influence on the test results. The freeze–break temperature of AC-13 asphalt mixture is lower than that of AC-16 asphalt mixture, because the former has smaller particle sizes and larger amounts of asphalt, thus delivering better low-temperature performance than the latter. The types of asphalt also have a certain influence on the test results. The freeze–break temperature of the matrix asphalt is higher than that of the modified asphalt, because the latter has a higher viscosity than the former and, furthermore, the toughness of the latter at a low temperature is better than that of the former.

4.2.2. Freeze–Thaw Splitting Test

According to the set plan, the freeze–thaw splitting test is carried out for three kinds of asphalt mixtures, namely matrix asphalt AC-13, SBS modified AC-13, and SBS modified AC-16. Two indicators, i.e., splitting strength and freeze–thaw splitting strength ratio, are obtained from the test calculation. The test results are shown in Table 7.

Table 7. Freeze–thaw splitting test results of asphalt mixture.

Asphalt Mixture	AC-13		SBS AC-13		SBS AC-16	
Freeze–Thaw Cycles	Strength (MPa)	Ratio (%)	Strength (MPa)	Ratio (%)	Strength (MPa)	Ratio (%)
0	7.39	100.00	8.65	100.00	8.73	100.00
10	6.14	83.09	7.88	91.10	7.85	89.92
20	5.66	76.59	7.46	86.24	7.54	86.37
30	5.42	73.34	7.13	82.43	7.12	81.56
50	5.16	69.82	6.89	79.65	6.83	78.24

It can be seen from the test results that the splitting strength of asphalt mixtures and the ratio of freeze–thaw splitting test strength are significantly decreased under the condition of frequent freeze–thaw cycles; with the increase in the number of freeze–thaw cycles, the splitting strength and the ratio of freeze–thaw splitting test strength would gradually go down. These results demonstrated that frequent freeze–thaw cycles would deliver a significant effect on the splitting tensile strength and water stability of asphalt mixtures. As the freeze–thaw splitting test is carried out in a water-containing state, when the temperature drops below 0 °C, the water in the asphalt mixtures will begin to freeze and expand, causing cracks, or the original cracks will keep expanding. When the temperature rises above 0 °C, the ice within the sample would be re-thawed, and the water will continuously infiltrate into the new cracks. Then, the process of freezing and expanding will begin again in the next freeze–thaw cycle, which can deliver a significant impact on the low-temperature crack resistance and water stability of the asphalt mixtures.

5. Conclusions

This study explores the freeze–thaw cycles' characteristics and the cracking problems of asphalt pavements in high-cold and high-altitude areas. Laboratory tests are made to evaluate the pavements' performance of asphalt binder as well as the performance of low-temperature crack resistance of asphalt mixtures under frequent freeze–thaw cycles. The main conclusions are as follows.

(1) There are basically more than 120 freeze–thaw cycles per year in the high-cold and high-altitude areas, or even more than 200 cycles. With the increase in altitudes, the number of freeze–thaw cycles shows an up-ticking trend. Frequent freeze–thaw cycles have a significant impact on asphalt pavements, the cracks in which have become a typical problem.
(2) The performances of the three different types of asphalt binders used in the test basically show no change after 50 freeze–thaw cycles, mainly because under such cycles, asphalt binders will only change its physical phases, but its internal chemical composition does not transfer or change, so neither the properties of the asphalt.
(3) The asphalt types have a significant effect on the low-temperature performance of asphalt mixtures. The modified asphalt shows a higher viscosity than the matrix asphalt, with better toughness than that of the matrix asphalt at low temperature. This result demonstrates that the freeze–break temperature of the matrix asphalt is higher than that of the modified asphalt.
(4) Frequent freeze–thaw cycles significantly influence the low-temperature splitting tensile strength and water stability of asphalt mixtures. With increased freeze–thaw cycles, the splitting strength and freeze–thaw splitting tensile strength ratio will gradually decrease to a significant level.

Author Contributions: H.C.: conceptualization, methodology, validation, investigation, writing—original draft. T.C.: conceptualization, writing—review and editing, supervision, project administration, resources, funding acquisition. H.Z.: methodology, writing—review and editing. H.R.: writing—review and editing, supervision. All authors have read and agreed to the published version of the manuscript.

Funding: This research was funded by National Natural Science Foundation of China (Grant No. 52178419), Transportation Industry Key Science and Technology Projects of China (Grant No. 2020-MS1-059) and Youth Science and Technology Innovation Fund project of FHCC (Grant No. YGY2019QC-02).

Institutional Review Board Statement: Not applicable.

Informed Consent Statement: Not applicable.

Data Availability Statement: Data will be made available on reasonable request.

Acknowledgments: Thanks to Zhendong Qian, who proposed the idea of the performance test in the current study.

Conflicts of Interest: The authors declare no conflict of interest.

References

1. Krcmarik, M.; Varma, S.; Kutay, M.E.; Jamrah, A. Development of Predictive Models for Low-Temperature Indirect Tensile Strength of Asphalt Mixtures. *J. Mater. Civ. Eng.* **2016**, *28*, 04016139. [CrossRef]
2. Kumar, A.; Choudhary, R. Characterisation of asphalt binder modified with ethylene–propylene–diene–monomer (EPDM) rubber waste from automobile industry. *J. Road Mater. Pavement Des.* **2021**, *22*, 2044–2068. [CrossRef]
3. Wen, H. Use of fracture work density obtained from indirect tensile testing for the mix design and development of a fatigue model. *Int. J. Pavement Eng.* **2013**, *14*, 561–568. [CrossRef]

4. Peiwen, H.; Dengliang, Z. Research on Evaluation Target for the Cracking Resistance Property of Bituminous Mixture in Low Temperature. *J. Highw.* **2000**, *45*, 63–67.
5. Weiwei, G.; Zutang, Z. Influence of ray aging on low temperature performance of asphalt mixture in plateau-cold region. *J. Chongqing Jiaotong Univ. Nat. Sci.* **2012**, *31*, 51–53.
6. Zhang, Y. Study on Low Temperature Crack Resistance and Mix Design Method of Asphalt Mixture in Permafrost Regions. Ph.D. Thesis, Chang'an University, Xi'an, China, 2004.
7. Russo, F.; Veropalumbo, R.; Oreto, C.; Biancardo, S.A.; Abbondati, F.; Viscione, N. Verifying the Mechanical Performance of Cold and Hot Asphalt Mastics Containing Jet Grouting Waste as a Filler. *J. Coat.* **2021**, *11*, 751. [CrossRef]
8. Little, D.N.; Mahboub, K. Engineering properties of first generation plasticized sulfur binders and low temperature fracture evaluation of plasticized sulfur paving mixtures. *Transp. Res. Rec.* **1985**, *1034*, 103–111.
9. Luo, X.; Luo, R.; LLytton, R. Characterization of fatigue damage in asphalt mixtures using pseudostrain energy. *J. Mater. Civ. Eng.* **2013**, *25*, 208–218. [CrossRef]
10. Xiao, F.; Zhao, W.; Amirkhanian, S.N. Laboratory investigation of fatigue characteristics of rubberized asphalt mixtures containing warm asphalt additives at a low temperature. *J. Test. Eval.* **2011**, *39*, 1.
11. Van, D.W. Practical Fatigue Characterization of Bituminous Mixes. *J. Assoc. Asph. Paving Technol.* **1975**, *44*, 38–72.
12. Edward, T. *Fatigue Response of Asphalt-Aggregate Mixes*; Strategic Highway Research Program (SHRP-A-404); National Research Council: Washington, DC, USA, 1999; pp. 38–42.
13. Songtao, L.; Xiaoge, T.; Jianlong, Z. Simulation Research on Thermal Fatigue Damage to Bituminous Mixture. *J. Highw.* **2005**, *50*, 116–118.
14. Xiaoge, T.; Jianlong, Z.; Zhihong, X. Research on calculation method of dissipated energy during temperature fatigue of asphalt mixture. In Proceedings of the Fourth International Road and Airport Pavement Technology Conference of China Highway Society, Kunming, China, 23 April 2002; pp. 450–454.
15. Qiang, F. Analysis of temperature fatigue expansion life of asphalt pavement cracks. In Proceedings of the 10th China Association for Science and Technology Annual Conference of China Association for Science and Technology and Henan Provincial People's Government, Zhengzhou, China, 1 September 2008; pp. 503–511.
16. Yongxiang, Z.; Hailin, Y.; Zheng, L. Fracture mechanic analysis of the asphalt concrete pavement structure with surface crack under low-temperature shrinkage. *J. Transp. Sci. Eng.* **2010**, *26*, 21–24.
17. Youpo, W.; Hao, H.; Kai, C. Asphalt mixture's flexural-tensile strength's influencing factor in plateau area. *J. West. China Commun. Sci. Technol.* **2009**, *2*, 5–7.
18. Kliewer, J.E.; Zeng Huayang Vinson, T.S. Aging and low-temperature cracking of asphalt concrete mixture. *J. Cold Reg. Eng.* **1996**, *10*, 134–148. [CrossRef]
19. Bo, G.; Ni, G. Effect of component factors of asphalt mixture on pavement performance in areas with big temperature difference. *J. Road Mach. Constr. Mech.* **2017**, *34*, 81–85.
20. Dongqing, L.; Qingzhou, M.; Jianhong, F. Test and Research on performance of Frost Thawing Resistance of Asphalt Mixture. *J. Highw.* **2007**, *52*, 145–147.
21. Biao, M.; Xueyan, Z.; Wei, S.; Ning, L.; Shuigen, P. Study on the water stability and high temperature performance of asphalt mixtures in Qinghai-Tibet cold regions. *J. Glaciol. Geocryol.* **2015**, *37*, 175–182.
22. Youmian, T.; Tao, L.; Shujun, L.; Hongliang, L.; Jie, Z. Static and dynamic laboratory performance study on asphalt binder under large temperature difference. *J. Pet. Asph.* **2017**, *31*, 17–23.
23. Yong, L.; Yang, L.; Peng, Y. Mix design and discussion of HRA in alpine region. *J. China Foreign Highw.* **2011**, *31*, 252–255.
24. Shuigen, P. Study on Pavement Performance of Asphalt Mixture of Lhasa-Gongga Airport Expressway. Ph.D. Thesis, Chang'an University, Xi'an, China, 2013.
25. Xuesong, M.; Zhe, H.; Fengjie, Z. Adaptability of pavement typical structure in high altitude cold area. *J. Chongqing Jiaotong Univ. Nat. Sci.* **2017**, *36*, 23–29.
26. *JTG F40-2004*; Technical Specifications for Construction of Highway Asphalt Pavements. Ministry of Transportation Highway Research Institute: Beijing, China, 2004.
27. *JTG E20-2011*; Standard Test Methods of Bitumen and Bituminous Mixtures for Highway Engineering. Ministry of Transport: Beijing, China, 2011.

coatings

Article

Effect of Fine Aggregate Gradation on Macro and Micro Properties of Cold Recycling Mixture Using Emulsified Asphalt

Zhigang Li [1,*], Kexin Li [1], Jianmin Zhang [2], Ruibo Ren [1,*], Pinru Du [3], Pinhui Zhao [1], Quanman Zhao [1] and Litao Geng [1]

[1] School of Transportation Engineering, Shandong Jianzhu University, Jinan 250101, China; likexin5386@163.com (K.L.); zhaopinhui08@163.com (P.Z.); zhaoquanman@sdjzu.edu.cn (Q.Z.); glt@sdjzu.edu.cn (L.G.)
[2] Shandong Luqiao Group Co., Ltd., Jinan 250101, China; zhangjm@sdluqiao.com
[3] Shaanxi Transportation Holding Group Co., Ltd., Xi'an 710048, China; dupinru@126.com
* Correspondence: lizhigang@chd.edu.cn (Z.L.); rrbgq@sdjzu.edu.cn (R.R.)

Abstract: In order to explore the influence of fine aggregate on the macro and micro properties of cold recycling mixture using emulsified asphalt (CRME), mechanical and microscopic property tests were carried out. The indirect tensile strength (ITS), unconfined compressive strength (UCS) and triaxial shear strength of different fine aggregate gradation was measured for analyzing the effects of fine aggregate on the mechanical strength, triaxial shear resistance and fracture energy of CRME. Meanwhile, the surface morphologies and air voids distribution of different CRME were observed by scanning electron microscopy (SEM) and X-ray computed tomography (X-ray CT). The results show that fine aggregate has a significant effect on the mechanical strength and shear resistance of CRME. With the same water and asphalt content, the fracture energy and failure strain of the mixture with less fine aggregate (G3-2) decreased by 16.2% and 18.2%, respectively. The less content of powder there was, the fewer cement hydration products there were due to some cement being coated by emulsified asphalt and the "cement hydration products fiber" length being shorter. Approximately 70% of the AFt hydration products in the G3-2 mixture were in the range of 1–4 μm, while those in the G1 mixture were in the range of 4–8 μm. With the increase in filler content, the number of air voids in the volume range of 0.5 $mm^3 \leq V < 5$ mm^3 in CRME decreased, and the number of air voids in the volume range of $V < 0.5$ mm^3 significantly increased, while the equivalent radius of air voids decreased slightly with the increase in filler content.

Keywords: cold recycling mixture using emulsified asphalt; fine aggregate; macro and micro properties; surface morphologies; air voids distribution

Citation: Li, Z.; Li, K.; Zhang, J.; Ren, R.; Du, P.; Zhao, P.; Zhao, Q.; Geng, L. Effect of Fine Aggregate Gradation on Macro and Micro Properties of Cold Recycling Mixture Using Emulsified Asphalt. *Coatings* **2022**, *12*, 674. https://doi.org/10.3390/coatings12050674

Academic Editor: Valeria Vignali

Received: 14 April 2022
Accepted: 12 May 2022
Published: 14 May 2022

Publisher's Note: MDPI stays neutral with regard to jurisdictional claims in published maps and institutional affiliations.

Copyright: © 2022 by the authors. Licensee MDPI, Basel, Switzerland. This article is an open access article distributed under the terms and conditions of the Creative Commons Attribution (CC BY) license (https://creativecommons.org/licenses/by/4.0/).

1. Introduction

The recycling of waste materials has become the predominant way to realize green highways. Cold/hot recycling technology, rubber asphalt technology and the reutilization of construction waste or recycled plastic is applied and researched [1–4]. Due to the lack of aggregates, asphalt and other resources, and considering the energy conservation and environmental sustainability of pavement maintenance, emulsified asphalt cold recycling technology has been widely applied [5–7]. The void volume (VV) of CRME is generally 8%–13%, and few pavement diseases caused by air voids have been found when the VV is less than 8%. Early and long-term permanent deformation, loosening, water damage and other diseases are easy to occur when the VV is larger than 13%. Fine aggregate mainly fills the forming air voids between coarse aggregate, which is one of the most important factors affecting the VV of CRME. At the same time, fine aggregate gradation has a significant effect on the formation and distribution of emulsified asphalt mortar, thus affecting the macro performance of CRME. Therefore, it is of great significance to explore the influence of fine aggregate gradation on the macro and micro properties of CRME for its material composition design and the improvement of its mechanical properties.

As machine-made sand content increases, the mechanical properties of CRME are also enhanced [8]. The contact angles and adhesive energy between the aggregates and asphalt significantly impact the water stability and fatigue performance of CRME [9], and the emulsified asphalt type has an appreciable effect on the adhesion between aggregates in a mixture [10]. In addition, intergranular mode fracture is the most common failure of mixtures for the weakness of the asphalt-aggregate interface [11]. Therefore, how to characterize the mortar interfacial characteristics of CRME is particularly important. Research work [12] studied the factors that affected the performance of CRME by macro and micro test methods. The stress dispersion state and cracking resistance mechanisms of mixtures were revealed through the micro interface. Generally, 1 wt.% ~1.5 wt.% cement is added to improve the early strength of CRME [13]. Meanwhile, high-temperature stability and water stability are also enhanced with increased cement addition [14]. With the increase in cement content, the low-temperature cracking resistance performance of cold recycled mixture did not decrease monotonically but presented a parabolic law [15]. That is, while the cement content was lower than 2%, the resistance to low-temperature cracking of the cold recycled mixture was enhanced as cement content increased. Pavement fracture toughness usually decreased with increasing percentages of RAP or decreasing cement content [16]. Cement that is not covered by emulsified asphalt completely hydrates with water in the mixture and forms a "three-dimensional space" structure by interweaving emulsified asphalt and cement hydration products. Furthermore, this three-dimensional space structure plays a "reinforcement" and "crack resistance" role in a mixture. The formed structure not only improved the strength of the mixture but also divided the air voids in the mixture, resulting in an increase in the number of air voids [17]. However, if part of the cement is covered by emulsified asphalt, the hydration reaction does not fully occur. Thus, this part of the cement is consistent with that of mineral powder [18,19]. Research pointed out that both low porosity and better coverage of aggregates could make the mixture show better performance [20]. Therefore, researchers had carried out a lot of research on the micro interface and void structure of CRME. The micro morphology and failure interface characteristics of CRME have been determined by X-ray CT scanning, SEM scanning and other micro testing methods combined with image processing, and it was found that the homogeneity of aggregate distribution, air void distribution and cement hydration products distribution is closely related to the macro mechanical strength of CRME [21–23]. The distributions of the air void parameters of cold recycled mixture and hot mix asphalt specimens along the specimen height are compared. It is found that the cold recycled mixture has more air voids and smaller diameters, while the HMA specimen has fewer voids and larger diameters. The change in the air void structure of cold recycled mixture is the fundamental reason for its strength attenuation [19]. There is a specific correlation between the distribution of air voids and the macro performance. How to evaluate the distribution characteristics of air voids in CRME has also attracted much attention. The Weibull statistical model can better reflect the distribution characteristics of air voids in cold recycled mixture [24]. At the same time, the performance of CRME can be prominently enhanced by adding fiber or waste materials [25].

At present, researchers mainly focus on the performance of CRME and have studied the influence of cement content, emulsified asphalt content, active filler type and fiber on the macro and micro performance of CRME. However, there is a lack of knowledge of the influence of fine aggregate gradation on the microstructure of a mixture, which could effectively provide theoretical support for the material composition design of CRME. It is necessary to know whether cement acts as an ordinary filler when ordinary fillers are insufficient. In this case, cement may not hydrate completely. Quantitative analysis of the cement hydration degree is also a difficult problem when ordinary filler is insufficient. Thus, this paper explores the influence of fine aggregate gradation on the macro mechanics, micro interface morphology and air void structure characteristics of CRME through laboratory tests, providing theoretical support for the material design of CRME.

2. Materials and Methods

2.1. Materials

2.1.1. RAP and Raw Aggregates

In the design process of CRME, reclaimed asphalt pavement (RAP) aggregates are usually regarded as "black stone" without considering the influence of aged asphalt. The RAP used in this study was collected from the Qingdao Jiaozhou Bay Highway, and the sieving results are shown in Table 1. The asphalt was reclaimed by the Aberson method, and the testing results are listed in Table 2. The apparent relative density of manufactured sand was 2.721, the sand equivalent was 83% and the methylene blue value was 1.7 g/kg. The raw aggregates and RAP were sieved for subsequent testing.

Table 1. Results of sieving for RAP.

Sieve Size/mm	26.5	19	16	13.2	9.5	4.75	2.36	1.18	0.6	0.3	0.15	0.075
Passing Percentage/%	100	95.4	83.2	66.7	60.8	32.2	17.9	13.5	7.2	5.9	3.7	2.5

Table 2. Testing results of recycled asphalt from RAP.

Testing Index	Results
Penetration (25 °C; 0.1 mm)	32
Ductility (25 °C; cm)	61
Softening point (°C)	39.5

2.1.2. Emulsified Asphalt and Cement

Slow cracking and setting emulsified asphalt was prepared by using an Akzo Nobel emulsifier and Shell 70# asphalt. The main technical performance test results of emulsified asphalt and matrix asphalt are shown in Tables 3 and 4. In order to promote the demulsification of emulsified asphalt and increase the early strength of CRME, cement was added to CRME. However, cement has little effect on the long-term performance of a mixture [26,27]. Ordinary Portland cement was added with 1.5% content by mass according to the "Technical Specifications for Highway Asphalt Pavement Recycling" (JTG/T 5521-2019) [28], and all the technical specifications of the cement met the requirements of Specification JTG/T 5521-2019 [28].

Table 3. Test results of the technical performance of matrix asphalt.

Test	Results	Specification
25 °C penetration/0.1 mm	72	60~80
Softening point/°C	47	≥46
60 °C Dynamic viscosity/Pa·s	243	≥180
10 °C ductility/cm	63	≥20
15 °C Density/(g/cm^3)	1.012	Report
Wax content/%	1.5	≤2.2

Table 4. Test results of the technical performance of emulsified asphalt.

Test		Results	Specification
Emulsifying speed		Slow-Breaking	Slow or median Breaking
Ionic charge		Cationic (+)	Cationic (+)
Residue of 1.18 mm sieves/%		0.05	≤0.1
Evaporative residues	Residual content/%	62.8	≥62
	25 °C penetration/0.1 mm	83.2	50~300
	15 °C ductility/cm	47	≥40
Storage stability	1 day/%	0.6	≤1
	5 days/%	2.7	≤5

2.1.3. Mixture Design

In the gradation design, 80 wt.% RAP and 20 wt.% raw aggregate were used. In order to reduce the influence of material variation, the gradation of mixtures was accurately controlled by sieving. A mixture with fine aggregate grading coefficients (n_{FA}) of 0.4, 0.6 and 0.8 (G1, G2, G3 and G3-2), respectively, was designed, as shown in Table 5, based on the N method grading theory revised by A.N. Talbot. CRME was designed by the modified Marshall method according to Specification JTG/T 5521-2019 [28]. The optimum water content (OWC) of CRME was determined by the maximum dry density of the heavy hammer test with 4.0 wt.% emulsified asphalt content. The optimum emulsified asphalt content (mass ratio of emulsified asphalt to aggregates) was determined according to the wet and dry ITS (ITS_{dry}). The Marshall design results are shown in Table 6. In order to eliminate the influence of emulsified asphalt content on the test results, the control group G3-2 was set. Its grading was the same as that of G3, and other design information was identical to that of the G1 mixture, which is shown in Table 6.

Table 5. Gradations of the three types of CRME.

Sieve Size/mm	Passing Percent/%			Specification
	G1	G2	G3	
26.5	100	100	100	100
19	92.7	92.7	92.7	90~100
16	79.2	79.2	79.2	-
13.2	72.4	72.4	72.4	-
9.5	65.3	65.3	65.3	60~80
4.75	42.6	42.6	42.6	35~65
2.36	32.3	27.4	24.1	20~50
1.18	24.2	18.3	13.7	-
0.6	18.4	12.3	8.0	-
0.3	14.2	8.2	4.9	3~21
0.15	10.5	5.4	2.7	-
0.075	8.2	3.6	1.6	2~8
n_{FA}	0.4	0.6	0.8	-

Table 6. Mix design results of CRME.

Type of Mixture	Optimum Emulsified Asphalt Content/wt.%	Optimum Water Content/wt.%	Maximum Dry Density/(g/cm^3)	ITS_{dry} /MPa	Requirements of ITS_{dry}/MPa
G1	4.7	4.2	2.068	0.624	≥0.4
G2	3.9	3.8	2.010	0.563	≥0.4
G3	3.6	3.6	1.923	0.487	≥0.4
G3-2	4.7	4.2	1.907	0.537	≥0.4

2.2. Preparation of CRME Specimens

RAP, cement, water, emulsified asphalt and raw aggregate were prepared in a certain proportion, as mentioned above. Standard Marshall specimens were adopted for the ITS, UCS test and X-ray CT scanning test. Gyratory Compacting specimens of 200 mm × 200 mm were made, and all the specimens were cured at 40 °C for 72 h in a constant temperature oven [29]. All of the specimens remained stationary at room temperature for 12 h after curing.

2.3. Experimental Method

2.3.1. Indirect Tensile Test

ITS was used for the mechanical properties of CRME, which reflected the cohesive properties of mixtures to a certain extent. At least four specimens were tested in each ITS test group, and the average value was calculated by the ITS results of four specimens. Before the test, the specimen was kept in a 15 °C constant-temperature water tank for 2 h, and the test loading rate was 50 mm/min.

2.3.2. Unconfined Compressive Strength

Due to the viscoelastic properties of asphalt mixture, it has significant temperature sensitivity [30]. In this study, UCS was adopted to characterize the carrying capacity of the mixtures. At least 6 h of heat preservation was required in a 25 °C incubator, with at least four specimens in each group. Testing was carried out by UTM-100 with a loading rate of 1 mm/min according to the "Standard Test Methods of Bitumen and Bituminous Mixtures for Highway Engineering" (JTG E20-2011) [31].

2.3.3. Triaxial Test

Specimens that were 100 mm in diameter and 200 mm in height were obtained by the drilling core test, and the top and bottom surfaces of the specimens were polished for the triaxial test. Before the triaxial test, the specimens were kept in a 25 °C incubator for at least 6 h. UTM-100 was used to carry out the triaxial test with a 3 mm/min loading rate [32]. Then, 50, 100 and 200 kPa were chosen as the confining pressures for the triaxial test, and at least three specimens were measured under each confining pressure. According to the calculation method of reference [32], the triaxial shear parameters of CRME were calculated. According to the stress-strain curve during the triaxial loading process, the fracture energy was calculated by integration. A schematic diagram of the calculation of the fracture energy and failure strain is shown in Figure 1.

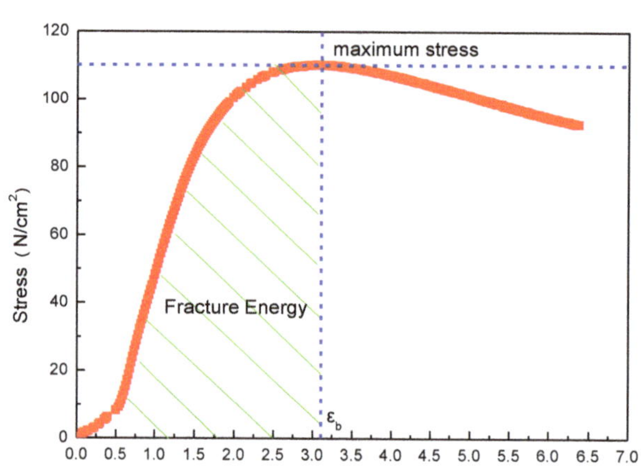

Figure 1. Calculation of fracture energy and failure strain.

2.3.4. Scanning Electron Microscopy Test

PHILIPS-FEI Quanta 200 SEM equipment was used to observe the micro-structure of the RAP aggregate at the ITS fracture interface. The micro-structure was observed to explore the influence of fine aggregate gradation on the spatial structure, which was an interweaving of emulsified asphalt mortar and cement hydration products in the mixture. The hydration products of AFt were quantitatively analyzed using Nano Measurer analysis software, and the hydration process of cement in the mixtures with different fine aggregate gradations was characterized by the length of AFt. Firstly, the standard length in the SEM image was calibrated, and then the length of the AFt needle was measured, as shown in Figure 2. After that, the distribution range of the AFt length was analyzed.

Figure 2. Nano Measurer statistical analysis for CRME (G1 mixture).

2.3.5. X-ray Computed Tomography Test

Y. Precision 225 kV X-ray CT with a scanning accuracy of two microns was adopted to determine the air void structure of the mixtures with different fine aggregate gradations. The optimum threshold value was determined by Otsu's algorithm, which is described in detail in article [24]. After three-dimensional scanning, VG software was used to reconstruct the air voids. The volume, surface area and position parameters of each air void in the mixtures were obtained, and all the information was used to explore the air void characteristics. Three image views and the three-dimensional reconstruction of the Marshall specimen after scanning are shown in Figure 3.

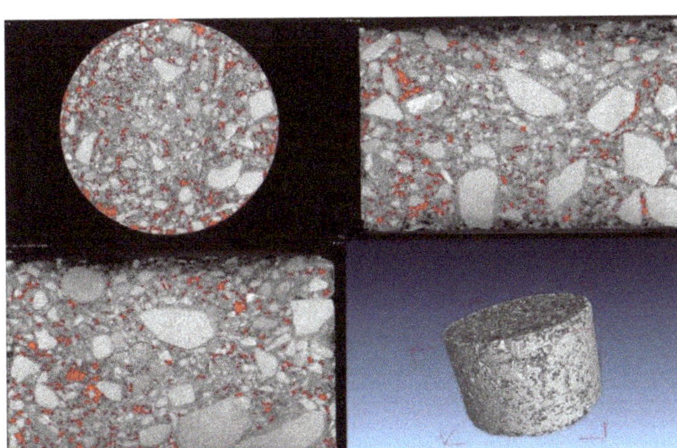

Figure 3. Three views and three-dimensional reconstruction image of a specimen.

3. Results and Discussion

3.1. Macro-Mechanical Properties of CRME

The results of the ITS and UCS are listed in Figure 4. The ITS and UCS of CRME decreased as the fine aggregate gradation coarsened. The UCS characterized the load-carrying capacity of CRME, which was related to the "inlay condition" of coarse aggregate. The greater the fine aggregate content was, the fuller the filling of air voids in the mixture was, and the mixture became denser, while the coarse aggregate had the same "inlay condition". The distribution of air voids was more uniform, and the stress concentration was low, which showed that the ITS and UCS were relatively large. At the same time, the powder fillers used to form emulsified asphalt mortar were sufficient, and the cement was not required to act as a powder filler. In contrast, when the powder fillers were insufficient, some cement needed to be consumed to form emulsified asphalt mortar, which led to the cement hydration products decreasing. Meanwhile, the "reticulate spatial structure", an interweaving of the cement hydration products and emulsified asphalt, became weakened, which resulted in the reduction of the macro-mechanical strength of CRME.

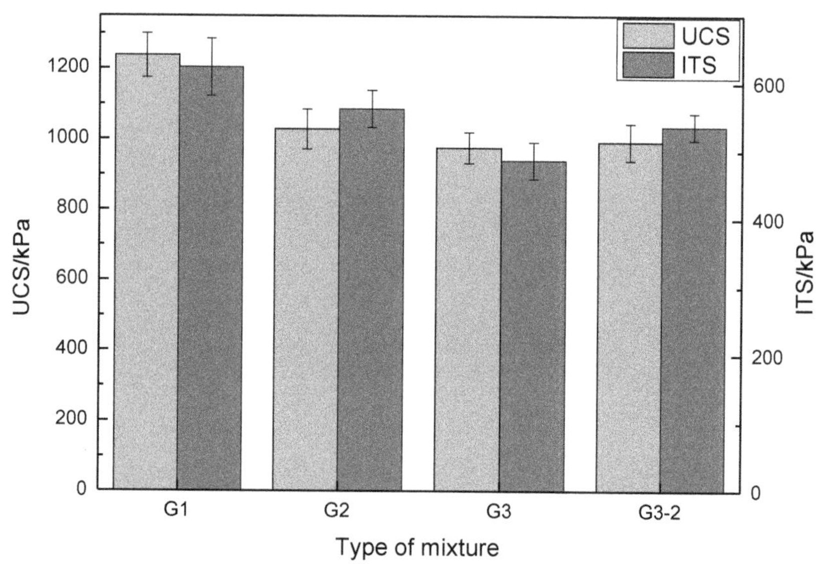

Figure 4. UCS and ITS results for the mixtures with different fine aggregate gradations.

The calculated results of the triaxial shear parameters, fracture energy and maximum failure strain are shown in Figure 5 and Table 7. According to the triaxial test results in Figure 5, as the fine aggregate gradation coarsened, the cohesion of CRME also decreased and the internal friction angle increased. Less effective filling occurred when the fine aggregate gradation coarsened. The adhesive effect of the emulsified asphalt mortar between the coarse aggregates decreased, which led to a decrease in the internal cohesion of CRME. On the contrary, due to the decreasing filler content, the emulsified asphalt content was reduced. This led to a strengthening of the "interlocked effect" between aggregates and a weakening of the "lubrication effect" of the emulsified asphalt. As a result, the internal friction angle of CRME increased.

In addition, the powder filler contents of the three gradations were quite different. Part of the cement was wrapped by emulsified asphalt due to the lower powder filler content, and the cement could not be fully used for hydration. The adhesion of the "network structure", an interweaving of the cement hydration products and emulsified asphalt, decreased and resulted in the cohesion decrease of CRME. However, the cement hydration products had little effect on improving the "interlocked effect" between coarse aggregates, and the internal friction angle of CRME mainly depended on the "interlocked condition" of the coarse aggregates themselves.

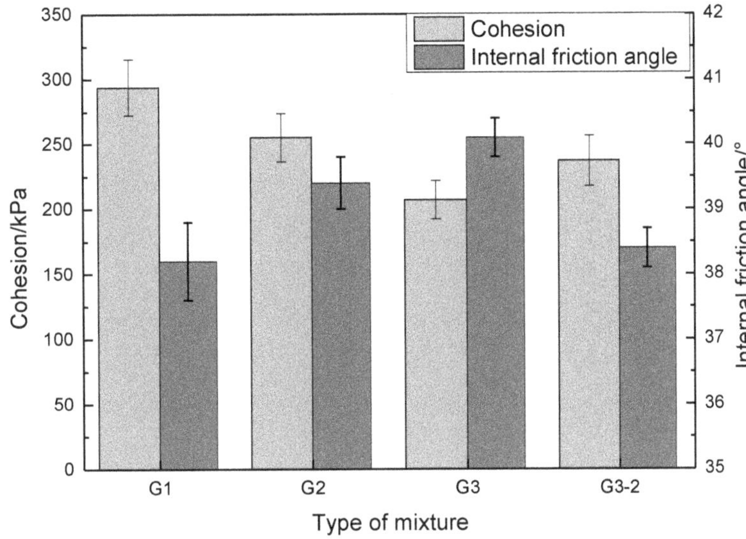

Figure 5. Triaxial test results for the mixtures with different fine aggregate gradations.

Table 7. Results of fracture energy and failure strain for the mixtures.

Type of Mixture	Confining Pressure 50 kPa		Confining Pressure 100 kPa		Confining Pressure 200 kPa	
	Fracture Energy/(N/cm^2)	Failure Strain/%	Fracture Energy/(N/cm^2)	Failure Strain/%	Fracture Energy/(N/cm^2)	Failure Strain/%
G1	247.6	4.6	322.7	4.4	462.3	3.8
G2	212.3	3.8	276.8	3.5	392.1	3.1
G3	173.4	3.0	242.5	2.6	354.2	2.3
G3-2	197.5	4.1	270.4	3.6	372.7	2.9

According to the test results in Table 7, as confining pressure increased, the fracture energy of CRME increased and the failure strain decreased. This indicated that confining pressure significantly improved the load-carrying capacity of CRME, and the triaxial test can better reflect the actual service state of pavement. With a decrease in filler content, both the fracture energy and failure strain of CRME decreased. Taking the test results of 100 kPa confining pressure as an example, compared with the G1 mixture, the fracture energy and failure strain of the G3 mixture decreased by 24.9% and 40.9%, respectively. Also compared with the G1 mixture, the fracture energy and failure strain of the G3-2 mixture decreased by 16.2% and 18.2%, respectively. It can be seen that the content of fine aggregate had a significant impact on the performance of CRME when compared with G1 and G3-2. The reason for this was that the lower the content of fine aggregates was, the lower the emulsified asphalt content was in the mixture, and the adhesion between the aggregates was weaker. Thus, less energy was needed to destroy the mixture. Similarly, as the emulsified asphalt mortar content decreased, the brittleness of CRME increased. Macroscopically, the failure strain was relatively small.

3.2. Microstructure Characteristics of the RAP Interface

Ordinary Portland cement is composed of gypsum and ground cement clinker, in which the cement clinker consists of C_3S, C_2S, C_3A, C_4AF, etc. The hydration process begins after the cement comes into contact with water. Then, hydration products such as calcium silicate hydrate (C-S-H), calcium hydroxide (CH) and AFt appeared. The air voids of CRME were filled and bridged by hydration products. Thus, a three-dimensional structure formed as the hydration products interweaved with emulsified asphalt mortar. Therefore, the quantity, shape and distribution of the hydration products have a great impact on the three-dimensional structure. However, C-S-H is mostly amorphous, and CH is in a layered structure. Both of them were difficult to quantify. Hence, in this paper, AFt was used for quantitative analysis to quantitatively describe the hydration product amounts.

As shown in Figure 6, there were a lot of cement hydration products and emulsified asphalt mortar on the RAP surface of the G1 mixture. The relatively smooth part in Figure 6 was emulsified asphalt mortar, while the needle-like and cluster structures were cement hydration products, namely, AFt and C-S-H, respectively. AFt and C-S-H interlaced with emulsified asphalt mortar resulted in the formation of a "network structure". Furthermore, the "network structure" acted as a filler between the air voids in CRME. AFt and C-S-H extended into the emulsified asphalt mortar, forming an "anchorage structure". The "bonding", "reinforcement" and "crack resistance" effects between the aggregates were strengthened. The air void structure was segmented and filled due to the distribution of cement hydration products. This was the main reason that the strength of CRME was changed. In order to reveal the influence of fine aggregate gradation on the micro-structure of CRME, the differences in the micro-morphology of the aggregate surfaces of the G1, G3 and G3-2 mixtures were analyzed by SEM, as shown in Figure 7.

Figure 7a–c show the surface morphologies of the G1, G3 and G3-2 mixtures, respectively. Compared with Figure 7b, many more cluster C-S-H and needle-like AFt cement hydration products can be found in Figure 7a, and there were significantly fewer in Figure 7c. Owing to the large amount of fine aggregate in the G1 mixture, the powder filler content was relatively abundant. Sufficient powder filler was coated by emulsified asphalt forming mortar. By contrast, the amount of powder filler coated with emulsified asphalt forming mortar was insufficient. Emulsified asphalt needs to be further coated with cement to form a mortar. As a result, part of the cement played the role of powder filler and could not be completely hydrated due to asphalt coating. Hence, relatively few cement hydration products were observed.

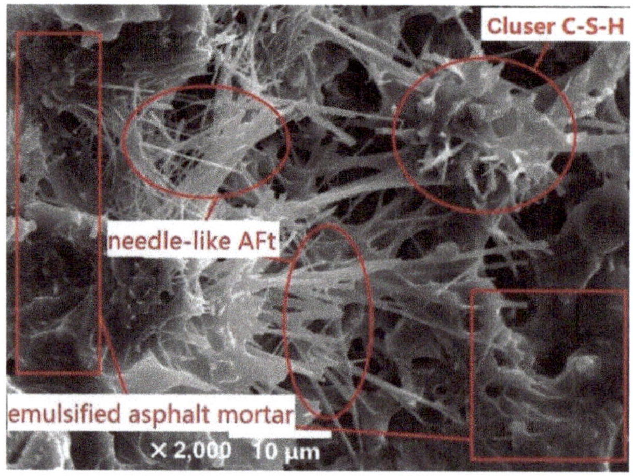

Figure 6. Interaction between the hydration products and asphalt mortar of the G1 mixture.

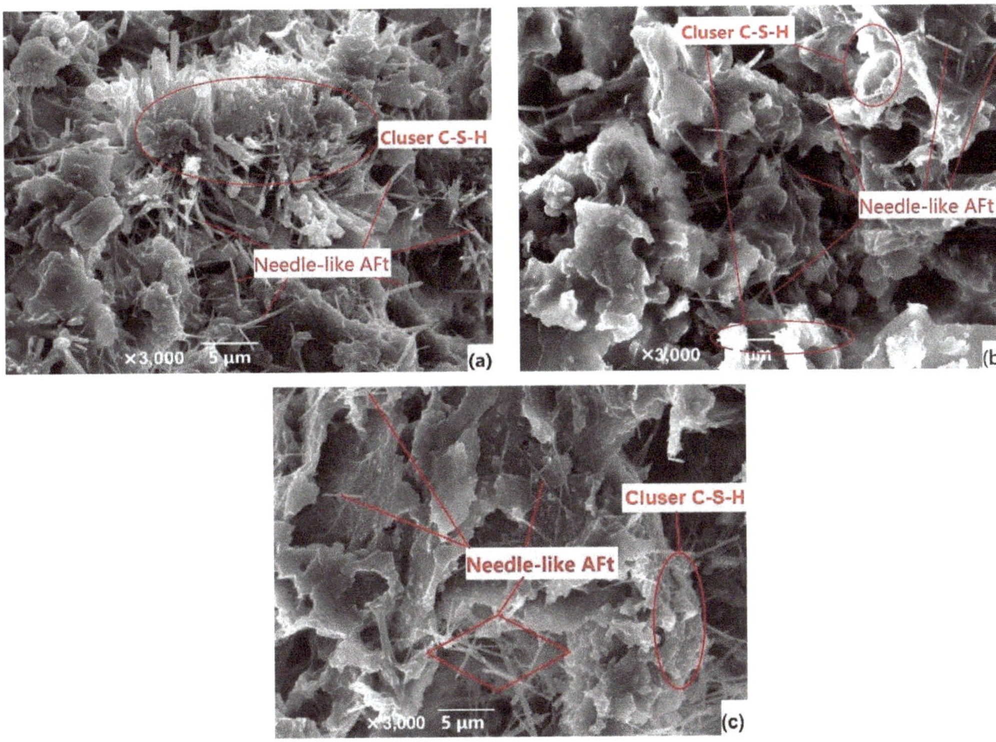

Figure 7. Morphology of aggregate surfaces of mixtures. (**a**) G1 mixture. (**b**) G3 mixture. (**c**) G3-2 mixture.

The length of the AFt hydration products in the G1, G3 and G3-2 mixtures was quantitatively analyzed using Nano Measurer software to evaluate the cement hydration degree.

According to the statistical results in Figure 8, 68.7% of the AFt hydration products in the G1 mixture were in the range of 4–8 μm, and 69.4% of the AFt hydration products in the G3 mixture were in the range of 2–5 μm, while 71.6% of the AFt hydration products in the G3-2 mixture were in the range of 1–4 μm. This indirectly showed that there were relatively more cement hydration products in the G1 mixture. It also was found that, as the cement content increased in CRME, the length of the AFt hydration products increased [18]. Therefore, this indicated that part of the cement in the G3 and G3-2 mixtures could not fully participate in the hydration reaction owing to the emulsified asphalt coating with cement (especially G3-2). At the same time, the longer the "AFt fiber" length of the hydration products was, the larger the interweaving area of the hydration products with emulsified asphalt was, and the "overlap effect" between the hydration products, the "embedding effect" into the aggregate surface and the "interweaving effect" with emulsified asphalt were all enhanced. The research results indicated that the quantity of cement hydration products and the interweaving state of asphalt and hydration products are the decisive factors of mixture performance [14]. However, fine aggregate gradation, especially the filler content, affected the composition of mortar in CRME. With the same emulsified asphalt content, the larger the proportion of the ordinary filler was, the less cement filler was used as ordinary filler, leading to an increase in the quantity of hydration products. Thus, the interfacial strength between the aggregate and emulsified asphalt mortar was enhanced, and the macroscopic strength of CRME was increased.

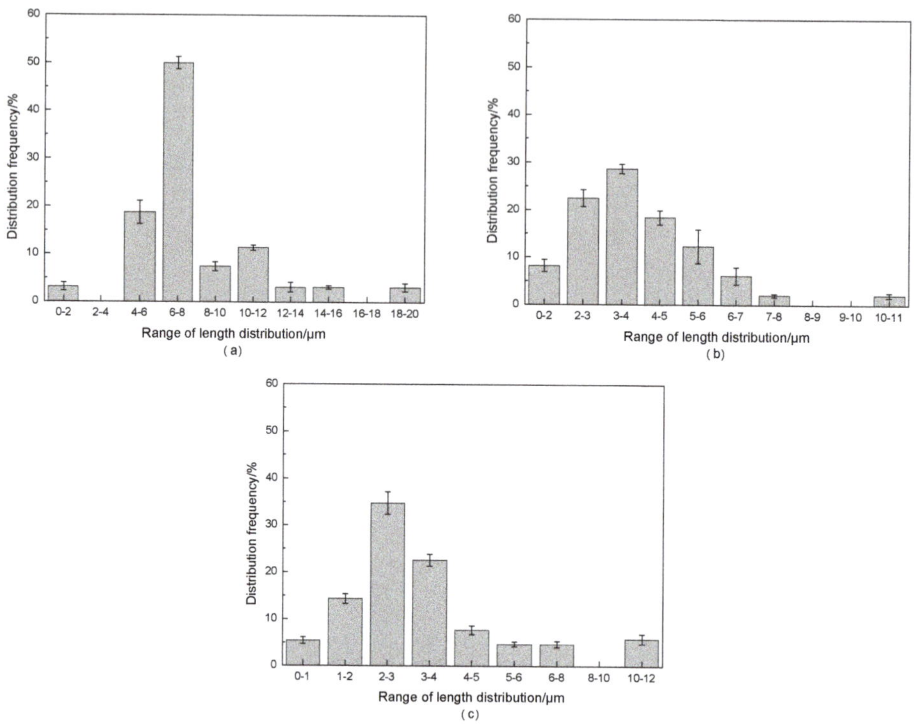

Figure 8. Quantitative analysis of cement hydration products distribution. (**a**) G1 mixture. (**b**) G3 mixture. (**c**) G3-2 mixture.

3.3. *Air Void Structure Characteristics of CRME*

3.3.1. Air Void Amount Distribution of CRME

According to the X-ray CT test results, the air void information (such as the amount, surface area, volume and three-dimensional position of each single air void) of CRME was obtained. The numbered percentage of air void amount was defined as the ratio of air voids in a certain volume range to the total air voids in CRME. The statistical results of the air void cumulative distribution in different mixtures are shown in Figure 9.

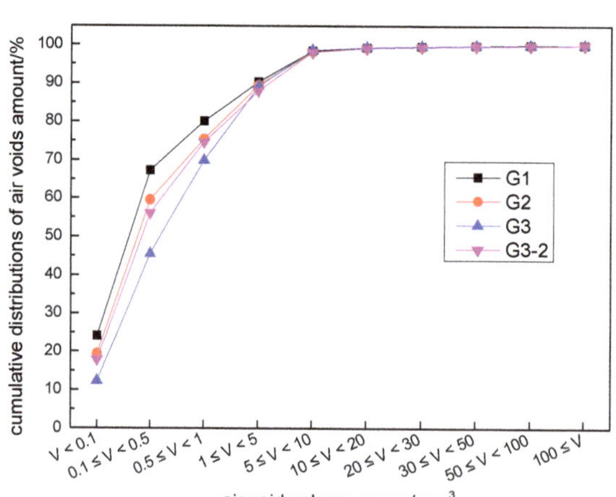

Figure 9. Cumulative distributions of air void amounts with different mixtures.

According to Figure 9, there was little difference in the distribution of air voids at a volume V < 5 mm^3. However, with an increase in the filler content, the number of air voids in the volume range of 0.5 mm^3 $\leq$ V < 1 mm^3 and 1 mm^3 $\leq$ V < 5 mm^3 was reduced. On the contrary, the number of air voids increased in the volume range of V < 0.5 mm^3. Compared with the G3 mixture, the number of air voids in the G1 mixture in the volume range of 0.5 mm^3 $\leq$ V < 5 mm^3 in CRME decreased by 47.4%; and that of G1 in the volume range of V < 0.5 mm^3 increased by 48.0%. However, when compared with the G3-2 mixture, the number of air voids in the G1 mixture in the volume range of 0.5 mm^3 $\leq$ V < 5 mm^3 in CRME decreased by 27.9%; and that of G1 in the volume range of V < 0.5 mm^3 increased by 20.1% (G1 and G3-2 had the same emulsified asphalt content). In other words, emulsified asphalt content and fine aggregate gradation both have significant effects on the distribution of air voids in CRME. On the one hand, the air voids of CRME were filled by redundant fine aggregate, and the original larger air voids were divided into several small-volume air voids. On the other hand, the higher the content of fine aggregate, the higher the content of emulsified asphalt required for coating aggregates. Larger volume air voids were filled and segmented by emulsified asphalt, which also led to a decrease in the number of large-volume air voids and an increase in the number of small-volume air voids in the mixture.

3.3.2. Equivalent Average Radius of Air Voids in CRME

Air voids were equivalent to a sphere of the same volume, and the radius of the sphere corresponding to the same volume as that of the air void was defined as the equivalent radius of the air voids. All equivalent radii of each air void in the mixture were averaged, and this was called the equivalent average radius. The equivalent average radius was used to describe the distribution characteristics of the air void size. The calculation method of the equivalent average radius is shown in Equation (1).

$$r = \frac{\sum_{i=1}^{n} \sqrt[3]{\frac{3V_i}{4\pi}}}{n} \tag{1}$$

where r is the equivalent average radius, mm; V_i is the volume of the air void numbered i, mm^3; n is the total number of air voids in the mixture.

All of the air voids of CRME were equivalent to spheres with the same volume. According to the CT scanning results and the calculation method, the equivalent average radius of the different mixtures was calculated, as shown in Table 8. The air voids were filled and divided into several small-volume voids as a result of a high content of fine aggregate and asphalt. Thus, the void volume and equivalent average radius were both small.

Table 8. Air void equivalent average radius of CRME with different gradations.

Type of Mixture	G1	G2	G3	G3-2
Equivalent average radius/mm	0.712	0.786	0.975	0.824

4. Conclusions

The influence of fine aggregate gradation on the mechanical properties, hydration products and air void structures were studied by macro and micro tests. The ultimate cause of the influence of fine aggregate gradation on the strength of CRME was revealed. However, the research will deeply explore the influence of air void morphology to find a relationship between the air void morphology parameters and mixture strength.

(1) The content of fine aggregate had a significant effect on the ITS, UCS and shear strength of CRME. As the gradation of fine aggregate becomes finer, the emulsified asphalt content increased obviously. As a result, compared with the G1 mixture, the fracture energy and failure strain of the G3-2 mixture decreased by 16.2% and 18.2%, respectively (100 kPa).

(2) Emulsified asphalt mortar and cement hydration products interweaved to form a "spatial network structure", which had "reinforcement", "anchorage", "filling" and "crack resistance" effects. It was one of the most important factors for the strength formation of CRME.

(3) While the powder filler was insufficient, part of the cement was coated with emulsified asphalt, which resulted in the cement failing to hydrate completely. Microscopically, fewer hydration products on the aggregate surface could be found with sufficient powder filler, while relatively more hydration products on the aggregate surface could be found with insufficient powder filler.

(4) Quantitative analysis of the length of AFt hydration products was carried out. At about 70% distribution frequency, the length of the hydration products of the G1 mixture ranged from 4 to 8 μm, while that of the G3 and G3-2 mixtures ranged from 2 to 5 μm and 1 to 4 μm, respectively. Part of the cement in the G3 and G3-2 mixtures was not completely hydrated due to being coated with emulsified asphalt.

(5) An increase in the filler content in fine aggregate resulted in an increase in the emulsified asphalt content in CRME. With the same emulsified asphalt content, when compared with the G3-2 mixture, the number of air voids in the G1 mixture in the volume range of $0.5\ mm^3 \leq V < 5\ mm^3$ in CRME decreased by 27.9%; and that of G1 in the volume range of $V < 0.5\ mm^3$ increased by 20.1%. Compared to the G3-2 and G1 mixtures, the equivalent radius of the air voids decreased by 13.4% with the increase in filler content.

The quantity, size and formation of the interweaving structure of cement hydration products and the number, distribution and size of the air voids were key to determining the performance of CRME. Therefore, in order to improve the performance of CRME, it is necessary to ensure the hydration quality of cement, increase the proportion of ordinary filler and improve the compaction effect to reduce the number of large volume air voids as much as possible.

Author Contributions: Conceptualization, Z.L. and R.R.; methodology, J.Z., K.L. and Z.L.; software, K.L.; formal analysis, K.L. and Q.Z; investigation, K.L., P.Z. and Z.L.; resources, Z.L., P.D. and Q.Z.; data curation, Z.L. and K.L.; writing—original draft preparation, Z.L.; writing—review and editing, Z.L. and R.R.; visualization, K.L. and J.Z.; supervision, Z.L. and L.G.; project administration, Z.L.; funding acquisition, Z.L. and L.G. All authors have read and agreed to the published version of the manuscript.

Funding: This research was financially supported by the project ZR2020QE273 supported by the Shandong Provincial Natural Science Foundation, project ZR2020KE007 supported by the Key Program of Shandong Provincial Natural Science and project 2019KJG004 supported by the Shandong Provincial Young Scholars Innovative Research Team Development Program in Colleges and Universities.

Institutional Review Board Statement: Not applicable.

Informed Consent Statement: Not applicable.

Data Availability Statement: Using the data of this article requires approval from the authors.

Acknowledgments: We express our sincere gratitude to the experts, teachers and students who have provided help for this article.

Conflicts of Interest: The authors declare no conflict of interest.

References

1. Hamed, A.; Ahmad, A.B. Life cycle assessment of incorporating recycled materials in pavement design. *J. King Saud Univ. Eng Sci.* 2022, *in press*.
2. Fawaz, A.; Fahad, A.; Meshal, A.; Husnain, H.; Ahmed, E.; Sherif, E. Sustainability Evaluation of Cold In-Place Recycling and Hot Mix Asphalt Pavements: A Case of Qassim, Saudi Arabia. *Coatings* **2022**, *12*, 50.
3. Ruviaro, S.A.; Silvestro, L.; Pelisser, F.; Azevedo, G.R.A.; Matos, R.P.; Gastaldini, G.L.A. Long-term effect of recycled aggregate on microstructure, mechanical properties, and CO_2 sequestration of rendering mortars. *Constr. Build. Mater.* **2022**, *321*, 126357. [CrossRef]

4. Ren, J.L.; Zhang, L.; Zhao, H.B.; Zhao, Z.D.; Wang, S.Y. Determination of the fatigue equation for the cement-stabilized cold recycled mixtures with road construction waste materials based on data-driven. *Int. J. Fatigue* **2022**, *158*, 106765. [CrossRef]
5. Turk, J.; Mauko, P.A.; Mladenovi, A. Cotič, Z.; Jurjavčič, P. Environmental comparison of two alternative road pavement rehabilitation techniques: Cold-in-place-recycling versus traditional reconstruction. *J. Clean. Prod.* **2016**, *121*, 45–55. [CrossRef]
6. Xu, J.Z.; Hao, P.W.; Zhang, D.P.; Yuan, G.A. Investigation of reclaimed asphalt pavement blending efficiency based on micro-mechanical properties of layered asphalt binders. *Constr. Build. Mater.* **2018**, *163*, 390–401. [CrossRef]
7. Iglesias, P.O.; Pasandín, A.M.R.; Pérez, I.P. Compaction and volumetric analysis of cold in-place recycled asphalt mixtures prepared using gyratory, static, and impact procedures. *Constr. Build. Mater.* **2021**, *296*, 123620.
8. Zarei, W.; Ouyang, J.; Yang, W.T.; Zhao, Y.Q. Experimental analysis of semi-flexible pavement by using an appropriate cement asphalt emulsion paste. *Constr. Build. Mater.* **2020**, *230*, 116994. [CrossRef]
9. Hou, Y.Q.; Ji, X.P.; Li, J.; Li, X.H. Adhesion between Asphalt and Recycled Concrete Aggregate and Its Impact on the Properties of Asphalt Mixture. *Materials* **2018**, *11*, 2528. [CrossRef]
10. Pan, C.L.; Liang, D.Q.; Mo, L.T.; Riara, M.; Lin, J.T. Influence of Different Modifiers on Bonding Strength and Rheological Performance of Bitumen Emulsion. *Materials* **2019**, *12*, 2414. [CrossRef]
11. Mubaraki, M.; Osman, S.; Sallam, H.E.M. Effect of RAP content on flexural behavior and fracture toughness of flexible pavement. *Lat. Am. J. Solids Struct.* **2019**, *16*, e177. [CrossRef]
12. Lv, Z.H.; Shen, A.Q.; Qin, X.; Guo, Y.C.; Ruan, C.H. Performance Optimization and Mechanism of Emulsified Asphalt Cold Recycled Mixtures. *J. Build. Mater.* **2018**, *21*, 90–95.
13. Lin, J.T.; Wei, T.Z.; Hong, J.X.; Zhao, Y.L.; Liu, J.P. Research on development mechanism of early-stage strength for cold recycled asphalt mixture using emulsion asphalt. *Constr. Build. Mater.* **2015**, *99*, 137–142. [CrossRef]
14. Yang, Y.H.; Yang, Y.; Qian, B.T. Performance and Microstructure of Cold Recycled Mixes Using Asphalt Emulsion with Different Contents of Cement. *Materials* **2019**, *12*, 2548. [CrossRef] [PubMed]
15. Xu, J.Z.; Hao, P.W.; Wang, H.; He, L.J. Influences of two binder materials on performance of cold recycled mixtures stabilized with foamed asphalt. *Acta Mater. Compos. Sinica* **2017**, *34*, 687–693.
16. Mubaraki, M.; Sallam, H.E.M. Reliability study on fracture and fatigue behavior of pavement materials using SCB specimen. *Int. J. Pavement Eng.* **2018**, *21*, 1563–1575. [CrossRef]
17. Wang, H. Meso-microscopic Void Distribution Characteristics of Emulsified Asphalt Cold Recycled Mixture with Different Cement Contents. *J. Highw. Transp. Res. Dev.* **2016**, *33*, 27–34.
18. Li, Z.G.; Hao, P.W.; Xu, J.Z. Study on Impacts of Freeze-thaw Cycles on the Shear Performances of Emulsified Asphalt Cold Recycle Mixture. *Mater. Rev.* **2016**, *30*, 121–125.
19. Li, Z.G.; Hao, P.W.; Liu, H.Y.; Xu, J.Z. Effect of cement on the strength and microcosmic characteristics of cold recycled mixtures using foamed asphalt. *J. Cleaner Prod.* **2019**, *230*, 956–965. [CrossRef]
20. Bahiense, A.V.; Alexandre, J.; Xavier, G.; Azevedo, G.R.A.; Monteiro, N.S. Dosage of interlocking paving with ornamental rock waste: An experimental design approach, particle packing and polluting potential. *Case Stud. Constr. Mater.* **2021**, *15*, e00596.
21. Lyu, Z.H.; Shen, A.Q.; Qin, X.; Yang, X.L.; Li, Y. Grey target optimization and the mechanism of cold recycled asphalt mixture with comprehensive performance. *Constr. Build. Mater.* **2019**, *198*, 269–277. [CrossRef]
22. Omrani, M.A.; Modarres, A. Stiffness and Fatigue Behavior of Emulsified Cold Recycled Mixture Containing Waste Powder Additives: Mechanical and Microstructural Analysis. *J. Mater. Civ. Eng.* **2019**, *3*, 04019061. [CrossRef]
23. Gao, L.; Wang, Z.Q.; Liu, Y.P.; Zheng, J.Q.; Li, H. Influence of Binder Property and Mortar Thickness on High-Temperature Performance of Cold Recycled Mixtures with Asphalt Emulsion. *Materials* **2019**, *12*, 2718. [CrossRef] [PubMed]
24. Gao, L.; Ni, F.; Luo, H.L.; Charmot, S. Characterization of air voids in cold in-place recycling mixtures using X-ray computed tomography. *Constr. Build. Mater.* **2015**, *84*, 429–436. [CrossRef]
25. Sun, J.X.; Liu, L.P.; Sun, L.J. Effect of fiber on fatigue performance of emulsified asphalt cold recycled mixture. *J. Wuhan Univ. Technol. (Transp. Sci. Eng.)* **2019**, *43*, 97–101.
26. Li, Z.G.; Hao, P.W.; Liu, H.Y.; Xu, J.Z. Investigation of early-stage strength for cold recycled asphalt mixture using foamed asphalt. *Constr. Build. Mater.* **2016**, *127*, 410–417. [CrossRef]
27. Lin, J.T.; Huo, L.; Xu, F.; Xiao, Y.; Hong, J.X. Development of microstructure and early-stage strength for 100% cold recycled asphalt mixture treated with emulsion and cement. *Constr. Build. Mater.* **2018**, *189*, 924–933. [CrossRef]
28. *JTG/T 5521-2019*; Technical Specifications for Highway Asphalt Pavement Recycling. Ministry of Transport of the People's Republic of China: Beijing, China, 2019.
29. Li, Z.G.; Hao, P.W. Effect of curing procedure to foamed asphalt cold recycled mixture properties. *J. Beijing Univ. Technol.* **2016**, *42*, 79–85.
30. Lytton, R.L.; Zhang, Y.; Gu, F.; Luo, X. Characteristics of damaged asphalt mixtures in tension and compression. *Int. J. Pavement Eng.* **2018**, *19*, 292–306. [CrossRef]
31. *JTG E20-2011*; Standard Test Methods of Bitumen and Bituminous Mixtures for Highway Engineering. Ministry of Transport of the People's Republic of China: Beijing, China, 2011.
32. Jenkins, K.J.; Collings, D.C. Mix design of bitumen-stabilised materials—South Africa and abroad. *Road Mater. Pavement Des.* **2017**, *18*, 331–349. [CrossRef]

coatings

MDPI

Article

Effect of Solid-Solid Phase Change Material's Direct Interaction on Physical and Rheological Properties of Asphalt

Haisheng Zhao [1,2], Jianmin Guo [3], Shijie Ma [2,*], Huan Zhang [2], Chunhua Su [2], Xiaoyan Wang [2], Zengguang Li [2], Jincheng Wei [2] and Shiping Cui [2]

1 School of Highway, Chang'an University, Xi'an 710064, China; zhaohaisheng@sdjtky.cn
2 Key Laboratory of Highway Maintain Technology Ministry of Communication, Jinan 250102, China; zhanghuan@sdjtky.cn (H.Z.); suchunhua@sdjtky.cn (C.S.); wangxiaoyan@sdjtky.cn (X.W.); lizengguang@sdjtky.cn (Z.L.); weijincheng@sdjtky.cn (J.W.); cuishiping@sdjtky.cn (S.C.)
3 Shandong Hi-Speed Company Limited, Jinan 250014, China; 18653160085@126.com
* Correspondence: mashijie@sdjtky.cn; Tel.: +86-18660163082

Abstract: Asphalt pavement is a temperature−sensitive structure that is prone to temperature-related diseases. Phase change material (PCM) is an excellent candidate for mitigating these diseases. This paper looked into the effects of indirect composite shape-stabilized PCM incorporation on the characteristics of asphalt. The compatibility, physical properties, and rheological properties of asphalt with various PCM content before and after aging were thoroughly investigated. No phase separation and no chemical reaction occurred between PCM and asphalt. The physical properties improved with the addition of PCM, and the high−temperature performance indexes improved while the low−temperature performance indexes decreased as the aging process progressed. The effects of PCM on the rheological properties of the matrix and SBS−modified asphalt was distinct. PCM was added to improve the high−temperature rheological characteristics of the matrix asphalt when the temperature was higher than 52 °C, while PCM reduced the high−temperature rheological properties of the SBS−modified asphalt. The aging process has an impact on the high−temperature rutting factor of asphalt with a high PCM content. The low−temperature creep behavior and PG grade of asphalt were both improved. The implication of PCM is that it cannot increase the thermoregulation of asphalt pavement without the cost of scarifying the performance of the asphalt or mixture.

Keywords: phase change material; physical property; rheological property; aging; direct interaction

Citation: Zhao, H.; Guo, J.; Ma, S.; Zhang, H.; Su, C.; Wang, X.; Li, Z.; Wei, J.; Cui, S. Effect of Solid-Solid Phase Change Material's Direct Interaction on Physical and Rheological Properties of Asphalt. *Coatings* **2022**, *12*, 625. https://doi.org/10.3390/coatings12050625

Academic Editor: Andrea Simone

Received: 23 March 2022
Accepted: 29 April 2022
Published: 3 May 2022

Publisher's Note: MDPI stays neutral with regard to jurisdictional claims in published maps and institutional affiliations.

Copyright: © 2022 by the authors. Licensee MDPI, Basel, Switzerland. This article is an open access article distributed under the terms and conditions of the Creative Commons Attribution (CC BY) license (https://creativecommons.org/licenses/by/4.0/).

1. Introduction

Asphalt pavement is widely used in highway engineering as it can provide passengers with a high degree of comfort. Asphalt has low−temperature elasticity and high−temperature viscosity, which can be attributed to its time and temperature sensitivity properties [1–4]. Thermal fatigue cracking, low−temperature cracking, and freeze−thaw cycling have a significant impact on the service performance and service life of asphalt pavement [5]. Repeated temperature cycles are the main trigger for the temperature sensitivity of asphalt, high tensile stress, weakened tensile strength, and repeated high/low stress cycling produced by temperature fluctuations and loading [6]. Thus, significant research efforts have focused on developing technologies and materials to prevent temperature defects in asphalt pavement [7–10].

Phase change energy storage technology, which has been proven to effectively reduce the magnitude of extreme temperature variations and extend the asphalt pavement service life, is one potential approach to solving the aforementioned temperature−related damages. Phase change material (PCM) can change its phase state, e.g., solid to solid, solid to liquid, solid to gas, liquid to gas, or vice−versa [11–13]. PCM has a small volume change throughout the phase change process and essentially keeps constant temperature by absorbing or transmitting energy (often known as latent heat) to the surrounding environment during melting or crystallization [14,15]. Phase change only depends on temperature; therefore, PCM is an excellent candidate for the thermoregulation of asphalt pavement and the reduction of temperature–related damages. PCM has various advantages, including high energy storage density, approximate isothermal process, a broad selection range of phase transition temperature, easy to control, and large heat–storage capacity [14,16].

The PCM is divided into different groupings based on the nature of the materials, such as organic, inorganic, and eutectic. Fatty acids, glycerine, phenol, caprylone, and other non–paraffin PCM subgroups are among the organic materials that are further classified as paraffin and non–paraffin. The inorganic materials are further classified as salt hydrate and metallics.

Direct incorporation [17], micro–encapsulate, macro–encapsulate [18,19], and aggregate impregnation methods are some of the PCM application methods that could be employed as modifiers for asphalt or fillers for asphalt mixtures. Each of these approaches has its advantages, as well as obvious limitations. Although direct incorporation is easy to carry out, the chemical compositions change significantly [17,19–22]. Although the micro–encapsulation method could protect PCM leaks from micro–capsules, the heat–transmission efficiency is insufficient to meet the requirements of temperature regulation of asphalt pavement [23–25]. The aggregate impregnation method could provide adequate protection from leakage, but its absorption capacities and temperature regulation performance are limited [26–29]. Challenges and solutions associated with the incorporation of PCM in asphalt or asphalt mixtures must be considered, and more research is required to find a reasonable approach to incorporate PCM in asphalt or asphalt mixtures.

Direct contact with asphalt or PCM leakage from micro–encapsulation would increase the aromatic and saturate fractions of asphalt and affect the colloidal structure of the asphalt binder [17,22], which would be of no benefit for storing or releasing energy in asphalts. The direct interaction of PCM with asphalt, as well as its impact on the physical and rheological properties of asphalt, must be investigated. Kakar et al. [30] studied the effect of asphalt's direct interaction with tetradecane on the properties of asphalt and found that the thermal energy released by PCM crystallization during cooling improved the rheological properties. Du et al. [31] investigated the effect of polyethylene glycol on direct interactions with asphalt. Tan et al. [22] found that the physical properties of asphalt that had been directly affected by organic phase change materials had been seriously impacted. The properties of asphalt modified with organic PCM cannot match the standards [32]. Zhang et al. [15] demonstrated that PCM enhances the rutting resistance of asphalt at high temperatures while also increasing the risk of fatigue and cracking.

Shape–stabilized technology, such as shape–stabilization [15], micro–encapsulation [33], and macro–encapsulation [34] is used to create composite PCM, which could withstand (a) the external force during mixing, paving, and compacting activities of asphalt pavement, and (b) high temperature to prevent PCM leakage. Zhang et al. [35] fabricated shape–stabilized composite PCM to conquer PCM leakage. Ma et al. [36] used an organic phase separation model to generate composite–modified PCMs with ethyl cellulose as the wall material and modified PCM as the core material. Composite shape–stabilized PCM was blended in asphalt mixtures to prevent the direct interaction of PCM with asphalt. Solid–solid PCM show clear advantages over solid–liquid PCM, including no liquid leakage problem, adjustable temperature region, good thermal stability, excellent durability, and high storage density [37–40].

Most research has focused on the application of PCM in asphalt mixture, with little attention paid to the impact of PCM on the properties of the asphalt binder. Wei [41,42], Zhang [15], Bueno [43] investigated the physical properties, rheological properties, and temperature adjustable performance of asphalt modified with polyurethane solid–solid PCM, expanded graphite (EG)/polyethylene glycol (PEG) composite PCM, and micro–encapsulated tetradecane PCM, respectively.

This paper deals with the influence of solid–solid composite shape–stabilized PCM, which are directly added to asphalt, on the properties of matrix and SBS–modified asphalt.

The investigated properties include physical properties (e.g., penetration, softening point, ductility, rotational viscosity) before and after aging, and rheological properties (e.g., high− and low−temperature rheological properties) before and after aging. Moreover, the morphology and reaction between PCM and asphalt will also be examined. The feasibility of the method of directly incorporating solid–solid composite shape−stabilized PCM in asphalt should be verified. This work will provide an insight into how solid−solid PCM affects the physical and rheological properties of asphalt before and after aging and attempts to attain a selection principle for the implication of PCM materials in asphalt pavement.

2. Materials and Methods

2.1. Materials

Matrix asphalt and SBS−modified asphalt were tested in this paper; the physical properties are listed in Table 1.

Table 1. The physical parameters matrix and SBS−modified asphalt.

Index	Matrix Asphalt	SBS−Modified Asphalt
Penetration (0.1 mm)	69.4	65.9
Softening Point (°C)	51.5	46.7
Ductility (cm)	64	93
Rotary Viscosity (135 °C, MPa·s)	0.881	2.797

Solid−solid composite shape−stabilized phase change materials (namely PCM) used in this paper were provided by a commercial company, the physical and chemical parameters of PCM are listed in Table 2.

Table 2. The physical and chemical parameters of solid−solid phase change material (PCM).

Parameter	Test Result
Apparent density (kg/m^3)	840
Latent heat value (Melting enthalpy value) (J/g)	67
Phase change point (°C)	17
Phase change interval (°C)	0–36
Phase change type	solid−solid
The decay rate of latent heat value (Melting enthalpy value) after 20 times phase change (%)	2

To investigate the influence of PCM on the physical and rheological properties of matrix and SBS−modified asphalt, the PCM was added to the matrix and SBS−modified, and the concentration of PCM added was 4% and 8% by the weight of the asphalt. Before testing, different contents of PCM were blended with asphalt in the laboratory. The blending processes were conducted using a high−speed mixer at 3000 rpm for 60 min. For matrix and SBS−modified asphalt, the blending temperatures were 140 °C and 170 °C, respectively.

2.2. Methods

2.2.1. Physical Property Test

Before modification, the physical properties of the control pure asphalt binders were tested to provide a basis to understand the impacts of PCM on asphalt binder properties.

The physical properties (e.g., penetration degree, softening point, ductility, rotational viscosity) of asphalts with various contents of PCM were tested according to the requirements of the Standard Test Methods of Bitumen and Bituminous Mixtures for Highway Engineering (JTG E20−2011) T0604, T0605, T0606, T0620, T0625 [44]. The physical properties are related to the high−temperature and low−temperature performance of asphalt.

2.2.2. Pressurized Aging Vessel (PAV) Test

Unmodified and PCM modified asphalts were subjected to the pressurized aging vessel (PAV) test following the Standard Test Methods of Bitumen and Bituminous Mixtures for Highway Engineering (JTG E20−2011) T0630 [44] to verify the influence of aging. The physical properties of asphalts were examined after artificial aging procedures.

2.2.3. Fluorescence Microscope

The phase composition and morphology were important for confirming the fusion status of the asphalt and PCM. The fluorescence microscope (LW2000C−LY, CEWEI, Shanghai, China) was employed in this investigation to observe the distribution of PCM in the asphalt immediately after modification.

2.2.4. Fourier Transform Infrared Spectroscopy (FTIR)

Fourier transform infrared spectroscopy (FTIR, Bruker TENSO II, Bruker, Karlsruhe, Germany) with wavenumbers ranging from 400 to 4000 cm^{-1} and KBr as the dispersing phase was used in this paper, the functional groups of the control asphalts and PCM−modified asphalts were analyzed to verify whether the PCM reacts with asphalt.

2.2.5. Rheological Property Tests

Dynamic shear rheometers (DSR, TA, New Castle, DE, USA) were used to examine the high−temperature rheological properties of the control asphalts and asphalts with various contents of PCM. The temperature sweep in DSR was measured in the strain−controlled mode at a constant frequency (10 rad/s). The frequency sweep test was performed at a constant strain of 9%. For the matrix asphalt, test temperatures ranged from 40 °C to 76 °C with 12 °C intervals. For the SBS−modified asphalt, temperatures ranged from 52 °C to 88 °C with 12 °C intervals. The frequency range was from 0.1–10 Hz, which corresponded to the traditional vehicle speed. To simulate short−term aging, the asphalt samples used in the DSR test were artificially aged by using the Rolling Thin Film Oven test (RTFOT).

The creep behaviors of the asphalt samples at the low temperature were investigated using a bending beam rheometer (BBR, CANNON, Centre County, PA, USA) test, with temperatures of −6 °C, −12 °C, and −18 °C. The Rolling Thin Film Oven test (RTFOT) was utilized to artificially age the asphalt samples used in the BBR test, and the RTFOT residues were subjected to the pressure aging vessel (PAV) to simulate the long−term aging of the asphalt samples.

The Standard Test Methods of Bitumen and Bituminous Mixtures for Highway Engineering (JTG E20−2011) T0628, T0627 [44] were used to conduct the rheological property testing.

3. Results

3.1. Physical Property Test Results

The effects of PCM on asphalt penetration, softening point, ductility, and rotary viscosity are shown in Figure 1.

As seen in Figure 1a, the penetration at 25 °C increased after the addition of PCM, but gradually reduced as the PCM content increased. When the PCM content was 4 wt% and 8 wt%, penetration values increased for the matrix asphalt by 16.1% and 12.2% and for the SBS−modified asphalt by 27.6% and −4.6%, respectively. The addition of PCM had a greater effect on the penetration of SBS−modified asphalt than matrix asphalt.

An asphalt binder with a lower penetration degree is more resistant to rutting in high temperatures. Comparing the penetration values in Figure 1a, all combinations had a higher penetration degree except SBS−modified asphalt with 8 wt% PCM. Furthermore, 4 wt% PCM had a stronger effect on increasing the penetration degree than 8 wt%. The addition of PCM could weaken the matrix asphalt's high−temperature resistance.

It can be noted from Figure 1b that with the addition of PCM, the softening point of matrix asphalts reduced slightly but increased as the PCM content increased. The softening point increased by 5.7 °C. For SBS−modified asphalt, the softening point increased by 7.3 °C.

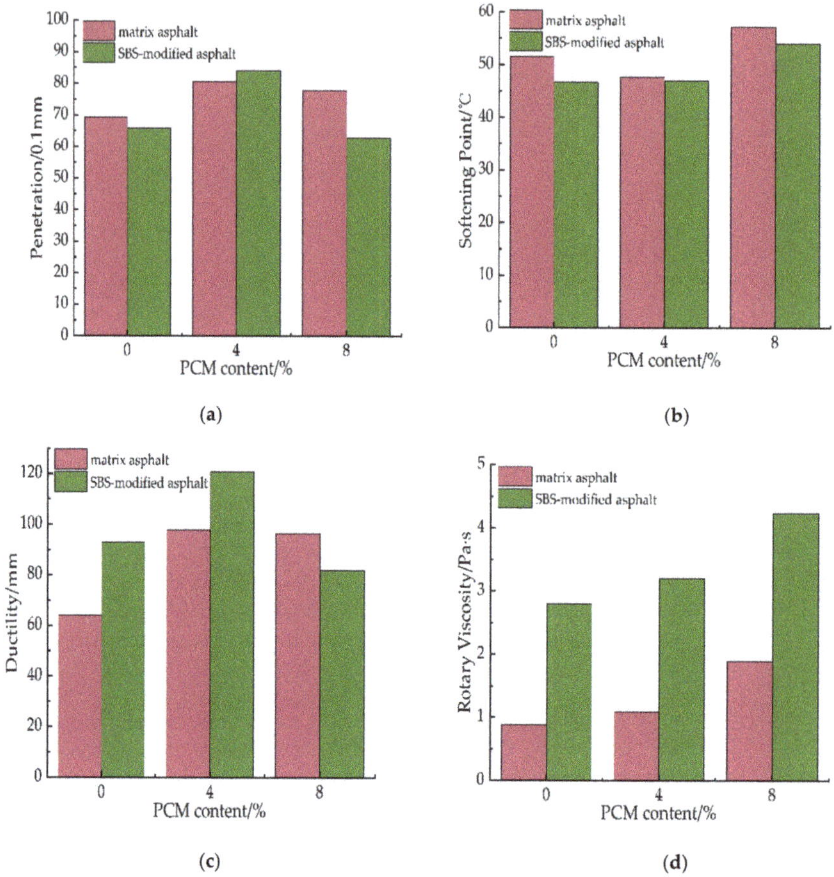

Figure 1. Physical property test results of asphalt samples with and without PCM. (**a**) penetration test results; (**b**) softening point test results; (**c**) ductility test results; (**d**) rotary viscosity test results.

The effect of PCM on the ductility was similar to the penetration index, as illustrated in Figure 1c. When the PCM addition was 4 wt%, the ductility values of the matrix asphalt (at 10 °C) and SBS−modified asphalt (at 5 °C) increased by 53.1% and 30.1%; when PCM was 8 wt%, the ductility values increased by 50.8% and −11.8%. Therefore, the high content of PCM reduces the cracking resistance of SBS−modified asphalt at low temperatures.

Figure 1d shows the rotary viscosity values of matrix and SBS−modified asphalt samples at 135 °C. As the PCM concentration increased, the rotary viscosity increased steadily. For 4 wt% and 8 wt% PCM, the increasing ratio of matrix asphalt was 23.1% and 114.5%, and for the SBS−modified asphalt, the increasing ratio was 14.4% and 51.4%, respectively.

3.2. Physical Property Test Results after Ageing

Penetration, softening point, ductility, and mass loss tests were performed to analyze the changes in the conventional physical properties of asphalts before and after PAV aging, the results are plotted in Figure 2.

During the asphalt pavement service period, the characteristics of asphalt vary with time and temperature (asphalt aging). Short−term aging during mixing and storage, as well as long−term aging throughout service, are both simulated at the laboratory scale employing thermal and oxidative treatment. In this study, the RTFOT residues were subjected to a pressure aging vessel (PAV) to simulate the long−term aging of asphalt.

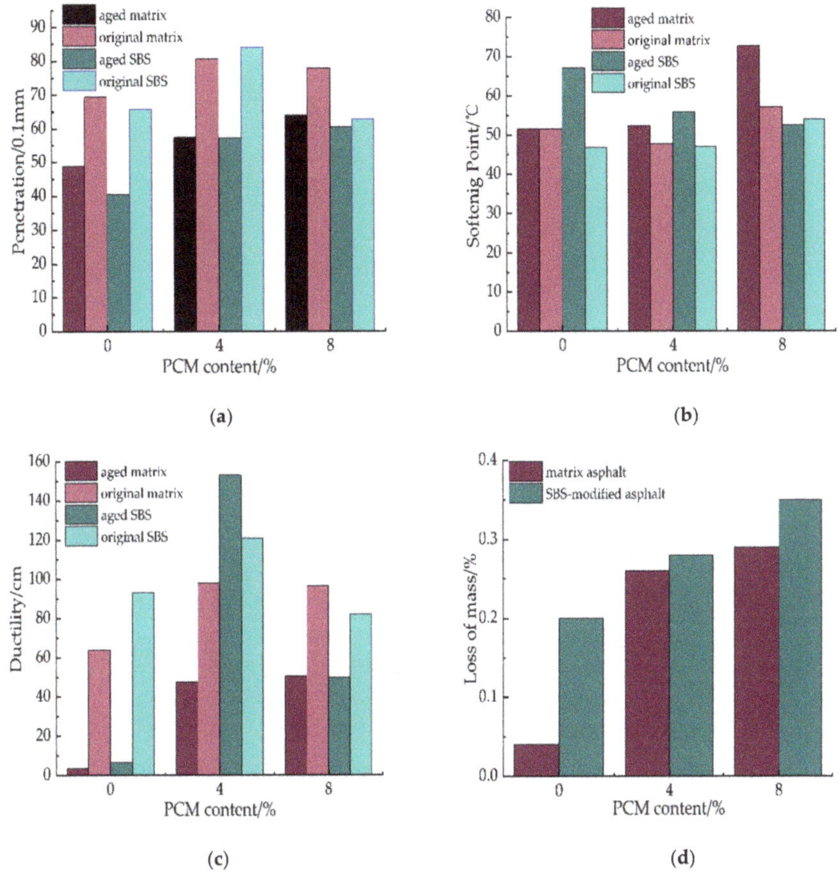

Figure 2. Physical property test results of asphalt samples before and after PAV aging. (**a**) penetration test results; (**b**) softening point test results; (**c**) ductility test results; (**d**) loss of mass test results.

Figure 2a illustrates how the penetration values of the asphalt samples before and after modification reduced after the PAV aging process. The decreasing trend in penetration values before and after PAV aging was reduced with the addition of PCM to asphalt, implying that adding PCM could postpone the effect of aging on the penetration index.

The softening point of asphalt with various contents of PCM before and after PAV aging increased, except for SBS−modified asphalt with 8% PCM, as shown in Figure 2b. However, the effect ratios for matrix and SBS−modified asphalt with varied PCM concentrations were incompatible. As PCM content grew, the difference between the matrix asphalt samples' softening points before and after PAV aging increased.

From Figure 2c, it can be noticed that, except for SBS−modified asphalt with 4 wt% PCM, asphalt samples before PAV aging had higher ductility. In comparison to the original asphalt samples, adding PCM potentially improved the cracking resistance after aging.

The loss of mass values after the PAV aging grew following the PCM content increasing, as shown in Figure 2d. The growth was not substantial, indicating that asphalts mixed with PCM had good thermal stability.

3.3. *Morphology of the Virgin Asphalt and PCM-Modified Asphalt*

Figure 3 displayed fluorescence images (magnified by 400 times) of asphalt samples with varying PCM content. Fluorescence images were used to investigate the distribution of PCM in asphalt.

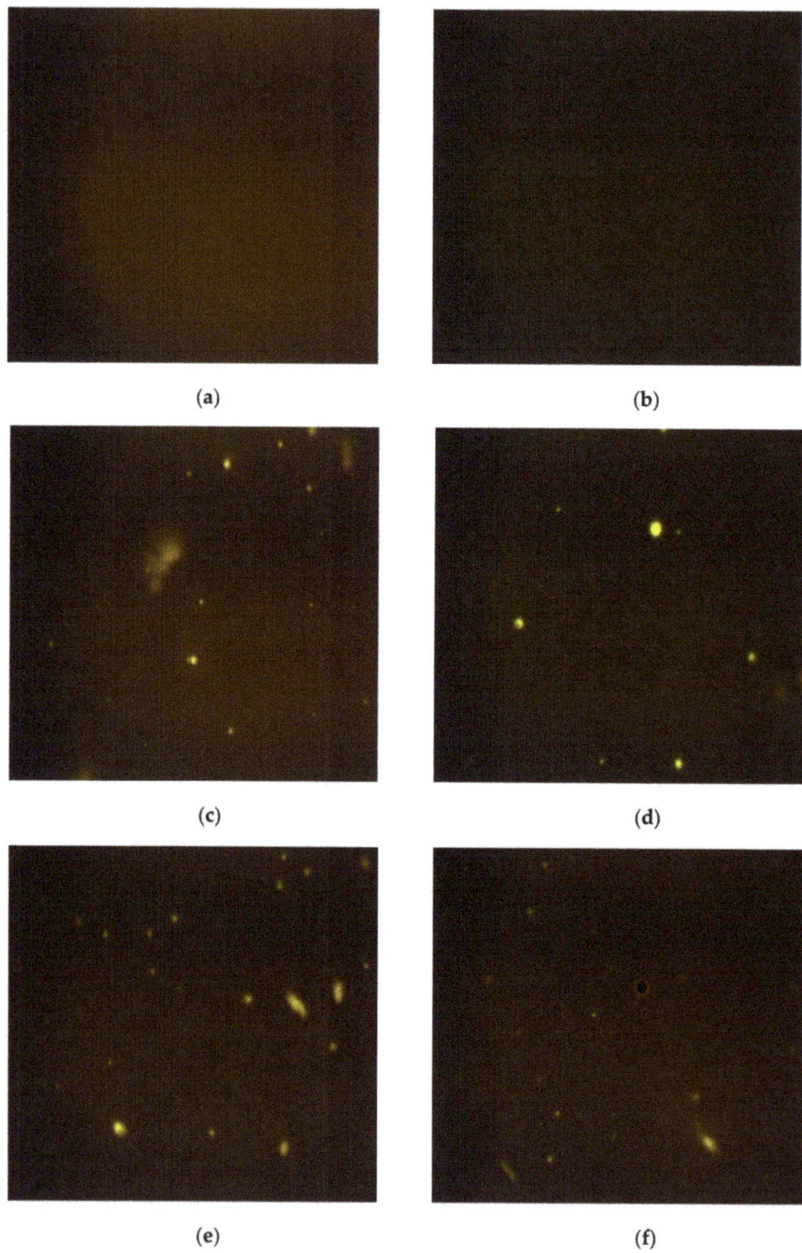

Figure 3. The morphology of asphalt samples before and after modification (magnified by 400 times): (**a**) pure matrix asphalt; (**b**) pure SBS−modified asphalt; (**c**) matrix asphalt with 4 wt% PCM; (**d**) SBS−modified asphalt with 4 wt% PCM; (**e**) matrix asphalt with 8 wt% PCM; (**f**) SBS−modified asphalt with 8 wt% PCM.

The virgin asphalt exhibits a flat and smooth fracture surface with no gully or layer structure, as illustrated in Figure 3a,b. It can be found that PCM was randomly distributed in the form of very small particles and these particles were always separated. When the PEG content increased, the particle number was multiplied. The phase interface between PCM particles and asphalt indicated that the PCM particles did not react with the asphalt.

3.4. FTIR Analysis

The FTIR (Bruker TENSO II, Bruker, Karlsruhe, Germany) analysis was used to determine the chemical changes in the asphalt samples. Each specific chemical bond can be defined by its specific wave−number range (cm^{-1}).

The FTIR spectra of matrix and SBS−modified asphalt samples before and after modification are depicted in Figure 4. The infrared spectrum of asphalt samples with different PCM content was highly similar to that of pure asphalt. In the spectra of asphalt samples with 4 wt% and 8 wt% of PCM, a new absorption peak at 1737 cm^{-1} (−C=O functional group) was found. The absorption peak increased as the PCM content increased, implying that the absorption peak occurred as a result of the addition of PCM. It could be attributed to the residual PCM that remained on the surface after the PCM microcapsules were prepared.

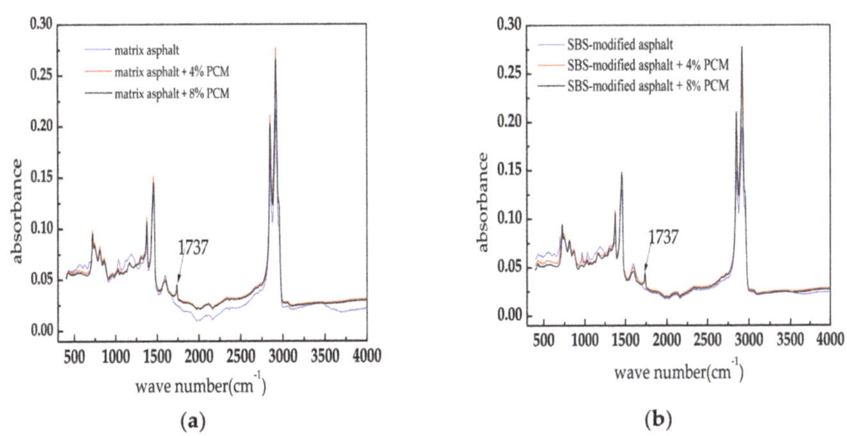

Figure 4. FTIR spectra of asphalt samples before and after modification. (**a**) matrix asphalt samples; (**b**) pure SBS−modified asphalt samples.

3.5. Effect of Aging and PCM Modification on Rheological Properties of Asphalt

3.5.1. High−Temperature Rheological Properties

High−Temperature Rheological Properties of Original Asphalt Samples

The frequency sweep test, the complex shear modulus, phase angle, and corresponding rutting factor are plotted in Figure 5.

It is shown in Figure 5a, that the complex shear modulus of the unaged matrix asphalt samples increased as the loading frequencies increased. The addition of PCM did not affect this trend. Under 40 °C, the complex shear modulus of the matrix asphalt with no PCM was the largest. When the test temperature was raised to 52 °C, the complex shear modulus of the matrix asphalt with no PCM and 8 wt% PCM was close and the complex shear modulus at 4 wt% was the smallest. The matrix asphalt with 8 wt% PCM had a higher complex shear modulus when the test temperature was raised to 64 °C and 76 °C. The differences between the three kinds of matrix asphalt samples grew as the test temperature was raised.

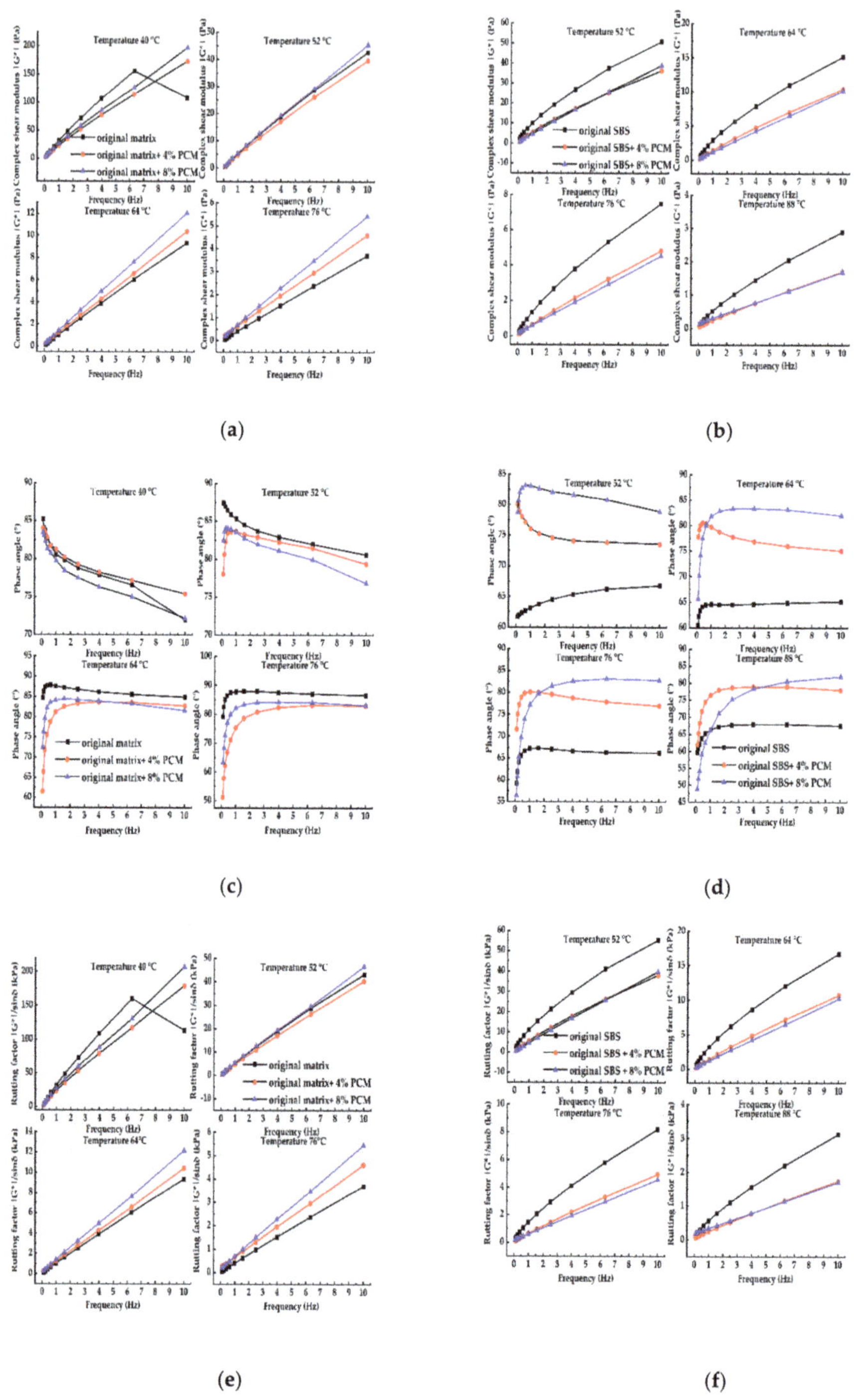

Figure 5. DSR frequency sweep test results of original asphalt samples with PCM under different test temperatures. (**a**) complex modulus of matrix asphalt samples; (**b**) complex modulus of SBS−modified asphalt samples; (**c**) phase angle of matrix asphalt samples; (**d**) phase angle of SBS−modified asphalt samples; (**e**) rutting factor of matrix asphalt samples; (**f**) rutting factor of SBS−modified asphalt samples.

The matrix asphalt without PCM had the highest complex shear modulus when the test temperature was lower. With the test temperature raised, the effect of PCM on the complex shear modulus became more significant, indicating that the solid–solid PCM started to absorb the heat. The inner heat was absorbed by the added PCM, which enhanced the complex shear modulus. The amount of PCM added had a significant impact on the complex shear modulus.

It can also be noted, from Figure 5b, that the complex shear modulus of SBS–modified asphalt becomes bigger with PCM added. The amount of PCM added did not affect the complex shear modulus of the matrix asphalt with the raising of the test temperature.

According to Figure 5c, the phase angle of the matrix asphalt with various PCM concentrations increased with the raising of the test temperature. At low loading frequencies, the phase angle of the matrix asphalt with PCM rose dramatically.

The effect of adding PCM on the phase angle of the SBS–modified asphalt was more complex than that of the SBS–modified asphalt without PCM, as shown in Figure 5d.

In comparison to the complex shear modulus, the rutting factors of the original matrix and SBS–modified asphalt with varied PCM content exhibited comparable trends, as seen in Figure 5e,f.

High–Temperature Rheological Properties of Aged Asphalt Samples

Figure 6a shows that the complex shear modulus of the aged matrix asphalt with 8 wt% PCM was the highest. The difference between the complex shear modulus of the aged matrix asphalt with 4 wt% and with 0 wt% PCM becomes insignificant with the test temperature raised. While the gap of the complex shear modulus between the aged matrix asphalt with 8 wt% PCM and other asphalts grows. The complex shear modulus of the aged matrix asphalt with various PCM concentrations was significantly higher than the unaged.

As can be observed from Figure 6b, under test temperatures of 52 °C and 64 °C, the complex shear modulus difference between aged the SBS–modified asphalt with 8 wt% and with 0 wt% PCM was insignificant. However, when the test temperature was raised to 76 °C and 88 °C, the complex shear modulus of the aged SBS–modified asphalt without PCM became the highest. The complex shear modulus of the aged SBS–modified asphalt with various PCM concentrations was significantly higher than the unaged.

The RTFOT aging process had a greater impact on the phase angle of the matrix and SBS–modified asphalt with 8 wt% PCM, as seen in Figure 6c,d.

In comparison to the complex shear modulus, the rutting factors of the aged matrix and SBS–modified asphalt with varied PCM content exhibited similar trends. Rutting factors under all test temperatures were bigger than that of original asphalt samples, as shown in Figure 6e,f.

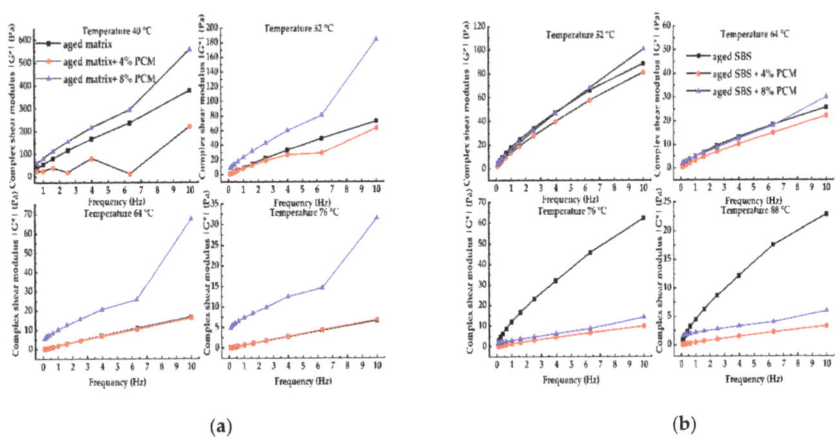

Figure 6. *Cont.*

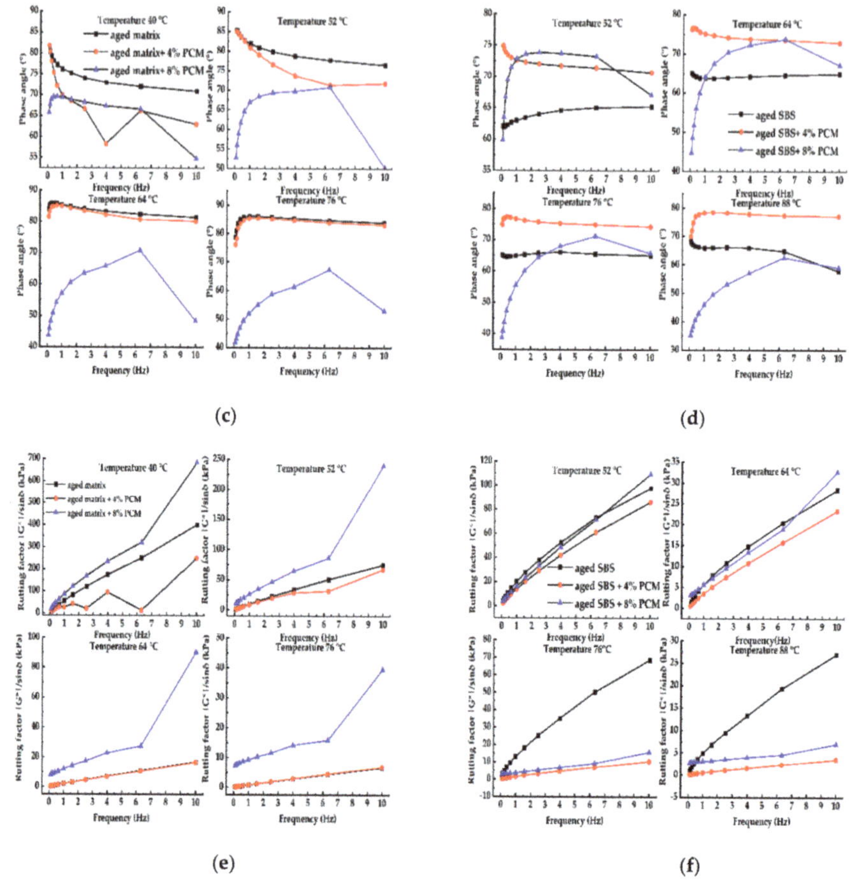

Figure 6. DSR frequency sweep test results of RTFOT aged asphalt samples with PCM under different test temperatures. (**a**) complex modulus of matrix asphalt samples; (**b**) complex modulus of SBS–modified asphalt samples; (**c**) phase angle of matrix asphalt samples; (**d**) phase angle of SBS–modified asphalt samples; (**e**) rutting factor of matrix asphalt samples; (**f**) rutting factor of SBS–modified asphalt samples.

3.5.2. Low–Temperature Rheological Properties

Figure 7 depicts the effect of PCM on the creep stiffness (donated as St) and creep curve slope (donated as m–value) of the matrix and SBS–modified asphalt with different PCM content.

It can be seen from Figure 7a, that the St of the matrix asphalt decreased as the PCM content and test temperature increased. The addition of PCM would improve the matrix asphalt's low–temperature crack resistance.

Adding PCM to the SBS–modified asphalt reduced St at all test temperatures, with 4 wt% PCM having the greatest effect, as shown in Figure 7b. Low–temperature resistance was reduced in the SBS–modified asphalt with PCM.

It can be observed, from Figure 7c,d, that the increase in PCM did not affect the m–value. Only 4 wt% PCM could increase the m–value when compared to asphalts without PCM.

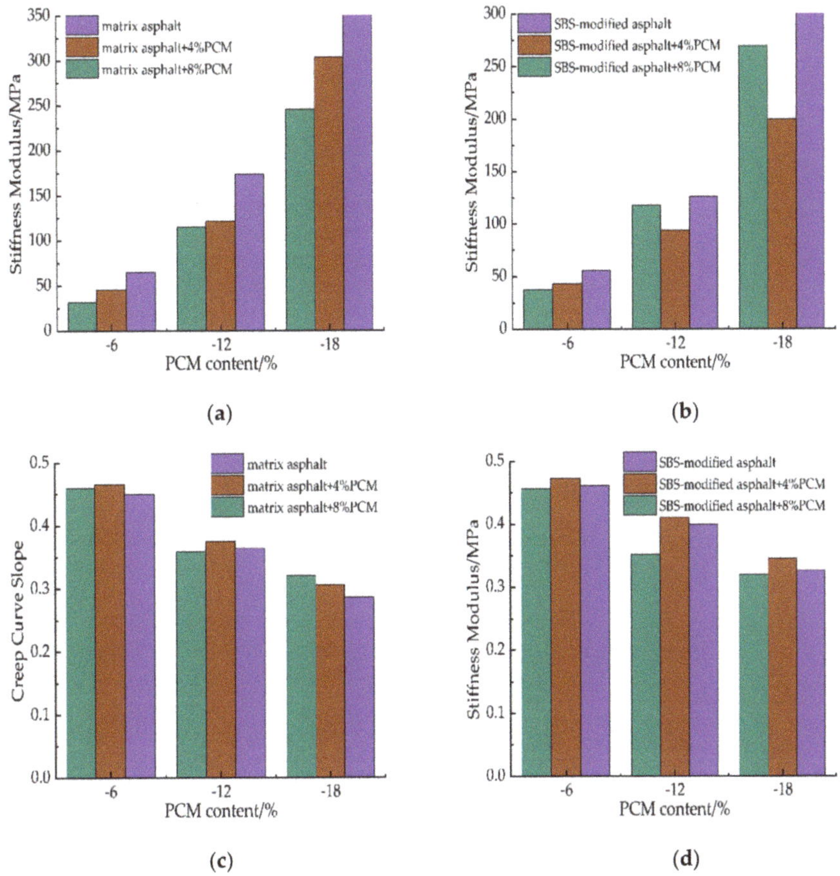

Figure 7. BBR test results of asphalt samples with PCM under different test temperatures. (**a**) stiffness modulus of matrix asphalt samples; (**b**) stiffness modulus of SBS−modified asphalt samples; (**c**) creep curve slope of matrix asphalt samples; (**d**) creep curve slope of SBS−modified asphalt samples.

4. Discussion

4.1. Effects of PCM Modification on Physical Properties of Asphalt

Previous research suggests that the incorporation of PCM in asphalt is not a feasible method for PCM application [22]. The impact of PCM on asphalt must be investigated as PCM particles would inevitably have contact with the asphalt binder during mixing and paving activities.

At high temperatures, asphalt with a lower penetration degree or a greater softening point is more resistant to rutting. Considering the penetration and softening point values in Figure 1, the rutting resistance at high temperatures was reduced with 4 wt% PCM. The effects of 8 wt% PCM on penetration and softening point values were adverse, indicating that the addition of PCM at different contents had a complex impact on the sensitivity to high temperature when compared to neat asphalt. Kakar et al. [30] indicate that the direct integration of PCM (tetradecane, Tm 6C) enhances the penetration, which is consistent with the results of Figure 1a. This issue could be due to the leakage of PCM microcapsules. During the preparation of PCM−modified asphalt, a limited number of micro−capsules may break due to high temperature and external force, thereby weakening the bitumen binder and increasing penetration.

The PCM went through a phase change during the heating process, which resulted in a higher softening point. These findings were in line with what was expected [42]. However, the softening point was slightly enhanced due to the effective micro−capsule protection against PCM.

The asphalt with higher ductility is more resistant to cracking at low temperatures. All combinations, except for the SBS−modified asphalt with 8 wt% PCM, were less sensitive to low temperatures than the neat asphalt binders in terms of ductility.

The workability of HMA is highly dependent on the viscosity of asphalt at high temperatures. Regarding the rotary viscosity values in Figure 1, the viscosity of PCM−modified asphalt rose as the PCM content increased. Furthermore, the rotary viscosity values of the matrix asphalt modified with PCM were smaller than the maximum acceptable (3 Pa·s). The rotary viscosity values of the SBS−modified asphalt mixed with various concentrations of PCM exceeded the maximum acceptable (3 Pa·s). Moreover, mixing PCM with SBS−modified asphalt increases asphalt viscosity, which indicates that mixing and compaction temperatures should be increased. It is a challenge to combine PCM with SBS−modified asphalt.

4.2. Aging of PCM−Modified Asphalt

Because of time, temperature, and ultraviolet rays, the asphalt binder ages during service life, resulting in changes in the physical behavior (e.g., stiffening) of asphalt pavement. The asphalt aging process is simulated in the laboratory for (a) short−term aging during mixing and storage; (b) the long−term aging during service life by applying thermal and oxidative treatment. It can be seen that adding PCM to asphalt reduces penetration, increases the softening point (except in SBS−modified asphalt with 8 wt% PCM), decreases ductility (except in SBS−modified asphalt with 4 wt% PCM), and increases the loss of mass.

When compared to the unaged asphalt samples, the high−temperature performance of the matrix and SBS−modified asphalt with PCM added was improved, while the low−temperature performance was lowered. In conclusion, adding PCM to asphalt is beneficial to the stability of asphalt at high temperatures, as well as its aging resistance and durability.

4.3. Morphology of the Original Asphalt and PCM−Modified Asphalt

It is noteworthy that the PCM was blended with asphalt in the form of small particles based on the fluorescence image results of the matrix and SBS−modified asphalt with varying PCM concentrations. PCM particles were visible, as was the interface between PCM and asphalt, indicating poor compatibility. When the PCM content was 4 wt%, phase separation emerged. The poor compatibility of PCM and asphalt would damage the durability and reliability of as−prepared blends, as well as prevent the use of higher PCM content.

When the PCM content reached 8 wt%, the phase separation between PCM and SBS−modified asphalt became lower than that between PCM and matrix asphalt (see Figure 3). This phenomenon could be explained by the styrene−butadiene−styrene network generated in the SBS−modified asphalt, which can settle the PCM particles, and so prevent phase segregation to some extent.

According to the test results, there was an obvious phase interface between PCM particles and asphalt. The PCM particles were in a dispersed phase. There was no dissolution reaction between PCM particles and asphalt. The spatial mesh structure of the co−blending system in the SBS−modified asphalt was not affected by the addition of PCM.

4.4. FTIR Analysis

In the spectra of asphalts modified with PCM, only a new absorption peak at 1737 cm^{-1} belonging to the stretching vibration peaks of the −C=O functional group was discovered. According to FTIR analysis, other peaks revealed the specific absorption characteristic of the matrix and SBS−modified asphalt. This new peak was attributed to the typical absorption of residual PCM. It can be concluded that no chemical reactions occurred between the asphalt binders and additives. The PCM was physically integrated into the asphalt without causing a chemical reaction.

4.5. Effect of Aging and PCM Modification on Rheological Properties of Asphalt

4.5.1. High−Temperature Rheological Properties

Frequency sweep tests were performed at four different test temperatures to evaluate the PCM phase state effect on the viscosity components of the asphalt. High values of complex shear modulus (G*) and low values of phase angle (δ) are desirable for improving asphalt fatigue resistance. Asphalts with a bigger phase angle (δ) are more sensitive to viscous deformation than asphalts with the same complex shear modulus (G*). According to Superpave specifications, G*/sin(δ) is a rutting factor that represents a measure of asphalt binder rutting resistance [45].

According to the results obtained from the high−temperature rheological properties analysis, the rheological measurements show a combined effect of the steep drop in complex modulus (G*) and increase in phase angle (δ) with temperature change. The phase angle (δ) did not increase consistently with the decrease in complex shear modulus (G*). At the high temperature, the phase angle (δ) would grow at low frequency and then reduce at high frequency. At high temperatures, the phase angle (δ) did not change much. This is a complicated topic that is partly explained by the phase state of PCM and its impact on the rheological properties of asphalt.

The complex shear modulus (G*) of the matrix asphalt modified with PCM reduced as the PCM content increased in the whole sweep frequency when the test temperature was less than 52 °C; when the test temperature was greater than 52 °C, the trend was reversed. As a result, integrating PCM into asphalt was beneficial for its complex shear modulus (G*) at high temperatures. Furthermore, the rutting factor (G*/sin(δ)) of the matrix asphalt with various PCM content exhibited remarkably comparable variation trends with the complex shear modulus (G*), implying that PCM may improve the high−temperature deformation resistance of matrix asphalt. Adding PCM to SBS−modified asphalt did not improve its rutting resistance under all contents and temperatures. SBS−modified with varied PCM content showed no phase change behavior at test temperatures above 52 °C.

The RTFOT aging process altered the rheological properties of asphalt with PCM added. The rutting factor (G*/sin(δ)) of asphalt with variable PCM concentrations was higher after aging than that of unaged, indicating that the aging process may improve the resisting deformation at high temperatures.

After aging, the matrix asphalt with 8 wt% PCM had the highest complex shear modulus (G*) under all test temperatures and loading frequency. As the temperature rose, the complex shear modulus (G*) of asphalt with 4 wt% and no PCM are simliar. At 52 °C and 64 °C, the complex shear modulus (G*) and rutting factor (G*/sin(δ)) of aged SBS−modified asphalt with 8 wt% and no PCM were close and slightly higher than those of aged SBS−modified asphalt with 4 wt% PCM. The complex shear modulus (G*) and rutting factor (G*/sin(δ)) of aged SBS−modified asphalt with PCM added decreased significantly with an increase in temperature and were subsequently smaller than those of SBS−modified asphalt without PCM. According to the rheological analysis of aged asphalt, the high content of PCM improves matrix asphalt's high−temperature deformation resistance. Adding PCM could reduce rutting resistance at high temperatures.

By combining variations of the complex shear modulus (G*) and phase angle (δ) at different temperatures, it is difficult to explicitly present the influence of PCM addition on the high−temperature rheological property of asphalt.

4.5.2. Low−Temperature Rheological Properties

BBR tests were conducted at low temperatures to analyze the creep behavior of asphalts with various PCM concentrations. The St and m−value indices were used in BBR testing to assess creep behavior. The ability of asphalt to withstand a constant load is measured in St, and asphalts with high St values are more brittle and prone to cracking. The m−value depicts how the stiffness of the asphalt changes as the loads are applied. The binder stiffness changes quickly with a high m−value, the accumulated thermal stress in asphalt disperses, and low−temperature cracking is prevented.

With the increase in PCM content, the creep stiffness modulus reduced, and the influence rate of different PCM content under all test temperatures was different, indicating that the addition was beneficial for asphalt's low temperature cracking resistance. Only the St of the matrix and SBS−modified asphalt without PCM failed to meet the specification of 300 MPa under all test temperatures, as shown in Figure 7.

The m−value of asphalts with varying PCM concentrations differed marginally from that of asphalt with no PCM added at all test temperatures. At all test temperatures, only the m−values of the matrix and SBS−modified asphalt without PCM were less than 0.3.

There is always a conflict between St and m−values during the evaluation of the low−temperature creep behavior of asphalt. Several researchers have chosen the St/m ratio as a new measurement for analyzing low temperature creep behavior. The St/m ratio is the ratio of the St and m−value at 60 s of test time [46]. A low St/m value is desired. Table 3 indicates that adding PCM to asphalt can improve its resistance to low−temperature cracking.

Table 3. The results of St/m at 60 s of test time.

Temperature/°C	Matrix	Matrix + 4% PCM	Matrix + 8% PCM	SBS	SBS + 4% PCM	SBS + 8% PCM
−6	144.3	98.2	69.6	120.1	90.7	81.7
−12	478.7	322.7	318.9	315.8	228.9	335.2
−18	1250.4	991.8	764.8	964.7	578.3	843.5

5. Conclusions

PCM is wildly used to regulate the inner temperature of pavement to prevent high− and low−temperature damage. PCM could be classified based on the phase change mechanism, e.g., solid to solid, solid to liquid. In this paper, the effect of solid−solid composite shape−stabilized PCM on the characteristics of matrix and SBS−modified asphalt was investigated. The physical property test, PAV test, Fluorescence Microscope test, FTIR test, and rheological test were employed to confirm the feasibility of directly incorporating solid−solid composite shape−stabilized PCM into the asphalt.

(1) The PCM was physically blended with asphalt in the form of small particles without chemical reaction. Phase separation was obvious. The compatibility between PCM and asphalt was poor;

(2) The addition of PCM could improve the physical properties of asphalt, the PCM concentrations had different influence degrees. The viscosity of SBS-modified asphalt including PCM cannot meet the requirements of the specification. After aging, the high−temperature performance was improved, and low−temperature performance was decreased compared to the asphalt without PCM;

(3) PCM could enhance the high−temperature rutting resistance of the matrix asphalt at a temperature higher than 52 °C at high concentration and reduce the high−temperature rutting resistance of the SBS−modified asphalt to some extent. The high−temperature rutting performance of HMA with PCM compared with HMA without PCM should be evaluated. The procedure of RTOF aging affects the high−temperature rutting resistance of asphalt, especially when PCM concentration is high;

(4) The low−temperature creep behavior and the PG grade of asphalt could be enhanced with the addition of PCM. The St/m ratio could be adopted to evaluate the influence of PCM on the low−temperature cracking resistance of asphalt.

The test results of this study lead to the conclusion that directly incorporating solid−solid composite shape−stabilized PCM into asphalt could alter the physical and rheological properties. The physical property of asphalt is insufficient to assess the impact of PCM, the rheological test should be adopted for evaluation. It is not a feasible method to directly incorporate PCM into asphalt; hence, the performance of HMA with PCM added should be carefully verified compared with HMA without PCM.

According to this study, the implicating criteria of PCM is that the addition of PCM does not cause deleterious effects on the road performance of the asphalt or asphalt mixture as it enhances the thermoregulation of the asphalt mixture. The addition of PCM cannot increase asphalt pavement performance without incurring the cost of scarifying the performance of the asphalt pavement. The influence of PCM on the physical and rheological characteristics of asphalt or asphalt mixture should be investigated and analyzed before application. The desired application approaches of PCM without scarifying the mechanical performance of asphalt or asphalt mixture should be explored.

In this paper, the effect of solid−solid PCM of various concentrations on asphalt was investigated. The method of using solid−solid composite shape−stabilized PCM in asphalt mixture, e.g., PCM blend with aggregates or filler, could be researched further. The effect of solid−solid PCM on the performance of asphalt mixture will be fully investigated in future studies.

Author Contributions: Conceptualization, H.Z. (Haisheng Zhao); methodology, S.M.; software, X.W.; validation, H.Z. (Haisheng Zhao), X.W., H.Z. (Huan Zhang) and J.W.; formal analysis, H.Z. (Haisheng Zhao) and Z.L.; investigation, H.Z. (Haisheng Zhao), Z.L. and S.C.; resources, J.G.; data curation, H.Z. (Haisheng Zhao); writing—original draft preparation, H.Z. (Haisheng Zhao); writing—review and editing, S.M.; visualization, H.Z. (Haisheng Zhao), H.Z. (Huan Zhang), C.S., J.W. and S.C.; supervision, S.M.; project administration, J.G. and S.M.; funding acquisition, J.G. All authors have read and agreed to the published version of the manuscript.

Funding: This research received no external funding.

Institutional Review Board Statement: Not applicable.

Informed Consent Statement: Not applicable.

Data Availability Statement: Not applicable.

Acknowledgments: We thank Zhaodi Yuan and Zhaojie Zhang for their assistance with experiments and valuable discussion.

Conflicts of Interest: The authors declare no conflict of interest.

References

1. Mondal, S. Phase change materials for smart textiles—An overview. *Appl. Therm. Eng.* **2008**, *28*, 1536–1550. [CrossRef]
2. Mitchell, M.R.; Link, R.E.; Amirkhanian, A.N.; Xiao, F.; Amirkhanian, S.N. Evaluation of High Temperature Rheological Characteristics of Asphalt Binder with Carbon Nano Particles. *J. Test. Eval.* **2011**, *39*, JTE103133. [CrossRef]
3. Behnia, B.; Buttlar, W.G.; Reis, H. Nondestructive Low-Temperature Cracking Characterization of Asphalt Materials. *J. Mater. Civ. Eng.* **2017**, *29*, 04016294. [CrossRef]
4. Huang, B.; Mohammad, L.N.; Wathugala, G.W. Application of a Temperature Dependent Viscoplastic Hierarchical Single Surface Model for Asphalt Mixtures. *J. Mater. Civ. Eng.* **2004**, *16*, 147–154. [CrossRef]
5. Zhang, C.; Wang, H.; You, Z.; Ma, B. Sensitivity analysis of longitudinal cracking on asphalt pavement using MEPDG in permafrost region. *J. Traffic Transp. Eng. (Engl. Ed.)* **2015**, *2*, 40–47. [CrossRef]
6. Mallick, R.B.; El-Korchi, T. *Pavement Engineering: Principles and Practice*, 3rd ed.; CRC Press: Boca Raton, FL, USA, 2017.
7. Lei, Y.; Wang, H.; Fini, E.H.; You, Z.; Yang, X.; Gao, J.; Dong, S.; Jiang, G. Evaluation of the effect of bio-oil on the high-temperature performance of rubber modified asphalt. *Constr. Build. Mater.* **2018**, *191*, 692–701. [CrossRef]
8. Jin, J.; Xiao, T.; Tan, Y.; Zheng, J.; Liu, R.; Qian, G.; Wei, H.; Zhang, J. Effects of TiO_2 pillared montmorillonite nanocomposites on the properties of asphalt with exhaust catalytic capacity. *J. Clean. Prod.* **2018**, *205*, 339–349. [CrossRef]
9. Jin, J.; Tan, Y.; Liu, R.; Lin, F.; Wu, Y.; Qian, G.; Wei, H.; Zheng, J. Structure characteristics of organic bentonite and the effects on rheological and ageing properties of asphalt. *Powder Technol.* **2018**, *329*, 107–114. [CrossRef]
10. Jin, J.; Tan, Y.; Liu, R.; Zheng, J.; Zhang, J. Synergy Effect of Attapulgite, Rubber, and Diatomite on Organic Montmorillonite-Modified Asphalt. *J. Mater. Civ. Eng.* **2019**, *31*, 04018388. [CrossRef]
11. Ma, B.; Si, W.; Ren, J.; Wang, H.-N.; Liu, F.-W.; Li, J. Exploration of road temperature-adjustment material in asphalt mixture. *Road Mater. Pavement Des.* **2014**, *15*, 659–673. [CrossRef]

12. Sharma, A.; Tyagi, V.V.; Chen, C.R.; Buddhi, D. Review on thermal energy storage with phase change materials and applications. *Renew. Sustain. Energy Rev.* **2009**, *13*, 318–345. [CrossRef]
13. Liston, L.; Krafcik, M.J.; Farnam, Y.; Tao, B.; Erk, K.A.; Weiss, J. Toward the use of phase change materials (PCM) in concrete pavements: Evaluation of thermal properties of PCM. In Proceedings of the 2014 FAA Worldwide Airport Technology Transfer Conference, Galloway, NJ, USA, 5–7 August 2014.
14. Guo, M.; Liang, M.C.; Jiao, Y.B.; Zhao, W.; Duan, Y.X.; Liu, H.Q. A review of phase change materials in asphalt binder and asphalt mixture. *Constr. Build. Mater.* **2020**, *258*, 119565. [CrossRef]
15. Zhang, D.; Chen, M.; Wu, S.; Riara, M.; Wan, J.; Li, Y. Thermal and rheological performance of asphalt binders modified with expanded graphite/polyethylene glycol composite phase change material (EP-CPCM). *Constr. Build. Mater.* **2019**, *194*, 83–91. [CrossRef]
16. Wei, K.; Ma, B.; Huang, X.; Xiao, Y.; Liu, H. Influence of NiTi alloy phase change heat-storage particles on thermophysical parameters, phase change heat-storage thermoregulation effect, and pavement performance of asphalt mixture. *Renew. Energy* **2019**, *141*, 431–443. [CrossRef]
17. Kong, W.; Liu, Z.; Yang, Y.; Zhou, C.; Lei, J. Preparation and characterizations of asphalt/lauric acid blends phase change materials for potential building materials. *Constr. Build. Mater.* **2017**, *152*, 568–575. [CrossRef]
18. He, L.; Li, J.; Zhou, C.; Zhu, H.; Cao, X.; Tang, B. Phase change characteristics of shape-stabilized PEG/SiO2 composites using calcium chloride-assisted and temperature-assisted sol gel methods. *Sol. Energy* **2014**, *103*, 448–455. [CrossRef]
19. Ryms, M.; Lewandowski, W.; Klugmann-Radziemska, E.; Denda, H.; Wcislo, P. The use of lightweight aggregate saturated with PCM as a temperature stabilizing material for road surfaces. *Appl. Therm. Eng.* **2015**, *81*, 313–324. [CrossRef]
20. Memon, S.A.; Cui, H.; Lo, T.Y.; Li, Q. Development of structural–functional integrated concrete with macro-encapsulated PCM for thermal energy storage. *Appl. Energy* **2015**, *150*, 245–257. [CrossRef]
21. Athukorallage, B.; Dissanayaka, T.; Senadheera, S.; James, D. Performance analysis of incorporating phase change materials in asphalt concrete pavements. *Constr. Build. Mater.* **2018**, *164*, 419–432. [CrossRef]
22. Bian, X.; Tan, Y.Q.; Lv, J.F.; Shan, L.Y. Preparation of Latent Heat Materials Used in Asphalt Pavement and Theirs' Controlling Temperature Performance. *Adv. Eng. Forum* **2012**, *5*, 322–327. [CrossRef]
23. Ren, J.; Ma, B.; Si, W.; Zhou, X.-Y.; Li, C. Preparation and analysis of composite phase change material used in asphalt mixture by sol–gel method. *Constr. Build. Mater.* **2014**, *71*, 53–62. [CrossRef]
24. Zhao, C.Y.; Zhang, G.H. Review on microencapsulated phase change materials (MEPCMs): Fabrication, characterization and applications. *Renew. Sustain. Energy Rev.* **2011**, *15*, 3813–3832. [CrossRef]
25. Jamekhorshid, A.; Sadrameli, S.M.; Farid, M. A review of microencapsulation methods of phase change materials (PCMs) as a thermal energy storage (TES) medium. *Renew. Sustain. Energy Rev.* **2014**, *31*, 531–542. [CrossRef]
26. Kheradmand, M.; Castro-Gomes, J.; Azenha, M.; Silva, P.D.; de Aguiar, J.L.B.; Zoorob, S.E. Assessing the feasibility of impregnating phase change materials in lightweight aggregate for development of thermal energy storage systems. *Constr. Build. Mater.* **2015**, *89*, 48–59. [CrossRef]
27. Manning, B.; Bender, P.R.; Cote, S.; Lewis, R.; Sakulich, A.; Mallick, R.B. Assessing the feasibility of incorporating phase change material in hot mix asphalt. *Sustain. Cities Soc.* **2015**, *19*, 11–16. [CrossRef]
28. Sakulich, A.R.; Bentz, D.P. Incorporation of phase change materials in cementitious systems via fine lightweight aggregate. *Constr. Build. Mater.* **2012**, *35*, 483–490. [CrossRef]
29. Nepomuceno, M.C.S.; Silva, P.D. Experimental evaluation of cement mortars with phase change material incorporated via lightweight expanded clay aggregate. *Constr. Build. Mater.* **2014**, *63*, 89–96. [CrossRef]
30. Kakar, M.R.; Refaa, Z.; Bueno, M.; Worlitschek, J.; Stamatiou, A.; Partl, M.N. Investigating bitumen's direct interaction with Tetradecane as potential phase change material for low temperature applications. *Road Mater. Pavement Des.* **2019**, *21*, 2356–2363. [CrossRef]
31. Du, Y.; Pu-sheng, L.; Wang, J.; Wang, H.; Hu, S.; Tian, J.; Li, Y. Laboratory investigation of phase change effect of polyethylene glycolon on asphalt binder and mixture performance. *Constr. Build. Mater.* **2019**, *212*, 1–9. [CrossRef]
32. Levy, F.L. The thermal conductivity of commercial brines and seawater in the freezing range. *Int. J. Refrig.* **1982**, *5*, 155–159. [CrossRef]
33. Kakar, M.R.; Refaa, Z.; Worlitschek, J.; Stamatiou, A.; Partl, M.; Bueno, M. Thermal and rheological characterization of bitumen modified with microencapsulated phase change materials. *Constr. Build. Mater.* **2019**, *215*, 171–179. [CrossRef]
34. Zhou, X.; Kastiukas, G.; Lantieri, C.; Tataranni, P.; Vaiana, R.; Sangiorgi, C. Mechanical and Thermal Performance of Macro-Encapsulated Phase Change Materials for Pavement Application. *Materials* **2018**, *11*, 1398. [CrossRef]
35. Zhang, X.; Yin, Z.; Meng, D.; Huang, Z.; Wen, R.; Huang, Y.; Min, X.; Liu, Y.; Fang, M.; Wu, X. Shape-stabilized composite phase change materials with high thermal conductivity based on stearic acid and modified expanded vermiculite. *Renew. Energy* **2017**, *112*, 113–123. [CrossRef]
36. Ma, B.; Adhikari, S.; Chang, Y.-J.; Ren, J.; Liu, J.; You, Z. Preparation of composite shape-stabilized phase change materials for highway pavements. *Constr. Build. Mater.* **2013**, *42*, 114–121. [CrossRef]
37. Chen, K.; Yu, X.; Tian, C.; Wang, J. Preparation and characterization of form-stable paraffin/polyurethane composites as phase change materials for thermal energy storage. *Energy Convers. Manag.* **2014**, *77*, 13–21. [CrossRef]

38. Du, X.; Wang, H.; Cheng, X.; Du, Z. Synthesis and thermal energy storage properties of a solid–solid phase change material with a novel comb-polyurethane block copolymer structure. *RSC Adv.* **2016**, *6*, 42643–42648. [CrossRef]
39. Sarı, A.; Alkan, C.; Biçer, A. Synthesis and thermal properties of polystyrene-graft-PEG copolymers as new kinds of solid–solid phase change materials for thermal energy storage. *Mater. Chem. Phys.* **2012**, *133*, 87–94. [CrossRef]
40. Xi, P.; Zhao, F.; Fu, P.; Wang, X.; Cheng, B. Synthesis, characterization, and thermal energy storage properties of a novel thermoplastic polyurethane phase change material. *Mater. Lett.* **2014**, *121*, 15–18. [CrossRef]
41. Wei, K.; Wang, Y.; Ma, B. Effects of microencapsulated phase change materials on the performance of asphalt binders. *Renew. Energy* **2019**, *132*, 931–940. [CrossRef]
42. Wei, K.; Ma, B.; Duan, S.Y. Preparation and Properties of Bitumen-Modified Polyurethane Solid–Solid Phase Change Materials. *J. Mater. Civ. Eng.* **2019**, *31*, 04019139. [CrossRef]
43. Bueno, M.; Kakar, M.R.; Refaa, Z.; Worlitschek, J.; Stamatiou, A.; Partl, M.N. Modification of asphalt mixtures for cold regions using microencapsulated phase change materials. *Sci. Rep.* **2019**, *9*, 20342. [CrossRef] [PubMed]
44. *JTG E20-2011*; Standard Test Methods of Bitumen and Bituminous Mixtures for Highway Engineering. Ministry of Transport of the People's Republic of China: Beijing, China, 2011.
45. Wong, W.-G.; Li, G. Analysis of the effect of wax content on bitumen under performance grade classification. *Constr. Build. Mater.* **2009**, *23*, 2504–2510. [CrossRef]
46. Wang, H.; Dang, Z.; You, Z.; Hao, P.; Huang, X. *Analysis of the Low-Temperature Rheological Properties of Rubberized Warm Mix Asphalt Binders*; ASTM International: West Conshohocken, PA, USA, 2012.

coatings

Article

Research on the Preparation of Graphene Quantum Dots/SBS Composite-Modified Asphalt and Its Application Performance

Youfu Lu [1], Nan Shi [2], Mingming Wang [2], Xinyang Wang [3], Liyang Yin [4], Qiang Xu [4] and Pinhui Zhao [4,*]

1 Shandong Hi-Speed Group, Jinan 250098, China; luyoufu1111@163.com
2 Shandong Hi-Speed Construction Management Group Co., Ltd., Jinan 250001, China; 0531shinan@163.com (N.S.); wangmm@lreis.ac.cn (M.W.)
3 Shandong Hi-Speed Testing Engineering Co., Ltd., Jinan 250001, China; wangxinyang1986@sina.com
4 School of Transportation Engineering, Shandong Jianzhu University, Jinan 250101, China; yinliyang01@163.com (L.Y.); longqiang1986@sdjzu.edu.cn (Q.X.)
* Correspondence: zhaopinhui08@163.com

Abstract: This study aims to prepare a graphene quantum dots (GQDs)/styrene-butadiene segmented copolymer composite (GQDs/SBS) as an asphalt modifier using the Pickering emulsion polymerization method. The physicochemical properties of the GQDs/SBS modifier and their effects on asphalt modification were investigated. In addition, the GQDs/SBS modifier was compared with the pure SBS modifier. Research results demonstrated that GQDs could be evenly dispersed into the SBS phase to form a uniform composite. Adding GQDs brings more oxygen-containing functional groups into the GQDs/SBS modifier, thus strengthening the polarity and making it disperse into the asphalt better. Compared with the SBS modifier, the GQDs/SBS modifier presents better thermostability. Moreover, GQDs/SBS composite-modified asphalt achieves better high-temperature performance than SBS-modified asphalt, which is manifested by the increased softening points, complex shear modulus and rutting factors. However, the low-temperature performance decreases, which is manifested by reductions in cone penetration, viscosity and ductility as well as the increased ratio between creep stiffness (S) and creep rate (m), that is, S/m. Furthermore, adding GQDs can improve the high-temperature performance of asphalt mixture, but it influences low-temperature and water stability slightly. GQDs/SBS also have the advantages of simple preparation techniques, low cost and are environmentally friendly. Therefore, they have become a beneficial choice as asphalt cementing material modifiers.

Keywords: graphene quantum dots (GQDs); GQDs/SBS composite-modified asphalt; pavement performances; rheology

Citation: Lu, Y.; Shi, N.; Wang, M.; Wang, X.; Yin, L.; Xu, Q.; Zhao, P. Research on the Preparation of Graphene Quantum Dots/SBS Composite-Modified Asphalt and Its Application Performance. *Coatings* **2022**, *12*, 515. https://doi.org/10.3390/coatings12040515

Academic Editor: Valeria Vignali

Received: 1 March 2022
Accepted: 7 April 2022
Published: 11 April 2022

Publisher's Note: MDPI stays neutral with regard to jurisdictional claims in published maps and institutional affiliations.

Copyright: © 2022 by the authors. Licensee MDPI, Basel, Switzerland. This article is an open access article distributed under the terms and conditions of the Creative Commons Attribution (CC BY) license (https://creativecommons.org/licenses/by/4.0/).

1. Introduction

In all pavement forms, asphalt pavement accounts for a very high proportion of road engineering around the world due to its remarkable advantages such as high riding comfort and convenient maintenance. Recently, transportation industries in countries around the world are booming. Increasing traffic loads, especially the growth in heavy-loaded and overloaded vehicles, intensifies damage to original roads. As a result, asphalt pavements may develop different types of early diseases soon after opening to traffic, such as ruts, pavement subsidence, upheaval, etc. [1,2]. These diseases affect the performance and service life of pavements significantly. Nowadays, improving durability and prolonging the service life of asphalt pavement is a key problem that has to be solved in the road field at present. Furthermore, developers often choose superior performance materials for asphalt pavements since excellent pavement structural performances are closely related to material performances. In particular, it is very important to choose a good performance asphalt binder because its quality is directly related to the performance of the asphalt pavement [3]. To obtain ideal performances from asphalt under various climatic and traffic load conditions, researchers tried adding modifiers such as rubber, resin, polymer and other fillers [4]. Recently, applications of new materials and new technologies achieved great progress. In particular, with the continuous development of nanotechnology, adding nanomaterials into polymer-modified asphalt materials, such as applications of nanoclay, nano-silica, nano-ore and nano-metal [5–7], occurs frequently. This is due to the superior high-temperature and low-temperature stability, durability and water stability of the nanomaterial/polymer composite-modified asphalt. Bhat et al. [8] found that the storage stability and the aging resistance of SBS-modified asphalt binder after adding Al_2O_3 were improved significantly. Martínez-Anzures et al. [9] pointed out that the softening point and viscosity of SBS-modified asphalt binder after adding Cloisite 15 A was improved under high temperatures. Golestani et al. [10] investigated the effect of organic montmorillonite (OMMT) in improving the storage stability and physical and rheological performances of SBS-modified asphalt binder. The asphalt mixture prepared by composite materials presented better flexural–tensile strength, elasticity modulus and smaller rutting depth. Bala et al. [11] found that mixing nano-silica could improve the compatibility of polypropylene modified asphalt, and the aging resistance and high-temperature performance of modified asphalt were improved significantly. Alhamali et al. [12] found that the viscosity and ductility of the polymer-modified asphalt binder were increased after nano-silica was added, accompanied by a significant improvement in high-temperature and storage stability. Zhang et al. [13] investigated the influences of nano TiO_2, nano-Zn and nano-$CaCO_3$ on the high-temperature and low-temperature performance of SBS and SBR polymer-modified asphalt binders. Results showed that nanomaterials could improve the dispersion of polymers in base asphalt and improve the compatibility between polymers and base asphalt. Therefore, the high-temperature and low-temperature performance of polymer-modified asphalt were improved.

Recently, another type of nanomaterials, that is, carbon nanomaterials, has attracted more and more research attention. Specifically, there are abundant products for the nano-modification of binding materials (referring to asphalt and concrete), including the nanomaterials of the graphene family (e.g., original graphene, monolayer graphene, multi-layer graphene and graphene nanosheets), graphene oxide (GO), single-wall carbon nanotube (CNT) and multi-wall CNTs [14–16]. Based on experimental studies, Liu [17] and Zhu [18] pointed out that GO could promote high-temperature viscoelasticity, strengthen humidity resistance and promote the internal healing ability of asphalt binder. Stratiev [19,20] found that the addition of H-Oil hydrocracked vacuum residual oils (H-Oil VTB) to the straight run vacuum residual oils (SRVRO) increases the oxidation rate of the SRVRO, which leads to a higher rate of asphaltenes formation, and, subsequently, a higher rate of softening point increasing. Santagata [21] stated after a survey that CNT could promote the high-temperature and low-temperature performance of asphalt binders and decrease the influence of the external climate on binder aging. Through a nano-scale study on an atomic force microscope (AFM), Mamun [16] pointed out that the combination of polymer and single-wall CNT composite improved the humidity resistance of the asphalt binder. Furthermore, Goli et al. [22] discussed the influences of CNT as a binder modifier on the performance storage stability of SBS asphalt. They found that the physical, rheological and storage stability of the SBS-modified asphalt binder were improved significantly after adding CNT. In recent years, graphene quantum dots (GQDs) have attracted strong attention for their unique structures, excellent performances and promising application prospects [23–26]. In particular, GQDs, which were prepared by asphaltene, were formed by the connection of abundant carboxyl and hydroxyl functional groups surrounding the 2–5 nm aromatic nucleus. Since GQDs are derived from asphalt, it can be reasonably speculated that GQDs have very good compatibility with asphalt [27]. Compared with other nanomaterials, GQDs contain a lot of carboxyl and hydroxyl on the surface and possess some acidity and surface activity. They are expected to develop some physical or chemical reactions with unsaturated dilute bonds of SBS, thus forming stable chemical crosslinking and developing the synergistic effect between GQDs and SBS effectively.

Nevertheless, carbon nanomaterials must overcome considerable surface tension in order to disperse into asphalt since they have great specific surface area [4]. As a result, the dispersion problem of carbon nanomaterials in SBS-modified asphalt is a key constraint against their development at present. With abundant carboxyl and hydroxyl functional groups on the surface, GQDs show some surface activity and they can be used to prepare Pickering emulsion as nano surfactant [28]. Further, the polymerization of Pickering emulsion is expected to be a new choice to prepare GQDs/SBS composite-modified materials [4,29].

This study aims to discuss the applications of GQDs as an asphalt modifier. For this purpose, the asphalt-based GQDs and SBS were used as the main raw material, and a new GQDs/SBS composite material prepared by the Pickering emulsion polymerization method was used as the modifier of asphalt. A series of chemical analyzers were used to analyze the functional group structures and thermostability of the GQDs and SBS modifier. On this basis, the conventional physical properties and rheological properties of GQDs/SBS composite-modified asphalt and SBS-modified asphalt were compared. Moreover, pavement performances of asphalt mixtures that were prepared using GQDs/SBS composite-modified asphalt and SBS-modified asphalt as binders, were characterized.

2. Materials and Methods

2.1. Materials

GQDs/SBS modifier was prepared using asphalt-based GQDs and linear SBSas the raw materials. Modified asphalt was prepared using Qinhuangdao AH-70 (PetroChina fuel asphalt Co., Ltd., Qinhuangdao, China)asphalt and GQDs/SBS modifier as raw materials. The conventional performances and four-component compositions of Qinhuangdao AH-70 asphalt are listed in Table 1. SBS, white fluffy rod-like solids, was provided by Yueyang Baling Petrochemical Corp (Sinopec, Yueyang, China). SBS is a linear molecule with an average molecular weight of 100,000 g/mol.

Table 1. Conventional Index of Qinhuangdao AH-70 asphalt.

Index		Units	Requirement	Results
Penetration (25 °C, 5 s, 100 g)		0.1 mm	60–80	71
Softening point (R&B)		°C	≥45	50.2
Ductility (10 °C)		cm	≮15	35.7
Dynamic viscosity at 60 °C		Pa·s	≮160	212
Four component	Saturates	%	–	17.8
	Aromatics	%	–	42.7
	Resins	%	–	25.2
	Asphaltenes	%	–	14.1

The deoiling asphalt (DOA, Asphaltenes content is 20%) from SINOPEC Jiujiang company, Jiujiang, China, was used as the raw material, and GQDs with asphaltene polycyclic aromatic hydrocarbon nucleus were prepared by nitric acid oxidation. The specific manufacturing technology was introduced as follows: 10 g DOA powder were added into a 250 mL flask, and 150 mL 65% concentrated nitric acid were added in slowly and stirred continuously. Under the strong stirring, the temperature increased gradually, and the reflux was heated. The reaction lasted for 4 h under 90 °C. After finishing the reaction, it was cooled to room temperature and then diluted with distilled water. It was filtered by a 0.2 um Millipore filter directly rather than neutralized by sodium hydroxide. The residual nitric acids in the filtrate were eliminated through reduced pressure distillation and then dried, thus obtaining nitric acid oxidized GQDs.

2.2. Preparation of GQDs/SBS Modifier

The GQDs/SBS modifier was prepared by the Pickering emulsion polymerization method. The preparation process is shown in Figure 1. Firstly, a certain mass of GQDs was dispersed into pure water, and a 5% (mass concentration) GQDs solution was gained through ultrasonic dispersion for 30 min. Meanwhile, SBS particles were dissolved into methylbenzene, and a 20 wt% (mass concentration) SBS methylbenzene solution was prepared. The SBS methylbenzene solution was added to the GQDs solution at a mass ratio of 1:1, and Pickering emulsion was acquired through 5 min high-speed shearing using a BME shearing machine under 4000 r/min. The Pickering emulsion was poured into a clean glass tray with a flat bottom. Finally, the tray was put in a vacuum drying box under 80 °C for 12 h. The Pickering emulsion developed auto polymerization under these conditions, thus obtaining a GQDs/SBS modifier. It can be seen from Figure 2 that the GQDs/SBS modifier is a black solid under room temperature.

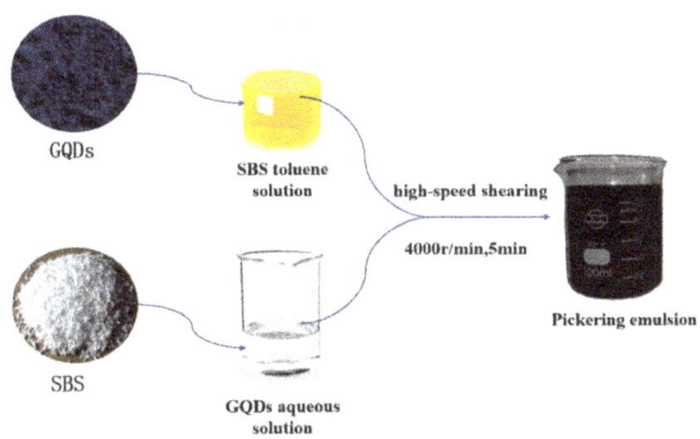

Figure 1. The preparation process of GQDs/SBS composite modifier Pickering emulsion.

Figure 2. GQDs/SBS composite modifier.

2.3. Structural Analysis of GQDs/SBS Modifier

2.3.1. FT-IR Spectral Analysis

The functional groups and material structures of the SBS modifier and the GQDs/SBS modifier were characterized using a Fourier infrared microscopic analysis spectrometer. Meanwhile, their chemical compositions were analyzed. A Nicolet IS 5-type infrared spectrometer (Thermo Science, Waltham, MA, USA) was used in the experiment. All tests were performed at room temperature. The resolution was 4 cm^{-1}, the scanning frequency was 32 times/min and the spectral wavenumber ranged between 4000 and 500 cm^{-1}. The samples were prepared by casting a film onto a potassium bromide (KBr) window from a 5% by weight solution in carbon tetrachloride (CCl_4).

2.3.2. Thermogravimetric Analysis (TGA)

The TGA-100 A (Shanghai all Instrument Equipment Co., Ltd., Shanghai, China) thermal gravimetric analyzer was applied in the experiment for the TGA of the SBS modifier and GQDs/SBS modifier. Under the nitrogen atmosphere, about 7 mg samples were collected and heated from 30 to 600 °C at the constant temperature rising rate of 10 °C/min. The thermostability performances of the two modifiers were evaluated by TG and DTG curves. To assure accuracy and decrease errors, all experiments were performed three times.

2.4. Preparation of Modified Asphalt

This study prepared GQDs/SBS composite-modified asphalt and SBS-modified asphalt (control group) using the melting–thawing mixed method. The melted–thawed AH-70 asphalt was collected and then poured into a cylinder container, which was then heated to 180 °C. Subsequently, 3 wt% (asphalt mass) compatilizer (extract oil) and 4 wt% modifier were added successively. Next, the mixture was processed by high-speed shearing for 30 min at the rate of 4000 r/min. Later, the temperature was lowered to 170 °C, and the stirring rate was 750 r/min. In total,0.25 wt% stabilizer was added and stirred continuously for 3 h. After full development, modified asphalt with stable performance was obtained.

2.5. Performance Characterization of Modified Asphalt

2.5.1. Characterization of Physical Properties of Modified Asphalt

With reference to Test Regulations on Highway Engineering and Asphalt Mixture (JTG E 20-2011), various physical properties of modified asphalt, including cone penetration, softening point, ductility and viscosity were characterized.

2.5.2. Rheological Test

The rheological properties of modified asphalt samples were characterized using the dynamic shear rheometer (DSR, TA, New Castle, DE, USA). The clamp chose the parallel plates with diameters of 8 mm and 25 mm, respectively. Firstly, the linear viscoelastic interval of the samples was determined through a stress and strain scan. Secondly, a small-angle vibration shearing test was carried out within the determined linear viscoelastic interval. The scanning results of isothermal frequencies (0.1–50 mads) were acquired under 30, 45, 60, and 75 °C. The specific operation process was introduced as follows: first, put about 0.1 g of the samples on the lower plate of the parallel plates. Second, install the parallel plates on the rheometer, and set the initial temperature. After the samples are softened, lower the upper plate to squeeze some samples. Finally, set the interval between the parallel plates to 1 mm (25 mm plate) or 1.5 mm (8 mm plate). The temperature scanning ranged from 58 to 95 °C. The temperature rising rate was 1 °C/min, and the frequency was 10 rad/s. The multi-stress repetitive creeping test was carried out under 100 and 3200 Pa. Each stress cycle number was set to 10, and each circle had 1 s stress loading and 9 s relaxation.

The bending dye rheometer (BBR, ATS, Butler, PA, USA) was used to measure the creep properties of the asphalt under low temperatures. The combination of BBR and DSR can present relatively comprehensive rheological information on asphalt under the used temperature. BBR uses the small beam principle in engineering to characterize the cracking trend of the asphalt upon temperature drop, through which two indexes could be gained: creep stiffness (S) and variation rate of stiffness with time (m). To avoid the cracking phenomenon of the asphalt under low temperatures, Peformance Grade (PG) classification norms require that the S for 60 s loading of BBR should be no higher than 300 MPa and the m value should be no smaller than 0.3. The temperature of the BBR test ranged between −18 and −24 °C.

In this study, the viscoelasticity within a wide-frequency and wide-temperature range was gained by the time–temperature equivalence principle. Such viscoelasticity with a very large span in orders of magnitude can hardly be measured directly. The time–temperature equivalence principle elaborates that influences of extended time (or decreased frequency on mechanical properties of materials are equivalent to temperature rise. Under conditions meeting the time–temperature equivalence principle, various viscoelastic parameters measured by experiment can be used to synthesize curves using translocation factors.

2.6. Performance Characterization of Asphalt Mixture

The SBS-modified asphalt (control group) and the prepared GQDs/SBS composite-modified asphalt were used as binders, respectively. The AC-20 asphalt mixture, which is commonly used in asphalt pavement surfaces, was chosen to design asphalt mixture by the Marshall Design method according to China's Construction Technological Norms on Highway Asphalt Pavement (JTG F40-2004). The grading curve is shown in Figure 3. Combining with engineering experiences, the optimal oil–stone ratio was determined to be 4.5 based on the target voidage of 4.0%. In this study, all evaluated asphalt mixtures had the same grading and optimal asphalt content.

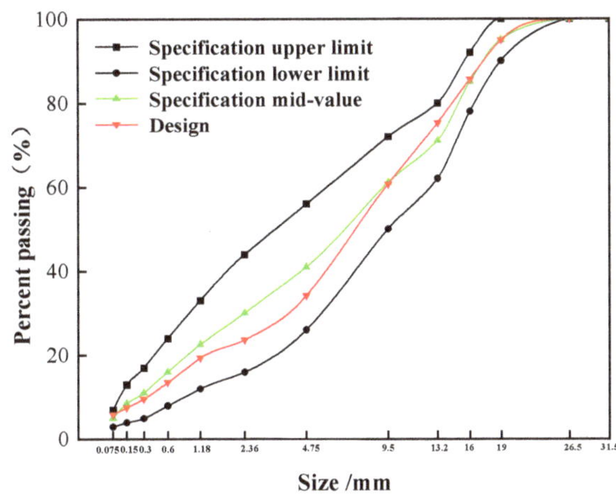

Figure 3. AC20 aggregate gradation curves.

Two asphalt mixtures were molded into specimens according to Test Regulations on Highway Engineering and Asphalt Mixture (JTG E 20-2011) of China. The properties of the asphalt mixtures, including high-temperature performance, low-temperature performance and water stability, were analyzed. Since the grading and asphalt consumptions of the two asphalt mixtures were the same, the volume indexes were similar, and their differences in performance indexes were mainly determined by the different performances of the asphalt cements.

3. Results and Discussions

3.1. Chemical Properties of Modifiers

3.1.1. FTIR Functional Group Analysis

The FT-IR spectra of the SBS modifier and GQDs/SBS composite modifier are shown in Figure 4. The IR region (wavenumber from 4000 to 400 cm^{-1}) was divided into a functional group zone (wavenumber from 4000 to 1330 cm^{-1}) and fingerprint zone (wavenumber from 1330 to 400 cm^{-1}) [30]. It can be seen from Figure 4 that SBS shows obvious methylene C-H asymmetric and symmetric stretching vibration peaks at 2917 and 2848 cm^{-1} as well as multiple absorption peaks between 3100–2950 cm^{-1}, which were stretching vibration absorption peaks of unsaturated hydrocarbons. The absorption peaks occurring simultaneously at 1630, 1600, 1560, and 1422 cm^{-1} corresponded to the stretching vibration of the aromatic ring skeleton ($-CH_2-$). The vibration within 1390~1000 cm^{-1} is the stretching vibration of the –C–O bond and single-bond skeleton vibration of C–C. In addition, absorption peaks near 697, 730, and 749 cm^{-1} were caused by the vibration absorption of single substituted benzene. The absorption peak near 972 cm^{-1} was caused by the twisting vibration of the C=C bond, while the absorption peak near 915 cm^{-1} is the infrared characteristic absorption peak of polybutadiene caused by out-of-plan swinging and vibration of $=CH_2$. As the petroleum asphalt-based GQDs are added, the peaks of the GQDs/SBS composite modifiers at these points are all strengthened. Moreover, a wide adsorption peak occurs at 3307 cm^{-1}, which is a combined peak of hydroxyl and amidogen stretching vibrations of petroleum asphalt-based GQDs. Meanwhile, there are obvious shoulder peaks at 1650–1580 cm^{-1}, which are stretching vibration peaks of the benzene ring. This reflects that GQDs and SBS develop polymerization reactions to form stable covalent bonds. These covalent bonds are enough to avoid phase separation which might occur during the simple physical mixing manufacturing of nanocomposites. In addition, modified asphalt contains more oxygen-containing functional groups (e.g., –C=O and –C–O). Furthermore, these functional groups mainly come from GQDs, and the increased oxygen content can also improve the polarity of GQDs/SBS composite modifiers, thus increasing their compatibility with asphalt [31].

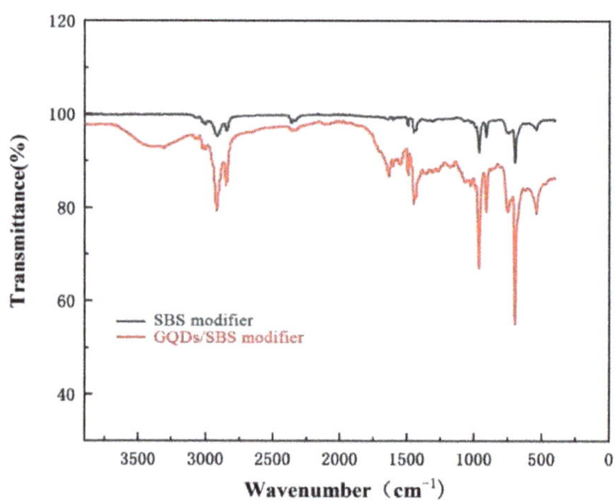

Figure 4. FTIR spectrum of GQDs/SBS composite modifier and SBS modifier.

3.1.2. TGA

The thermostability of the modifier is an important property that has to be considered when analyzing the structural characteristics of asphalt binders. In this study, the thermal stability of GQDs/SBS composite modifiers and the SBS modifier were discussed by TGA. It can be seen from Figure 5 and Table 2 that the TGA curves of the GQDs/SBS composite modifier and SBS modifier present the same trend, and they both experienced two major stages of mass loss. However, the thermodynamic behaviors of the GQDs/SBS composite modifier and SBS modifier are significantly different. The GQDs/SBS composite modifier and SBS modifier both enter the first stage of mass loss before 340 °C. In this stage, mass loss is mainly attributed to volatilization of crystal water adsorbed onto the sample surface as well as decomposition of some oxygen-containing functional groups in molecules (–OH and –COOH). Since the GQDs surface has a lot of oxygen-containing functional groups, the mass loss rate of the GQDs/SBS composite modifier is far higher than that of the SBS modifier in the first stage. The second stage of mass loss occurs in the temperature range of 340~490 °C. The mass loss of modifiers is mainly attributed to the decomposition of SBS into small molecules and volatilization. This is the major stage of mass loss. It can be seen from the TG curve that the initial decomposition temperatures of the GQDs/SBS composite modifier and SBS modifier are at about 416 °C (the tangent initial point of TG is about 416 °C), and the pyrolysis termination temperature is about 480 °C. The pyrolysis termination temperature of the GQDs/SBS composite modifier (478.8 °C) is slightly lower than that of the SBS modifier (479.9 °C), showing a very small difference. After finishing the pyrolysis, the residual mass of the GQDs/SBS composite modifier is 1.78%, and the mass change is 98.22%. The residual mass of the SBS modifier is only 0.05%, and the mass changes are 99.95%. This demonstrates that within this temperature range, the SBS modifier almost loses weight completely under the N_2 atmosphere, and it is decomposed into small molecules and then volatilized without producing any residual carbons. However, the GQDs/SBS composite modifier is decomposed incompletely under an N_2 atmosphere, and it will produce some residual carbons. This is because the carbon nucleus is the major structural unit of GQDs in the GQDs/SBS composite modifier. After surface oxygen-containing functional groups are lost in the first stage of mass loss, the residual carbon nucleus has very good thermostability, and it will not be decomposed again, thus resulting in a high residual mass. The maximum mass loss rate points (DTG peak value) of the GQDs/SBS composite modifier and SBS modifier are at 460 °C. The mass-loss rate of the SBS modifier (17.08%/min) is 0.85%/min higher than that of the GQDs/SBS composite modifier (16.23%/min). To sum up, the GQDs/SBS composite modifier has better thermostability than the SBS modifier. In other words, adding GQDs improves the thermostability of the SBS modifier.

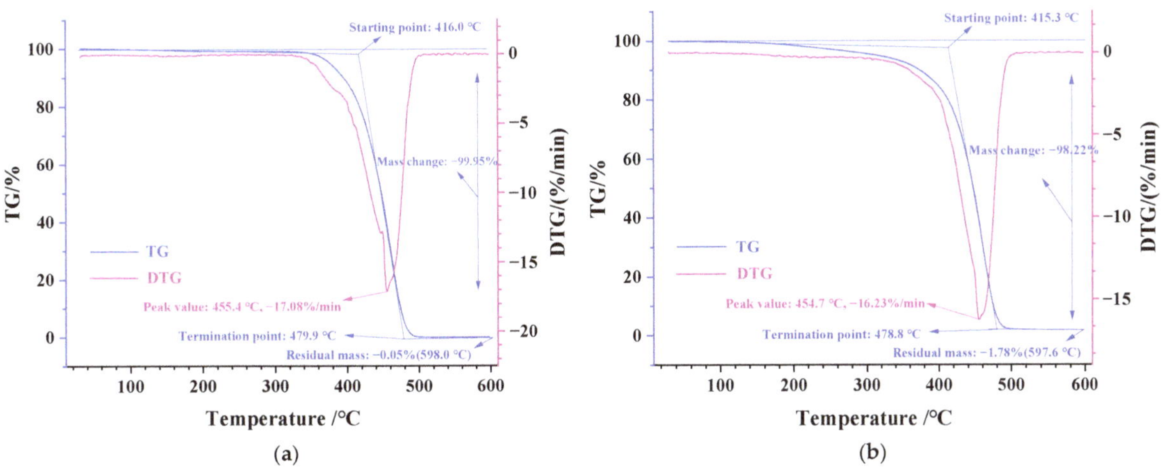

Figure 5. Thermal stability analysis of (**a**) SBS modifier and (**b**) GQDs/SBS composite modifier.

Table 2. TG and DTG parameters of SBS modifier and GQDs/SBS composite modifier.

Modifier Type	TG			DTG	
	Starting Temperature/°C	Termination Temperature/°C	Residual Mass/%	Peak Temperature/°C	Pyrolysis Rate/%/min
SBS	416.0	479.9	0.05	455.4	−17.08
GQDs/SBS	415.3	478.8	1.78	454.7	−16.23

3.2. Conventional Physical Properties of Modified Asphalt

The physical properties of the GQDs/SBS composite-modified asphalt and SBS-modified asphalt are listed in Table 3. Clearly, the cone penetration and ductility of the GQDs/SBS composite-modified asphalt decrease, while the softening point increases compared with those of the SBS-modified asphalt. This implies that adding GQDs can improve the high-temperature performance of SBS-modified asphalt but decrease the low-temperature performance to some extent.

Table 3. Physical property of SBS-modified asphalt and GQDs/SBS composite-modified asphalt.

Index	SBS	GQDs/SBS	Test Method
Softening point/°C	65.6	71.1	T0606
Penetration/0.1 mm	51.6	47.9	T0604
Ductility(5 °C)/cm	24.8	24.5	T0605

In addition, temperature influences the high-temperature flowing characteristics of asphalt. Flow characteristics of different samples show different degrees of sensitivity to temperature changes. In other words, asphalts have different temperature sensitivities. There are high-temperature zones and low-temperature zones. The temperature sensitivity of the high-temperature zone is closely related to the construction of asphalt mixture, the pumping of asphalt and other construction characteristics. In this study, a Brookfield rotary viscosimeter was applied, and the 27 # rotors were applied to the viscosity of the GQDs/SBS composite-modified asphalt and SBS-modified asphalt within the temperature range of 110–175 °C. The variation curves of viscosity with temperature are shown in Figure 6a. With the increase in temperature, the viscosity values of both the GQDs/SBS composite-modified asphalt and SBS-modified asphalt decline sharply in the beginning and then become stable. This is because modified asphalt changes from a non-Newtonian body to a Newtonian body gradually under high temperatures. Given the same temperature, the viscosity of GQDs/SBS composite-modified asphalt is lower than SBS-modified asphalt. With the increase in temperature, differences between the GQDs/SBS composite-modified asphalt and SBS-modified asphalt decrease. This reflects that chemical crosslinking between GQDs and SBS is disadvantageous to the strength of polymers.

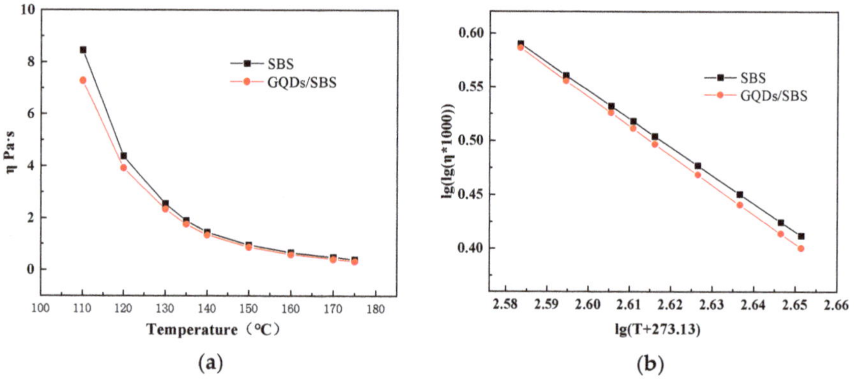

Figure 6. Viscosity–temperature performance of SBS and GQDs/SBS composite-modified asphalt. (**a**) Viscosity -temperature curve; (**b**)Viscosity temperature curve after fitting with Saal formula.

The Saal model (Equation (1)) proposed by ASTM D2493 was further applied to process the viscosity–temperature curves. It can characterize the temperature sensitivities of the GQDs/SBS composite-modified asphalt and SBS-modified asphalt.

$$\lg(\lg\eta * 1000) = n + m\cdot\lg\,(T + 273.13) \quad (1)$$

where m refers to the slope of the regression line; n denotes the intercept of the regression line on the lg(lgη*1000) axis; η is the viscosity (Pa·s); T is the temperature (°C). Saal fitting curves of the viscosity–temperature curves of the GQDs/SBS composite-modified asphalt and SBS-modified asphalt are shown in Figure 6b. The parameters of the corresponding Saal model are listed in Table 4. Moreover, m in the Saal model was defined as viscosity-temperature sensitivity (VTS). The smaller absolute value of VTS indicates that viscosity changes more slowly with temperature, and the temperature sensitivity is better.

Table 4. Viscosity–temperature curve fitting equation of SBS and GQDs/SBS composite-modified asphalt.

Asphalt Type	Viscosity–Temperature Curve Fitting Equation	VTS	R^2
SBS	$\lg(\lg\eta*1000) = -2.6184 * (T + 273.13) + 7.3543$	−2.6184	0.99732
GQDs/SBS	$\lg(\lg\eta*1000) = -2.7364 * (T + 273.13) + 7.6555$	−2.7364	0.99959

It can be seen from Figure 6 and Table 4 that the absolute value of VTS of the GQDs/SBS composite-modified asphalt is higher than that of the SBS-modified asphalt, indicating that adding GQDs is disadvantageous for the temperature sensitivity of SBS-modified asphalt.

3.3. Rheological Properties of Modified Asphalt

The usability of asphalt pavement is determined, to a very large extent, by the viscoelastic properties of the modified asphalt binder. The linear viscoelasticity of modified asphalt is very sensitive to the motion and interaction of polymer molecular chains. Moreover, the complexity of different high-molecular polymer modification systems may influence the internal structure of modified asphalt, thus influencing the rheological characteristics of asphalt. Rheological parameters in the linear viscoelasticity interval are independent of changes regarding stress and strain, and they are only related to the properties of the materials [32]. Therefore, linear viscoelasticity and dynamic rheological tests are very effective methods to elaborate on the influences of modifiers on the performances of modified asphalt and to study the influences of polymers on the viscoelasticity of asphalt.

3.3.1. Frequency Scanning under Middle and High Temperature

The major curve of the GQDs/SBS composite-modified asphalt and SBS-modified asphalt at 30 °C is shown in Figure 7. This curve was obtained from translocations of frequency scanning curves at 30, 45, 60, and 75 °C. It can be seen from Figure 7 that within the whole frequency scanning range, given the same frequency, the modulus of a complex number (G*) of the GQDs/SBS composite-modified asphalt is higher than that of the SBS-modified asphalt. Moreover, the major curves of the GQDs/SBS composite-modified asphalt and SBS-modified asphalt differ significantly in the low-ω zone. However, such difference decreases with the increase of frequency. According to the time–temperature equivalence principle, the low-frequency zone corresponds to the high-temperature zone. Hence, the GQDs/SBS composite-modified asphalt has a better high-temperature performance than SBS-modified asphalt. In other words, adding GQDs improves the rutting resistance of the SBS-modified asphalt. Additionally, the G* values of the GQDs/SBS and SBS-modified asphalt in the high-ω zone are close to the same value, indicating that GQDs influence the viscoelasticity performance of the SBS-modified asphalt in the high-ω zone.

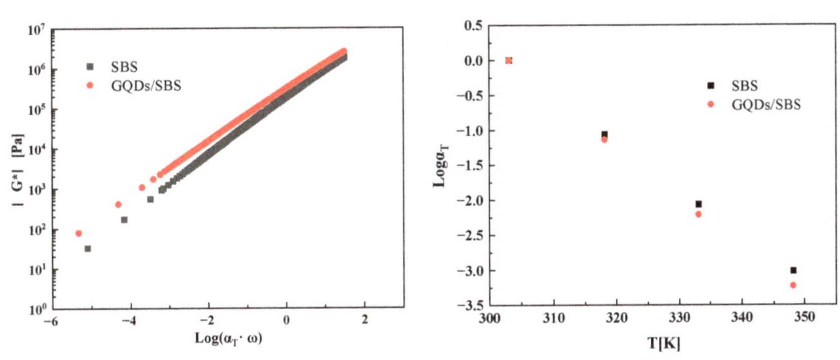

Figure 7. (**a**) Master curves for GQDs/SBS composite-modified asphalt and SBS-modified asphalt at the reference temperature of 30 °C and (**b**) Variations of the translocation factor with temperature.

It can be seen from major curves in Figure 7a that the time–temperature equivalence principle is highly applicable to GQDs/SBS composite-modified asphalt and SBS-modified asphalt. Variations of the translocation factor with temperature are shown in Figure 7b. Obviously, the translocation factors of GQDs/SBS composite-modified asphalt and SBS-modified asphalt are significantly different. The variations of translocation factor with temperature were fit using the Arrhenius-like equation (Figure 8). Differences in the translocation factors of different samples can be distinguished quantitatively by the activation energy of the Arrhenius-like equation. The activation energy is related to the temperature sensitivity of materials. This further proves that adding GQDs improves the high-temperature performance of the SBS-modified asphalt.

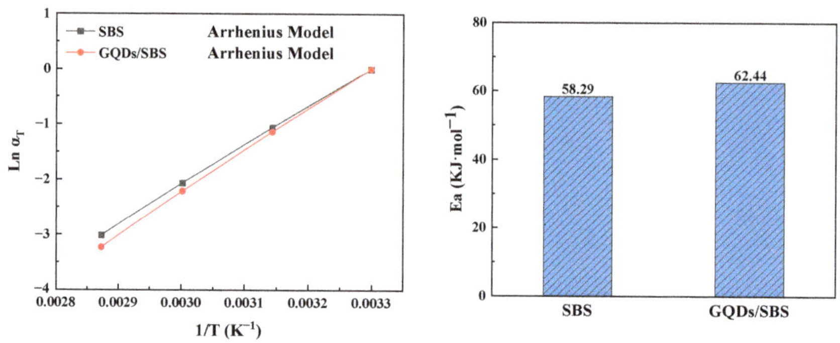

(a) (b)

Figure 8. (**a**) Shifting factor versus temperature and (**b**) activation energy for GQDs/SBS composite-modified asphalt and SBS-modified asphalt.

3.3.2. Temperature Scanning

In this study, temperatures of asphalt samples within a wide range (58–95 °C) were scanned. The variations of storage modulus (G′) and loss modulus (G″) with temperature are shown in Figure 9. Both G′ and G″ decrease dramatically with the increase in temperature. The reduction rates of the GQDs/SBS composite-modified asphalt and SBS-modified asphalt are different and finally, tend to be stable. Within a wide temperature range, the reduction rates of the G′ and G″ of the GQDs/SBS composite-modified asphalt with the temperature rise are lower than those of SBS-modified asphalt. This reflects that compared to SBS-modified asphalt, the GQDs/SBS composite-modified asphalt has better temperature sensitivity within a wide range.

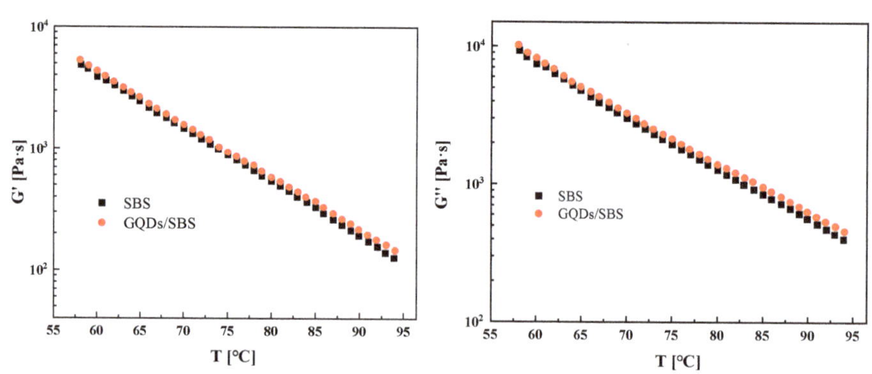

(a) (b)

Figure 9. (**a**) Variation of storage modulus G′ with temperature and (**b**) Variation of loss modulus G″ with temperature.

The rutting resistance of asphalt can be characterized by the rutting factor G*/sinδ and the failure temperature which is gained when G*/sinδ = 1.0 kPa. The higher the G*/sinδ and failure temperature, the better the high-temperature stability of asphalt. It can be seen from Figure 10 that in the middle-temperature and high-temperature intervals, asphalt is mainly in the sticky flow state. The elasticity and strength of the system are provided by polymers. The variation laws of the G*/sinδ of the GQDs/SBS composite-modified asphalt and SBS-modified asphalt with temperature are consistent with the variation laws of G′ and G″. Their temperatures at G*/sinδ = 1.0 Kpa are 84.82 and 86.20 °C, respectively. This further demonstrates that GQDs bring higher hardness of SBS-modified asphalt so that SBS-modified asphalt presents better mechanical properties and better resistance to deformation.

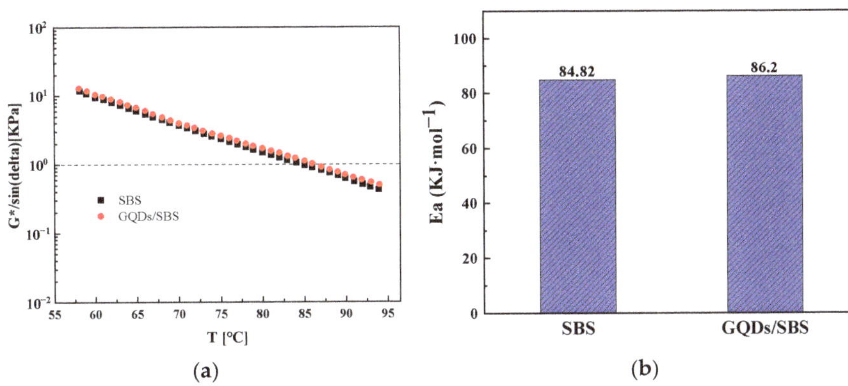

Figure 10. (**a**) Evolutions with temperature of the G*/sinδ and (**b**) Temperature calculated at the point of G*/sinδ = 1.0 KPa for SBS-modified asphalt and GQDs/SBS composite-modified asphalt.

3.3.3. MSCR

After the reciprocal action of vehicle loads for a long period, asphalt pavement may develop shear creep deformation and form ruts. A multi-stress cyclic creep (MSCR) test is an index used to evaluate the high-temperature performance of modified asphalt in recent years. MSCR usually provides 10 loading cycles to samples. In each cycle, loads are applied for 1s, and then the stress is eliminated for resilience for 9 s. In this study, MSCR tests were carried out at 60 °C under two stress levels (100 Pa and 3200 Pa). In MSCR tests, the recovery rate (R) and unrecoverable compliance (Jnr) could be calculated from the recoverable and unrecoverable strains, respectively.

The strain responses of the GQDs/SBS composite-modified asphalt and SBS-modified asphalt after 10 cycles at 60 °C and two stress levels (100 Pa and 3200 Pa) are shown in Figure 11. For one creep–recovery cycle, the strain of the GQDs/SBS composite-modified asphalt at the end of the creep stage and its strain at the end of the recovery stage are smaller than those of SBS-modified asphalt, which are attributed to the added GQDs.

For the quantitative comparison of high-temperature performance between the GQDs/SBS composite-modified asphalt and SBS-modified asphalt, the parameters of R and Jnr of the two samples at 60 °C under two stress levels (100 Pa and 3200 Pa) are shown in Figure 12. With the addition of GQDs, R increases while Jnr decreases. Given the same conditions, the R of the GQDs/SBS composite-modified asphalt is higher than that of the SBS-modified asphalt, while Jnr is smaller. This implies that adding GQDs increases the high-temperature rutting resistance of asphalt. In addition, the R values of both the SBS-modified asphalt and GQDs/GQDs composite-modified asphalt decrease with the increase of stress. Meanwhile, the Jnr of the two samples increases to some extent. In a word, increasing vehicle loads may weaken the recoverable capacity of asphalt pavement significantly under high temperatures in summer, thus causing rutting damages.

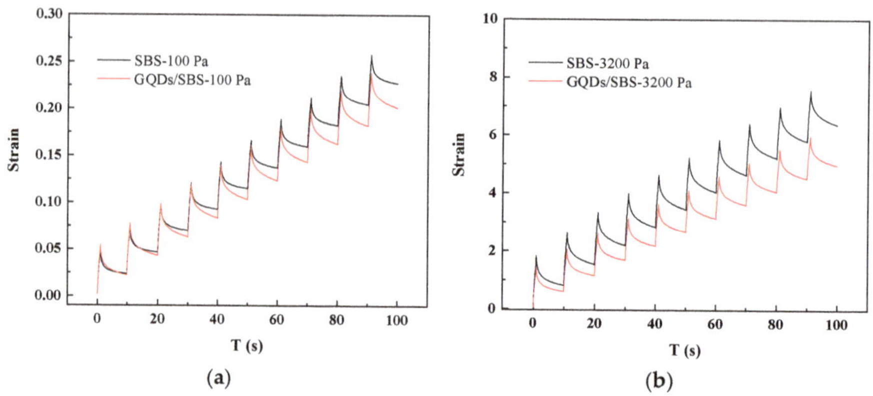

Figure 11. Strain response of SBS-modified asphalt and GQDs/SBS composite-modified asphalt (**a**) in 10 cycles at 60 °C 100 Pa and (**b**) in 10 cycles at 60 °C 3200 Pa.

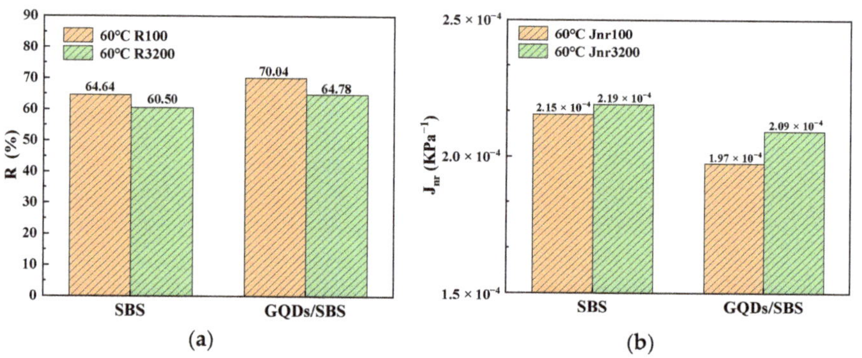

Figure 12. (**a**) The percent recovery and (**b**) Non-recoverable creep compliance calculated from MSCR test at 60 °C and two stress levels for SBS-modified asphalt and GQDs/SBS composite-modified asphalt.

3.3.4. Low-Temperature Creep Properties

Asphalt pavement may crack under low temperatures. Since there is a binding force between the asphalt mixture layer and the lower layer, it will hinder shrinkage and produce translocations, thus generating tensile stress. Cracks occur when the tensile stress exceeds the tensile strength of the asphalt mixture. This requires the asphalt to have a high creep rate to release stresses generated under low temperatures or in the cooling process.

Since DSR cannot test asphalt which has considerable hardness under low temperatures, BBR is usually applied to measure the creep properties of asphalt when the temperature is very low. BBR uses the small beam principle to characterize the cracking trend of asphalt when temperature declines. Tow indexes can be gained from BBR: creep stiffness (S) and creep rate (m). These two indexes are used to characterize the load resistance and relaxation ability of asphalt. If S is too large, the possibility of cracking is high. If m is relatively low, the relaxation ability is insufficient to release stress produced by the reduction of temperature and the probability of cracking increases.

Variations of the S and m of the GQDs/SBS composite-modified asphalt and SBS-modified asphalt at −18 and −24 °C, which are measured by BBR with time, are shown in Figure 13. S drops quickly with the increase of loading time, while m increases significantly. The variable rates of S and m are different. Given the same loading time, the m of the GQDs/SBS composite-modified asphalt at −18 °C is smaller than that of SBS-modified asphalt. However, the m of the GQDs/SBS composite-modified asphalt at −24 °C is higher. S presents the opposite variation trend. To further compare the low-temperature performance of the GQDs/SBS composite-modified asphalt and SBS-modified asphalt, S/m at 60 s was used to characterize the low-temperature crack resistance of asphalt. The lower S/m indicates the stronger crack resistance of asphalt and better low-temperature performance. It can be seen from Figure 14 that given the same temperature, the S/m of the GQDs/SBS composite-modified asphalt is higher than that of the SBS-modified asphalt, indicating that the GQDs/SBS composite-modified asphalt has poor low-temperature crack resistance.

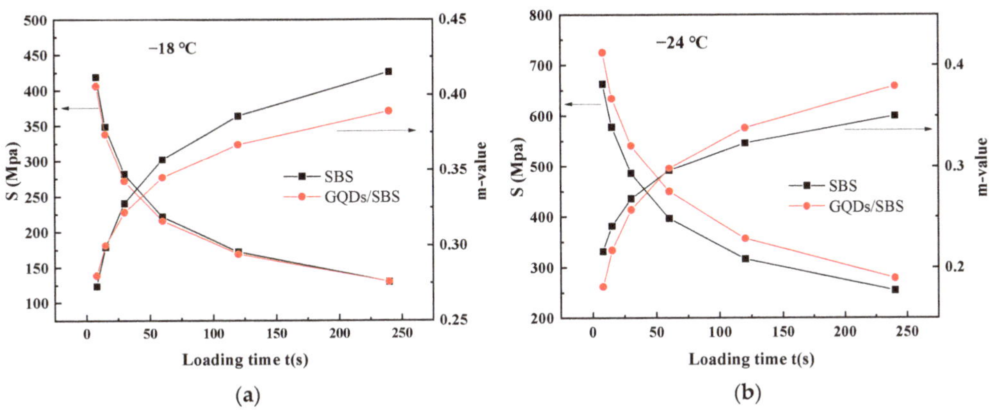

Figure 13. Evolution of S and m-value versus loading time for GQDs/SBS composite-modified asphalt and SBS-modified asphalt at (**a**) −18 and (**b**) −24 °C.

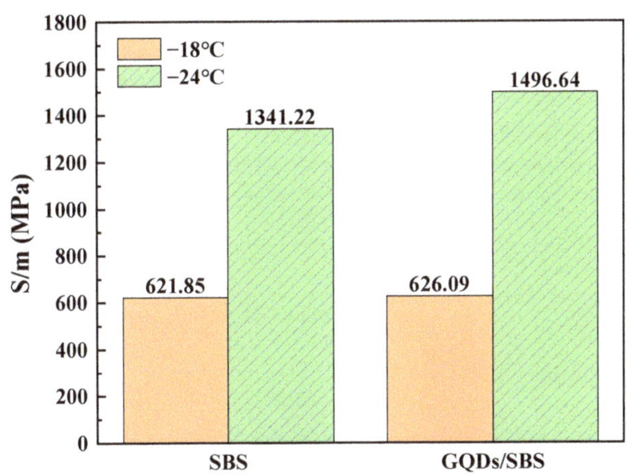

Figure 14. The ratio of creep stiffness S and m of SBS-modified asphalt and GQDs/SBS composite-modified asphalt.

4. Pavement Performance Test Analysis of Mixture

4.1. High-Temperature Stability

In the present study, the high-temperature stabilities of the SBS-modified asphalt mixture and the GQDs/SBS composite-modified asphalt mixture were evaluated by the dynamic stability in the high-temperature (60 °C) rutting test. The high-temperature stability test results are shown in Figure 15.

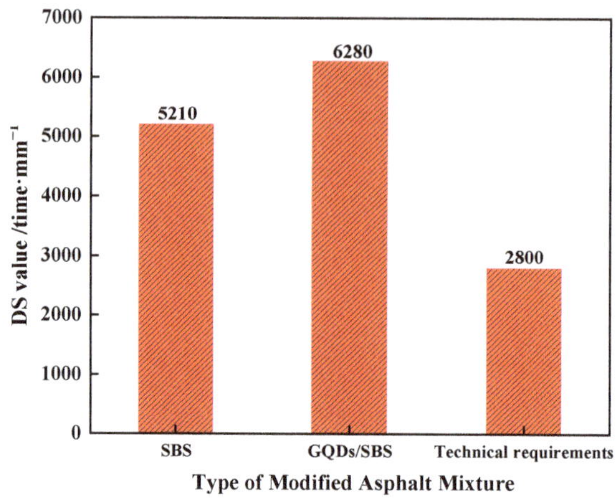

Figure 15. Results of wheel tracking test.

It can be seen from Figure 15 that the dynamic stability of the GQDs/SBS composite-modified asphalt mixture is significantly higher than that of the SBS-modified asphalt mixture, indicating that adding GQDs increases the cohesive force of asphalt. Moreover, more compact structures are formed by adjusting the skeleton of asphalt mixtures in the compacting process, thus increasing the internal friction angle. With the increase of cohesive force and internal friction angle, the shear strength of the asphalt mixture increases, thus making it equipped with good high-temperature stability.

4.2. Low-Temperature Crack Resistance

The resistance of the asphalt mixture to low-temperature cracking performance was evaluated through a low-temperature small beam bending test. Small beam specimens (250 mm (Length)*30 mm (Width)*35 mm (Height)) were used in the test, and the loading rate and temperature were set to 50 mm/min and −10 °C, respectively. The low-temperature crack resistance test results are shown in Figure 16.

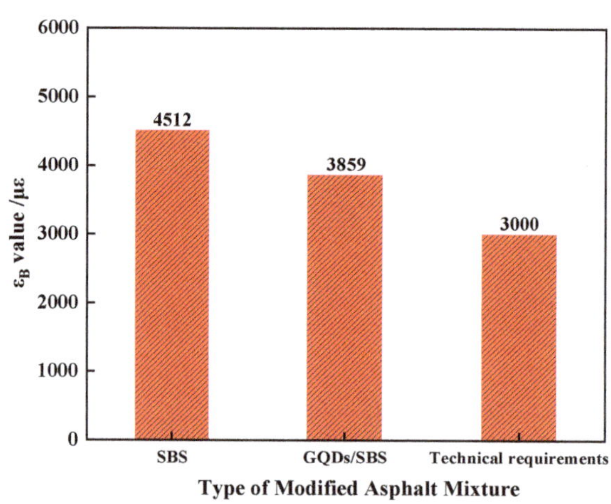

Figure 16. Results of beam bending test.

It can be seen from Figure 16 that the maximum bending strain of the GQDs/SBS composite-modified asphalt mixture at low-temperature failures is 14.5% lower than that of the SBS-modified asphalt mixture. However, all test results meet the standard requirements, indicating that adding GQDs decreases the tenacity and temperature sensitivity of the asphalt mixture under a low-temperature state. As a result, the low-temperature crack resistance declines accordingly.

4.3. Water Stability

The water stabilities of two modified asphalt mixtures were evaluated by the freeze–thaw splitting test. After freeze–thaw cycles of specimens based on the Marshall test, freeze–thaw splitting residual strength was tested, thus enabling us to analyze the resistance of the asphalt to water damage under tough environments. The water stability test results are shown in Figure 17.

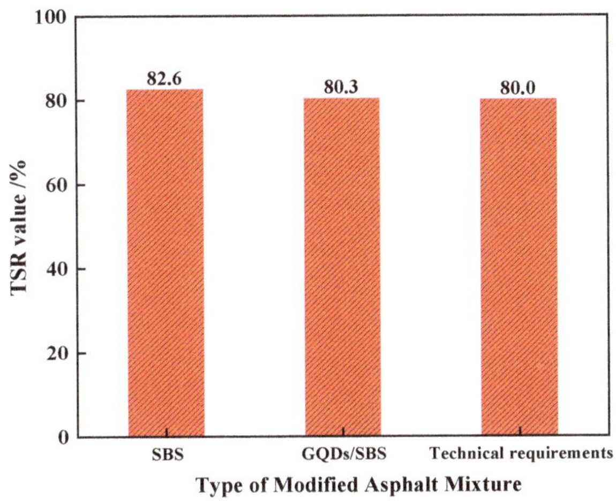

Figure 17. Results of immersion Marshall test.

It can be seen from Figure 17 that the residual strengths of the GQDs/SBS composite-modified asphalt mixture before and after freezing and thawing decrease compared with those of the SBS-modified asphalt mixture. This reveals that adding GQDs decreases the adhesion between asphalt and aggregate, thus decreasing the resistance of the asphalt mixture to water damage. However, the residual strength meets the requirements of technical specifications. This implies that the prepared GQDs/SBS composite modifier influences the water stability of the mixture slightly.

5. Conclusions

In this study, the GQDs/SBS composite modifier was prepared using the Pickering emulsion polymerization method. In addition, the physical and chemical properties of the GQDs/SBS composite modifier, physical and rheological properties of the binders, as well as pavement performances of the GQDs/SBS composite-modified asphalt mixture were investigated. According to results and discussions, some conclusions could be drawn:

(1) The GQDs/SBS composite modifier is prepared by the simple Pickering emulsion polymerization method. GQDs can evenly disperse into the SBS modifier to form a uniform composite. The GQDs/SBS composite modifier contains more oxygen-containing functional groups than the SBS modifier. Furthermore, the pyrolysis rate of the GQDs/SBS composite modifier is lower than the SBS modifier, and its residual mass is higher, thus showing better thermostability.

(2) The conventional physical properties and rheological properties of the GQDs/SBS composite-modified asphalt and SBS-modified asphalt are compared. The GQDs/SBS composite-modified asphalt shows a higher softening point, complex shear modulus, activation energy, rutting factor and recovery rate than the SBS-modified asphalt, thus

showing better high-temperature performance. However, the cone penetration and ductility of the GQDs/SBS composite-modified asphalt decrease while S/m increases, indicating that its low-temperature performance is worsened.

(3) The pavement performance of the GQDs/SBS composite-modified asphalt mixture and SBS-modified asphalt mixture are compared. The high-temperature stability of the GQDs/SBS composite-modified asphalt mixture is improved to some extent compared to that of the SBS-modified asphalt mixture, while its water stability changes slightly and the low-temperature performance declines to some extent.

Author Contributions: Conceptualization, M.W.; methodology, X.W.; validation, Q.X.; data curation, L.Y.; writing—original draft preparation, Y.L. and N.S.; writing—review and editing, P.Z.; supervision, P.Z.; project administration, P.Z.; funding acquisition, P.Z. All authors have read and agreed to the published version of the manuscript.

Funding: This study was monetarily bolstered by the Natural Science Fund project in Shandong province: ZR2021ME189, Project of science and technology support for youth entrepreneurship in Colleges and universities of Shandong Province (2019KJG004).

Institutional Review Board Statement: Not applicable.

Informed Consent Statement: Not applicable.

Data Availability Statement: The information used to bolster the discoveries of this research are from prior detailed researches cited before. This original copy does not include distributed Figures, Tables, and Charts before; thus, all Figures, Tables and Charts of this original copy are unique.

Acknowledgments: We would like to recognize numerous co-workers, students and research facility associates for offering specialized assistance on instrument examination.

Conflicts of Interest: The authors have no conflict of interest.

References

1. Gong, X.; Shao, C. The Gray Superiority Analysis of Asphalt Pavement Performance Evaluation Index and Main Diseases. In Proceedings of the 2009 Second International Conference on Intelligent Computation Technology and Automation, Washington, DC, USA, 10–11 October 2009; pp. 806–808.
2. Zhao, P.; Dong, M.; Yang, Y.; Shi, J.; Wang, J.; Wu, W.; Zhao, X.; Zhou, X.; Wang, C. Research on the Mechanism of Surfactant Warm Mix Asphalt Additive-Based on Molecular Dynamics Simulation. *Coatings* **2021**, *11*, 1303. [CrossRef]
3. Zhao, P.; Ren, R.; Zhou, H.; Ouyang, J.; Li, Z.; Sun, D. Preparation and properties of imidazoline surfactant as additive for warm mix asphalt. *Constr. Build. Mater.* **2021**, *273*, 121692. [CrossRef]
4. Zhao, P.; Song, X.; Dong, M.; Sun, H.; Wu, W.; Zhang, R.; Sun, M.; Zhao, X. Preparation and characterization of CQDs/SBS composites and its application performance as asphalt modifier. *Constr. Build. Mater.* **2022**, *320*, 126312. [CrossRef]
5. Zhu, C.; Zhang, H.; Zhang, Y. Influence of layered silicate types on physical, rheological and aging properties of SBS modified asphalt with multi-dimensional nanomaterials. *Constr. Build. Mater.* **2019**, *228*, 116735. [CrossRef]
6. Sun, L.; Xin, X.; Ren, J. Asphalt modification using nano-materials and polymers composite considering high and low temperature performance. *Constr. Build. Mater.* **2017**, *133*, 358–366. [CrossRef]
7. Li, X.K.; Chen, G.S.; Duan, M.W.; Yang, W.C.; Tang, S.C.; Cao, Y.D.; Luo, Y. Branched Hydroxyl Modification of SBS Using Thiol-Ene Reaction and Its Subsequent Application in Modified Asphalt. *Ind. Eng. Chem. Res.* **2017**, *56*, 10354–10365. [CrossRef]
8. Bhat, F.S.; Mir, M.S. A study investigating the influence of nano Al_2O_3 on the performance of SBS modified asphalt binder. *Constr. Build. Mater.* **2021**, *271*, 121499. [CrossRef]
9. Martínez-Anzures, J.D.; Zapién-Castillo, S.; Salazar-Cruz, B.A.; Rivera-Armenta, J.L.; Antonio-Cruz RD, C.; Hernández-Zamora, G.; Méndez-Hernández, M.L. Preparation and properties of modified asphalt using branch SBS/nanoclay nanocomposite as a modifier. *Road Mater. Pavement Des.* **2019**, *20*, 1275–1290. [CrossRef]
10. Golestani, B.; Nam, B.H.; Nejad, F.M.; Fallah, S. Nanoclay application to asphalt concrete: Characterization of polymer and linear nanocomposite-modified asphalt binder and mixture. *Constr. Build. Mater.* **2015**, *91*, 32–38. [CrossRef]
11. Nura, B.; Ibrahim, K.; Madzlan, N.; Hartadi, S.M. Polymer Nanocomposite-Modified Asphalt: Characterisation and Optimisation Using Response Surface Methodology. *Arab. J. Sci. Eng.* **2018**, *44*, 4233–4243.
12. Alhamali, D.I.; Wu, J.; Liu, Q.; Hassan, N.A.; Yusoff NI, M.; Ali SI, A. Physical and rheological characteristics of polymer modified bitumen with nanosilica particles. *Arab. J. Sci. Eng.* **2016**, *41*, 1521–1530. [CrossRef]
13. Zhang, H.-L.; Su, M.-M.; Zhao, S.-F.; Zhang, Y.-P.; Zhang, Z.-P. High and low temperature properties of nano-particles/polymer modified asphalt. *Constr. Build. Mater.* **2016**, *114*, 323–332. [CrossRef]

14. Kh, A.; Cy, A.; Qy, B.; Yc, A.; Gc, A.; Rong, M.C. Multi–scale enhancement mechanisms of graphene oxide on styrene–butadiene–styrene modified asphalt: An exploration from molecular dynamics simulations. *Mater. Des.* **2021**, *208*, 109901.
15. Wang, R.; Yue, M.; Xiong, Y.; Yue, J. Experimental study on mechanism, aging, rheology and fatigue performance of carbon nanomaterial/SBS-modified asphalt binders. *Constr. Build. Mater.* **2020**, *268*, 121189. [CrossRef]
16. Mamun, A.A.; Arifuzzaman, M. Nano-scale moisture damage evaluation of carbon nanotube-modified asphalt. *Constr. Build. Mater.* **2018**, *193*, 268–275. [CrossRef]
17. Liu, K.; Zhu, J.; Zhang, K.; Wu, J.; Yin, J.; Shi, X. Effects of mixing sequence on mechanical properties of graphene oxide and warm mix additive composite modified asphalt binder. *Constr. Build. Mater.* **2019**, *217*, 301–309. [CrossRef]
18. Zhu, J.; Zhang, K.; Liu, K.; Shi, X. Performance of hot and warm mix asphalt mixtures enhanced by nano-sized graphene oxide. *Constr. Build. Mater.* **2019**, *217*, 273–282. [CrossRef]
19. Stratiev, D.; Shishkova, I.; Dinkov, R.; Kirilov, K.; Yordanov, D.; Nikolova, R.; Veli, A.; Tavlieva, M.; Vasilev, S.; Suyunov, R.R. Variation of oxidation reactivity of straight run and H-Oil hydrocracked vacuum residual oils in the process of road asphalt production. *Road Mater. Pavement Des.* **2021**. [CrossRef]
20. Stratiev, D.; Nenov, S.; Nedanovski, D.; Shishkova, I.; Dinkov, R.; Stratiev, D.D.; Stratiev, D.D.; Sotirov, S.; Sotirova, E.; Atanassova, V.; et al. Empirical Modeling of Viscosities and Softening Points of Straight-Run Vacuum Residues from Different Origins and of Hydrocracked Unconverted Vacuum Residues Obtained in Different Conversions. *Energies* **2022**, *15*, 1755. [CrossRef]
21. Santagata, E.; Baglieri, O.; Tsantilis, L.; Dalmazzo, D. Rheological characterization of bituminous binders modified with carbon nanotubes. *Procedia-Soc. Behav. Sci.* **2012**, *53*, 546–555. [CrossRef]
22. Goli, A.; Ziari, H.; Amini, A. Influence of carbon nanotubes on performance properties and storage stability of SBS modified asphalt binders. *J. Mater. Civ. Eng.* **2017**, *29*, 04017070. [CrossRef]
23. Chakraborty, A.K.; Coleman, K.S.; Dhanak, V.R. The electronic fine structure of 4-nitrophenyl functionalized single-walled carbon nanotubes. *Nanotechnology* **2009**, *20*, 155704. [CrossRef] [PubMed]
24. Li, L.; Wu, G.; Yang, G.; Peng, J.; Zhao, J.; Zhu, J.-J. Focusing on luminescent graphene quantum dots: Current status and future perspectives. *Nanoscale* **2013**, *5*, 4015–4039. [CrossRef] [PubMed]
25. Ponomarenko, L.A.; Schedin, F.; Katsnelson, M.I.; Yang, R.; Hill, E.W.; Novoselov, K.S.; Geim, A.K. Chaotic Dirac billiard in graphene quantum dots. *Science* **2008**, *320*, 356–358. [CrossRef]
26. Zheng, X.T.; Ananthanarayanan, A.; Luo, K.Q.; Chen, P. Glowing graphene quantum dots and carbon dots: Properties, syntheses, and biological applications. *Small* **2015**, *11*, 1620–1636. [CrossRef]
27. Zhao, P.; Yang, M.; Fan, W.; Wang, X.; Tang, F.; Yang, C.; Dou, X.; Li, S.; Wang, Y.; Cao, Y. Facile One-Pot Conversion of Petroleum Asphaltene to High Quality Green Fluorescent Graphene Quantum Dots and Their Application in Cell Imaging. *Part. Part. Syst. Charact.* **2016**, *33*, 635–644. [CrossRef]
28. Zhang, L.; Yu, J.; Yang, M.; Xie, Q.; Peng, H.; Liu, Z. Janus graphene from asymmetric two-dimensional chemistry. *Nat. Commun.* **2013**, *4*, 1–7. [CrossRef]
29. Ren, R.; Zhang, Z.; Zhao, P.; Shi, J.; Han, K.; Yang, Z.; Gao, D.; Bi, F. Facile and one-step preparation carbon quantum dots from biomass residue and their applications as efficient surfactants. *J. Dispers. Sci. Technol.* **2019**, *40*, 627–633. [CrossRef]
30. Ren, R.; Han, K.; Zhao, P.; Shi, J.; Zhao, L.; Gao, D.; Zhang, Z.; Yang, Z. Identification of asphalt fingerprints based on ATR-FTIR spectroscopy and principal component-linear discriminant analysis. *Constr. Build. Mater.* **2019**, *198*, 662–668. [CrossRef]
31. Shi, J.T.; Zhao, P.H.; Fan, W.Y.; Yang, Z.Q.; Lin, Y.; Ouyang, J. Facile preparation and application performance evaluation of SBS/C-9 petroleum resin blends as modifier for high viscosity asphalt. *Constr. Build. Mater.* **2020**, *262*, 120073. [CrossRef]
32. Zhang, Q.; Gu, X.; Yu, Z.; Liang, J.; Dong, Q. Viscoelastic Damage Characteristics of Asphalt Mixtures Using Fractional Rheology. *Materials* **2021**, *14*, 5892. [CrossRef] [PubMed]

coatings

MDPI

Article

The Direct and Oblique Shear Bond Strength of Geogrid-Reinforced Asphalt

Qiaoyi Li [1], Guangqing Yang [2,*], He Wang [2] and Zhijie Yue [2]

1 School of Traffic and Transportation, Shijiazhuang Tiedao University, Shijiazhuang 050043, China; liqiaoyi@stdu.edu.cn
2 School of Civil Engineering, Shijiazhuang Tiedao University, Shijiazhuang 050043, China; wanghe@stdu.edu.cn (H.W.); zi.yue.mce@foxmail.com (Z.Y.)
* Correspondence: yanggq@stdu.edu.cn

Abstract: The interlayer bonding strength is an essential property of geogrid-reinforced asphalt. To study the interlayer bonding characteristics of geogrid-reinforced asphalt, direct shear and oblique shear tests were carried out in the laboratory. The direct interlaminar shear strength of geogrid-reinforced asphalt was lower than that of unreinforced asphalt. The oblique shear strength of the carbon–carbon geogrid-reinforced sample was the highest, the unreinforced sample was second, and the carbon–glass geogrid-reinforced sample was the lowest. The stiffness of the geogrid affects the oblique shear strength. The interlayer direct shear strengths of AC-20C asphalt samples were higher than AC-13C asphalt samples. The oblique shear strengths of AC-20C asphalt samples were almost the same as the AC-13C asphalt samples. Normal stress made the double-layered sample tend to behave as a homogeneous granular material. The direct shear strength vs. shear displacement curves showed an area of oscillation, but the oblique shear curves were smooth throughout the process.

Keywords: geogrid; reinforcement; asphalt; direct shear strength; oblique shear strength

Citation: Li, Q.; Yang, G.; Wang, H.; Yue, Z. The Direct and Oblique Shear Bond Strength of Geogrid-Reinforced Asphalt. *Coatings* **2022**, *12*, 514. https://doi.org/10.3390/coatings12040514

Academic Editors: Leilei Chen and Andrea Simone

Received: 7 March 2022
Accepted: 6 April 2022
Published: 11 April 2022

Publisher's Note: MDPI stays neutral with regard to jurisdictional claims in published maps and institutional affiliations.

Copyright: © 2022 by the authors. Licensee MDPI, Basel, Switzerland. This article is an open access article distributed under the terms and conditions of the Creative Commons Attribution (CC BY) license (https://creativecommons.org/licenses/by/4.0/).

1. Introduction

Using a geogrid in asphalt pavement effectively reduces reflection cracks and extends the service life of the pavement [1–3]. However, the presence of the geogrid introduces a fragile surface at the points of contact between asphalt layers and the geogrid, which can cause problems such as slips and detachments [4]. It is crucial to study the interlayer shear properties of the geogrid-reinforced asphalt. Interface shear bond tests have been considered a good and effective method to test the interlayer bond characteristics of geogrid-reinforced asphalt [5,6]. In recent years, many scholars have studied the interlayer bond characteristics of reinforced asphalt layers through shear tests.

N.S. Correia et al. [7] evaluated the effect of the binder rate and geogrid characteristics on the bond strength of reinforced asphalt interfaces. Results showed that a stiffer geogrid provided lower interface shear strength and interface shear stiffness. Sagnol et al. [8] pointed out that the shear strength of reinforced specimens was reduced by 40% compared with unreinforced specimens. They also suggested that using a non-woven layer between the geogrid strands is not conducive to interlocking. Pasetto et al. [9] conducted pure shear tests on asphalt reinforced with three innovative composite materials. Results showed that the unreinforced interface exhibited brittle failure with a very high interlayer shear strength. In contrast, the presence of the composite reinforcement led to a sensible reduction of interface shear resistance without physical failure. Sudarsanan et al. [10] studied the effect of temperature on the bond strength of reinforced asphalt and found an increase in strength of 10–15% with a change in temperature from −10 to 10 °C, followed by an 80% reduction in strength up to 30 °C. Walubita et al. [11] evaluated the impact of polyester (FA) and fiberglass (FG) geogrids on the HMA-overlays interlayer bond strength. Their research found HMA samples reinforced with polyester-based geogrids performed better than fiberglass-based, geogrid-reinforced HMA samples. Francesco et al. [12] evaluated interlayer bonding by means of shear tests, pointing out that the presence of a geogrid leads to a peak resistance reduction, whereas the residual (post-peak) interlayer friction is not influenced by reinforcement.

A study of the literature finds that shear tests are mainly plane-direct shear tests. The plane-direct shear test reflects the shear mode of the road surface when a vehicle is driving at a constant peed. However, the vehicle will accelerate or decelerate, including emergency braking on steep slopes. In either situation, oblique shear occurs between asphalt pavement layers. It is pointed out that, when the vehicle is in the most unfavorable state of emergency braking on a downhill section, the ratio of interlayer normal stress to interlayer shear stress is close to 1:1 [13]. At present, there is no general standardized test method to characterize and quantify the interlaminar, oblique shear characteristics of reinforced pavement. Reports of research on interlaminar, oblique shear are few.

In this paper, the oblique shear properties of geogrid-reinforced asphalt layers were studied through an oblique shear test device built in-house to obtain a more comprehensive understanding of geogrid-reinforced asphalt. At the same time, pure direct shear tests were carried out. The difference between direct shear properties and oblique shear properties is considered and analyzed. Also, the internal friction angles of different asphalt samples are calculated. This paper can provide a reference for the application and design of geogrid-reinforced asphalt pavement.

2. Experimental Program

2.1. Geosynthetic Reinforcements

Two types of geogrids for asphalt reinforcement were used in this study. One type of geogrid (CC) had transverse and longitudinal ribs composed of 12K carbon fiber wires; the other geogrid (CG) had transverse ribs made of 12K carbon fiber and longitudinal ribs made of 1100 Tex glass fiber. One manufacturer provided both types of geogrids coated with bitumen. Table 1 shows the physical and mechanical properties of the geogrids used in this study.

Table 1. Characteristics of CC and CG geogrids.

Geogrid	Direction	Material	Grid Size (mm)	Grid Thickness (mm)	Elongation at Rupture (%)	Tensile Force (kN/m)
CC	Longitudinal	Carbon fiber	22	0.7	2–2.5	76.00
	Transversal	Carbon fiber	22	0.7	2–2.5	76.53
CG	Longitudinal	Glass fiber	22	0.7	3–4.5	45.00
	Transversal	Carbon fiber	22	0.7	2–2.5	76.92

2.2. Asphalt Mix Design

In this study, two types of double-layered hot mix asphalt slab specimens were made in the laboratory. For the first type, the lower and upper layers were both an AC-20C asphalt mixture composed of crushed limestone aggregates (19 mm nominal maximum size) and 4.3% bitumen content by weight. In the second type, the upper layer was an AC-13C asphalt mixture composed of crushed basaltic aggregates (13.2 mm nominal maximum size) and 4.8% bitumen content by weight, and the lower layer was the same as in the first type of asphalt. It is noted that the asphalt mix design was based on current Chinese guidelines (JTGF40-2004). The mix design sieve curve is presented in Figure 1, and bitumen properties are shown in Table 2.

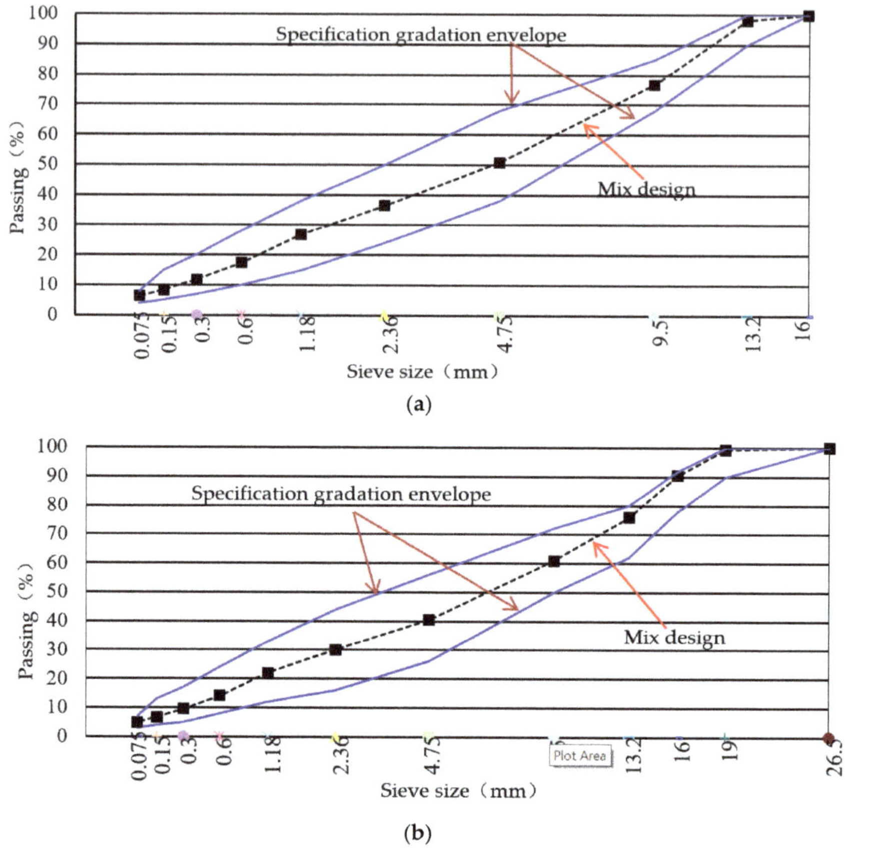

Figure 1. Curve and specification gradation envelope: (**a**) AC-13C mix design curve; (**b**) AC-20C mix design curve.

Table 2. Properties of bitumen.

	Properties	
SBS modified asphalt	**Penetration at 25 °C (0.1 mm)**	55
	Softening point, R&B (°C)	81
	Flashpoint (°C)	270
	Brookfield viscosity 135 °C (Pa/s)	1.9
	Ductility at 5 °C, 5 cm/min (cm)	29
	Penetration Index (Pen/R&B)	0.1
	Penetration (0.1 mm, after RTFOT)	39.6
	Penetration decrease at 25 °C after RTFOT	15.4
	Remaining penetration after RTFOT (%)	75
	Aging resistance (mass % change after RTFOT)	−0.1
	Ductility (cm, 5 °C, 5 cm/min, after RTFOT)	15.3

2.3. Sample Production

Preparation of the hot-mix asphalt used an automatic mixture device. The homogenous asphalt mixture was placed in a mold (300 mm × 300 mm × 50 mm). Then, compaction was carried out so as to form the bottom slab. When the bottom asphalt slab cooled to approx. 25 °C, a tack coat was applied on the surface using a bristle brush before the installation of the geogrid. After the tack coat break, a second layer of hot-mix asphalt was placed and compacted over the geogrid. During the whole preparation process, the mixing temperature of aggregate was not lower than 170 °C, and the tack coat was be painted at 0.5 L/m^2. Four core samples were extracted from the double-layered asphalt slab. Every core sample was 100 mm in diameter and 100 mm thick, with both lower and upper layers being 50 mm thick. The preparation of the test specimen is presented in Figure 2. Tack coat properties are presented in Table 3.

Figure 2. The preparation of test specimens.

Table 3. Tack coat properties.

Emulsion Type	**Cationic Rapid Setting**	
Identification of Cationic Property	**Positive**	
	Sieve test, 1.18 mm (%)	0.09
Viscosity	Angler viscosity E_{25}	10
	Standard viscosity $C_{25,3}$ (s)	25
Residual binder	Content (%)	50.1
	Penetration 100 g, 25 °C, 5 s (dmm)	55
	Softening point, R&B (°C)	63
	Ductility at 5 °C	25
	Solubility in trichloroethylene (%)	99.5
Storage stability	1 d (%)	<0.5
	5 d (%)	<2

2.4. *Interface Shear Tests*

In this study, a constant shear displacement rate of 2.54 mm/min was applied at an ambient temperature of 20 °C for both the direct shear tests and oblique shear tests. This shear displacement rate has been used in many research projects [14–16]. Tests were ended when the shear strength reached 60% of the maximum shear strengths. The direct shear and oblique shear apparatus are shown in Figure 3. The direct shear equipment has a gap width of 5 mm, while the oblique shear apparatus has a gap width of 10 mm.

(**a**) (**b**)

Figure 3. The direct shear and oblique shear apparatus: (**a**) the direct shear apparatus; (**b**) the oblique shear apparatus (Chinese in the figure: Changchun Kexin Company Test Instrument Research Institute, Chinese Academy of Sciences).

3. Results and Analysis

3.1. Interface Direct Shear Strength

Figures 4 and 5 show the results of the direct shear tests. From Figure 4, it can be seen that the ultimate direct shear strengths of the unreinforced asphalt were 0.436 MPa and 0.409 MPa, corresponding to AC-20C and AC-13C, respectively. The ultimate direct shear strengths of the CG geogrid-reinforced samples were 0.338 MPa and 0.290 MPa, while the CC geogrid-reinforced samples reached 0.369 MPa and 0.361 MPa, corresponding to AC-20C and AC-13C, respectively. The ultimate direct shear strengths of the geogrid-reinforced samples were lower compared with the unreinforced samples. This is in accordance with previous investigations [8,9,11,12] carried out with various experimental devices. The ultimate interface direct shear strengths of CC geogrid-reinforced samples were higher than the CG geogrid-reinforced samples. It is considered that the difference was caused by the stiffness of longitudinal ribs. Comparing the AC-20C and AC-13C asphalt samples, it can be found that the interlayer direct shear strengths of AC-20C asphalt samples were higher than AC-13C asphalt samples. The reason may be the maximum aggregate size in the AC-20C was larger than in the AC-13C.

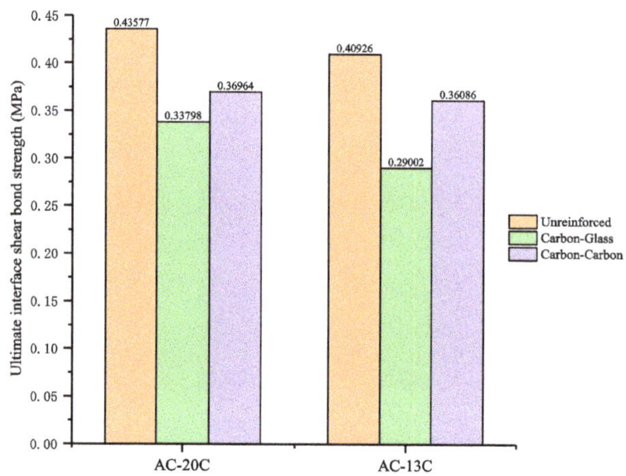

Figure 4. The results of the direct shear tests.

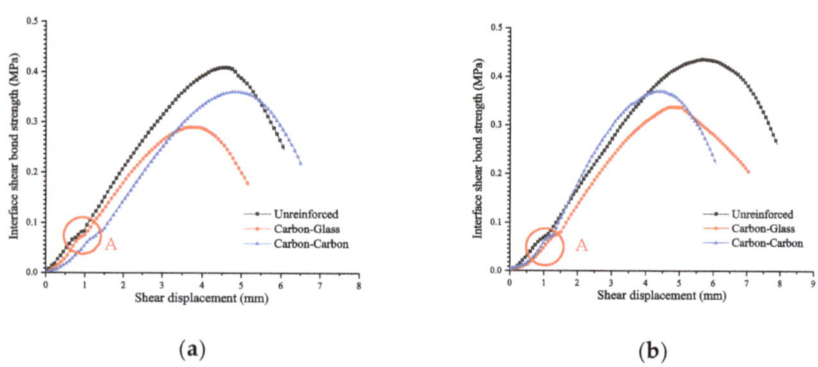

Figure 5. Interface shear strength vs. shear displacement curves in direct shear experiment. (**a**) AC-13C and (**b**) AC-20C.

The shear strength vs. shear displacement curves are shown in Figure 5. The trends of the six curves are similar. This indicates that the interlayer bonding properties of the asphalt do not change when a geogrid is added; the changes occur in the interlayer bonding strength.

3.2. Interface Oblique Shear Strength

The stress state of the specimen under oblique shear is shown in Figure 6.

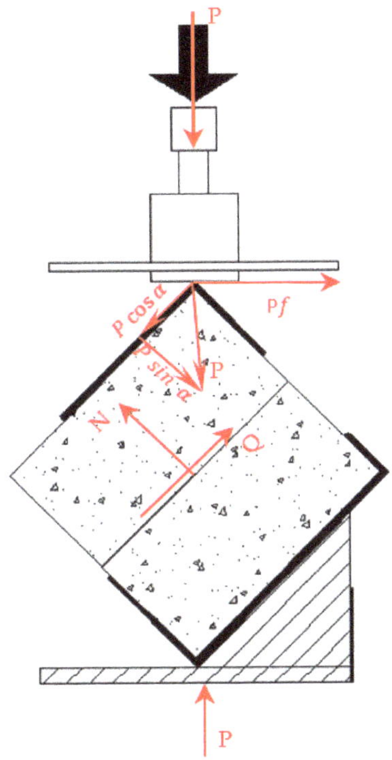

Figure 6. The stress state of the specimen under oblique shear.

P—Stress (The outputs of pressure senor), *N*
N—Normal force on shear plane, *N*
Q—Tangential force on shear plane, *N*
f—Friction coefficient, about 0.003
α—Shear angle, 45°
A—Shear area, mm

And,

N-axis: $N - P\cos\ \alpha - Pf\sin\ \alpha = 0$
Q-axis: $Q + Pf\cos\ \alpha - P\sin\ \alpha = 0$
Then, $\sigma_n = \frac{P}{A}(\cos\ \alpha + f\sin\ \alpha)$

$$\tau_n = \frac{P}{A}(\sin\ \alpha - f\cos\ \alpha)$$

σ_n—Normal stress, MPa
τ_n—Shear stress, MPa

Figure 7 shows the results of the oblique shear tests. They show that, among the three types of asphalt samples, the ultimate oblique shear strength of the CC geogrid-reinforced sample was the highest, with values of 1.441 MPa and 1.456 MPa, corresponding to AC-20C and AC-13C, respectively. The unreinforced sample was second, with values of the ultimate oblique strength of 1.368 MPa and 1.422 MPa, corresponding to AC-20C and AC-13C; respectively. The CG geogrid-reinforced sample was the lowest, with values of 1.300 MPa and 1.203 MPa for AC-20C and AC-13C, respectively.

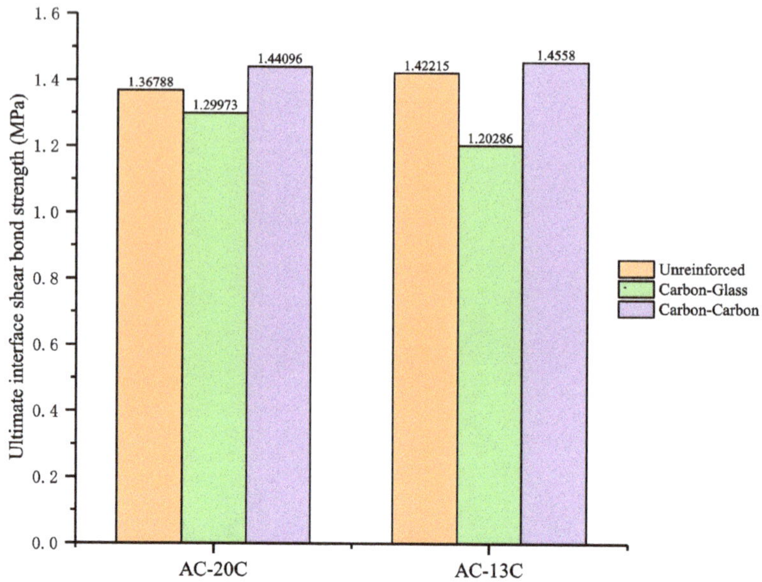

Figure 7. The results of the oblique shear tests.

Comparing the AC-20C and AC-13C asphalt samples, it can be seen that the ultimate oblique shear strengths of the AC-20C asphalt samples are almost the same as those of the AC-13C asphalt samples. In the oblique shear test, with the generation and increase of normal stress, the effect of the asphalt mixture gradation on interlayer shear strength decreases. The AC-20C and AC-13C samples had a tendency to behave the same as would a granular material.

Comparing the direct shear curves and the oblique shear curves, it can be seen that the direct shear curves show vibration when the shear displacements reach 1–2 mm, denoted by point "A". It is believed that the direct shear strength is mainly provided by cohesion and friction. When the shear displacement exceeds 2 mm, cohesion between asphalt layers becomes dominant. The oblique shear curves are smooth throughout the whole shear process. (See Figure 8).

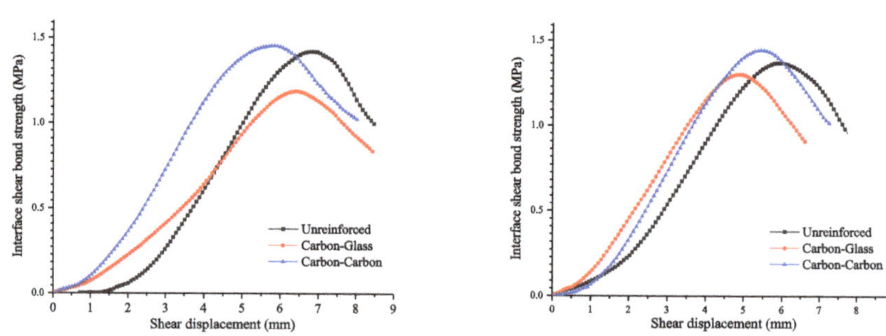

(a) **(b)**

Figure 8. Interface shear strength vs. shear displacement curves in the oblique shear experiment (**a**) AC-13C and (**b**) AC-20C.

3.3. *τ-σ Graph*

The direct shear test can be considered as an oblique shear test with a shear angle (α) of 90°. Then, normal stress $\sigma_n = \frac{P}{A} * f$, $\tau_n = \frac{P}{A}$; with f assumed to be about 0.003

$\sigma_n \approx 0$ MPa. The τ-σ graph is shown in Figure 10. It can be seen that with the normal stress σ_n increasing, the shear stress τ_n substantially converges to a certain value. It can be concluded that the geogrid-reinforced asphalt and unreinforced asphalt have the same interfacial bonding strengths under high normal stress, especially since the geogrid's transverse and longitudinal ribs are made of the same material have the same stiffness. From the data shown in Figures 4, 7, and 9, the yield surface can be drawn in Figure 10. The values of φ are shown in Table 4. The friction angle of the reinforced asphalt is obviously higher than that for the unreinforced asphalt, which means that the mesh structure of the geogrid enhances the friction between asphalt layers.

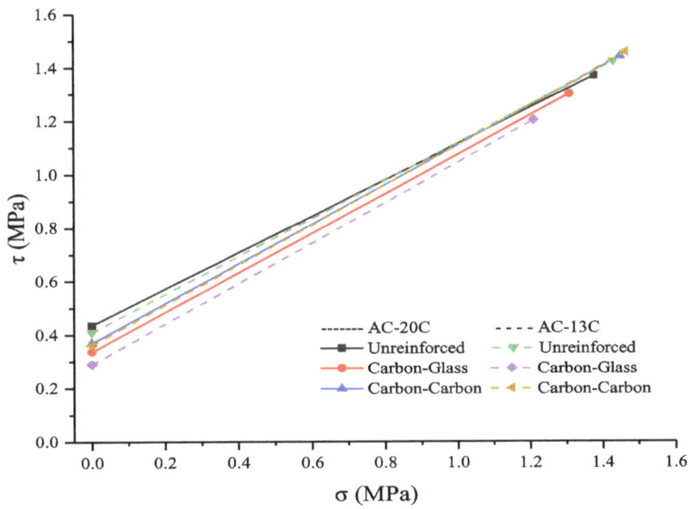

Figure 9. τ-σ image from the direct shear and oblique shear tests.

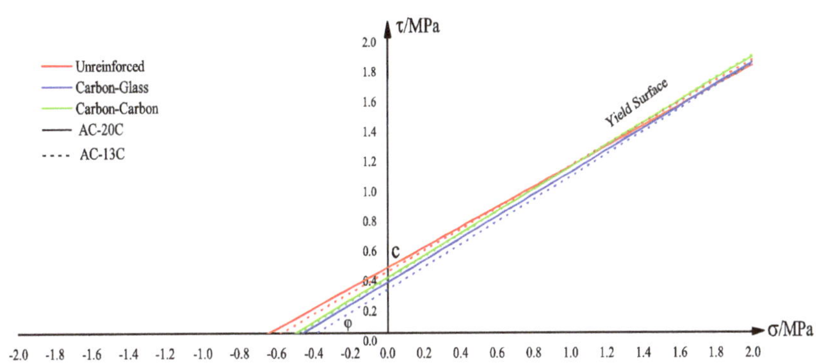

Figure 10. The yield surface.

Table 4. The values of φ of different asphalts.

Friction Angle (°)	Asphalt Mix	Unreinforced	Carbon–Glass	Carbon–Carbon
φ	AC-20C	34.12	36.31	36.47
	AC-13C	35.31	37.04	36.79

4. Summary

This article has discussed the interlaminar shear characteristics of geogrid-reinforced asphalt and unreinforced asphalt through direct shear and oblique shear tests. The following conclusions can be drawn regarding the results presented:

1. The interlaminar direct shear strength of geogrid-reinforced asphalt was lower than that of unreinforced asphalt. The direct shear strength of carbon–carbon geogrid-reinforced asphalt was reduced by 4% and that of carbon–glass geogrid-reinforced asphalt was reduced by 10%.
2. For the oblique shear test, the carbon–carbon geogrid-reinforced asphalt had the highest shear strength of the three asphalts. The oblique shear strength was about four times the direct shear strength.
3. The geogrid-reinforced asphalt and unreinforced asphalt had similar strength vs. shear displacement curves for both the direct shear and oblique shear tests. The oblique shear curves were smooth throughout the whole process, but the direct shear curves showed an area of oscillation.
4. The interlayer direct shear strengths of the AC-20C asphalt samples were higher than those of the AC-13C asphalt samples. The oblique shear strengths of the AC-20C asphalt samples were almost the same as the AC-13C asphalt samples.
5. With the normal stress increasing, the shear stress gradually converged to a certain value. The friction angle of the reinforced asphalt was obviously higher than that of the unreinforced asphalt.

When a vehicle is driving on asphalt pavement, it will shear the asphalt pavement obliquely at different angles. In this paper, due to the limitations of the test equipment, oblique shear tests at various angles were not carried out, but rather we only performed the test at 45°. In future work, the modeling of other oblique angles will be developed so as to more fully understand the bonding characteristics of geogrid-reinforced asphalt.

Author Contributions: Conceptualization, Q.L.; formal analysis, H.W.; investigation, Q.L.; data curation, Z.Y.; writing—original draft preparation, Q.L.; writing—review and editing, G.Y.; supervision, G.Y.; project administration, Q.L.; funding acquisition, G.Y. All authors have read and agreed to the published version of the manuscript.

Funding: This research was funded by the Innovation Funding Project for Postgraduates in Hebei Province (Grant no. CXZZBS2021113).

Institutional Review Board Statement: Not applicable.

Informed Consent Statement: Not applicable.

Data Availability Statement: Not applicable.

Conflicts of Interest: The authors declare no conflict of interest.

References

1. Zofka, A.; Maliszewski, M.; Maliszewska, D. Glass and carbon geogrid reinforcement of asphalt mixtures. *Road Mater. Pavement Des.* **2017**, *18*, 471–490. [CrossRef]
2. Lee, J.H.; Baek, S.B.; Lee, K.H.; Kim, J.S.; Jeong, J.H. Long-term performance of fiber-grid-reinforced asphalt overlay pavements: A case study of Korean national highways. *J. Traffic Transp. Eng. (Engl. Ed.)* **2019**, *6*, 366–382. [CrossRef]
3. Kim, H.; Sokolov, K.; Poulikakos, L.D.; Partl, M.N. Fatigue evaluation of porous asphalt composites with carbon fiber reinforcement polymer grids. *Transp. Res. Rec.* **2009**, *2116*, 108–117. [CrossRef]
4. Zamora-Barraza, D.; Calzada-Peréz, M.; Castro-Fresno, D.; Vega-Zamanillo, A. New procedure for measuring adherence between a geosynthetic material and a bituminous mixture. *Geotext. Geomembr.* **2010**, *28*, 483–489. [CrossRef]
5. Raab, C.; Partl, M.N. Interlayer bonding of binder, base and subbase layers of asphalt pavements: Long-term performance. *Constr. Build. Mater.* **2009**, *23*, 2926–2931. [CrossRef]
6. Petit, C.; Chabot, A.; Destrée, A.; Raab, C. Recommendation of RILEM TC 241-MCD on interface debonding testing in pavements. *Mater. Struct. Constr.* **2018**, *51*, 96. [CrossRef]
7. Correia, N.S.; Mugayar, A.N. Effect of binder rates and geogrid characteristics on the shear bond strength of reinforced asphalt interfaces. *Constr. Build. Mater.* **2021**, *269*, 121292. [CrossRef]

8. Sagnol, L.; Quezada, J.C.; Chazallon, C.; Stöckner, M. Effect of glass fibre grids on the bonding strength between two asphalt layers and its Contact Dynamics method modelling. *Road Mater. Pavement Des.* **2019**, *20*, 1164–1181. [CrossRef]
9. Pasetto, M.; Pasquini, E.; Giacomello, G.; Baliello, A. Innovative composite materials as reinforcing interlayer systems for asphalt pavements: An experimental study. *Road Mater. Pavement Des.* **2019**, *20*, S617–S631. [CrossRef]
10. Sudarsanan, N.; Karpurapu, R.; Amrithalingam, V. An investigation on the interface bond strength of geosynthetic-reinforced asphalt concrete using Leutner shear test. *Constr. Build. Mater.* **2018**, *186*, 423–437. [CrossRef]
11. Walubita, L.F.; Nyamuhokya, T.P.; Komba, J.J.; Ahmed Tanvir, H.; Souliman, M.I.; Naik, B. Comparative assessment of the interlayer shear-bond strength of geogrid reinforcements in hot-mix asphalt. *Constr. Build. Mater.* **2018**, *191*, 726–735. [CrossRef]
12. Francesco, C.; Leonello, B.; Gilda, F.; Andrea, G. Shear and flexural characterization of grid-reinforced asphalt pavements and relation with field distress evolution. *Mater. Struct.* **2015**, *48*, 959–979. [CrossRef]
13. Hu, T.; Zi, J.-M.; Jin, S.-J. Shear Performance of Bridge Deck Pavement Based on Multi-angle Shear Test. *J. Civ. Eng. Manag.* **2019**, *36*, 104–108. [CrossRef]
14. Ferrotti, G.; Canestrari, F.; Virgili, A.; Grilli, A. A strategic laboratory approach for the performance investigation of geogrids in flexible pavements. *Constr. Build. Mater.* **2011**, *25*, 2343–2348. [CrossRef]
15. Ferrotti, G.; Canestrari, F.; Pasquini, E.; Virgili, A. Experimental evaluation of the influence of surface coating on fiberglass geogrid performance in asphalt pavements. *Geotext. Geomembr.* **2012**, *34*, 11–18. [CrossRef]
16. Mohammad, L.N.; Raqib, M.A.; Huang, B.; Mohammad, L.N. Influence of Asphalt Tack Coat Materials on Interface Shear Strength. *Transp. Res. Record.* **2002**, *2*, 56–65. [CrossRef]

coatings

MDPI

Article

Preparation and Temperature Susceptibility Evaluation of Crumb Rubber Modified Asphalt Applied in Alpine Regions

Youjie Zong, Rui Xiong *, Yaogang Tian, Mingfeng Chang, Xiaowen Wang, Jiahao Yu, Yixing Zhang, Baozhu Feng, Haoyu Wang and Chuang Li

School of Materials Science and Engineering, Chang'an University, Xi'an 710061, China; 2020031003@chd.edu.cn (Y.Z.); ygtian@chd.edu.cn (Y.T.); mfchang99@126.com (M.C.); wangxiaowen20211@163.com (X.W.); jiahao17yu@163.com (J.Y.); 2020131065@chd.edu.cn (Y.Z.); 2019131026@chd.edu.cn (B.F.); 2020231070@chd.edu.cn (H.W.); 2021131065@chd.edu.cn (C.L.)
* Correspondence: xiongr61@126.com

Abstract: In alpine regions, the durability of asphalt pavement is worse due to the harsh climate. However, crumb rubber modified asphalt has the potential to improve the durability of pavement. Based on matrix asphalt, crumb rubber, Trans-Polyoctenamer Rubber Reactive Modifier (TOR), and an orthogonal test, the preparation scheme of crumb rubber modified asphalt suitable for alpine regions was obtained. The crumb rubber was selected 30 mesh, and the content of crumb rubber was 24% of the quality of matrix asphalt. Shearing time was 60 min, and preparation temperature was 205 °C. Shearing rate was 5000 r/min, and optimum TOR content was set as 4.5% of the quality of crumb rubber. The temperature susceptibility of matrix asphalt (JZ), crumb rubber modified asphalt (AR), crumb rubber modified asphalt mixed with TOR (TAR) was investigated. The high temperature performance indexes, including phase angle, the complex modulus index, and rutting factor, and the low temperature performance indexes including creep rate and stiffness modulus, the ratio of creep rate and stiffness modulus, Burgers model parameters, dissipated energy ratio, and the derivative of creep compliance were analyzed in depth with multiple parameters. Meanwhile, the mechanisms of prepared AR and TAR were explored. The results indicate that the PG grading is PG64-22 of JZ, PG70-34 of AR, and PG82-28 of TAR. The deformation resistance of TAR at high temperature is superior to AR. The addition of crumb rubber not only improves the temperature susceptibility, but also enhances its viscoelasticity at low temperatures. After the crumb rubber swell in the asphalt system, the network structures are crosslinked and the physical and chemical effects are produced, such as cracking and repolymerization in the asphalt system. TOR can further enhance the swelling and network crosslinking effects of rubber-asphalt. The TAR has the strongest weather resistance. The study provides guidance for AR and TAR to be used in alpine regions.

Keywords: road engineering; alpine regions; crumb rubber modified asphalt; TOR; temperature susceptibility; microscopic mechanism analysis

Citation: Zong, Y.; Xiong, R.; Tian, Y.; Chang, M.; Wang, X.; Yu, J.; Zhang, Y.; Feng, B.; Wang, H.; Li, C. Preparation and Temperature Susceptibility Evaluation of Crumb Rubber Modified Asphalt Applied in Alpine Regions. *Coatings* **2022**, *12*, 496. https://doi.org/10.3390/coatings12040496

Academic Editor: Valeria Vignali

Received: 3 March 2022
Accepted: 4 April 2022
Published: 7 April 2022

Publisher's Note: MDPI stays neutral with regard to jurisdictional claims in published maps and institutional affiliations.

Copyright: © 2022 by the authors. Licensee MDPI, Basel, Switzerland. This article is an open access article distributed under the terms and conditions of the Creative Commons Attribution (CC BY) license (https://creativecommons.org/licenses/by/4.0/).

1. Introduction

The alpine region is characterized by low temperature, high altitude, large temperature difference, and intense ultraviolet radiation. The daily temperature difference can reach more than 35 °C. Winds in the mountain pass areas can reach force 10 to 12 [1]. The number of freeze–thaw cycles is as high as 80 times per year. The alternating time of positive and negative temperature is more than 6 months per year [2]. The average relative humidity is only 35 percent, far less than 65 percent in eastern China [3]. Asphalt pavement is rapidly aged under strong ultraviolet radiation. Continuous low temperature action, high and low temperature cycle change, and damage of freeze–thaw and frost heaving make asphalt surfaces crack, and the pavement performances are significantly reduced [2,3]. With the extension of service life, the expressways with excellent pavement performance may generate various diseases [4,5]. The coupling effects of multiple factors in the environment lead to the aggravation of temperature fatigue of pavement structures and accelerate the fatigue damage and performance degradation of pavement structures and materials. It brings about frequent pavement diseases in early service [6–9]. Meanwhile, the increasing traffic volume proposes a severe test to the stability and durability of the pavement surface in alpine regions [10–12].

With the development and improvement of crumb rubber modified asphalt technology, the application and system are gradually being perfected. Generally, when the crumb rubber content is about 10%, modified asphalt has relatively weak performance improvement. The main purpose of its use is to improve the adhesion of asphalt and aggregate and reduce the temperature sensitivity of asphalt pavement. The crumb rubber modified asphalt is mostly used for asphalt pavements with low traffic volume. When the crumb rubber content is about 15%, the performance of crumb rubber modified asphalt has been significantly improved [13]. The use of crumb rubber modified asphalt can slow down the occurrence of road diseases and improve pavement durability, and the advantages of anti-skid and wear resistance are obtained. It is used for the pavement surface layers of heavy traffic and worse original road condition [13]. The plant-produced crumb rubber asphalt shows good storage stability and satisfying road properties compared to other binders, while the asphalt mixture prepared with plant-produced crumb rubber asphalt shows satisfying road performances. In general, the plant-produced crumb rubber asphalt could be a promising replacement for SBS modified asphalt based on the mixture type evaluated [14,15]. State departments of transportations are currently experimenting with use of ground tire rubber (crumb rubber) in bituminous construction and as a crack sealer [16]. The performances of the recycled asphalt concrete with stable crumb rubber asphalt binder were investigated. Both the normal recycled asphalt mixtures and the stable crumb rubber asphalt recycled asphalt mixtures were prepared with 0%, 30%, or 50% content of the reclaimed asphalt pavement. The stable crumb rubber asphalt is much better than the virgin asphalt in recycling the aged asphalt mixtures, with large reclaimed asphalt pavement content which could reach 50% [17–19]. The microstructure and performance of crumb rubber modified asphalt were explored. The crumb rubber modified asphalt have better performance compared with matrix asphalt. The properties of crumb rubber display significant effects on the performance of crumb rubber modified asphalt. The performance of crumb rubber modified asphalt are improved with the decrease in ash content and the increase in acetone extract. Micrographs of crumb rubber modified asphalt prepared under different conditions showed that, compared with crumb rubber modified asphalt prepared by the traditional preparation process, the scattering state of rubber powder and matrix asphalt in crumb rubber modified asphalt prepared by matrix asphalt preblending process is excellent and even, and a perfect polymer reticular structure is formed in the sample system [20–22].

Crumb rubber modified asphalt mixture can reduce the thickness of pavement surface, the incidences of reflective cracks and temperature shrinkage cracks [14,15]. Meanwhile, crumb rubber comes from waste tire containing high polymers such as natural rubber, which is beneficial for enhancing asphalt performance [23,24]. The use of waste tire is also an environmentally friendly waste material recycling technology, and reduces carbon emissions [25,26]. The mechanical response of two gap-graded asphalt rubber mixtures manufactured by the dry process (AR_{dry}) were evaluated. The observed behavior was compared with that of a similar gap-graded mixture without rubber granulate, used as reference. The results were also compared with analogous asphalt rubber mixes produced elsewhere by the wet process (AR_{wet}). The laboratory results and behavior concluded that mechanical performance of the tested AR_{dry} is better than that measured for the reference blend, and was at the same level of performance as AR_{wet}, provided that a proper mixture design and some construction directives are used [27].

The test sections and physical projects using crumb rubber modified asphalt are involved in south, southwest, and light freezing regions of China. However, research on the specific preparation process and temperature susceptibility of crumb rubber modified asphalt suitable for alpine regions is scarce. Meanwhile, the final performances of crumb rubber modified asphalt are closely related to the chemical composition of the matrix asphalt, physical and chemical properties of the crumb rubber, such as type, size, amount, etc., and preparation parameters, such as the preparation of temperature, shearing rate, mixing time, etc. [28,29]. Raw materials and preparation process are strictly controlled. The crumb rubber modified asphalt of high quality and best performance is prepared. In addition, the technical application system of crumb rubber modified asphalt in alpine regions is still incomplete. In view of this, preparation and temperature susceptibility evaluations of crumb rubber modified asphalt applied in alpine regions were investigated, and modified mechanisms were analyzed.

2. Materials and Methods

2.1. Materials

Crumb rubber modified asphalt is a complex combination system consisting of SK-90# matrix asphalt, crumb rubber, and admixtures. The performance of each raw material has important influences on the performance indexes of crumb rubber modified asphalt. This section mainly studies the properties of SK-90# matrix asphalt, crumb rubber, and admixtures.

2.1.1. Matrix Asphalt

The SK-90# matrix asphalt was used in alpine regions. Table 1 shows its technical indicators, and Figure 1 shows the diagrams of atomic force microscope, which was made by Park Systems in Suwon, Korea. The dispersion phases with peak structure are evenly distributed on the asphalt surface. The structures are thought to be the rise and fall of asphalt in a flat area, which are represented by "spikes" in the three-dimensional diagram. The "bee-structure" of SK-90# matrix asphalt mainly comes from asphaltene and some colloid in the asphalt. The wax elements in asphalt will crystallize at the tip of the "bee-structure", which is not conducive to the adhesion of asphalt and aggregates. The R_q of root mean square roughness is 6.73 nm calculated.

Table 1. Technical indicators of SK-90# matrix asphalt.

Test Items		Unit	Results	Technical Requirements	Methods
Penetration (25 °C, 100 g, 5 s)		0.1 mm	94.5	80~100	T0604
Ductility (15 °C, 5 cm/min)		cm	>150	≮100	T0605
Softening Point (R&B)		°C	46.0	≮44	T0606
Solubility		%	99.6	≮99.5	T0607
Brookfield viscosity (135 °C)		Pa·s	1.885	-	T0625
Flash point		°C	260	≮245	T0611
Specific gravity		g/cm^3	1.030	Measured value	T0603
RTFOT (163 °C, 75 min)	Mass loss	%	+0.4	±0.8	T0610
	Ductility (10 °C)	cm	12	≮8	T0605
	Penetration ratio (25 °C)	%	57.8	≮57	T0604

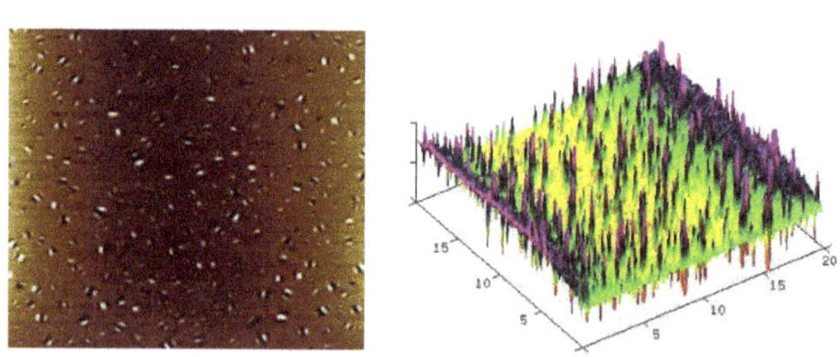

Figure 1. The atomic force microscope images of SK-90# matrix asphalt.

2.1.2. Crumb Rubber

Waste truck radial tire was used as raw material to prepare crumb rubber by normal temperature grinding process. Figure 2 shows the structure of load radial tire, and Table 2 shows the differences in material compositions of different tires by test analysis.

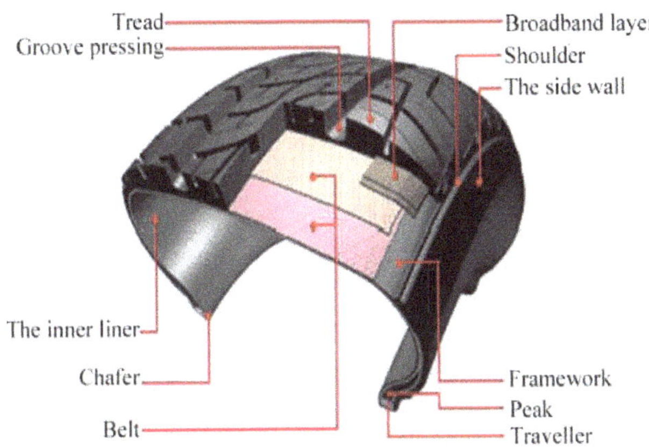

Figure 2. The structure of load radial tire.

Table 2. Comparison of tire composition between car and truck.

Tire Type	Material Content (%)						
	Rubber	Black Carbon	Metal	Fabric	Zinc Oxide	Sulphur	Additive
Car	48	22	15	5	5	1	4
Truck	43	21	27	-	2	1	6

Figure 3 shows the grinding process at normal temperature. As the crumb rubber particles have an irregular shape and large relative specific surface area by the grinding process. It is conducive to full interaction with asphalt, and the cost is the lowest [30,31].

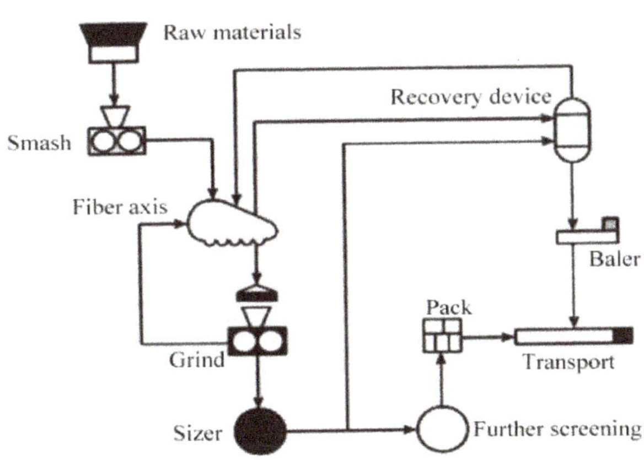

Figure 3. The grinding process at normal temperature.

The crumb rubber images of a scanning electron microscope (SEM) and an infrared spectrum are shown in Figures 4 and 5. Hitachi S-4800 cold field emission scanning electron microscope was made by Japanese high-tech manufacturer, and Fourier transform infrared spectrometer (BRUKER TENSOR II) was made by Germany. Wave numbers of 2958 and 2917 cm^{-1} are the results of stretching vibration of C–H bond in $–CH_3$ and $–CH_2–$. In total,

2848 cm^{-1} is produced by symmetric expansion of $–CH_2–$; 1655 cm^{-1} is produced by the vibration of C=O double bond in the carboxy group; 1448 cm^{-1} is produced by the variable angle vibration of $–CH_2–$; and 1092 cm^{-1} is produced by the scaling of C–O. The wave number 1539 cm^{-1} is the stretching vibration result of NO_2 in the $R–NO_2$. The wave numbers of 743, 834, and 1375 cm^{-1} are generated by out-of-plane and in-plane bending vibration of the benzene torus. The wave number of 965 cm^{-1} is produced by the out-of-plane bending vibration of the C–H bond in the trans unsaturated group of butadiene. The main components of crumb rubber are natural rubber and styrene-butadiene rubber. The technical specifications are shown in Tables 3 and 4.

Figure 4. The SEM image of crumb rubber (×1000).

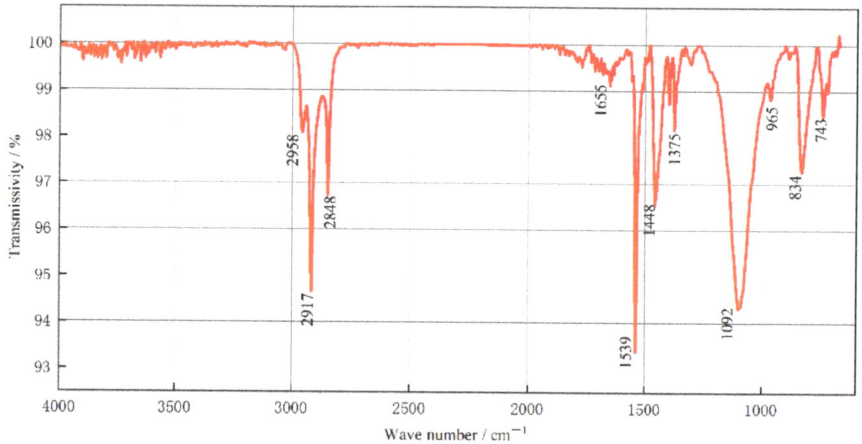

Figure 5. The infrared spectrum diagram of crumb rubber.

Table 3. Physical indicators of crumb rubber.

Index	Metal Content (%)	Fiber Content (%)	Moisture Content (%)	Relative Density
Technical requirements ①	≤0.05	≤1	≤1	1.10~1.30
Test results	0.023	0	0.3	1.20

① Reference standard <Ground vulcanized rubber of scrap tires for highway engineering> (JT/T 797-2011).

Table 4. Chemical indicators of crumb rubber.

Index	Rubber Hydrocarbon Content (%)	Ash Content (%)	Acetone Extract (%)	Carbon Black Content (%)
Technical requirements ①	≥42	≤8	≤22	≥28
Test results	57.5	6.9	12.1	31.5

① Reference standard <Ground vulcanized rubber of scrap tires for highway engineering> (JT/T 797-2011).

2.1.3. Trans-Polyoctenamer Rubber Reactive Modifier

Trans-Polyoctenamer Rubber Reactive Modifier (TOR) is a polymer with a double bond structure. As shown in Figure 6, TOR is translucent crystalline particles. TOR crosslinks the sulfur in the asphalt to the sulfur on the crumb rubber surface. The network structures of rings and chains of polymers are formed.

Figure 6. The macro morphology of TOR.

Figure 7 shows the infrared spectrum diagram of TOR. TOR is mainly composed of saturated and unsaturated carbon chains. The absorption peak of TOR is produced by the antisymmetric absorption vibration of C–H bond in methylene CH_2– at 2917 cm^{-1}. In total, 2849 cm^{-1} is the symmetric vibration effect of C–H in alkane and cycloalkane. The absorption peak of 1467 cm^{-1} is the bending vibration in C–H bond plane. The wave number of 965 cm^{-1} is produced by the out-of-plane bending vibration of the C–H bond in the trans unsaturated group of butadiene. The in-plane wobble vibration produced by saturated methylene is approximately 719 cm^{-1}. The technical specifications are shown in Table 5.

Table 5. Technical indicators of TOR.

Test Item	Unit	Result	Method
Molecular weight (GPC)	/	90,000	Acc-DIN55672-1
Glass-transition temperature (Tg)	°C	−65	ISO 11357
Crystallinity (23 °C)	%	−30	ISO 11357
Melting point	°C	54	ISO 11357
Thermal decomposition (TGA)	°C	275	ISO 11357
Cis/trans double bond ratio	%	20/80	SOP 0188
Mooney viscosity	/	<10	ASTM D1646
Ash content	%	≤0.1	DIN535568.Part1
Volatile substance (1 h, 105 °C)	%	≤0.5	ISO 248
Density (23 °C)	g/cm^3	0.910	ISO 1183

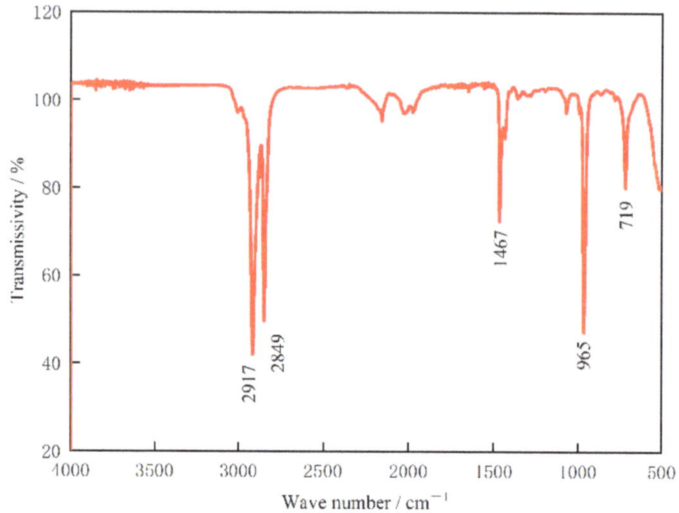

Figure 7. The infrared spectrum diagram of TOR.

2.2. Methods

2.2.1. Preparation Method of Crumb Rubber Modified Asphalt

The matrix asphalt selected SK-90#. It was referred to as JZ. Crumb rubber modified asphalt was referred as AR. When TOR was added, crumb rubber modified asphalt was referred as TAR.

In order to prepare crumb rubber modified asphalt suitable for pavement in alpine regions, the orthogonal test method was adopted. Penetration (25 °C, 100 g, 5 s), softening point, ductility (5 °C), segregation softening point difference (48 h, 163 °C), elastic recovery rate (25 °C), the |B| value of viscosity–temperature regression coefficient, and shearing strength (25 °C) were selected as evaluation indexes. According to Equation (1), the shearing strength was obtained through cone penetration test. The cone penetration value of crumb rubber modified asphalt was tested by cone penetration tester, as shown in Figure 8. |B| values were acquired by Equation (2) and regression analysis of crumb rubber modified asphalt viscosity with temperatures of 135, 150, 165, 180, and 195 °C.

$$\tau = \frac{981Q\cos^2\left(\frac{\alpha}{2}\right)}{\pi h^2 \tan\left(\frac{\alpha}{2}\right)} \quad (1)$$

in which τ is shearing strength, kPa; Q is total mass of conical needle, connecting rod and weights, 195 g; h is cone penetration, mm; and α is the angle of cone tip, 30°.

$$\eta = Ae^{B/T} \quad (2)$$

in which η is viscosity, Pa·s; T is the absolute temperature, K; and A and B are constants. |B| is viscosity–temperature regression coefficient.

The crumb rubber mesh, crumb rubber content, shear time, preparation temperature, and shear rate were taken as the main influencing factors. A four-level table of five factors (A, B, C, D, and E) was listed, as shown in Table 6. The orthogonal table L16 (45) and test scheme are listed in Table 7.

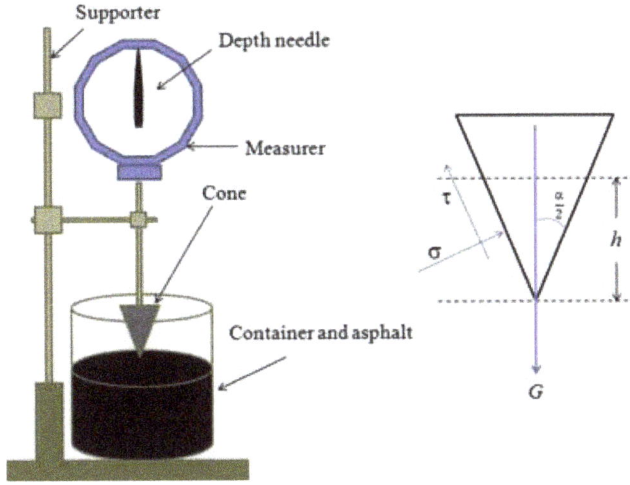

Figure 8. Cone penetration tester.

Table 6. Factor and level of orthogonal test.

Level	Factor				
	A/Crumb Rubber (Mesh)	B/Crumb Rubber Content (%)	C/Shear Time (min)	D/Preparation Temperature (°C)	E/Shear Rate (r/min)
1	25	12	45	175	3500
2	40	16	60	190	4000
3	60	20	75	205	4500
4	80	24	90	220	5000

Table 7. Orthogonal test scheme.

Scheme	Crumb Rubber (Mesh)	Content (%)	Shear Time (min)	Preparation Temperature (°C)	Shear Rate (r/min)
1	25	12	45	175	3500
2	25	16	60	190	4000
3	25	20	75	205	4500
4	25	24	90	220	5000
5	30	12	60	205	5000
6	30	16	45	220	4500
7	30	20	90	175	4000
8	30	24	75	190	3500
9	40	12	75	220	4000
10	40	16	90	205	3500
11	40	20	45	190	5000
12	40	24	60	175	4500
13	60	12	90	190	4500
14	60	16	75	175	5000
15	60	20	60	220	3500
16	60	24	45	205	4000

According to the orthogonal test scheme, the preparation process of crumb rubber modified asphalt is shown in Figure 9. On the basis of Standard Test Methods of Bitumen and Bituminous Mixtures for Highway Engineering (JTG E20—2011) and test results, the best preparation process and parameters were determined.

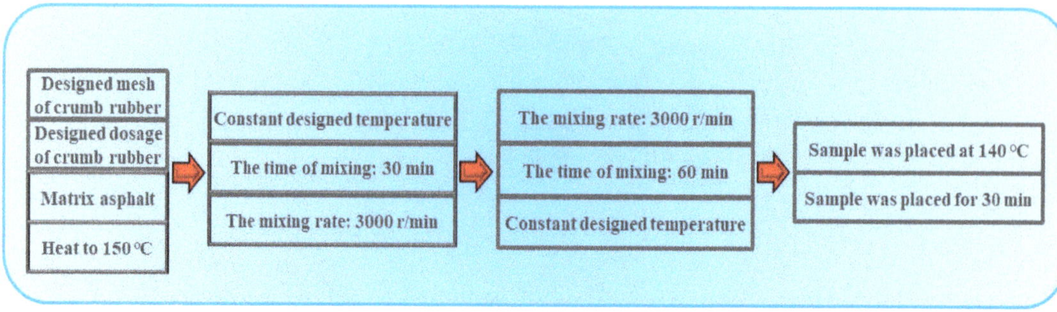

Figure 9. The preparation process of crumb rubber modified asphalt.

2.2.2. Determination of TOR Content

TOR was added to improve pavement performance of crumb rubber modified asphalt. TOR was 0, 4.0%, 4.5%, or 5.0% of the content of crumb rubber. This is represented as W_0, $W_{4\cdot0}$, $W_{4\cdot5}$, or $W_{5\cdot0}$, respectively. According to Standard Test Methods of Bitumen and Bituminous Mixtures for Highway Engineering (JTG E20-2011), the experiments of penetration, softening point, ductility, elastic recovery rate, cone penetration, segregation softening point difference, and Brinell viscosity at 135, 150, 165, 180, and 195 °C were conducted. The optimal content of TOR was determined in crumb rubber modified asphalt. Meanwhile, shear strength is converted from the cone penetration, according to Equation (1). The Viscosity–Temperature-Susceptibility was calculated by Equation (2).

2.2.3. Dynamic Shear Rheological Test

Referring to AASHTO-TP5 test method, the dynamic shear rheological test was carried out by DHR-1 dynamic shear rheometer. The strain control mode was selected. The strain value was 8%. The frequency was 10 rad/s. The temperature scanning test was carried out in the temperature range of 46–82 °C. Through comparative analysis of phase angle, the complex modulus index GTS, and rutting factor, the effects of crumb rubber and TOR on the high temperature performance of crumb rubber modified asphalt were investigated.

2.2.4. Bending Beam Rheometer Test

The bending beam rheometer is used to test the low temperature performance of asphalt samples. After the three kinds of asphalt samples were aged by TFOT and PAV, the beam samples were prepared. Its length, width, and thickness were 127 ± 2.0, 12.7 ± 0.1, and 6.4 ± 0.1 mm, respectively. The test temperature of JZ was set to 6, −12, and −18 °C. The test temperature of AR/TAR were set to −12, −18, −24, and −30 °C. The stiffness modulus and creep rates of different asphalt were obtained by testing under constant stress of 240 s and set temperature. The stiffness modulus and creep rate were calculated by Equations (3) and (4), respectively [32–34].

$$S(t) = \frac{Pl^3}{4bh^3v(t)} \tag{3}$$

$$m = d \times \log(S(t))/d \times \log(t) \tag{4}$$

in which S is stiffness modulus, MPa; m is creep rate; l is the span of the beam, mm; v(t) is deformation of beam midpoint at some moment, mm; h is the height of the beam, mm; p is the constant load, N; and b is the width of the beam, mm.

2.2.5. Microcosmic Test of Crumb Rubber Modified Asphalt

The prepared JZ, AR, and TAR were placed under a fluorescence microscope, scanning electron microscope, and Fourier infrared spectrometer, respectively. The magnifications of the scanning electron microscope and Fourier microscope were selected 1000 and 5000 times, respectively. The magnification of fluorescence microscope is 100 times. The microstructures of three kinds of asphalts were observed, and mechanisms were analyzed. Figure 10 shows flow chart of material preparations and experimental programs.

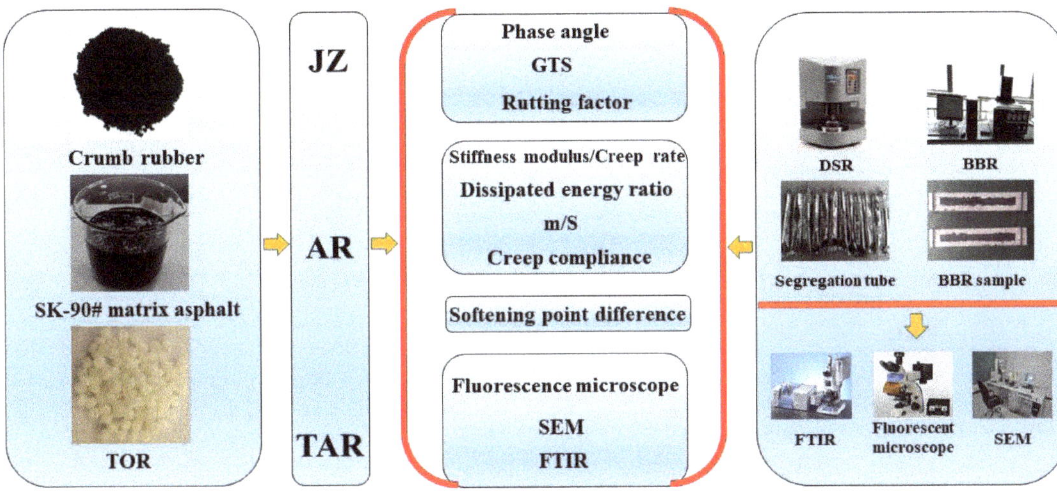

Figure 10. Flow chart of material preparations and experimental programs.

3. Results and Discussion

3.1. Preparation of Crumb Rubber Modified Asphalt

3.1.1. Analysis of Orthogonal Test Results

According to the designed orthogonal test scheme, 16 groups of crumb rubber modified asphalt were prepared. Test results are shown in Table 8.

Table 8. Orthogonal test results.

Number	Penetration (25 °C, 100 g, 5 s) (0.1 mm)	Ductility (15 °C, 5 cm/min) (cm)	Softening Point (°C)	Elastic Recovery Rate (%)	Cone Penetration of 25 °C (cm)	Segregation Softening Point Difference (°C)	Viscosity of 180 °C (cp)	∣B∣ Value	Shearing Strength (kPa)
1	53.2	8.9	57.4	56	48.3	3.6	537.5	788.4	70.3
2	54.2	11.5	59.8	71	53.2	5.3	1040.0	772.2	58.4
3	77.3	13.2	54.2	43	71.6	3.2	830.0	862.2	32.8
4	85.4	16.9	53.7	38	82.6	3.0	675.0	725.6	24.4
5	67.7	9.6	54.2	56	58.5	0.6	395.0	673.4	48.2
6	73.4	11.2	54.4	71	64.2	8.7	520.0	766.7	40.9
7	52.7	12.0	62.8	68	48.9	6.7	1815.0	896.3	68.4
8	58.9	13.8	63.6	78	50.3	3.7	1775.0	814.0	64.1
9	78.1	12.2	52.6	66	67.4	4.8	342.5	761.0	36.0
10	81	14.2	52.3	54	73.2	3.1	307.5	814.9	30.5
11	84.1	16.6	52.7	58	76.9	5.3	530.0	821.6	28.4
12	46.5	12.5	66.4	82	37.1	2.6	2160.0	749.1	118.3
13	74.4	10.6	50.3	42	64.5	1.3	172.5	872.1	40.7
14	85.3	12.0	50.9	78	12.2	3.1	307.5	761.1	27.3
15	84.2	14.2	51.2	64	76.1	5.4	390.0	798.1	28.2
16	87.4	17.7	54.0	70	72.9	4.4	852.5	800.1	31.1

The range values of penetration, softening point, ductility, segregation softening point difference, elastic recovery rate, |B| value, and shearing strength were analyzed. The range values of various factors and the effects of factor level on the corresponding index were calculated, with the results shown in Table 9.

Table 9. Range values of corresponding indexes under various experimental factors.

Factor	Penetration (25 °C, 100 g, 5 s) (0.1 mm)	Ductility (15 °C, 5 cm/min) (cm)	Softening Point (°C)	Elastic Recovery Rate (%)	Viscosity (cp)	Segregation Softening Point Difference (°C)	|B| Value	Shearing Strength (KPa)
A/Powder mesh	19.7	2.2	7.2	16.3	695.6	1.4	21.2	23.5
B/Powder content	6.4	5.0	5.1	13.5	1003.8	2.6	72.3	20.5
C/shear time	11.6	1.7	3.3	17.8	386.3	2.0	79.0	23.3
D/Preparation temperature	20.7	1.8	6.4	15.2	653.9	2.7	57.1	38.8
E/Shear rate	12.8	1.9	3.5	11.3	535.6	2.3	67.2	25.5

The performances of crumb rubber modified asphalt prepared under different combination schemes are different. The crumb rubber mesh has the most significant effect on the softening point value. The softening point values of crumb rubber modified asphalt prepared by 30 mesh and 40 mesh crumb rubber are higher. However, the softening point value of crumb rubber modified asphalt prepared by 60 mesh crumb rubber is minimal. The content of crumb rubber has a significant effect on the low temperature ductility and viscosity of crumb rubber modified asphalt. When the content of crumb rubber is 24%, the ductility and viscosity are about 1.5 times and 3 times higher, respectively, than those of the 12% sample. Therefore, the high content of crumb rubber is beneficial to improve the low temperature performance. Shear time has the most effect on elastic recovery rate and |B| value. With the extension of shear time, elastic recovery rate first increases and then decreases, while the |B| value is the opposite. When shear time is 60 min, elastic recovery rate is the strongest and temperature sensitivity is the lowest. Preparation temperature significantly affects the penetration, segregation softening point difference, and shear strength of the crumb rubber modified asphalt. With the increase in preparation temperature, the penetration gradually increases, and it is also affected by the crumb rubber mesh. The segregation softening point difference first decreases and then increases with the increase in the preparation temperature, and the minimum value is obtained when the preparation temperature is 205 °C. The shear strength decreases significantly with the increase in the preparation temperature. The reason is that the effect of temperature will lead to desulfurization of crumb rubber in asphalt. The shear rate has no significant effect on the performances of crumb rubber modified asphalt.

According to performance requirements of crumb rubber modified asphalt in alpine regions, proper analysis were conducted for penetration, softening point, ductility and elastic recovery rate, segregation softening point difference, shearing strength, and |B| value. The optimal combination schemes required by composite corresponding indexes were selected, as shown in Table 10.

The optimal combination schemes for the preparation of crumb rubber modified asphalt is A_2-B_4-C_2-D_3-E_4, that is, the crumb rubber mesh is 30. The content of crumb rubber is 24%. The shear time is 60 min. The preparation temperature is 205 °C, and the shear rate is 5000 r/min. The confirmed process flow charts of preparation of crumb rubber modified asphalt are shown in Figure 11.

Table 10. Analysis and summary of orthogonal test results.

Index	Factors Influence the Order of Significance	Optimal Portfolio
Penetration	D/Preparation temperature > A/Powder mesh > E/Shear rate > C/shear time > B/Powder content	A_4-B_3-C_3-D_4-E_4
Ductility	B/Powder content > A/Powder mesh > E/Shear rate > D/Preparation temperature > C/shear time	A_3-B_4-C_1-D_3-E_4
Softening point	A/Powder mesh > D/Preparation temperature > B/Powder content > E/Shear rate > C/shear time	A_2-B_4-C_2-D_1-E_2
Elastic recovery rate	C/shear time >A/Powder mesh > D/Preparation temperature > B/Powder content > E/Shear rate	A_2-B_4-C_2-D_2-E_2
Viscosity	B/Powder content > A/Powder mesh > D/Preparation temperature > E/Shear rate > C/shear time	A_4-B_1-C_1-D_4-E_4
Segregation softening point difference	D/Preparation temperature > B/Powder content > E/Shear rate > C/shear time > A/Powder mesh	A_4-B_1-C_2-D_3-E_4
∣B∣ value	C/shear time > B/Powder content > D/Preparation temperature > E/Shear rate > A/Powder mesh	A_1-B_4-C_2-D_3-E_4
Shearing strength	D/Preparation temperature > E/Shear rate > A/Powder mesh > C/shear time > B/Powder content	A_2-B_4-C_2-D_1-E_3

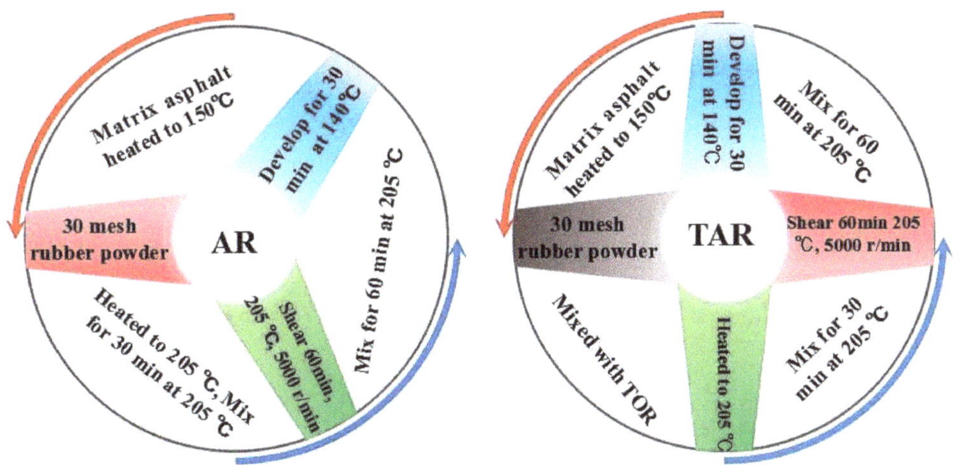

Figure 11. Flow charts of AR and TAR preparation process. (**a**) AR, (**b**) TAR.

3.1.2. Determination of TOR Content

When the content of TOR is different, the performances of crumb rubber modified asphalt is shown in Figures 12 and 13. In Figure 12, the softening point, elastic recovery rate, and shear strength of crumb rubber modified asphalt enhance with the increase in TOR, while the penetration, low temperature ductility, segregation softening point difference, and VTS decrease, indicating that TOR significantly improved the flexibility, temperature susceptibility, and storage stability of crumb rubber modified asphalt at low temperatures. With the increase in TOR content, the elastic recovery rate increases by 1%, 7%, and 10%, respectively. The shear strength increases by 5%, 51%, and 79%, respectively. The VTS decreases by 1%, 6%, and 9%, respectively. The difference of segregation softening point decreases by 9%, 8%, and 7%, respectively. When the content of TOR is 4.5%, all the evaluation indexes change significantly. When the content of TOR increases from 4.5% to 5%, the difference of the segregation softening point increases from 0.7 to 1.2, and the viscosity at 150 °C increases from 4700 to 6075 cp, as shown in Figure 13. Considering the high viscosity, poor storage stability of crumb rubber modified asphalt, the climate of alpine regions, and economical analysis, the appropriate TOR content is determined to be 4.5%.

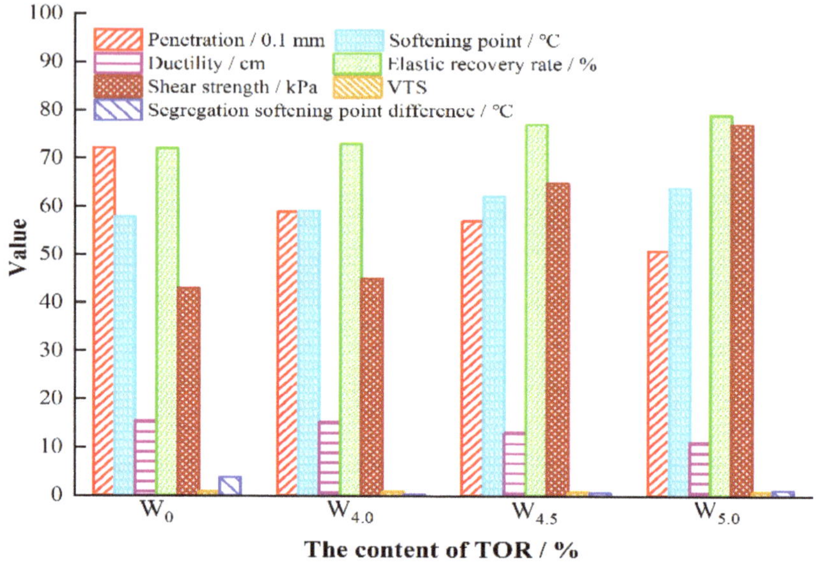

Figure 12. Performance indexes of crumb rubber modified asphalt with different TOR content.

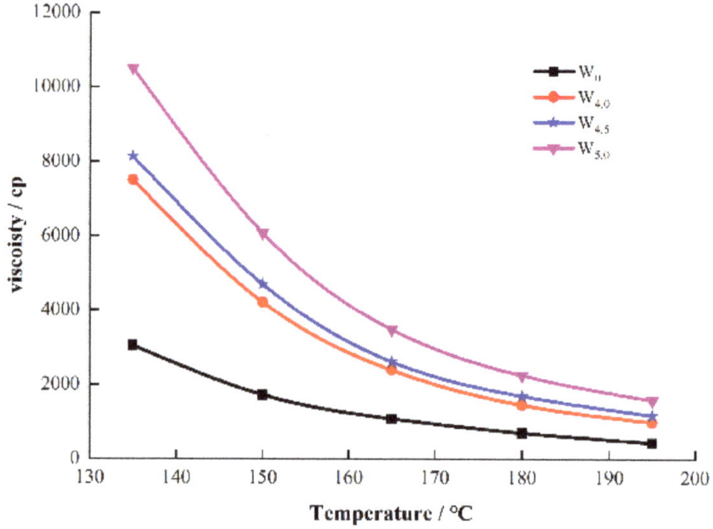

Figure 13. Viscosity of crumb rubber modified asphalt with different TOR content.

3.2. *Investigation on High Temperature Performance of Asphalt*

3.2.1. Phase Angle

The parameters of dynamic shear rheological test were performed in the temperature range of 46–82 °C. The strain value was 8%, and the frequency was 10 rad/s. The results are shown in Figure 14.

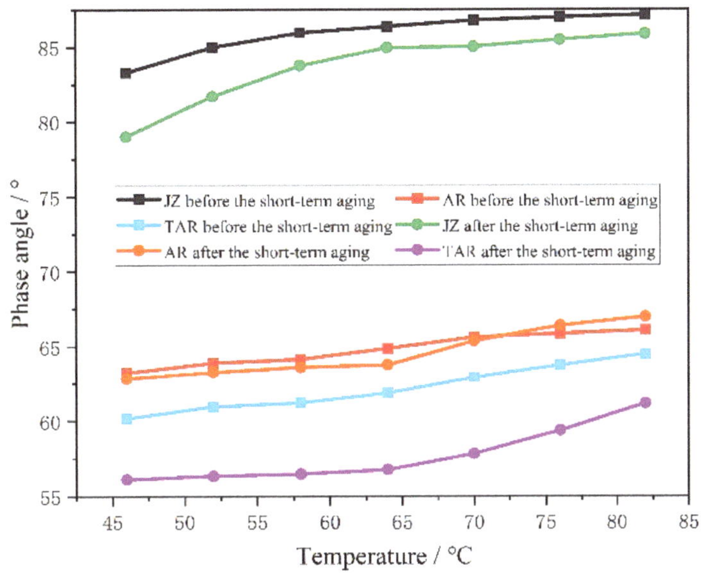

Figure 14. Phase angle.

As shown in Figure 14, the values of phase angle gradually increase as the temperature increases. Under the same temperature, the values of phase angle are JZ > AR > TAR. The delay time of asphalt response to stress is affected by asphalt type and temperature condition. When the temperature is 64 °C, the δ_{AR}/δ_{JZ} and δ_{TAR}/δ_{JZ} are 0.76 and 0.72, respectively, before the short-term aging. The reason is that the incorporation of crumb rubber increases the elastic components in the complex modulus of asphalt. The high temperature deformation resistance of asphalt under load is enhanced. The addition of TOR further enhances the effect. After short-term aging, the phase angles of the three kinds of asphalts decrease; δ_{AR}/δ_{JZ} and δ_{TAR}/δ_{JZ} are 0.75 and 0.67, respectively. Meanwhile, the light components of asphalt volatilize. The asphalts become hardened by oxidation. The elastic components of asphalt increase, and the phase angles decrease. The significant order of short-term aging effect on the three kinds of asphalts is TAR > AR = JZ.

3.2.2. Complex Modulus Exponent

The complex modulus exponent method is used to characterize the temperature sensitivity of medium and high temperature regions of asphalt by dynamic shear rheometer test. The complex modulus exponent was calculated by Equation (5) [35–37].

$$\lg(\lg G^*) = GTS \times \lg T + C \quad (5)$$

in which G* is complex modulus, Pa; GTS is the complex modulus exponent; C is constant; and T is the test temperature, K.

Dynamic shear rheometer was used to test the rheological properties of asphalt samples at 58, 64, 70, 76, and 82 °C. Figure 15 shows the regression equations of GTS and regression coefficient results. The greater the absolute value of GTS, the greater the temperature sensitivity of asphalt. The complex modulus G* double logarithms of the three kinds of asphalts have good linear relationships with the temperature logarithm.

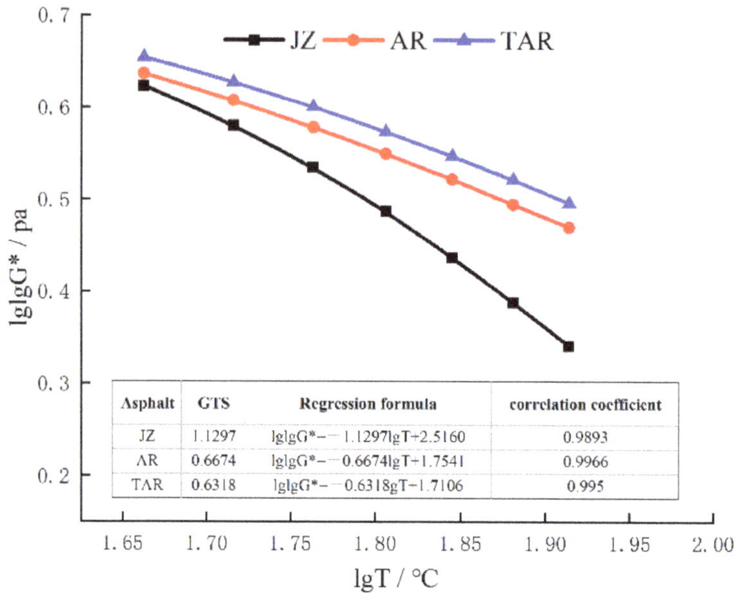

Figure 15. The variations of complex modulus G* double logarithm with temperature logarithm.

3.2.3. Rutting Factor

In Figure 16, as the temperature increases, the rutting factors of the three kinds of asphalts exhibit decreasing trends. The reason is that asphalts behave as an elastic state at low temperature. As the temperature increases, the asphalt changes into a viscous flow state, and the strain caused by the shear force decreases gradually. Under the same temperature, the rutting factors are TAR > AR > JZ. When temperature is 64 °C, before the short-term aging, the rutting factors of AR and TAR increase by 2.3 and 0.6 times compared with JZ. The adding of crumb rubber enhances the rutting resistance of asphalts at high temperature. The reason is that the swelling effect of the compounds of asphalt and crumb rubber increases the elastic components of asphalt. Moreover, the addition of TOR enhances the interactions between asphalt and crumb rubber, so TOR enhances the deformation resistance of TAR at high temperature. At 58 °C, after the short-term aging, the rutting factors of the three kinds of asphalts increase by 0.8, 0.3, and 0.7 times, respectively. The reason is that the rutting factors increase after short-term aging, while AR is the least affected by short-term aging. Meanwhile, after short-term aging, the PG grading of AR is reduced by one grade. The final PG grading temperature of high temperatures are 64 °C (JZ), 70 °C (AR), and 82 °C (TAR).

3.3. *Low Temperature Performance Evaluation of Asphalt*

3.3.1. Low Temperature Creep Test

According to the SHRP evaluation methods, the BBR test results of three kinds of asphalts are shown in Tables 11 and 12. When the temperature is −12 °C, compared with m_{JZ} and S_{JZ}, m_{AR} increases by 19% and S_{AR} decreases by 73%. The results show that crumb rubber can enhance the relaxation ability at low temperature and crack resistance of asphalt. Under the same temperature, compared with AR and TAR, the m_{JZ} is slightly lower, and the S_{JZ} is slightly higher. In conclusion, the low temperature performances of the three kinds of asphalts are AR > TAR > JZ. According to the specification, the slope m of the creep curve is not less than 0.3, and the stiffness modulus S is not more than 300 MPa. The PG grading temperatures of low temperature can be obtained as −22 °C (JZ), −34 °C (AR), and −28 °C (TAR).

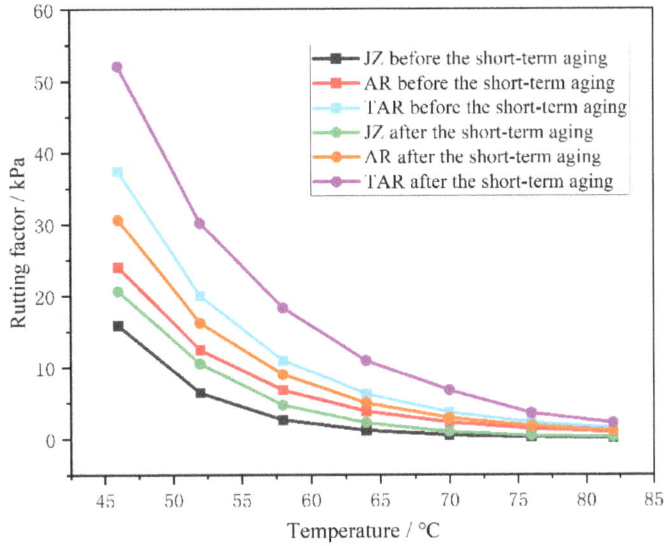

Figure 16. The effects of short-term aging on rutting factor.

Table 11. Low temperature bending test results of JZ.

Asphalt	−6 °C		−12 °C		−18 °C	
	m	S/MPa	m	S/MPa	m	S/MPa
JZ	0.449	47.6	0.365	123	0.286	296

Table 12. Low temperature bending test results of AR and TAR.

Asphalt	−12 °C		−18 °C		−24 °C		−30 °C	
	m	S/MPa	m	S/MPa	m	S/MPa	m	S/MPa
AR	0.436	33.7	0.378	80.6	0.358	235	0.232	455
TAR	0.42	43.1	0.343	106	0.297	253	0.236	430

Figure 17 shows the variations of creep rate and stiffness modulus changing with time. The creep rate decreases with the decrease in temperature, while the stiffness modulus increases. At the 87th load action, the creep rate curves of AR intersect at −18 and −24 °C, and then continue to accumulate with the number of load actions. However, $m_{-24°C}$ is larger than $m_{-18°C}$. It shows that PG grading cannot completely distinguish the low temperature performance of asphalt. It is limited to reflect the low temperature performance of asphalt only by m or S. Therefore, multiple processing and calculation methods should be used to evaluate [38–40].

3.3.2. m/S Value in Creep Compliance

The m/S ratio was adopted to evaluate the low temperature performance of asphalt. As the m/S ratio increases, the low temperature performance of asphalt is better [41,42].

Figure 18 shows the variations of m/S values with loading time. With the increase in loading time, the m/S values increase gradually and eventually tend to be flat. These show that the low temperature performances of three kinds of asphalts tend to be stable with the increase in loading time. At −12 °C, the m/S values of AR and TAR are significantly higher than JZ, indicating that the addition of crumb rubber improves the low temperature performance of asphalt. The m/S differences between AR and TAR decrease with decreasing temperature. The results show that the effects of crumb rubber and TOR on the low temperature performance of asphalt are reduced at lower temperature.

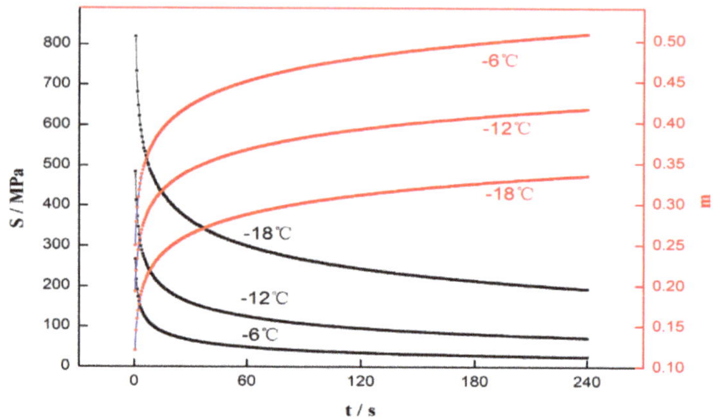

(**a**)

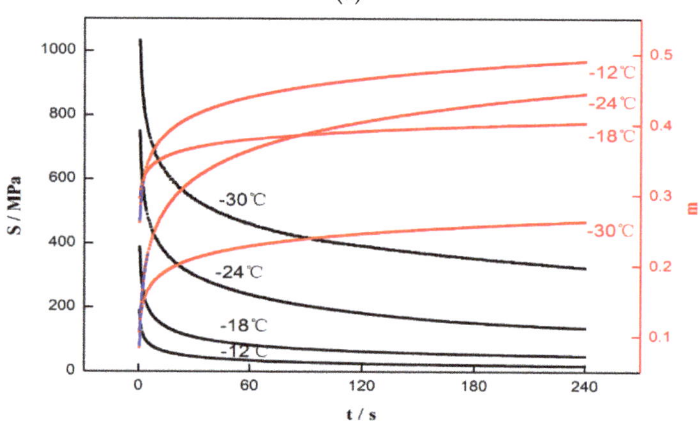

(**b**)

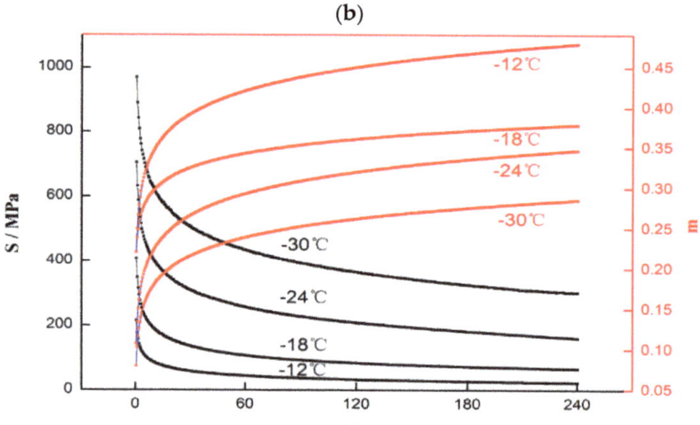

(**c**)

Figure 17. The variations of creep rate and stiffness modulus changing with time. (**a**) JZ, (**b**) AR, and (**c**) TAR.

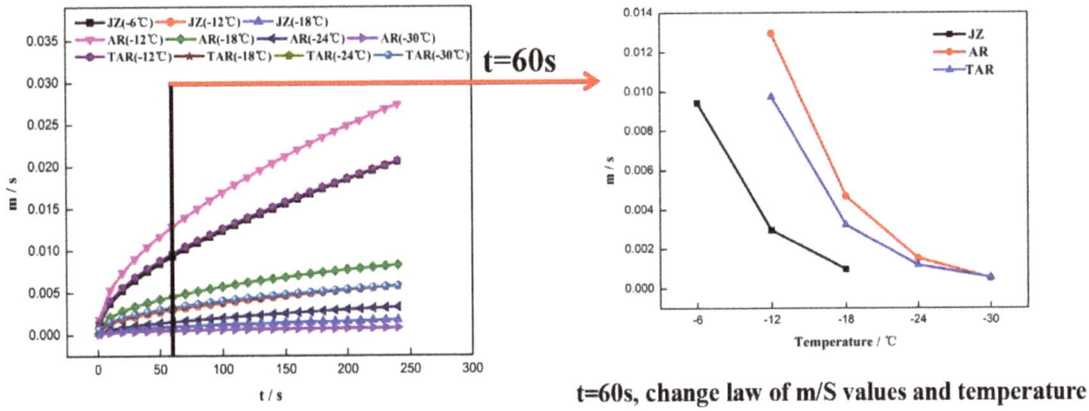

Figure 18. The variations of m/S values with loading time.

3.3.3. Determine Burgers Model Parameters

According to the creep compliance (Equation (6)) of Burgess model and the Levenberg–Marquardt (LM) method, the asphalt creep data were fitted. The elastic modulus (E_1, E_2) and viscosity coefficient (η_1, η_2) were obtained by fitting. The relaxation time (λ) and delay time (τ) were obtained according to Equations (7) and (8) [43–45]. Table 13 shows the Burgers model parameter results at different temperatures.

$$J(t) = \frac{1}{E_1} + \frac{1}{\eta_1}t + \frac{1}{E_1}\left(1 - e^{\frac{-E_2}{\eta_2}t}\right) \tag{6}$$

$$\lambda = \frac{\eta_1}{E_1} \tag{7}$$

$$\tau = \frac{\eta_2}{E_2} \tag{8}$$

in which J(t) is creep compliance of Burgess model, Pa; E_1 is the elastic coefficient of Maxwell model, Pa; η_1 is the damping coefficient of Maxwell model, Pa; E_2 is the elastic coefficient of Kelvin model, Pa; η_2 is the damping coefficient of Kelvin model, Pa; and t is the load time, s.

Table 13. Burgers model parameter results at different temperatures.

Asphalt	Temperature (°C)	R^2	E_1	E_2	N_1	N_2	λ	τ
JZ	−6	0.9944	280.857	280.857	7461.986	1135.144	26.5686	4.0417
	−12	0.9928	285.224	2169.703	26,020.424	3365.985	91.2280	1.5514
	−18	0.9934	789.674	789.674	92,339.996	10,941.991	116.9343	13.8563
AR	−12	0.9934	199.717	199.717	5398.507	756.450	27.0308	3.7876
	−18	0.9915	385.665	385.665	17,037.055	2070.255	44.1758	5.3680
	−24	0.9952	723.123	723.123	48,590.296	7549.043	67.1951	10.4395
	−30	0.9369	956.527	956.527	191,557.173	19,670.318	200.2632	20.5643
TAR	−12	0.9939	222.573	222.573	7354.134	987.512	33.0414	4.4368
	−18	0.9913	388.469	388.469	25,389.389	3049.974	65.3576	7.8513
	−24	0.9945	667.333	667.333	72,246.635	8309.587	108.2617	12.4519
	−30	0.9940	905.090	905.090	159,510.036	17,540.130	176.2367	19.3794

Figures 19 and 20 show the relaxation time (λ) and delay time (τ) changes of the three kinds of asphalts at different temperatures. The relaxation time can represent the variation of stress in the material with time. The larger λ is, the lower the stress relaxation ratio is, which is not conducive to rapid stress dissipation. Under the same temperature, the relaxation time of asphalts is $\lambda_{JZ} > \lambda_{TAR} > \lambda_{AR}$. The low temperature relaxation capacities of AR and TAR are much higher than JZ, because the blend of crumb rubber increases the viscosity of asphalt. TOR causes crumb rubber and asphalt to form network structures of large rings and chains polymers. At low temperatures, the intermolecular motion is difficult. The stress relaxation time is long.

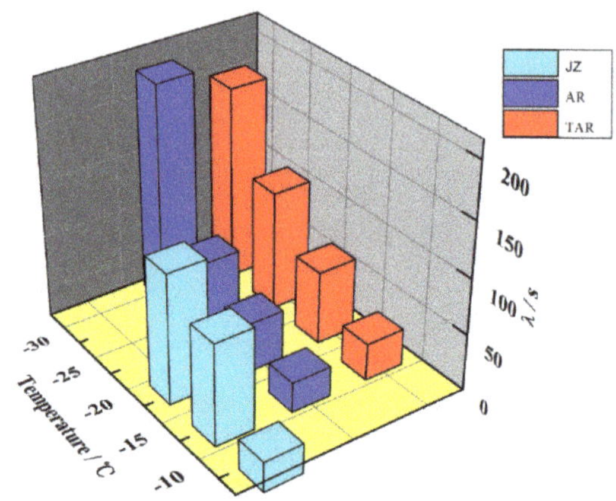

Figure 19. The variations of relaxation time (λ).

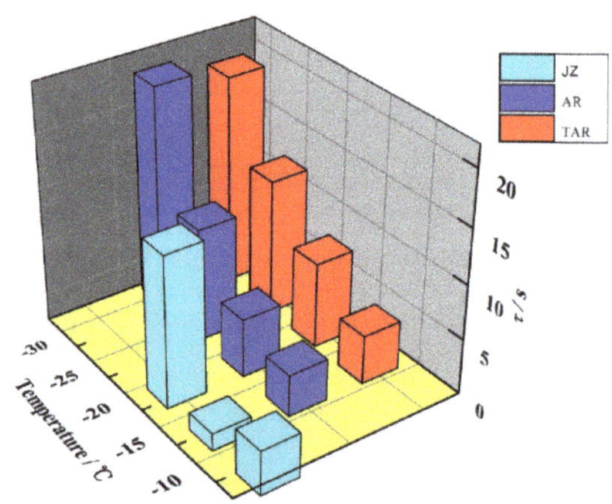

Figure 20. The variations of delay time (τ).

The relaxation time of asphalt increases with decreasing temperature. On one hand, with the decrease in temperature, the ratio of elastic component to viscous component in asphalt gradually increases. The energy consumption of asphalt under stress decreases. The relaxation capacity decreases, and the relaxation time increases. On the other hand, the molecular motion of the swelling crumb rubber polymer in asphalt is temperature dependent. At low temperatures, the internal friction of relative motion between molecular chains is larger, which leads to a longer stress relaxation time.

3.3.4. Dissipated Energy Ratio

Dissipated energy ratio is the ratio of dissipated energy to stored energy, which can reflect the stress relaxation capacity. The higher the ratio of dissipated energy, the better the stress relaxation capacity and low temperature performance will be. The dissipated energy ratio varies with time. According to the established storage energy (Equation (9)) and dissipated energy (Equation (10)), the dissipated energy ratio at t = 60 s was explored to analyze the low temperature performances of the three kinds of asphalts [46–48].

$$G_s(t) = \sigma_0^2\left[\frac{1}{E_1} + \frac{1}{2E_2}(1 - 2e^{\frac{-E_2}{\eta_2}t} + e^{\frac{-2E_2}{\eta_2}t})\right] \tag{9}$$

$$G_d(t) = \sigma_0^2\left[\frac{1}{\eta_1}t + \frac{1}{2E_2}(1 - e^{\frac{-2E_2}{\eta_2}t})\right] \tag{10}$$

in which $G_s(t)$ is storage energy; $G_d(t)$ is dissipated energy; E_1 is the elastic coefficient of Maxwell model, Pa; η_1 is the damping coefficient of Maxwell model, Pa; E_2 is the elastic coefficient of Kelvin model, Pa; η_2 is the damping coefficient of Kelvin model, Pa; t is the load time, s; and σ_0 is the stress, MPa.

Figure 21 is the variations of dissipated energy ratio at different temperatures. When the temperature decreases, the dissipated energy ratios of the three kinds of asphalts decrease gradually, and the reduction amplitudes also decrease gradually. These show that the elastic component of asphalt increases as the temperature decreases. When temperature is low, to a certain extent, the asphalt behaves as an elastomer. The stored energy increases in the asphalt, but the dissipated energy decreases. The stress relaxation capacity declines, which is not good for the crack resistance of AR and TAR at low temperature. Under the same temperature, the dissipated energy ratios of AR and TAR are much higher than JZ. The reason is that the adding of crumb rubber can enhance the energy dissipation in asphalt, and reduce the energy storage. Therefore, the stress relaxation capacity is enhanced. With the continuous decrease in temperature, the dissipated energy ratios of AR and TAR are close to the same, indicating that the influences of crumb rubber and TOR on the energy dissipation and storage capacity are much less than temperature.

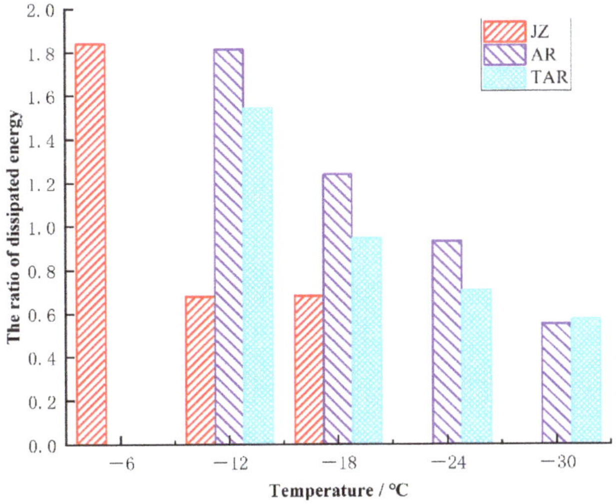

Figure 21. The variations of dissipated energy ratio at different temperatures.

3.3.5. Derivative of Creep Compliance

The derivative of creep compliance with respect to time reflects the change rate of creep compliance with time. The higher the derivative of creep compliance, the better the crack resistance of asphalt at low temperature. Equation (11) is the derivative of Equation (6), which is $J'(t)$ [48].

$$J'(t) = \frac{1}{\eta_1} + \frac{1}{\eta_2}e^{\frac{-E_2}{\eta_2}t} \tag{11}$$

According to Equation (11), J′(t) values at different temperatures within 240 s were obtained. Its curves, changing with time, are shown in Figure 22.

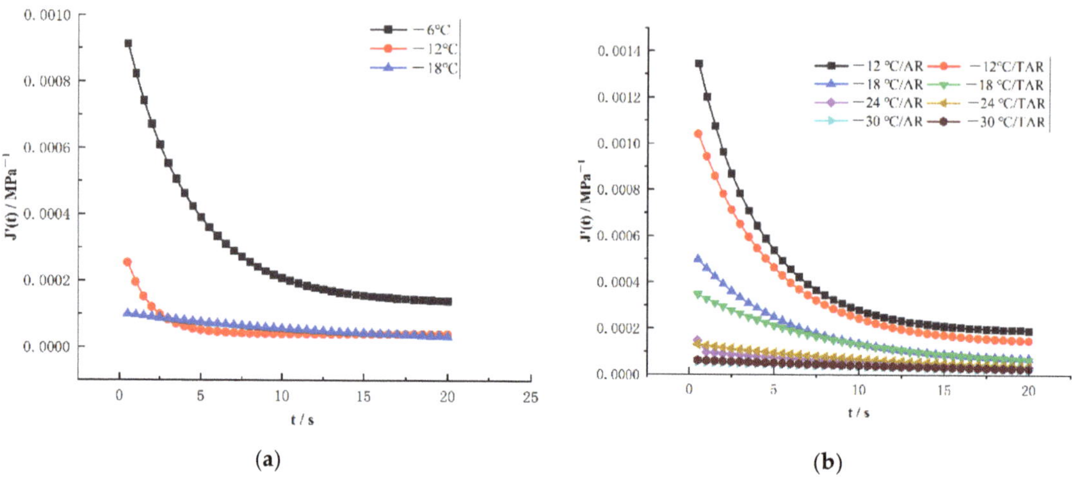

Figure 22. The variations of J′(t) at different temperatures. (**a**) JZ (**b**) AR and TAR.

In Figure 22, with the increase in loading time, J′(t) values gradually decrease and eventually tend to be stable. Therefore, the longtime load makes the low temperature crack resistances of asphalts tend to be stable, and the low temperature performances of three kinds of asphalts are no longer affected by the load action time. When the loading time is short, the J″(t) values of the same asphalt decrease with the decrease in temperature. When the temperature is low, J′(t) values tend to be stable, indicating that temperature has the most significant effect on the low temperature performances of three kinds of asphalts.

Based on viscoelastic theory and BBR creep test, the low temperature performances of JZ, AR, and TAR were evaluated and analyzed by m and S values, dissipated energy ratio, m/s value, and creep compliance derivative. The results of the four evaluation methods are consistent. The low temperature performances of the three kinds of asphalts are AR > TAR > JZ.

4. Microscopic Mechanism Analysis

Figure 23 shows fluorescence microscope images of three kinds of asphalts. The JZ is green-yellow. The additions of crumb rubber and TOR cause the color to deepen, so AR and TAR are dark green. The crumb rubber particles are black without fluorescence, and evenly disperse in the asphalt. TOR particles have no fluorescence effect and appear white. TOR particles evenly disperse in black crumb rubber particles–asphalt system. By comparing the colors of the fluorescent microscope images, it can be seen that the surface of crumb rubber particles absorbs the light components of asphalt, and the swelling effect is caused The swelling effect promotes the degradation of crumb rubber. Degraded crumb rubber cross-links to form uniform network structures. Based on the characteristics of TOR, it can further enhance the swelling and network crosslinking effects of rubber-asphalt blend.

Figure 24 shows the scanning electron microscope images of AR and TAR. The crumb rubber particles are evenly dispersed in the asphalt system, but there is obvious difference in the interface between the asphalt and crumb rubber particles. In AR, the crumb rubbers are angular, and there are obvious gaps in the stacking place. In TAR, the swelling effects of asphalt and crumb rubbers are more sufficient. The interfaces of asphalt and crumb rubbers are seamless and integrated, indicating that TOR significantly enhances the compatibility of asphalt and crumb rubbers blending system.

(a) (b) (c)

Figure 23. Fluorescent microscope images of asphalts. (**a**) JZ, (**b**) AR, and (**c**) TAR.

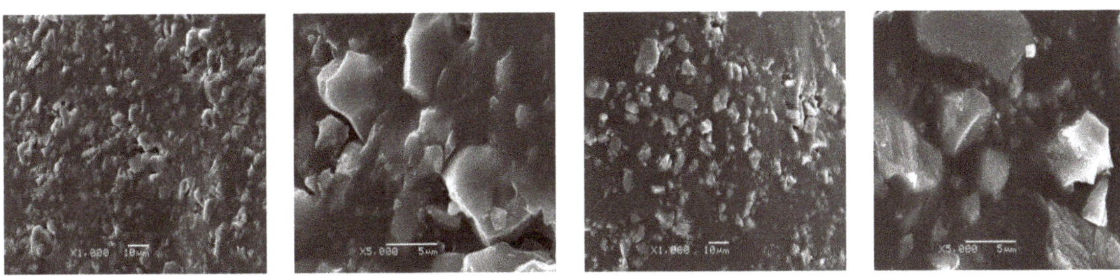

(**a**) ×1000 (**b**) ×5000 (**c**) ×1000 (**d**) ×5000

Figure 24. Scanning electron microscope images of the AR and TAR. ((**a**,**b**): AR. (**c**,**d**): TAR).

Figure 25 shows infrared spectrum diagrams of three kinds of asphalts. The infrared spectra diagrams of JZ, AR, and TAR are similar, indicating that the modification process is mainly physical modification. 1599 cm^{-1} is stretching vibration of C=C bond; 1456 cm^{-1} is the vibrational effect of the methylene $-CH_2-$ in-plane surface; and 1376 cm^{-1} is flexural vibration of C–H bond in $-CH_3-$. The benzene-ring substitution region of 1000 to 650 cm^{-1} is the vibrational effect of C–H and C–C bond of the benzene ring. Based on the infrared spectrum of crumb rubber and TOR, comparing the areas at the peak of curves, the absorption peak areas of AR and TAR at 1599 cm^{-1} decrease relatively, caused by the skeletal vibration of the conjugate double bond C=C of benzene ring. The main reason is the bending vibration of C–H bond of crumb rubber at 1400 cm^{-1}, indicating that there are weak chemical modification effects inside crumb rubber modified asphalt. The addition of TOR enhances the weak chemical interactions between crumb rubber and the asphalt interface to a certain extent.

Figure 26 shows the schematic diagram of AR and TAR asphalt molecular structure. Asphalt is composed of light components and asphaltene. When crumb rubber and asphalt are mixed at high temperatures, crumb rubber absorbs part of the light components in asphalt and expands in volume. The gel structure is formed around it. When TOR is added, the circular macromolecules and linear molecules of TOR intertwine with each other on the surface of crumb rubber, which gives TOR great deformation resistance and improved resilience after crosslinking.

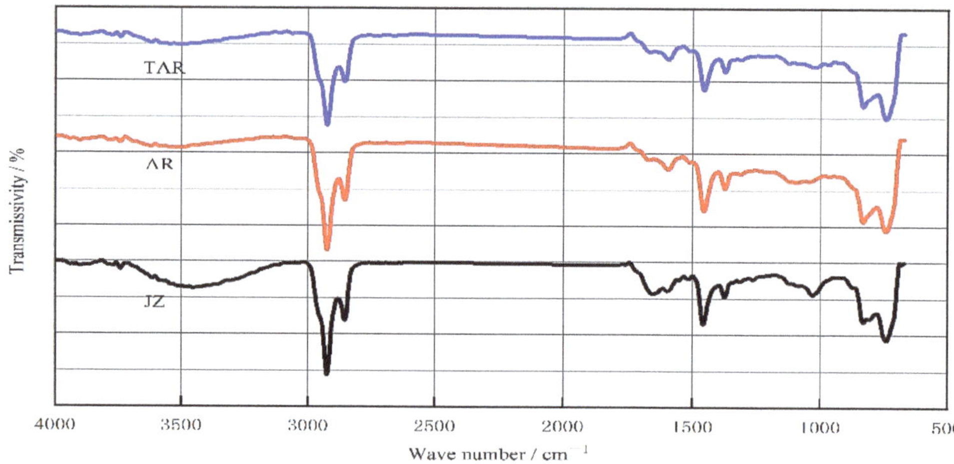

Figure 25. Infrared spectrum diagram of three kinds of asphalts.

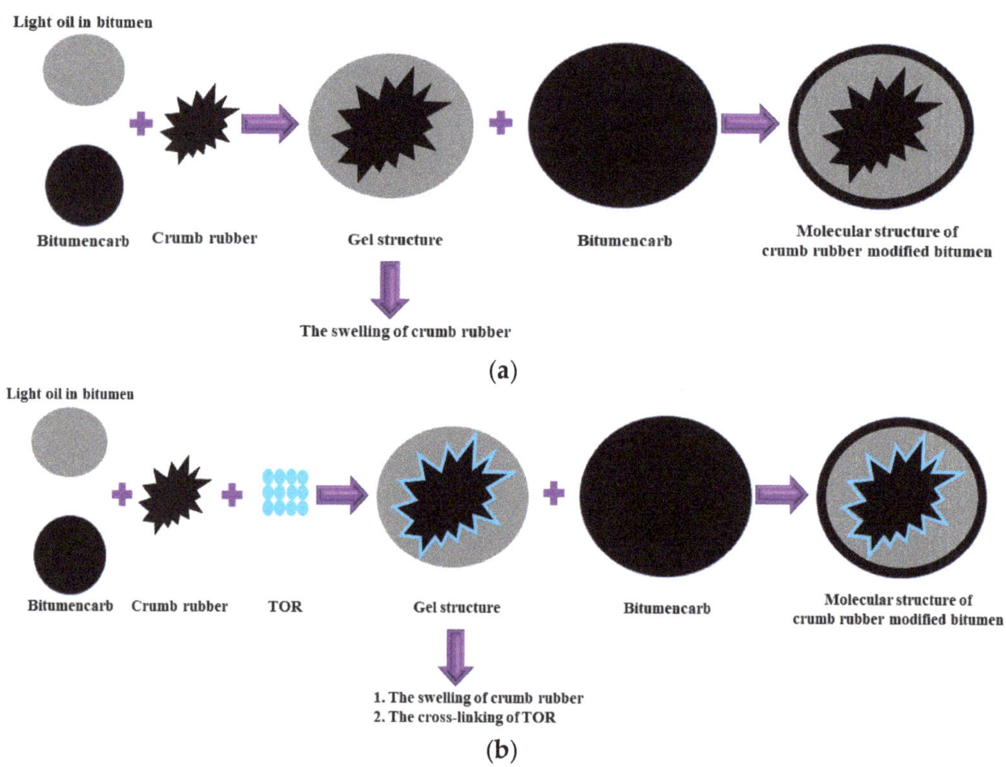

Figure 26. Schematic diagram of crumb rubber modified asphalt. (**a**) Schematic diagram of AR asphalt molecular structure. (**b**) Schematic diagram of TAR asphalt molecular structure.

5. Conclusions

In this study, the preparation process and parameters of crumb rubber modified asphalt suitable for alpine regions were proposed in detail. Meanwhile, the temperature susceptibility evaluations of matrix asphalt (JZ), crumb rubber modified asphalt (AR), and crumb rubber modified asphalt blended with TOR (TAR) were investigated by multi-parameter The following conclusions were drawn.

- The preparation process of crumb rubber modified asphalt (AR and TAR) suitable for alpine region was determined. Crumb rubber is selected 30 mesh, and the content is 24%. The preparation temperature and shearing rate are 205 °C and 5000 r/min, respectively. Shearing time is confirmed to be 60 min. TOR content is 4.5% of the quality of crumb rubber.
- PG grading is PG64-22 of JZ, PG70-34 of AR, and PG82-28 of TAR. The deformation resistance of TAR at high temperatures is superior to AR, but its crack resistance at low temperatures is relatively poor.
- The crumb rubber can improve the temperature susceptibility, and enhance its viscoelasticity at low temperatures. Meanwhile, the relaxation time is reduced. The dissipated energy and stress relaxation capacity are increased. Under the same temperature, the velocity of asphalt entering the creep stable phase decreases. The lower the temperature is, the less time it takes for asphalt to reach creep stability.
- When the crumb rubber is dispersed in the asphalt system, the surface of the crumb rubber absorbs the light components to produce swelling, and the network structures are cross-linked inside. If TOR is added, it can further enhance the swelling and network crosslinking effects of rubber–asphalt. In the asphalt system, the amount of precipitation of crumb rubber is reduced in a certain time. TOR, and the effects of desulfurization and depolymerization, enhance the storage stability in high temperature conditions.
- Based on the research results of temperature susceptibility of crumb rubber modified asphalt in alpine regions, its storage stability and weather resistance under the coupling effect of multiple factors should be further explored. The references for the design and construction of crumb rubber modified asphalt pavement are provided in alpine regions.

Author Contributions: Data curation, Y.T.; Formal analysis, M.C.; Investigation, Y.Z. (Youjie Zong); Methodology, R.X. and X.W.; Project administration, J.Y.; Software, Y.Z. (Yixing Zhang), B.F. and C.L.; Writing–original draft, H.W. All authors have read and agreed to the published version of the manuscript.

Funding: The authors wish to thank the financial support from the program of Scientific Innovation Practice Project of Postgraduates of Chang'an University in 2021, CHD. (No. 300103714063), and Provincial Undergraduate Innovation and Entrepreneurship Training Program in 2021, CHD. (No. S202110710332).

Institutional Review Board Statement: Not applicable.

Informed Consent Statement: Not applicable.

Data Availability Statement: The authors confirm that the data supporting the findings of this study are available within the article.

Conflicts of Interest: The authors declare no conflict of interest.

References

1. Wei, S.I.; Biao, M.A.; Nan, X.I.; Zeren, G.E. Compression characteristics of asphalt mixture under freeze-thaw cycles in cold plateau region. *J. Highw. Transp. Res. Dev.* **2013**, *30*, 6–10.
2. Gao, W.; Dong, N.; Wang, X.; Zeng, M. Test Analysis on Drainage Base Mixture of Asphalt Pavement in Cold Area. In *IOP Conference Series: Earth and Environmental Science*; IOP Publishing: Beijing, China, 2020; Volume 510, p. 052006.
3. Behnia, B.; Buttlar, W.G.; Reis, H. Cooling cycle effects on low temperature cracking characteristics of asphalt concrete mixture. *Mater. Struct.* **2014**, *47*, 1359–1371. [CrossRef]
4. Palit, S.K.; Reddy, K.S.; Pandey, B.B. Laboratory evaluation of crumb rubber modified asphalt mixes. *J. Mater. Civ. Eng.* **2004**, *16*, 45–53. [CrossRef]
5. Li, H.; Li, W.; Sheng, Y.; Lv, H. Influence of compound action of rubber powder and SBS on high-temperature performance of asphalt pavement surface. *J. Mater. Civ. Eng.* **2021**, 33, 04021126. [CrossRef]

6. Ai, C.-F. Characteristics and Design Methods of Asphalt Pavement in Plateau-Cold Region. Diploma Thesis, Southwest Jiaotong University, Chengdu, China, 2008.
7. Liu, J.N.; Qi, L.; Wang, X.F.; Li, M.; Wang, Z.J. Influence of aging induced by mutation in temperature on property and microstructure development of asphalt binders. *Constr. Build. Mater.* **2022**, *319*, 126083. [CrossRef]
8. Liu, J.N.; Zhang, T.H.; Guo, H.Y.; Wang, Z.J.; Wang, X.F. Evaluation of self-healing properties of asphalt mixture containing steel slag under microwave heating: Mechanical, thermal transfer and voids microstructural characteristics. *J. Clean. Prod.* **2022**, *342*, 130932. [CrossRef]
9. Li, P.; Jiang, X.; Ding, Z.; Zhao, J.; Shen, M. Analysis of viscosity and composition properties for crumb rubber modified asphalt. *Constr. Build. Mater.* **2018**, *169*, 638–647. [CrossRef]
10. Wang, Q.Z.; Wang, N.N.; Tseng, M.L.; Huang, Y.M.; Li, N.L. Waste tire recycling assessment: Road application potential and carbon emissions reduction analysis of crumb rubber modified asphalt in China. *J. Clean. Prod.* **2020**, *249*, 119411. [CrossRef]
11. Erpeng, D.; Songlin, M.A.; Haimin, J. Asphalt pavement performance prediction model based on gray system theory. *J. Tongji Univ.* **2010**, *38*, 1161–1164.
12. Liu, W.; Xu, Y.; Wang, H.; Shu, B.; Barbieri, D.M.; Norambuena-Contreras, J. Enhanced Storage Stability and Rheological Properties of Asphalt Modified by Activated Waste Rubber Powder. *Materials* **2021**, *14*, 2693. [CrossRef]
13. Tao, M.A.; Chen, C.-L.; Zhang, Y.; Zhang, W.-G. Development of Using Crumb Rubber in Asphalt Modification: A Review. *China J. Highw. Transp.* **2021**, *34*, 1.
14. Picado-Santos, L.G.; Capitão, S.D.; Neves, J.M.C. Crumb rubber asphalt mixtures: A literature review. *Constr. Build. Mater.* **2020**, *247*, 118577. [CrossRef]
15. Ding, X.; Ma, T.; Zhang, W.; Zhang, D. Experimental study of stable crumb rubber asphalt and asphalt mixture. *Constr. Build. Mater.* **2017**, *157*, 975–981. [CrossRef]
16. Azizian, M.F.; Nelson, P.O.; Thayumanavan, P.; Williamson, K.J. Environmental impact of highway construction and repair materials on surface and ground waters: Case study: Crumb rubber asphalt concrete. *Waste Manag.* **2003**, *23*, 719–728. [CrossRef]
17. Shill, S.K.; Al-Deen, S.; Ashraf, M. Saponification and scaling in ordinary concrete exposed to hydrocarbon fluids and high temperature at military airbases. *Constr. Build. Mater.* **2019**, *215*, 765–776. [CrossRef]
18. Rocha Segundo, I.G.; Dias, E.A.; Fernandes, F.D.; Freitas, E.F.; Costa, M.F.; Carneiro, J.O. Photocatalytic asphalt pavement: The physicochemical and rheological impact of TiO2 nano/microparticles and ZnO microparticles onto the bitumen. *Road Mater. Pavement Des.* **2019**, *20*, 1452–1467. [CrossRef]
19. Ding, X.; Chen, L.; Ma, T.; Ma, H.; Gu, L.; Chen, T.; Ma, Y. Laboratory investigation of the recycled asphalt concrete with stable crumb rubber asphalt binder. *Constr. Build. Mater.* **2019**, *203*, 552–557. [CrossRef]
20. Xiang, L.; Cheng, J.; Que, G. Microstructure and performance of crumb rubber modified asphalt. *Constr. Build. Mater.* **2009**, *23*, 3586–3590. [CrossRef]
21. Rocha Segundo, I.; Landi, S.; Margaritis, A.; Pipintakos, G.; Freitas, E.; Vuye, C.; Blom, J.; Tytgat, T.; Denys, S.; Carneiro, J. Physicochemical and rheological properties of a transparent asphalt binder modified with nano-TiO_2. *Nanomaterials* **2020**, *10*, 2152. [CrossRef]
22. Picado-Santos, L.G.; Capitão, S.D.; Dias, J.L.F. Crumb rubber asphalt mixtures by dry process: Assessment after eight years of use on a low/medium trafficked pavement. *Constr. Build. Mater.* **2019**, *215*, 9–21. [CrossRef]
23. Jamal, M.; Giustozzi, F. Low-content crumb rubber modified asphalt for improving Australian local roads condition. *J. Clean. Prod.* **2020**, *271*, 122484. [CrossRef]
24. Torretta, V.; Rada, E.C.; Ragazzi, M.; Trulli, E.; Istrate, I.A.; Cioca, L.I. Treatment and disposal of tyres: Two EU approaches. A review. *Waste Manag.* **2015**, *45*, 152–160. [CrossRef] [PubMed]
25. Li, J.; Zhu, Y.; Wang, H.; Wang, S.; Zhang, Y.; Zhang, Y. High temperature storage stability of asphalt modified with crumb rubber. *China Synth. Rubber Ind.* **2009**, *32*, 259–263.
26. Singh, D.; Sawant, D.; Xiao, F. High and Intermediate Temperature Performance Evaluation of Crumb Rubber Modified Binders with RAP. *Transp. Geotech.* **2016**, *10*, 13–21. [CrossRef]
27. Dias, J.L.F.; Picado-Santos, L.G.; Capitão, S.D. Mechanical performance of dry process fine crumb rubber asphalt mixtures placed on the Portuguese road network. *Constr. Build. Mater.* **2014**, *73*, 247–254. [CrossRef]
28. Yu, H.; Chen, Y.; Wu, Q.; Zhang, L.; Zhang, Z.; Zhang, J.; Miljković, M.; Oeser, M. Decision support for selecting optimal method of recycling waste tire rubber into wax-based warm mix asphalt based on fuzzy comprehensive evaluation. *J. Clean. Prod.* **2020**, *265*, 121781. [CrossRef]
29. Zhang, L.; Xing, C.; Gao, F.; Li, T.S.; Tan, Y.Q. Using DSR and MSCR tests to characterize high temperature performance of different rubber modified asphalt. *Constr. Build. Mater.* **2016**, *127*, 466–474. [CrossRef]
30. Chen, Z.; Wang, T.; Pei, J.; Amirkhanian, S.; Xiao, F.; Ye, Q.; Fan, Z. Low temperature and fatigue characteristics of treated crumb rubber modified asphalt after a long-term aging procedure. *J. Clean. Prod.* **2019**, *234*, 1262–1274. [CrossRef]
31. Chen, S.; Gong, F.; Ge, D.; You, Z.; Sousa, J.B. Use of reacted and activated rubber in ultra-thin hot mixture asphalt overlay for wet-freeze climates. *J. Clean. Prod.* **2019**, *232*, 369–378. [CrossRef]
32. Fiore, N.; Caro, S.; D'Andrea, A.; Scarsella, M. Evaluation of asphalt modification with crumb rubber obtained through a high pressure water jet (HPWJ) process. *Constr. Build. Mater.* **2017**, *151*, 682–691. [CrossRef]

33. Ragab, M.; Abdelrahman, M. Enhancing the Crumb Rubber Modified Asphalt's Storage Stability through the Control of its Internal Network Structure. *Int. J. Pavement Res. Technol.* **2017**, *11*, 13–27. [CrossRef]
34. Chang, M.; Zhang, Y.; Pei, J.; Zhang, J.; Wang, M.; Ha, F. Low-Temperature Rheological Properties and Microscopic Characterization of Asphalt Rubbers Containing Heterogeneous Crumb Rubbers. *Materials* **2020**, *13*, 4120. [CrossRef] [PubMed]
35. Lu, Z.; Li, B.; Yang, X.; Wei, Y.; Zhang, Z. Production Process Parameter Optimization for Crumb Rubber Modified Asphalt Based on TOPSIS. *J. Highw. Transp. Res. Dev.* **2017**, *34*, 1–7.
36. Liang, M.; Xin, X.; Fan, W.; Sun, H.; Yao, Y.; Xing, B. Viscous properties, storage stability and their relationships with microstructure of tire scrap rubber modified asphalt. *Constr. Build. Mater.* **2015**, *74*, 124–131. [CrossRef]
37. Xiang, L.; Cheng, J.; Kang, S. Thermal oxidative aging mechanism of crumb rubber/SBS composite modified asphalt. *Constr. Build. Mater.* **2015**, *75*, 169–175. [CrossRef]
38. Wang, L.; Wang, Z.; Li, C. Low temperature performance of polyphosphoric acid asphalt based on viscoelastic theory. *Acta Mater. Compos. Sin.* **2017**, *34*, 322–328.
39. Hosseinnezhad, S.; Kabir, S.F.; Oldham, D.; Mousavi, M.; Fini, E.H. Surface functionalization of rubber particles to reduce phase separation in rubberized asphalt for sustainable construction. *J. Clean. Prod.* **2019**, *225*, 82–89. [CrossRef]
40. Yang, X.; You, Z.; Hasan, M.R.; Diab, A.; Shao, H.; Chen, S.; Ge, D. Environmental and mechanical performance of crumb rubber modified warm mix asphalt using Evotherm. *J. Clean. Prod.* **2017**, *159*, 346–358. [CrossRef]
41. Liu, S.; Cao, W.; Shang, S.; Qi, H.; Fang, J. Analysis and application of relationships between low-temperature rheological performance parameters of asphalt binders. *Constr. Build. Mater.* **2010**, *24*, 471–478. [CrossRef]
42. Wang, S.; Guo, C.; Peng, F. Research on modified mechanism of asphalt with crumb tire rubber. *J. Chang. Univ.* **2010**, *4*, 34–38.
43. Wang, K.; Hao, P. Analysis of asphalt low temperature performance and viscoelasticity based on BBR test. *J. Liaoning Tech. Univ.* **2016**, *10*, 1138–1143.
44. Mashaan, N.S.; Ali, A.H.; Karim, M.R.; Abdelaziz, M. An overview of crumb rubber modified asphalt. *Int. J. Phys. Sci.* **2012**, *7*, 166–170.
45. Jeong, K.D.; Lee, S.J.; Amirkhanian, S.N.; Kim, K.W. Interaction effects of crumb rubber modified asphalt binders. *Constr. Build. Mater.* **2010**, *24*, 824–831. [CrossRef]
46. Bocci, E.; Prosperi, E. Recycling of reclaimed fibers from end-of-life tires in hot mix asphalt. *J. Traffic Transp. Eng.* **2020**, *7*, 678–687. [CrossRef]
47. Wang, H.; Ma, Z.; Chen, X.; Hasan, M.R.M. Preparation process of bio-oil and bio-asphalt, their performance, and the application of bio-asphalt: A comprehensive review. *J. Traffic Transp. Eng.* **2020**, *7*, 137–151. [CrossRef]
48. Liu, J.H. Experimental Study on Improving the Performance of Rubber with TOR. Diploma Thesis, Changsha University of Science and Technology, Changsha, China, 2019.

coatings MDPI

Article

Investigation of Tri-Axial Stress Sensing and Measuring Technology for Tire-Pavement Contact Surface

Jiaxi Guan [1], Xinglin Zhou [1,*], Lu Liu [1], Maoping Ran [2] and Yuan Yan [2]

1 School of Machinery and Automation, Wuhan University of Science and Technology, Wuhan 430081, China; guanjiaxi@wust.edu.cn (J.G.); liulu@wust.edu.cn (L.L.)
2 School of Automotive and Traffic Engineering, Wuhan University of Science and Technology, Wuhan 430065, China; ranmaoping@wust.edu.cn (M.R.); sahara1990@163.com (Y.Y.)
* Correspondence: zhouxinglin@wust.edu.cn

Abstract: A tri-axial stress sensor was designed to measure contact stresses in the tire–pavement contact patch. The shape and size of the sensor surface were designed considering both the asphalt pavement texture and the tire pattern. The top-down cracking mechanism was also taken into account, and the sensor was placed at the vertical crack depth. Temperature drifts and zero drifts were compensated for. The sensor had high structural strength and met the sensing requirements of specialized heavy vehicles. In a preliminary study, three sensors were fabricated and calibrated in three directions. Simulated measurements were performed using a tire–pavement surface contact test bench. Signals from the L-shaped sensor region were obtained for the upper, middle, and lower parts of the tire, and preliminary stress distributions were determined at different positions on the contact surface. This study has laid a foundation for the design and construction of a more precise test system in the future.

Keywords: tire–pavement contact; tri-axial stress; sensor array; sensing element; stress distribution

Citation: Guan, J.; Zhou, X.; Liu, L.; Ran, M.; Yan, Y. Investigation of Tri-Axial Stress Sensing and Measuring Technology for Tire-Pavement Contact Surface. *Coatings* **2022**, *12*, 491. https://doi.org/10.3390/coatings12040491

Academic Editor: Valeria Vignali

Received: 6 March 2022
Accepted: 31 March 2022
Published: 6 April 2022

Publisher's Note: MDPI stays neutral with regard to jurisdictional claims in published maps and institutional affiliations.

Copyright: © 2022 by the authors. Licensee MDPI, Basel, Switzerland. This article is an open access article distributed under the terms and conditions of the Creative Commons Attribution (CC BY) license (https://creativecommons.org/licenses/by/4.0/).

1. Introduction

The tire is the only part of an automobile in contact with the pavement. The forces exerted by the vehicle on the pavement are transmitted through the tires; therefore, the tire force is an important parameter in the study of pavement mechanical behavior and has always been a research hotspot in pavement engineering, vehicle engineering, tire science, and other related fields. Many high-grade asphalt pavements do not reach their designed service life due to pavement functional failures such as rutting, cracking, pitting, and raveling. Moreover, adverse weather conditions, in particular rain and snow, can exacerbate pavement damage and cause vehicle slippage, which can affect the driver's behavioral response and increase the number of road traffic accidents [1,2].

Tire–pavement contact stress is influenced by the tire tread pattern, pavement texture, vehicle load, and many other factors. Simplified models are often adopted in simulations. For instance, the continuum theory of elastic layers under double circular uniform vertical load has long been used as the gold standard in the design and construction of asphalt pavement structures [3]. However, as highlighted repeatedly in the literature [4], circular contact surfaces and uniformly distributed loads do not match the actual situation.

A tire–road contact patch can be obtained using the thermal paper method or paint footprint method [5]. The effective tire contact area should be further corrected for various tread patterns [6]. The contact patches of special tires, such as tractor tires, are more complex [7]. Tire type [8], tire inflation, and tire load directly affect the mechanical properties of the contact patch [9]. When the load is small, the contact patch is elliptical, but it becomes closer to a rectangle after a certain load is reached, and the characteristics of an approximate rectangle become more obvious as the load increases [10,11]. Different contact patch shapes lead to different stress distributions, and the various resulting damage phenomena are difficult to explain or predict. Therefore, an accurate tri-axial stress distribution cannot be obtained using only simulation methods. A sensor that measures actual stress distributions could be used to better analyze the mechanical characteristics of tire–pavement interaction.

The pavement materials determine the design method of sensors. The typical flexible base asphalt pavement structure in the middle section of the Sui-Yue Expressway is taken as an example [12]. As shown in Figure 1, the pavement consists of three layers of compacted materials. The total thickness of the pavement is 62 cm, comprised of the following structure: an upper layer of SMA-13 modified asphalt concrete with a thickness of 4 cm, a middle layer of AC-20I medium-grained asphalt concrete with a thickness of 6 cm, and a lower layer of AC-25 coarse-grained asphalt concrete with a thickness of 8 cm. The pavement sits atop a base layer of ATB-25 asphalt-stabilized gravel with a thickness of 24 cm and a bottom base layer of cement-stabilized gravel with a thickness of 20 cm. Soil and natural subgrade are located further below. The mechanical properties of asphalt pavements are completely different from those of concrete- and soil-based pavement structures. Under the tire rolling condition, in addition to the vertical force F_z applied to the pavement surface, the tire also creates lateral forces F_y, longitudinal forces F_x, and vibration impacts.

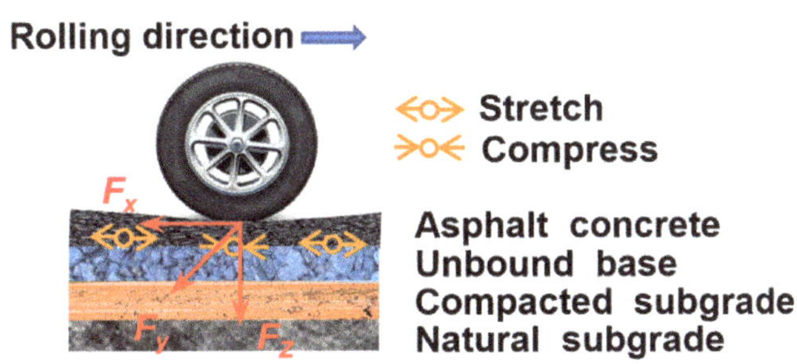

Figure 1. Schematic diagram of tire–pavement interaction.

The dynamic vehicle load is transmitted downward through each structural layer and decomposed. Due to an incomplete understanding of pavement disease mechanisms, in earlier studies, structural deformation under traffic loads was believed to be the main cause of the rutting and cracking of pavements, while wheel wear, compaction deformation, and plastic shear flow deformation were ignored [13]. Asphalt pavement is not only elastic but also clearly exhibits viscosity [14,15]. Non-uniform tri-axial forces exerted through the tires in the tire–pavement contact area lead to strain concentrations, which can lead to the top-down cracking of pavements [16,17].

Many factors influence tire–pavement tri-axial contact stress. To better determine the true stress distribution, in addition to modeling and simulation methods, many scholars have designed various types of sensors to directly measure tri-axial stresses through experiments. Howell et al. [18] measured the static contact stress of aircraft tires. Lippmann [19] studied contact patch stress based on tire deformation. To obtain better measurement results, new measurement methods were introduced. Marshek [20] used a photocell sensor that can measure the intensity of small grease stains left by ink bubbles after bursting due to tire pressure, which was used to determine the static contact stress. Smith et al. [21] used a water-pressure sensor to measure static stress in the tire–ground contact area. Cheli [22] used a thin polyvinylidene fluoride (PVDF) piezoelectric film as a sensor to dynamically measure contact stress, and many other scholars used the pressure plate method to study contact stresses.

Certain methods can only be used for static measurements, and others can only be used for vertical measurements. Owing to recent research developments, pressure sensors with a wide measuring range that can be used to dynamically measure force distributions in three directions have emerged. Anghelache [23] designed an L-shaped structure sensor. De Beer [24] introduced the stress-in-motion (SIM) system for the quantification of tri-axial (3D) tire contact force (or stress) distributions. Yi Xie and Qun Yang [25] embedded pressure sensors in rectangular asphalt concrete to measure the vertical pressure distribution of tires with a reduced diameter. The matrix array method uses multiple sensors and has the advantages of a wide measuring range and the ability to simultaneously measure large weights along multiple axes.

In this paper, the design of a new sensor array for measuring tri-axial stresses in the tire–pavement contact patch is presented. The sensor element considers several different factors, including the pavement texture, temperature compensation, and top-down crack mechanism. An array of sensor elements was designed, and the sensors were used to measure changes in the tri-axial stress distribution in the pavement throughout the whole tire rolling process; the test data are helpful for exploring the cause of road damage.

The remainder of this article is organized as follows. In Section 2, the sensor design and details of the finite element analysis are introduced. Zero-point temperature drift and temperature compensation were performed. In Section 3, we calibrate the sensor performance, verify the effectiveness of the sensor using the tire loading test platform, and discuss the results of the sensor performance. In Section 4, the main conclusions of this study and future directions are summarized.

2. Stress Sensor Design

2.1. Mechanical Structure

In China, most expressways have been designed with a high bearing capacity using semi-rigid base asphalt pavement. However, the shrinkage and cracking of the semi-rigid base material can lead to the reflection cracking of the asphalt overlay. Rigid watertight pavement with poor permeability can promote water damage. Moreover, wet–dry cycles, freeze–thaw cycles, thermal expansion, and cold weather contraction, as well as repeat loading, can reduce the strength of the base material. Thus, pavement surface defects cannot be avoided, and many pavements do not reach their designed lifespan [26].

Newly built high-grade expressways typically adopt an asphalt-mixed graded gravel or non-bonded graded gravel filling base. This flexible base asphalt pavement has been gradually replacing semi-rigid pavements. Despite the obvious advantages, top-down cracking can still occur in flexible pavements during service. Cracks can be divided into three main types [27]. Type I cracks are V-shaped fractures that arise due to external tensile stress. Fractures typically expand in the direction orthogonal to the fracture plane, and temperature change is an important factor. Type II cracks are caused by the sliding mode of fracture. Uneven vertical shear stress can trigger mutual slip, resulting in the formation concave and convex surface cracks. Type III cracks are caused by the tearing mode of fracture. Under the joint action of uneven parallel tensile stress and vertical shear stress, dislocation phenomena occur.

Type I cracks, mainly caused by temperature change, are the most dangerous and have the greatest possibility of breaking the pavement surface. Samples were taken by drilling cores into the pavement where cracks appeared. Top-down cracks mostly appeared on the surface pavement and extended a distance of 2 cm to 4 cm into the pavement. Early crack widths were between 6.25 and 12.5 mm. Throughout most of China's high-grade expressways, AC-13 coarse- and dense-graded asphalt concrete or SMA-13 modified asphalt concrete is used in the upper layer, with a particle size of about 13 mm and an average texture depth of 1 mm. The asphalt parameters are the key to designing a suitable transducer structure.

Tires are not in perfect contact with pavement surfaces, and the pavement texture and the tire pattern are the most important factors that contribute to the non-uniform distribution of tri-axial force. As shown in Figure 2, the tire and pavement are not in smooth contact; therefore, the tire is not perfectly parallel with the pavement surface, and surface gravel will enter the grooves in the tire pattern. In addition, the load on the contact surface is complex due to uneven vehicle loads, pavement slope, wheel steering, and other factors.

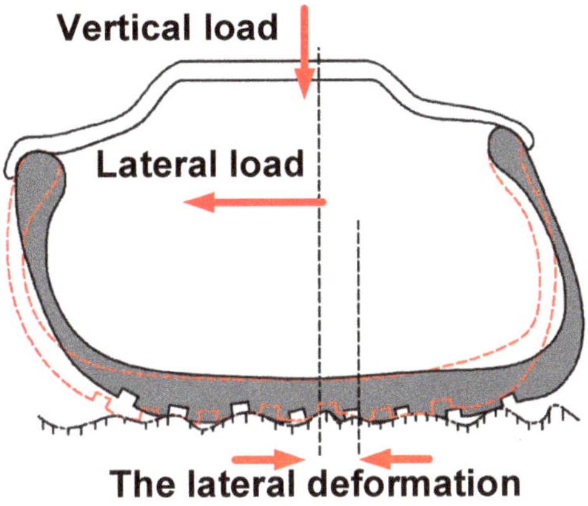

Figure 2. Diagram of tire–pavement contact patch.

Existing commercial sensors cannot meet the test requirements in terms of size, strain gauge position, and measurement range. Taking these factors into consideration, a basic finite element model of the sensor was established. The upper surface area of the sensor was 12 mm $\times$ 12 mm = 144 mm^2, which is slightly smaller than the particle size of the upper surface of asphalt concrete. The sensor surface was roughened to increase the friction force and to ensure adequate contact with the rough pavement surface. The sensor had a cuboidal geometry at a distance of 5 mm from the top to match the tire pattern and the local wedging of the pavement surface. When the sensor was partially wedged into the tire pattern, the tire rotated, and the forces acting on the wedged area of the sensor were more complicated, especially when the tire pressure was low, which created a larger contact area. The main part of the sensor was 8 $\times$ 8 $\times$ 15 mm^3. Compared with the upper surface, this section was slenderer, with obvious changes in stress distribution. A strain gauge was attached to obtain more accurate measurements. The strain gauge and measuring circuit were better protected by the concave structure of the sensor. The cable of the strain gauge circuit was fed into a through-hole directly below the sensor. The cylindrical pile head was fixed at the bottom, the side face was cut with a semicircular chamfer with a diameter of 3 mm, and the chamfer was fitted to the base bolt to ensure the sensor remained fixed to the base, with the ability to bear a certain impact without rotating or sliding.

2.2. Finite Element Analysis

Finite element analysis was performed in ANSYS 19.0 [28] using an elastomer, a sensor that did not have to be assembled, as the stainless steel structure was the elastic body. Stress and strain values were calculated, and the optimal patch position was obtained. The sensor and bottom plate were tightly fixed and seamless. To simplify the analysis, a fixed-constraint mode was used for the bottom of a single transducer. Static conditions were adopted, meshing included 13227 nodes and 7337 elements, and the analysis was carried out within the linear interval. The tire pressure, tire contact area, maximum load, and material strength of the sensor were considered. The sensor was also used to measure the contact stress of heavy vehicles and special equipment. The calibration forces in three directions were determined as 740 N, which is relatively conservative compared with the maximum sensor measurement limit. The sensor was made of 17-4PH martensitic stainless steel, which exhibits high strength, high hardness, good corrosion resistance, and excellent mechanical properties after heat treatment.

Since the transducer was symmetrical, F_X and F_Y were equivalent. As shown in Figure 3a, a lateral force $F_X(F_Y)$ of 740 N horizontal force was applied to the left side of the top of the sensor. The compressive strain in the orange region was about 0.0013869, and an equivalent tensile strain was generated in the symmetric region. Thus, the strain gauge measured strain in the sensitive area of the sensor to realize force measurements in the X or Y direction.

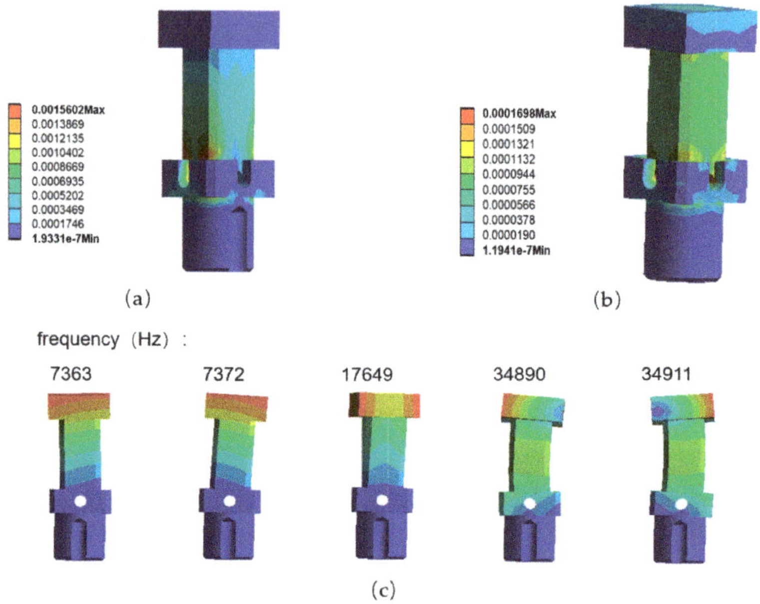

Figure 3. Finite element analysis results of the elastomer: (**a**) equivalent elastic strain in X/Y direction; (**b**) equivalent elastic strain in Z direction; (**c**) modal analysis of orders 1 to 5.

A vertical downward force F_Z of 740 N was applied in the Z direction at the center of the top surface, and the strains were found to be highest at the four vertices in the middle of the transducer; therefore, the transducer was most sensitive in this area. As shown in Figure 3b, the strain was approximately 0.00011321. An XY strain gauge was attached to the lower regions of the sensor, and a Z strain gauge was attached to the upper region; then, a modal analysis of orders 1 to 5 was performed, as shown in Figure 3c. The sensor frequency ranged from 7362.7 Hz to 34911 Hz and increased with the increasing tire crushing speed. Based on experience, for a 0.2 m contact length and a speed of 30 m/s (108 km/h), an excitation frequency of 75 Hz is much lower than the basic natural frequency. In this paper, quasi-static measurements were performed, and the natural frequency of the sensor was shown to easily meet the sensing requirements.

2.3. Strain Gauge Arrangement and Measuring Circuit

A strain gauge is a highly sensitive device. When attached to the surface of an elastomer, the strain gauge will produce the same strain as the component. As the strain gauge is stretched or compressed, the resistance value changes and can be converted into a voltage or current. Based on a stress and strain analysis of the elastic element, the appropriate strain gauge arrangement to form the measuring bridge was determined. The sensitivity of the sensor was improved, and the cross-link effect was reduced.

The results of the finite element analysis suggested that the proposed sensor structure had sufficient sensitivity. To test the strain gauge arrangement, strain gauges were arranged on the elastomer, as shown in Figure 4. The specific sticking position of the strain gauge is shown in Figure 5. R1 and R4 were on the same side, and R2 and R3 were on the opposite side. The other strain gauges were also arranged in the same way.

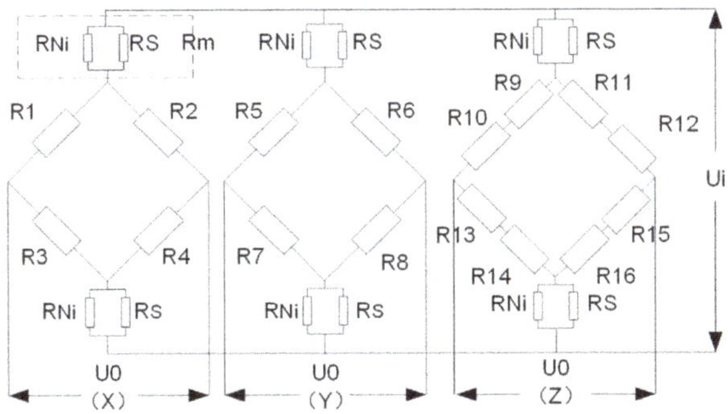

Figure 4. Strain gauge bridge and compensation circuit.

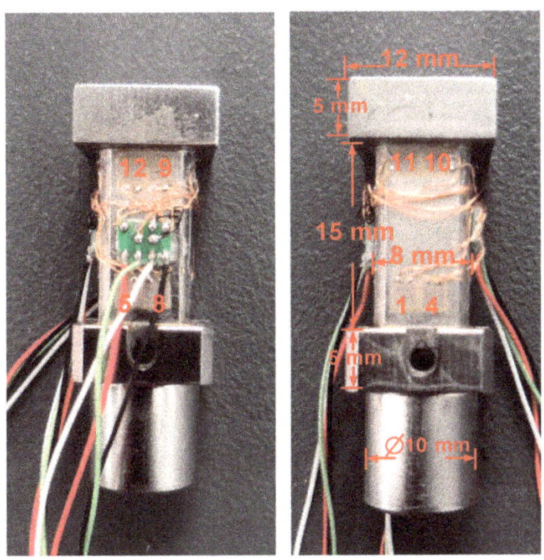

Figure 5. Physical dimensions of strain sensor.

Strain gauges R1 and R4 produced positive strains when stretched, and R2 and R3 produced negative strains when compressed, thus forming a Wheatstone bridge, which was used to measure strain in the X direction. Similarly, the same Wheatstone bridge could be used to measure strains in the Y direction. In this case, R5 and R8 produced positive strains when stretched, and R6 and R7 produced negative strains when compressed; however, the direction was orthogonal to the X direction. Two groups of strain gauges in series were selected to measure strains in the Z direction, forming a set of bridge arms: R9, R13, R12, and R14 were arranged vertically and compressed when stress was applied; R10, R11, R14, and R15 were arranged laterally and were stretched when stress was applied. Altogether, eight strain gauges were used to measure strain in the Z direction. The dimensions of the sensor structure are illustrated in Figure 5. After forming the bridge, the outer end of the sensor was sealed with silica gel to protect the circuit. We used a BEK 350 series strain gauge, thickness 2.5 μm, sensitive grid size 1.5 mm × 2.5 mm, base size 4.6 mm × 3.6 mm, resistance 350 Ω ± 0.5%.

2.4. Zero-Drift and Temperature Compensation

As mentioned above, temperature drift is an important factor in pavement diseases; therefore, sensor designs should consider temperature changes and ensure measurement accuracy under different temperature conditions. Zero-drift and temperature compensation are crucial. For a Wheatstone bridge composed of four strain gauges, the initial zero of a transducer will not be zero, due to differences between the strain gauge resistance and the resistance of the wire. The transducer signal is poor at zero resistance. The zero-point signal error is

$$S = \frac{U_0}{Ui} = \frac{1}{4\frac{R}{\Delta R} + 2} \times 10^3 (\mathrm{mV/V}) \tag{1}$$

When $S = A$ (mV/V), the difference in resistance must be compensated:

$$\Delta R = \frac{4A \times R}{1000 - 2A} \tag{2}$$

where U_i is the bridge supply voltage, R is the strain gauge resistance, and U_0 is the output voltage.

To rebalance the bridge, equivalent Constantan wire must be connected to the bridge arm using a reverse change for compensation. After zero compensation, since the printed circuit board (PCB) and bridge are connected by copper wire, the zero point will also change at high and low temperatures and the zero-point temperature coefficient $TC0$ must be compensated to zero. The sensor zero-point change due to temperature change is

$$TC0/_{R-H} = \frac{S_{0/H} - S_{0/R}}{2Cn_{/R}} \times 10^6 \tag{3}$$

The unit is ppm/10 °C.

Thus, the following resistance must be compensated:

$$\Delta R_T = \frac{8R \times Cn_{/R} \times TC0_{R-H}}{10^9 - 4Cn_{/R} \times TC0_{R-H}} \tag{4}$$

where $S_{0/H}$ is the sensor signal at a high temperature (greater than 40 °C) with zero output, $S_{0/R}$ is the sensor signal at room temperature (around 20 °C) with zero output, $C_{n/R}$ is the sensitivity of the sensor at room temperature (around 20 °C), and $TC0/_{R-H}$ is the sensor from room temperature to high temperature.

The sensor was an elastomer made of stainless steel. The elastic modulus (E) of stainless steel decreases as the temperature increases. When the temperature rises, the material softens; therefore, when the sensor was subjected to the same load, the strain would increase and the signal output would be larger. To compensate for the sensitivity to temperature increase, two nickel (Ni) resistors with resistance values that increased with increasing temperature were connected in series in the positive and negative electrodes of the power supply. Compensating resistors reduce the supply bridge voltage to offset the increase in signal caused by changes in the modulus of the elastomer. Since the resistance of Ni does not change linearly with temperature, a precision resistor of about four times the resistance of the resistor also had to be added in parallel to linearize the Ni resistor.

3. Results and Discussion of Sensor Performance

3.1. Calibration Test

The range, sensitivity, repeatability, creep, and other performance indicators directly affect measurement accuracy. For a tri-axial force transducer, calibration is a critical step The sensors are fabricated by cutting and processing the elastomer, and cross-coupling or interference may arise between the tri-axial channels. Since the relationship between the deformation of the elastomer and the circuit measurement is almost linear, partial cross-links can be eliminated by using a decoupling matrix.

The gravitational acceleration in the laboratory was 9.7941 m/s^2. As shown in Figure 6 calibration tests were carried out on the calibration test bench, which was comprised of a weight-loading table, control system, and data-acquisition system. The sensor was fixed on the bracket and placed directly below the loading hook to ensure close contact and no sliding. Then, weight was gradually added up to the maximum design weight. The inset in

Figure 6 shows the calibration of the sensor in the X direction on the left, which was rotated 90 degrees for the calibration, and the vertical Z-direction calibration is shown on the right. From the A/D data-acquisition circuit, the relationship between loading force and output voltage could be derived. The test results are presented in Figure 7.

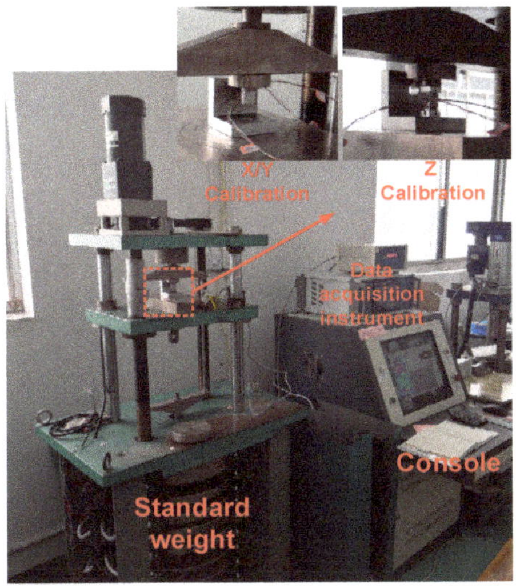

Figure 6. Calibration test bench.

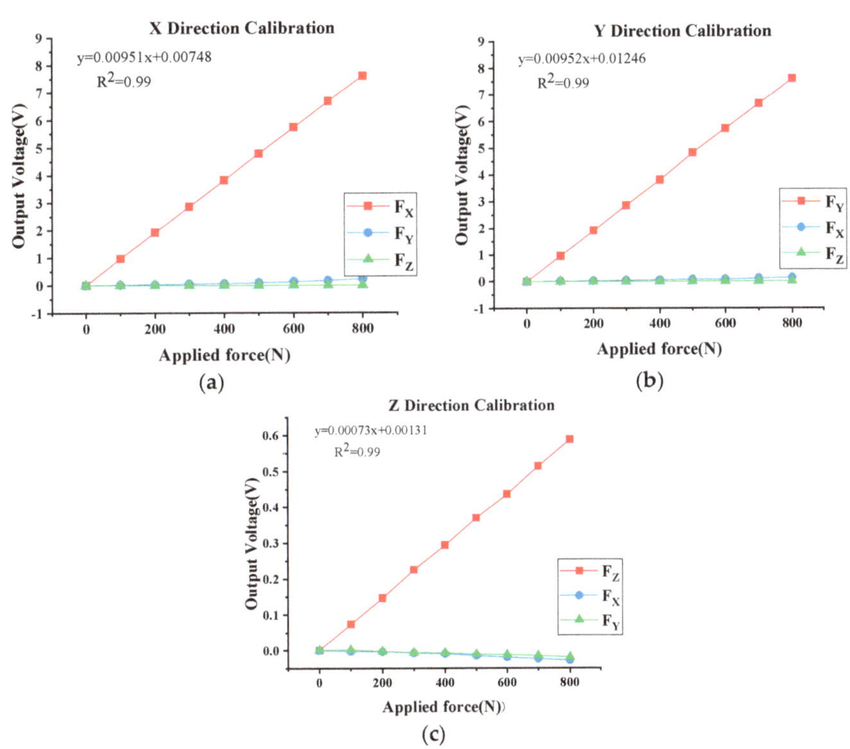

Figure 7. Calibration results: (**a**) X direction; (**b**) Y direction; (**c**) Z direction.

The X direction and Y direction were mutually perpendicular and symmetric, and the loading curves and output voltages were similar. The measurements were approximately linear over the tested range. The performance of the sensor was determined through multiple loading and unloading tests from 0 to 800 N in the three directions (Table 1). From the third column in Table 1, it can be seen that the maximum error value of repeatability was less than or equal to 0.356%. This indicates that the sensor has good repeatability in three directions.

Table 1. Main performance parameters of sensor 1 (see Figure 8).

Direction of the Force	Sensitivity (mV/V)	Linearity Error %F·S	Repetitive %F·S
X	1.36956	−0.499	0.123
Y	1.36527	−0.551	0.356
Z	0.11993	−0.804	0.250

Note: F·S means full scale.

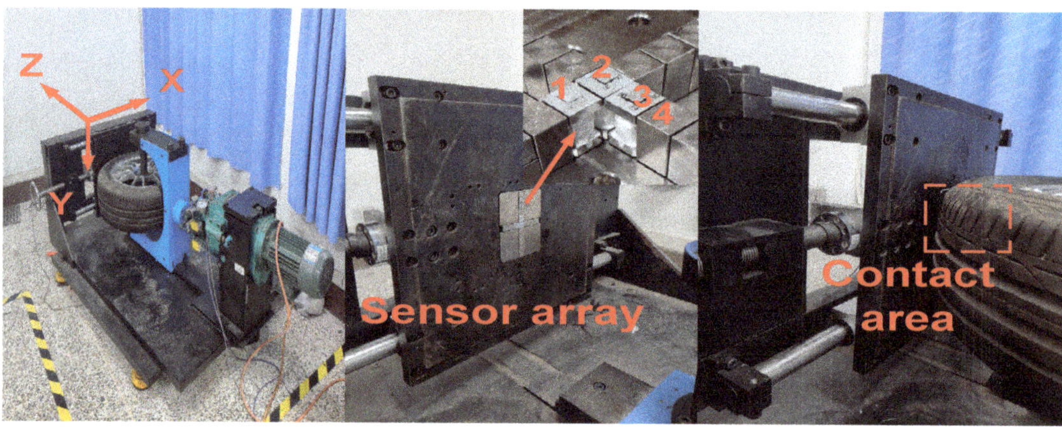

Figure 8. Simulated tire–pavement contact load test platform.

Creep is another factor that should be considered in the practical application of sensors, particularly those used for slow-loading and quasi-static measurements. The creep values of the sensor were recorded for 900 s as the constant load limit in the range was applied. The creep results are presented in (Table 2). The results suggest that the real-time tri-axial force output remained relatively stable, indicating good creep resistance.

Table 2. Creep resistance of sensor 1 (see Figure 8).

Direction of Applied Force	Maximum Measurement Variation (Mv/V)	Creep %F·S
X	−0.00442	−0.3192
Y	0.00716	0.4558
Z	−0.00242	−0.2483

3.2. Tire Loading Measurements

Although the sensors met the actual tri-axial stress measurement demands for tire–pavement contact, based on the calibration test results, there are differences between actual engineering applications and the theoretical test environment. To verify the effectiveness and stability of the sensor, the sensor was further tested in simulated tire–pavement contact load experiments. The test platform is shown in Figure 8.

The simulated loading platform was mainly comprised of an experimental bench, control cabinet, and data-acquisition system. A program was used to control the hydraulic push rod point dynamic loading or constant loading via the control cabinet, and the vertical force F_Z was the output. The bottom plate of the test bench could be moved left and right on the slide by manually rotating a screw, which simulated traction force F_X in the driving direction. The tire could move up and down within a certain range, and stress distributions at different positions on the tire could be obtained by moving the tire latterly. The movement of a vehicle tire was studied under light load conditions, and the interaction between the tire and pavement was simulated more accurately. By setting certain tri-axial loading parameters, a constant load or reciprocating load with a frequency of 0 to 3 Hz was applied in the Z direction. The tires used in the experiment were 205/55R16 radial car tires with an air pressure of 240 kPa.

The measuring surface in contact with the tire was composed of a cross-shaped array with the same transverse width as the tire and could measure the changes in the tire contact stress across the entire cross-section. The longitudinal sensor measured the influence of the tire tread pattern on the stress distribution, and its longitudinal length met at least two pattern periods. Since only the sensor performance was tested here, the sensors were not used in an array. Instead, three sensors were tested in an L-shaped arrangement in the center of the array—sensors 1, 2, and 3, as shown in Figure 8. Sensor 4 and other surrounding bolts of equivalent volume were substituted with metal blocks. The sensors and replacement blocks were close to the gravel size of pavement, resulting in measured values closer to the real situation. Nonetheless, the pressure produced by the tires of actual trucks and cars would be far greater than those generated by the simulated test platform. To have a better sensing effect, the sensor and an equivalent metal block were used to construct a cross-array plane, which was slightly higher (0.3 mm) than the bottom plane. In this way, the tire and the sensor were in closer contact, and the force was greater, which overcame the problem of insufficient pressure on the test platform.

In the first step, the sensor was fixed to an area above the tread center and F_Z was slowly applied up to 4000 N, then held constant. An upward thrust was applied in the vertical direction, while ensuring no relative sliding between the tire and the sensor. Then, the handle was slowly rotated at a uniform speed along the needle, moving the entire bottom plate to the right as the tire rolled from the right edge of sensors 2 and 3 toward sensor 1. At this point, the tire was in contact with approximately 12 sensors and equivalent replacement blocks. From Figure 9d, it can be seen that the stress distributions of the three sensors in the three directions were not uniform. Sensor 2 was located at a gap in the tire pattern. During the rolling process, sensor 1 had a delay of about 0.5 s; however, the measured values of all three sensors showed similar trends, with a maximum value of 23 kPa in the X direction, 41 kPa in the Y direction, and 208 kPa in the Z direction. It can be seen from Figure 9a that, as the tire rolled in the X direction and approached the sensor, the forward thrust of the tire was exerted on the sensors. After the tire had completely rolled over the top of the sensor, the force tended to zero. During the second half of the tire rolling process, the direction of the force changed and the magnitude rapidly declined after reaching a peak value. The variation of force during a single rolling cycle resembled a sine function. This trend was almost the same in the Z and Y directions. However, the force in the Z direction was five times larger. It should be noted that there was a certain difference between the magnitude of the force in the Y direction and the real measured force of the tire in the direction perpendicular to the ground obtained from the loading experiment, but the simulated test rig met the preliminary test of the sensor performance in the early stage.

After completing the test in the upper tire region (left edge of the tire in the longitudinal view), the tire position was changed to ensure contact with sensors located in the middle and lower regions. The handle was reset to the initial position, the bottom plate was reversed, and the tire was returned to its initial position. Stress values in the middle and lower regions were subsequently determined. The strain signals from sensors 1 and 2 were obtained through the A/D data-acquisition system and converted into equivalent stresses. In order to improve the repeatability of the test, all equipment was reset to its original position in each test. However, the vibration of the electric motor and hydraulic push rod, as well as the irregularity of the tire pattern, would have had a certain influence on the test results. Nevertheless, the overall trend of the measured curve was consistent.

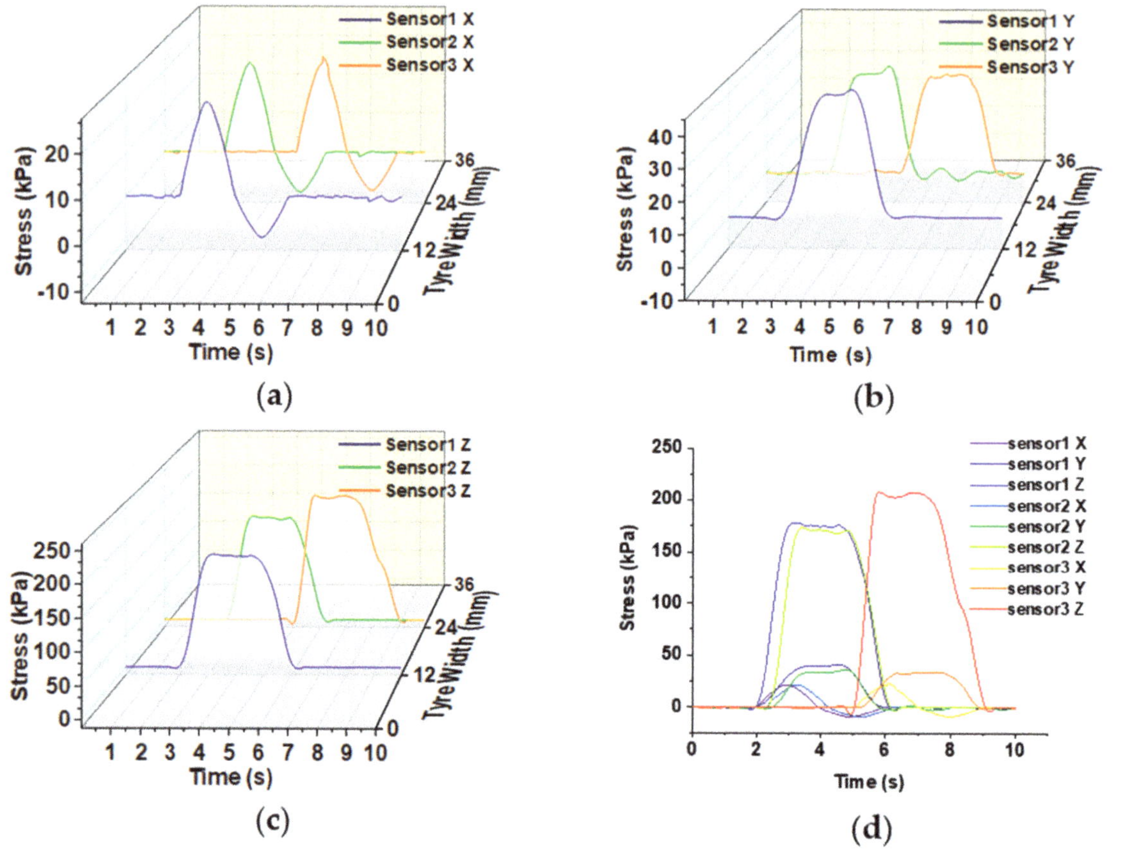

Figure 9. Relationship between time and stress: (**a**) variation of stress in X direction with time; (**b**) variation of stress in Y direction with time; (**c**) variation of stress in Z direction with time; (**d**) time-dependent stress of all transducers.

Sensors 1 and 2 were in the same plane, which allowed trends in the stress distribution to be obtained at the same horizontal position. The test was performed three times to determine changes in stress distributions in the upper, middle, and lower regions of the tire during a complete rolling process. In Figure 10, the horizontal axis is the tire width and the vertical axis is the tire rolling length. Figure 10 is a heat map composed of three test results using the test method in Figure 9. This figure can clearly reflect the stress change of the contact patch between the tire and pavement surface in a complete rolling process. The stress distribution between the tire and the sensor was not uniform during the contact process. Under a small force, the stress was greater in the middle compared with the surrounding areas, regardless of the direction (X, Y, or Z). The tire pattern and pavement texture were important factors leading to the uneven distribution of stress. The stresses in the X direction in Figure 10a are consistent with those in Figure 9a, exhibiting a typical sinusoidal trend. The stresses in the Y and Z directions shown in Figure 10b,c are also consistent with those in Figure 9b,c. There is only one peak, and the stresses decrease rapidly after reaching the maximum value.

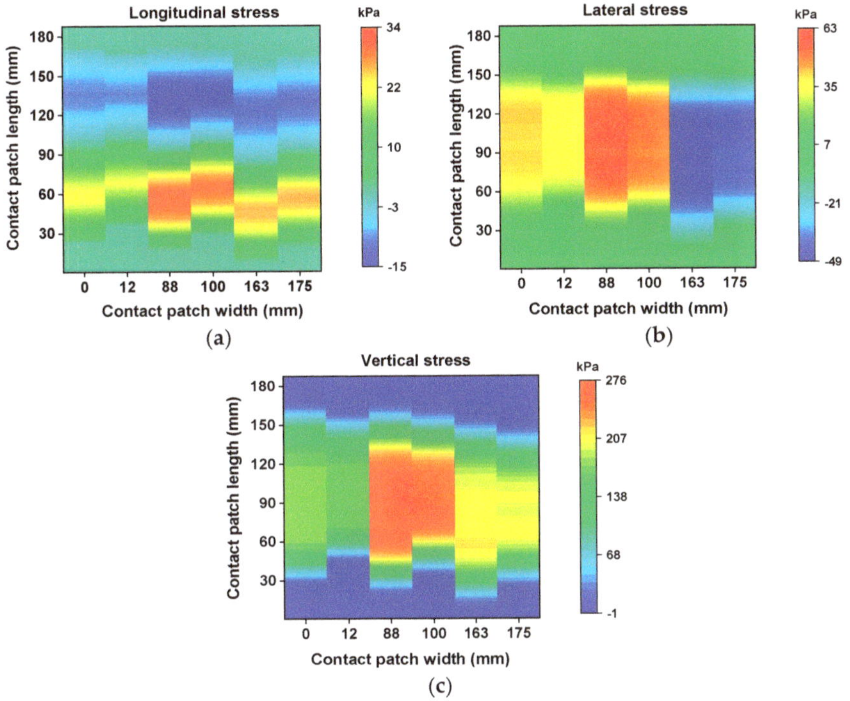

Figure 10. (**a**) Longitudinal (X) stress distribution; (**b**) lateral (Y) stress distribution; (**c**) vertical (Z) stress distribution.

4. Conclusions

A new tire–pavement tri-axial-stress-measuring sensor with complete performance compensation was designed in this study. The main conclusions are summarized below:

(1) Considering the asphalt pavement condition, a tire–pavement tri-axial stress sensor was designed. A surface width of 12 mm was selected by taking into account the aggregate size of typical asphalt pavement used in China. The strain gauges were located at the same depth as the top-down crack formation mechanism.

(2) A basic cross-shaped sensing array was formed in a cross-configuration to measure the contact stress distribution. The transverse sensors were used to measure the variation of stress across the whole contact surface during tire rolling. The longitudinal sensor could measure changes in stress before and after the tire pattern had rolled along with the contact patch. At present, only three sensors have been tested in the system. Other sensor positions were replaced with metal blocks of a similar structure. The sensors in this structure could be used for measurements under larger loads and higher velocities.

(3) The designed sensor has good linearity (maximum linearity error $\leq$ 0.804%), repeatability (maximum error $\leq$ 0.356%), and creep resistance (creep $\leq$ 0.4558%). The simulation test platform can reflect the characteristics of tire–pavement contact stress distribution in a complete rolling process.

(4) The stress distribution between the tire and the sensor was not uniform during the contact process. Under a small force, the stress was greater in the middle compared with the surrounding areas, regardless of the direction (X, Y, or Z). The tire pattern and pavement texture were important factors leading to the uneven distribution of stress.

In the future, we will continue to optimize the sensor's performance and build additional sensors to form a complete sensor array for more accurate measurement of tri-axial stress distribution in tire–pavement contact patches. This will be helpful for further study on the load transfer effect of wheels on asphalt pavement and the cause of pavement damage.

Author Contributions: Conceptualization, X.Z.; Formal analysis, M.R.; Investigation, L.L.; Methodology, J.G.; Writing—review & editing, Y.Y. All authors have read and agreed to the published version of the manuscript.

Funding: This work is supported by a Project Fund provided by the Special Fund for Research on National Major Research Instruments of China (51827812), National Natural Science Foundation of China (52172392), Key Research and Development Project of Hubei Province (2021BAA180), to which the authors are grateful.

Institutional Review Board Statement: Not applicable.

Informed Consent Statement: Not applicable.

Data Availability Statement: Not applicable.

Conflicts of Interest: The authors declare that they have no known competing financial interests or personal relationships that could have appeared to influence the work reported in this paper.

References

1. Chan, C.Y.; Huang, B.S.; Yan, X.D.; Richards, S. Investigating effects of asphalt pavement conditions on traffic accidents in Tennessee based on the pavement management system (PMS). *J. Adv. Transp.* **2010**, *44*, 150–161. [CrossRef]
2. Mamlouk, M.; Vinayakamurthy, M.; Underwood, B.S.; Kaloush, K.E. Effects of the International Roughness Index and Rut Depth on Crash Rates. *Transp. Res. Rec* **2018**, *2672*, 418–429. [CrossRef]
3. *JTGD50-2017*; Code for Design of Asphalt Pavement for Highway, Industry Standard—Traffic. China Communications Highway Planning and Design Institute: Beijing, China, 2017; 143p.
4. Grecenko, A. Tyre footprint area on hard ground computed from catalogue values. *J. Terramechanics* **1995**, *32*, 325–333. [CrossRef]
5. Els, P.S.; Stallmann, M.J.; Botha, T.R.; Guthrie, A.G.; Jimenez, E. Comparison of Tire Footprint Measurement Techniques. In Proceedings of the ASME 2016 International Design Engineering Technical Conferences and Computers and Information in Engineering Conference, Charlotte, NC, USA, 21–24 August 2016.
6. Muniandy, R.; Moazami, D.; Hamid, H.; Hassim, S. Characterization of Effective Tire Contact Area for Various Tread Patterns. *Instrum. Sci. Technol.* **2014**, *42*, 15–26. [CrossRef]
7. Kenarsari, A.E.; Vitton, S.J.; Beard, J.E. Creating 3D models of tractor tire footprints using close-range digital photogrammetry. *J. Terramechanics* **2017**, *74*, 1–11. [CrossRef]
8. Myers, L.; Roque, R.; Ruth, B.; Drakos, C. Measurement of Contact Stresses for Different Truck Tire Types to Evaluate Their Influence on Near-Surface Cracking and Rutting. *Transp. Res. Rec. J. Transp. Res. Board* **1999**, *1655*, 175–184. [CrossRef]
9. Tielking, J.T.; Abraham, M.A. Measurement of truck tire footprint pressures. *Transp. Res. Rec. J. Transp. Res. Board* **1994**, *1435*, 92–99.
10. Xu, T. Research on a Semi-Physical Tire Model Considering the Characteristics of the Contact Patch. Ph.D. Thesis, Jilin University, Changchun, China, 2018.
11. de Beer, M.; Kannemeyer, L.; Fisher, C. Towards improved mechanistic design of thin asphalt layer surfacings based on actual tyre/pavement contact Stress-In-Motion (SIM) data in South Africa. In Proceedings of the 7th Conference on Asphalt Pavements forSouthern Africa (CAPSA '99), Theme 5: Innovation in Asphalt Design, Victoria Falls, Zimbabwe, 29 August–2 September 1999.
12. Zeng, C. Mechanical Behaviors of Flexible Base Asphalt Pavement Structure under Nonuniform Vertical Contact Stresses. Master's thesis, Southwest Jiaotong University, Chengdu, China, 2015.
13. Peng, W.; Liu, M.; Liu, S. Analysis of Shear Dynamic Response of Asphalt Pavement under Repeated Brake Loads by 3D Finite Element Method. *J. Highw. Transp. Res. Dev.* **2009**, *9*, 46–52.
14. Blab, R.; Harvey, J.T. Modeling Measured 3D Tire Contact Stresses in a Viscoelastic FE Pavement Model. *Int. J. Geomech.* **2002**, *2*, 271–290. [CrossRef]
15. Chou, Y.T. Stresses and Displacements in Viscoelastic Pavement Systems under a Moving Load. *Field Tests.* 1969, pp. 25–40. Available online: https://trid.trb.org/view/100156 (accessed on 5 March 2022).
16. Miao, Y.; He, T.G.; Yang, Q.; Zheng, J.J. Multi-domain hybrid boundary node method for evaluating top-down crack in Asphalt pavements. *Eng. Anal. Bound. Elem.* **2010**, *34*, 755–760. [CrossRef]
17. Svasdisant, T.; Schorsch, M.; Baladi, G.; Pinyosunun, S. Mechanistic Analysis of Top-Down Cracks in Asphalt Pavements. *Transp. Res. Rec. J. Transp. Res. Board* **2002**, *1809*, 126–136. [CrossRef]
18. Howell, W.E.; Perez, S.E.; Vogler, W.A. Aircraft tire footprint forces. In *The Tire Pavement Interface, ASTM ASP 929*; Pottinger, M.G., Yager, T.J., Eds.; American Society for Testing and Materials: Philadelphia, PA, USA, 1986; pp. 110–124.

19. Lippmann, S.A. Effects of tire structure and operating conditions on the distribution of stress between the tread and the road. In *The Tire Pavement Interface, ASTM ASP 929*; Pottinger, M.G., Yager, T.J., Eds.; American Society for Testing and Materials: Philadelphia, PA, USA, 1986; pp. 91–109.
20. Marshek, K.M.; Chen, H.H.; Connell, R.B.; Hudson, R.W. Experimental Determination of Pressure Distribution of Truck Tire-Pavement Contact. *Transp. Res. Rec.* **1986**, *1070*, 9–14.
21. Smith, B.E.; Matthes, R.K.; Watson, W.F. Measurements of maximum pressure at the soil-tyre interface of a rubber-tyred skidder. *Pap.-Am. Soc. Agric. Eng.* **1990**, 107815091.
22. Cheli, F.; Braghin, F.; Brusarosco, M.; Mancosu, F.; Sabbioni, E. Design and testing of an innovative measurement device for tyre–road contact forces. *Mech. Syst. Signal. Process.* **2011**, *25*, 1956–1972. [CrossRef]
23. Anghelache, G.; Moisescu, R.; Sorohan, S.; Buretea, D. Measuring system for investigation of tri-axial stress distribution across the tyre-road contact patch. *Measurement* **2011**, *44*, 559–568. [CrossRef]
24. De Beer, M.; Fisher, C. Stress-In-Motion (SIM) system for capturing tri-axial tyre-road interaction in the contact patch. *Measurement* **2013**, *46*, 2155–2173. [CrossRef]
25. Xie, Y.; Yang, Q. Tyre-pavement contact stress distribution considering tyre types. *Road Mater. Pavement* **2019**, *20*, 1899–1911. [CrossRef]
26. Braunfelds, J.; Senkans, U.; Skels, P.; Janeliukstis, R.; Bobrovs, V. FBG-Based Sensing for Structural Health Monitoring of Road Infrastructure. *J. Sens.* **2021**, *2021*, 8850368. [CrossRef]
27. Canestrari, F.; Ingrassia, L.P. A review of top-down cracking in asphalt pavements: Causes, models, experimental tools and future challenges. *J. Traffic Transp. Eng.* **2020**, *7*, 541–572. [CrossRef]
28. Nayal, R.; Sharma, A. Mechanism-Driven Piezoelectric Energy Harvester. In *Advances in Engineering Design*; Springer: Singapore, 2021.

coatings

MDPI

Article

Utilization of Modified Red Mud Waste from the Bayer Process as Subgrade and Its Performance Assessment in a Large-Sale Application

Shijie Ma, Zhaoyun Sun *, Jincheng Wei, Xiaomeng Zhang and Lei Zhang

Shandong Transportation Research Institute, Ji'nan 250102, China; mashijie@sdjtky.cn (S.M.); weijincheng@sdjtky.cn (J.W.); zhangxiaomeng@sdjtky.cn (X.Z.); realchungsdu@163.com (L.Z.)
* Correspondence: andersontwo@163.com

Abstract: The utilization of red mud waste discharged from the Bayer production process used for extracting alumina from bauxite presents a pressing demand in the aluminum industry. This study aims to adopt a chemical modifier to solidify the Bayer red mud for its application in highway subgrade. The mechanism and properties of the modified red mud using a modifier composed of cement, phosphogypsum and organic polymer, were analyzed and investigated. It was found that the optimal modifier dosage of the solidified modifier was 8%. The three-day unconfined compressive strength of the modified Bayer red mud could reach up to 3 MPa and its strength loss when immersed in water at 7 days and 28 days measured less than 20%. For its real application as subgrade, its road performance could be achieved with good bearing capacity, including a resilient modulus value greater than 90 MPa, a dynamic deformation modulus reaching up to 140 MPa and the Falling Weight Deflectometer (FWD) value measuring less than 100 (0.01 mm). Compared with traditional lime or cement stabilized soil, using locally modified Bayer red mud for subgrade filling can reduce the project cost, minimize the consumption of non-renewable resources and reduce the emission of environmental hazards, thus providing an engineering reference for large-scale and resource-based road applications.

Keywords: red mud waste; subgrade; modification; road performance

Citation: Ma, S.; Sun, Z.; Wei, J.; Zhang, X.; Zhang, L. Utilization of Modified Red Mud Waste from the Bayer Process as Subgrade and Its Performance Assessment in a Large-Sale Application. *Coatings* **2022**, *12*, 471. https://doi.org/10.3390/coatings12040471

Academic Editor: Valeria Vignali

Received: 6 March 2022
Accepted: 28 March 2022
Published: 30 March 2022

Publisher's Note: MDPI stays neutral with regard to jurisdictional claims in published maps and institutional affiliations.

Copyright: © 2022 by the authors. Licensee MDPI, Basel, Switzerland. This article is an open access article distributed under the terms and conditions of the Creative Commons Attribution (CC BY) license (https://creativecommons.org/licenses/by/4.0/).

1. Introduction

Red mud is an industrial solid waste discharged from the production process of extracting alumina from bauxite, with the Bayer process representing a widely used alkaline alumina production process [1,2]. Approximately 1.0 to 1.8 tons of waste residue is discharged per ton of alumina produced as a byproduct of the Bayer process, with the waste referred to as Bayer red mud. With the rapid development of the aluminum industry in China, the amount of discharged Bayer red mud continues to increase. At present, the alumina production industry generally stores Bayer red mud through a dry-heaping embankment technology [3]. However, its open-air storage occupies a large amount of land, which continuously consumes management and maintenance costs, and more importantly brings risks and hidden dangers to the surrounding ecological environment [4,5].

Many studies have attempted to implement the comprehensive utilization of Bayer red mud and offer technical solutions for recycling of the production of industrial building materials [6–8], extraction and recovery of valuable metals [9–11], production of new materials [12,13], and environmental restoration [14–16]. However, Bayer red mud still cannot be used in large-scale operations thus far, due to mutual constraints of various factors such as technical and economic feasibility, and secondary pollution. Therefore the cumulative stock of red mud in China has exceeded 350 million tons, but its overall comprehensive utilization rate remains less than 4% [17].

In practice, it is acknowledged that the construction of roads, railways and other infrastructure projects require large amounts of soil and rock for subgrade filling, which not only wastes a large amount of non-renewable land resources, but also causes increasingly significant damage to the environment and ecology. Using Bayer red mud as a roadbed filling material for engineering construction through systematic research can alleviate the shortage of soil sources during the construction process and realize the large-scale consumption and utilization of bulk solid waste in the aluminum smelting industry, which has important practical significance.

Bayer red mud is alkaline, contains trace amounts of chromium, fluoride and other harmful components, and has characteristics of high natural moisture content, fine particles, and poor water stability. Therefore, it cannot be directly used as subgrade filling material from the perspective of environmental protection and engineering applications. Over the years, extensive work has been conducted by researchers worldwide to develop various utilization methods of red mud in road engineering. E Mukiza et al. prepared an eco-friendly road base material using mainly red mud and characterized its mechanical properties, hydration and leaching characteristics [18]. CVH Rao carried out unconfined compressive strength, splitting tensile strength and California bearing ratio tests on different amounts of GGBS stabilized red mud, and concluded that red mud stabilized with GGBS can be used as subbase, base and subgrade [19]. V Mymrin et al. used red mud, clay slate mining wastes and polishing sludge to prepare composite materials, resulting in the improvement of all its mechanical properties, and which could be used in road foundations, airport runways and building foundations [20]. J Qi et al. carried out the preparation and curing test of red mud road base material based on the analysis of the physical and chemical properties of red mud. Subsequent test results showed that the red mud base exhibited better compressive strength and resilience modulus than the traditional road base material [21]. Previous laboratory studies indicated that red mud was feasible for use as a raw material in road base or subgrade by adopting advanced technology/treatment methods. These studies mainly focused on the selection of red mud curing materials, mix design and improvement of mechanical properties, but research on improving the utilization rate of red mud, the control of leaching pollutants and real road performance remain inconclusive.

In this study, based on the analysis of the production process, the chemical composition and engineering and environmental properties of original Bayer red mud, its pollution source, environmental risks and road characteristics were first assessed. Furthermore, a chemical modifier was adopted to solidify the Bayer red mud, attain effective inhibition of the leaching of harmful components in red mud and to achieve the utilization rate of red mud above 90%. In a real highway project, a large-scale engineering application and road performance of the modified Bayer red mud as subgrade filling were evaluated. In this paper, the key technologies of road performance improvement and environmental risk control were comprehensively considered for the modification treatment of red mud, and the modified material with red mud as the main body was used for expressway subgrade filling for the first time. The road and environmental performance were monitored and evaluated by the physical project, providing a valuable reference for the large-scale application of red mud in road engineering.

2. Materials and Methodology

2.1. Original Red Mud Waste from the Bayer Process

2.1.1. Chemical Composition of Bayer Red Mud Waste

The chemical composition of Bayer red mud waste, collected from a local aluminum industry, is shown in Table 1. The Bayer process uses strong alkaline to dissolve bauxite, and the resulting red mud waste contains a high content of iron, and its mineral composition mainly includes hematite, perovskite, goethite, calcite, calcite, diaspore and so on [22]. The chemical composition of Bayer red mud mainly depends on the composition of bauxite [23], the production process of alumina and the additives mixed in the production process, which include mainly iron oxide, alumina, silicon dioxide, calcium oxide, in addition to some trace non-ferrous metals. It was thus found that there were higher contents of SiO_2, Al_2O_3 and Fe_2O_3 than that of other oxides.

Table 1. Chemical constituents of Bayer red mud.

Chemical Composition	Fe_2O_3	Al_2O_3	SiO_2	CaO	MgO	TiO_2	Na_2O	K_2O	Loss
Content/%	34.3	21.4	20.1	3.2	0.3	2.0	8.1	0.2	9.7

2.1.2. Engineering Properties of Red Mud Waste

The Bayer red mud used in this study was sampled on site and the related physical and mechanical properties were tested in laboratory.

- The specific gravity of Bayer red mud ranged between 2.70–3.10;
- The measured value of compression coefficient $a_{1\text{-}2}$ ranged between 0.15–0.35 MPa^{-1} which has medium compressibility in accordance with Code for design of building foundation (GB 50007-2011);
- The liquid limit (WL) was more than 50%, and the plasticity index IP was less than 17, which is similar to the engineering characteristics of high liquid limit silty clay in accordance with Test methods of soils for highway engineering (JTG 3430-2020);
- The unconfined compressive strength under the optimal water content state was between 400–600 kPa, and its water stability was poor, leading to disintegration immediately after immersion.

Figure 1 shows the particle size distribution of Bayer red mud. Its particles were very fine, in which the particle content of 1.0–10 μm accounted for more than 90%. The mineral composition of red mud particles with negative charges bears strong hydrophilicity, which makes the thickness of particle water film thick, and the water content can still remain between 35–40% after pressure filtration.

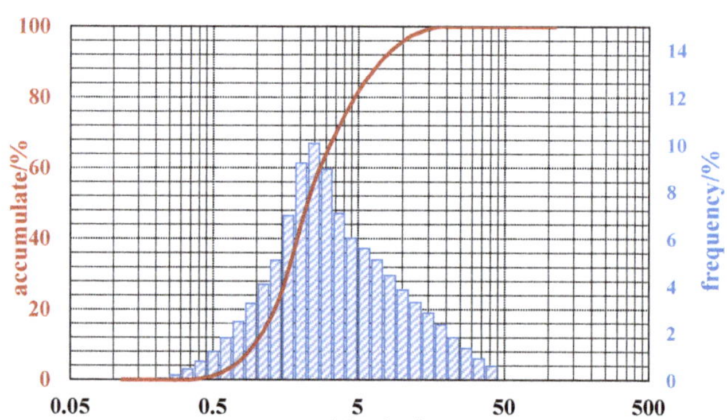

Figure 1. Particle distribution curve of Bayer red mud.

It can be concluded that the engineering characteristics of Bayer red mud used exhibited high natural water content, high liquid limit, low strength and poor water stability, which cannot be directly used for road subgrade filling.

2.1.3. Leaching of Hazardous Substances of Red Mud Waste

Solid waste substances from production, living and other activities can be divided into solid waste and hazardous waste [24]. According to the test items and methods specified in the General Principles for Identification of Hazardous Wastes (GB 5085.7-2019), the Bayer red mud was tested. It was found that the pH test value ranged from 11.4 to 12.2. The leaching concentrations of detectable hazardous substances are shown in Table 2. The results showed that leaching concentration of certain harmful elements of Bayer red mud were lower than that of the concentration limit specified in the standard, and should thus not be listed in the national hazardous waste list.

Table 2. Leaching concentrations of some harmful elements from Bayer red mud.

Project	Total Cr	As	Se	Mo	Sb	V	Cr^{6+}	F^-
Leaching concentration (mg/L)	1.5	0.0097	0.01	0.16	0.03	1.10	1.45	16.1
Limit value of hazardous waste (mg/L)	15	5	1	—	—	—	5	100

2.2. *Modification of Bayer Red Mud*

2.2.1. Modifier used in this Study

Due to the poor engineering properties measured above, Bayer red mud requires modification before application. In this study, the solidification and modification treatment adopted a modifier to improve the physical and mechanical properties, to reduce pH value, to inhibit the effect of metal ion leaching, through the action of charge neutralization, adsorption bridging and surface adsorption.

The main components of the modifier comprised cement, phosphogypsum and polymer stabilizer, each component fully mixed in certain mass proportions under normal temperatures. The modifier was then used to treat the Bayer red mud. The cement used was a hydraulic cementing material with 42.5 grade ordinary Portland cement employed. Phosphogypsum is a solid waste produced by a wet-process phosphoric acid process, which was dried and ground to below 0.15 mm before use. The polymer stabilizer is a high molecular stabilizer and water-soluble polymer. This modifier material is a powdery solid at room temperature. After homogenous mixing with Bayer red mud and water, it can be chemically solidified. Cement forms C–S–H gel, $Ca(OH)_2$, Aft and Afm. The pH value of the modified red mud can be adjusted to about 10, which can effectively inhibit the leaching of lead and cadmium. The polymer stabilizer in the modifier can reduce the distance between red mud particles through charge neutralization and form colloidal particle aggregation structure. At the same time, the strong adsorption capacity of the polymer stabilizer effectively connects the dispersed particles to form a whole through "bridging", which not only forms the strength through the crystal network structure skeleton, but also stabilizes the harmful metal ions in the red mud through physical wrapping and chemical adsorption solidification. Compared with traditional lime, cement and other curing materials, addition of modifier can not only improve the strength of red mud, but also improve the water stability of solidified red mud and the diffusion and migration of some metal ions, making it better in actual road performance and environmental sustainability.

2.2.2. Mixture Design and Sample Preparation

The modified red mud mixtures were prepared with five different amounts of modifier at 4%, 6%, 8%, 10% and 12% by mass, respectively. If the content of modifier is lower than 4%, mixing with red mud may not be uniform enough to affect the later modification effect. If the content of modifier is higher than 12%, it will reduce the economic value.

A set of standard Proctor compaction tests in accordance with Test methods of soils for highway engineering (JTG 3430-2020) were conducted to determine optimum moisture content (ω) and maximum dry density (ρ_d). The cylinder specimen with a diameter of 39.1 mm and a height of 80 mm was compacted for each condition, and its degree of compaction was 96%. After molding, it was sealed with a sealing bag and placed in a standard curing box at a temperature of 20 ± 1 °C and a relative humidity of $\geq$90% for a specific number of curing days in the laboratory.

2.3. Case Study of a Practical Application of Modified Bayer Red Mud as Subgrade

In order to realize large-scale utilization of modified Bayer red mud in real-world applications, a demonstration road trial was built in a new Ji-Qing Highway Reconstruction and Expansion Project in 2015 in China. This highway was completed in early 1994 and bears a total length of 318 km. In recent years, with the rapid development of regional economy and society and the rapid growth of traffic volume, its actual traffic volume has far exceeded the design traffic volume, thus it is urgent to implement the reconstruction and expansion of engineering projects. In 2015, the Ji-Qing Highway Reconstruction and Expansion Project was officially constructed, from two-way four lanes to two-way eight lanes.

Locally, a number of large alumina enterprises are distributed in Binzhou and Zibo along the reconstruction and expansion project of Jinan Qingdao Highway. For effective usage of locally available resources to minimize cost, a demonstration road trial using the modified red mud was concerned and located on the right side of the stake number (k289 + 090.5~k294 + 385). The subgrade length was about 4900 m, the subgrade width was 11 m, and the filling thickness was 0.2 m. The modified Bayer red mud was used to replace 0.2 m lime-stabilized soil on the upper part of roadbed as shown in Figure 2. The Bayer red mud was evenly mixed with the solidified modifier in the mixing plant, and then transported to the subgrade filling site for paving and compaction. The mass ratio of Bayer red mud and solidified modifier by mass was 92:8. The demonstration road consumed about 17,000 tons of Bayer red mud and saved about 14,800 m^3 of natural soil.

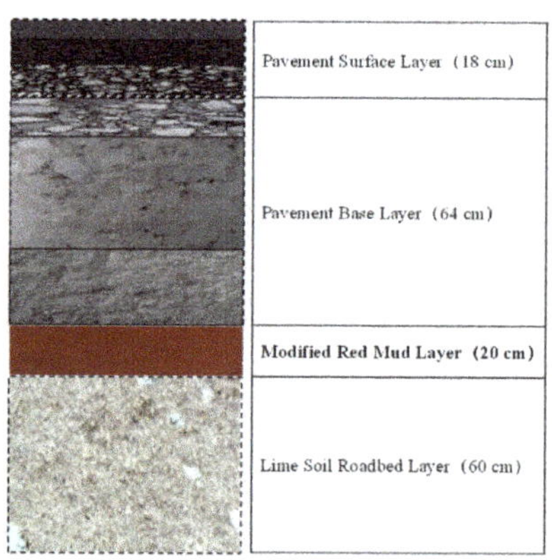

Figure 2. Pavement structure diagram of demonstration road.

3. Results and Discussion

3.1. Properties of Solidified Red Mud in the Laboratory

3.1.1. Effect of Modifier on Optimum Moisture Content and Maximum Dry Density

Figure 3 shows the influence of modifier dosage on the optimal moisture content and maximum dry density of the modified red mud. It was found that the modifier content had a great influence on the maximum dry density of Bayer red mud, while it had less influence on the optimum moisture content. This could be due to a chemical reaction that occurred when the solidified modifier made contact with the Bayer red mud, which would consume a certain amount of water. At the same time, the modifier material allowed red mud particles to more easily form dense agglomerative structures, and its dry density improved from a macroscopic point of view.

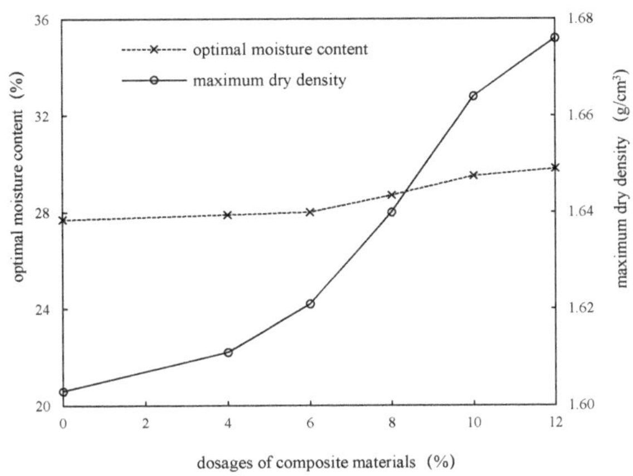

Figure 3. Influence of modifier on optimum moisture content and maximum dry density.

3.1.2. Unconfined Compressive Strength of Solidified Red Mud

Figure 4 illustrates the unconfined compressive strength of the modified red mud cured for 3 days, 7 days and 28 days. The unconfined compressive strength test was carried out under the condition of zero lateral pressure, with the application of an axial pressure with an equal axial deformation, until sample failure. It can be observed that with the curing days and the modifier dosage, the strength continued increasing. However, compared to the 28-day strength, the influence of the modifier dosage became insignificant after 8% and the change of strength was slight. This implies that an optimum modifier dosage to treat the red mud waste would be 8%.

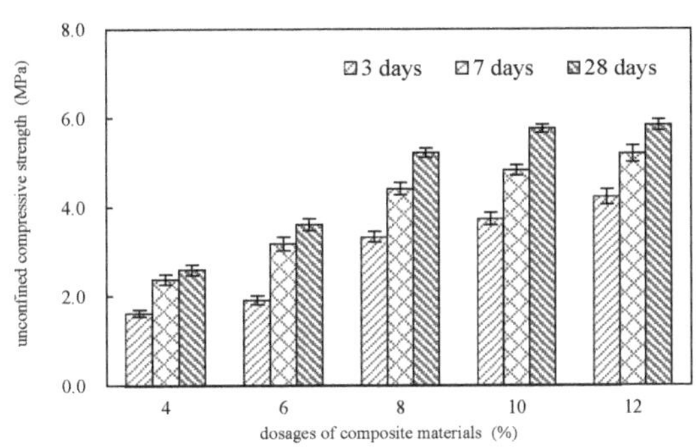

Figure 4. Unconfined compressive strength of solidified red mud at different dosages.

After immersion in water for one day, the unconfined compressive strength of the specimens cured for 7 days, and 28 days were tested as well, and the test results are shown in Figure 5. For this immersion test, the specimens were removed from the curing box and soaked in a 20 °C water bath for 24 h. Then, the compressive test was carried out according to the standard method. It was found that after one-day immersion in water, the strength of all specimens had decreased. The level of decrease for the 28 days cured mixture prepared with 8% of modifier was less than 20%.

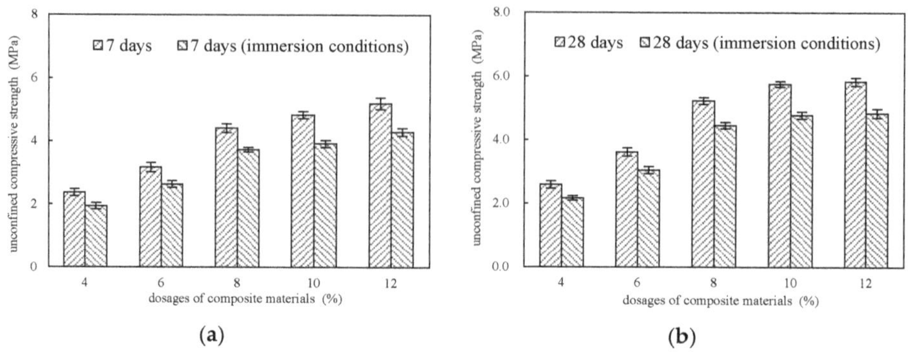

Figure 5. Unconfined compressive strength of solidified red mud after immersion in water. (**a**) is strength comparison of 7 days, (**b**) is strength comparison of 28 days.

3.1.3. Road Performance of Bayer Red Mud before and after Curing Modification

By comparing the results of unconfined compressive strength and strength loss of the modified red mud under standard curing and immersion conditions, the optimal mixing ratio of the modified modifier was determined as 8%. The solidification of Bayer red mud containing 8% modifier was carried out in the laboratory, and the solidification effects on road performance were compared and analyzed. The test results are listed in Table 3. In Table 3, ω denotes optimum moisture content, ρ_d denotes maximum dry density, w_L denotes liquid limit index, w_p denotes plastic limit index, I_P denotes plasticity index, φ denotes internal friction angle, c denotes cohesion, E_0 denotes modulus of resilience, CBR denotes California Bearing Ratio.

Table 3. Comparison of road performance of Bayer red mud before and after curing modification.

Category	ω (%)	ρ_d (g/cm^3)	w_L (%)	w_P (%)	I_P	φ (°)	c (kPa)	E_0 (MPa)	CBR (%)
Original red mud	27.7	1.603	51.4	38.1	13.3	24.2	31.0	27.3	4.3
Modified red mud	28.7	1.660	37.2	22.1	15.1	36.8	143.2	705.4	124

Through the solidification and modification treatment, the liquid limit of the Bayer red mud reduced from 51.4% to 37.2%, the plasticity index increased from 13.3 to 15.1, and the physical properties were improved. At the same time, the shear strength, modulus of resilience and CBR of modified Bayer red mud were significantly improved, and the results showed that the modification treatment could effectively improve the mechanical and road performance of Bayer red mud.

3.1.4. Leaching of Hazardous Substances in Solidified Red Mud

Table 4 lists the leaching concentration and reduction ratio of some hazardous substances in the solidified red mud compared with the original one in Table 2. Comparative analysis showed that the pH value of leaching solution decreased from 10–12 to 9–10. The leaching concentrations of total chromium, arsenic, selenium, molybdenum, antimony, vanadium, hexavalent chromium and fluoride decreased significantly, and the reduction ratio of hexavalent chromium, fluoride, selenium, arsenic and vanadium was more than 70%.

Table 4. Leaching concentration and reduction ratio of some hazardous substances in solidified red mud.

Category	Total Cr	As	Se	Mo	Sb	V	Cr^{6+}	F^-
Leaching concentration(mg/L)	1.21	0.0028	0.0013	0.12	0.02	0.34	0.031	2.68
Percentage reduction (%)	19.3	71.1	87.0	37.5	33.3	69.1	97.9	83.4

In general, the modified Bayer red mud had a significant effect on inhibiting the leaching of hazardous substances. The modified Bayer red mud can meet the standard requirements of subgrade filling and environmental protection in accordance with specifications for design of highway subgrades (JTG D30-2015) and technical specification of red mud (Bayer) subgrade application for highway engineering (DB 37/T 3559-2019).

3.2. Performance of Modified Bayer Red Mud as Subgrade in Practical Application

3.2.1. Construction Method and Quality Control of Subgrade with Modified Bayer Red Mud

In a real project, a centralized plant mixing method was adopted to produce the modified Bayer red mud and to construct the subgrade. The main technological process is given below: preparation before construction → Bayer red mud storage and transportation → preparation of modified materials → construction site leveling and distribution → centralized mixing at mixing station → transportation of modified red mud to site → paving of modified red mud → vibration compaction (strong vibration twice and weak vibration twice) → final compaction → quality inspection → maintenance → next layer construction. This process is used for paving and rolling on the solidified lime soil, which will not adversely affect the lower bearing layer. The construction site of mixing, paving, compacting and curing is shown in Figure 6.

(**a**) Mixing (**b**) Paving

(**c**) Compaction (**d**) Curing

Figure 6. Plant mixing construction of Bayer red mud in a project.

Regarding the construction, the modified Bayer red mud represents a kind of non-traditional roadbed filling material. Due to its own physical properties and the particularity of the solidification process, some targeted controls on technological aspects were implemented:

The main physical and mechanical properties shall be tested before construction. When the source of red mud waste or the production process changes, these tests must be conducted again:

- The moisture content of the modified Bayer red mud should be tested after mixing in the plant and spreading on site, and the measured moisture content should be controlled at 2–3 percentage points higher than the optimal moisture content.
- The dosage of the solidified modifier should be controlled accurately in the mixing station, and the error should not be more than ±0.5%. After mixing, the surface color of the modified red mud is even and consistent, and there is no ash mass and gray strip. The maximum particle size should not exceed 5 mm.
- In a single-layer construction, the on-site compaction should be completed within 4 h after mixing. In the multi-layer continuous construction, the next filling layer should be compacted within 4 h of the previous filling layer.
- After compaction, the surface should be covered with felt or watered for curing, and the curing period should not be less than 2 days.

3.2.2. Environmental Monitoring of Subgrade Filled with the Modified Red Mud

In the Bayer red mud demonstration road, four sampling wells for groundwater quality monitoring were buried on both sides of the subgrade. The bottom of the sampling wells was below the groundwater level, and the groundwater was sampled regularly [25].

The monitoring well was buried before the Bayer red mud was filled as subgrade, and the groundwater was sampled for testing as a control. After the construction of Bayer red mud, the groundwater in the well was collected regularly at certain interval periods [26]. Figure 7 shows typical indexes of pH, Cr^{6+} and F^- in groundwater samples as a function of time. Through continuous tracking collection and testing, it was found that the pH value and hazardous substance concentrations of groundwater samples in four sampling wells presented different variations. The pH value of groundwater in four places decreased slightly after construction, and finally stabilized in the range of 7.0–8.0. The concentration of hexavalent chromium showed a downward trend from the first blank sampling to 28 days after the completion of the test road, and then measured lower than the limits. The concentration of fluoride showed an upward trend from the first blank sampling to 28 days after the completion of the test road. The concentrations of other metals (total mercury, total chromium, cadmium, lead, beryllium, selenium, silver, copper, antimony, vanadium and cyanide) were lower than that of the detection limits.

According to the comparative analysis of long-term test results above, the pH values and concentrations of hazardous substances in the continuous sampling of groundwater in the test road were lower than that of the limit values of class III groundwater. At the same time, the monitoring results of the test road and the conventional road were consistent overall, and there were no abnormal changes in the concentration of one or more hazardous substances caused by the red mud test road.

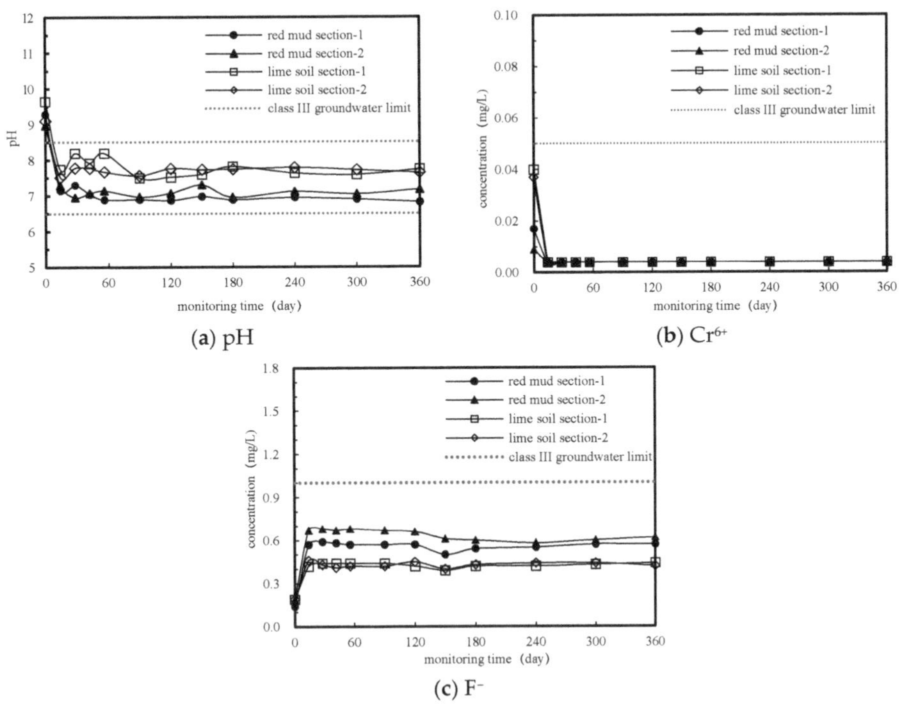

Figure 7. Typical indexes pH, Cr^{6+} and F^{-} in groundwater samples as a function of time.

3.2.3. Road Performance of Subgrade Filled with the Modified Red Mud

After seven days of moisturizing and curing in the test road, the compacted Bayer red mud subgrade in the demonstration road was tested on site for road performance. The resilient modulus was measured using a bearing plate method, dynamic modulus of deformation was measured by the portable deflectometer method (PFWD) and the deflection value was measured by the falling weight deflectometer method (FWD) [27–29]. In order to analyze the overall uniformity of subgrade and the correlation of each test index [30], 200 m was selected as the fixed-point test on site, and points were arranged at 20 m intervals in the two lanes of the widening part. The test results are listed in Table 5.

Table 5. Performance test results of fixed point test sections.

Measuring Points	E/MPa	E_{vd}/MPa	l/0.01 mm	Measuring Point	E/MPa	E_{vd}/MPa	l/0.01 mm
1-1	79.4	98.2	138.9	2-1	76.9	86.0	116.2
1-2	108.3	166.0	89.5	2-2	143.9	217.1	49.6
1-3	88.9	118.1	111.2	2-3	164.7	229.5	54.4
1-4	100.7	152.3	96.9	2-4	125.2	152.0	63.1
1-5	110.7	156.9	81.6	2-5	174.0	260.6	50.6
1-6	90.6	139.5	101.7	2-6	151.1	209.6	56.2
1-7	112.3	143.5	75.4	2-7	115.4	201.1	74.0
1-8	134.2	187.6	74.5	2-8	169.8	199.8	62.2
1-9	123.1	185.0	69.9	2-9	185.3	267.2	51.4
1-10	142.3	184.4	63.7	2-10	145.7	197.9	62.5

It can be seen from Table 5 that the resilient modulus of most measuring points in the test section was greater than 90 MPa, the dynamic deformation modulus of PFWD was greater than 140 MPa, and the FWD deflection value was less than 100 (0.01 mm). The overall bearing capacity of the subgrade was good and met the road requirements.

Figure 8 further shows correlations between different test indices. These correlations were good enough to provide reference for the quality control and evaluation of the modified Bayer red mud subgrade.

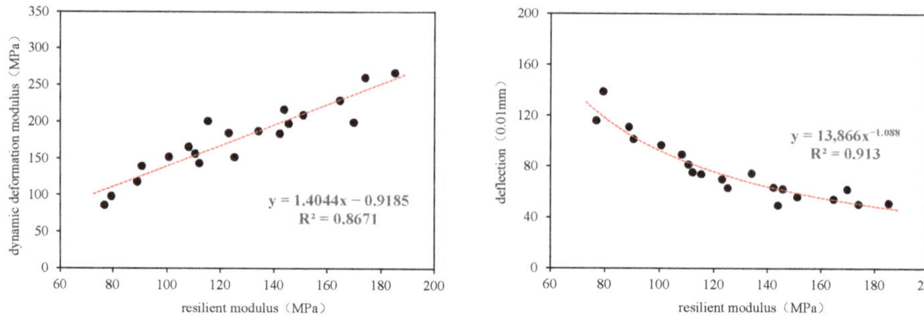

Figure 8. Correlation between different performance indexes of modified red mud subgrade. (**a**) is correlation between dynamic deformation modulus and resilience modulus, (**b**) is correlation between deflection and resilience modulus.

Figure 9 shows the cored cylindrical subgrade samples (Φ150 mm × 150 mm) on site for the unconfined compressive strength test. After testing, the average value of the unconfined compressive strength was 2.4 MPa, which met the specification requirements in this project. In the 95% confidence interval, the unconfined compressive strength of the cored specimen was 2.06–2.77 MPa, and the standard deviation was 0.50. In addition, the white substance on the top surface of the on-site coring specimen comprised Na_2CO_3, $CaCO_3$ and other salt substances, which represent the mineral precipitation caused by the humidity change on the top surface of the structural layer exposed to the air.

Figure 9. Site coring and testing of Bayer red mud Subgrade. (**a**) is field coring sample, (**b**) is unconfined compressive strength test of core sample.

3.2.4. Benefit Analysis of Subgrade Filled with the Modified Red Mud

In this practical project, Bayer red mud was used as a local waste resource and its utilization did not require purchasing of materials, with only a certain amount of transportation costs involved. Compared with traditional cement or lime stabilized soil (6%), when the transportation distance was between 15 and 20 km, the modified Bayer red mud needed to fill the subgrade could save CNY 10 per ton compacted earthwork. Through the field test, the representative deflection value of modified Bayer red mud subgrade was less than the representative deflection value of lime stabilized soil, and the modulus was greater than the modulus of lime stabilized soil. This showed that the overall bearing capacity and road performance of Bayer red mud subgrade were better than that of lime stabilized soil subgrade. Therefore, from the analysis of the economic benefits and utilization of wastes, the use of modified Bayer red mud instead of cement or lime stabilized subgrade is more effective.

Meanwhile, Bayer red mud is now stored in the form of dry-heaping embankment technology, and the enterprise needs to invest approximately CNY 35 per ton in the maintenance and management of the yard every year in China. Among them, the energy cost (gasoline or diesel) of vehicles for loading, unloading, transportation, transportation and dust suppression by watering accounts for about 55%, approximately CNY 19.25 per ton, equivalent to 2.85 L per ton of energy (as calculated by comparison with diesel). Additionally, using one ton of Bayer red mud can directly save approximately 3.49 kg of standard coal per year.

4. Conclusions

This study presented the properties of the modified Bayer red mud waste and its large-scale application as subgrade in a real project. The main findings are provided below:

Regarding the leaching results of Bayer red mud, it was identified as a general industrial solid waste by environmental protection standards. However, its engineering characteristics exhibited a high natural water content, high liquid limit, low strength and poor water stability, and thus cannot be directly used for road subgrade filling.

The modifier of Bayer red mud adjusted charge neutralization, adsorption bridging and surface adsorption, and pH value of the microenvironment. Subsequently, a network aggregation structure was formed to reduce the leaching concentration of harmful substances and to improve the mechanical and water stability of the modified Bayer red mud.

Through laboratory testing, the optimal modifier dosage was identified at 8%. The road performance of the modified Bayer red mud revealed a great improvement. The pH value of the leaching solution reduced from 10–12 to 9–10, and the three-day unconfined compressive strength could reach up to 3 MPa. Moreover, the leaching concentration of typical harmful components was reduced by more than 70%.

In the demonstrated engineering application, the resilient modulus of the filled subgrade measured more than 90 MPa, the dynamic deformation modulus more than 140 MPa, and the FWD deflection value less than 100 (0.01 mm). The pH value and hazardous substance concentrations of long-term monitoring groundwater quality samples in the test section were consistent with the monitoring results in the conventional sections and had no adverse effects on groundwater quality.

Author Contributions: Data curation, J.W.; Formal analysis, X.Z. and L.Z.; Writing—original draft, S.M.; Writing—review & editing, Z.S. All authors have read and agreed to the published version of the manuscript.

Funding: The authors acknowledge the financial support of the National Key R&D Program of China (2018YFB1600103), Shandong Transportation Science and Technology Innovation Project (2016B43) and project ZR2020QE271 supported by Shandong Provincial Natural Science Foundation.

Institutional Review Board Statement: Not applicable.

Informed Consent Statement: Not applicable.

Data Availability Statement: The authors confirm that the data supporting the findings of this study are available within the article.

Conflicts of Interest: The authors declare no conflict of interest.

References

1. Hind, A.R.; Bhargava, S.K.; Grocott, S.C. The surface chemistry of Bayer process solid: A review. *Colloids Surf. A Physicochem. Eng. Asp.* **1999**, *146*, 359–374. [CrossRef]
2. Brunori, C.; Cremisini, C.; Massanisso, P.; Pinto, V.; Torricelli, L. Reuse of a treated red mud bauxite waste: Studies on environmental compatibility. *J. Hazard. Mater.* **2005**, *117*, 55–63. [CrossRef]
3. Xue, S.; Li, Y.; Guo, Y. Environmental impact of bauxite residue: A comprehensive review. *J. Univ. Chin. Acad. Sci.* **2017**, *34*, 401–412.

4. Gräfe, G.P.; Klauber, C. Bauxite residue issues: I. Current management, disposal and storage practices. *Hydrometallurgy* **2011**, *108*, 33–45.
5. Liu, S.; Guan, X.; Feng, C.H.; Qiu, M. Progress on red mud stockpiling and utilization. *Bull. Chin. Ceram. Soc.* **2015**, *34*, 2194–2200.
6. Linora, M.D.; Selvamony, C.; Anandakumar, R.; Seeni, A. Investigations on optimum possibility of replacing cement partially by redmud in concrete. *Sci. Res. Essays* **2015**, *10*, 137–143.
7. Tsakiridis, P.E.; Agatzini-Leonardou, S.; Oustadakis, P. Red mud addition in the raw meal for the production of Portland cement clinker. *J. Hazard. Mater.* **2004**, *116*, 103–110. [CrossRef] [PubMed]
8. Pontikes, Y.; Rathossi, C.; Nikolopoulos, P.; Angelopoulos, G.N.; Jayaseelan, D.D.; Lee, W.E. Effect of firing temperature and atmosphere on sintering of ceramics made from Bayer process bauxite residue. *Ceram. Int.* **2009**, *35*, 401–407. [CrossRef]
9. Liu, W.; Yang, J.; Xiao, B. Application of Bayer red mud for iron recovery and building material production from alumosilicate residues. *J. Hazard. Mater.* **2009**, *161*, 474–478. [CrossRef] [PubMed]
10. Piga, L.; Pochetti, F.; Stoppa, L. Recovering metals from red mud generated during alumina production. *JOM* **1993**, *45*, 54–59. [CrossRef]
11. Li, Y.; Wang, J.; Wang, X.; Wang, B.; Luan, Z. Feasibility study of iron mineral separation from red mud by high gradient superconducting magnetic separation. *Phys. C Supercond. Appl.* **2011**, *471*, 91–96. [CrossRef]
12. Hairi, S.N.M.; Jameson, G.N.; Rogers, J.J.; MacKenzie, K.J. Synthesis and properties of inorganic polymers (geopolymers) derived from Bayer process residue (red mud) and bauxite. *J. Mater. Sci.* **2015**, *50*, 7713–7724. [CrossRef]
13. Wei, H.; Ma, X.; Guan, X.; Zhang, W.; Huang, A.; Li, S. Preparation of Bayer red mud light-weight thermal insulation ceramics. *Bull. Chin. Ceram. Soc.* **2019**, *38*, 749–751, 761.
14. Altundogan, H.S.; Altundogan, S.; Tumen, F.; Bildik, M. Arsenic adsorption from aqueous solutions by activated red mud. *Waste Manag.* **2002**, *22*, 357–363. [CrossRef]
15. Bhatnagar, A.; Vilar, V.J.; Botelho, C.M.; Boaventura, R.A. A review of the use of red mud as adsorbent for the removal of toxic pollutants from water and wastewater. *Environ. Technol.* **2011**, *32*, 231–249. [CrossRef]
16. Yang, H.Z.; Cheng, Q.K.; Wang, X.Z.; Chen, H.K.; Zhang, H.Y.; Huang, X.F.; Yang, X.F.; Tang, Y.T.; Shi, T.H. Preparation of modified Bayer process red mud granules and stabilization treatment of lead in soil. *Chin. J. Environ. Eng.* **2018**, *12*, 1171–1181.
17. Sun, Z.; Cheng, Y. Present situation of red mud stockpiling and comprehensive utilization in Shandong Province. *Shandong Jiaotong Keji* **2017**, *2*, 99–100, 107.
18. Mukiza, E.; Liu, X.; Zhang, L.; Zhang, N. Preparation and characterization of a red mud-based road base material: Strength formation mechanism and leaching characteristics. *Constr. Build. Mater.* **2019**, *220*, 297–307. [CrossRef]
19. Rao, C.H.V.H.; Naidu, P.G.; Satyanyarana, P.V.V.; Adiseshu, S. Application of GGBS stabilized redmud in road construction. *IOSR J. Eng.* **2012**, *2*, 14–20.
20. Mymrin, V.; Aibuldinov, E.K.; Alekseev, K.; Avanci, M.A.; Rolim, P.H.; Catai, R.E.; Carvalho, K.Q. Sustainable material manufacturing from hazardous bauxite red mud in composites with clay slate waste and lime production waste. *Int. J. Adv. Manuf. Technol.* **2022**, *111*, 1375–1385. [CrossRef]
21. Qi, J.; Yang, J.; Wang, M. Experiment research on road base material of red mud. *J. Highw. Transp. Res. Dev.* **2005**, *22*, 30–33.
22. Snars, K.; Gilkes, R.J. Evaluation of bauxite residues (red muds) of different origins for environmental applications. *Appl. Clay Sci.* **2009**, *46*, 13–20. [CrossRef]
23. Bogatyrev, B.A.; Zhukov, V.V.; Tsekhovsky, Y.G. Formation conditions and regularities of the distribution of large and superlarge bauxite deposits. *Lithol. Miner. Resour.* **2009**, *44*, 135–151. [CrossRef]
24. Duan, H. *Study on Identification Theory and Method for Extraction Toxicity of Hazardous Waste*; Chinese Research Academy of Environmental Sciences: Beijing, China, 2006.
25. Yao, Y. Summarization on monitoring methods & instrument for underground water. *Autom. Water Resour. Hydrol.* **2010**, *1*, 6–13.
26. Zhou, Y.; Li, W. Groundwater quality monitoring and assessment. *Hydrogeol. Eng. Geol.* **2008**, *1*, 1–11.
27. Kim, J.R.; Kang, H.B.; Kim, D.; Park, D.S.; Kim, W.J. Evaluation of in situ modulus of compacted subgrades using portable falling weight deflectometer and plate-bearing load test. *J. Mater. Civ. Eng.* **2006**, *19*, 492–499. [CrossRef]
28. Shi, H.; Gu, Z.; Wang, X. Research on test method of resilient modulus bearing plate for highway subgrade. *Highway* **2008**, *9*, 118–123.
29. Sun, L.; Wang, D. Nondestructive detection and evaluation of subgrade compaction using data from portable falling deflectometer. *J. Highw. Transp. Res.* **2012**, *29*, 41–47.
30. Sun, Z.; Wei, J.; Cheng, Y. Study on modulus characteristics of modified Bayer red mud subgrade. *J. China Foreign Highw.* **2018**, *38*, 54–57.

coatings

MDPI

Article

Properties of SBS/MCF-Modified Asphalts Mixtures Used for Ultra-Thin Overlays

Quanman Zhao [1,*,†], Xiaojin Lu [1,†], Shuo Jing [1], Yao Liu [1], Wenjun Hu [1,*], Manman Su [2], Peng Wang [1], Jifa Liu [3], Min Sun [1] and Zhigang Li [1]

1 School of Transportation Engineering, Shandong Jianzhu University, Jinan 250101, China; luxiaojin0329@163.com (X.L.); jingshuo99@163.com (S.J.); liuyao199704@163.com (Y.L.); wangpeng@sdjzu.edu.cn (P.W.); sunmin@sdjzu.edu.cn (M.S.); lizhigang@sdjzu.edu.cn (Z.L.)

2 School of Civil Engineering, Yantai University, Yantai 264005, China; ldusuman@126.com

3 Tai'an Highway Planning and Design Institute, Tai'an 271000, China; believevae@163.com

* Correspondence: zhaoquanman@sdjzu.edu.cn or bestcupid@163.com (Q.Z.); huwenjun@sdjzu.edu.cn (W.H.)

† These authors contributed equally to this work.

Abstract: In order to produce high-viscosity and high-toughness asphalt for ultra-thin overlays, the conventional asphalt cement was modified with high-content SBS and micro carbon fiber (MCF). The performances of the modified asphalt were studied by tests of penetration, softening point, ductility, kinematic viscosity, multiple stress creep recovery (MSCR), and by dynamic shear rheometer (DSR) and bending beam rheometer (BBR). Mixture properties were studied by tests of rutting, low-temperature bending, freeze–thaw splitting, four-point bending fatigue and dynamic modulus. The results reflect that the addition of MCF could enormously improve the high- and low-temperature properties, increase the viscosity of asphalt, and improve the toughness of asphalt. When SBS content was 6%, with the increase of MCF content, $G^*/\sin\delta$ and R values first increased and then decreased, and the J_{nr} value first decreased and then increased. When MCF content was 0.8%, the overall performance was best. Adding MCF into an asphalt mixture or increasing the content of SBS improved the rutting resistance, low-temperature crack resistance, water stability, and fatigue performance of the asphalt mixture. At the same temperature and frequency, there was little difference in phase angle between the 6%SBS + 0.8%MCF and 5%SBS + 0.0%MCF modified asphalt mixtures, and the dynamic modulus was slightly higher over the whole range. It can be concluded that the addition of SBS and MCF can enormously enhance the viscosity and toughness of asphalt. The viscosity of the 6%SBS + 0.8%MCF modified asphalt met the performance requirements of high-viscosity asphalt. When used for ultra-thin overlays, it had great road service performance and met the application requirements.

Keywords: road engineering; ultra-thin overlays; micro carbon fiber; modified asphalt; road performance

Citation: Zhao, Q.; Lu, X.; Jing, S.; Liu, Y.; Hu, W.; Su, M.; Wang, P.; Liu, J.; Sun, M.; Li, Z. Properties of SBS/MCF-Modified Asphalts Mixtures Used for Ultra-Thin Overlays. *Coatings* **2022**, *12*, 432. https://doi.org/10.3390/coatings12040432

Academic Editor: Valeria Vignali

Received: 3 March 2022
Accepted: 21 March 2022
Published: 23 March 2022

Publisher's Note: MDPI stays neutral with regard to jurisdictional claims in published maps and institutional affiliations.

Copyright: © 2022 by the authors. Licensee MDPI, Basel, Switzerland. This article is an open access article distributed under the terms and conditions of the Creative Commons Attribution (CC BY) license (https://creativecommons.org/licenses/by/4.0/).

1. Introduction

As of 2020, the total highway mileage in China measured at up 5.20 million kilometers, including 0.16 million kilometers of expressway. A total of 98.96% of highway is maintained, and the maintenance task is arduous. In order to restore the pavement function, extend the pavement's life, and save building materials [1], ultra-thin overlays are often used as a new technology for preventive maintenance [2]. It is suggested in the "Technical Specifications for Maintenance of Highway Asphalt Pavement" (JTG 5142-2019) that high-viscosity modified asphalt, rubber asphalt, and high-content SBS-modified asphalt can be used for ultra-thin overlays [3]. Therefore, so as to promote the use of ultra-thin overlays in preventive maintenance of asphalt pavement, new high-performance modified asphalt needs to be researched and developed.

In the 1970s, France first proposed the use of thin overlays and ultra-thin overlays in the maintenance of asphalt pavement, and Nova Chip with a skeletal void structure was used for ultra-thin overlays in the 1980s [4]. In recent years, so as to heighten the performance of ultra-thin overlays, many scholars have begun to focus on improving the performance of asphalt [5–8]. Chen [9] and Zhou et al. [10] added SBS and a tackifier into a conventional asphalt cement and found that the modified asphalt had higher viscosity and better high-temperature performance than the conventional asphalt cement. A new SMA-5 high-viscosity asphalt mixture was developed by Ren et al. [11]; its performance is better than SBS-modified asphalt mixture, particularly in high-temperature stability and fatigue life. Liu et al. [12] appraised the high-temperature property of polyphosphoric acid–modified asphalt and proposed that the high-temperature property could be accurately evaluated by softening point and cumulative strain. Li et al. [13] used small penetration-grade asphalt to produce high-viscosity asphalt and found that the modification process could greatly improve the rutting resistance and reduce the sensitivity to loading frequency. Zhang et al. [14] developed a new type of SBS-PU high-viscosity asphalt and compared its properties with those of two kinds of high-viscosity modified asphalt, SINOTPS and TPS. Other scholars prepared high-viscosity bitumen by adding rubber powder into SBS [15]; Zhou [16] and Yang et al. [17] analyzed its thermal storage stability and viscoelasticity, and concluded that it has the best performance when the content of rubber powder is 30%. Ming et al. [18] evaluated the rheological properties of several kinds of high-viscosity modified asphalt and proposed that the appropriate modifier should be selected according to the actual needs.

As the most widely used asphalt admixture, fiber materials [19] have the functions of absorbing oil, increasing viscosity, and enhancing toughness. Their type and content directly affect the asphalt properties, for the sake of affecting the high- and low-temperature performance, fatigue durability, and water stability of an asphalt mixture. Micro carbon fiber (MCF) is a kind of crystalline graphite material with more than 95% carbon content and a 0.007–0.044 mm fineness. Thanks to its excellent properties [20–24], it is a widely used material. Its excellent electrochemical properties allow it to be commonly used in the energy field [25]. It is often added to composites to improve the interlaminar properties because of its good toughening effect [26]. It is widely used as a reinforcement material [27] in view of its low density, high modulus [28], and high strength. After mixing it into asphalt, it produces a bridging effect and forms a unique interface construction with the asphalt, effectively limiting the drift of the asphalt and enhancing the high-temperature rheological properties of the asphalt so as to reduce rutting diseases, reduce the temperature reactivity of the asphalt, and improve its low-temperature crack resistance [29]. Dong et al. [30] added biochar into asphalt and found that its aging performance, rutting resistance, and low-temperature performance were improved. Liu et al. [31] studied carbon fiber–graphite-modified asphalt and found that alterations of internal resistance of the pavement can be determined by its conductivity, so as to predict pavement cracking. Vo et al. [32] verified that the tensile strength of an asphalt mixture was significantly improved after adding carbon fiber. Hasan et al. [33] studied the property of an asphalt modified with carbon nanofibers and it was found that all of the performance properties were improved, especially the adhesion of the asphalt.

In short, there are many mediums for ameliorating the performance of asphalt but few studies on adding MCF into asphalt. In this paper, SBS and MCF were added to the conventional asphalt cement for the purpose of achieving modification. The characteristics of the modified asphalt were appraised by tests of penetration, softening point, ductility, kinematic viscosity, multiple stress creep recovery (MSCR) and by dynamic shear rheometer (DSR) and bending beam rheometer (BBR). Mixture properties were studied by tests of rutting, low-temperature bending, freeze–thaw splitting, four-point bending fatigue, and dynamic modulus. The influence of MCF and SBS dosage on the performance of the modified asphalts was analyzed, and the optimal dosages of SBS and MCF were determined. Then, by studying the pavement performance of the SBS/MCF composite modified asphalt mixtures, their applicability in ultra-thin overlays was analyzed.

2. Materials and Methodology

2.1. Materials

2.1.1. Raw Materials

The raw materials required for the article are MCF, styrene butadiene styrene (SBS), Qinhuangdao 70# conventional asphalt cement, additive (including stabilizer and extracted oil), and aggregates. The major technical indexes of the new type of fiber material MCF are presented in Table 1. SBS is the linear T6302 produced by Dushanzi Petrochemical. WDJ4H stabilizer and Iranian extracted oil were selected. The aggregates are basalt gravel with particle sizes of 5–10 mm and 0–3 mm. Table 2 presents the relevant technical indicators of the aggregate and mineral powder. Table 3 presents the basic indexes of the conventional asphalt cement.

Table 1. Technical indexes of MCF.

Index	Data Results
Specification type(mm)	0.017
Density (g/cm^3)	1.8
Tensile strength (MPa)	4900
Modulus of elasticity (GPa)	230

Table 2. Technical indexes of basalt aggregate and mineral powder.

Material Type	Relative Apparent Density	Relative Gross Volume Density	Water Content (%)	Asphalt Absorption Coefficient	Relative Effective Density (g/cm^3)
Basalt 5–10	2.9307	2.8194	1.3511	0.60	2.8859
Basalt 0–3	2.8689	2.8689	-	0.93	2.8689
Mineral Powder	2.6335	2.6335	-	-	2.6335

Table 3. Indexes of conventional asphalt cement.

Technical Property	Unit	Test Results	Specification Requirement
Penetration (25 °C)	0.1 mm	63.8	60~80
Penetration index (PI)	-	−1.47	−1.5~+1.0
Ductility (10 °C)	cm	52	≮20
Softening point	°C	47	≮46
Flash point (opening)	°C	320	≮260
Kinematic viscosity (60 °C)	Pa. s	198	≮180
Density (15 °C)	g/cm^3	1.032	≮1.01
After Thin-Film Oven Test (TFOT)			
Mass Loss	%	0.11	≯±0.8
Penetration (25 °C)	0.1 mm	42.3	-
Residual penetration ratio	%	66.3	≥61
Residual ductility	cm	10.5	≮6

2.1.2. Mixture Composition and Production

1. Modified Asphalt

The modified asphalt was prepared according to a specific process [12]. The content of SBS was 5%, 6%, and 7%, the content of MCF was 0.0%, 0.4%, 0.8%, and 1.2%, the content of extracted oil was 2%, and the content of stabilizer was 0.2%.

2 Modified Asphalt Mixture

Owing to the requirements of ultra-thin overlays, SMA-10 was selected as the grading type. According to the maximum and minimum values of gradation specified in the specification [34], within the scope of the design specification [35]. Figure 1 displays the gradation of SMA-10.

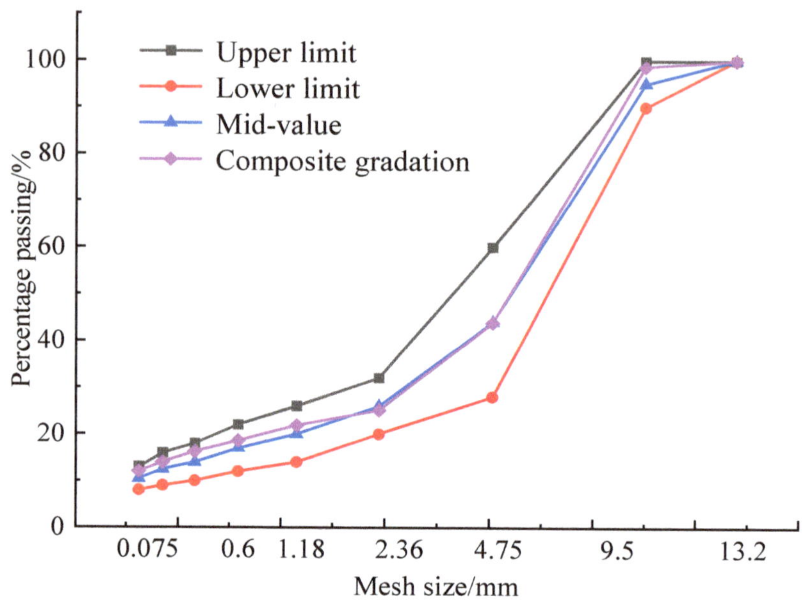

Figure 1. Designed gradation curve.

After the gradation of SMA-10 was designed, Marshall specimens were geared up and tested in light of the specification requirements [35], and the most appropriate asphalt content was 6.8%. Table 4 presents the Marshall indexes of SMA-10 obtained through the Marshall test, which all meet the specification requirements [35].

Table 4. SMA-10 Marshall specimen test results.

Marshall Indexes	Gross Volume Density (g/cm^3)	Maximum Theoretical Density (g/cm^3)	Volume of Air Void (VV/%)	Voids in Mineral Aggregate (VMA/%)	Voids Filled with Asphalt (VFA/%)	Stability (kN)	Flow Value (mm)
Test Results	2.452	2.555	4.0	20.3	80.1	16.5	4.9
Specification Requirements	-	-	3~4	≥17.0	75~85	≥6.0	-

2.2. *Methodology*

2.2.1. Tests for Modified Asphalt

(1) Conventional tests

Penetration, softening point, ductility, 60 °C kinematic viscosity, and viscosity toughness tests were conducted in accordance with "Standard Test Methods of Bitumen and Bituminous Mixtures for Highway Engineering" (JTG E20-2011) [35].

(2) Rotary viscosity

The rotary viscosity test was performed in keeping with the specification [35]. The measured temperatures were 100 °C, 120 °C, 135 °C, 155 °C, and 175 °C, and the rotor model selected was 27#.

(3) DSR

The temperature range of asphalt sample scanning was 52~88 °C, the heating rate of DSR was 2 °C/min, the strain of 27# rotor was 0.5%, and the frequency was 1 Hz. The variation law of rutting factor $G^*/\sin\delta$ at different temperatures was measured.

(4) MSCR

MSCR was carried out continuously at 64 °C and the stress levels were 0.1 kPa and 3.2 kPa [13]. Each stress level was maintained for 10 cycles, and each cycle [36] lasted 10 s (1 s loading creep stage and 9 s unloading recovery stage).

(5) BBR

Preparation and test operation of BBR were carried out according to specification [35] and American AASHTO M320-10. The test temperatures were −12 °C, −18 °C, and −24 °C [30].

2.2.2. Tests for Asphalt Mixture

(1) Rutting test

Rutting test was conducted at 60 °C according to the specification [35].

(2) Low-temperature bending test

According to the specification [35], beams with 250 mm × 30 mm × 35 mm were prepared as the test specimens, and the single-point loading was implemented at the temperature of −10 °C with the loading speed of 50 mm/min.

(3) Freeze–thaw splitting test

Freeze–thaw splitting test was conducted according to the specification [35].

(4) Four-point bending fatigue test

Stress control mode was chosen to conduct the four-point bending fatigue test. Therefore, the load was applied at the speed of 0.01 mm/s and the test would not stop until the specimen broke in the static load test. The stress levels were selected at the stress ratios of 0.3, 0.4, and 0.5, a non-discontinuous asymmetric equal amplitude sine wave was used as the loading waveform for this experiment, and the loading frequency was 10 Hz [37].

(5) Dynamic modulus test

Dynamic modulus test specimens were prepared according to the specification [35], and the test parameters of the gyratory compactor (Cangzhou Huayun Experimental Instrument Co., Ltd., Cangzhou, China) were set as follows: rotation angle 1.25°, rotation rate 30 r/min, and vertical pressure 600 kPa. The dynamic modulus test of the modified asphalt mixture was carried out at 5 °C, 20 °C, 30 °C, 45 °C and the load frequencies were 0.5 Hz, 1 Hz, 5 Hz, 10 Hz, and 20 Hz, respectively. In order to reduce the test error, the loading was applied in the order of load frequency and test temperature, from low to high. Figure 2 presents the complete experimental design of the study.

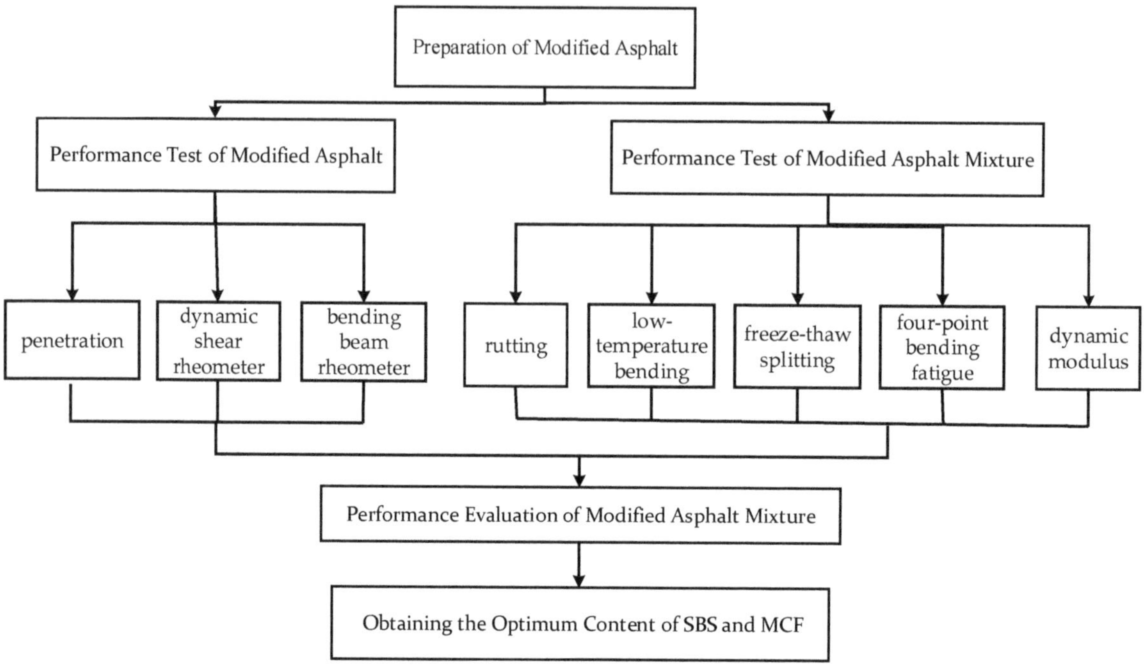

Figure 2. The experimental design.

3. Results and Discussion

3.1. Properties of Modified Asphalt

Table 5 presents the basic index tests results of the modified asphalts.

Table 5. Test results of basic indexes of modified asphalts.

Modified Asphalt Type	Penetration (mm)	Softening Point (°C)	Ductility (mm)	60 °C Kinematic Viscosity (kPa·s)	Viscosity Toughness (N·m)	Toughness (N·m)
5%SBS + 0.0%MCF	53.77	74.70	29.5	16.00	24.6	16.0
5%SBS + 0.4%MCF	52.77	76.90	30.2	19.00	27.0	21.0
5%SBS + 0.8%MCF	52.40	77.70	33.5	22.00	30.0	22.0
5%SBS + 1.2%MCF	53.60	74.45	34.6	20.15	30.2	23.0
6%SBS + 0.0%MCF	52.83	77.70	31.2	17.00	27.2	18.1
6%SBS + 0.4%MCF	51.33	78.50	32.4	23.00	29.8	22.0
6%SBS + 0.8%MCF	50.77	79.10	34.2	25.00	32.0	23.7
6%SBS + 1.2%MCF	51.70	79.02	35.1	22.20	34.2	24.0
7%SBS + 0.0%MCF	48.80	81.05	32.6	21.45	32.7	20.6
7%SBS + 0.4%MCF	46.40	81.55	33.0	29.75	34.0	24.0
7%SBS + 0.8%MCF	46.30	82.55	34.8	31.25	34.6	24.6
7%SBS + 1.2%MCF	47.90	81.05	35.5	28.84	35.1	25.2

According to the results, it can be summarized that increasing the content of MCF while keeping the content of SBS unchanged, with the increase in MCF content, the penetration first decreased and then heightened, the softening point and 60 °C kinematic viscosity first increased and then decreased, and the ductility, viscosity toughness, and toughness gradually increased. When the content of MCF remained unchanged, increasing the content of SBS, the penetration decreased gradually, and the softening point, ductility, 60 °C kinematic viscosity, viscosity toughness, and toughness increased continuously. When the SBS content was 5%, 6%, and 7%, the best content of MCF was 0.8%, in which the penetration was the minimum, the softening point was the highest and the viscosity toughness was the maximum. The data depict that the addition of MCF could notably advance the high- and low-temperature performance of conventional asphalt cement, significantly enhance the kinematic viscosity, and effectively improve the viscosity toughness. When SBS content was 6% and MCF content was 0.8%, the kinematic viscosity of the modified asphalt was in line with the requirements of ≥20,000 Pa·s for high-viscosity asphalt.

Based on the above analysis, 5%SBS + 0.0%MCF, SBS (content: 5%, 6%, 7%) + 0.8%MCF and 6%SBS + MCF (content: 0.0%, 0.4%, 0.8%, 1.2%) were selected as the optimum asphalt content to research the rheological performances of modified asphalt, and then determine the optimal content of SBS and MCF for ultra-thin overlays.

3.2. Rheological Properties of Asphalt

3.2.1. Rotational Viscosity Analysis

Figure 3 presents the test results.

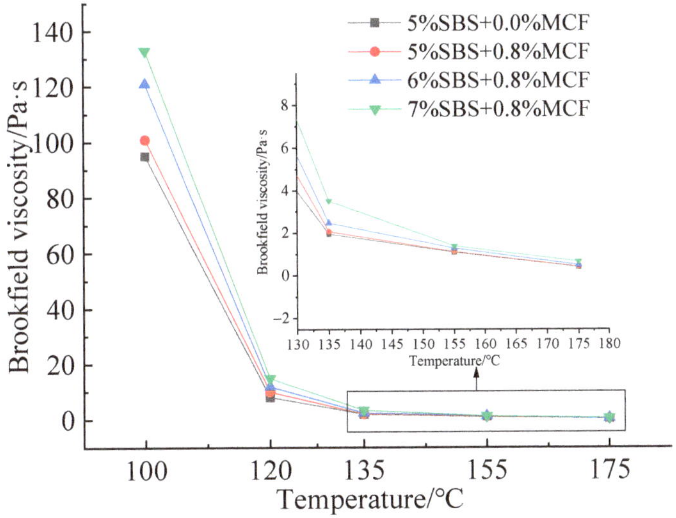

Figure 3. Rotational viscosity test result for the modified asphalt.

It can be concluded from Figure 3 that when the content of MCF was constant, the viscosity increased slowly with the increase of SBS content. The viscosity raise significantly after adding MCF, when the content of SBS was 5%, indicating that the viscosity of the modified asphalt was notably heightened by adding SBS and MCF. At all temperatures, the viscosity relationship of asphalt was the following: 7%SBS + 0.8%MCF > 6%SBS + 0.8%MCF > 5%SBS + 0.8%MCF > 5%SBS + 0.0%MCF. At 135 °C, the viscosity of 7%SBS + 0.8%MCF was 3.517 Pa·s (>3 Pa·s), the specification requirements were not met [34]. At 135 °C, the viscosity of 6%SBS + 0.8% MCF was 2.475 Pa·s, and the construction work ability of the modified asphalt met the specification requirements [34].

3.2.2. DSR

DSR test was used to appraise the high-temperature performance of the modified asphalts [14]. Figure 4 reveals the variation law of rutting factor $G^*/\sin\delta$.

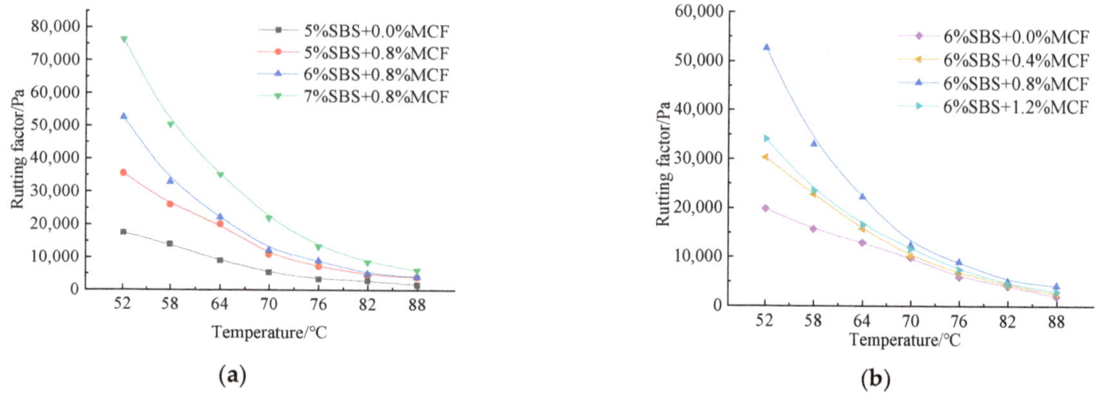

Figure 4. Rutting factors for different modified asphalts. (**a**) Rutting factors for the modified asphalts with different SBS content; (**b**) Rutting factors for modified asphalts with different MCF content.

The conclusion can be drawn from Figure 4 that G*/sinδ is coversely associated with the change of temperature and gradually decreases with the expand of temperature, and the higher the temperature, the smaller the difference. Figure 4a shows that G*/sinδ gradually heightened with the increase of SBS content. Figure 4b depicts that when the SBS content was 6%, G*/sinδ increased first and then decreased with the increase of MCF content. When the MCF content was 0.8%, it reached the maximum, and the high-temperature deformation resistance was the best. Taking 64 °C as an example, when the content of MCF was 0.8%, G*/sinδ was increased by 108.5%, 21.6%, and 6.3%, respectively, compared with when the MCF content was 0.0%, 0.4%, and 1.2%.

3.2.3. MSCR

Table 6 presents repeated cumulative strain (S), deformation recovery rate (R), and average unrecoverable creep compliance (J_{nr}), as obtained from the MSCR test.

Table 6. MSCR test results for the modified asphalts.

Asphalt Type	5%SBS + 0.0%MCF	5%SBS + 0.8%MCF	6%SBS + 0.8%MCF	7%SBS + 0.8%MCF	6%SBS + 0.0%MCF	6%SBS + 0.4%MCF	6%SBS + 1.2%MCF
$S_{0.1kPa}$	0.25	0.10	0.04	0.05	0.06	0.09	0.11
$S_{3.2kPa}$	8.67	4.32	1.30	1.90	2.12	3.00	2.48
$J_{nr0.1}$ (kPa^{-1})	0.25	0.10	0.04	0.05	0.06	0.09	0.08
$J_{nr3.2}$ (kPa^{-1})	0.27	0.13	0.04	0.06	0.07	0.09	0.11
$R_{0.1}$ (%)	46.91	75.10	85.68	85.17	82.69	72.25	64.69
$R_{3.2}$ (%)	47.07	70.38	84.54	82.81	79.07	70.42	82.02

The data from Table 6 support that the variation law of cumulative strain with time under the two stress levels was overall the same, and the strain of 6%SBS + 0.8%MCF was not greater than for other modified asphalts, indicating that adding MCF to SBS-modified asphalt improved the deformation resistance, and the performance was the best when the content was 6%SBS + 0.8%MCF. When the SBS content was unchanged, adding MCF could effectively reduce the J_{nr} value of the asphalt binder. When the content of MCF remained unchanged, enhancing the content of SBS, the J_{nr} value under the two stress levels first dropped and then was increased, indicating that increasing the content of SBS could improve the rutting resistance of asphalt, and the performance of 6%SBS + 0.8%MCF was the best.

Under the two stress levels, although the variation law of R was inconsistent at SBS content of 5%, 6% and 7%, R was the largest when the content of MCF was 0.8%. Keeping the content of MCF unchanged, the value of R under the two stress levels increased gradually with the increase of SBS content. This showed that adding the suggested amount of MCF to the modified asphalt could heighten the deformation resistance of the asphalt binder, and the optimum content was 0.8%.

3.2.4. Low-Temperature Bending Test

Low-temperature crack resistance of the modified asphalts was evaluated by gauging the stiffness modulus (S) and the creep rate (m) of the modified asphalts by BBR. Figure 5 shows the test results.

The conclusion can be concluded from Figure 5 that when the temperature was −12 °C and −18 °C, the values of S and m met the requirements of the specification ($S \leq 300$ Mpa and $m \geq 0.3$). All of the modified asphalts had the same low-temperature grade which was PG-28. When the temperature was −24 °C, the modified asphalts had little difference in m value, but there was a lot of difference in S value. Therefore, it was one-sided to use S or m value to dissect the low-temperature property of asphalt [38].Therefore, the Burgers model was used to evaluate the low-temperature performance of asphalt with m/S values, as shown in Figure 6.

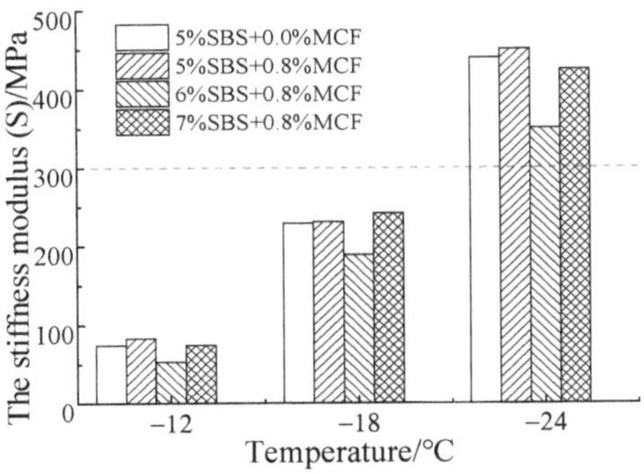

(a)

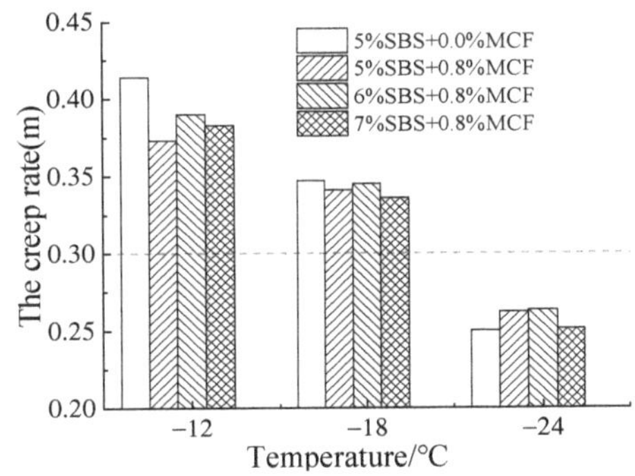

(b)

Figure 5. BBR results of modified asphalts at different temperatures. (**a**) Test results for S value; (**b**) Test results for m value.

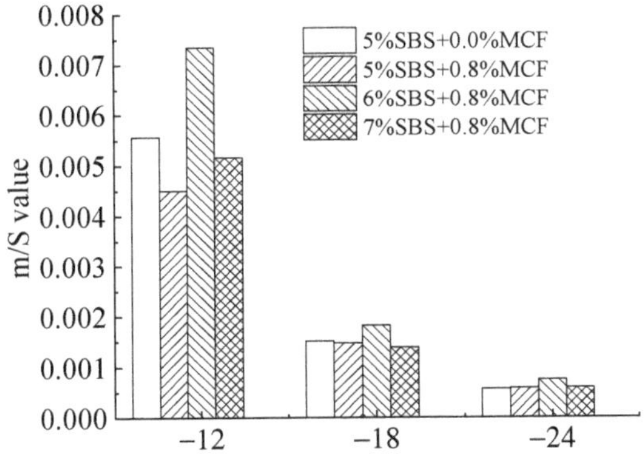

Figure 6. Values of m/S for different modified asphalt.

It is clear in Figure 6 that the m/S value of modified asphalt are positively correlated with temperature, and the decline range from −12 °C to −18 °C was greater than that from

−18 °C to −24 °C. The differences of m/S values for modified asphalts became smaller when the temperature decreased. The m/S value of 6%SBS + 0.8%MCF was the largest, indicating that its low-temperature performance was the best.

3.3. Performance of the Modified Asphalt Mixtures

3.3.1. High-Temperature Stability

Table 7 presents the dynamic stability (DS) index of rutting test [35].

Table 7. Results of the rutting test.

Mixture Type	$d_{45min/mm}$	$d_{60min/mm}$	DS (time/mm^{-1})	Specification Requirement (time/mm^{-1})
5%SBS + 0.0%MCF	1.706	1.830	5045	≥2500
5%SBS + 0.8%MCF	1.728	1.840	5625	
6%SBS + 0.8%MCF	1.903	1.996	6774	
7%SBS + 0.8%MCF	1.323	1.398	8362	

Table 7 shows that DS values of 5%SBS + 0.8%MCF, 6%SBS + 0.8%MCF, and 7%SBS + 0.8%MCF increased by 11.5%, 20.4%, and 48.7%, respectively, compared with 5%SBS + 0.0%MCF, indicating that enhancing the content of SBS and MCF can effectively increase the high-temperature stability of an asphalt mixture.

3.3.2. Low-Temperature Crack Resistance

Table 8 presents the test outcomes of the low-temperature bending tests and specification [34] requirements.

Table 8. Results of low-temperature bending tests.

Mixture Type	Flexural Tensile Strength R_B (MPa)	Failure Strain ε_B (με)	Bending Stiffness Modulus SB (MPa)	Specification Requirement (MPa)
5%SBS + 0.0%MCF	7.532	2867	2627.0	≥2500
5%SBS + 0.8%MCF	9.343	3226	2896.2	
6%SBS + 0.8%MCF	10.506	3578	2936.2	
7%SBS + 0.8%MCF	11.375	3447	3299.9	

These results of Table 8 revealed that the failure strain of 6%SBS + 0.8%MCF was the largest, indicating that after the crack appeared, the crack development speed of the specimen was the slowest, and its fracture toughness was better than that of other specimens. For ultra-thin overlays, modified asphalt with strong strain resistance is preferred, so 6%SBS + 0.8%MCF modified asphalt would be preferred for ultra-thin overlays.

3.3.3. Water Stability

Table 9 presents the freeze–thaw splitting test results and from it, the order of freeze–thaw splitting strength ratio was established as: 7%SBS + 0.8%MCF > 6%SBS + 0.8%MCF > 5%SBS + 0.8%MCF > 5%SBS + 0.0%MCF, indicating that adding MCF or increasing the content of SBS can improve the water stability of an asphalt mixture.

Table 9. Results of the freeze–thaw splitting tests.

Mixture Type	$\overline{R}_{T1}$(MPa)	$\overline{R}_{T2}$(MPa)	TSR (%)	Specification Requirement (%)
5%SBS + 0.0%MCF	1.492	1.352	90.6	TSR ≥ 80
5%SBS + 0.8%MCF	1.805	1.650	91.3	
6%SBS + 0.8%MCF	1.695	1.585	93.5	
7%SBS + 0.8%MCF	1.697	1.617	95.3	

3.3.4. Fatigue Performance

Through the previous research, it was found that the performance of the 6%SBS + 0.8%MCF modified asphalt mixture was excellent. Therefore, fatigue and dynamic modulus tests were tested on it and it was in conflict with the conventional 5%SBS + 0.0%MCF modified asphalt mixture. A four-point bending test [24,25] was used to test the fatigue property of the asphalt mixture. Figure 7 presents the bending stress–deflection curve for bending tensile strength test. It can be concluded, as summarized in Figure 7, that under static loading conditions, there was little difference between 6%SBS + 0.8%MCF and 5%SBS + 0.0%MCF for the properties of stress deflection. So as to comprehensively judge the fatigue performance of the asphalt mixtures under distinct stress standards, fatigue tests at the stress ratio levels of 0.3, 0.4, and 0.5 were implemented. Figure 8 presents the results of the fatigue test and it can be concluded that the fatigue life decreased significantly with the enhancement of the stress ratio. With a stress ratio of 0.3, 0.4, and 0.5, respectively, the fatigue life of the 6%SBS + 0.8%MCF modified asphalt mixture was 1.17, 1.13, and 1.32 times that of the 5%SBS + 0.0%MCF modified asphalt mixture, respectively, and the fatigue performance of the 6%SBS + 0.8%MCF modified asphalt mixture was excellent.

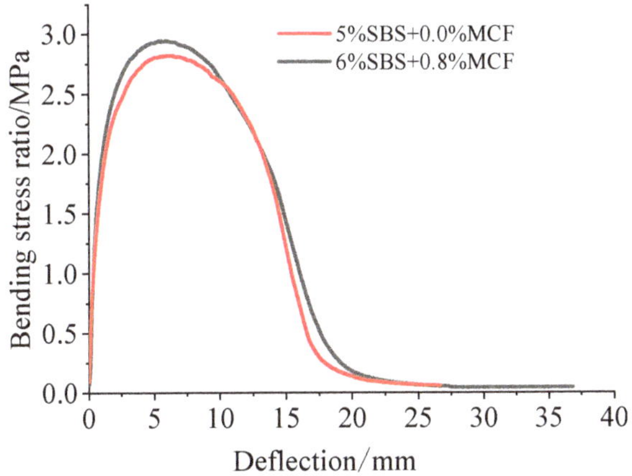

Figure 7. Bending stress–deflection curve.

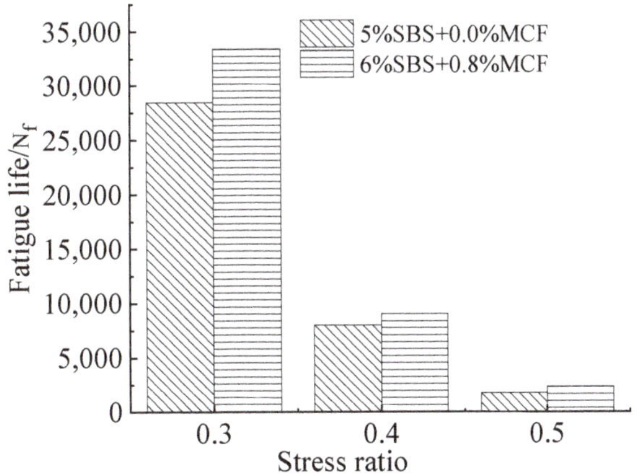

Figure 8. Results of fatigue tests for different stress levels.

3.3.5. Dynamic Modulus

Tables 10 and 11 present separately the data of the dynamic modulus and phase angle obtained from the dynamic modulus test.

Table 10. Results of the dynamic modulus test for modified asphalt mixtures.

Frequency/Hz	Dynamic Modulus (MPa)							
	5%SBS + 0.0%MCF				6%SBS + 0.8%MCF			
	5 °C	20 °C	30 °C	45 °C	5 °C	20 °C	30 °C	45 °C
0.1	5526.5	1341.0	596.6	307.0	6133.0	1486.0	646.4	339.5
0.5	7906.0	2284.0	919.5	393.3	8781.0	2529.5	992.7	427.9
1	9085.5	2918.5	1185.0	458.9	9923.0	3227.5	1279.5	492.1
5	12007.0	4775.0	2118.5	793.1	13014.0	5251.0	2295.0	832.5
10	13264.5	5795.5	2713.0	995.8	14387.5	6352.0	2947.0	1047.5
20	14359.0	6915.0	3488.0	1293.5	15814.5	7555.0	3776.5	1377.5

Table 11. Results of phase angle for modified asphalt mixtures.

Frequency/Hz	Phase Angle (°)							
	5%SBS + 0.0%MCF				6%SBS + 0.8%MCF			
	5 °C	20 °C	30 °C	45 °C	5 °C	20 °C	30 °C	45 °C
0.1	24.260	30.130	25.395	18.600	23.085	29.950	25.155	17.830
0.5	20.435	30.640	29.120	21.775	19.555	30.150	29.270	22.160
1	18.805	29.700	30.015	23.910	17.760	29.090	30.275	24.630
5	14.835	26.985	31.195	26.565	13.930	25.960	31.245	27.880
10	13.310	25.370	30.910	28.610	12.520	24.270	30.800	29.965
20	11.955	23.515	29.740	30.005	11.245	22.420	29.550	31.315

(1) The relationship between dynamic modulus and temperature

As is shown in Table 10, during high-frequency loading, the reduction range of dynamic modulus gradually decreased as the temperature rose. On the contrary, it increased with the decrease of temperature during low-frequency loading. Under the same conditions frequency or temperature, comparing the dynamic modulus of the 6%SBS + 0.8%MCF and 5%SBS + 0.0%MCF, it was found that the former had a larger value. This result indicates that 6%SBS + 0.8%MCF could maintain a higher modulus at higher temperatures, and had stronger resistance to high-temperature rutting than that of 5%SBS + 0.0%MCF modified asphalt pavement.

(2) The connection between phase angle and temperature

Table 11 presents that, under the same temperature and frequency, comparing the phase angle of the 6%SBS + 0.8%MCF and the 5%SBS + 0.0%MCF, the value of the former is slightly lower. This indicates that 6%SBS + 0.8%MCF had better elasticity, and better resistance to high-temperature rutting.

(3) Establishment of master curve

For viscoelastic materials, the time–temperature conversion principle was available for use in studying the mechanical properties under different temperatures and load frequencies. The master curve was established through Formula (1) [39], and the obtained master curve parameter values are shown in Table 12. The main curve parameters were brought into Equations (2) and (3) to obtain the master curve equation for dynamic modulus and shift factor at different temperatures, so as to draw the master curve of dynamic modulus and master curve of phase angle [40], as presented in Figure 9.

$$\log|E*| = \delta + \frac{Max - \delta}{1 + e^{\beta + \gamma \log t_r}} \quad (1)$$

$$\log[a(T)] = \frac{\Delta E_a}{19.14714}\left(\frac{1}{T} - \frac{1}{T_r}\right) \tag{2}$$

$$\log|E*| = \delta + \frac{Max - \delta}{1 + e^{\beta+\gamma\{\log f + \frac{\Delta E_a}{19.14714}[(\frac{1}{T})-(\frac{1}{T_r})]\}}} \tag{3}$$

where $|E^*|$—Dynamic modulus (MPa); *Max*—Logarithm of ultimate maximum dynamic modulus; t_r—Reduction time at reference temperature; β, γ, δ—Fitting parameters, and the initial value is $\delta = 0.5$, $\beta = -1.0$, $\gamma = -0.5$.

Table 12. Parameter values for master curve.

Mixture Type	δ	ΔE_a	β	γ	S_e/S_y	R^2
5%SBS + 0.0%MCF	4.3693	200000	0.4020	0.5976	0.0044	0.999
6%SBS + 0.8%MCF	4.4550	200000	0.4134	0.6377	0.0039	0.999

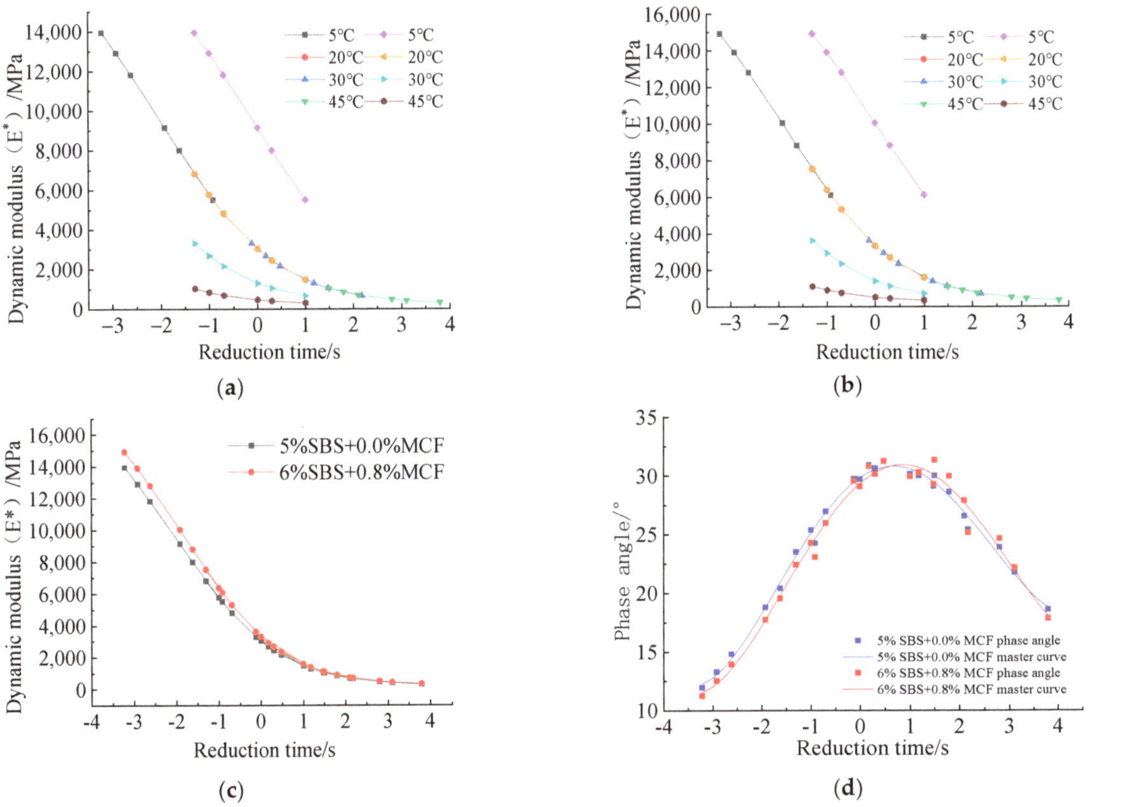

Figure 9. Master curve for modified asphalt mixtures. (**a**) Master curve for 5%SBS + 0.0%MCF; (**b**) Master curve for 6%SBS + 0.8%MCF; (**c**) Comparison of master curves for dynamic modulus; (**d**) Comparison of master curves for phase angle.

As indicated in Figure 9a,b, the dynamic modulus decreased with the increase of reduction time, with a slower trend at both ends of the curve and an accelerated decline in the middle of the curve. Figure 9c shows that the influence of frequency on the 6%SBS + 0.8%MCF and 5%SBS + 0.0%MCF asphalt mixtures was overall the same, compared with 5%SBS + 0.0%MCF asphalt mixture, 6%SBS + 0.8%MCF mixture has a higher dynamic modulus, indicating that the deformation resistance of the 6%SBS + 0.8%MCF modified asphalt mixture was better compared to the 5%SBS + 0.0%MCF asphalt mixture. In Figure 9d it can be seen that the phase angle of the two asphalt mixtures had a tiny gap, indicating that the content of SBS and MCF had no significant effect on the phase angle.

4. Conclusions

The major conclusions are as follows:

(1) Addition of MCF can obviously heighten the high- and low-temperature properties of asphalt, significantly increase the kinematic viscosity, and effectively improve the viscosity and toughness of asphalt. The addition of SBS and MCF can lead to high viscosity and toughness in asphalt. The kinematic viscosity of the 6%SBS + 0.8%MCF modified asphalt met the criteria for high-viscosity asphalt ($\geq$20,000 Pa·s);

(2) DSR and MSCR experiments showed that when the SBS content was 6%, with the increase in MCF content, the $G^*/\sin\delta$ and R values first increased and then decreased, and the J_{nr} value first decreased and then increased. The optimum content of MCF was 0.8%. Adding an appropriate amount of MCF into high-content SBS-modified asphalt can significantly heighten the high-temperature deformation resistance. BBR tests found that a high content of SBS and an appropriate content of MCF can enhance the low-temperature performance of asphalt, and the best high- and low-temperature performance of modified asphalt was achieved when the SBS content was 6% and the MCF content was 0.8%;

(3) It was found that adding MCF or increasing the content of SBS in modified asphalt can increase the rutting resistance, low-temperature crack resistance, water stability, and fatigue performance of the asphalt mixture, and the performance of the 6%SBS + 0.8%MCF modified asphalt mixture was the best, which is feasible for ultra-thin overlays.

(4) The dynamic modulus test indicated that for the phase angle of the 6%SBS + 0.8%MCF and 5%SBS + 0.0%MCF modified asphalt mixture there was little difference at equal temperature and frequency. The dynamic modulus of the 6%SBS + 0.8%MCF asphalt mixture was superior to that of the 5%SBS + 0.0%MCF asphalt mixture. The modified asphalt mixture with 6%SBS + 0.8%MCF had better deformation resistance than the 5%SBS + 0.0%MCF mixture.

The modified asphalt mixture of 6%SBS + 0.8%MCF not only had good viscosity and toughness, but also its road performance was better than that of general modified asphalt. Therefore, it can be used in road maintenance technology as ultra-thin overlays.

Author Contributions: Conceptualization, Q.Z. and W.H.; methodology, Q.Z., M.S. (Manman Su) and M.S. (Min Sun); software, X.L.; validation, Q.Z. and Y.L.; formal analysis, X.L. and W.H.; investigation, Y.L., S.J. and Z.L.; resources, Q.Z. and P.W.; data curation, X.L.; writing—original draft preparation, X.L.; writing—review and editing, Q.Z. and W.H.; visualization, X.L.; supervision, Q.Z. and W.H.; project administration, Q.Z.; funding acquisition, Q.Z. and J.L. All authors have read and agreed to the published version of the manuscript.

Funding: This research was financially supported by the project 2021B24 supported by Shandong Transportation Science and Technology Plan, and project ZR2020QE273 supported by Shandong Provincial Natural Science Foundation.

Institutional Review Board Statement: Not applicable.

Informed Consent Statement: Not applicable.

Data Availability Statement: Using the data of this article requires application from the author.

Acknowledgments: We express our sincere gratitude to the experts, teachers, and students who have provided help for this paper.

Conflicts of Interest: The authors proclaim no conflict of profit.

References

1. Mokhtari, A.; Nejad, F.M. Mechanistic Approach for Fiber and Polymer Modified SMA Mixtures. *Constr. Build. Mater.* **2012**, *36*, 381–390. [CrossRef]
2. Li, X.; Ye, J.; Badjona, Y.; Chen, Y.; Luo, S.; Song, X.; Zhang, H.; Yao, H.; Yang, L.; You, L.; et al. Preparation and performance of colored Ultra-Thin overlay for preventive maintenance. *Constr. Build. Mater.* **2020**, *249*, 118619. [CrossRef]
3. *JTG 5142-01-2021*; Technical Specifications for Preventive Maintenance of Highway Asphalt Pavement. Ministry of Transport of the People's Republic of China: Beijing, China, 2021.
4. Yu, J.; Yang, N.; Yu, H. Research and Application Status of High-Performance Asphalt Ultra-Thin Wearing Layer Technology. *J. Cent. South Univ. Sci. Technol.* **2021**, *52*, 2287–2298. (In Chinese)
5. Wang, S.; Zhao, S.; Al-Qadi, I.L. Real-Time Density and Thickness Estimation of Thin Asphalt Pavement Overlay During Compaction Using Ground Penetrating Radar Data. *Surv. Geophys.* **2020**, *41*, 431–445. [CrossRef]
6. Jiang, W.; Yuan, D.; Shan, J.; Ye, W.; Lu, H.; Sha, A. Experimental study of the performance of porous ultra-thin asphalt overlay. *Int. J. Pavement Eng.* **2020**. [CrossRef]
7. Wang, K.; Wang, K.; Zhu, Y.; Bao, Y.; Fang, Z. Study of Structural Characteristics of Asphalt Overlays on Airport Pavement with Damaged Load Transfer Efficiency of Joints. *Transp. Res. Rec.* **2021**, *39*, 57–60. [CrossRef]
8. Yu, J.; Yang, N.; Chen, F.; Chen, Y.; Lin, Z.; Yu, H. Design of Cold-Mixed High-Toughness Ultra-Thin Asphalt Layer towards Sustainable Pavement Construction. *Buildings* **2021**, *11*, 619. [CrossRef]
9. Chen, X.; Li, C.; Jiang, Y.; Zhang, W.; Xu, G. Comparisons with high viscosity additive effects on base and modified asphalt. *Pet. Sci. Technol.* **2019**, *37*, 1331–1337. [CrossRef]
10. Zhou, Z.; Chen, G. Preparation, Performance, and modification mechanism of high viscosity modified asphalt. *Constr. Build. Mater.* **2021**, *310*, 125007. [CrossRef]
11. Ren, R.; Geng, L.; Wang, L.; Wang, P. Design and Performance Evaluation of A SMA-5 High Viscosity Asphalt Mixture. *J. Build. Mater.* **2016**, *19*, 762–766. (In Chinese)
12. Liu, H.; Xu, J.; Zhang, Z.; Chang, R.; Hao, P. Investigation Into High Temperature Evaluation Indexes of Polyphosphoric Acid-Modified Asphalt. *J. South China Univ. Technol. (Nat. Sci. Ed.)* **2016**, *44*, 98–105. (In Chinese)
13. Li, M.; Zeng, F.; Xu, R.; Cao, D.; Li, J. Study on Compatibility and Rheological Properties of High-Viscosity Modified Asphalt Prepared from Low-Grade. Asphalt. *Materials* **2019**, *12*, 3776. [CrossRef] [PubMed]
14. Zhang, Z.; Tian, Z.; Huang, S.; Zhao, F. Composition and Process Parameters of SBS-PU Modified High-Viscosity and High-Elasticity Asphalt. *J. Build. Mater.* **2020**, *23*, 100–107. (In Chinese)
15. Hu, X.; Wang, H.; Wang, H. The Performance of Nanometer $CaCO_3$ High-Viscosity Compound Modified Asphalt. *Adv. Transp. Infrastruct. Mater.* **2016**, *1*, 210–218. [CrossRef]
16. Zhou, Z.G.; Chen, G.H.; Zhang, H.B.; Ling, Y.Y. Study On the Preparation and Properties of Modified Asphalt By Rubber Power/SBS and High Viscosity Modifier Composite. *Mater. Rep.* **2021**, *35*, 6093–6099. (In Chinese)
17. Yang, S.; Sun, S.; Qin, L.; Li, Q. Microstructure and Meso-Mechanical Properties of Asphalt Mixture Modified by Rubber Powder under a Multi-Scale Effect. *Coatings* **2021**, *11*, 1321. [CrossRef]
18. Liang, M.; Xin, X.; Fan, W.; Zhang, J.; Jiang, H.; Yao, Z. Comparison of rheological properties and compatibility of asphalt modified with various polyethylene. *Int. J. Pavement Eng.* **2021**, *22*, 11–20. [CrossRef]
19. Mateos, A.; Harvey, J.; Paniagua, J.; Paniagua, F.; Fan Liu, A. Mechanical characterisation of concrete-asphalt interface in bonded concrete overlays of asphalt pavements. *Eur. J. Environ. Civ. Eng.* **2017**, *21*, 43–53. [CrossRef]
20. Bu, A.; Zhang, Y.; Zhang, Y.; Chen, W.; Cheng, H.; Wang, L.; Wang, Y. A Novel Electrolytic Plasma Spraying Preparation SiO_2/SiC Coating on Carbon Fiber Fabric. *Coatings* **2018**, *8*, 344. [CrossRef]
21. Nijs, A.; Selezneva, M.; Swolfs, Y.; Hirano, N.; Taketa, I.; Karaki, T.; Verpoest, I.; Gorbatikh, L. Notch-sensitivity of hybrid carbon-fibre/self-reinforced polypropylene composites. *Compos. Sci. Technol.* **2020**, *200*, 108422. [CrossRef]
22. Sergei, B.; Sapozhnikov, Y.; Stepan, V. Mode I and II interlaminar critical energy release rates in all-carbon interlayer unidirectional fibre-hybrids based on ultrahigh-modulus and high-strength fibres. *Compos. Struct.* **2020**, *236*, 111886.
23. Zhang, S.; Hao, A.; Nguyen, N.; Oluwalowo, A.; Liu, Z.; Dessureault, Y.; Park, J.G.; Liang, R. Carbon nanotube/carbon composite fiber with improved strength and electrical conductivity via interface engineering. *Carbon* **2019**, *144*, 628–638. [CrossRef]
24. Kandola, B.; Sarker, F.; Luangtriratana, P.; Myler, P. Thermal Protection of Carbon Fiber-Reinforced Composites by Ceramic Particles. *Coatings* **2016**, *6*, 22. [CrossRef]
25. Jiang, X.; Wang, J.; Qiu, L.; Bai, B.; Jin, Z.; Mei, D.; Du, P. Applications of Carbon Based Fiber Materials in Energy Field. *Mater. Rep.* **2021**, *35*, 470–478.
26. Jiang, C.; Che, Z.; Xing, F.; Wang, S.; Li, M. Interlaminar Property of Carbon Nanotube-Continuous Fiber Reinforced Resin Matrix Composites a Review. *Acta Mater. Compos. Sin.* **2022**, 1–20. [CrossRef]
27. Newcomb, B.A. Processing, structure, and properties of carbon fibers. *Compos. Part A Appl. Sci. Manuf.* **2016**, *91*, 262–282. [CrossRef]
28. Chen, S.; Qiu, L.; Cheng, H.M. Carbon-Based Fibers for Advanced Electrochemical Energy Storage Devices. *Chem. Rev.* **2020**, *120*, 2811–2878. [CrossRef]
29. Yu, H.Y.; Ma, T.; Wang, D.W.; Wang, C.H.; Lu, S.T.; Zhu, X.Y. Review on China's Pavement Engineering Research. *China J. Highw. Transp.* **2020**, *33*, 1–66. (In Chinese)

30. Dong, W.; Ma, F.; Li, C.; Fu, Z.; Huang, Y.; Liu, J. Evaluation of Anti-Aging Performance of Biochar Modified Asphalt Binder. *Coatings* **2020**, *10*, 1037. [CrossRef]
31. Liu, L.; Zhang, X.; Xu, L.; Zhang, H.; Liu, Z. Investigation on the piezoresistive response of carbon fiber-graphite modified asphalt mixtures. *Constr. Build. Mater.* **2021**, *301*, 124140. [CrossRef]
32. Vo, H.V.; Park, D.W.; Seo, W.J.; Yoo, B.S. Evaluation of Asphalt Mixture Modified with Graphite and Carbon Fibers for Winter Adaptation: Thermal Conductivity Improvement. *J. Mater. Civ. Eng.* **2017**, *29*. [CrossRef]
33. Hasan, R.; Ali, A.; Decarlo, C.; Elshaer, M.; Mehta, Y. Laboratory Evaluation of Electrically Conductive Asphalt Mixtures for Snow and Ice Removal Applications. *Transp. Res. Rec.* **2021**, *2675*, 48–62. [CrossRef]
34. *JTG F40-2004*; Technical Specifications for Construction of Highway Asphalt Pavements. Ministry of Transport of the People's Republic of China: Beijing, China, 2004.
35. *JTG E20-2011*; Standard Test Methods of Bitumen and Bituminous Mixtures for Highway Engineering. Ministry of Transport of the People's Republic of China: Beijing, China, 2011.
36. Yan, Y.; Yang, Y.; Ran, M.; Zhou, X.; Zou, L.; Guo, M. Application of Infrared Spectroscopy in Prediction of Asphalt Aging Time History and Fatigue Life. *Coatings* **2020**, *10*, 959. [CrossRef]
37. Wang, L.; Shan, M.; Chang, C.; Zhou, X. The macro-and meso-cracking characteristics of warm mix crumb rubber asphalt mastics before and after aging. *Constr. Build. Mater.* **2020**, *262*. [CrossRef]
38. *JTG F80/1-2017*; Inspection and Evaluation Quality Standards for Highway Engineering. Ministry of Transport of the People's Republic of China: Beijing, China, 2017.
39. You, Y. The Research on the Dynamic Modulus of Asphalt Mixture. Master's Thesis, Shandong Jianzhu University, Jinan, China, 2017. (In Chinese).
40. Liu, F.; Zhou, Z.; Zhang, X. Construction of complex shear modulus and phase angle master curves for aging asphalt binders. *Int. J. Pavement Eng.* **2022**, *23*, 536–544. [CrossRef]

coatings

MDPI

Article

Interlaminar Shear Characteristics, Energy Consumption, and Carbon Emissions of Polyurethane Mixtures

Yufeng Bi [1], Min Sun [2,*], Shuo Jing [2], Derui Hou [2], Wei Zhuang [1], Sai Chen [1], Xuwang Jiao [1] and Quanman Zhao [2,*]

1 Shandong Provincial Communications Planning and Design Institute Co., Ltd., Jinan 250031, China; 18866130036@163.com (Y.B.); zhuangweijtky@163.com (W.Z.); chensaiyile@hotmail.com (S.C.); xuwang.jiao@outlook.com (X.J.)
2 School of Transportation Engineering, Shandong Jianzhu University, Jinan 250101, China; jingshuo99@163.com (S.J.); hdr09191210@163.com (D.H.)
* Correspondence: 15253170143@163.com (M.S.); bestcupid@163.com (Q.Z.)

Abstract: The interlaminar shear characteristics of a polyurethane (PU) mixture composite structure, quantitatively calculating its energy consumption and carbon emissions were analyzed in this study. Inclined shear tests were carried out on thirteen structures without interlaminar treatment, and high-temperature water bath accelerated loading tests were conducted on three composite structures; further, the interlaminar shear strength of the tire trace position after the accelerated loading test was tested to analyze the influence of both the high-temperature water bath environment and loading on the structure. In addition, based on the medium repair project of the Qingdao-Yinchuan expressway, the construction log of PU mixture pavement and asphalt pavement was investigated. Combined with the calculation parameters provided by the United Nations Intergovernmental Panel on Climate Change (IPCC), the consumption and carbon emissions of the two types of mixtures were calculated and compared quantitatively. The results showed that the shear strength between layers of asphalt mixtures, PU mixture and asphalt mixture was less than 2 MPa; however, the shear strength between PU mixture–cement-stabilized macadam and PU mixture–PU mixture was greater than 3 MPa. Therefore, it was recommended to spread a 0.4 L/m^2 two-component PU adhesive layer as the interlayer treatment scheme for the structure of asphalt mixture-PU mixture and PU mixture-asphalt mixture; the high-temperature water area and accelerated loading had different effects on the interlaminar shear strength of the three structures. The PUM-16 mixture could effectively reduce energy consumption by 88.3 and 87.2%, carbon emissions by 81.1% and 79.1% in comparison to Stone Matrix asphalt with Nominal maximum particle size of 13.2 mm (SMA-13) and Stone Matrix asphalt with Nominal maximum particle size of 19.0 mm (SMA-20). Thus, the PU mixture was confirmed to be an environmentally friendly road pavement material.

Keywords: polyurethane; mixture; shear strength; energy consumption; carbon emission

Citation: Bi, Y.; Sun, M.; Jing, S.; Hou, D.; Zhuang, W.; Chen, S.; Jiao, X.; Zhao, Q. Interlaminar Shear Characteristics, Energy Consumption, and Carbon Emissions of Polyurethane Mixtures. *Coatings* **2022**, *12*, 400. https://doi.org/10.3390/coatings12030400

Academic Editor: Valeria Vignali

Received: 28 January 2022
Accepted: 11 March 2022
Published: 17 March 2022

Publisher's Note: MDPI stays neutral with regard to jurisdictional claims in published maps and institutional affiliations.

Copyright: © 2022 by the authors. Licensee MDPI, Basel, Switzerland. This article is an open access article distributed under the terms and conditions of the Creative Commons Attribution (CC BY) license (https://creativecommons.org/licenses/by/4.0/).

1. Introduction

Asphalt pavement is widely used all over the world; however, its mixing, transportation, paving, and rolling machinery consumes significant amounts of oil and energy, emitting high levels of greenhouse gases such as CO_2. Further, asphalt pavement is prone to rutting, looseness, and other issues [1,2]. To solve the aforementioned problems, high molecular polymers polyurethane (PU) is used as binder, the PU mixture pavement is intended to replace asphalt pavement [3]. The entire construction process of PU mixture production, transportation, and compaction is conducted at room temperature, which greatly reduces the consumption of energy consumption and carbon emissions. Further, PU mixtures have better high- and low-temperature stability, fatigue resistance, and dynamic mechanical properties, which can effectively improve pavement service life and greatly reduce the maintenance frequency and cost [4,5].

Over the past decade, many researchers have conducted studies on composition design, reaction mechanism, and performance characteristics of PU mixtures [6–11]. Wang et al. tested the strength and modulus of Porous Polyurethane Mixture (PPM) by using the unconfined compression test on cement concrete cubes and the flexural test on rectangular beams [6]. Based on the entire load stress test, the strength characteristics and failure mechanism of PPM were studied. Wang et al. prepared a PERS mixture with a porosity of 35% using a two-component PU material provided by BASF Co., Ltd. [7,8]. The invention of a new surface layer material with PU was proposed by Schacht et al., and the vehicle pavement interaction facilities of the Federal Highway Research Institute (BAST) was conducted, it was determined that the surface system has the highest acoustic performance possible. Cong et al. used two-component PU as a binder to replace traditional asphalt binders, and the prepared PPM overcame the limitations of traditional OGFC mixtures [10,11]. However, the curing reaction of PU binders involves a chemical condensation reaction process. Through the reaction of -NCO in the binder with water in the air and active hydrogen on the substrate surface, a cross-linked structure—including a urea bond, amino formate bond, and urea formate macromolecular network—is generated; finally, a three-dimensional network structure is formed, which bonds the substrate together [12].

The strength formation mechanism of PU mixtures differs from that of asphalt mixture and cement concrete. When it is used in conjunction with other road building materials to form a composite pavement structure, a large number of studies on polyurethane mixture composite pavement structure show that the interlayer shear resistance is the key factor affecting its service performance. Therefore, it is necessary to study the interlaminar shear characteristics of composite specimens and further clarify the appropriate interlaminar treatment scheme. Chen et al. carried out interlaminar inclined shear tests to detect the interface shear performance between porous polyurethane mixtures and asphalt sublayers [13]. Zhang et al. studied the effects of waterproof bonding materials, the base surface roughening mode, the upper mixture type, and test conditions on the peak interlaminar shear strength of composite specimens [14]. Mohammad et al. evaluated the effects of waterproof bonding materials, interface types, dry and wet conditions of the base surface, and specimen preparation methods on the peak value of interlaminar shear strength [15]. Luo et al. compared the effects of interface pollution and water saturation conditions on interlayer bonding characteristics [16]. Cao et al. studied the effects of waterproof bonding materials, the base surface roughening mode, the forming mode, the shear mode, and the temperature on the interlayer shear strength [17]. Current research mainly focuses on the influence of different factors (such as the interface roughness mode, waterproof bonding materials, temperature, immersion, freeze–thaw cycles, and interlayer pollution degree) on the interlayer shear strength of asphalt pavement; further, there are some studies on the interlayer shear failure characteristics or mechanism of PU mixture composite pavement structures [18–21].

The greenhouse effect caused by road construction has recently attracted increasing research attention [22]. As a low-carbon and environmentally friendly road building material, the energy conservation and emission reduction effect of PU mixtures in road construction have not been evaluated quantitatively, to the best of our knowledge. Therefore, the energy-saving and emission reduction effect of PU mixture cannot be highlighted, which limits the popularization and application speed of PU mixture pavement. However, the quantitative calculation method for the energy consumption and emissions of road construction based on engineering practice is a relatively mature field of study. Kim et al. established a framework for estimating greenhouse gas emissions [23]. White et al. calculated the average CO_2 value of a road by using parameters such as the thickness and characteristics of materials [24]. Cass et al. established a comprehensive life cycle analysis (LCA) model according to an actual project [25]. Wang et al. estimated the carbon emissions of roads, bridges, and other construction structures in Southwest China [26]. Thives et al. evaluated the carbon dioxide emission and energy consumption generated by road pavement through a literature review [27]. The relationship between energy demand and the environmental impact of pavement in the construction cycle can be analyzed through life cycle analysis (LCA), and the corresponding environmental impact factors have been determined in the past [28–30]. Therefore, these mature methods can be used to quantitatively calculate the energy conservation and emission reduction effect of polyurethane mixture pavement in combination with physical engineering.

In summary, it is necessary to carry out interlaminar shear tests on PU mixture composite specimens, analyze the influence of environmental factors and the load on the interlaminar shear strength, and further clarify the appropriate interlaminar treatment scheme. At the same time, based on the medium repair project of Qingdao–Yinchuan expressway, the construction log of PU mixture pavement and asphalt pavement was investigated. The construction process was divided into three stages: mixture production, mixture transportation, and mixture site construction, which include seven constituent processes. In combination with the calculation parameters provided by the United Nations Intergovernmental Panel on Climate Change (IPCC), a quantitative model of the energy consumption and carbon emission of mixtures was established, which were subsequently compared quantitatively.

2. Experiment Scheme

2.1. Materials

2.1.1. Raw Material

The materials mainly include a one-component PU binder, two-component PU binder, 70# base asphalt, styrene butadiene styrene (SBS) modified asphalt, and aggregates. The technical indexes of asphalts meet the requirements of JTG F40-2004. One-component PU was used as the binder for the PU mixture, and two-component PU was used as the interlayer bonding material, which are prepared by Wanhua Chemical Co., Ltd. (Yantai, China) [12]. The PU material used is synthesized from 4,4-diphenylmethane diisocyanate (MDI) modified by carbodiimide, polyether polyol, and other additives. It is almost non-toxic and can be used in pavement engineering. The main reaction in the PU synthesis process is the reaction of polyol and diisocyanate to prepare long-chain prepolymer with -NCO end group. The chemical reaction equation is shown in Figure 1. The specific technical indexes of asphalt are given in Table 1 and the indexes of PU binders are given in Table 2.

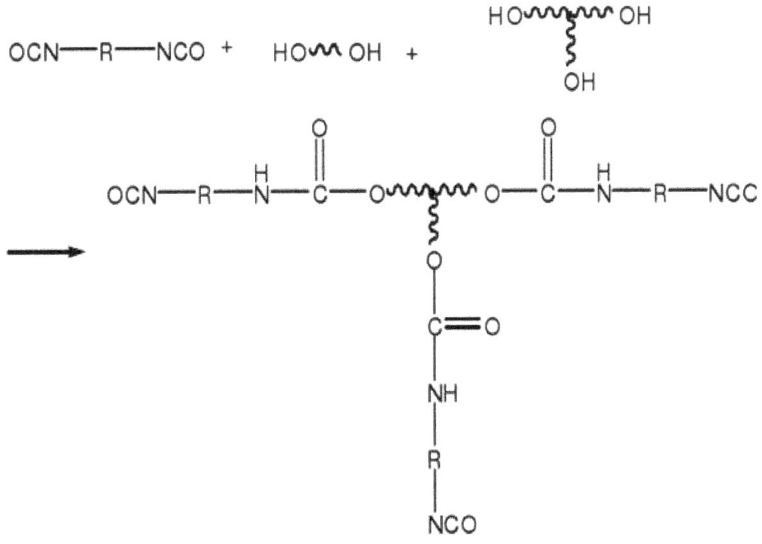

Figure 1. The chemical reaction equation of synthesis process.

Table 1. Technical indexes of PU asphalt.

Technical Indicators	Unit	Technical Requirement	
		70# Base Asphalt	SBS Modified Asphalt
Penetration	0.1 mm	71	38
Ductility	cm	>100 (15 °C)	25.7 (5 °C)
Softening point	°C	48.2	81.6
60 °C dynamic viscosity	Pa.s	158.7	6350.5

Table 2. Technical indexes of PU binder.

Technical Indicators	Unit	Technical Requirement	
		One-Component PU	Two-Component PU
Surface drying time	min	40 ± 10	20 ± 5
Tensile strength	MPa	≥15.0	≥20.0
Fracture elongation	%	≥100	≥100
Molecular weight	/	13,000–17,000	15,000–20,000

2.1.2. Mixture Material Composition

The technical indexes of the asphalt concrete (AC) and Stone Matrix asphalt (SMA) meet the provisions of JTG F40-2004. The technical indexes of the cement-stabilized macadam mixture meet the provisions of JTJ 034-2018. The Nominal maximum particle sizes of 13.2, 16.0, 19.0 and 26.5 mm are symbolized as 13, 16, 20, and 25, respectively. The mineral aggregate gradation of multi gravel PU concrete (PUM) is designed according to the maximum density theory and the optimum dosage of PU binder is determined by Marshall mix design method. The composition design results of PUM with different aggregate nominal maximum sizes are shown in Table 3.

Table 3. Material composition design results of PU mixtures.

Sieve Size (mm)	Cumulative Passing Percentage of Each Sieve (mm)/%		
	PUM-13	PUM-16	PUM-20
26.5	100	100	100
19	100	100	99.7
16	100	91.8	90.3
13.2	95.6	78.4	79.4
9.5	66.8	60.5	63.2
4.75	31.2	39.2	40.3
2.36	21.5	28.4	29.5
1.18	16.7	19.8	21.4
0.6	11.5	13.1	13.8
0.3	8.6	11.3	10.4
0.15	6.3	7.3	7.5
0.075	2.5	3.1	3.2
Binder content (%)	5.1	5.0	5.0

2.1.3. Mixture Preparation

The SMA-13, AC-20 and AC-25 mixtures were prepared according to the regulations of "Standard Test Methods of Bitumen and Bituminous Mixtures for Highway Engineering" (E20-2011). The PUM-13, PUM-16, and PUM-20 mixture was prepared using an asphalt mixture mixer; however, the aggregates and binder were at room temperature. First, the specified proportion of aggregates were mixed in the mixing pot for 20–30 s. Next, the corresponding proportion of PU binder was added and mixed for 30–50 s. Finally, the corresponding proportion of mineral powder was added and mixed for 30–50 s, and the smoothness of the mixture was checked.

2.2. *Inclined Shear Test*

2.2.1. Inclined Shear Test Device

The interlaminar inclined shear test device was used to obtain the interlaminar shear strength of different types of composite specimens. The interlaminar shear test device is designed with reference to the relevant provisions in "Specifications for Design and Construction of Pavement on Highway Steel Deck Bridge" (JTG/T 3364-02-2019) [31]. The inclination angle between the structure layer and the horizontal direction was set to 45°, such that the interlayer had an equivalent shear and normal stress. The controlled strain mode was adopted, with a loading rate of 5 mm/min [10]. The test device used for the interlaminar shear process is shown in Figure 2, and the shear stress corresponding to the peak value of the stress–strain curve was the maximum vertical load. There were five test pieces in each group. The calculation formula for the inclined shear strength is shown in Equation (1), where τ represents the interlaminar shear strength, F represents the maximum vertical load, and S represents the interface contact area between the two layers.

$$\tau = \frac{F \cdot sin\, 45^{\circ}}{S} \tag{1}$$

Figure 2. Inclined shear test device.

2.2.2. Preparation of the Interlaminar Shear Specimens

A 100 mm thick rutting test mold was used to form the composite rutting plate, and the interlayer shear specimens were then cut and prepared [15]. The specific manufacturing method was as follows: (1) prepare the lower layer mixture rutting test piece with a thickness of 50 mm; (2) install the lower layer test piece onto the rutting test mold with a thickness of 100 mm; (3) carry out the interlayer treatment on the surface of the lower layer according to the treatment scheme; (4) prepare the upper layer mixture and put it into the rutting test mold; (5) form the composite rutting specimen using the wheel rolling method, and remove the formwork after curing; and (6) cut the test piece, remove a 3 cm width from each side (the size of the test pieces after cutting was 50 mm × 50 mm × 100 mm). After the composite specimens were formed, they were cured at room temperature for four days to ensure the complete curing of the PU binder [12]. As far as the PU adhesive layer was concerned, the upper layer mixture was compacted within 0.5–1 h after the two-component PU adhesive layer was spread.

2.2.3. Interlaminar Shear Test

Two types of interlaminar shear tests on different composite structures were carried out. Table 4 shows the combination of specimens without interlaminar treatment, whereas Table 5 shows the combination of specimens with interlaminar treatment.

Table 4. Composite specimens without adhesive layer.

Serial Number	I-1	I-2	I-3	I-4	I-5	I-6	I-7
Upper layer	SMA-13	SMA-13	PUM-13	PUM-16	PUM-16	PUM-13	PUM-20
Lower layer	AC-20	PUM-20	AC-20	AC-25	PUM-16	PUM-20	Cement stabilized macadam

Table 5. Composite specimens with adhesive layer.

Serial Number	II-1	II-2	II-3	II-4	II-3	II-5	II-6	II-7	II-6
Upper layer	SMA-13				PUM-13				
Adhesive layer	emulsified asphalt	PU	PU	PU	PU + 70% macadam (3~5 mm)	PU	PU	PU	PU + 70% macadam (3~5 mm)
Content/(L/m^2)	0.5	0.2	0.4	0.6	0.4	0.2	0.4	0.6	0.4
Lower layer	AC-20	PUM-20				AC-20			

2.3. *Accelerated Loading Test*

In order to clarify the change of interlayer shear strength of composite pavement structure under long-term water temperature and load, the accelerated loading test with high-temperature water area was carried out to simulate the coupling effect of water, temperature and load. On the basis of the interlaminar shear test, structures of I-6, II-1, and II-3 were selected for the accelerated loading test. The forming method of the double-layer rutting plate specimens was the same as that of the inclined shear test [20–22]. The all-environment loading system (ALT-S100) developed by Shandong Jiaotong University was used for the accelerated loading tests (Figure 3a). The test speed was 4.5 km/h, the dynamic load was 1000 kg, the loading speed was 4000 times/h, and the effective working length of the equipment was 1 m. The device shown in Figure 3b was used to arrange the test specimens, where the middle three plates were effective test specimens and the two plates on both sides were cushion blocks.

(**a**) (**b**) (**c**)

Figure 3. Test equipment and specific working conditions. (**a**) Accelerated loading test device; (**b**) Specimen layout; (**c**) High-temperature water area test.

High-temperature water area loading tests of the specimens with the three structures were carried out, and three plates were formed for each structure. In the high-temperature water area test, 50 °C water was injected into the test tank, the liquid level was flush with the top of the test piece (Figure 3c), the test specimens were immersed for 8 h in the test environment, and the accelerated loading test was carried out in a circulating water bath environment. The test rolling times were set as 300,000 times.

2.4. Energy Consumption and Emission Calculation Method

2.4.1. Project Introduction

Based on the construction log of the medium maintenance project of the Xiajin section of the Qingdao–Yinchuan expressway, data regarding the various equipment and transportation were obtained, and the energy consumption and carbon emission of the mixtures were calculated. The implementation scheme of the project involved removing the floating dust and slag after milling the upper and middle surface of the carriageway, followed by repaving the upper and middle layers. The PU mixture pavement was located in the downward direction of Qingyin–Yinchuan expressway, with a total length of 400 m, from K451 + 600~k451 + 820 to K451 + 840~k452 + 000. The original pavement structure, main line pavement structure, and PU mixture pavement structure are shown in Figure 4.

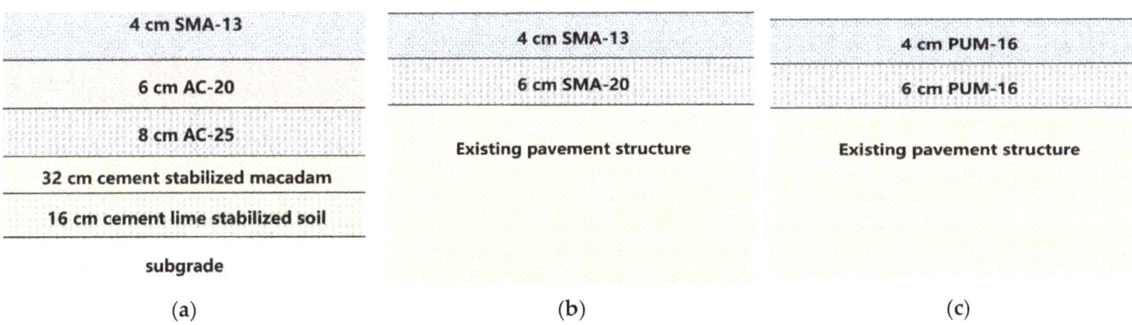

Figure 4. Composite pavement structure. (**a**) Old pavement structure; (**b**) Main line pavement structure; (**c**) Polyurethane mixture pavement structure.

2.4.2. Calculation of the Energy Consumption and Carbon Emission during Construction

To calculate the energy consumption and carbon emissions during the mixture construction, the process was divided into three stages: mixture production, mixture transportation, and mixture site construction, which include seven constituent processes. The energy of the mechanical equipment during the construction process was mainly provided by diesel, electricity, and natural gas [27,28]. The carbon emissions generated through the energy consumption and high-temperature volatilization of hot mixtures, the proportions of energy consumption, and the amount of carbon emissions during mixture construction are shown in Figure 5.

The carbon emission factor (CEF), energy calorific value (Q), and global warming potential value (GWP) provided by the United Nations IPCC were used as the calculation parameters of the energy consumption and carbon emissions [29,30]. The Q values at various energies are shown in Table 6; the CEF and GWP are shown in Table 7.

Table 6. Energy calorific value of energy consumption.

Energy Type	Unit	Heating Coefficient
diesel oil	MJ/kg	42.652
natural gas	MJ/m^3	38.931
electric power	MJ/(kW·h)	3.6

Table 7. CEF and GWP at different energy consumptions.

Energy Type		Unit	CO_2	CH_4	N_2O
CEF	diesel oil	Mg/MJ	74,100.0	3.0	0.6
	natural gas	Mg/MJ	56,100.0	1.0	0.1
	electric power	Mg/MJ	317,000.0	/	/
GWP		/	1	28	265

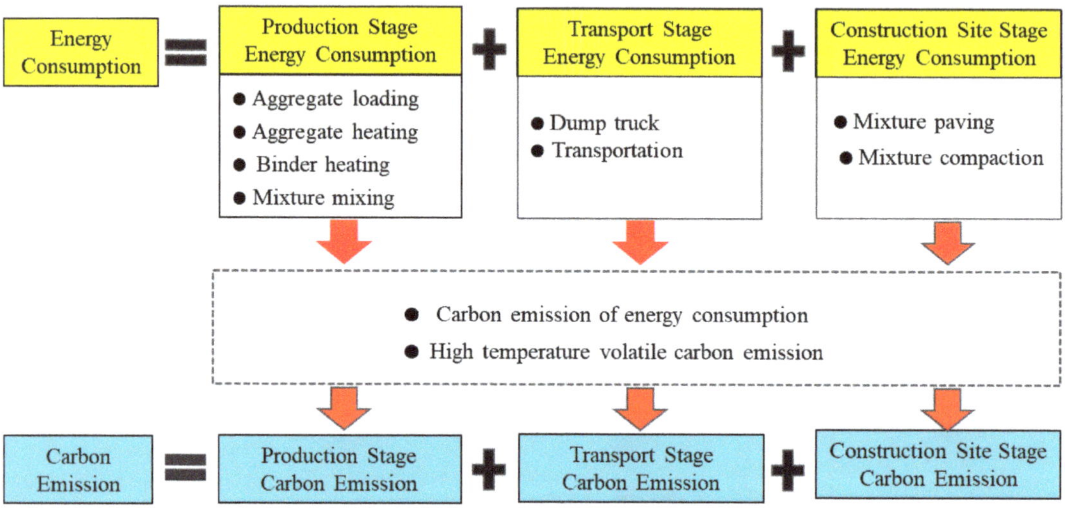

Figure 5. Energy consumption and carbon emission of the mixtures.

According to the Q and energy consumption survey data, a quantitative model of the energy consumption during mixture construction was established, as shown in Equation (2). According to the CEF, GWP, and energy consumption survey data, the calculation model of energy consumption and carbon emissions during mixture construction was established [28], as shown in Equations (3)–(5).

$$QC = \sum_{i=1}^{3}\sum_{j=1}^{7} QC_{ij} = \sum_{i=1}^{3}\sum_{j=1}^{7}(M_{ij}Q_{ij} + V_{ij}Q_{ij} + W_{ij}Q_{ij}) \quad (2)$$

$$EC_{ijk} = M_{ij}Q_{ij}C_{ijk} + V_{ij}Q_{ij}C_{ijk} + W_{ij}Q_{ij}C_{ijk} \quad (3)$$

$$EC_{ij} = \sum_{k=1}^{3}\left(EC_{ijk}p_k\right)/10^6 \quad (4)$$

$$EC = \sum_{i=1}^{3}\sum_{j=1}^{7} EC_{ij} = \sum_{i=1}^{3}\sum_{j=1}^{7}\sum_{k=1}^{3}\left(EC_{ijk}p_k\right)/10^6 \quad (5)$$

where QC is the total energy consumption during pavement construction (MJ/t); EC is the energy consumption emission during pavement construction (kg/t); i represents the three stages of pavement construction, including production, transportation, and construction; j refers to the seven processes during pavement construction, including aggregate supply, aggregate heating, binder heating, mixing, mixture transportation, mixture paving, and mixture compaction; K represents three types of greenhouse gases, including CO_2, CH_4, and N_2O; EC_{ijk} is the emission of K kinds of greenhouse gases corresponding to the jth process of the ith stage (mg/t); EC_{ij} is the energy consumption of carbon emissions corresponding to the jth process of the ith stage (kg/t); M_{ij} is the diesel consumption corresponding to the *j*th process of the *i*th stage (kg/t); V_{ij} is the natural gas consumption corresponding to the *j*th process of the *i*th stage (m^3/t); W_{ij} is the electric energy consumption corresponding to the *j*th process of the *i*th stage (kW·h/t); Q_{ij} is the energy calorific value corresponding to the *j*th process of the *i*th stage (MJ); C_{ijk} is the energy consumption emission factor of *K* greenhouse gases corresponding to the *j*th process of the *i*th stage (mg/MJ); and p_k is the global warming potential of K greenhouse gases [30].

A large amount of asphalt smoke and dust were produced during the mixing, transportation, paving, and rolling of hot mixtures, which contained gases with significantly high greenhouse effects, including CO_2, CH_4, and N_2O. Therefore, the ZR-3110 multi-gas monitor was used to detect CO_2, CH_4, and N_2O in asphalt smoke and dust when the pavement was constructed [32,33].

2.4.3. Carbon Emission Calculation of the Curing Stage

The strength of the PU mixture was ensured by the curing reaction of the PU binder; water was the key factor affecting the strength of the PU mixture [34]. Further, water in the air could react with the PU binder to form a urea bond, promoting cross-linking and curing.

Measurement of the NCO Content

The PU binder was dissolved in toluene solution, and the isocyanate in the PU binder reacted with excess di-n-butylamine to generate the corresponding substituted urea. When the reaction was complete, the remaining di-n-butylamine was titrated with a hydrochloric acid standard solution. The reaction equation is shown in Equations (6) and (7).

$$\text{R-NCO} + (\text{C}_4\text{H}_9)_2\text{NH} \rightarrow \text{RNHCON}(\text{C}_4\text{H}_9)_2 \quad (6)$$

$$(\text{C}_4\text{H}_9)_2\text{NH} + \text{HCL} \rightarrow (\text{C}_4\text{H}_9)_2\text{NH}\bullet\text{HCL} \quad (7)$$

The specific steps were as follows:

First, about 4.62 g of the weighed PU binder sample was put into a 250 mL conical flask with a stopper; next, 25 mL of anhydrous toluene was added, the bottle stopper was covered and shaken to completely dissolve the sample [5–8]. Then, 25 mL of 0.1 mol/L di-n-butylamine toluene solution was added with a pipette, the bottle covered with a stopper, shaken for 15 min, and—finally—100 mL of isopropanol and 4–6 drops of bromophenol blue indicator solution were added. The solution was then titrated with 0.1 mol/L hydrochloric acid standard solution. The end point of the reaction was when the solution turned from blue to yellow. The volume of hydrochloric acid used was recorded at this time. Further, blank tests were conducted without adding samples. The mass fraction of isocyanate was calculated according to Equation (8).

$$\omega_{\text{NCO}} = \frac{(V_0 - V_s) \times C \times 42}{1000m} \times 100\% \quad (8)$$

where ω_{NCO} is the mass fraction of NCO (%); V_0 is the volume of hydrochloric acid standard solution consumed by blank titration (mL); V_s is the volume of hydrochloric acid standard solution consumed by sample titration (mL); *C* is the concentration of the hydrochloric acid solution (mol/L); *m* is the sampling quantity (g); and 42 is the molar mass of NCO (g/mol).

Calculation of CO_2 Emissions

After determining the mass fraction of isocyanate, the CO_2 emissions per ton of the PU mixture curing reaction can be calculated according to the reaction equation between isocyanate and water (shown in Equation (9)), according to Equation (10):

$$2\text{RNCO} + \text{H}_2\text{O} \rightarrow \text{RNHCONHR} + \text{CO}_2 \uparrow \quad (9)$$

$$m_{\text{CO}_2} = \frac{\omega_{\text{NCO}} \times A}{42} \times \frac{1}{2} \times 44 \times 10^3 \quad (10)$$

where m_{CO_2} is the CO_2 emission (kg); A is the PU content of the PU mixture (%); and 44 is the molar mass of CO_2 (g/mol).

3. Results and Discussion

3.1. Interlaminar Shear Properties of the Polyurethane Mixture

3.1.1. Interlaminar Shear Properties of Composite Specimens without Interlaminar Treatment

The inclined shear test results of seven composite specimens without interlayer treatment are shown in Figure 6.

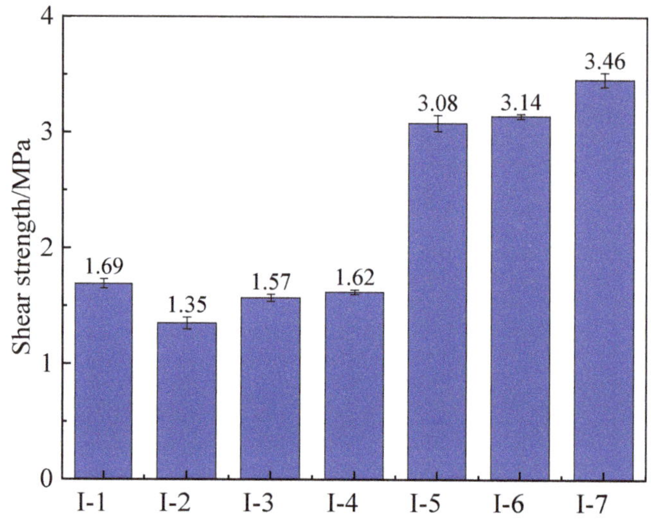

Figure 6. Shear strength of specimens without interlaminar treatment.

The data of Figure 6 showed that the shear strengths between layers of asphalt mixture–asphalt mixture and PU mixture–asphalt mixture were less than 2 MPa, and the shear failure mostly occurred at the interlayer. The shear failure form of specimens was shown in Figure 7a–c. The shear strength of I-3 was 0.22 MPa greater than that of I-2, indicating that when the gradations of the upper and lower layers were similar, the interlayer shear strength was higher when the PU mixture used in the upper layer. The shear strength of specimen I-1 was greater than those of I-2, I-3, and I-4, indicating that the interlaminar shear strength between asphalt mixtures was larger than that between PU mixture and asphalt mixture. Therefore, the shear strengths of these structures with different interlaminar treatment schemes are analyzed in the next section.

The interlayer bonding effect of PU mixtures, PU mixture, and cement-stabilized macadam were better than that of other structures. The interlayer shear strength was greater than 3 MPa, and until shear failure occurred, there was no obvious damage between the layers of these structures, and the interlayer was in a good condition. However, the upper and lower materials were damaged. The shear failure form of specimens without interlaminar treatment was shown in Figure 7d–f. The reason may be that the macromolecular chain of the PU binder between the two layers of PU mixtures produced an interweaving effect, and the macromolecular chain segment of the PU binder between the PU mixture layer and the cement-stabilized macadam layer interacted with the inorganic materials; thus, the interlayer stability was good, and treatment was not required between the layers.

Figure 7. Shear failure form of specimens without interlaminar treatment. (**a**) I-1; (**b**) I-2; (**c**) I-3; (**d**) I-5; (**e**) I-6; (**f**) I-7.

3.1.2. Interlaminar Shear Properties of Composite Specimens with Different Treatments

The shear test results for SMA-13 + AC-20, SMA-13 + PUM-20, and PUM-13 + AC-20 with different interlayer treatment schemes are shown in Figure 8.

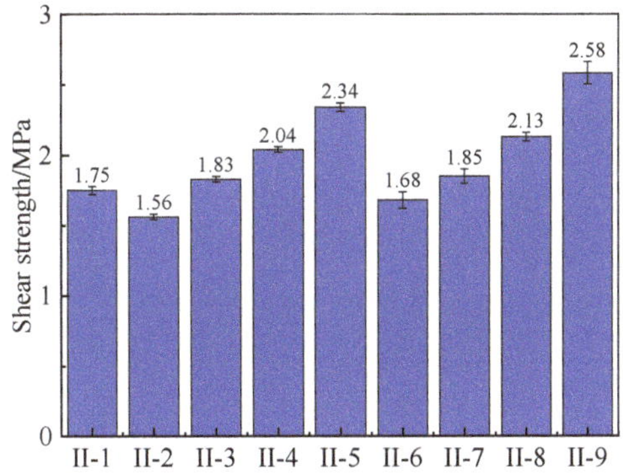

Figure 8. Shear strength of specimens.

According to Figure 8, for the SMA-13 + PUM-20 and PUM-13 + AC-20 structures, the interlayer shear strength increased with increases in the distribution amount of the two-component PU adhesive layer. The variation trend of interlayer shear strength with the distribution amount is shown in Figure 9. When the distribution amount of the adhesive layer was 0.4 L/m^2, the interlayer shear strength of the SMA-13 + PUM-20 structure and PUM-13 + AC-20 structure was greater than that of II-1 with the structure of SMA-13 + AC-20 with an adhesive layer. Therefore, when the composite pavement used an SMA-13 + PUM-20 structure and PUM-13 + AC-20 structure, it was recommended to spread 0.4 L/m^2 two-component PU adhesive layer as the interlayer treatment scheme.

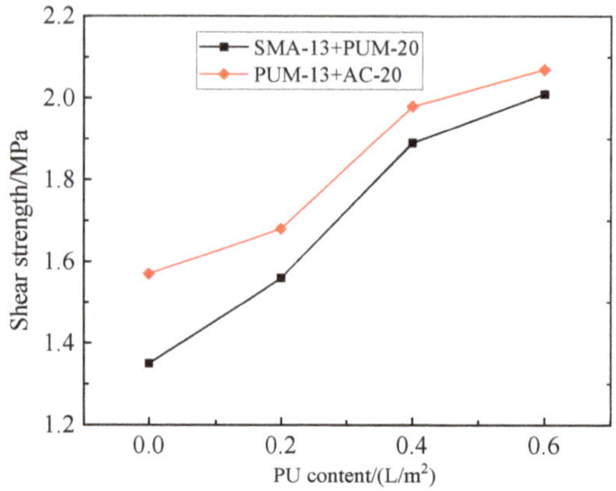

Figure 9. Specimens with a PU adhesive layer.

When the two-component PU macadam seal was used in the treatment, the interlayer shear strength was the highest; the interlayer shear strength of the SMA-13 + PUM-20 structure was 2.34 MPa, and the interlayer shear strength of the PUM-13 + AC-20 structure was 2.58 MPa. When the two structures are used for special road conditions, such as heavy loads and long longitudinal slopes, the technology of the two-component PU macadam seal can be used as the interlayer treatment.

3.2. *Long-Term Shear Properties of Composite Specimens*

3.2.1. Surface Properties after Long-Term Accelerated Loading

The tire trace position of the test piece is cut after accelerated loading to a size of 50 mm × 50 mm × 10 mm, and the cut specimens are shown in Figure 10.

(**a**) (**b**) (**c**)

Figure 10. Specimen surfaces after the accelerated loading test. (**a**) I-6; (**b**) II-1; (**c**) II-3.

It can be seen that, after the accelerated loading test, the fine surface aggregate of II-1 and II-3 was mostly missing, while the surface of test piece I-6 basically maintained integrity, although a small amount of fine aggregate was missing. This shows that, under the dual action of long-term high-temperature water and loading, asphalt mixtures are prone to particle loosening and breaking, while PU mixtures had relatively good stability.

3.2.2. Long-Term Interlaminar Shear Property

The shear strength of the three structures before and after accelerated loading tests are shown in Figure 10. The shear strength ratio was calculated by dividing the original shear strength by the shear strength after the accelerated loading test; the results are shown in Figure 11.

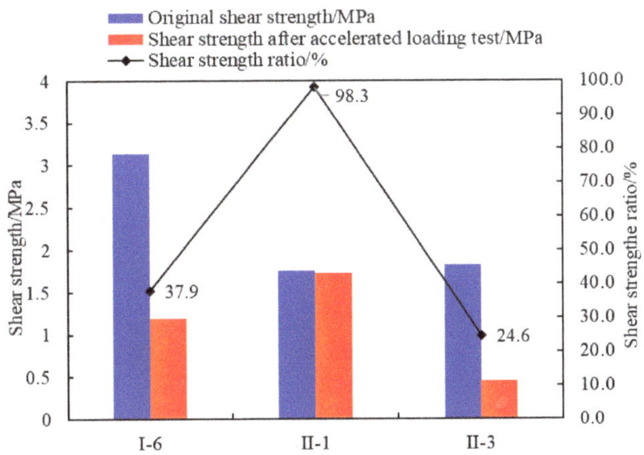

Figure 11. Shear strength and shear strength ratio of the three structures.

From Figure 11, it can be seen that the accelerated loading test had different effects on the interlaminar shear strength of the three structures. Among them, the interlaminar shear strength of II-3 decreases the most, and the shear strength ratio after the accelerated loading test was only 24.6%. The specimen after the shear test is shown in Figure 12c, and the failure surface between the two layers was found to be very smooth. This indicates that the high-temperature water area and loading greatly reduced the interlayer shear strength of the composite structure of the PU mixture–asphalt mixture; therefore, this structure was not suitable for high-temperature and wet environments.

Figure 12. Failure form of the shear test after accelerated loading test. (**a**) I-6; (**b**) II-1; (**c**) II-3.

The interlaminar shear strength of the double PU mixture layer structure (I-6) was greater than 3 MPa; however, the shear strength ratio after the accelerated loading test was only 37.9%. The sheared specimen is shown in Figure 12a, indicating that the high-temperature water area and load have a great impact on the interlaminar shear strength of the structure. A large number of studies show that when the interlaminar shear strength of pavement structure is greater than 1MPa, it can meet the basic service requirements of pavement [28]. The interlaminar shear strength after accelerated loading was greater than 1 MPa, thus it can meet the basic requirements of pavement structures.

The interlaminar shear strength of the double asphalt mixture layer structure specimen (II-1) changed slightly before and after accelerated loading, and the shear strength ratio after accelerated loading was 98.3%. The sheared specimen is shown in Figure 12b, indicating that the high-temperature water area and loading have little impact on the interlaminar shear characteristics of the double-layer asphalt mixture structure.

3.3. *Evaluation of the Energy-Saving and Emission-Reduction Effects of Composite Pavement*

3.3.1. Energy Consumption during Pavement Construction

According to the site construction log, the data of various equipment were obtained for energy consumption calculation. The energy consumption of each construction stage and the process of the three mixtures are shown in Table 8.

Table 8. Energy consumption list of mixtures at each stage.

Material Type	Production Stage				Transportation Stage	Construction Site Stage	
	Aggregate Feed (Diesel)/(L·t^{-1})	Aggregate Heating (Natural Gas)/(m^3·t^{-1})	Binder Heating (Natural Gas)/(m^3·t^{-1})	Mixture Mixing (Electric Energy)/(kW·h)	Mixture Transportation (Diesel)/(L·t^{-1})	Mixture Paving (Diesel)/(L·t^{-1})	Mixture Rolling (Diesel)/(L·t^{-1})
SMA-13	0.195	8148	1353	3200	0.27	0.182	0.277
SMA-20	0.142	7276	1279	3011	0.43	0.155	0.231
PUM-16	0.158	0	0	3342	0.36	0.167	0.188

The energy consumption and carbon emissions during the construction process per ton of the mixture were calculated using Equations (2)–(5). The results are shown in Figure 13, and the proportions of energy consumption of different construction stages of the three mixtures are shown in Figure 14.

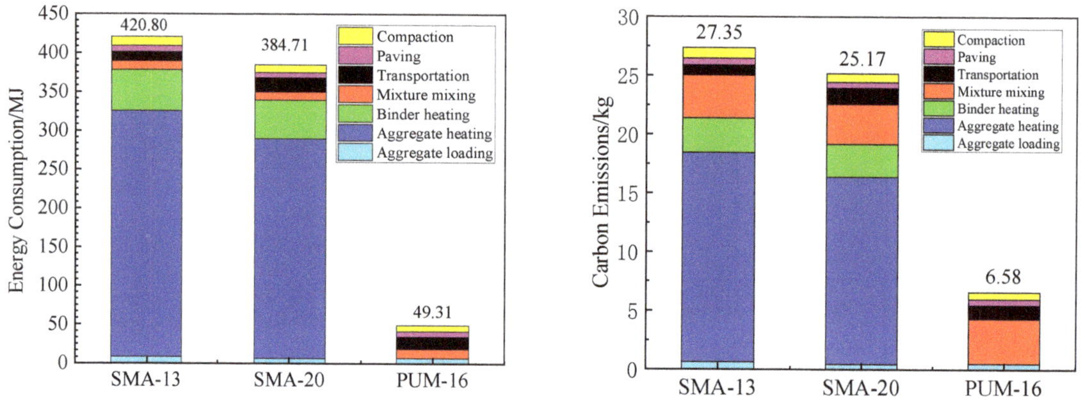

(a) (b)

Figure 13. Energy consumption and carbon emissions of pavement construction. (**a**) Energy consumption; (**b**) Carbon emissions.

According to Figures 13 and 14, the energy consumption of the PUM-16 mixture in the construction stage was 11.7% and 12.8% of that of SMA-13 and SMA-20, respectively, and the carbon emissions were 24.2% and 26.1% of that of SMA-13 and SMA-20, respectively, indicating that the PUM-16 mixture could effectively reduce the energy consumption and carbon emissions of the construction process and meet the requirements of energy conservation and emission reduction.

For the asphalt mixture, the energy consumption and emission for aggregate heating accounted for the largest proportion of the whole, followed by asphalt heating and asphalt mixing. The energy consumption for aggregate heating, asphalt heating, and mixture mixing accounted for about 75%, 12%, and 3% of the total energy consumption, respectively. For the PU mixture, the energy consumption of mixture transportation and mixing was the largest, accounting for about 26.1% and 24% of the entire construction process, respectively, followed by that for compaction, paving, and aggregate feed.

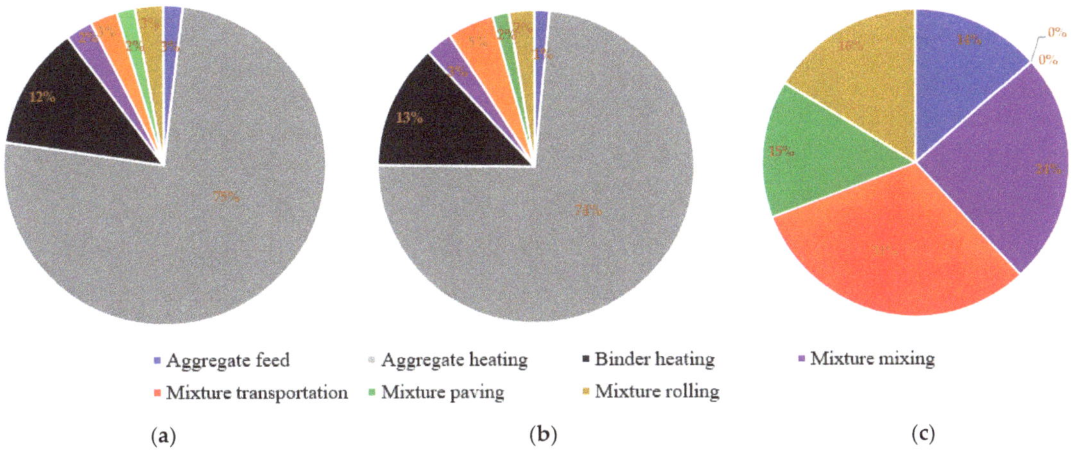

Figure 14. Energy consumption of pavement construction. (**a**) SMA-13; (**b**) SMA-20; (**c**) PUM-16.

3.3.2. Carbon Emission during the High-Temperature Volatilization and Curing Stage

The high-temperature volatile carbon emissions of the SMA-13 and SMA-20 mixtures during each construction process were detected. Owing to the normal temperature construction of the PUM-13 mixture, there were no high-temperature volatile carbon emissions. The results are shown in Table 9.

Table 9. High-temperature volatile carbon emissions from hot-mix asphalt mixture kg·t^{-1}.

Mixture Type	Mixing/kg·t^{-1}	Transportation/kg·t^{-1}	Paving/kg·t^{-1}	Compaction/kg·t^{-1}	Sum/kg·t^{-1}
SMA-13	0.040	0.247	1633	19.468	21.388
SMA-20	0.035	0.1888	1509	17.085	18.818
PUM-16	0	0	0	0	0

The NCO content of the PU mixture was calculated as 10%, and the CO_2 produced during the curing reaction of 1t of the PUM-13 mixture was 2.62 kg.

3.3.3. Total Energy Consumption and Carbon Emissions of the Three Mixtures

The carbon emissions of the three mixtures during the construction stage, high-temperature volatilization stage, and curing stage can be summed, as shown in Figure 15. It can be seen that the PUM-16 mixture can effectively reduce the energy consumption by 88.3% and 87.2% and carbon emissions by 81.1% and 79.1%, respectively, in comparison to SMA-13 and SMA-20. From the perspective of the medium maintenance project of Qingdao–Yinchuan expressway, the pavement structure of PUM-16 + PUM-16 was compared with the structure of SMA-20 + SMA-13; it was found that the energy consumption was reduced by 87.8%, the carbon emissions were reduced by 80.2%, and the energy conservation and emission reduction effects of the PUM-16 + PUM-16 structure were remarkable.

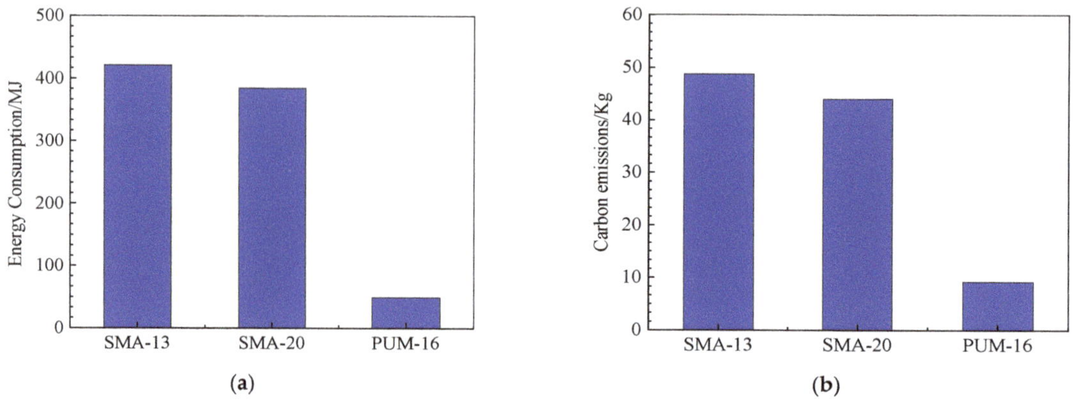

Figure 15. Energy consumption and carbon emissions of the three mixtures. (**a**) Energy consumption; (**b**) Carbon emissions.

4. Conclusions

The following conclusions can be drawn:

1. The shear strength between the layers of asphalt mixtures, PU mixture, and asphalt mixture was less than that of the PU mixture–cement-stabilized macadam and PU mixture–PU mixture structures.
2. It is recommended to spread 0.4 L/m^2 of two-component PU adhesive layer as the interlayer treatment scheme for the SMA-13 + PUM-20 structure and PUM-13 + AC-20 structure.
3. The long-term coupling action of water, temperature, and load has different effects on different composite pavement structures. Under the coupling action, the shear strength of the double PU mixture layer structure, the PU mixture, and asphalt mixture composite structure decreased greatly, while the shear strength of the double asphalt mixture layer decreased slightly.
4. In comparison to asphalt mixtures, the energy conservation and emission reduction effect of the PU mixture was remarkable.

Author Contributions: Conceptualization, Y.B.; Methodology, S.C.; Validation, M.S. and S.J.; Formal analysis, M.S. and Q.Z.; Investigation, Y.B.; Resources, W.Z. and X.J.; Data curation, S.J. and D.H.; Writing—original draft preparation, M.S. and X.J.; Writing—review and editing, W.Z.; Supervision, S.C. and D.H.; Project administration, Q.Z. and W.Z. All authors have read and agreed to the published version of the manuscript.

Funding: This work was financially supported by Key Scientific Research Projects in the Transportation Industry of the Ministry of Transport (2019MS2028), Shandong Expressway Group Project (HSB 2021-72).

Institutional Review Board Statement: Not applicable.

Informed Consent Statement: Not applicable.

Data Availability Statement: This original copy does not include distributed figures, tables, and charts before, thus all figures, tables, and charts of this original copy are unique.

Acknowledgments: We would like to recognize numerous co-workers, students, and research facility associates for giving specialized assistance on instrument examination.

Conflicts of Interest: The authors declare no conflict of interest.

References

1. Xiao, F.; Yao, S.; Wang, J.; Wei, J.; Amirkhanian, S. Physical and chemical properties of plasma treated crumb rubbers and high temperature characteristics of their rubberised asphalt binders. *Road Mater. Pavement Des.* **2020**, *21*, 587–606. [CrossRef]
2. Jiang, W.; Xiao, J.; Yuan, D.; Lu, H.; Xu, S.; Huang, Y. Design and experiment of thermoelectric asphalt pavements with power-generation and temperature-reduction functions. *Energy Build.* **2018**, *169*, 39–47. [CrossRef]
3. Min, S.; Bi, Y.; Zheng, M.; Chen, S.; Li, J. Evaluation of a cold-mixed high-performance polyurethane mixture. *Adv. Mater. Sci. Eng.* **2019**, *2019*, 1507971. [CrossRef]
4. Sun, M.; Bi, Y.; Zheng, M.; Wang, J.; Wang, L. Performance of polyurethane mixtures with skeleton-interlocking structure. *J. Mater. Civ. Eng.* **2020**, *32*, 04019358. [CrossRef]
5. Xu, S.; Zhang, Y.; Guo, Y.; Ma, C.; Gao, D.; Peng, G. Determination of polyurethane concrete compaction timing based on penetration resistance test system. *China J. Highw. Transp.* **2021**, *34*, 226–235. (In Chinese)
6. Wang, H.; Li, R.; Wang, X.; Ling, T.; Zhou, G. Strength and road performance for porous polyurethane mixture. *China J. Highw. Transp.* **2014**, *27*, 24–30.
7. Wang, D.; Schacht, A.; Leng, Z.; Leng, C.; Kollmann, J.; Oeser, M. Effects of materials composition on mechanical and acoustic performance of poroelastic road surface (PERS). *Constr. Build. Mater.* **2017**, *135*, 352–360. [CrossRef]
8. Wang, D.; Liu, P.; Leng, Z.; Leng, C.; Lu, G.; Buch, M.; Oeser, M. Suitability of poroelatic road surface (PERS) for urban roads in cold regions: Mechanical and functional performance assessment. *J. Clean. Prod.* **2017**, *165*, 1340–1350. [CrossRef]
9. Lu, G.; Renken, L.; Li, T.; Wang, D.; Li, H.; Oeser, M. Experimental study on the polyurethane-bound pervious mixtures in the application of permeable pavements. *Constr. Build. Mater.* **2019**, *202*, 838–850. [CrossRef]
10. Cong, L.; Wang, T.; Tan, L.; Yuan, J.; Shi, J. Laboratory evaluation on performance of porous polyurethane mixtures and OGFC. *Constr. Build. Mater.* **2018**, *169*, 436–442. [CrossRef]
11. Guo, G.; Cong, L.; Yang, F.; Tan, L. Application progress of polyurethane material in pavement engineering. *J. Highw. Transp. Res. Dev.* **2020**, *37*, 1–9. (In Chinese)
12. Sun, M.; Bi, Y.; Zhuang, W.; Chen, S.; Zhao, P.; Pang, D.; Zhang, W. Mechanism of polyurethane binder curing reaction and evaluation of polyurethane mixture properties. *Coatings* **2021**, *11*, 1454. [CrossRef]
13. Chen, J.; Yao, C.; Wang, H.; Huang, W.; Ma, X.; Qian, J. Interface shear performance between porous polyurethane mixture and asphalt sublayer. *Appl. Sci.* **2018**, *8*, 623. [CrossRef]
14. Juan, Z. *Study on the Waterproofing and Boding Layer of Concrete Bridge Deck*; Chang'an University: Xi'an, China, 2008. (In Chinese)
15. Mohammad, L.N.; Bae, A.; Elseifi, M.A.; Button, J.; Patel, N. Effects of pavement surface type and sample preparation method on tack coat interface shear strength. *Transp. Res. Rec.* **2020**, *2180*, 93–101. [CrossRef]
16. Luo, Z.G.; Zeng, J.; Wang, S.Y. Experiment and study on influence of interface pollution on interlaminar bonding performance between semi-rigid base and asphalt concrete course. *Highway* **2011**, *1*, 109–113. (In Chinese)
17. Cao, M.M.; Huang, W.Q.; Lu, Y.; Tan, Q.Q. Test and evaluation method of interlaminar shear property of composite pavement. *J. Highw. Transp. Res. Dev.* **2018**, *35*, 40–48. (In Chinese)
18. Leng, Z.; Al-Qadi, I.L.; Carpenter, S.H.; Ozer, H. Interface bonding between hot-mix asphalt and various Portland cement concrete surfaces: Assessment of accelerated pavement testing and measurement of interface strain. *Transp. Res. Rec.* **2009**, *2127*, 20–28. [CrossRef]
19. Tian, S.; Zhang, T.Z.; Li, Y.; Wang, K.M. Research on shear test of the interface between bridge deck wit carbon fiber and deck pavement. *J. Wuhan Univ. Technol.* **2011**, *33*, 80–84.
20. Sheng, L.I.; Zhao-hui, L.I.U.; Zhang, J.Y. Test evaluation of pavement performance of interlaminar asphalt material for rigid-flexible composite pavement. *J. Highw. Transp. Res. Dev.* **2013**, *30*, 7–12. (In Chinese)
21. Lun, J.I.; Yun-liang, L.I.; Jun-da, R.E.N. Method fo determining the spraying amount of waterproof binder for bridge deck pavement. *J. Harbin Inst. Technol.* **2014**, *46*, 57–62. (In Chinese)
22. Zhu, C.; Gao, D. A research on the factors influencing carbon emission of transportation industry in "the belt and road initiative" countries based on panel data. *Energies* **2019**, *12*, 2405. [CrossRef]
23. Kim, B.; Lee, H.; Park, H.; Kim, H. Framework for estimating greenhouse gas emissions due to asphalt pavement construction. *J. Constr. Eng. Manag.* **2012**, *138*, 1312–1321. [CrossRef]
24. White, P.; Golden, J.S.; Biligiri, K.P.; Kaloush, K. Modeling climate change impacts of pavement production and construction. *Resour. Conserv. Recycl.* **2010**, *54*, 776–782. [CrossRef]
25. Cass, D.; Mukherjee, A. Calculation of greenhouse gas emissions for highway construction operations by using a hybrid life-cycle assessment approach: Case study for pavement operations. *J. Constr. Eng. Manag.* **2011**, *137*, 1015–1025. [CrossRef]
26. Wang, X.; Duan, Z.; Wu, L.; Yang, D. Estimation of carbon dioxide emission in highway construction: A case study in southwest region of China. *J. Clean. Prod.* **2015**, *103*, 705–714. [CrossRef]
27. Thives, L.P.; Ghisi, E. Asphalt mixtures emission and energy consumption: A review. *Renew. Sustain. Energy Rev.* **2017**, *72*, 473–484. [CrossRef]
28. Ma, F.; Sha, A.; Lin, R.; Huang, Y.; Wang, C. Greenhouse gas emissions from asphalt pavement construction: A case study in China. *Int. J. Environ. Res. Public Health* **2016**, *13*, 351. [CrossRef]
29. Peng, B.; Cai, C.; Yin, G.; Li, W.; Zhan, Y. Evaluation system for CO_2 emission of hot asphalt mixture. *J. Traffic Transp. Eng.* **2015**, *2*, 116–124. [CrossRef]

30. Peng, B.; Deng, H.L.; Cao, S.J.; Li, W.Y. Carbon emission quantification and evaluation system of hot mix asphalt mixture. *J. Chang'an Univ. Nat. Sci. Ed.* **2019**, *39*, 1–9. (In Chinese)
31. *Specifications for Design and Construction of Pavement on Highway Steel Deck Bridge*; JTG /T3364-02-2019; People's Communications Publishing House Co., Ltd.: Beijing, China, 2019.
32. Li, Q.; He, D.; Ding, H. Effects of tack coat on mechanical properties of pavement interface with combination of different types. *China J. Highw. Transp.* **2015**, *28*, 22–30.
33. Fakhri, M.; Shahryari, E.; Ahmadi, T. Investigate the use of recycled polyvinyl chloride (PVC) particles in improving the mechanical properties of stone mastic asphalt (SMA). *Constr. Build. Mater.* **2022**, *36*, 126780. [CrossRef]
34. Fakhri, M. The effects of nano zinc oxide (ZnO) and nano reduced graphene oxide (RGO) on moisture susceptibility property of stone mastic asphalt (SMA). *Case Stud. Constr. Mater.* **2021**, *15*, e00655. [CrossRef]

coatings

MDPI

Article

Research on the Quality of Asphalt Pavement Construction Based on Nondestructive Testing Technology

Wei Chen [1], Guiling Hu [1,*], Wenyang Han [2], Xiaomeng Zhang [2], Jincheng Wei [1,2], Xizhong Xu [2] and Xiangpeng Yan [2]

1 School of Transportation Engineering, Shandong Jianzhu University, Jinan 250101, China; chenwei13572468@163.com (W.C.); sdjtky@126.com (J.W.)
2 Science and Technology Innovation Center, Shandong Transportation Institute, Jinan 250102, China; hanwenyang@sdjtky.cn (W.H.); zhangxiaomeng@sdjtky.cn (X.Z.); xxz137152@163.com (X.X.); yanxiangpeng@sdjtky.cn (X.Y.)
* Correspondence: huguilingtech@foxmail.com

Abstract: In order to better evaluate the construction quality of asphalt pavement, nondestructive testing techniques are used to inspect newly paved asphalt mixture pavement. The proposed system for the evaluation of asphalt pavement construction quality uses three-dimensional ground-penetrating radar (GPR) and a non-nuclear density gauge. The GPR and the non-nuclear density gauge test results were used to establish a dielectric constant–porosity model by fitting. This approach can more accurately determine the dielectric constant selection scheme of the GPR based on the average value of every 10 dielectric constant data points in the length direction of the radar antenna and every three data channels in the width direction. The GPR collected the dielectric constants of the road surface based on the total reflection method and used the average value of the local dielectric constant to evaluate the construction quality of the road. The non-nuclear density gauge used the local porosity to assess the construction quality of the road. It is recommended that the two testing schemes described above be used to evaluate the quality of asphalt pavement construction. They can provide theoretical guidance for future applications in practical processes.

Keywords: nondestructive testing; construction quality evaluation system; dielectric constant–porosity prediction model; average of local dielectric constants; proportion of local severe segregation

Citation: Chen, W.; Hu, G.; Han, W.; Zhang, X.; Wei, J.; Xu, X.; Yan, X. Research on the Quality of Asphalt Pavement Construction Based on Nondestructive Testing Technology. *Coatings* **2022**, *12*, 379. https://doi.org/10.3390/coatings12030379

Academic Editor: Valeria Vignali

Received: 22 February 2022
Accepted: 10 March 2022
Published: 14 March 2022

Publisher's Note: MDPI stays neutral with regard to jurisdictional claims in published maps and institutional affiliations.

Copyright: © 2022 by the authors. Licensee MDPI, Basel, Switzerland. This article is an open access article distributed under the terms and conditions of the Creative Commons Attribution (CC BY) license (https://creativecommons.org/licenses/by/4.0/).

1. Introduction

By the end of 2020, China had 160,000 km of motorways open to traffic, putting it in first place globally; the highway coverage of cities with a population of 200,000 or more exceeded 98% [1]. For highways with design lives of 20 years, asphalt mixture pavements are used, but they are susceptible to heavy loads, high temperatures, rain, and other adverse coupled factors [2]. Furthermore, they are prone to early severe damage, such as wheel ruts, potholes, looseness, flooding oil, and swelling within the first 1–2 years of use [3]. As these phenomena are not evident early on in use, the development of early failure is difficult to control, increasing the maintenance and repair costs of the road at later stages, affecting the comfort and safety of vehicle travel, and reducing the normal service life of the road. The quality control problems contribute to early failure during the construction of new asphalt mixture pavement [4], which is mainly reflected by the porosity of the asphalt mixture. Thus, the porosity of the asphalt mixture should be controlled to within a reasonable range. Excessive porosity in asphalt pavement can cause water and air to enter the pavement, leading to water damage, aging, spalling, and cracking [5]. Therefore, accurate and efficient assessment of the quality of asphalt pavement construction is one of the most important means of preventing the development of early asphalt pavement distress.

Current three-dimensional ground-penetrating radar technology is relatively mature for detecting road thickness [6,7], and electromagnetic pulses are mainly transmitted through an antenna to penetrate the road surface. The reflected echoes generated by the surface and internal unevenness of the road surface are recorded to measure the penetration depth [8]. Baltruaitis et al. used an existing mathematical model to determine the density of asphalt pavement by drilling a core and using ground-penetrating radar (GPR) [9]. However, the construction quality evaluation system for asphalt mixtures using GPR is still in the developmental research stage [10,11]. Related research has shown that GPR can optimize asphalt pavement design and identify the exact location of internal damage to the pavement [6]. Zhang et al. used GPR to accurately locate and detect wet damage on asphalt pavement, proving that GPR has excellent detection performance [12]. Ma et al. found that GPR is more effective at detecting stripping damage on flexible pavement. In order to achieve this goal, a finite-difference time-domain simulation program was used to study the propagation of GPR signals in a stripped pavement [13]. Although the traditional core drilling and sampling method can accurately measure the density of the structural layer of asphalt pavement [14,15], it is destructive to the pavement, as well as complex and inefficient [15–17]. A non-nuclear density gauge can accurately measure the density of asphalt pavement and other indicators through calibration before the measurement [18]. Wim et al. studied the parameters of the non-nuclear density gauge, and the use of PQI is recommended as part of the standard quality control process by Flanders [19], but this required a single point of detection, which was inefficient [20]. Both existing nondestructive testing methods are under-representative and so cannot accurately and efficiently evaluate the quality of asphalt pavement construction [21]. Therefore, it is necessary to evaluate the construction quality of asphalt pavement comprehensively to improve the construction process of asphalt mixtures and reduce the influence of porosity on the service life of asphalt pavement.

The purpose of this research was to combine three-dimensional GPR with a non-nuclear density gauge. Based on two nondestructive testing techniques, two concepts, the average of local dielectric constant values and the proportion of local porosity, are proposed to establish an asphalt pavement construction quality evaluation system to provide a theoretical basis for later application and development in actual projects.

2. Nondestructive Testing Equipment and Principles

2.1. Three-Dimensional Ground-Penetrating Radar (GPR)

In this test, the Geo Scope ^{MK} IV GPR system produced by 3D-Radar was used to study the dielectric constant of asphalt pavement. The system was equipped with a DX1821 air-coupled antenna, distance measuring wheel (DMI), computer, and Geo Scope radar mainframe, as shown in Figure 1. The vehicle-mounted radar system could obtain dielectric constant gray images to reflect the condition of the road in real time. According to the different reflection effects of high-frequency electromagnetic waves in mixtures of different media [22], GPR transmits pulsed high-frequency electromagnetic waves to the asphalt pavement via a butterfly monopole antenna. The reflected electromagnetic wave signals are processed by software to understand the reflection of different media in the pavement's structural layer and obtain the dielectric constant of the road section.

2.2. Non-Nuclear Density Gauge

A new generation of non-nuclear density gauge (PQI 380) from Trans Tech, Adamstown, MD, USA, was used in this study, as shown in Figure 2. Because asphalt mixtures are prepared from asphalt and aggregate, the dielectric constants of these materials vary considerably, and air can be present in the interstitial spaces. The PQI 380 emits a circular electromagnetic wave through the bottom to the asphalt pavement. As the material composition of the mixture changes, the transmission direction and propagation speed of the electromagnetic wave will change accordingly. This causes a difference in the signal received at the receiving end of the instrument, and an empirical algorithm is built into the device for density detection [23]. However, for different types of asphalt mixes with different testing depths, calibration must be carried out before using the PQI 380, as follows

First, an area of approximately 1.5 m × 0.75 m was selected in the calibration area, and three circles were drawn in this area. The PQI 380 measured the density data at each of the five locations of each circle, as shown in Figure 3, then the height and density of five sample cores were measured in the laboratory. The ratio of the actual densities of the core samples to the density measured by the PQI 380 was calculated, and the density correction coefficient was obtained so that the density of the asphalt pavement measured by the PQI 380 at the same location was consistent with the actual density value.

Figure 1. Three-dimensional ground-penetrating radar (GPR) inspection vehicle.

Figure 2. Non-nuclear density gauge (PQI 380).

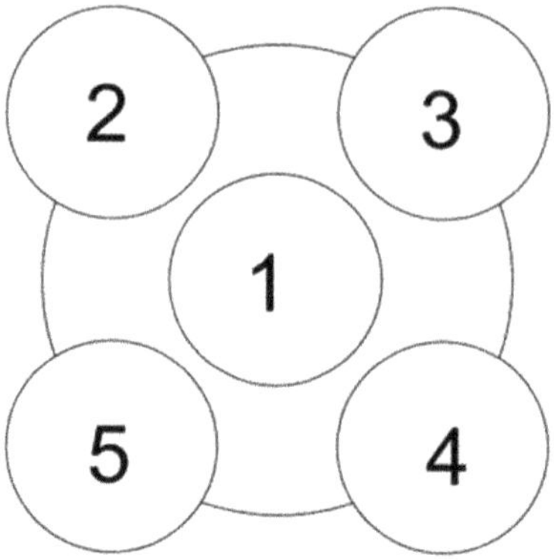

Figure 3. Calibration method of PQI 380.

2.3. Total Reflection Method

The total reflection method was as follows. First, a standard iron plate was placed under the air-coupled antenna, the total reflection signal generated by the GPR electromagnetic wave was collected through the iron plate, and then the electromagnetic wave was transmitted to the asphalt pavement [24,25]. When the electromagnetic wave encountered the layered road surface, a small part was scattered, and a part was received by the receiving antenna. The GPR antenna measured the peak-to-peak amplitude A_0 of the reflection on the road surface and the peak-to-peak amplitude A_p of the reflection from the iron plate, which are defined in Figure 4, and the dielectric constant of the first layer $\varepsilon_{r,1}$ was calculated as follows:

$$\varepsilon_{r,1} = \left(\frac{1 + \frac{A_0}{A_P}}{1 - \frac{A_0}{A_P}} \right)^2. \tag{1}$$

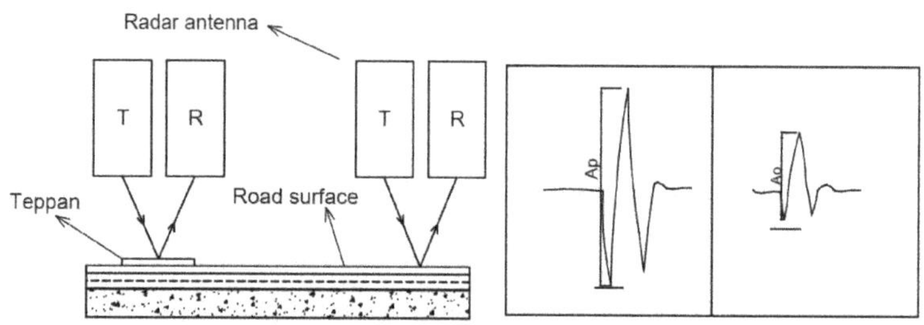

Figure 4. Schematic diagram of total reflection detection method of three-dimensional ground-penetrating radar.

3. Selection of Representative Values for Ground-Penetrating Radar Dielectric Constants

Asphalt mixes are made up of asphalt, aggregate, and filler; the dielectric constant of asphalt is generally in the range of 2.6–2.8, the dielectric constant of aggregate is generally in the range of 4.5–6.4, and the dielectric constant of air is 1. A dielectric constant of the asphalt mixture pavement needs to be assumed to be an overall dielectric constant [26]. The optimal dielectric constant value method can accurately determine the porosity of the pavement, better predict early issues with the road, and enable the asphalt mixture pavement to have a longer service life.

3.1. Design of Scheme for Value of Dielectric Constant for Asphalt Mixture Pavements

The air-coupled antenna used in this study had 21 channels, with 7.5 cm between them, as shown in Figure 5. Each of the 21 channels could collect the corresponding dielectric constants. Therefore, choosing the average value of the appropriate number of channel data as the representative value of the dielectric constant of the pavement at this time is of great significance for evaluating the construction quality of asphalt pavement. The asphalt layer was examined using 3dr-Examiner data analysis software, and the reflection amplitude of the asphalt layer and the reflection amplitude of the iron plate were obtained. The amplitude total reflection method [25] was used to calculate the dielectric constant of the asphalt pavement and to measure the porosity of the asphalt pavement at this time. By fitting the representative values of the dielectric constant and porosity, a more accurate method of determining the dielectric constant was obtained, and a corresponding dielectric constant–porosity prediction model was established.

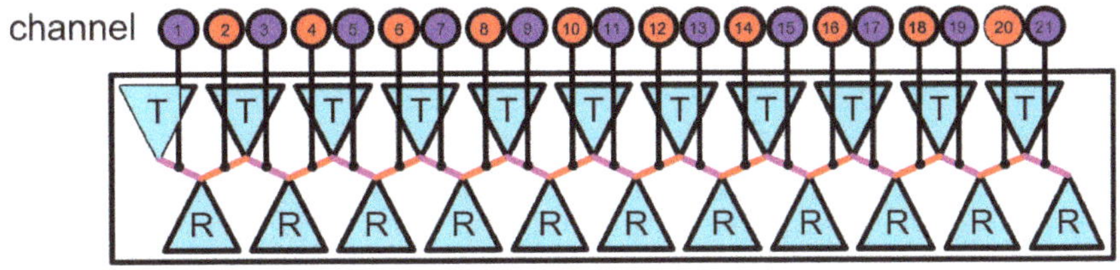

Figure 5. Internal vibrator arrangement of air-coupled antenna of three-dimensional ground-penetrating radar.

In this study, a test Section 5 m in length and 1.5 m in width was designed in the front of the test section, and a grid of squares with an interval of 0.5 m was drawn in the middle section of this road. First, the non-nuclear density gauge was calibrated, and then one data point was collected in an area of 0.5 m × 0.5 m. Since the acquisition interval in the radar length direction was 0.05 m, every set of 10 data points was averaged in the radar length direction. Three methods for determining the dielectric constant in the radar width direction were examined: Method 1: the dielectric constant of the middle channel of every seven channels in the radar width direction was determined, and a total of 10 data points were averaged; Method 2: the average value of the dielectric constant of the middle three channels of every seven channels was calculated in the radar width direction, and a total of 30 data points were averaged; Method 3: the average value of the dielectric constant for every seven channels was taken in the radar width direction, so a total of 70 data points were averaged, as shown in Figure 6. The dielectric constants measured by the three methods were fitted to the porosity measured by the non-nuclear density gauge to determine which method gave the best fit.

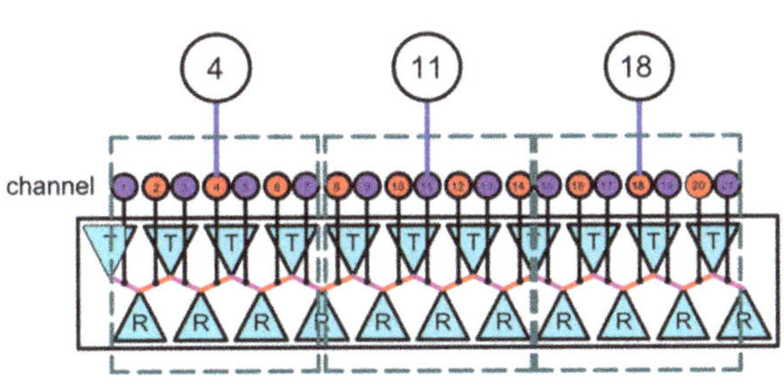

(a)

Figure 6. *Cont.*

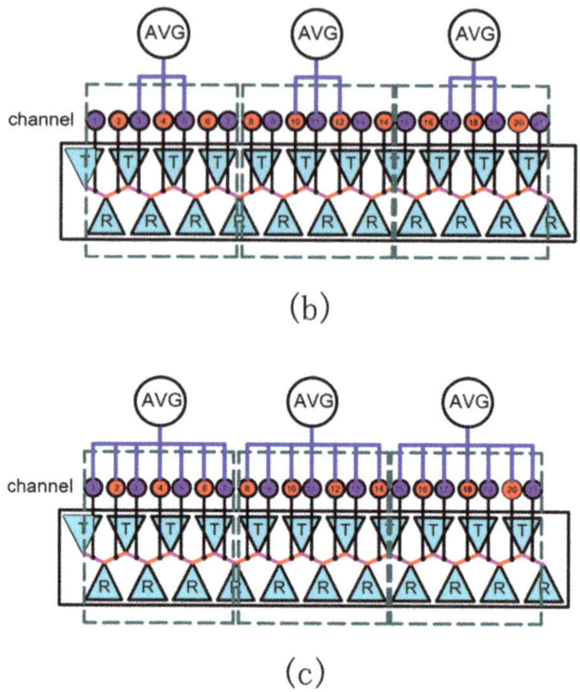

Figure 6. Experimental images of asphalt mixture tests used in this research: (**a**) Method 1, (**b**) Method 2, and (**c**) Method 3.

3.1.1. Method 1

Based on the calculated average of the 10 dielectric constant data points in Method 1 and the one porosity value measured by the non-nuclear density gauge, the best fit was obtained. The specific fit is shown in Figure 7.

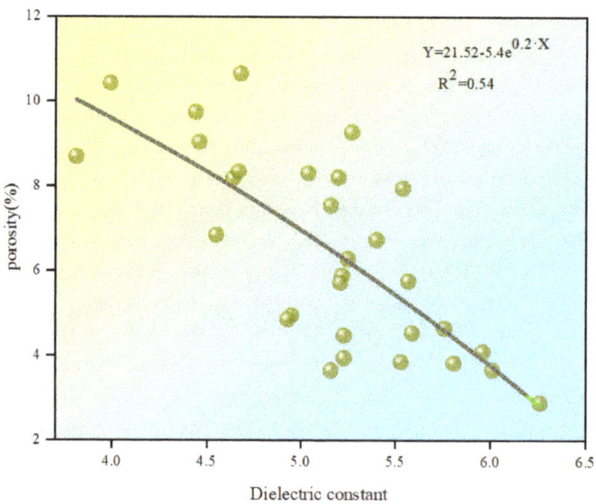

Figure 7. Fitting of the representative values of the dielectric constant obtained by Method 1 and the porosity.

The fitting equation was $Y = 21.52 - 5.4e^{0.2 \cdot X}$, with $R^2 = 0.54$. Based on the low correlation coefficient, it was determined that the fit of the dielectric constant calculated by Method 1 and the porosity was poor. This was mainly because the effective detection width of the radar antenna was 1.5 m, which was equally divided by 21 channels, and the average detection width per channel was 7.14 cm. If only the dielectric constants of the four, 11, and 18 channels were used as representative values, the average detection width per channel was 50 cm. Its effective detection width was diluted by 700.28%, so the dielectric constant–porosity fit was poor.

3.1.2. Method 2

The diameter of the non-nuclear density gauge test chassis was approximately the length of three channels, so a second method of taking the dielectric constant was developed. Data were collected every 10 channels along the length of the test section and the middle three channels of every seven channels in the width direction. A total of 30 data points were averaged, and the average values obtained and the porosity values measured by the non-nuclear density gauge were fitted, as shown in Figure 8.

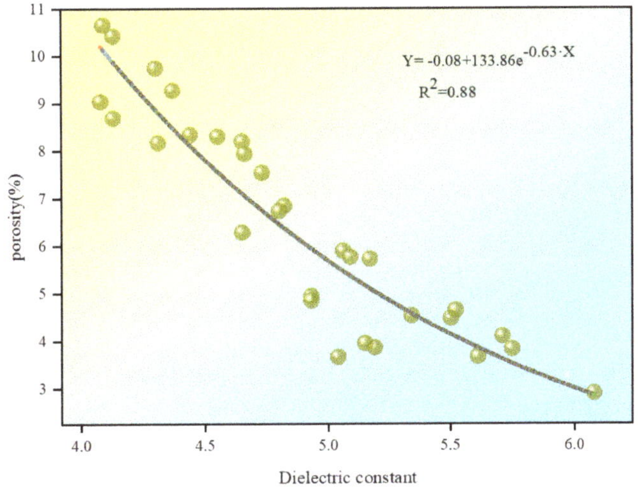

Figure 8. Fitting of the representative values of the dielectric constant obtained by Method 2 and the porosity.

As shown in Figure 8, the fitted model showed a high correlation with the average values of the dielectric constants collected from channels 3, 4, 5, 10, 11, 12, 17, 18, and 19 and the porosity data. The fitted equation was $Y = -0.08 + 133.86e^{-0.63 \cdot X}$, with $R^2 = 0.88$. Mainly during the process of GPR data being collected by the radar antenna and sent to the radar computer for analysis and interpretation, propagation parameters such as the wave velocity, attenuation coefficient, and wave impedance were collected from the scattering data to obtain the quantitative information required for detection. The close spacing of the transmitting and receiving antennas resulted in the unique oscillation characteristics of the GPR data. The area close to the transmitting antenna, where a rapidly decaying low-frequency energy field was generated due to the presence of an electrostatic field and induced electric field (e.g., channels 8 and 9 and passages 13 and 14, channels 6 and 7 and passages 15 and 16), also produced similar oscillations, but relatively weakly. This low-frequency energy field often resulted in a torsional component in the received signal, making the echo signal more variable, and the dielectric constant obtained after unified processing could exhibit large fluctuations and reduced accuracy. This is why the dielectric constants of the middle three channels were chosen to be averaged to avoid this issue.

3.1.3. Method 3

In the third approach, the average value of all the data collected by the GPR in an area of 0.5 m × 0.5 m with a total of 70 dielectric constants was calculated. The fit of the average value of the dielectric constant to the corresponding porosity data is shown in Figure 9.

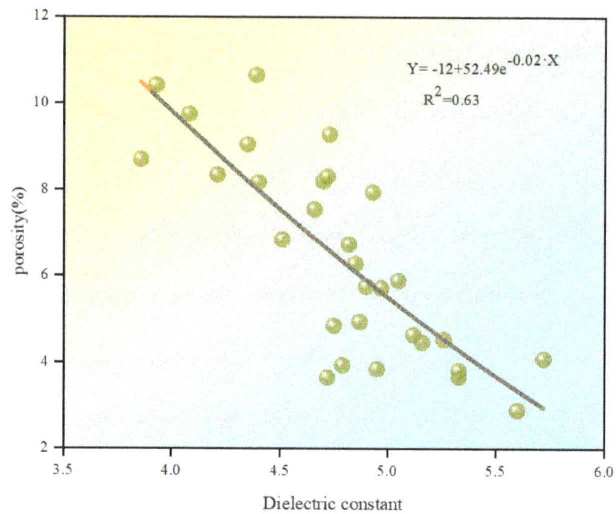

Figure 9. Fitting of the representative values of the dielectric constant obtained by Method 3 and the porosity.

The fitting equation of the average value of the dielectric constant of each seven channels and the porosity measured by the PQI 380 was $Y = -12 + 52.49e^{-0.02 \cdot X}$, with $R^2 = 0.63$. Theoretically, denser test points would better reflect the actual dielectric constant of the asphalt pavement, but this was not observed. The analysis showed that the GPR ground detection exhibited a rapid attenuation of the radar signal as the electromagnetic signal emitted downward from the radar antenna entered the asphalt pavement. As the depth of detection increased, the signal decayed more rapidly. These reflected signals were transmitted back to the radar host via the receiving antenna and needed to be corrected before they could be output. Hence, compensation processing was needed. The general decay of the signal amplitude over time is called the time gain. The time gain process is nonlinear, and a uniform physical model function is required to determine the time gain function. Thus, the transmission times of the radar signals on channels 1, 2, 20, and 21 and channels 10–13 were not the same, and the filtering treatment before and after the time gain treatment changed accordingly, resulting in a large error between the calculated dielectric constant after averaging and the actual dielectric constant of the pavement. Thus, the fitted function correlated poorly with the data.

3.2. *Determination of Dielectric Constant Value Method of Asphalt Mixture Pavement*

Table 1 can be obtained by summarizing the dielectric constant–porosity formulas obtained from the above test schemes. After a comprehensive analysis, due to the oscillations and time gain processing during the GPR data processing analysis, the correlation between the empirical prediction model relating the dielectric constant and porosity did not become more significant as more dielectric constant data were collected. In this test, the average dielectric constant measured by Method 2 was better correlated with the porosity.

Table 1. Statistical table of dielectric constant–porosity fit data.

	Method 1	Method 2	Method 3
Fitting equation	$Y = 21.52 - 5.4e^{0.2 \cdot X}$	$Y = -0.08 + 133.86e^{-0.63 \cdot X}$	$Y = -12 + 52.49e^{-0.02 \cdot X}$
R^2	0.54	0.88	0.63

4. Asphalt Pavement Construction Quality Evaluation Study

4.1. Engineering Background

To better evaluate the construction quality of asphalt pavement, it is necessary to obtain a large amount of dielectric constant and porosity data. The test was carried out on the intermediate layer of an AC-20 asphalt mixture on a new test road from K12+600 to K12+650.

4.1.1. Materials

The test road section used styrene–butadiene–styrene (SBS, I-D)-modified asphalt according to the Chinese standard JTG E20-2011 [27]. The technical indicators of the asphalt were tested, and the specific technical indicators are shown in Table 2.

Table 2. Styrene–butadiene–styrene (SBS, I-D)-modified asphalt technical specifications.

Items	Test Values	Specification [28]
Penetration (25 °C, 0.1 mm)	52	40–60
Ductility (5 °C, cm)	29.3	≥20
Softening point (°C)	74	≥60
Kinematic viscosity (135 °C, Pa.s)	1.80	≤3

The coarse and fine aggregates were made from high-quality limestone and tested for each functional index according to the Chinese standard JTG F40-2004 [28]. The specific indicators are shown in Tables 3 and 4.

Table 3. Properties of coarse aggregate.

Items	Unit	Limestone Test Results	Specification [28]
Apparent relative density	g/cm^3	2.686	≥2.50
Water absorption	%	0.97	≤3.0
Crush value	%	23.1	≤28
Abrasion value	%	21.6	≤30
Soundness	%	7.0	≤12

Table 4. Properties of fine aggregate.

Items	Unit	Test Values	Specification [28]
Apparent relative density	g/cm^3	2.726	≥2.5
Sand equivalent	%	78	≥60
Soundness	%	14	≥12
Angularity	s	49	≥30

4.1.2. Grading Design of Asphalt Mixes

The optimum amount of bitumen for the intermediate layer in the asphalt mixture for this section was determined by Marshall testing to be 4.3%. The maximum theoretical relative density was 2.539 g/cm^3, and the gradation composition is shown in Table 5.

Table 5. AC-20 asphalt mixture grading design.

Items	Percentage of Mass Passing the following Sieve Holes (%)										
	19 mm	16 mm	13.2 mm	9.5 mm	4.75 mm	2.36 mm	1.18 mm	0.6 mm	0.3 mm	0.15 mm	0.075 mm
Upper limit of gradation	100	92	80	72	56	44	33	24	17	13	7
Lower limit of gradation	90	78	62	50	26	16	12	8	5	4	3
Production mixture ratio	94.6	84.4	74.4	59.2	37.6	25.2	17.8	13.0	9.8	7.5	5.4

4.2. Experimental Design

The length of the designed test section was 50 m, and the width was 7.5 m. The test section was divided into five measurement lanes, each 50 m long and 1.5 m wide, and a grid was drawn with an interval of 0.5 m. In the length direction of the test road, the measurement lane could be divided into 100 grids, and 100 data points could be collected. In the direction of the width, the test section was divided into 15 square grids of 0.5 m × 0.5 m, and 15 data points were collected. The specific detection path of the PQI 380 is shown in Figure 10. According to the design method, 1500 data points were collected on the test section, and the detection interval was 0.5 m.

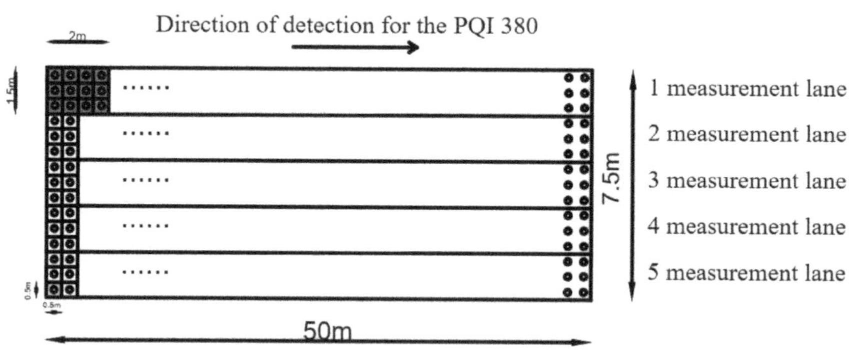

Figure 10. PQI 380 detection distribution map.

The detection of the GPR is shown in Figure 11. The best detection width of the GPR was 1.5 m, the collection interval was 5 cm, and five measurement lanes were designed. Each measuring channel was 50 m long and 1.5 m wide. In the detection width direction, there were five measurement lanes and 21 channels per lane, and the total width was 7.5 m. Thus, 105 data points could be collected. According to the optimum value scheme for the dielectric constant of the asphalt pavement, in a range of 0.5 m × 0.5 m, the radar could collect 3 × 10 valid data points. This method is accurate but requires selecting the required channel data from a large number of GPR data, which is too inefficient. In order to evaluate the construction quality of asphalt pavement more accurately and efficiently, each measurement lane was divided into 25 partial areas, each with a size of 2 m × 1.5 m. The construction quality of the asphalt pavement was mainly evaluated from the local heavy segregation of the asphalt mixture pavement.

Based on related research and actual engineering experience [29], the area where the porosity of the intermediate layer of the test engineering section was more than 6.5% was defined as severe segregation. Only severe segregation of the asphalt pavement was considered. The 0.5 m × 0.5 m grid drawn during the detection by the PQI 380 was combined into a 2 m × 1.5 m local cell, which corresponded to the local area measured by the GPR. The quality of the construction of the asphalt pavement was first evaluated using data collected by a PQI 380, and the GPR collected values according to Method 2. The fitted dielectric constant–porosity model was obtained, and the quality of the asphalt pavement construction was evaluated. If the findings from the two types of nondestructive testing (NDT) methods were the same, then the above evaluation system could be used to evaluate the quality of the construction of asphalt pavement.

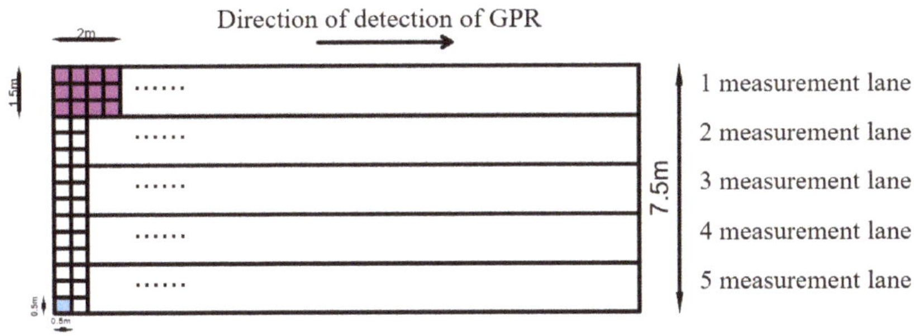

Figure 11. GPR detection distribution map.

4.3. Evaluation of Asphalt Pavement Construction Quality Based on PQI 380

Calibration must be carried out before using the PQI 380 to detect the porosity of the intermediate layer from K12+600 to K12+650. First, a dry and clean area of approximately 1.5 m × 0.75 m was selected on the test section. The area was divided equally into three circles, and data were measured at five locations around each circle using the five-point method of the PQI 380. Then, the height of the core sample at the center of each circle and the laboratory density were measured, and the measured density of the core sample was divided by the density measured by the PQI 380 to obtain the correction coefficient, as shown in Table 6. The 1500 data points collected were corrected by a correction factor, and 12 small grids with lengths of 0.5 m and widths of 0.5 m were combined based on the design plan into a local cell that was 2 m long and 1.5 m wide. It was suspected that severe segregation occurred when the porosity of the intermediate layer of this test section was greater than 6.5%, at which time there would be a certain percentage of detection points with porosities greater than 6.5% at the corresponding position on the asphalt pavement. If the locations of these inspection points were in the same area, it could indicate that the construction quality problem in this area was more severe. The ratio obtained by dividing the number of small grids with porosities greater than 6.5% in a local cell by 12 was defined as the proportion of local severe segregation. The smaller the value of local severe segregation, the better the construction quality became, and vice versa.

Table 6. PQI 380 calibration.

	Core Sample 1	Core Sample 2	Core Sample 3
Center Point 1	2.3725	2.3651	2.3656
Center Point 2	2.3967	2.3854	2.3895
Center Point 3	2.4661	2.3962	2.3864
Center Point 4	2.3732	2.3979	2.3997
Center Point 5	2.3822	2.3921	2.4123
Average	2.3981	2.3873	2.3907
Measured core density	2.396	2.413	2.446
Correction factor	0.9991	1.0107	1.0231
Average of correction factors		1.011	

As shown in Figure 12, when the distance was 0–26 m, most of the local severe segregation in measurement lane 1 was above 0.5, showing a higher horizontal fluctuation trend. The distance was 26–28 m, with a decreasing trend. In the interval between 28 and 50 m, the proportion of local severe segregation fluctuated at a relatively low level. The mean value was 0.39, the extreme difference was 0.67, the maximum value was 0.75

(the horizontal range was 24–26 m), and the minimum value was 0.08 (the horizontal range was 32–34 m). The analysis showed that the pavement had a large porosity at distances of 8–12 m, 16–20 m, and 22–26 m, which may have been due to construction quality problems caused by uneven paver discharge and requires attention when adding the upper surface layer.

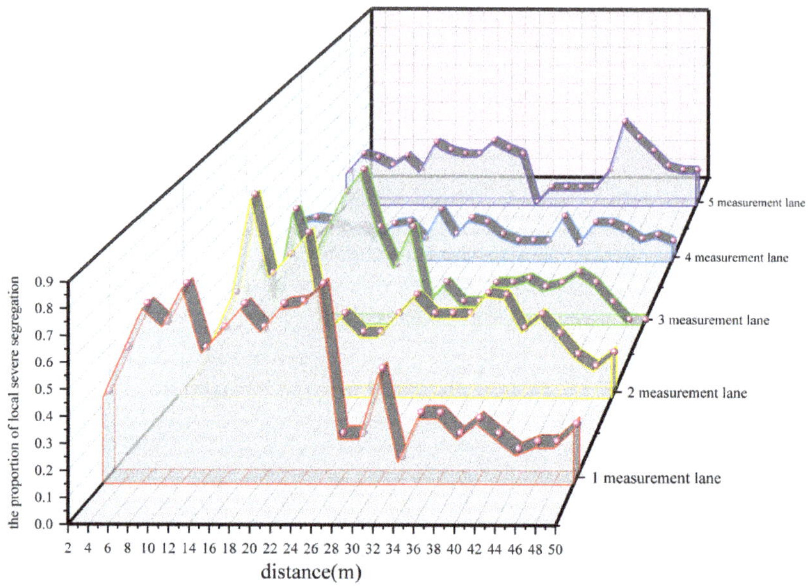

Figure 12. Proportion of local severe segregation on asphalt pavement.

In measurement lane 2, the proportion of local severe segregation at a distance of 0–10 m showed an upward trend, and the peak value was higher. In the range of 10–18 m, there was a step-down trend. Within a distance of 18–50 m, there was a fluctuating trend at lower levels. The mean value was 0.33, and the extreme difference was 0.83, with a maximum value of 0.83 (the horizontal range was 8–10 m) and a minimum value of 0 (the horizontal range was 0–2 m). The analysis showed that at distances of 8–12 m and 14–16 m, the pavement had a large porosity and heavy segregation, and the construction quality of the pavement was poor.

In measurement lane 3, the proportion of local severe segregation at a distance of 0–16 m underwent a step-like upward trend. At a distance of 16–24 m, there was a step-down trend. At distances of 24–42 m, the trend fluctuated at lower levels. There was a decreasing trend in the distance range of 42–50 m. The mean value was 0.23, and the extreme difference was 0.68, with a maximum value of 0.68 (in the horizontal range of 14–16 m) and a minimum value of 0 (in the horizontal range of 46–50 m). The analysis showed that, at distances of 8–12 m and 14–16 m, the pavement had a large porosity and heavy segregation, and the construction quality of the pavement was poor.

In measurement lane 4, the overall detection of the proportion of local severe segregation in the region showed a steady fluctuating trend with no particularly prominent points. The mean value was 0.14, the extreme difference was 0.16, the maximum value was 0.23 (the horizontal range was 34–36 m), and the minimum value was 0.07 (the horizontal range was 36–38 m). The overall porosity of measurement lane 4 was under control, and there were few areas where heavy segregation occurred, so the construction quality was good.

In measurement lane 5, the proportion of local severe segregation at a distance of 0–26 m fluctuated at a level near 0.2. When the distance was 26–28 m, it showed a sudden downward trend. A rising trend appeared at distances of 28–40 m. In the range of 40–50 m, there was a downward trend. The mean value was 0.21, and the extreme difference was

0.42, with a maximum value of 0.42 (in the horizontal range of 38–40 m) and a minimum value of 0 (in the horizontal range of 26–28 m). The porosity of the asphalt pavement was greater at distances of 0–26 m and 38–46 m than in the corresponding areas of measurement lane 4, which had a greater degree of heavy segregation. Thus, the construction quality was worse than that in measurement lane 4.

The mean values of the proportion of local severe segregation for the above five measurement lanes were in the order of measurement: lane 4 < measurement lane 5 < measurement lane 3 < measurement lane 2 < measurement lane 1. Most of the data from channel 4 were smaller than the those in the other four channels and more evenly distributed throughout the channel, with less severe segregation. The extreme differences in the proportion of local severe segregation for the measurement lanes were in the order of measurement: lane 4 < measurement lane 5 < measurement lane 1 < measurement lane 3 < measurement lane 2. The range of fluctuations in the proportion of local severe segregation was minimal for measurement lane 4 and was more evenly distributed throughout the measurement lane. A comprehensive analysis of the results of the five test tracks showed that the construction quality of measurement lane 4 was the best and that the construction quality of measurement lanes 1 and 2 was poor.

4.4. Evaluation of Asphalt Pavement Construction Quality Based on GPR

The object of the GPR study was also the intermediate layer from K12+600 to K12+650. The optimal detection width for GPR is 1.5 m, and the optimal acquisition interval is 0.05 m. The designed detection method is shown in Figure 11. On the test road, the GPR detected a total of five measurement lanes, of which 21 data points could be detected in the direction of the width of each measurement lane. One measurement lane of the GPR could collect up to 1000 data points in each measurement lane length. Within the detection range of 0.5 m × 0.5 m of the PQI 380, the GPR could collect 30 data points. To evaluate the construction quality of the intermediate layer in asphalt pavement more accurately, a partial area of 1.5 m × 2 m was used as the study area for the dielectric constant of the asphalt pavement, which corresponded to the partial area merged with the PQI 380, and the average value of the local dielectric constant was used to evaluate the construction quality of the asphalt pavement.

In this study, areas with porosities greater than 6.5% were defined as severely segregated. By fitting the calculated dielectric constant–porosity prediction model based on the second method of taking the dielectric constant and converting it, it was concluded that severe segregation occurred when the dielectric constant of the asphalt pavement was greater than 5.167. If the data points with a dielectric constant greater than 5.167 were too concentrated in one area, it could be assumed that there was a serious quality problem.

According to Figure 13, the average value of the local dielectric constant for measurement lane 1 fluctuated relatively widely over a distance of 4–36 m. There was essentially no severe segregation in the range of 36–50 m. The overall mean value of the average value of the local dielectric constant values was 4.7596, with a maximum value of 6.59 and a minimum value of 3.5. The area of severe segregation detected by the GPR essentially covered the area of severe segregation detected by the PQI 380. At a distance of 34–36 m, the PQI 380 did not reflect the severe segregation in this area due to the limited number of data points that it collected.

In measurement lane 2, the average value of the local dielectric constant was relatively small over a distance of 0–8 m. The average values of the local dielectric constant values were relatively large (severe segregation) over distances of 8–16 m and 26–46 m. The overall mean of the average values of the local dielectric constants was 4.8584, the maximum value was 7.41, and the minimum value was 3.73. The large variations in the overall mean value of the dielectric constant could indicate that the construction quality of this test measurement lane was unstable. This was consistent with the determination of the PQI 380.

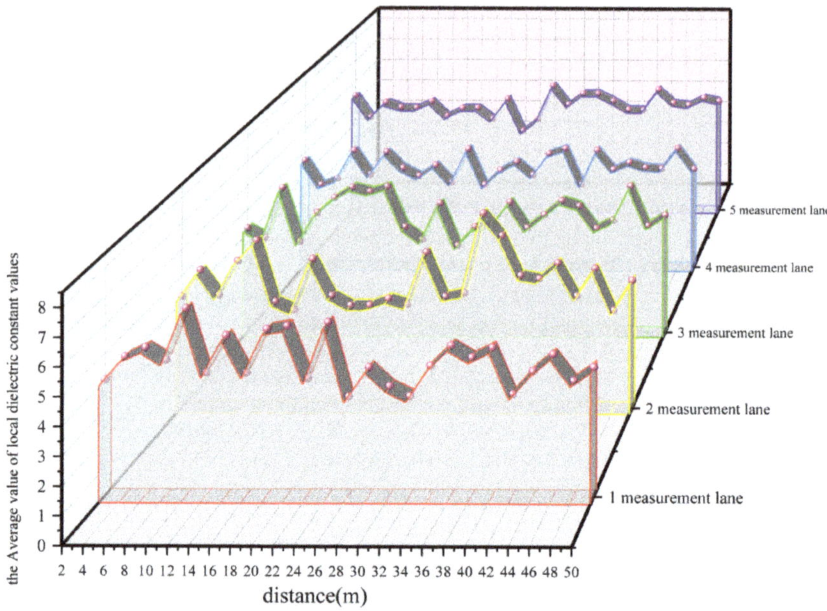

Figure 13. Distribution of the average values of the local dielectric constants in asphalt pavements.

In measurement lane 3, the average value of the local dielectric constant within the range of 6–20 m fluctuated significantly, and there were many severe segregation phenomena. In the range of 32–38 m and 42–46 m, the evaluation index of the PQI 380 did not reflect the severe segregation at this specific location. This may have been because the PQI 380 only measured data at the center of a 0.5 m × 0.5 m square grid, and edge locations were not examined. It is possible that the midpoint value was not representative of the entire small grid with severe segregation.

The average values of the local dielectric constant values for measurement lane 4 were generally stable with small fluctuations. The overall average was 4.43, the maximum value was 5.25, and the minimum value was 3.59. These are the same results as those obtained by the PQI 380.

The average values of the local dielectric constant values for measurement lane 5 were large. The average over a distance of 30–42 m was 5.15, which was close to the dielectric constant of 5.17 when severe coarse segregation occurred. The overall mean of the average values of the local dielectric constant values was 4.91, the maximum value was 5.66, and the minimum value was 3.7. Although some of the local mean values of the dielectric constants for measurement lane 5 exceeded 5.17, they were more stable overall, with some fluctuations. The data points were evenly distributed throughout the measurement lanes, and the overall construction quality was better than that for measurement lane 2.

Based on Figures 12 and 13, the test results of the nucleus-free density meter designed to evaluate the quality of asphalt pavement construction were compared to the test results obtained by GPR and verified, and the same test conclusions were found. This demonstrated that GPR and the PQI 380 could be used to evaluate the asphalt pavement construction quality more quickly, efficiently, and accurately. The GPR used a full detection range and was more efficient. The PQI 380 used a single point of detection; its detection data were accurate but inefficient. In an area of 0.5 m × 0.5 m, the PQI 380 could only collect data from one midpoint, and the edges could not be collected. For example, no construction quality problems were detected between 34 and 36 m on measurement lane 1. Furthermore, no quality problems were detected between 32 and 38 m and between 42 and

46 m on measurement lane 3. However, the actual pavement had a large porosity, and the construction quality was unstable.

When carrying out actual asphalt pavement construction quality inspection, if the inspection area is small, the construction quality of the asphalt pavement can be evaluated directly using a PQI 380 based on the proportion of local severe segregation. When the area of the pavement to be tested is large, first, a 5-m test section is selected and porosity data are collected and analyzed using the PQI380. GPR is used to collect the dielectric constant data based on the second method of taking the dielectric constant and establishing a dielectric constant–porosity empirical prediction model. The GPR is then used to detect the whole road surface, and finally, the average value of the local dielectric constant values is used in the data processing process to evaluate the construction quality of the asphalt pavement.

5. Conclusions

In this paper, we proposed the use of a PQI 380 with GPR as a nondestructive testing tool for more efficient and accurate inspection of asphalt pavement construction quality. A nondestructive testing evaluation system for the quality of asphalt pavement construction was designed and validated, allowing the following conclusions to be drawn.

(1) The optimization of the dielectric constants collected by the GPR led to the identification of a new choice of values for the dielectric constants of the radar antenna channels. Every 10 data points along the length of the test were collected, and the average value of the dielectric constant of the middle three channels of every seven channels in the radar width direction was computed. A total of 30 data points were averaged.

(2) The data measured by the PQI 380 and GPR were analyzed and fitted to the dielectric constant and porosity data, and a corresponding dielectric constant–porosity prediction model was developed: $Y = -0.08 + 133.86e^{-0.63 \cdot X}$, with $R^2 = 0.86$. By analyzing and comparing different methods of obtaining the dielectric constant, this prediction model was found to provide the best fit.

(3) We proposed using the average value of the local dielectric constant to evaluate the construction quality of asphalt pavement in the results measured by GPR. For the detection of the PQI 380, we proposed using the proportion of local severe segregation to determine the problem of road segregation and further evaluate the construction quality of asphalt pavement. The results of the two NDT methods were identical: the best construction quality was found on measurement lane 4, and the worst construction quality was found on measurement lane 2.

Author Contributions: Writing—original draft, W.C.; data curation, G.H. and W.C.; methodology, X.X., W.H. and G.H.; project administration, W.H., X.Y., X.X. and X.Z.; Investigation, J.W., X.Y. and X.Z.; Supervision, J.W. All authors have read and agreed to the published version of the manuscript.

Funding: This research was funded by the Shandong Provincial Natural Science Foundation (grant number ZR2020QE272), the Shandong Jianzhu University Doctoral research foundation (grant number X18073Z), the National Key R&D Program of China (grant number 2018YFB1600100), and the Shandong Natural Science Foundation Committee (ZR2020QE271).

Institutional Review Board Statement: Not applicable.

Informed Consent Statement: Not applicable.

Data Availability Statement: Not applicable.

Acknowledgments: The authors would like to thank the Shandong Transportation Institute for their support.

Conflicts of Interest: The authors declare no conflict of interest.

References

1. Wang, L. Significance analysis of influencing factors of highway freight transportation in China and multi-variable grey prediction for its development. *J. Intell. Fuzzy Syst.* **2021**, *41*, 1237–1246. [CrossRef]
2. Sun, L.; Li, X.; Wang, X. Study on the damage of semi rigid base asphalt pavement under high temperature condition. In Proceedings of the 4th International Conference on Mechatronics, Materials, Chemistry and Computer Engineering 2015, Xi'an, China, 12–13 December 2015.
3. Tutu, K.A.; Timm, D.H. A Recursive Pseudo Fatigue Cracking Damage Model for Asphalt Pavements. *Int. J. Pavement Eng.* **2018**, 1–21. [CrossRef]
4. Zhi, Z.; Dai, C.S.; Guo, S.Y. Study on risk early warning system of asphalt concrete pavement design. In Proceedings of the 2013 International Conference on Management Science and Engineering (ICMSE), Harbin, China, 17–19 July 2013.
5. Nobakht, M.; Zhang, D.; Sakhaeifar, M.S.; Lytton, R.L. Characterization of the adhesive and cohesive moisture damage for asphalt concrete. *Constr. Build. Mater.* **2020**, *247*, 118616. [CrossRef]
6. Scullion, S.T. Road evaluation with ground penetrating radar. *J. Appl. Geophys.* **2000**, *43*, 119–138.
7. Solla, M.; González-Jorge, H.; Lorenzo, H.; Arias, P. Uncertainty evaluation of the 1GHz GPR antenna for the estimation of concrete asphalt thickness. *Measurement* **2013**, *46*, 3032–3040. [CrossRef]
8. Chamberlain, A.T.; Sellers, W.; Proctor, C.; Coard, R. Cave Detection in Limestone using Ground Penetrating Radar. *J. Archaeol. Sci.* **2000**, *27*, 957–964. [CrossRef]
9. Baltruaitis, A.; Vaitkus, A.; Smirnovs, J. Asphalt Layer Density and Air Voids Content: GPR and Laboratory Testing Data Reliance. *Balt. J. Road Bridge Eng.* **2020**, *15*, 93–110. [CrossRef]
10. Loizos, A.; Georgiou, P.; Plati, C. A comprehensive approach for the assessment of HMA compactability using GPR technique. *Near Surf. Geophys.* **2016**, *14*, 117–126.
11. Benedetto, A.; Tosti, F.; Ciampoli, L.B.; D'Amico, F. An overview of ground-penetrating radar signal processing techniques for road inspections. *Signal Process.* **2017**, *132*, 201–209. [CrossRef]
12. Zhang, J.; Yang, X.; Li, W.; Zhang, S.; Jia, Y. Automatic detection of moisture damages in asphalt pavements from GPR data with deep CNN and IRS method. *Autom. Constr.* **2020**, *113*, 103119. [CrossRef]
13. Ma, Y.; Elseifi, M.A.; Dhakal, N.; Bashar, M.Z.; Zhang, Z. Non-Destructive Detection of Asphalt Concrete Stripping Damage using Ground Penetrating Radar. *Transp. Res. Rec. J. Transp. Res. Board* **2021**, *2675*, 938–947. [CrossRef]
14. Loizos, A.; Plati, C. Accuracy of pavement thicknesses estimation using different ground penetrating radar analysis approaches. *NDT E Int.* **2007**, *40*, 147–157. [CrossRef]
15. Shangguan, P.; Alqadi, I.L.; Leng, Z.; Schmitt, R.; Faheem, A. An Innovative Approach for Asphalt Pavement Compaction Monitoring Using Ground Penetrating Radar. *Transp. Res. Rec. J. Transp. Res. Board* **2013**, *2347*, 79–87. [CrossRef]
16. Plati, C.; Loizos, A. Using ground-penetrating radar for assessing the structural needs of asphalt pavements. *Nondestruct. Test. Eval.* **2012**, *27*, 273–284. [CrossRef]
17. Shangguan, P.; Al-Qadi, I.L. Calibration of FDTD Simulation of GPR Signal for Asphalt Pavement Compaction Monitoring. *IEEE Trans. Geosci. Remote Sens.* **2014**, *53*, 1538–1548. [CrossRef]
18. Praticò, F.G.; Moro, A.; Ammendola, R. Factors Affecting Variance and Bias of Non-nuclear Density Gauges for Porous European Mixes and Dense-Graded Friction Courses. *Balt. J. Road Bridge Eng.* **2009**, *4*, 99–107. [CrossRef]
19. Van den Bergh, V.; Vuye, C.; Kara, P.; Couscheir, K.; Blom, J.; Van Bouwel, P. The use of a non-nuclear density gauge for monitoring the compaction process of asphalt pavement. *IOP Conf. Ser. Mater. Sci. Eng.* **2017**, *236*, 012014. [CrossRef]
20. Fernandes, F.M.; Pais, J.C. Virtual Special Issue Ground-Penetrating Radar and Complementary Non-Destructive Testing Techniques in Civil Engineering Laboratory observation of cracks in road pavements with GPR. *Constr. Build. Mater.* **2017**, *154*, 1130–1138. [CrossRef]
21. Keefe, B.J. Permanent International Association of Road Congresses (PIARC): Report on the XVII World Road Congress Held in Sydney, October 1983. *Qld. Roads* **1983**, *22*, 44.
22. Olhoeft, G.R. Applications and frustrations in using ground penetrating radar. *IEEE Aerosp. Electron. Syst. Mag.* **2002**, *17*, 12–20. [CrossRef]
23. Cho, Y.; Wang, C.; Zhuang, Z. Framework for Empirical Process to Improve Nonnuclear Gauge Performance in Hot-Mix Asphalt Pavement Construction. *J. Constr. Eng. Manag. ASCE* **2013**, *139*, 601–610. [CrossRef]
24. Sihvola, A.; Lindell, I.V. Polarizability and Effective Permittivity of Layered and Continuously Inhomogeneous Dielectric Spheres. *J. Electromagn. Waves Appl.* **1988**, *4*, 1–26. [CrossRef]
25. Al-Qadi, I.; Lahouar, S.; Loulizi, A. Successful Application of Ground-Penetrating Radar for Quality Assurance-Quality Control of New Pavements. *Transp. Res. Rec. J. Transp. Res. Board* **2003**, *1861*, 86–97. [CrossRef]
26. Wang, L.; Wang, Y.; Mohammad, L.; Harman, T. Voids Distribution and Performance of Asphalt Concrete. *Int. J. Pavements* **2002**, *1*, 22–33.
27. *JTG E20-2011*; Standard Test Methods of Bitumen and Bituminous Mixtures for Highway Engineering. Ministry of Transport, China: Beijing, China, 2011.
28. *JTG F40-2004*; Technical Specifications for Construction of Highway Asphalt Pavements. Ministry of Transport, China: Beijing, China, 2004.
29. Li, X.; Zhou, Z.; Lv, X.; Xiong, K.; Wang, X.; You, Z. Temperature segregation of warm mix asphalt pavement: Laboratory and field evaluations. *Constr. Build. Mater.* **2017**, *136*, 436–445. [CrossRef]

coatings

MDPI

Article

Properties of Different Waterproof Bonding Layer Systems for Cement Concrete Bridge Deck Pavement

Jiancun Fu [1,2,*], Aiqin Shen [1] and Zhaodi Yuan [2]

1 School of Highway, Chang'an University, Xi'an 710064, China; saq6305@163.com
2 Shandong Transportation Institute, Jinan 250002, China; yuanzhaodi365@163.com
* Correspondence: fujiancunhit@163.com

Abstract: The performance of waterproof bonding layer systems significantly affects the service life of concrete bridge deck pavement. This article aims to compare the bonding properties and composite structure mechanical properties of different waterproof bonding layer systems in bridge deck pavement to recommend appropriate structures based on service conditions. To fulfill this objective, four kinds of waterproof bonding material composite systems were designed, and test samples were fabricated. Then, the waterproof bonding layer properties were studied by penetration resistance tests, scalding resistance tests, and corrosion resistance tests. Pull-off tests were conducted to compare the bonding properties. Then, dynamic modulus tests, Hamburg wheel tracking tests, and static load creep tests of composite structure specimens were implemented to evaluate the dynamic and static mechanical performances and water-temperature stability. The findings showed that the four schemes had good impermeability. The pull-off strengths of scheme III at 25 °C were about 2.5-, 2.5-, and 4-times those of schemes I, II, and IV, respectively, and the pull-off strengths at 50 °C were about 1.6-, 2-, and 1.6-times those of schemes I, II, and IV, respectively, and the bonding performances and dynamic modulus of different schemes were ranked as follows: III > I > II > IV. The results of the Hamburg wheel tracking test were consistent with the static load creep test. The high-temperature stability of scheme IV was the best, while the stabilities of scheme III and I were better than that of scheme II. In conclusion, the scheme of concrete bridge deck pavement can be selected based on the environmental conditions and traffic load, and schemes III and I should be considered first. The findings of this research could provide technical support for the future design of bridge deck waterproof bonding systems.

Keywords: cement concrete; bridge deck pavement; waterproof bonding layer; properties; impermeability; dynamic modulus

Citation: Fu, J.; Shen, A.; Yuan, Z. Properties of Different Waterproof Bonding Layer Systems for Cement Concrete Bridge Deck Pavement. *Coatings* **2022**, *12*, 308. https://doi.org/10.3390/coatings12030308

Academic Editor: Andrea Nobili

Received: 25 December 2021
Accepted: 21 February 2022
Published: 24 February 2022

Publisher's Note: MDPI stays neutral with regard to jurisdictional claims in published maps and institutional affiliations.

Copyright: © 2022 by the authors. Licensee MDPI, Basel, Switzerland. This article is an open access article distributed under the terms and conditions of the Creative Commons Attribution (CC BY) license (https://creativecommons.org/licenses/by/4.0/).

1. Introduction

With increasing traffic volumes and axle loads and increasingly harsh weather conditions, serious early disease appears on bridge deck pavement, which significantly affects the function of expressways and induces traffic accidents [1–5]. The performance differences between asphalt concrete and cement concrete are significant, so the interface is often considered to be vulnerable zone. There will inevitably be an insufficient bond strength between the interlayer if there are no waterproof bonding layers between the asphalt concrete and the cement concrete. This will cause delamination, rutting, cracking, and other diseases on the bridge deck pavement, which will affect the safety of traffic [6–8]. According to the research and analysis of multi objective grey target decision theory, some scholars have obtained the priority relationship among shear performance, bond performance, and cost [9]. Therefore, this paper mainly analyzes the four schemes from the bonding properties and composite structure mechanical properties.

To strengthen the overall bonding performance of the bridge deck and pavement materials, a considerable amount of research has been conducted on the performances of waterproof bonding materials around the world [10–18]. Large numbers of investigations have shown that the bonding effect of the waterproof system under low and normal temperatures is the main factor affecting the layer performance. Under high temperatures, the working temperature of the bridge deck is close to or even reaches the softening point of the asphalt-based waterproof material. Therefore, during the construction process, the rolling action of the roller and the high temperature generated during paving will cause the waterproof layer to fail to varying degrees [19–22]. Kruntcheva et al. [23] showed that the type of contact between the layers has a significant influence on the performance of the interface through dynamic and static tests. It is recommended to introduce vertical and horizontal shear response modulus to describe the stiffness and strength of the waterproof bonding layer. Xu et al. [24] used laboratory direct shear and pull-out tests to test the interface bonding strength of the waterproof bonding layer. Guo et al. [25,26] selected three different waterproof bonding layers to do waterproof and shear strength tests, and the results showed that SBS-modified asphalt was better. Zhang et al. [27] found that modified asphalt had the highest bonding strength to concrete slabs through experiments, followed by ordinary asphalt and then emulsified asphalt. Guo et al. [28] studied the influence of factors such as the drawing rate, roughness, thickness, temperature, and humidity on the drawing strength and shear strength of waterproof bonding layers. The research results showed that the construction quality of the waterproof bond system was greatly affected by temperature and humidity. The construction temperature should be 25 °C, and the relative humidity should not exceed 60%. Zhang et al. [10] analyzed the interlaminar performance of waterproof and cohesive materials for concrete bridge deck via oblique shear tests, pull-off tests, and shear tests after loading under specific test conditions. The results showed that the order of the shear failure life of the selected waterproof and cohesive materials is SBS-modified asphalt were superior to those of asphalt rubber (19%). For the bridge deck paving–waterproof bonding layer, not only the performance of the waterproof adhesive material itself needs to be tested, but the composite structure of the bridge deck paving system also needs to be tested and evaluated, which will help to determine the overall anti-destructive ability of the pavement. Ai et al. [29] carried out a direct tensile, direct shear, and 45-degree skew shear tests at different temperatures (0, 25, and 70 °C) on specimens to compare the bonding properties of different waterproof bonding material composites. The results showed that the composite specimens with the best direct tensile strength and fatigue life used CILBOND®55C as the upper bonding materials and CILBOND®62 W as the lower bonding materials. Qiu et al. [9] used a waterproof bonding system consisting of two layers, a waterproof layer and a bonding layer, to ensure that the deck pavement and the steel panel could form a monolithic structure. They conducted a series of laboratory experiments that demonstrated that applying a waterproof bonding system improved the fatigue life of the bridge deck pavement significantly.

In summary, there are many types of waterproof bonding layers on the market at present. Although there have been a large number of experiments to detect the waterproof bonding layer [30–35], most of the relevant research has focused on optimizing a specific aspect of its performance, such as its shear strength or fatigue performance. Waterproof bonding layers have different influences on the upper paving structure, so the whole bridge deck pavement system's mechanical properties and temperature stability are changed However, the previous research on the properties of different cement concrete bridge deck pavement waterproof bonding layer systems was not comprehensive, which limits the structural design of waterproof bonding layer of concrete bridge deck. Therefore, four commonly used cement concrete bridge-deck waterproof-bonding-material composite systems were designed and evaluated comprehensively in this paper, and the research framework is shown in Figure 1. The waterproof bonding layer properties were studied by impermeability tests, scalding resistance tests, and corrosion resistance tests. The interface bonding performances were studied by pull-out tests at two different test temperatures (25 and 50 °C). The properties of the composite bridge deck pavement structure were tested by Hamburg wheel track tests, dynamic modulus tests, and static load creep tests. Based on this study, the scheme of concrete bridge deck pavement can be selected based on the environmental conditions and traffic load, which is of great significance for the structure design of cement concrete bridge deck pavement.

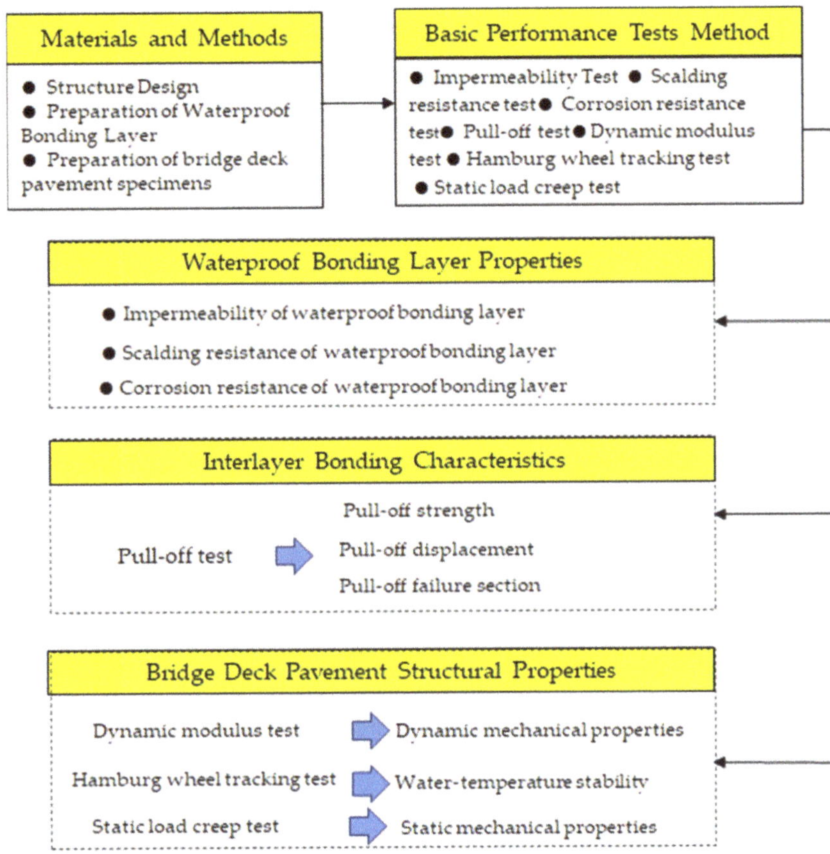

Figure 1. The research flowchart.

2. Experiment Program

2.1. Materials and Methods

In order to compare the bonding properties and composite structure mechanical properties of different waterproof bonding layer systems of bridge deck pavement, a deep research study was conducted. Firstly, the four waterproof bonding layers and composite bridge deck pavement schemes were designed. Secondly, the preparation of waterproof bonding layer and the technical indices of component materials were introduced. Then, the four composite bridge deck pavement specimens were prepared based on the cement prefabricated slabs with dimensions of 32 cm × 26 cm. The details are described in the following chapter.

2.1.1. Structure Design of Bridge Deck Pavement

The four commonly used waterproof bonding layers and composite bridge deck pavement schemes were designed and are shown in Table 1. About 10 kinds of materials were examined, including styrene–butadiene–styrene (SBS) modified asphalt, modified emulsified asphalt, and protective board. Asphalt sand (AC–5), pouring asphalt concrete (GA–10), and stone matrix asphalt (SMA–13) were used in the different schemes, the aggregate gradation of the three mixtures is shown in Table 2. The technical indexes of

AC-5, SMA-13 meet the requirements of JTG F40-2004, technical indexes of GA-5 comply with standard technical regulations of "Design and Construction of Pavement on Highway Steel Deck Bridge" (JTG/T 3364-02-2019).

Table 1. Cement concrete bridge deck systems.

Scheme	Waterproof Bonding Layer	Upper Bridge Deck Pavement Structure
I	Gravel asphalt sealing layer + 20 mm AC–5 + modified emulsified asphalt	60 mm SMA–13
II	5 mm rubber asphalt + 3.6 mm protective board + modified emulsified asphalt	70 mm SMA–13
III	GIS–I primer + 30 mm GA–10	50 mm SMA–13
IV	1.2 mm fiber-reinforced waterproof coating (five–layer spraying)	80 mm SMA–13

Table 2. Aggregate gradation of the three mixtures.

Sieve Size (mm)	16.0	13.2	9.50	4.75	2.36	1.18	0.60	0.30	0.15	0.075	Binder Content/%
AC–5	100	100	100	95	65	45	30	20	12.5	7.5	6.1
GA–10	100	100	90	71.5	55.5	45	39	33.5	30	25	7.3
SMA–13	100	81.6	61.2	24.2	19.9	16.8	14.5	12.4	10.9	9.8	6.2

2.1.2. Preparation of Waterproof Bonding Layer

(1) Scheme I

First, the SBS-modified asphalt was sprayed with the amount of 1.2 L/m^2. Second, a layer with a thickness of 5–10 mm was spread with 3–5‰ asphalt ready-mixed crushed stone added. Third, when the sealing layer was complete, the AC–5 mixture was constructed by a linear rubbing test machine. Finally, the SBS–modified emulsified asphalt bonding layer was sprayed with a distribution amount of 1.0 kg/m^2.

The technical indices of the SBS–modified asphalt complied with the requirements of the I–D standard JTG F40–2004. The modified emulsified asphalt met the technical requirements of the specifications of JTG–2004.

(2) Scheme II

First, rubber asphalt was coated with a controlled thickness of 5 mm. The rubber asphalt was different from the conventional rubber asphalt, which was processed through the composite modification process and had a high elasticity, viscosity, and toughness [36]. The test indices of rubber asphalt are shown in Table 3.

Table 3. Technical indices of rubber asphalt.

Project	Measured Value	Test Method
Penetration at 25 °C,100 g, and 5 s (0.1 mm)	47.5	T0604
Fluidity at 60 °C (Pa·s)	2.3	ASTM D5329
Softening point (°C)	89.5	T0606
G*/sinδ (with the pressure of 4800 Pa and rotation speed of 8 rad/s)	18.7	T0628
G*/sinδ (with the pressure of 1600 Pa and rotation speed of 10 rad/s)	4.95	T0628

The protective plate impregnated with asphalt was placed, which mainly protected the rubber asphalt. The protective plate with the size of 30 mm × 30 mm × 4 mm was used in the laboratory. Finally, the modified emulsified asphalt bonding layer was sprayed with a distribution of 1.0 kg/m^2.

(3) Scheme III

The waterproof bonding layer was mainly composed of a GIS–I solvent-based adhesive coating and GA–10 waterproof layer. The GIS–I was coated first and then the GA–10 was constructed. The pouring asphalt concrete had a small porosity and a good waterproofing function. Due to the high asphalt content, it exhibited excellent interface contact performances. The GIS–I solvent-based adhesive was a single-component solvent-based waterproof adhesive formed by melting asphalt and a variety of polymer-resin-based additives in organic solvents through special processing technology. Table 4 shows the drying and curing times of the GIS–I adhesive at different temperatures.

Table 4. Hard surface time and curing time of GIS–I adhesives at different temperatures.

Temperature (°C)	Hard Surface Time (h)	Curing Time (h)
5	8	48
15	5	36
25	3	24
35	2	18
45	1	12

(4) Scheme IV

The waterproof bonding layer was a mechanically sprayed waterproof coating film. Different from the conventional flexible waterproof film, a polymer-modified asphalt waterproof coating was sprayed as five layers, and alkali-free glass was sprayed simultaneously when spraying the 2nd and 4th layers were completed. The technical indicators of the bridge deck bonding waterproof coating are shown in Table 5. The fiber that was sprayed simultaneously with the waterproof coating was alkali-free glass-fiber non-twisted coarse-sand-glass type-E fiber, with a single-fiber diameter of 13 μm and the technical indicators shown in Table 6.

Table 5. Technical indices of polymer-modified asphalt waterproof coating.

Number	Project		Type II
1	Solid content		≥50
2	Hard surface time, h		<2
3	Actual dry time, h		4
4	Heat resistance, °C		180
5	Impermeable, 0.3 MPa, 30 min		Impermeable
6	Low-temperature flexibility, °C		−20
7	Tensile strength, MPa		1.0
8	Elongation at break, %		≥800
9	Salt treatment	Tensile strength retention rate, %	≥80
		Elongation at break, %	≥800
		Low-temperature flexibility, °C	−15
		Quality increase, %	≤2.0
10	Hot aging	Tensile strength retention rate, %	≥80
		Elongation at break, %	≥600
		Low-temperature flexibility, °C	−15
		Heating expansion rate, %	≤1.0
		Quality loss, %	≤1.0

Table 6. Technical indicators of alkali-free glass fiber.

Burn Loss	Water Content	Stiffness	Dispersion
0.8	0.1	≥140	≥95

2.1.3. Preparation of Bridge Deck Pavement Specimens

For the specimen preparation, the following steps were taken:

1. The cement prefabricated slabs with dimensions of 32 cm × 26 cm were prepared. To approximate the actual site conditions, the bridge-cement concrete-box-girder construction ratio and materials were used, which were formed and cured simultaneously. The box girder was cured for 28 days to meet the strength requirements. Then, cutting

and surface grinding were carried out successively to remove the laitance on the surface and increase the surface roughness, as shown in Figure 2.

2. The four waterproof bonding layers and composite bridge deck pavement structures were constructed on the bridge deck according to schemes I–IV. The four bridge deck pavement specimens are shown in Figure 3.
3. According to the demands of the test, the bridge deck pavement specimens could be cut or drilled with a specific size. Figure 4 shows the drilling of cylindrical specimens.

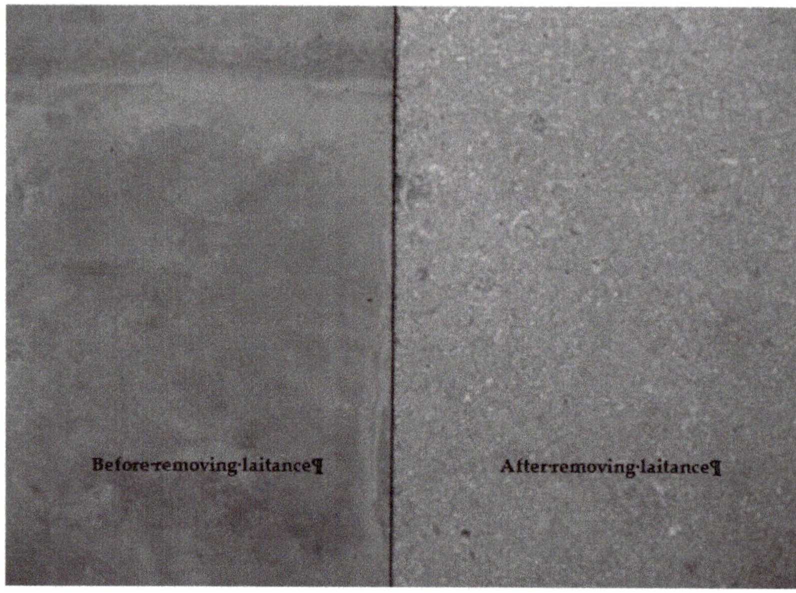

Figure 2. Cement concrete deck.

Figure 3. Formed bridge deck pavement specimens.

Figure 4. Drilling cylindrical specimen.

2.2. Basic Performance Tests Method

2.2.1. Waterproof Bonding Layer Property Tests

(1) Impermeability Test

The composite specimens composed of cement concrete, asphalt mixture, and four kinds of waterproof and cohesive materials were formed. The impermeability was tested according to the regulations of "Standard Test Methods of Bitumen and Bituminous Mixtures for Highway Engineering" (JTG E20–2011) T 0730–2011. The impermeability test of a composite specimen was carried out by using the material permeameter for the structural layer, which is shown in Figure 5. The number of replicates was no less than three. The material permeameter for the structural layer was composed of a barometer, a pressure cylinder with a plate, a clamping bolt, and a base [37]. The water seepage on the side of the rutting plate was observed, and the starting time of the water seepage was recorded.

Figure 5. ZHY permeability instrument.

(2) Scalding resistance test

This scalding resistance test refers to "Standard Test Methods of Bitumen and Bituminous Mixtures for Highway Engineering" (JTG E20–2011) T 0755–2011. This test mainly verified whether the crack resistance of each waterproof layer material decreased or if the layer lost its waterproof effect after the surface layer was constructed. A release agent was coated on a glass plate, and the waterproof layer film was completely peeled off after forming. The waterproof layer film was covered with a cement concrete slab, and the asphalt mixture was overlaid via rotary compaction or rolling with a Hamburg linear kneading machine. The temperature was the on-site construction control temperature. After simulating the on-site rolling and forming, it was determined whether the bottom of the waterproof membrane was damaged.

(3) Corrosion resistance test

Corrosion resistance test refers to "Corrosion tests in artificial atmospheres. Salt spray tests" (GB/T 10125–2012). The purpose of this experiment was to qualitatively simulate the durability, corrosion resistance, and protection of the bridge deck when chlorine salt seeped into the waterproof layer with water after repeated high-temperature and low-temperature cycles [38]. The specimens were obtained by cutting the bridge deck pavement specimens with dimensions of 10 mm × 30 mm × 10 mm. There were three replicates for every scheme. Figure 6 shows the device for simulating chloride penetration into the waterproof layer specimens.

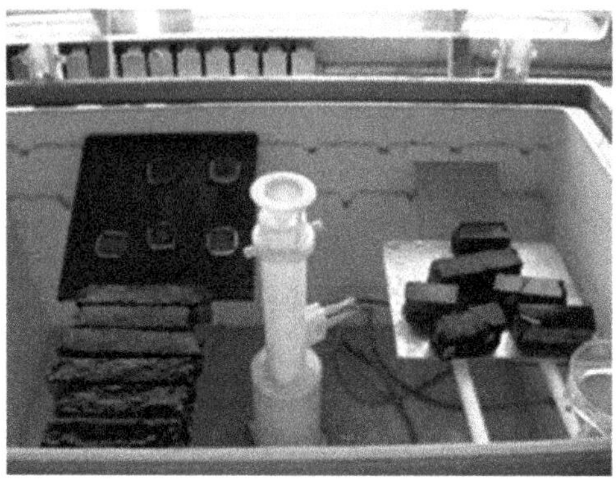

Figure 6. Device for simulating chloride penetration.

The test procedures were as follows:

1. When the specimen is formed, it was vacuumed and immersed in water at room temperature for 7 days. It was divided into two groups: ordinary fresh water and seawater;
2. The specimens were saturated in a 60 °C constant-temperature water bath for 7 days, and then they were put in air for 7 days. Finally, they were placed into a −10 °C freezer for 7 days, and the process was repeated twice;
3. The test specimens were removed and placed into a salt spray corrosion box with 5% salt spray at 35 °C. The test was stopped after 4 days;
4. The damage of the waterproof bonding layer at any time during the period was observed.

2.2.2. Interlayer Bonding Characteristics Test Method

In the bridge deck pavement structure, the interface bonding properties of the waterproof bonding materials are very important to the overall performance of the composite structure. To evaluate the tensile strength of different waterproof bonding materials, pull-off tests were conducted with the LGZ–1 structural layer material strength drawing instrument (Figure 7). The test temperatures were 25 and 50 °C, and the drawing head diameter was 100 mm. Two tests were carried out for each structure, and the average was taken as the result. The test results were processed, and the interlayer adhesion force can be calculated as follows:

$$C = \frac{F}{S}, \quad (1)$$

where C is the interlayer pull strength (MPa), F is the pulling force measured by the drawing instrument (kN), and S is the shear area between layers (m^2).

Figure 7. Pull-off test instrument.

2.2.3. Bridge Deck Pavement Structure Test Method

(1) Dynamic modulus test

The dynamic modulus was tested according to the regulations of American Association of State Highway and Transportation Officials (AASHTO) T312. The specimens were obtained by coring the test board. Each group had three samples. The test temperatures were 5, 25, and 50 °C, and the load frequencies were 0.1 and 10 Hz.

(2) Hamburg wheel tracking test

The test was conducted according to T324–2014 of AASHTO. The Hamburg wheel tracking device was from Helmut-wind company in Hamburg, Germany. There were two specimens in each group. The device is shown in Figure 8.

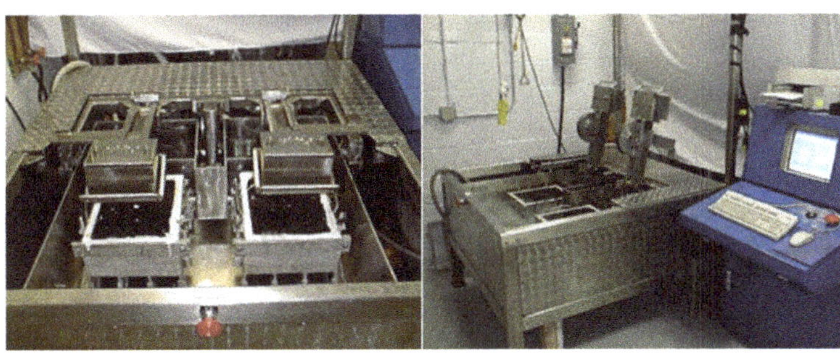

Figure 8. Hamburg test instrument.

(3) Static load creep test

The MTS material testing machine was used for uniaxial, static-load, high-temperature creep tests. The axial stress was 0.2 MPa and the loading time was 3600 s. To eliminate the contact gap, the sample was preloaded with a 200-N load for 60 s before the formal test. The test temperature was strictly controlled at 50 °C, and the test piece was made by drilling cores of the composite rutting plate with dimensions of Ø100 mm × 100 mm.

3. Test Conclusions and Analysis

3.1. Waterproof Bonding Layer Properties

(1) Impermeability of waterproof bonding layer

There was no water seepage in the four waterproof bonding layer structures after water injection for 24 and 48 h. This indicated that the four kinds of waterproof bonding systems had good permeability resistances.

(2) Scalding resistance of waterproof bonding layer

The surface of the waterproof bonding layer had various degrees of damage after rolling. However, the damage of scheme II was the greatest. In the process of rolling, the board that played a protective role was more complicated. When the lower rubber asphalt was less and the concrete slab was relatively flat, the underside of the protective board became scalded as shown in Figure 9a, and there was a layer of non-sticky fiber plant between the rubber asphalt and the concrete, as shown in Figure 9b. In most cases, the protective plate was crushed, resulting in local enrichment. The protective plate was much less viscous than rubber asphalt, which could affect the bonding strength between the layers.

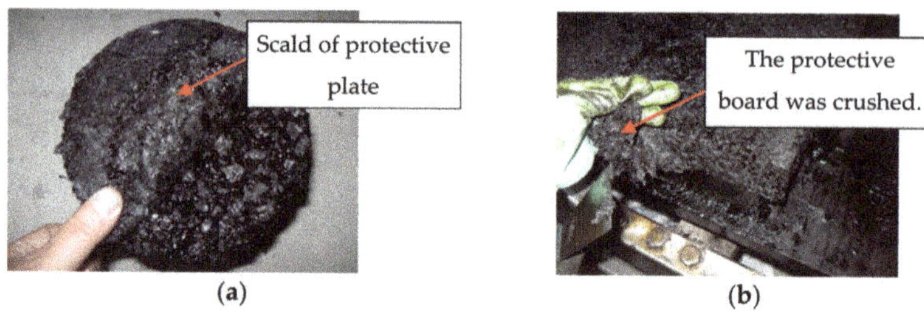

Figure 9. Burst resistance test phenomena of scheme II. (**a**) Scald of protective plate. (**b**) Protective board was crushed.

(3) Corrosion resistance of waterproof bonding layer
The test results of the corrosion resistance test are shown in Figure 10.

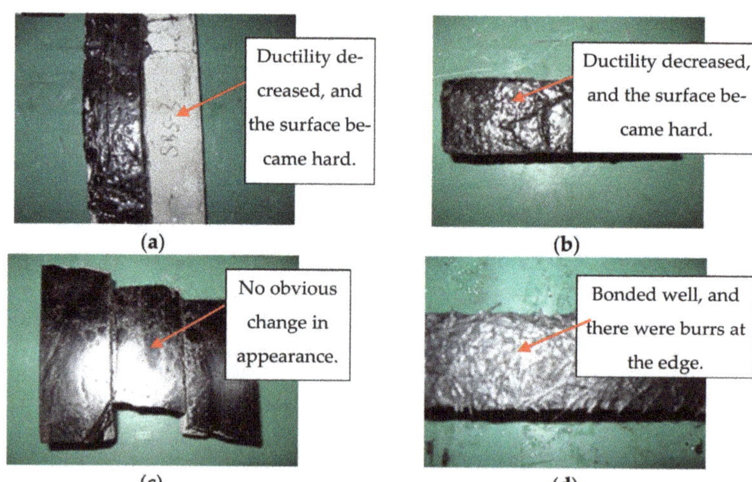

Figure 10. Specimen after corrosion resistance test. (**a**) Results of scheme I. (**b**) Results of scheme II. (**c**) Results of scheme III. (**d**) Results of scheme IV.

As shown in Figure 10, after repeated freezing and thawing cycles and salt spray corrosion, the four waterproof layers or undercoats were all well connected to the cement concrete bridge deck. However, the surfaces of schemes I and II became hard when pressed with a finger. The reason may have been that the chemical properties and physical composition of SBS–modified asphalt and rubber asphalt changed after the high-temperature and low-temperature cycles, and the viscosity and toughness decreased. There was no significant change of scheme III, indicating that the GIS–I–type primer had strong durability and corrosion resistance. Scheme IV bonded well, and there were burrs at the edge. Therefore the long-term coupling effects of water, temperature, and salinity would accelerate the deterioration of waterproof layer materials in a short time. Thus, the influence of these factors should be considered in the design of bridge deck pavement structures.

3.2. Interlayer Bonding Characteristics

The pull-off test results of the four schemes are shown in Figure 11, and the section forms after the pull-out test are shown in Figure 12.

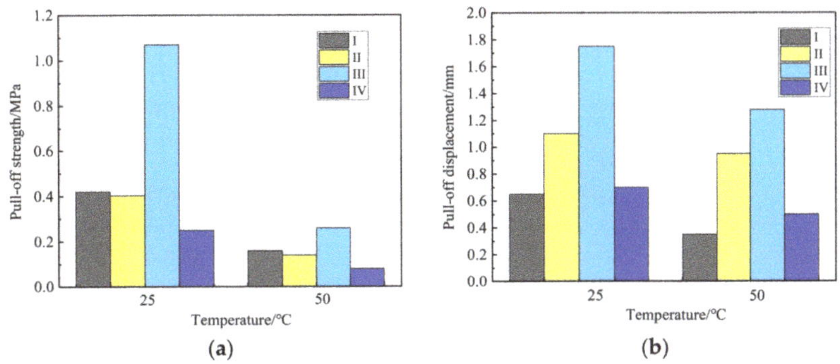

Figure 11. Test results of pull-off test. (**a**) Pull-off strength. (**b**) Pull-off displacement.

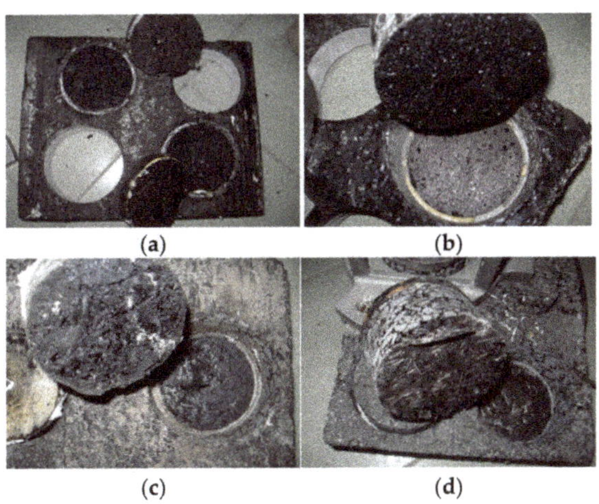

Figure 12. Sections of specimens after pull-off test. (**a**) Scheme I. (**b**) Scheme II. (**c**) Scheme III. (**d**) Scheme IV.

(1) Pull-off strength and deformation

The pull-off strength and deformation of bridge deck pavement scheme III were the largest at 25 and 50 °C. The pull-off strengths of scheme III at 25 °C were about 2.5-, 2.5-, and 4-times those of schemes I, II, and IV, respectively, and the pull-off strengths at 50 °C were about 1.6-, 2-, and 1.6-times those of schemes I, II, and IV, respectively. Hence, the order of the pull-off strengths of the bonding materials in the different schemes was as follows: III > I > II > IV, indicating that the order of the bonding performances between different waterproof bonding composite systems and cement concrete slab was as follows: III > I > II > IV.

The pull-off strengths of schemes I and II were relatively close, and the pull-off strength at 25 °C was greater than that at 0.35 MPa, which could meet the technical requirements of bridge deck pavement in "Technical Specifications for Construction of Highway Asphalt Pavement" (JTG F40–2004). The bonding principles of the two pavement structures were the same. The modified asphalt waterproof membrane achieved a good connection with the cement concrete bridge deck, and the bonding with the upper SMA–13 pavement layer was enhanced by the modified emulsified asphalt layer. However, the high-temperature pull-off test results showed the bonding strength of scheme II decreased rapidly, indicating that the thickness and temperature stability of the asphalt film determined the stability of the bonding layer. Therefore, it is necessary to optimize the distribution of modified asphalt and the appropriate thickness of rubber asphalt during the design of the bridge deck pavement structure.

The pull-off strength of scheme IV was not ideal. The reason may have been that the manual distribution method was used in the indoor tests, which could not accurately simulate the roller brush construction process. This affected the distribution uniformity of alkali-free glass fibers, resulting in partial thickening and strong plate properties and affecting the connection with the upper layer. Otherwise, based on the damage condition after the pull-off tests, the waterproof coating was well connected with the bridge deck. It cracked at the connection with the upper paving layer or between the waterproof layer. This indicated that scheme IV overemphasized the connection with the bridge deck and the overall anti-cracking capacity, while the current distribution amount and number of glass fibers made the waterproof coating too hard, and the contact area with the mixture of the upper paving layer was small. This reduced the pull-off strength between layers.

The pull-off deformation amounts and pull-off strengths of the four schemes at 50 °C were less than those at 25 °C. The variation range of scheme IV was small, which was mainly due to the low-temperature sensitivity of the cold-constructed GIS–I–type primer, and the temperature had little effect on its resistance to pull-off deformation.

(2) Failure section after pull-off test

The pull-off failure section of scheme III appeared between the pull-off head and the epoxy layer, as shown in Figure 12c. The reason was that, on the one hand, the GIS–I organic solvent was used as the bonding layer material, which had a strong permeability and improved the surface characteristics of the cement concrete bridge deck. On the other hand, the pouring asphalt concrete had a small porosity, good durability, strong adhesion, and good adaptability with deformation. Thus, it increased the interlayer bonding strength.

3.3. Bridge Deck Pavement Structural Properties

(1) Dynamic mechanical properties

The dynamic modulus test results are shown in Figure 13.

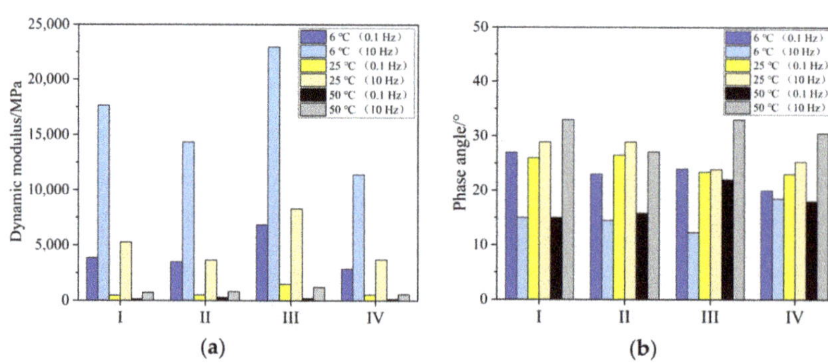

Figure 13. Test results of dynamic modulus. (**a**) Dynamic modulus. (**b**) Phase angle.

For the same bridge deck pavement structure, the higher the test temperature was, the smaller the measured value of dynamic modulus was. The higher the test frequency was, the larger the measured value of dynamic modulus was. Hence, the order of the dynamic modulus of the different schemes was as follows: III > I > II > IV. The dynamic modulus of scheme III was higher than those of the other three schemes. The dynamic modulus of scheme III at 6, 25, and 50 °C were, respectively, 1.5–2.4 times, 1.5–3.0 times, and 1–2.2 times those of schemes I, II, and IV. This indicated that scheme III exhibited good dynamic mechanical properties.

With the loading frequencies of 0.1 and 10 Hz, the phase angles of the four bridge deck pavement systems changed between 15–30°, indicating that the viscoelastic properties of the four composite structures were stable in the service temperature range of 6–50 °C, which was very beneficial to the stress of the whole bridge deck pavement structure.

To analyze the influence of the temperature on the dynamic modulus of different bridge deck pavement structures, the temperature-sensitive values (TSVs) of each bridge deck pavement structure with different temperature ranges were calculated. The formula is as follows:

$$TSV_{ij} = \frac{D_i - D_j}{i - j}, \tag{2}$$

where D_i and D_j are the dynamic modulus of the pavement structure at test temperatures i and j (°C). The calculation results are shown in Figure 14.

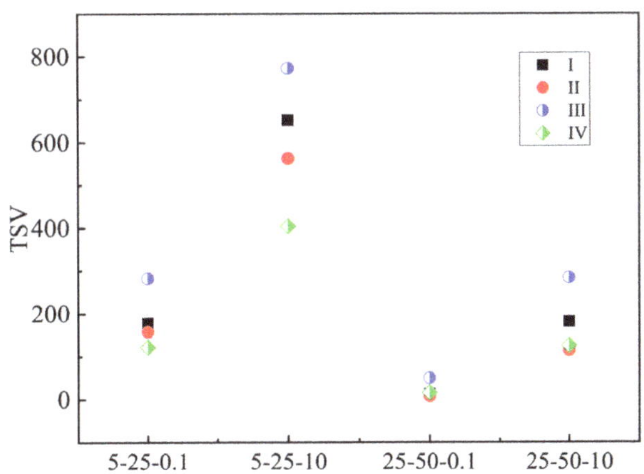

Figure 14. Temperature sensitive value.

The TSVs of scheme III in the low- and high-temperature ranges were greater than those of the other three schemes, indicating that scheme III's dynamic mechanical properties were greatly affected by the temperature. Although scheme III had good comprehensive properties in the medium-temperature region, it was still possible to be brittle at low temperatures and creep at high temperatures. This may have been caused by the high asphalt content of GA–10. The TSVs of schemes I and IV were less than those of the other three schemes, indicating that schemes I and IV had good temperature stability.

(2) Water-temperature stability

Until the Hamburg wheel tracking test stopped after 20,000 loading cycles, inflection points of the four bridge deck pavement systems did not appear, so the peeling inflection points were not obtained. The Hamburg wheel tracking test system automatically recorded the deformation of 12 points of the left and right wheel ruts, and the maximum deformation generally occurred near point 6. Therefore, the rutting depth and maximum rutting depth of each structure at points 3, 6, and 9 were collected, and the data are shown in Figure 15.

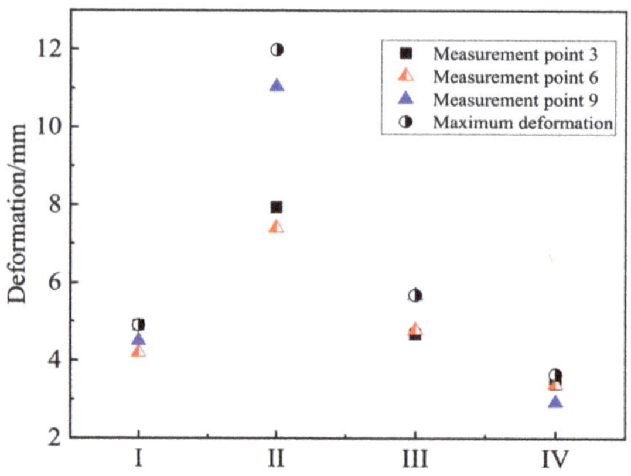

Figure 15. Hamburg wheel tracking deformation.

Figure 15 shows that the maximum deformation and the deformation at measurement point 9 of scheme II were larger than 10 mm, while the deformation values of the other three schemes did not exceed 8 mm, indicating that the water-temperature stability of scheme II was poor. Thus, it was not suitable for high-temperature environments.

The order of the deformation amounts of the different schemes were as follows: II > III > I > IV, indicating that the order of the water stabilities of the different waterproof bonding composite systems were as follows: IV > I > III > II. The deformation of scheme I was between 4–5 mm, that of scheme III was between 4.5–6 mm, and that of scheme IV was the smallest, between 2.5–3.5 mm. These results indicated that the water temperature stabilities of schemes I, III, and IV were good. Scheme IV showed an excellent water temperature stability. The reason may be that the fibers played a reinforcement role, effectively improving the overall strength of the waterproof bonding layer and the puncture resistance during construction. Furthermore, it could adapt to the dynamic load and the compression and tensile characteristics of the bridge, and it exhibited good waterproof performances and corrosion resistance.

(3) Static mechanical properties

Figure 16 shows the creep deformation and time curves of the four schemes measured by static load creep test. Although the creep deformation increases of different schemes were different at 50 °C, the variations were consistent. If the relationship between the creep deformation and time before rheology was plotted on semi-logarithmic coordinates, there was a good linear relationship between them, the linear correlation formulas were shown in Figure 17. Thus, according to the linear correlation formulas the creep deformation of the four structures can be predicted easily.

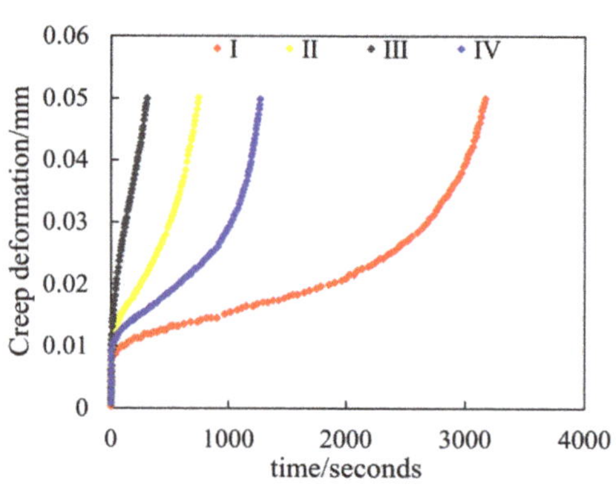

Figure 16. Results of static load creep test.

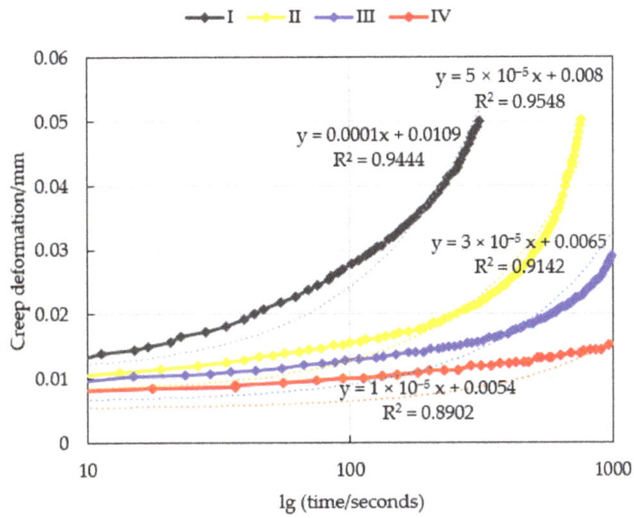

Figure 17. Deformation–time semilog curves.

The creep deformation amounts of the four schemes under load went through three stages. First, in the migration period, the deformation increased quickly with time. Second, in the stable period, the deformation increased steadily. Third, the deformation increased rapidly with time until failure. The rheological time was defined as the time when the total volume did not change, that is, the beginning of the third stage was the rheological point, and the corresponding time was the rheological time (Ft). Thus, the Ft values of the four schemes were calculated, as shown in Table 7.

Table 7. Rheological time of four schemes.

Scheme	I	II	III	IV
Rheological time/seconds	364	298	905	1636

The rheological time of scheme IV was the longest, and it was 4.5-, 5.5-, and 1.8-times those of schemes I, II, and III, respectively. The rheological time of scheme III was about 2.5- and 3.0-times greater than that of schemes I and II, respectively, indicating that the schemes III and IV had strong high-temperature deformation resistances. Scheme II had the weakest high-temperature deformation resistance. This result was consistent with the test conclusion of the Hamburg rutting test. Therefore, when the ambient temperature was high, scheme III is recommended.

4. Conclusions

In this study, a series of tests were carried out to evaluate the waterproof bonding material composite systems of cement concrete bridge deck pavement. The following conclusions can be drawn from the tests:

(1) The four waterproof bonding systems exhibited good permeability resistances. In the process of rolling, scheme II was easily damaged because the protective board was scalded and crushed. According to the corrosion resistance test, the stability of scheme III was the best.

(2) The pull-off strengths of scheme III at 25 °C were about 2.5-, 2.5-, and 4-times those of schemes I, II, and IV, respectively, and the pull-off strengths at 50 °C were about 1.6-, 2-, and 1.6-times those of schemes I, II, and IV, respectively. The pull-off tests indicated that the bonding performances of different waterproof bonding composite

systems and cement concrete slabs ranked as follows: III > I > II > IV, and only scheme IV could not meet the requirements of the standard.

(3) The dynamic modulus of scheme III was higher than those of the other three schemes. The dynamic modulus of scheme III at 6, 25, and 50 °C were respectively 1.5–2.4 times, 1.5–3.0-times, and 1–2.2-times those of schemes I, II, and IV. This indicated that scheme III exhibited good dynamic mechanical properties. The results of dynamic modulus tests showed that the dynamic modulus of the different schemes ranked as follows: III > I > II > IV. However, the TSV of scheme III in the low-temperature and high-temperature range was greater than those of the other three schemes, indicating that the scheme III dynamic mechanical properties were greatly affected by temperature.

(4) The results of the Hamburg wheel tracking test were consistent with the static load creep test results. The high-temperature stability of scheme IV was the best, while the stability of schemes III and I were better than that of scheme II. In conclusion, the scheme of concrete bridge deck pavement can be selected according to the environmental conditions and traffic loads, and schemes III and I are recommended.

(5) In this paper, the performances of four waterproof bonding layer systems for cement concrete bridge deck pavement are tested and evaluated. However, there are many different systems in practical engineering. Therefore, it is necessary to carry out more extensive research on the concrete bridge deck pavement structure, evaluate the properties of more structures, to serve the concrete bridge deck pavement structure design.

Author Contributions: Methodology, J.F. and A.S.; data curation, J.F.; writing—original draft preparation, J.F. and Z.Y.; writing—review and editing, A.S. and Z.Y.; funding acquisition, J.F. All authors have read and agreed to the published version of the manuscript.

Funding: This research was funded by Science and Technology Projects of Inner Mongolia Department of Transportation, (Contract number: NJ-2013-30).

Institutional Review Board Statement: Not applicable.

Informed Consent Statement: Not applicable.

Data Availability Statement: The data that support the findings of this study are available from the corresponding author upon reasonable request.

Conflicts of Interest: The authors declare no conflict of interest. The funders had no role in the design of the study; in the collection, analyses, or interpretation of data; in the writing of the manuscript, or in the decision to publish the results.

References

1. Manning, D.G. *Waterproofing Membranes for Concrete Bridge Decks*; Transportation Research Board: Washington, DC, USA, 1995; pp. 55–56.
2. Van Til, C.J.; Carr, B.J.; Vallerga, B.A. Waterproof membranes for protection of concrete bridge decks-laboratory phase. *NCHRP* **1976**, *165*, 66–67.
3. Zhou, Q.; Xu, Q. Experimental study of waterproof membranes on concrete deck: Interface adhesion under influences of critical factors. *Mater. Des.* **2009**, *30*, 1161–1168. [CrossRef]
4. PRIC, A.R. Waterproofing of Concrete Bridge Decks Site Practice and Failures. *TRRL* **1991**, *317*, 57–60.
5. PRICE, A.R. Laboratory Tests on Waterproofing Systems for Concrete Bridge Decks. *TRRL* **1990**, *248*, 35–39.
6. Chen, X.; Zhu, Y.; Cai, D.; Xu, G.; Dong, T. Investigation on Interface Damage between Cement Concrete Base Plate and Asphalt Concrete Waterproofing Layer under Temperature Load in Ballastless Track. *Appl. Sci.* **2020**, *10*, 2654. [CrossRef]
7. Qian, G.; Li, S.; Yu, H.; Gong, X. Interlaminar Bonding Properties on Cement Concrete Deck and Phosphorous Slag Asphalt Pavement. *Materials* **2019**, *12*, 1427. [CrossRef]
8. Haynes, M.A.; Coleri, E.; Obaid, I. Performance of Waterproofing Membranes to Protect Concrete Bridge Decks. *Transp. Res. Rec.* **2021**, *2675*, 1693–1706. [CrossRef]
9. Qiu, Y.J.; Shaoke, A.; Ali, R.; Ai, C.F. Evaluation and optimization of bridge deck waterproof bonding system using multi-objective grey target decision method. *Road Mater. Pavement Des.* **2020**, *21*, 1844–1858. [CrossRef]
10. Zhang, K.; Luo, Y.F. Interlaminar Performance of Waterproof and Cohesive Materials for Concrete Bridge Deck under Specific Test Conditions. *J. Mater. Civ. Eng.* **2018**, *30*, 04018161. [CrossRef]

11. Ye, F.; Zhou, K.; Jia, X. Evaluation of shear performance of flexible waterproof-adhesive layer in concrete bridge pavement based on grey correlation analysis. *Road Mater. Pavement Des.* **2009**, *10*, 349–360.
12. Xie, H.; Zhao, R.; Wang, R.; Xi, Z.; Yuan, Z.; Zhang, J.; Wang, Q. Influence of thermal shock on the performance of B-staged epoxy bond coat for orthotropic steel bridge pavements. *Constr. Build. Mater.* **2021**, *294*, 123598. [CrossRef]
13. Wang, H.; Jin, C.; Liu, H.; Xue, Z. Rubber asphalt waterproof adhesive layer for steel bridge gussasphalt pavement. *Int. J. Struct. Integr.* **2020**, *12*, 261–270. [CrossRef]
14. Li, Y.; Li, S.; Lv, R.; Zhang, P.; Xu, Y.; Hou, G.; Cui, C. Research on failure mode and mechanism of different types of waterproof adhesive materials for bridge deck. *Int. J. Pavement Eng.* **2015**, *16*, 602–608. [CrossRef]
15. Feng, D.C.; Xu, M.; Wei, W.D. Analysis of the Influence of Cement Concrete Deck Moisture Content on the Bonding Performance of Waterproof Adhesion Layer. *J. Highw. Transp. Res. Dev.* **2014**, *8*, 31–36. [CrossRef]
16. Mazzotta, F.; Lantieri, C.; Vignali, V.; Simone, A.; Dondi, G.; Sangiorgi, C. Performance evaluation of recycled rubber waterproofing bituminous membranes for concrete bridge decks and other surfaces. *Constr. Build. Mater.* **2017**, *136*, 524–532. [CrossRef]
17. Xu, G.; Zhou, J.; Chen, X.; Liang, Y.; Cai, D.; Lou, L. Temperature features of the asphalt concrete waterproofing layer on high-speed railway in cold regions. *Constr. Build. Mater.* **2021**, *305*, 124665. [CrossRef]
18. Shi, X.; Zhang, H.; Bu, X.; Zhang, G.; Zhang, H.; Kang, H. Performance evaluation of BDM/unsaturated polyester resin-modified asphalt mixture for application in bridge deck pavement. *Road Mater. Pavement Des.* **2020**, *23*, 684–700. [CrossRef]
19. Zhang, M.; Hao, P.; Men, G.; Liu, N.; Yuan, G. Research on the compatibility of waterproof layer materials and asphalt mixture for steel bridge deck. *Constr. Build. Mater.* **2021**, *269*, 121346. [CrossRef]
20. Xu, Y.; Lv, X.; Ma, C.; Liang, F.; Qi, J.; Chou, Z.; Xu, S. Shear Fatigue Performance of Epoxy Resin Waterproof Adhesive Layer on Steel Bridge Deck Pavement. *Front. Mater.* **2021**, *7*, 469. [CrossRef]
21. Cao, M.-M.; Huang, W.-Q.; Lu, Y.; Tan, Q.-Q. Test and Evaluation Method of Interlaminar Shear Performance of Composite Pavement. *J. Highw. Transp. Res. Dev.* **2018**, *12*, 33–43. [CrossRef]
22. Liu, Y.; Wu, J.T.; Chen, J. Mechanical properties of a waterproofing adhesive layer used on concrete bridges under heavy traffic and temperature loading. *Int. J. Adhes. Adhes.* **2014**, *48*, 102–109. [CrossRef]
23. Kruntcheva, M.R.; Collop, A.C.; Thom, N.H. Properties of asphalt concrete layer interfaces. *J. Mater. Civ. Eng.* **2006**, *18*, 467–471. [CrossRef]
24. Xu, Q.; Zhou, Q.; Medina, C.; Chang, G.K.; Rozycki, D.K. Experimental and numerical analysis of a waterproofing adhesive layer used on concrete-bridge decks. *Int. J. Adhes. Adhes.* **2009**, *29*, 525–534. [CrossRef]
25. Guo, M.; Tan, Y.; Wang, L.; Ye, Z.; Hou, Y.; Wu, J.; Yang, H. Study on water permeability, shear and pull-off performance of waterproof bonding layer for highway bridge. *Int. J. Pavement Res. Technol.* **2018**, *11*, 396–400. [CrossRef]
26. Jiao, Y.; Zhang, Y.; Fu, L.; Guo, M.; Zhang, L. Influence of crumb rubber and tafpack super on performances of SBS modified porous asphalt mixtures. *Road Mater. Pavement Des.* **2019**, *20*, S196–S216. [CrossRef]
27. Zhang, H.; Gao, P.; Zhang, Z.; Pan, Y. Experimental study of the performance of a stress-absorbing waterproof layer for use in asphalt pavements on bridge decks. *Constr. Build. Mater.* **2020**, *254*, 119290. [CrossRef]
28. Guo, L.C.; Zeng, G.D. Study on mechanical properties of typical steel bridge deck pavement waterproof bonding system. *J. Phys. Conf. Ser.* **2021**, *1802*, 022018. [CrossRef]
29. Ai, C.F.; Huang, H.W.; Ali, R.; Shaoke, A. Establishment of a new approach to optimized selection of steel bridge deck waterproof bonding materials composite system. *Constr. Build. Mater.* **2020**, *264*, 120269. [CrossRef]
30. AASHTO. *LRFD Bridge Design Specifications*, 3rd ed.; American Association of State Highway and Transportation of Officials: Washington, DC, USA, 2004.
31. AASHTO. *AASHTO Guide Specifications for Fatigue Evaluation of Existing Steel Bridges*; American Association of State Highway and Transportation of Officials: Washington, DC, USA, 1990.
32. *BS5400 Steel, Concrete and Composite Bridges*; British Standard Institute: London, UK, 2000.
33. AASHTO. *AASHTO Guide Manual for Condition Evaluation and Load and Resistance Factor Rating (LRFR) of Highway Bridges*; American Association of State Highway and Transportation of Officials: Washington, DC, USA, 2002.
34. Wang, T.L.; Liu, C.; Huang, D.; Shahawy, M. Truck loading and fatigue damage analysis for girder bridges based on weightin-motion data. *J. Bridge Eng.* **2005**, *10*, 12–20. [CrossRef]
35. Zhao, P.; Dong, M.; Yang, Y.; Shi, J.; Wang, J.; Wu, W.; Zhao, X.; Zhou, X.; Wang, C. Research on the Mechanism of Surfactant Warm Mix Asphalt Additive-Based on Molecular Dynamics Simulation. *Coatings* **2021**, *11*, 1303. [CrossRef]
36. Ma, J.; Hu, M.; Sun, D.; Lu, T.; Sun, G.; Ling, S.; Xu, L. Understanding the role of waste cooking oil residue during the preparation of rubber asphalt. *Resour. Conserv. Recycl.* **2020**, *167*, 105235. [CrossRef]
37. Xiao, X.; Wang, J.; Cai, D.; Lou, L.; Xiao, F. A novel application of thermoplastic polyurethane/waste rubber powder blend for waterproof seal layer in high-speed railway. *Transp. Geotech.* **2021**, *27*, 100503. [CrossRef]
38. Wang, F.; Lei, S.; Ou, J.; Li, W. Effect of PDMS on the waterproofing performance and corrosion resistance of cement mortar. *Appl. Surf. Sci.* **2020**, *507*, 145016. [CrossRef]

coatings

MDPI

Article

Comparative Study on Road Performance of Low-Grade Hard Asphalt and Mixture in China and France

Guodong Zeng [1], Wenjuan Wu [2,*], Juechi Li [1], Qinsheng Xu [2], Xianghang Li [1], Xiangpeng Yan [2,*], Ye Han [2] and Jincheng Wei [2]

1 Foshan Transportation Science and Technology Co., Ltd., Foshan 528315, China; wutongshuxi_a@163.com (G.Z.); ljc20211013@163.com (J.L.); dfy8307@163.com (X.L.)
2 Shandong Transportation Institute, Jinan 250031, China; 15966652536@163.com (Q.X.); hanye@sdjtky.cn (Y.H.); weijincheng@sdjtky.cn (J.W.)
* Correspondence: wuwenjuan@sdjtky.cn (W.W.); yanxiangpeng@sdjtky.cn (X.Y.)

Abstract: The high-modulus asphalt mixture produced by hard asphalt has played a good role in reducing asphalt pavement rutting and improving pavement durability. It was widely used in the construction of various levels of road engineering in Europe. However, low-grade hard asphalt was rarely used in road engineering in China, and the use of hard asphalt for the design and construction of high modulus asphalt mixtures lacks sufficient engineering experience. Considering the above issues, comparative research on the performance of Chinese and French low-grade hard asphalt and mixture was carried out. In this paper, the performance of French 15# hard asphalt commonly used for EME (Enrobés à Module Elevé Class) high modulus asphalt mixture and China 15# hard asphalt was analyzed comparatively, where six typical high modulus asphalt mixtures corresponding to two low-grade hard asphalts were designed referring to the design requirements of French high modulus asphalt mixtures. The road performance of a high-modulus asphalt mixture was evaluated to verify the feasibility of its application in engineering. The research results show that the performance indicators of both Chinese and French asphalts meet the requirements of binder materials used in high modulus asphalt mixtures, and the performance of their corresponding mixtures of the two asphalts also meet the specifications of high modulus asphalt mixtures. However, the two asphalts and the performance of their corresponding mixture are slightly different. The high modulus asphalt mixture of Chinese low-grade hard bitumen can be used well in road engineering applications in China. It can strongly promote the wide application of high modulus asphalt mixture in China.

Keywords: low-grade asphalt; high-modulus asphalt mixture; road performance; road engineering application

Citation: Zeng, G.; Wu, W.; Li, J.; Xu, Q.; Li, X.; Yan, X.; Han, Y.; Wei, J. Comparative Study on Road Performance of Low-Grade Hard Asphalt and Mixture in China and France. *Coatings* **2022**, *12*, 270. https://doi.org/10.3390/coatings12020270

Academic Editor: Valeria Vignali

Received: 25 January 2022
Accepted: 15 February 2022
Published: 17 February 2022

Publisher's Note: MDPI stays neutral with regard to jurisdictional claims in published maps and institutional affiliations.

cc BY

Copyright: © 2022 by the authors. Licensee MDPI, Basel, Switzerland. This article is an open access article distributed under the terms and conditions of the Creative Commons Attribution (CC BY) license (https://creativecommons.org/licenses/by/4.0/).

1. Introduction

High modulus asphalt mixture refers to an asphalt mixture with a complex modulus ≥14,000 Mpa under the conditions of 15 °C and 10 Hz, which is derived from the French high modulus asphalt mixture EME (Enrobés à Module Elevé Class) and the high modulus asphalt mixture HMAC (High Modulus Asphalt Concrete) of the middle surface layer in the concept of permanent pavement (in the USA) [1–3]. Its core is to improve the modulus and high-temperature stability of asphalt concrete and reduce the strain of asphalt concrete under load to achieve the purpose of improving the anti-rutting ability of the pavement reducing the thickness of the pavement and improving the durability of the pavement High modulus asphalt mixture has become a research hotspot due to its excellent high temperature stability, strong resistance to water damage and good fatigue resistance [4–7] Some Chinese researchers have carried out research on high modulus asphalt mixture from different angles. However, most of the existing high-modulus asphalt mixtures in China have a completely different technical approach to production from French high modulus mixtures. The production of high modulus asphalt mixtures in France has two main technical methods: one is to use low-grade asphalt, mainly using 15# hard asphalt, that is, non-modified asphalt with a penetration of 10–20 (0.1 mm) at 25 °C; the other is the use of high modulus additives. The former accounted for about 70%, and the latter accounted for about 30%. However, in China, the exploration of high modulus asphalt mixtures is more inclined to the contribution of high modulus modifiers to its performance. French road workers [8–15] applied hard asphalt with a penetration of 151/10 mm to the high-modulus concrete EME structure of the bottom layer and increased the amount of asphalt to the same level as the surface layer at the same time and the results showed that the permanent deformation resistance of the high-modulus asphalt mixture produced by the hard asphalt with a degree of 151/10 mm is better than that of the SBS modified asphalt mixture, and the fatigue life is 30% longer than that of the original subsurface mixture. Newcombet al. [16] used low-grade hard asphalt and modified asphalt on the same road, respectively, and conducted a comparative test study. The test results show that low-grade hard asphalt has better road performance than modified asphalt. De Backer C [17] gave a detailed introduction to the use of hard asphalt in high-modulus asphalt concrete in Belgium at the European Transport Association. The test results show that the hard asphalt has good road performance. Visscher [18] and Backer [19] introduced the research progress of high-modulus asphalt concrete in Europe at the 4th European Conference on Asphalt and Asphalt Mixtures and clearly expressed the importance of using hard asphalt for high-modulus concrete. Rohde [20] found that the high-modulus asphalt concrete containing hard bitumen has good resistance to permanent deformation and good low-temperature performance by comparing and analyzing different types of high-modulus asphalt concrete. In China, Liaoning Transportation Research Institute Co., Ltd. carried out research on the application technology of high modulus asphalt concrete and adopted the method of adding high modulus admixtures to improve the high-temperature dynamic modulus of asphalt mixtures [2]. Linquan Guo et al. [21] carried out research on the application of high modulus asphalt concrete in road rutting treatment and studied the relationship between the modulus of asphalt mixture and rutting from the aspects of molecular weight and modification mechanism of SBS (styrene–butadiene–styrene block copolymer) modified asphalt. Qing Huazhou et al. [5] carried out a series of high modulus asphalt mixture performance tests and road engineering application research from the aspect of mixture admixtures. They used French PR series high modulus modifiers to increase the mixture modulus and adopted an SK-70 asphalt +High modulus modifier to produce AC (Asphalt Concrete)-20, AC-25, SUP-20, SUP(Superpave)5, EME-14, EME-20 and other gradation type mixtures for performance test research, and the test roads were paved. Although some progress has been made in the research of high modulus asphalt mixtures, the application of low-grade hard asphalt has been neglected, resulting in the high cost of high-modulus asphalt mixtures, which restricts its large-scale application in engineering [22].

Hard asphalt was first produced and used in France, mainly used in high modulus asphalt mixture EME. After nearly two decades of development, it has been widely used in European countries and was promoted in Australia, South Africa and other countries. Chinese road workers have also carried out some research on hard asphalt and its mixtures based on foreign experience. Zhaohui Liu et al. [5] found that 30# hard asphalt has good resistance to deformation at high temperatures through experimental research and road paving tests. Yuming Dong et al. [23] conducted dynamic shear rheological tests on 30# hard asphalt produced in China, finding that it exhibited good high-temperature performance. Chunyu Liang [24] and Weihong Xiong [25] conducted comparative experiments on the performance of low-grade asphalt with a penetration of 20–30, 70# asphalt, SBS modified asphalt and rubber asphalt and found that the low-grade hard asphalt mixture has excellent high-temperature performance and its dynamic stability is more than five times higher than that of 70# asphalt under the same test conditions. Zhao Lei et al. [26] carried out research on the application of hard asphalt mixture in rutting treatment engineering of trunk highways and found that the mixture has good compactability, and applying it to the lower layer of trunk highways can significantly improve the anti-rutting performance of the road surface. Some research has been performed on hard asphalt and its high modulus asphalt mixtures, but most of the hard asphalt performance studies simply evaluated whether it meets the requirements of the specifications and have not compared the performance of the 15# hard asphalt commonly used in France EME2. In the design and evaluation of high modulus asphalt mixtures, there is no systematic study in accordance with the French high modulus asphalt mixture EME system. Most studies only discuss the high-temperature anti-deformation ability and stiffness modulus of hard asphalt mixtures, without comprehensive analysis of its low-temperature anti-cracking performance, water stability, fatigue resistance and durability. Moreover, there are few engineering applications for low-grade hard asphalt, and there is little experience in the design and application of high-modulus asphalt mixtures based on low-grade hard asphalt.

In this paper, two kinds of low-grade hard asphalt from China and France were used, and three kinds of high modulus asphalt mixture, including EME-14 continuous, EME-14 discontinuous and HMAC-20, which are typical and commonly used in the middle and lower layers, were selected. The performance difference of the two low-grade hard asphalts from China and France was analyzed, and the road performance and mechanical properties of three typical high modulus asphalt mixtures based on the China and France hard asphalt were systematically evaluated. It is verified whether the performance of hard asphalt high modulus asphalt mixture produced in China based on limestone meets the requirements, and the feasibility of its application in road engineering was analyzed, which provides a reference for the application of hard asphalt high-modulus asphalt mixture in China.

2. Materials and Experimental

2.1. Materials

The crushed limestone and its ground ore powder were used as aggregates and filler, and the technical indicators meet the requirements of standard JTG-F40-2004 [27], respectively, as shown in Table 1.

Table 1. Properties of coarse and fine aggregate.

	Technical Index		Test Results	Index Requirements
Coarse aggregate	Apparent relative density	15~20 mm	2.762	≤2.50
		10~15 mm	2.732	
		5~10 mm	2.746	
		3~5 mm	2.758	
	Water absorption rate (%)	15~20 mm	0.45	≤3.0
		10~15 mm	0.32	
		5~10 mm	0.45	
		3~5 mm	0.63	
	Crushing value (%)	-	19.2	≤28
	Needle flake content (%)	15~20 mm	3.5	≤18
		10~15 mm	9.3	
		5~10 mm	10.7	
	Soft stone content (%)		0.55	≤5
	Washing method <0.075 mm particle content (%)	15~20 mm	0.3	≤1.0
		10~15 mm	0.6	
		5~10 mm	0.4	
		3~5 mm	0.9	
	Adhesion to asphalt, grade		5	≥4
Fine aggregate	Apparent relative density		2.644	≥2.5
	Sand equivalent (%)		67	≥60
	Angularity of fine aggregate (s)		36	≥30
	Methylene blue (g/Kg)		7	≤25

Two types of low-grade asphalt were used in the French high modulus asphalt mixture. One is the hard asphalt with the label of 10/20, namely France 15# hard asphalt, the other is China 15# hard asphalt produced in China, which has similar performance to France 15#. The technical indicators of the two types of hard bitumen meet the requirements of standard EN 13924 [28], and the test results are shown in Table 2.

Table 2. Asphalt conventional technical indicators.

Pilot Projects		China 15#	France 15#	EN 13924
Penetration (100 g, 5 s, 25 °C)/0.1 mm		16.5	16.8	10–20
Softening Point (5 °C)/°C		64.25	66.6	58–78
Ductility (5 cm/min, 15 °C)		Brittle	Brittle	-
Solubility/%		99.6	99.78	≥99
density (15 °C)/g/cm^3		1.029	1.033	-
Flash point/°C		280	344	≥245
dynamic viscosity/Pa·s		6076	7219	≥700
RTFOT	Quality change/%	−0.31	0.051	≤0.5
	Residual penetration ratio/%	78	71.6	≥55

2.2. Mix Design

EME-14 (continuous and intermittent) and AC-20, three types of typical high modulus asphalt mixture gradations for France and China, suitable for the middle and lower layers, were selected in this study. According to the aggregate screening results and the actual high-modulus road engineering application situation, the designed composite gradation is shown in Table 3 and the gradation curve is shown in Figure 1.

Table 3. Synthetic gradation design of high modulus asphalt mixture.

Sieve Size (mm)	26.5	19	16	13.2	9.5	4.75	2.36	1.18	0.6	0.3	0.15	0.075
EME-14 continuous	100	100	100	97	90.1	48.7	33.1	25.7	16.1	10.6	7.9	5.7
EME-14 intermittent	100	100	100	88.6	63.7	47.1	33.9	26.4	16.4	10.7	7.7	5.8
HMAC-20	100	96.2	86.2	75.4	56.2	34.6	24.3	19.3	12.6	8.9	7.1	5.7
Design upper limit	100	100	-	-	82.0	64.0	43.0	-	-	-	-	8.0
Design lower limit	100	90.0	-	-	66.0	41.0	28.0	-	-	-	-	6.0
Design median	100	95.0	-	-	74.0	52.5	35.5	-	-	-	-	7.0

The asphalt binder dosage is determined by the abundance coefficient *K*. The abundance coefficient *K* is a ratio of the conventional thickness of the asphalt film attached to the surface of the aggregate with an asphalt binder [29]. *K* has nothing to do with the density of gravel. According to LCPC (Laboratoire Central des Ponts et Chaussées) Bituminous Mixtures Design Guide, the amount of asphalt is calculated and estimated by the abundance coefficient *K*, which satisfies $K \geq 3.4$. The asphalt binder dosage of high modulus asphalt mixture was calculated according to Equations (1)–(3), as shown in Table 4.

The relationship between the abundance coefficient *K* and the oil–stone ratio:

$$TL_{ext} = K \times \alpha \sqrt[5]{\sum} \tag{1}$$

$$100\sum = 0.25G + 2.3S + 12s + 150f \tag{2}$$

$$\alpha = \frac{2.65}{\rho G}. \tag{3}$$

where:

TL_{ext} is the percentage of admixture (whetstone ratio), %;
K is the specific surface area, m^2/kg;
G is the ratio of aggregates with a particle size greater than 6.3 mm, %;
S is the ratio of aggregates with a particle size of 0.25~0.63 mm, %;
s is the aggregate ratio with a particle size of 0.063~0.25 mm, %;
f is the ratio of aggregates with a particle size of less than 0.063 mm, %;
a is the correlation coefficient related to aggregate density;
ρG is the density of aggregate, g/cm^3;

2.3. Test Methods

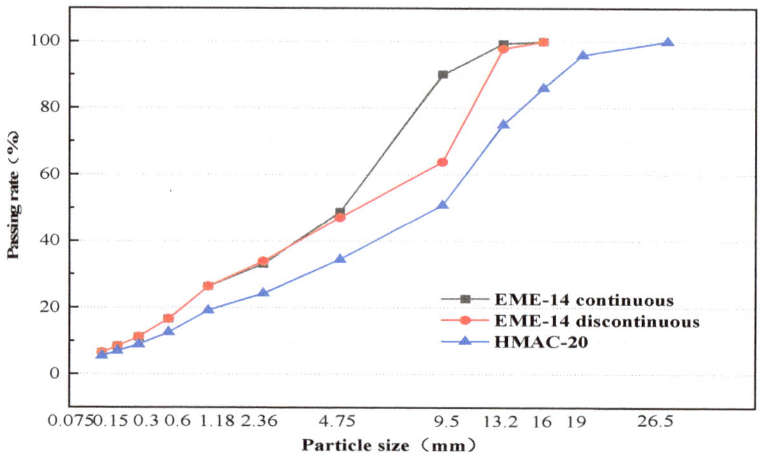

Figure 1. Synthetic gradation curves of three typical high modulus asphalt mixtures.

Table 4. Abundance coefficient calculation table.

Mixture Type	G(%)	S(%)	s(%)	f(%)	ρG	Oil–Stone Ratio/%	Abundance Coefficient
EME-14 (continuous and intermittent)	37	50.4	4.2	7.7	2.730	5.7	3.46
HMAC-20	58.6	33.2	3.4	4.8	2.730	5.3	3.44

2.3.1. Asphalt Binder Performance Test

1. High-temperature rheological properties test

The dynamic shear rheometer DSR (Dynamic Shear Rheometer) is usually used to evaluate the high-temperature stability of an asphalt binder. According to Chinese standard JTG E20-2011 [30], a fully automatic dynamic shear rheometer was used to determine the dynamic shear modulus and phase angle of asphalt.

2. Low-temperature rheological properties test

A Bending Beam Rheometer (BBR) is usually used to evaluate the low-temperature performance of asphalt materials. According to Chinese standard JTG E20-2011 [30], a fully automatic bending beam rheological tester was used to measure the flexural creep stiffness modulus S and creep rate m.

3. Pneumatic rheological rebound test

The maximum creep deformation when loaded and the ability to recover from deformation when unloaded are unique properties of each thermoplastic material. The LTI-210 asphalt quality rapid testing equipment can quickly evaluate the mechanical response and road performance of asphalt materials at a certain temperature, which mainly used nitrogen loading technology to measure the creep and creep recovery capacity of asphalt under single stress or multi-stress conditions.

2.3.2. High Modulus Asphalt Mixture Performance Test

Referring to the design concept of the French high modulus asphalt mixture EME2, the performance of six kinds of high modulus asphalt mixtures, EME-14 continuous + France 15#, EME-14 continuous + China 15#, EME-14 intermittent + France 15#, EME-14 intermittent + China 15#, HMAC-20 + France 15#, HMAC-20 + China 15#, under the test level 1 to 4 was evaluated according to LPC Bituminous Mixtures Design Guide [31]. The test methods are shown in Table 5. Moreover, the low-temperature crack resistance was also evaluated according to JTG E20-2011 [30].

Table 5. High modulus asphalt mixture performance index verification test plan.

Test Level	Test Item	Test Methods
Level 1	Rotational compaction void ratio (%)	EN 12697-31 [32]
	Water stability: Durize test (Compressive strength ratio, %)	EN 12697-12 (Method B) [33]
Level 2	High-temperature stability: French wheel rutting test (30,000 times, 60 °C)	EN 12697-22 [34]
Level 3	Stiffness modulus: complex modulus (MPa, 15 °C, 10 Hz/0.02 s)	EN 12697-26 [35]
Level 4	Fatigue life: two-point bending fatigue of trapezoidal beam (10 °C, 25 Hz, 106 cycles, 130 με)	EN 12697-24 [36]

3. Performance Evaluation of Asphalt Binder

3.1. Penetration, Softening Point, Dynamic Viscosity and Residual Penetration

The data comparison of the penetration, softening point, dynamic viscosity and residual penetration of China 15# and France 15# hard asphalt are shown in Table 2, respectively. The penetration, softening point and dynamic viscosity at 60 °C of France 15# hard asphalt are slightly higher than those of China 15#, indicating that the high-temperature stability of France 15# is relatively better. After short-term RTFOT aging, the residual penetration of China 15# is 8% higher than that of France 15#, indicating that the anti-aging performance of Chinese hard asphalt is better.

3.2. High-Temperature Rheological Properties of Asphalt Binder

The dynamic shear rheometer DSR (Dynamic Shear Rheometer) can be used to evaluate the high-temperature stability of the asphalt binder, and the viscoelastic properties of asphalt materials can be characterized by measuring the complex modulus G^* and phase angle δ of asphalt under different temperature and load conditions [5]. The rutting factor $G^*/\sin\delta$ is used as an index to evaluate the high-temperature rutting resistance of asphalt materials.

G^* represents the rutting resistance of the asphalt material, and δ represents the viscoelasticity of the asphalt material. The larger G^*, the smaller the δ, the better the resistance of the asphalt material to deformation under load. Figure 2 shows the variation of complex modulus G^* and phase angle δ with temperature. It can be seen that the complex modulus G^* and the phase angle δ of the two kinds of hard asphalts have the same trend of change with temperature, but the change range is slightly different. When the temperature is low, the two asphalts have higher complex modulus and smaller phase angle, the asphalt tends to be elastic, and its ability to resist deformation is better; with the increase in temperature, the complex modulus G^* decreases, the phase angle δ increases rapidly, the asphalt tends to be more viscous, and its ability to resist deformation becomes weaker. The complex modulus of France 15# hard asphalt is slightly higher than that of China 15#, and the phase angle is relatively small, indicating that the high-temperature stability of France 15# is better than that of China 15#.

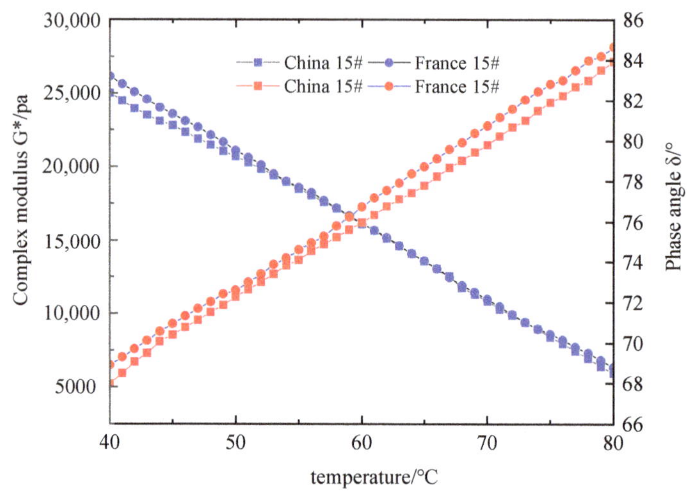

Figure 2. The complex modulus G* and phase angle δ.

Figure 3 shows the variation trend of the rutting factor with temperature. Studies have shown that the greater the anti-rutting factor G*/sinδ, the stronger the anti-rutting ability of the asphalt material and the better its high-temperature stability. It can be seen from the figure that the anti-rutting factor of the two low-grade hard asphalts has the same trend of change, and both gradually decrease with the increase in temperature. The G*/sinδ of the France 15# hard asphalt is larger, indicating that the rutting resistance of France 15# hard asphalt is better than that of China 15# hard asphalt.

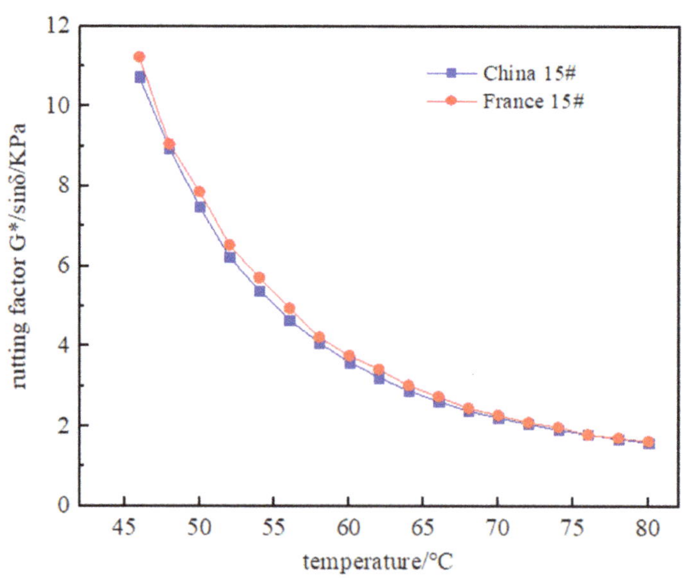

Figure 3. The rutting factor G*/sinδ.

3.3. Low-Temperature Rheological Properties of Asphalt Binder

The Bending Beam Rheometer (BBR) can evaluate the low-temperature performance of asphalt materials. Since most of the low-temperature cracking of asphalt pavement in China occurs in the early stage, the low-temperature rheological performance of asphalt materials after short-term RTFOT aging was analyzed [25,26]. By measuring the flexural creep stiffness modulus S and creep rate m of the two low-grade hard asphalt, the low-temperature crack resistance performance was evaluated. The larger the modulus of stiffness, the smaller the creep rate, and the easier it is to fracture for asphalt material when it is stretched at low temperature.

The deflection changes of the two hard asphalts with loading time are shown in Figure 4. The stiffness modulus and creep rate results are given respectively in Table 6. It can be seen that the stiffness modulus S of China 15# is slightly lower than that of France 15#, and the creep rate is lower than that of France 15#, indicating the crack resistance and stress relaxation performance of the French low-grade asphalt in the low-temperature environment are better than China 15# hard asphalt.

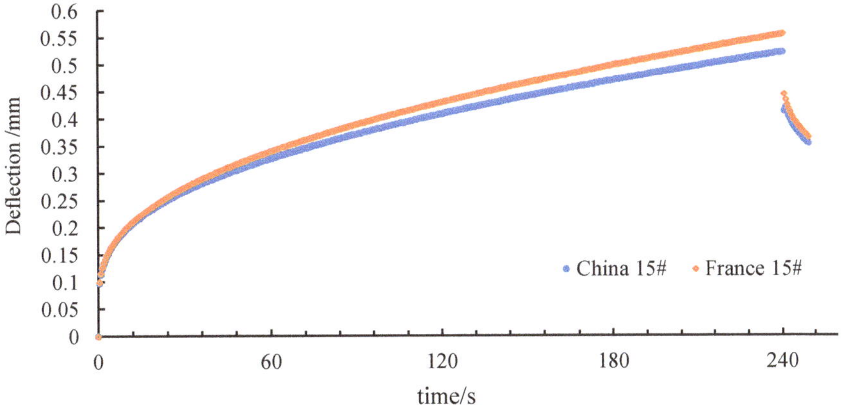

Figure 4. Deflection change of two hard bitumen with loading time.

Table 6. The values of creep stiffness modulus S and creep rate m of two hard bitumen.

Asphalt Type	Creep Stiffness Modulus S/−6 °C (Average Value)	Standard Deviation	Creep Rate m/−6 °C (Average Value)	Standard Deviation
France 15#	256	1.63	0.365	0.021
China 15#	264	2.06	0.36	0.016

3.4. Pneumatic Rheological Rebound Performance

The data from the rebound test of asphalt rheological properties in Figure 5 and Table 7 shows that the change trends of the two low-grade hard asphalts in China and France are the same, but under different stress conditions, the maximum deformation and the degree of deformation recovery when unloading are different. Under the standard conditions of 9.75 Psi, the maximum creep deflection of China 15# hard asphalt is small, and the deformation recovery rate when unloading is high, indicating that Chinese low-grade hard asphalt has high hardness (high modulus) and better viscoelasticity. However, as the pressure increases to 20 Psi, the maximum creep deflection of France 15# hard asphalt is smaller and the deformation recovery rate is higher than China 15#, indicating that the French low-grade hard asphalt is less sensitive to changes in stress levels.

Table 7. Maximum deformation and elastic recovery rate.

Asphalt Material	9.75 Psi		20 Psi	
	Maximum Creep Deflection (mm)	Elastic Recovery Rate (%)	Maximum Creep Deflection (mm)	Elastic Recovery Rate (%)
China 15#	0.0039	82.1	0.0128	61.7
France 15#	0.0046	79.5	0.0104	65.4

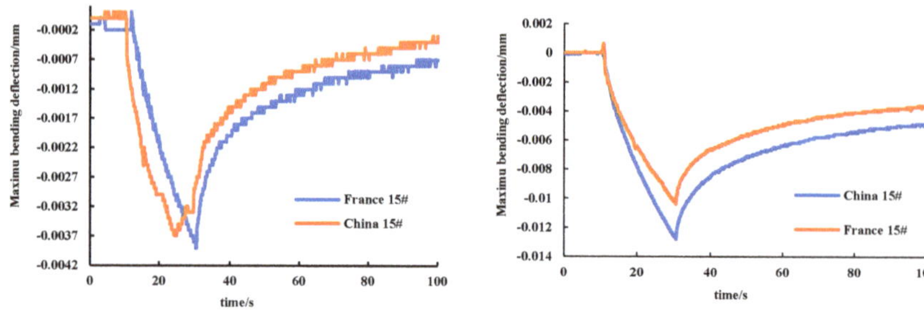

(a) (b)

Figure 5. The deformation and deformation recovery corresponding to (**a**) 9.75 and (**b**) 20 Psi.

4. Analysis and Evaluation of High Modulus Asphalt Mixture Performance

4.1. Porosity Verification and Water Sensitivity Evaluation

According to the French high modulus asphalt mixture design guide LCPC level 1 performance requirements [31], the PCG rotary compaction test and Durize test were carried out on the six designed high modulus asphalt mixtures, and the porosity and water sensitivity of the two low-grade high modulus asphalt mixtures of China and France were analyzed. The results of the void ratio and water sensitivity are shown in Tables 8 and 9, respectively.

Table 8. Porosity of different types of high modulus asphalt mixtures.

Gradation Type	Void Ratio/%	Standard Deviation
EME-14 continuous-France 15#	3.65	0.033
EME-14 discontinuous-France 15#	3.48	0.012
HMAC-20-France 15#	3.58	0.024
EME-14 continuous-China 15#	3.57	0.012
EME-14 discontinuous-China 15#	3.52	0.021
HMAC-20-China 15#	3.63	0.016

Table 9. The compressive strength ratio of different types of high modulus asphalt mixtures.

Gradation Type	Specimen of Immersion Group (Average Value)		Test Pieces of the Comparison Group (Average Value)		Compressive Strength Ratio	Standard Deviation
	Void Ratio (%)	Strength (MPa)	Void Ratio (%)	Strength (MPa)		
EME-14 continuous-France 15#	3.98	15.13	3.88	14.95	101.2	1.29
EME-14 discontinuous-France 15#	2.96	18.53	3.12	18.16	102.04	2.09
HMAC-20-France 15#	3.78	17.38	3.56	17.05	101.94	1.32
EME-14 continuous-China 15#	4.07	14.8	4.16	14.75	100.34	2.16
EME-14 discontinuous-China 15#	3.1	17.41	2.52	17.12	101.69	1.28
HMAC-20-China 15#	4.01	16.15	3.76	15.95	101.25	1.51

It can be seen from Table 8 that the porosity of asphalt mixture specimens of different gradation types obtained by the PCG rotary shear compaction all meet the design index requirements (<6%) of French high modulus asphalt mixture, indicating that the asphalt dosage determined by the calculation of the abundance coefficient formula meets the design requirements. In Table 9, the compressive strength ratios of these six kinds of high modulus asphalt mixtures are greater than 75%, and all meet the control threshold requirements of the level 1 test index. Moreover, the compressive strength ratio of the two low-grade asphalt mixture for the same gradation type is close to 1(1.01:1), indicating that the water stability of the high-modulus asphalt mixtures based on France 15# and China 15# low-grade asphalts is not very different.

4.2. High-Temperature Performanc

Referring to LCPC level 2 rutting performance requirements [34], the French large-scale rutting test was carried out on six kinds of high modulus asphalt mixture under 60 °C to analyze the high-temperature performance. Figure 6 shows the variation of the rutting deformation rate with the increase in rolling times. The rutting deformation rate curve of the six types of mixture has the same changing trend with the increase in the rolling times. In the initial stage, the rutting deformation rate increases rapidly, and with the increase in the rolling times, the rutting deformation rate increases smoothly. As specified in CSN EN 12697-22 [34] and Chinese standard DB 37/T 3564-2019 [37], the rutting deformation rate of the high modulus asphalt mixture should be less than 7.5% under the condition of 60 °C and 30,000 rolling times. It can be seen in Figure 5 that the rutting deformation rates of different types of high modulus asphalt mixtures are all less than 7.5% under the specified 30,000 rolling times, which meets the high modulus asphalt mixture level 2 test performance requirements.

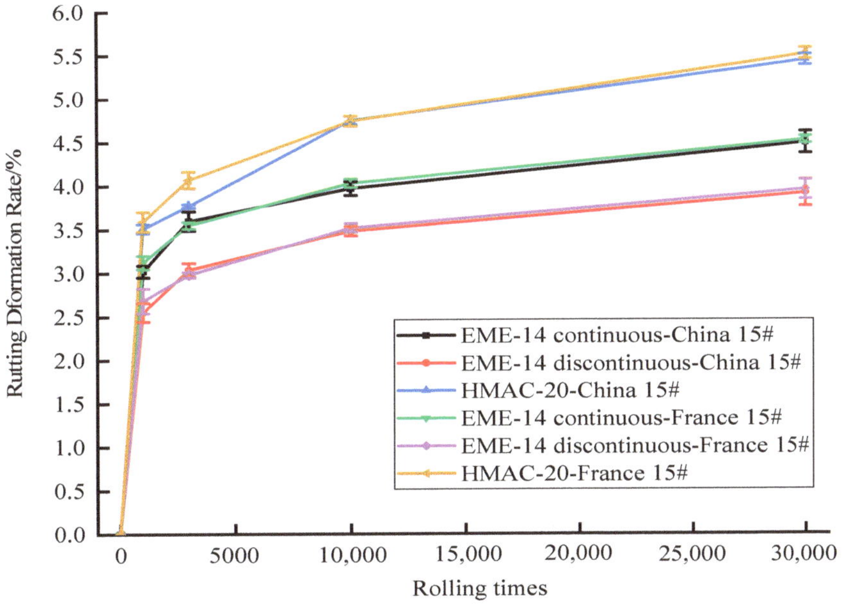

Figure 6. The rutting deformation rate curve of different types of high modulus asphalt mixtures.

According to the test results, the relationship between rutting deformation rate Pi and rolling number N was exported in the double logarithmic coordinate system, as shown in Figure 7, and the fitting curve relationship was exhibited in Table 10. Under the same test conditions, the rutting deformation rate of the three high modulus asphalt mixtures corresponding to China 15# asphalt is smaller than that of France 15# asphalt mixtures, and the rutting deformation rate curve parameter b values for China 15# asphalt mixtures are also smaller. It shows that the high-modulus asphalt mixture corresponding to the Chinese low-grade asphalt has stronger resistance to high-temperature deformation than French, and the high-temperature performance of the mixture is better. As to the rutting deformation rate slope, the values of curve parameter a for France 15# hard asphalt mixtures are smaller than those of China 15# hard asphalt mixtures, which indicated that the high modulus asphalt mixture based on the French low-grade asphalt is less sensitive to load changes.

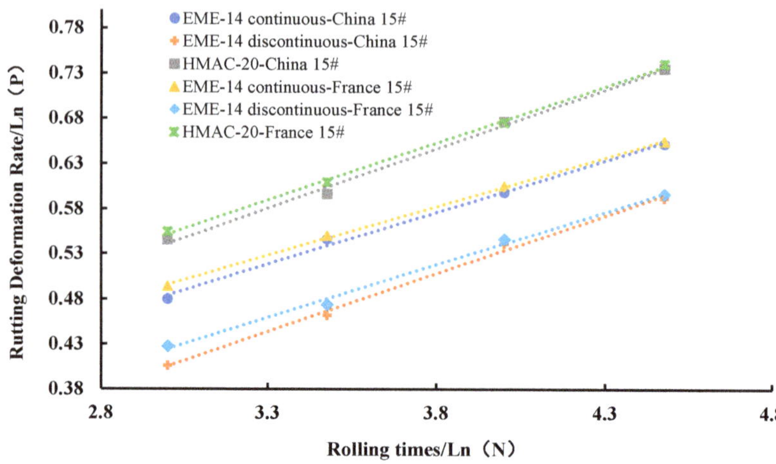

Figure 7. The relationship between rutting deformation rate Pi and rolling number N in the double logarithmic coordinate system.

Table 10. Rutting deformation rate curve.

Mixture Type	a	b	Curve Equation	R^2
EME 14 continuous-France 15#	0.1086	0.1703	$\text{Ln}(Pi) = 0.1086 \times \text{Ln}(N) + 0.1703$	0.9992
EME 14 continuous-China 15#	0.1148	0.1402	$\text{Ln}(Pi) = 0.1148 \times \text{Ln}(N) + 0.1402$	0.9955
EME 14 discontinuous-France 15#	0.1168	0.0749	$\text{Ln}(Pi) = 0.1168 \times \text{Ln}(N) + 0.0749$	0.9955
EME 14 discontinuous-China 15#	0.1285	0.0201	$\text{Ln}(Pi) = 0.1285 \times \text{Ln}(N) + 0.0201$	0.9957
HMAC 20-France 15#	0.126	0.1742	$\text{Ln}(Pi) = 0.126 \times \text{Ln}(N) + 0.1742$	0.9985
HMAC 20-China 15#	0.1316	0.1465	$\text{Ln}(Pi) = 0.1316 \times \text{Ln}(N) + 0.1465$	0.9958

4.3. Low-Temperature Performance

According to Chinese standard JTG E20-2011 [30], the low-temperature bending tests were conducted on six designed EME-14 (continuous and intermittent) and AC-20 asphalt mixtures to evaluate the low-temperature crack resistance performance of the two low-grade high-modulus asphalt mixtures. The maximum flexural strain was taken as the mixture low-temperature performance evaluation index in this paper and the result is shown in Figure 8. As seen from Figure 8, the strain values of the EME-14 continuous, EME-14 discontinuous and AC-20 high modulus asphalt mixtures based on the two low-grade asphalts of China and France are all greater than 2000 MPa and all meet the requirements of bending and tensile strain specified in Chinese standard GB/T36143-2018 [38]. The flexural strain values of France 15# asphalt mixture are about 10% larger than that of China 15#, which showed that the low-temperature crack resistance of high modulus asphalt mixture based on the French low-grade hard asphalt is better.

4.4. Stiffness Modulus

Through the complex modulus test of the CRT-2PT two-point trapezoidal beam, the result of the complex modulus at 15 °C and 10 Hz was exhibited in Figure 9, and the main curves of modulus and phase angle were established, as shown in Figure 10a,b, respectively.

At 15 °C and 10 Hz, the complex modulus values of high-modulus asphalt mixtures based on Chinese and French low-grade asphalt are > 14,000 MPa, which all meet the requirements of the level 3 modulus index (≥14,000 MPa). When the gradation types are the same, the ratios of the modulus values of the two low-grade hard asphalt mixtures are chosen to be 1, indicating that the complex modulus indexes of Chinese and French low-grade hard asphalt mixtures are almost equivalent.

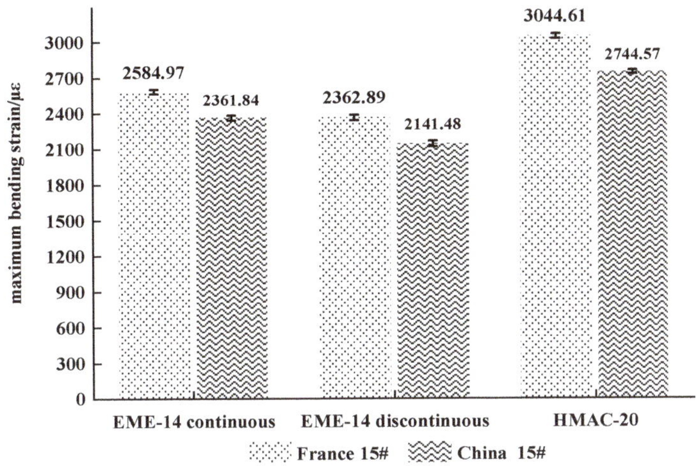

Figure 8. Flexural strain of different types of high modulus asphalt mixtures.

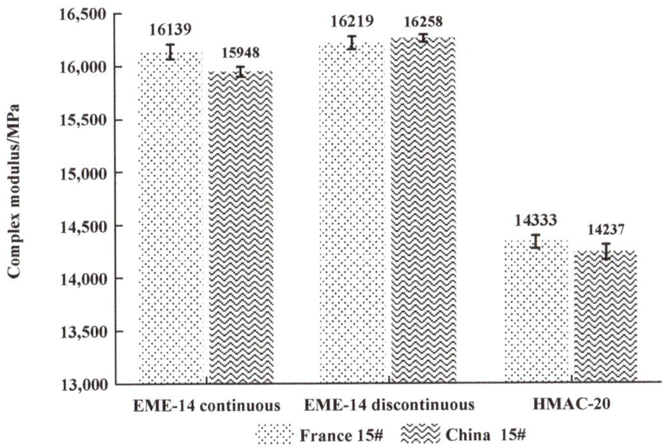

Figure 9. The complex modulus of high modulus asphalt mixtures.

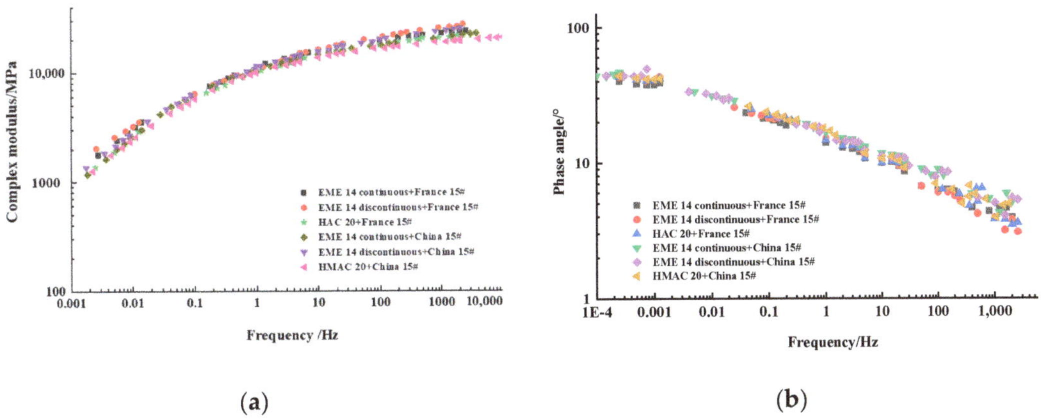

Figure 10. Modulus and phase angle master curves of high modulus asphalt mixtures: (**a**) Modulus; (**b**) Phase angle.

The research on the high and low-temperature performance of the asphalt mixture can also be realized by fitting the master curve. The modulus and phase angle main curve of the mixtures at different temperatures fitted with 15 °C as the reference temperature in Figure 10 shows that the modulus and phase angle of different types of high modulus asphalt mixtures change with the loading frequency in the same trend. As the frequency increased, the modulus of the mixture increased and the phase angle decreased. On the whole, the modulus and phase angle master curves of these two low-grade hard asphalt mixtures are close to overlap, indicating that the mixture modulus and phase angle are similar.

4.5. Fatigue Life

The fatigue test was performed on the designed low-grade high-modulus asphalt mixture referring to CRT-2PT two-point trapezoidal beam fatigue test. The test result is shown in Table 11, and the fatigue curve was drawn according to the test results, as shown in Figure 11.

Table 11. Fatigue life of low-grade high-modulus asphalt mixtures with different grading types.

Gradation Type	Average Void Ratio (%)	Strain Level (με)	Average Fatigue Life (Time)	Standard Deviation
EME-14 Continuous + France 15#	2.22	110 130 150 200	4,422,308 1,390,390 617,943 147,639	11,746.56 4874.22 2153.63 912.68
EME-14 Discontinuous + France 15#	2.02	110 130 150 200	3,765,624 1,129,163 570,676 130,954	16,056.46 5384.39 2062.17 852.25
HMAC-20 France + 15#	2.92	110 130 150 200	2,349,959 1,098,623 482,917 114,285	10,545.53 4472.47 1357.56 519.04
EME-14 Continuous + China 15#	2.46	110 130 150 200	4,638,924 1,490,490 605,541 142,573	10,318.48 3135.88 2196.78 889.59
EME-14 Discontinuous + China 15#	2.30	110 130 150 200	4,054,673 1,242,715 563,702 128,627	11,574.17 3633.73 1838.54 800.34
HMAC-20 + China 15#	3.01	110 130 150 200	2,674,851 1,181,271 472,865 101,752	10,018.67 4154.14 2398.79 838.11

According to the data in Table 8, under the specified test conditions of 10 °C and 25 Hz, the fatigue life of EME-14 continuous + China 15# is 1,490,490 times, the fatigue life of EME-14 discontinuous + China 15# is 1,242,715 times, and the fatigue life of HMAC-20 + China 15# is 1,181,271 times. The fatigue life of EME-14 continuous + France 15# is 1,390,390 times, the fatigue life of EME-14 discontinuous + France 15# is 1,129,163 times, and the fatigue life of HMAC-20 + France 15# is 1,098,623 times. The fatigue times of the above six low-grade high-modulus asphalt mixtures are $\geq 10^6$ times under the conditions of 10 °C, 25 Hz and 130 με, indicating that the fatigue performance of the mixture meets the fatigue performance index requirements of high-modulus asphalt mixtures under French standards.

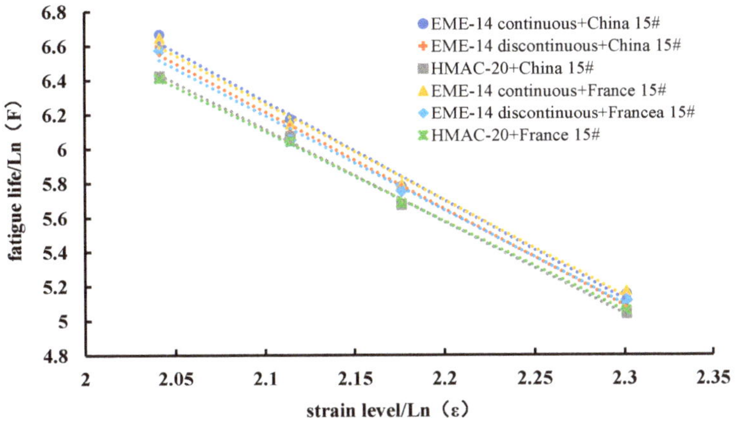

Figure 11. Fatigue curves of high modulus asphalt mixture.

It can be seen from Figure 11 that the fatigue life of different gradation types of high modulus asphalt mixtures based on the two low-grade asphalts of China and France have the same change trend. The fatigue life of the mixture decreases with the increase in strain level, but the degree of fatigue life decrease for different asphalt mixtures is different with the increase in strain.

According to Figure 11, the fatigue model $Ln(N) = a + (1/b) \times Ln(\varepsilon)$ of fatigue life N and strain level ε was established, and the fatigue parameters a and 1/b are shown in Table 12. The larger the value of a, the better the fatigue resistance of the mixture, and the larger the absolute value of 1/b, the more sensitive the fatigue life of the mixture is to changes in strain levels. When the gradation type is the same, the parameter a value of Chinese low-grade high modulus asphalt mixture is slightly larger than that of French low-grade high modulus asphalt mixture. It indicated that Chinese low-grade high-modulus asphalt mixtures showed greater fatigue life and better fatigue performance. As to the parameter 1/b, its absolute value of French low-grade high modulus asphalt mixture is slightly smaller than that of Chinese low-grade high-modulus asphalt mixture, showing that the fatigue performance of the French low-grade high-modulus asphalt mixture is less sensitive to strain changes.

Table 12. Fatigue equation parameters.

Mixture Type	a	1/b	Curve Equation	R^2
EME 14 continuous + France 15#	18.068	−5.6211	$Ln(N) = -5.6211 \times Ln(\varepsilon) + 18.068$	0.9932
EME 14 continuous + China 15#	18.436	−5.7886	$Ln(N) = -5.7886 \times Ln(\varepsilon) + 18.436$	0.9938
EME 14 discontinuous + France 15#	17.764	−5.5089	$Ln(N) = -5.5089 \times Ln(\varepsilon) + 17.764$	0.992
EME 14 discontinuous + China 15#	18.111	−5.6639	$Ln(N) = -5.6639 \times Ln(\varepsilon) + 18.111$	0.9945
HMAC 20 + France 15#	17.09	−5.232	$Ln(N) = -5.232 \times Ln(\varepsilon) + 17.09$	0.9994
HMAC 20 + China 15#	17.476	−5.4089	$Ln(N) = -5.4089 \times Ln(\varepsilon) + 17.476$	0.9981

5. Conclusions

This research assessed the mechanical performance of low-grade hard asphalt in China and France comparatively. The high and low-temperature performance, water stability, two-point bending complex modulus test and bending fatigue of trapezoidal test were performed to evaluate the road performance of high-modulus asphalt mixtures for Chinese and French hard asphalt, respectively. The following conclusions were given based on the results in this study:

1. The performance indicators of the China 15# hard asphalt meet the requirements of the binder used in the high modulus asphalt mixture. China 15# and France 15# hard

bitumen are similar in performance but slightly different. The French low-grade hard asphalt showed better high-temperature stability and less sensitivity to changes in the level of change, while the Chinese low-grade hard asphalt had better anti-aging performance and viscoelastic properties.
2. The high and low-temperature performance, water stability, stiffness modulus and fatigue life of the high-modulus asphalt mixture based on China 15# hard asphalt all meet the specification requirements, indicating that the application of the high modulus asphalt mixture of Chinese low-grade hard bitumen can be carried out in China.
3. The road performance of the Chinese and French 15# hard asphalt mixtures is similar but slightly different. Chinese low-grade hard asphalt mixtures showed better high-temperature performance and fatigue performance. The French 15# hard asphalt mixture was less sensitive to changes in load times and strain levels and performed better at low temperature, which can increase the low-temperature performance by about 10%. There is not much difference in water stability and stiffness modulus for the two low-grade hard asphalts mixtures.

The results of this study helped to evaluate whether the Chinese hard asphalt mixture can meet the performance index requirements of French high modulus asphalt mixture and verified the feasibility of the application of Chinese hard asphalt in road engineering. The results will also guide the construction of hard bitumen high modulus asphalt mixture pavement of the highway in China, promoting the wide application of high modulus asphalt mixture in China.

Author Contributions: G.Z., W.W. and Q.X. conceived and designed the experiments. Q.X., J.L. and X.L. performed the experiments. W.W., X.Y. and Y.H. analyzed the data. G.Z., Q.X. and J.W. contributed reagents/materials/analysis tools. W.W. and Q.X. wrote the paper. W.W., Q.X. and J.W. revised the paper. All authors have read and agreed to the published version of the manuscript.

Funding: This research was funded by the National Key R&D Program of China, grant No.2018YFB1600103, the National Natural Science Foundation of China, grant No. 42107213, Shandong Provincial Natural Science and Foundation, grant No. ZR2020QE271 and Shandong Provincial Key Research and Development Program, grant No. 2019GSF109020.

Institutional Review Board Statement: Not applicable.

Informed Consent Statement: Not applicable.

Data Availability Statement: Data sharing is not applicable to this article.

Conflicts of Interest: The authors declare no conflict of interest.

References

1. *NF P98-141*; Asphalt Mixture-Wearing Course and Base Course: High Modulus Asphalt Mixture-Definition-Classification-Characteristics-Production and Construction. French Standardization Association: Paris, France, 1999.
2. Xia, X.P.; Zhang, R.N.; Fu, H.W. Characteristics and new development of high modulus asphalt mixture. *Chin. For. Highw.* **2005**, 2, 123–127.
3. Serfass, P.; Bense, P.; Pellevoisin, P. Properties and New Developments of High Modulus Asphalt Concrete. In Proceedings of the Lecture Series 8th International Conference on Asphalt Pavements, Seattle, WA, USA, 10–14 August 1997.
4. Jean, F.C. *Development and Uses of Hard-Grade Asphalt and of High-Modulus Asphalt Mixes in France*; Transportation Research Board: Washington, DC, USA, 2001; pp. 12–31.
5. Zhou, Q.H.; Sha, A.M.; Yang, Q. Experimental study on mechanical properties of high modulus asphalt concrete. *J. Zhengzhou Univ.* **2008**, 3, 128–131.
6. Witczak, M. *Simple Performance Tests: Summary of Recommended Methods and Database*; NCHRP-547; Transportation Research Board: Washington, DC, USA, 2005; pp. 2–7.
7. Montanelli, E.F. Fiber/Polymeric Compound for High Modulus Polymer Modified Asphalt (PMA). *Procedia-Soc. Behav. Sci.* **2013**, *104*, 39–48. [CrossRef]
8. Gottam, S.R.; Adepu, R.; Vijayapuri, V.R.; Ramu, P. Laboratory evaluation of hard grade bitumen produced with PPA addition. *Int. J. Pavement Res. Technol.* **2020**, *14*, 505–512.

9. Maupin, G.W.; Diefenderfer, B.K. Design of a High-Binder-High-Modulus Asphalt Mixture. Virginia Transportation Research Council: Charlottesville, VA, USA, 2006; pp. 1–30.
10. Silvino, D.C.; Luis, P.S. Assessing permanent deformation resistance of high modulus asphalt mixtures. *J. Transp. Eng.* **2006**, *132*, 394–401.
11. Didier, D.; Philippe, G.; Patrice, O. Comparative experiment for PMB and hard bitumen on a thin anti rutting maintenance treatment. In Proceedings of the Vienna: 3rd Eurasphalt & Eurobitume Congress, Vienna, Australia, 12–14 May 2004.
12. Serfass, J.P.; Bauduin, A.; Garnier, J.F. High modulus asphalt mixes laboratory evaluation, practical aspects and structural design. In Proceedings of the 7th International Conference on Asphalt Pavements, Nottingham, UK, 16–20 August 1992; Volume 1, pp. 275–288.
13. Lee, H.J.; Lee, J.H.; Park, H.M. Performance evaluation of high modulus asphalt mixtures for long life asphalt pavements. *Constr. Build. Mater.* **2007**, *21*, 1079–1087.
14. Yang, J. Study on low temperature performance of gussasphalt on steel decks with hard bitumen. *J. Southeast Univ.* **2003**, *19*, 160–164.
15. Corte, J.F. Development and uses of hard grade asphalt and of high modulus asphalt mixes in France. *Transp. Res. Circ.* **2003**, *503*, 12–31.
16. Newcomb, D.E.; Buncher, M.; Huddleston, I.J. *Concepts of Perpetual Pavements*; Transportation Research Board: Washington, DC, USA, 2001; pp. 4–11.
17. Claude, B.; Joëlle, V.; Lieve, G.; Ann, V.; Stefan, V.; Lucien, H. A Comparative High-modulus Asphalt Experiment in Belgium. *Transp. Res. Arena Eur.* **2008**, *32*, 22–38.
18. De Visscher, J.; Vansteenkiste, S.; Vanelstraete, A. Test sections in high-modulus asphalt: Mix design and laboratory performance testing. In Proceedings of the 4th Eurasphalt and Eurpbitume Congress, Copenhagen, Denmark, 21–23 May 2008; pp. 1–11.
19. De Backer, C.; Glorie, L.; Reynaert, R. Test sections in high-modulus asphalt:a comparative experiment with ten variants. In Proceedings of the 4th Eurasphalt and Eurpbitume Congress, Copenhagen, Denmark, 21–23 May 2008; pp. 1–11.
20. Rohde, L.; Ceratti, J.A.P.; Nunez, W.P.; Vitorello, T. Using APT and laboratory testing to evaluate the performance of high modulus asphalt concrete for base courses in Brazil. In Proceedings of the Third International Conference Centro de Estudios y Experimentación de Obras Públicas (CEDEX) Transportation Research Board, Madrid, Spain, 1–3 October 2008.
21. Guo, Q.; Tan, J. Application of High Modulus Asphalt Concrete in Pavement Rutting Treatment. *J. Highw. Transp. Res. Dev.* **2006**, *7*, 59–62.
22. Han, C. Design and Application Research of Durable High Modulus Asphalt Mixture Based on Hard Asphalt. Master's Thesis, Southeast University, Nanjing, China, 2018.
23. Dong, Y.; Tan, Y. Research on high temperature performance of hard grade asphalt based on repeated shear creep test. *High* **2015**, *2*, 160–164.
24. Liang, C.; Liu, F. Research on properties of 30# hard asphalt and asphalt mixture. *J. China Foreign Highw.* **2006**, *6*, 185–188.
25. Xiong, W. Experimental study on road performance of low-grade asphalt mixture. *Highw. Automot. Appl.* **2008**, *1*, 80–82.
26. Zhao, L.; Han, C.; Zheng, G.; An, F.; Dong, S. Application of Domestic Hard Asphalt in Rutting Treatment Project of Trunk Highway. *Shanghai Highw.* **2019**, *4*, 6–8+64+3.
27. *JTG F40-2004*; Technical Specifications for Construction of Highway Asphalt Pavements. People's Communications Press: Beijing, China, 2005.
28. *BS EN 13924*; Bitumen and Bituminous Binders-Specifications for Hard Paving Grade Bitumens. British standard: London, UK, 2006.
29. Corté, J.-F.; Serfass, J.-P. The French approach to asphalt mixtures design: A performance-related system of specifications. *Assoc. Asph. Paving Technol.* **2000**, *69*, 794–834.
30. *JTG E20-2011*; Standard Test Methods of Bitumen and Bitumen Mixtures for Highway Engineering. People's Communications Press: Beijing, China, 2011.
31. France, R.S.T.; "Asphalt Mixture Design" Working Group. *LPC Bituminous Mixtures Design Guide*; Laboratoire Central des Ponts et Chaussées: Paris, France, 2010.
32. *CSN EN 12697-31*; Bituminous Mixtures-Test Methods for Hot Mix Asphalt-Part 31: Specimen Preparation by Gyratory Compactor. The Commonwealth Standards Network: London, UK, 2007.
33. *CSN EN 12697-12*; Bituminous Mixtures-Test Methods for Hot Mix Asphalt Part 12: Determination of the Water Sensitivity of Bituminous Specimens. The Commonwealth Standards Network: London, UK, 2008.
34. *CSN EN 12697-22*; Bituminous Mixtures-Test Methods for Hot Mix Asphalt-Part 22: Wheel Tracking. The Commonwealth Standards Network: London, UK, 2020.
35. *CSN EN 12697-26*; Bituminous Mixtures-Test Methods for Hot Mix Asphalt-Part 26: Stiffness. The Commonwealth Standards Network: London, UK, 2003.
36. *CSN EN 12697-24*; Bituminous Mixtures-Test Methods for Hot Mix Asphalt Part 24: Resistance to Fatigue. The Commonwealth Standards Network: London, UK, 2012.
37. *DB 37/T 3564-2019*; Anti-Rutting and Anti-Fatigue High Modulus Asphalt Mixture Design and Construction Technical Specifications. Shandong Provincial Market Supervision Administration: Jinan, China, 2019.
38. *GB/T 36143-2018*; High Modulus Anti-Fatigue Asphalt Mixture for Road. People's Communications Press: Beijing, China, 2018.

coatings

MDPI

Article

Experimental Study and Performance Characterization of Semi-Flexible Pavements

Guosheng Li [1], Huan Xiong [1,*], Qi Ren [2], Xiaoguang Zheng [2] and Libao Wu [2]

[1] Chengdu Xingcheng Construction Management Co., Ltd., Chengdu 610041, China; gsl11041@hotmail.com
[2] Shanghai Municipal Engineering Design Institute (Group) Co., Ltd., Shanghai 200092, China; qrendavis@gmail.com (Q.R.); zhengxiaoguang@smedi.com (X.Z.); wulibao@smedi.com (L.W.)
* Correspondence: hxiong9527@hotmail.com

Abstract: Semi-flexible pavement (SFP) is made up of a porous skeleton of asphalt mixture and cement grout. This special structure granted SFP superior strength and durability and made it a promising solution for the paving of heavy trafficked area. This paper performed in-depth study on the mechanistic behavior of SFP. Firstly, the volumetric mix design of SFP was introduced, and followed with strength, moisture susceptibility, viscoelastic behavior, fatigue life as well as rutting resistance through a series of laboratory tests. Marshall stability tests and dynamic stability tests suggested that SFP gained fair strength and rutting resistance from the curing of cement grout. Meanwhile, SFP was found not sensitive to freeze–thaw cycles through indirect tensile tests. In dynamic modulus tests, SFP exhibited significant viscoelastic behaviors. In four-point beam fatigue tests, the average fatigue lives of SFP reached 85.4 k loading repetition under 400 $\mu\varepsilon$ level. In Hamburg wheel tracking tests, the ultimate rutting depth of SFP was smaller than 2.5 mm. The viscoelastic behavior and rutting propagation of SFP was characterized with master curve and power function by fitting the test results. SFP was also compared with traditional asphalt mixtures in MMLS3 accelerated tests and its performance turned out to be prevailing.

Keywords: semi-flexible pavement; Marshall stability; dynamic modulus; four-point beam fatigue test; Hamburg wheel tracking; MMLS3

Citation: Li, G.; Xiong, H.; Ren, Q.; Zheng, X.; Wu, L. Experimental Study and Performance Characterization of Semi-Flexible Pavements. *Coatings* **2022**, *12*, 241. https://doi.org/10.3390/coatings12020241

Academic Editor: Andrea Simone

Received: 13 January 2022
Accepted: 10 February 2022
Published: 13 February 2022

Publisher's Note: MDPI stays neutral with regard to jurisdictional claims in published maps and institutional affiliations.

Copyright: © 2022 by the authors. Licensee MDPI, Basel, Switzerland. This article is an open access article distributed under the terms and conditions of the Creative Commons Attribution (CC BY) license (https://creativecommons.org/licenses/by/4.0/).

1. Introduction

Semi-flexible pavement (SFP) is a composite structure of open-graded asphalt mixture filled with cement grout. It was firstly proposed by French engineers in the 1960s, known as Salviacim [1]. Starting from 1970s, this technique was further utilized in airfields across Europe and North America with extensive studies on its mechanical properties [2]. During this period, researchers referred to this material as resin modified pavement [3] or Densiphalt [4]. With better knowledge of its performance, subsequent researchers in recent decades preferred the acronym "SFP" [5] to highlight its semi-flexible behavior or "grouted macadam" [6] to emphasize its composition. Despite its various names in history, the fundamental idea of filling the air voids in porous asphalt mixture with cement grout remained consistent [7].

SFP was initially designed as an alternative solution to concrete pavement due to its jointless feature as well as lower cost. Early applications in airfield pavements were positively reported [8]. The heavy and channelized load caused by aircraft wheels was challenging to most asphalt pavements, but SFP was able to maintain fair rutting resistance skid resistance and durability. The superior performance soon received worldwide recognition. The application of SFP was thus expanded to bridge deck pavements [9], heavy trafficked intersections [10], and bus rapid transit systems [11].

The material design of SFP included two key elements, open-graded asphalt mixture and cement grout. The open-graded asphalt mixture was supposed to possess a void ratio of at least 25% [12], in order to permit the cement infiltration. The design of cement grout was supposed to maintain a balance between fluidity and strength [13]. In recent years, innovative techniques and methodologies have been frequently used in SFP design. Husain et al. [14] studied the effects of aggregate gradation and concluded that air voids in asphalt matrix would positively contribute to the compressive strength of SFP. This behavior was contrary to traditional asphalt pavements. As for moisture susceptibility concerns, Fakhri and Mottahed [15] studied the performance of warm mixed asphalt mixture containing RAP and nanoclay, which employed a similar mechanism with SFP. The conclusion suggested that enhancing the strength of asphalt mixture could also result in better moisture performance. A comparative study of rutting resistance and moisture damage test methods was carried out by Fakhri et al. [16]. Fair correlation was found between deformation strength, dynamic creep test, and wheel tracking results. Wang et al. [17] added carboxyl latex into cement grout and used a warm-mix technique in open-graded asphalt mixture. Modified SFPs were evaluated with various tests and compared with unmodified specimens. The comparison suggested that carboxyl latex could effectively improve rutting resistance as well as fatigue lives. Cai et al. [18] used nanoindentation, scanning electron microscopy as well as energy dispersive spectroscopy to identify the microstructural characteristics of SFP. It was found that the increase in grout material would further enhance the bonding between asphalt and cement. Cai et al. [19] evaluated the interlocking of SFP with X-ray computed tomography. The work proposed a classification method by identifying the effective contacts between aggregates and cement grout, and concluded that only a limited number of contacts were helpful to the bonding.

Despite the extensive research on the material properties of SFP, its structural design has not been thoroughly studied yet. Oliveira et al. [20] investigated the fatigue life of SFP with two comparative approaches. The empirical fatigue threshold was found conservative and thus damage accumulation was necessary to make more accurate estimations. Bharath et al. [21] used cored specimens from SFP test sections to evaluate the mechanical properties. This study employed indirect tensile strength and resilient modulus tests, and highly commended the performance of SFP. Skid resistance and rutting depth were also recorded in subsequent observations and were found satisfactory. Zhang et al. [22] conducted performance tests with indirect tensile approach. The dynamic modulus of SFP was found higher than that of traditional asphalt mixture. Fair resistance to rutting and moisture damage was also observed. Despite the efforts, it should also be noted that a standard test-characterization-prediction procedure for SFP was still absent. Although Mechanistic-Empirical Pavement Design Guide (MEPDG) was introduced two decades ago [23], SFP has not been incorporated into the framework yet.

Based on previous research, there were still a few concerns regarding the SFP performance and modeling. This paper intended to perform an in-depth investigation in terms of the mechanistic behavior as well as model characterization of SFP. A series of laboratory tests were conducted to address multiple concerns as listed in Table 1. Tests 1 to 4 were strength/stability-based tests, which were used to validate the mix design and provide a fundamental understanding of the SFP material. Tests 5 to 8 were more advanced performance-oriented tests, aiming to quantitatively characterize the mechanistic behavior under simulated traffic loading. The study of this work was used to support the pavement construction of the latest Chengdu Bus Rapid Transit stations.

Table 1. Performance test of SFP and the major concerns.

No.	Test Names	Concerns
1	Marshall Stability	Influence of cement curing
2	Dynamic Stability	Rutting resistance
3	Indirect tensile strength	Freeze–thaw cycles
4	Flexural Strength	Flexural strength
5	Dynamic modulus	Viscoelastic behavior
6	Four-point beam fatigue	Fatigue life
7	Hamburg wheel tracking	Rutting propagation
8	MMLS3 accelerated wheel tracking	Comparison with asphalt mixtures

2. Materials Design

The mix design of SFP specimen consisted of 3 steps. Firstly, an open-graded asphalt mixture was designed using volumetric method with a gyratory compactor. Secondly, the water–cement ratio of cement grout was determined based on strength, fluidity, shrinkage rate as well as bleeding rate. Lastly, the strength and stability of SFP must be tested through a series of tests to validate the mix design.

2.1. Open-Graded Asphalt Mixture

Gradation was a primary concern in the mix design of asphalt mixture. In this study, 3 different gradations were prepared, as shown in Figure 1. The first and second gradation were close to the asphalt mixture commonly used in permeable pavements. The tested void ratios of the fabricated mix were 23% and 19%, respectively. Since the void ratios were not satisfactory for SFP mix, the final gradation was further adjusted to achieve a void ratio of 25%.

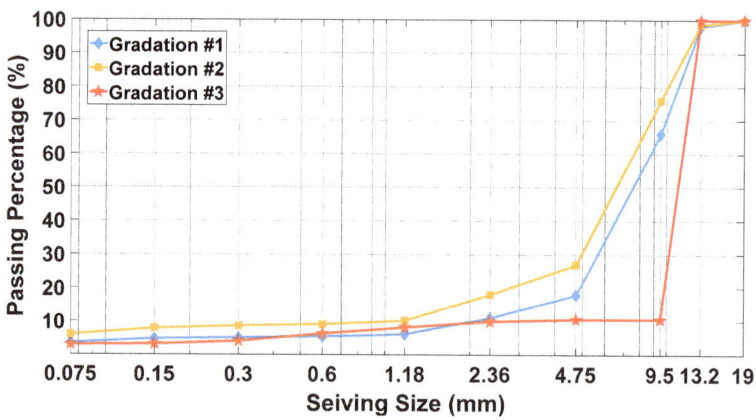

Figure 1. Gradation of open-graded asphalt mixture. Void ratios are 23%, 19% and 25% for #1, #2 and #3, respectively.

A high-viscosity modified asphalt was used to prepare the open-graded asphalt mixture using volumetric design method. The penetration was 68 (0.1 mm) and viscosity was 70,000 Pa·S based on vacuum capillary viscometer test. The asphalt content was determined by a group of Marshall stability tests and eventually adjusted to 2.5%. Cylinder specimens were prepared for dynamic modulus tests and Hamburg wheel tracking. As for four-point beam fatigue tests, slab specimens with the same gradation and asphalt content were compacted with a roller.

2.2. Cement Grout

For cement-based material, the water–cement ratio was a key parameter in the design. The rise of this ratio would result in a lower strength but better workability, i.e., easier to flow and more convenience in the construction phase. Thus, an ideal water–cement ratio should be able to balance the concerns of strength as well as workability. It was selected based on flexural strength, compressive strength, fluidity, shrinkage rate and bleeding rate. The strength test results were shown in Figure 2.

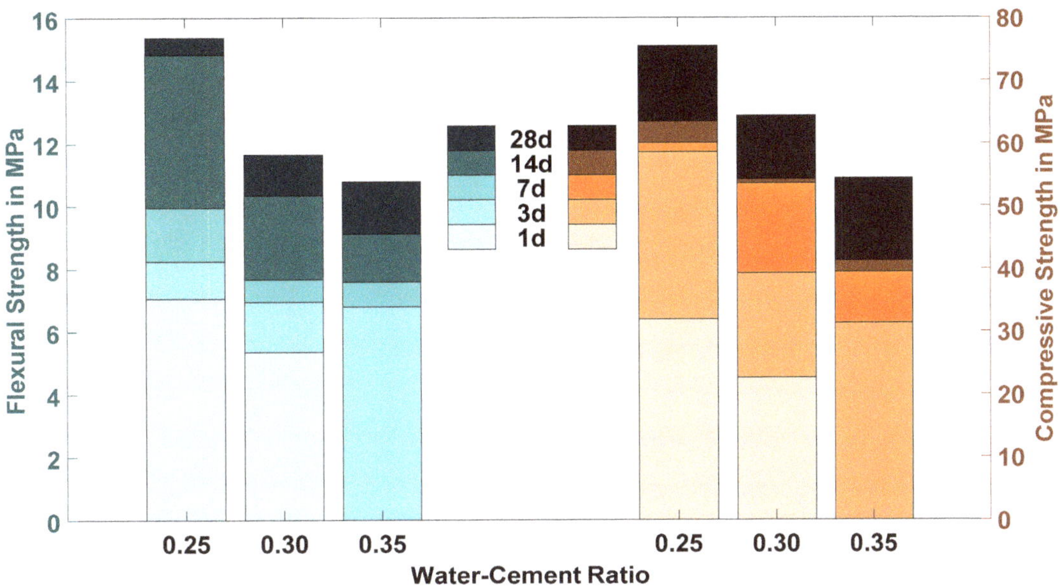

Figure 2. Flexural strength and compressive strength of cement grout given different curing times.

For the cases of 0.35 water–cement ratio, the specimen cannot cure in 1 d. It could be seen that the impact of water–cement ratio was obvious. Both flexural strength and compressive strength increased significantly with curing time. The difference increasing rate also suggested certain anisotropy. As water–cement ratio raised from 0.25 to 0.35, the 28 d-strength would be decreased by 30% and 28% percent in flexural mode and compressive mode, respectively. The tradeoff between strength and workability needed to be further investigated through a series of tests on fluidity, shrinkage rate and bleeding rate, as shown in Table 2.

Table 2. Fluidity, shrinkage rate and bleeding rate at different w/c ratios.

W/C	Fluidity in s	Shrinkage Rate in %	Bleeding Rate in %
0.25	11.2	0.10	1.8
0.30	9.3	0.15	2.1
0.35	8.5	0.28	2.5

The final water–cement ratio was set to 0.25 to balance strength and workability of cement grout. After the open-graded asphalt mixture was fabricated and cooled, the specimens were covered in a plastic membrane and placed in a cylinder or rectangular mold. The cement grout would be poured from the top surface to form SFP specimens, as shown in Figure 3.

(a)

(b)

Figure 3. SFP specimen fabrication: (**a**) filling cement grout; (**b**) cured SFP specimen.

2.3. Mix Design Validation

The designed SFP specimens were validated through a series of tests to investigate the compressive strength, flexural strength and stability, including Marshall stability tests, dynamic stability tests, freeze–thaw cycles and fracture strength tests.

Marshall stability test was a classical method to evaluate the stability of HMA specimens. A compressive load would be applied from the lateral direction of the cylinder at a constant loading rate of 50.8 mm/min till failure. For SFP specimens, this simple method could be used to quantify the effect of cement curing. Five groups of specimens were tested. The first group was tested before cement grout was filled. The second and third group was cured for 3 and 7 days, respectively. The fourth group was tested after 7 days curing and 48 h immersion to examine moisture susceptibility. The last group was tested after 28 days curing, but the strength exceeded the range of the equipment. The test results were plotted in Figure 4.

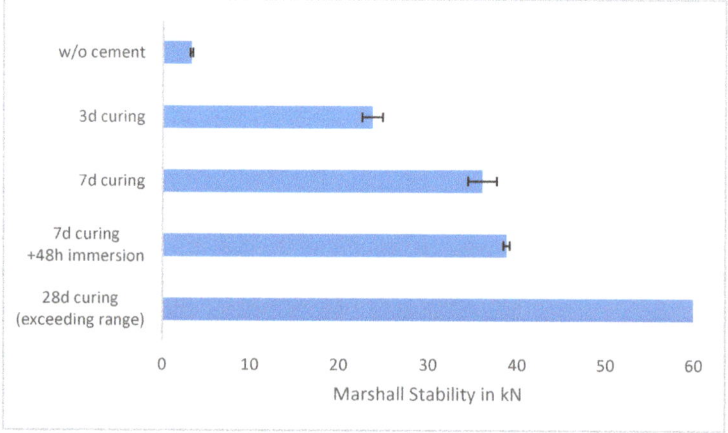

Figure 4. Marshall stability test results at different curing times.

The Marshall stability increased along with the curing time. After 3 d curing, it increased approximately 7 times compared with the original open-graded asphalt mixture. In the fourth group, the stability further increased after immersion in water, which suggested that the cement was further cured. The moisture susceptibility was proofed satisfactorily. As for the last group, the 60 kN result represented the maximum value the equipment could measure, so the exact stability remained unknown. In contrast, the Marshall stability of 2 commonly used HMA in China, SMA-13 and AC-20, were both smaller than 20 kN. It should also be noted that Marshall stability test was not a feasible test method for rigorous study of the mechanistic behavior of SFP. In this study, it served as a tentative attempt to capture the influence of cement curing.

Dynamic stability was an efficient approach to study the rutting resistance from qualitative level. A slab SFP specimen would be loaded with a rubber wheel. The tire pressure was set to 0.7 MPa to simulate traffic load. The loading times needed to cause 1 mm rut depth was recorded as dynamic stability. For SFP specimens, this method was used to study the effect of curing as well as loading temperature. Four groups of dynamic stability tests were conducted. The test results were shown in Figure 5.

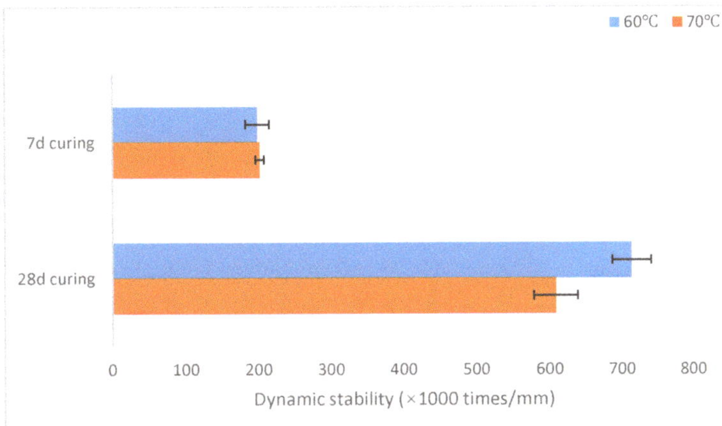

Figure 5. Dynamic stability test results at different test temperature and curing times.

It could be seen that the curing time could greatly affect the dynamic stability of SFP. The rutting resistance of 28 d-curing specimens were 259% and 202% percent higher than that of 7 d-curing specimens. The improvement of rutting resistance was supposed to be caused by the curing of cement grout. The dynamic stability of SFP material was about 10 times greater than commonly used asphalt mixtures.

As for indirect tension test, a vertical load would be applied from the top and cause tension in the center of the specimen. The test was in strain-controlled mode and the maximum tensile stress was recorded as the strength. The indirect tensile strength and void ratio variation were tested after freeze–thaw cycles. The test results were shown in Figure 6.

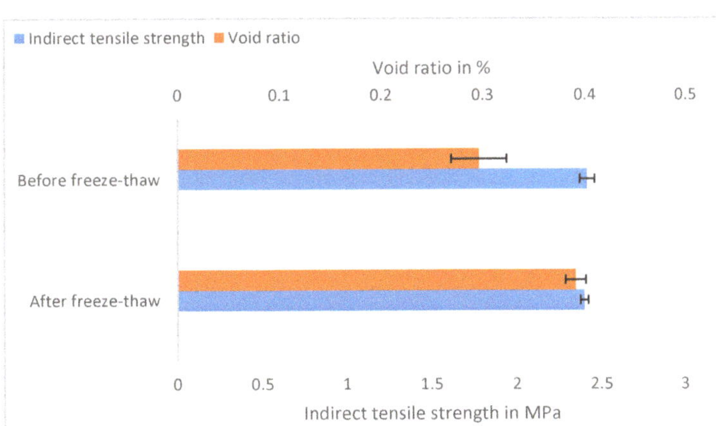

Figure 6. Indirect tensile strength and void ratio changes after freeze–thaw cycles.

The indirect tensile strength of SFP specimens after freeze–thaw cycles did not significantly decrease. The specimens maintained approximately 99.4% tensile strength. Meanwhile, the void ratio increased from 0.296% to 0.391%. The test results suggested that SFP material could be regarded as waterproof.

The flexural strength test was performed at −10 °C to study the influence of low temperature conditions. The size of the beam specimen was 250 mm × 35 mm × 30 mm. The specimen was supported at a span of 200 mm and loaded from the center at a constant rate of 50 mm/min till failure. In total 7 SFP specimens were tested. The flexural strength was tabulated along with their failure strains in Table 3.

Table 3. Statistical results of flexural strength tests.

-	Flexural Strength in MPa	Failure Strain in $\mu\varepsilon$
Average	11.18	2233
Median	11.16	2283
75% quantile	9.95	1983
Standard deviation	0.87	217

The average flexural strength and failure strain of SFP specimens reached 11.18 MPa and 2233 $\mu\varepsilon$, respectively. The flexural strength range of dense graded asphalt was between 4.5 and 7.5 MPa, according to Cui et al. [24], depending on the angularity and sphericity. Obviously, SFP possessed significantly greater strength.

3. Performance Tests

3.1. Performance Test Method

As a composite material of asphalt mixture and cement, the application of SFP in pavements involved the following concerns:

(1) Did viscoelasticity exist in SFP since open-graded asphalt mixture was part of its composition?
(2) Under repeated loading, what was the expected fatigue life and rutting depth of SFP?
(3) How good was the performance of SFP compared with other asphalt mixture?

The strength and stability tests conducted in Section 2 was not sufficient to address these issues, so a few more advanced test methods were needed. In this section, dynamic modulus test was used to study the viscoelastic behavior by applying multi-frequency loads at different temperatures, and a sigmoidal master curve would be fitted to characterize the modulus variation under these conditions. Fatigue life and rutting depth would be studied and characterized by four-point beam fatigue test and Hamburg wheel tracking test, respectively. In the end, a direct comparison between SFP and other commonly used asphalt materials was made through a one-third scale accelerated loading test using Model Mobile Load Simulator (MMLS3). The tests setup was shown in Figure 7.

A dynamic modulus test was conducted using a universal testing machine (UTM) in accordance with AASHTO T 342 [25], as shown in Figure 7a. A cylindrical SFP specimen with a diameter of 100 mm and height of 150 mm was subjected to a sinuous load with a series of frequencies of 0.1, 0.5, 1, 5, 10 and 25 Hz. The deformation of the central 70 mm was measured with three linear variable differential transformers (LVDTs). The test temperature was set to 20, 40 and 60 °C.

Four-point beam fatigue tests were employed to study the fatigue resistance of SFP material. The specimen preparation and test equipment were shown in Figure 7b. The size of the specimens was 380 mm × 50 mm × 63.5 mm. The curing time was 28 days. The test temperature was set to 15 °C and loading frequency was 10 Hz. The specimen was loaded at a constant strain of 400 $\mu\varepsilon$.

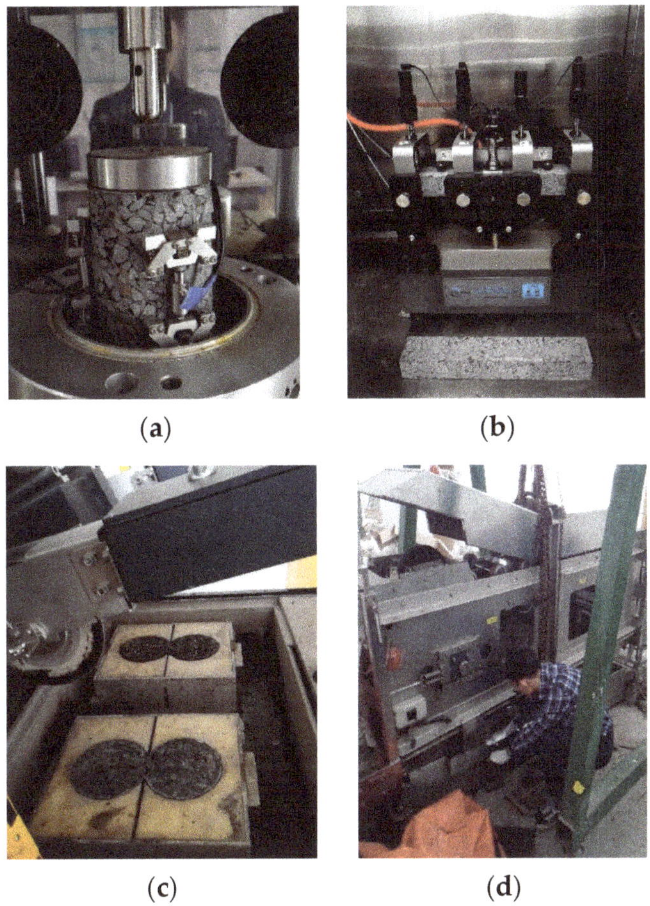

Figure 7. Performance tests of SFP. (a) Dynamic modulus test; (b) four-point beam fatigue test; (c) Hamburg wheel tracking test; (d) MMLS3 accelerated wheel tracking test.

The rutting resistance of SFP was evaluated using Hamburg wheel tracking (HWT) test. For HWT tests, 150 mm × 60 mm cylinder specimens were fabricated with gyratory compactor. The test condition was 50 °C and water bath. The load was applied with a steel wheel with a magnitude of 0.7 MPa pressure. The test would terminate when rutting depth was greater than 10 mm or the loading repetition reached 20,000 times The HWT apparatus and specimens were shown in Figure 7c.

MMLS3 accelerated test was used to compare the rutting propagation between SFP specimens and typical asphalt mixtures. The advantage of this device was validated by Epps et al. [26] and Lee et al. [27]. The MMLS3 device and test setup were shown in Figure 7d. The size of the specimens was 900 mm × 200 mm with a thickness of 100 mm. The specimen would be bathed in 20 °C air during the test period. The loading system would apply channelized loads with rubber tires at a magnitude of 50 kN and tire pressure of 700 kPa to simulate real traffic loads. The loading frequency was 5000 times/h to accelerate rutting propagation.

3.2. Test Results and Analysis

3.2.1. Dynamic Modulus Test

The test results of dynamic modulus tests were plotted in Figure 8.

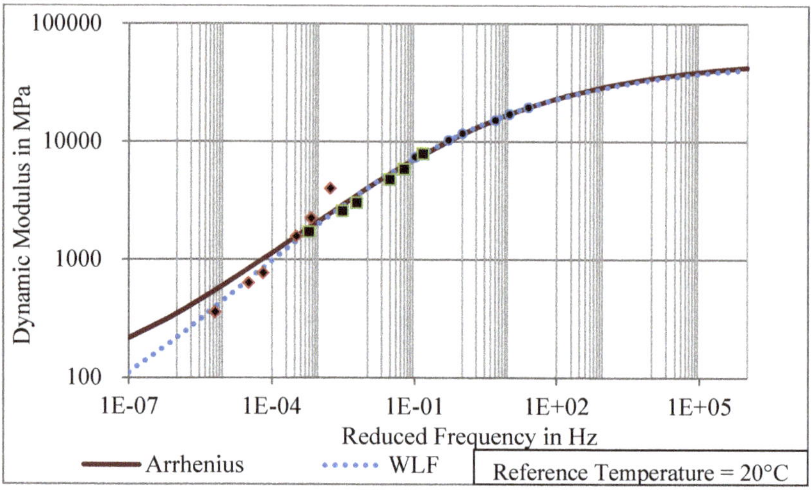

Figure 8. Dynamic modulus test results and fitted master curve with WLF and Arrhenius equation.

To characterize the viscoelastic behavior of SFP, a sigmoidal master curve was fitted using two approaches, known as the WLF method and the Arrhenius equation [28]. The master curve could be expressed as Equation (1).

$$\log E* = \delta + \frac{\alpha}{1 + \exp(\beta + \gamma \log(f_r))} \quad (1)$$

where, δ, α, β, γ were parameters from regression,

$E*$ was the dynamic modulus in MPa,

f_r was the reduced frequency, which could be calculated as Equation (2).

$$\log(f_r) = \log(f) + a_T \quad (2)$$

where, f was the loading frequency in dynamic modulus test,

a_T was the shift factor depending on the test temperatures, which could be calculated as Equations (3) and (4). For WLF method, the shift factor could be calculated as

$$\log(a_T) = \frac{-C_1(T - T_r)}{C_2 + T - T_r} \quad (3)$$

where, T was the test temperature in Celsius,

T_r was the reference temperature in Celsius, which was set to 20 °C,

C_1, C_2 were regression parameters.

Alternatively, the shift factor could be calculated using Arrhenius Equation as

$$\log(a_T) = \frac{E_a}{\ln(10) \times R}\left(\frac{1}{T} - \frac{1}{T_r}\right) \quad (4)$$

where, E_a and R could be acquired from regression,

T and T_r were the test temperature and reference temperature in Kelvin.

It could be inferred that SFP specimens exhibited obvious viscoelastic behavior. The dynamic modulus of SFP materials could be properly predicted by the master curve fitted either with the WLF method or Arrhenius equation. The master curves were close to each other in high frequency but differs in low frequency. The fitted parameters were tabulated in Table 4.

Table 4. Fitted parameters and adjust R^2 of master curves.

Shift Method	WLF		Arrhenius Equation	
Fitting Parameters	δ	0.9465	δ	1.6874
	α	3.7549	α	3.0311
	β	−1.5970	β	−1.2871
	γ	−0.3539	γ	−0.3698
	C_1	8.1560	E_a	68,415.9120
	C_2	48.6732	R	2.9042
Adjusted R^2	0.9973		0.9939	

The fitting results suggested that the adjusted R^2 was close to 1, and thus the master curves fitted from both methods were able to explain the modulus variation with time and loading frequency. The viscoelastic behavior of SFP material could also be validated by the modulus-phase angle plot shown in Figure 9, known as black plot.

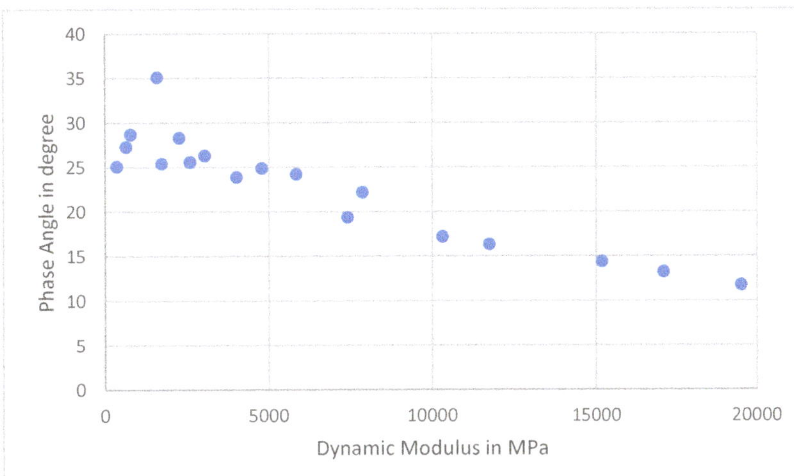

Figure 9. Black plot of SFP specimen (dynamic modulus vs. phase angle).

It could be seen that the phase angle would basically decrease with the increase in dynamic modulus. SFP tended to be more elastic at high frequencies and more viscous at low frequencies.

3.2.2. Four-Point Beam Fatigue Test

The test results of eight specimens were summarized in Table 5 along with certain statistical analysis. Despite the significant variance, the test results did not reject normality hypothesis, which suggested that the fatigue lives of SFP specimens could be described by a normal distribution. The test results suggested that the average fatigue lives of SFP reached 85.4 k loading repetition, which was obviously higher than open-graded asphalt mixture reported by literature [29].

However, the comparison between SFP and dense-graded asphalt mixture could be tricky. Due to the introduction of cement, the specimen became less flexible. Under strain-controlled mode, if the strain level was high enough, the fatigue life of SFP would be shorter than dense-graded asphalt mixture, which was quite misleading. Thus, to make this comparison justifiable, stress-controlled mode should be employed, which would be extremely time consuming. Alternatively, such comparison should be made through a wheel tracking test, which was discussed in Section 3.2.4.

Table 5. Four-point beam fatigue test results and statistical normality tests.

-	Average	25% Percentile	Median	75% Percentile	Standard Deviation
Fatigue life (×1000)	85.4	40.7	91.9	127.4	50.5
Normality test p value (5% significance level) Reject normality?		Anderson-Darling test 0.9026 No		Kolmogorov–Smirnov test 0.9552 No	

3.2.3. Hamburg Wheel Tracking (HWT) Test

The test results of HWT test results were shown in Figures 10 and 11 along with the fitting results.

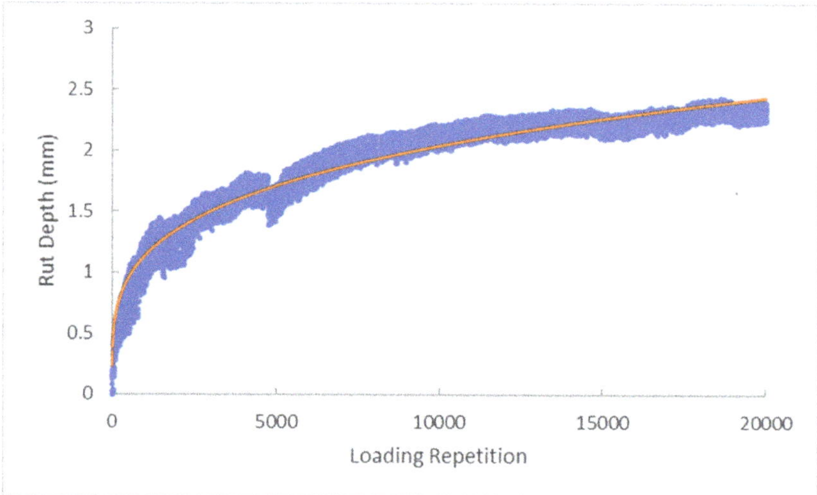

Figure 10. 1# HWT test results and fitted power model.

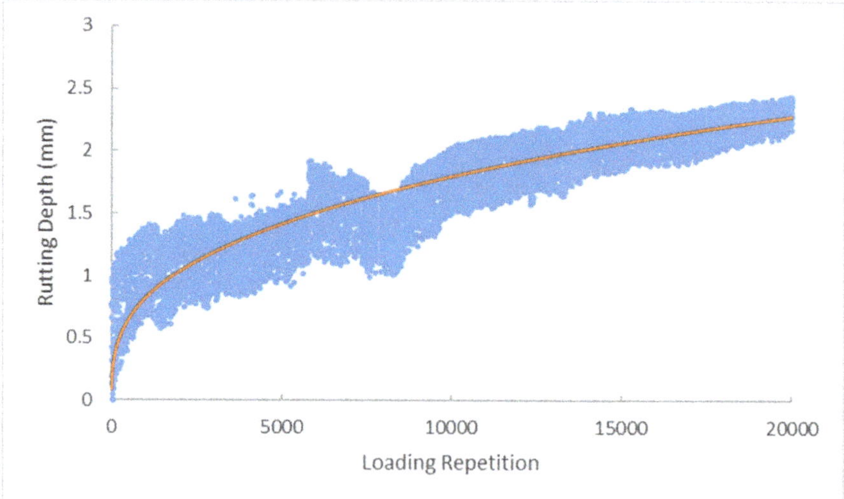

Figure 11. 2# HWT test results and fitted power model.

The correlation between rutting depth and loading repetition could be characterized with a power equation as

$$d_p = a \times N^b \tag{5}$$

where, d_p was the cumulative rutting depth in mm,

N was the number of loading repetitions,

a, b were regression parameters.

The fitting parameters were tabulated in Table 6.

Table 6. The regression results of HWT test.

Parameters	HWT Test 1#	HWT Test 2#
a	0.195	0.075
b	0.255	0.345
Adjusted R^2	0.94	0.82

It could be seen that the rutting depth was smaller than 2.5 mm at 20,000 loading repetition. Considering the fact that the specimens were water bathed at 50 °C, the moisture susceptibility and rutting resistance of SFP was satisfactory. Meanwhile, the fitted curves maintained R^2 of 0.94 and 0.82 for each HWT test results. It could be inferred that the power model could be used to predict the rutting propagation of SFP materials.

3.2.4. MMLS3 Accelerated Pavement Test

Four different structures were tested, as summarized in Table 7, along with their mix types and nominal maximum aggregate sizes. The first structure was composed of gap-graded stone mastic asphalt (SMA) and dense graded asphalt concrete (AC), which was a commonly used structure in pavement construction. The second structure was made up of two 5 cm-SMA layers, and high-viscosity modified asphalt was used to enhance its rutting resistance. To investigate the performance of SFP, the SMA used in the top layer of the first structure and bottom layer of the second structure were replaced with SFP.

Table 7. Four structures tested in MMLS3 accelerated wheel tracking test.

Structure No.	Top Layer	Bottom Layer
1	4 cm SMA-13	6 cm AC-20
2	5 cm SMA-13 (High viscosity modified asphalt)	5 cm SMA-13 (High viscosity modified asphalt)
3	4 cm SFP-13	6 cm AC-20
4	5 cm SMA-13 (High viscosity modified asphalt)	5 cm SFP-13

The test results of MMLS3 accelerated tests were plotted in Figure 12. No cracking or stripping was observed during the test. The rutting depths were recorded after 1 thousand, 5 thousand, 10 thousand, 100 thousand, 300 thousand, 500 thousand, 700 thousand and 1 million loading repetitions. It could be seen that the rutting propagation was greatly delayed when SFP was introduced in the 3rd and 4th structure. The ultimate rutting depth of the 3rd structure was only 1/3 of that of the 1st structure. Meanwhile, it could be inferred that using SFP as the surface layer would better contribute to the rutting performance by comparing the ultimate rutting depth of these structures.

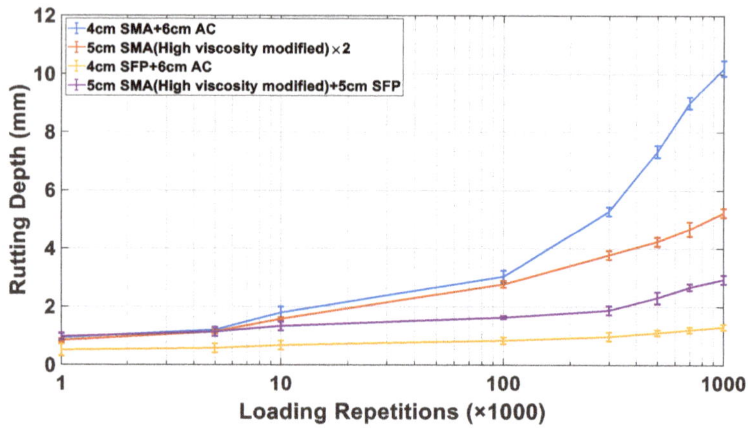

Figure 12. MMLS3 accelerated wheel tracking test results, loading repetitions are in log scale.

4. Summary

SFP material have been considered as a promising solution for heavily trafficked pavements since the last century. In this paper, a series of performance tests were conducted to study its mechanical strength as well as performance under simulated traffic loads. Compared with previous research, SFP was thoroughly studied in a quantitative manner. Mechanistic models were fitted based on test results to better characterize and predict the performance. The major findings were summarized as follows.

1. The SFP material consisted of two parts, open-graded asphalt mixture and cement grout. The mix design of open-graded asphalt mixture could be accomplished with volumetric method, and cement grout should be designed based on strength, fluidity, shrinkage rate and bleeding rate. Thus, two critical parameters in material design were target void ratio and water–cement ratio, which were 25% and 0.25 in this study, respectively. These two parameters needed to be chosen carefully to ensure the performance and durability of SFP material.
2. A series of strength and stability tests were conducted to validate the mix design of SFP. The Marshall stability test suggested that the strength of SFP specimens increased approximately 7 times compared with the original open-graded asphalt mixture. The Marshall stability could raise to 35 kN after 7 days curing, which was significantly higher than HMA. The dynamic stability test results suggested that the rutting resistance of SFP would be greatly affected by the curing time. The rutting resistance of 28 d-curing specimens would be 202% to 259% higher than that of 7 d-curing specimens. This implied that the rutting resistance was actually provided by the cement grout instead of asphalt mixture. The indirect tensile strength suggested that SFP specimens were able to maintain 99.4% tensile strength after freeze–thaw cycles. Meanwhile, the void ratio barely changed so it could be regarded as waterproof The flexural Strength of SFP could reach 10 MPa, which was believed to be satisfactory for heavy traffic.
3. Viscoelastic behavior of SFP was observed in dynamic modulus tests. Two master curves were fitted using WLF and Arrhenius equation, respectively. The master curves were mostly identical except for the low frequencies smaller than 10^{-4} Hz The adjusted R^2 suggested both of the master curves could characterize the modulus variation of SFP material under different temperatures and loading frequencies.
4. Strain-controlled four-point beam fatigue test was employed to study the fatigue resistance of SFP. The test results suggested that the average fatigue life of SFP under 400 $\mu\varepsilon$ reached 85.4 k loading repetitions. As for the characterization of rutting resistance, a Hamburg wheel tracking test was used and a power model was fitted

based on tested data. The ultimate rutting depth after 20,000 times loading repetition maintained below 2.5 mm and tended to be convergent.

5. Four different structures were tested with a MMLS3 accelerated testing device to compare the performance between SFP and other commonly used asphalt mixtures. It was found that the rutting propagation could be greatly stemmed if SFP was introduced into a typical HMA structure. Meanwhile, placing SFP on the top layer of certain structure turned out to be more effective in design.

To conclude, the overall performance of SFP material was believed to be superior, compared with traditional asphalt mixtures. Future research should focus on the engineering practice of SFP and its long-term performance in the field.

Author Contributions: Conceptualization, G.L. and H.X.; methodology, G.L.; validation, H.X., Q.R. and X.Z.; formal analysis, Q.R.; investigation, L.W.; resources, G.L.; data curation, X.Z.; writing—original draft preparation, G.L.; writing—review and editing, H.X.; visualization, Q.R.; supervision, X.Z.; project administration, G.L.; funding acquisition, G.L. All authors have read and agreed to the published version of the manuscript.

Funding: This research was funded by Chengdu Xingcheng Construction Management Co. Ltd., grant number K2020K180B-01. The APC was funded by Shanghai Municipal Engineering Design Institute (Group) Co. Ltd.

Institutional Review Board Statement: Not applicable.

Informed Consent Statement: Not applicable.

Data Availability Statement: The data presented in this study are available on request from the corresponding author.

Acknowledgments: The authors would acknowledge the Key laboratory of Road and Traffic Engineering of the Ministry of Education at Tongji University for the support in laboratory tests.

Conflicts of Interest: The authors declare no conflict of interest.

References

1. Blight, G.E. The properties of salviacim: A bitumen bound cemented paving material. In *Asphalt: Road to 2000, Proceedings of the 6th International Asphalt Conference, Sydney, Australia, 27–30 January 1986*; Australian Asphalt Pavement Association: Hawthorn, Australia, 1986.
2. Anderton, G.L. *Engineering Properties of Resin Modified Pavement (RMP) for Mechanistic Design*; U.S. Army Engineer Research and Development Center: Vicksburg, MS, USA, 2000.
3. Anderton, G.L. The resin modified pavement system: A new pavement surfacing technology. In *Airport Facilities: Innovations for the Next Century, Proceedings of the 25th International Air Transportation Conference, Austin, TX, USA, 14–17 June 1998*; American Society of Civil Engineers: Reston, VA, USA, 1998.
4. Collop, C.; Elliott, R.C. Assessing the mechanical performance of Densiphalt. In *Performance and Durability of Bituminous Materials and Hydraulic Stabilised Composites, Proceedings of the Third European Symposium, Nottingham, UK, 11–12 April 2002*; Aedificatio Publishers: Freiburg, Germany, 1999.
5. Mayer, J.; Thau, M. Jointless pavements for heavy-duty airport application: The semi-flexible approach. In *Advancing Airfield Pavements, Proceedings of the 27th International Air Transportation Conference, Chicago, IL, USA, 5–8 August 2001*; American Society of Civil Engineers: Reston, VA, USA, 2001; pp. 87–100.
6. Oliveira, J.R.M. *Grouted Macadam: Material Characterisation for Pavement Design*; University of Nottingham: Nottingham, England, 2006.
7. Hassani, A.; Taghipoor, M.; Karimi, M.M. A state of the art of semi-flexible pavements: Introduction, design, and performance. *Constr. Build. Mater.* **2020**, 253, 119196. [CrossRef]
8. Anderton, G.L. *User's Guide: Resin Modified Pavement*; U.S. Army Engineer Waterways Experiment Station: Vicksburg, MS, USA, 1996.
9. Al-Qadi, L.; Gouru, H.; Weyers, R.E. Asphalt portland cement concrete composite: Laboratory evaluation. *J. Transp. Eng.* **1994**, *120*, 94–108. [CrossRef]
10. Battey, R.L.; Whittington, J.S. *Construction, Testing and Performance Report on the Resin Modified Pavement Demonstration Project*; No. FHWA/MS-DOT-RD-07-137; Mississippi Department of Transportation: Jackson, MS, USA, 2007.
11. Wang, K.; Hong, J. *Hybrid Concrete for Advancing Pavement Performance*; No. IHRB Project TR-708B; Iowa State University: Ames, IA, USA, 2018.

12. Pei, J.; Cai, J.; Zou, D.; Zhang, J.; Li, R.; Chen, X.; Jin, L. Design and performance validation of high-performance cement paste as a grouting material for semi-flexible pavement. *Constr. Build. Mater.* **2016**, *126*, 206–217. [CrossRef]
13. Zhang, J.; Cai, J.; Pei, J.; Li, R.; Chen, X. Formulation and performance comparison of grouting materials for semi-flexible pavement. *Constr. Build. Mater.* **2016**, *115*, 582–592. [CrossRef]
14. Husain, N.; Karim, M.R.; Mahmud, H.B.; Koting, S. Effects of aggregate gradation on the physical properties of semiflexible pavement. *Adv. Mater. Sci. Eng.* **2014**, *2014*, 529305. [CrossRef]
15. Fakhri, M.; Mottahed, A.R. Improving moisture and fracture resistance of warm mix asphalt containing RAP and nanoclay additive. *Constr. Build. Mater.* **2021**, *272*, 121900. [CrossRef]
16. Fakhri, M.; Maleki, H.; Hosseini, S.A. Investigation of different test methods to quantify rutting resistance and moisture damage of GFM-WMA mixtures. *Constr. Build. Mater.* **2017**, *152*, 1027–1040. [CrossRef]
17. Wang, D.; Liang, X.; Jiang, C.; Pan, Y. Impact analysis of carboxyl latex on the performance of semi-flexible pavement using warm-mix technology. *Constr. Build. Mater.* **2018**, *179*, 566–575. [CrossRef]
18. Cai, X.; Zhang, J.; Zhang, H.; Yao, Z.; Chen, X.; Yang, J. Identification of microstructural characteristics in semi-flexiblepavement material using micromechanics and nano-techniques. *Constr. Build. Mater.* **2020**, *246*, 118426. [CrossRef]
19. Cai, X.; Yang, J.; Chen, X.; Zhang, J.; Zhang, H. Interlocking property evaluation of dual skeleton in semi-flexible pavement material by micromechanical model and X-ray computed tomography. *Constr. Build. Mater.* **2020**, *254*, 118934. [CrossRef]
20. Oliveira, J.R.; Thim, N.H.; Zoorob, S.E. Design of pavements incorporating grouted macadams. *J. Transp. Eng.* **2008**, *134*, 7–14. [CrossRef]
21. Bharaath, G.; Shukla, M.; Nagabushana, M.N.; Chandra, S.; Shaw, A. Laboratory and field evaluation of cement grouted bituminous mixes. *Road Mater. Pavement Des.* **2019**, *21*, 1694–1712. [CrossRef]
22. Zhang, W.; Shen, S.; Goodwin, R.D.; Wang, D.; Zhong, J. Performance characterization of semi-flexible composite mixture. *Materials* **2020**, *13*, 342. [CrossRef]
23. AASHTO. *Mechanistic-Empirical Pavement Design Guide: A Manual of Practice*; American Association of State Highway and Transportation Officials: Washington, DC, USA, 2008.
24. Cui, P.; Xiao, Y.; Yan, B.; Li, M.; Wu, S. Morphological characteristics of aggregates and their influence on the performance of asphalt mixture. *Constr. Build. Mater.* **2008**, *186*, 303–312. [CrossRef]
25. AASHTO. *Standard Method of Test for Determining Dynamic Modulus of Hot-Mix Asphalt (HMA)*; American Association of State Highway and Transportation Officials: Washington, DC, USA, 2011.
26. Epps, L.; Ahmed, T.; Little, D.C.; Mikhail, M.Y.; Hugo, F. Performance assessment with the MMLS3 at westrack (with discussion). *J. Assoc. Asph. Paving Technol.* **2001**, *70*, 509–542.
27. Lee, J.; Kim, Y.R.; Lee, J. Rutting performance evaluation of asphalt mix with different types of geosynthetics using MMLS3. *Int. J. Pavement Eng.* **2015**, *16*, 894–905. [CrossRef]
28. Pellinen, T.K.; Witczak, M.W.; Bonaquist, R.F. Asphalt mix master curve construction using sigmoidal fitting function with non-linear least squares optimization. In *Recent Advances in Materials Characterization and Modeling of Pavement Systems, Proceedings of the 15th Engineering Mechanics Division Conference, New York, NY, USA, 2–5 June 2002*; American Society of Civil Engineers: Reston, VA, USA, 2004; pp. 83–101.
29. Liu, Q.; Schlangen, E.; Van De Ven, M.; Van Bochove, G.; Van Montfort, J. Evaluation of the induction healing effect of porous asphalt concrete through four point bending fatigue test. *Constr. Build. Mater.* **2012**, *29*, 403–409. [CrossRef]

coatings

MDPI

Article

Assessment of Modulus Attenuation of Cement and Lime-Fly Ash Semi-Rigid Road Base Materials

Luchuan Chen [1], Sixin Yu [2], Ying Zhu [1], Xiaomeng Zhang [3,*], Wenjuan Wu [3], Qiang Sun [3], Tingting Chen [3], Xiaoyan Wang [3] and Jincheng Wei [3]

1 Shandong Hi-Speed Group Co., Ltd., Jinan 250101, China; chenluchuan01@163.com (L.C.); zhuying2005@126.com (Y.Z.)
2 Shandong Hi-Speed Transportation Construction Group Co., Ltd., Jinan 250101, China; chd008@126.com
3 Shandong Transportation Institute, Jinan 250101, China; wuwenjuan@sdjtky.cn (W.W.); sunqiang@sdjtky.cn (Q.S.); chentingting@sdjtky.cn (T.C.); wangxiaoyan@sdjtky.cn (X.W.); weijincheng@sdjtky.cn (J.W.)
* Correspondence: zhangxiaomeng@sdjtky.cn

Abstract: For asphalt pavement structures, semi-rigid road base course has to sustain repeated high-axle load during its service life and the performance of semi-rigid road base materials directly influences the durability of pavement structures. The dynamic compressive resilience modulus of two commonly used semi-rigid road base materials, cement stabilized aggregates (CSG) and lime-fly ash stabilized aggregates (LFSG) were evaluated at different frequencies using a Universal Testing Machine (UTM). The results showed that LFSG had higher dynamic modulus than that of CSG and the load frequency had less influence on the dynamic modulus of these two semi-rigid road base materials. The four-point bending test was applied to measure the flexural-bending strength and the fatigue life of these two semi-rigid materials. A higher flexural-bending strength of LFSG indicated its better bearing capacity than that of CSG. The fatigue life of LFSG and CSG decreased with the increase of stress–strength ratio and the LFSG performed better in terms of fatigue resistance. The fatigue damage models of CSG and LFSG based on Stress-Life (*S-N*) curve are established. As per incremental-recursive mechanics, a general modulus degradation model was established and verified by the results of full-scale accelerate loading test. This model cannot only be used to predict the fatigue deterioration of semi-rigid road base materials under different stress levels, but is also able to calculate the current bending elastic modulus based on its initial modulus value.

Keywords: cement stabilized aggregates; lime-fly ash stabilized aggregates; fatigue; modulus degradation model

Citation: Chen, L.; Yu, S.; Zhu, Y.; Zhang, X.; Wu, W.; Sun, Q.; Chen, T.; Wang, X.; Wei, J. Assessment of Modulus Attenuation of Cement and Lime-Fly Ash Semi-Rigid Road Base Materials. *Coatings* **2022**, *12*, 216. https://doi.org/10.3390/coatings12020216

Academic Editor: Claudio Lantieri

Received: 18 January 2022
Accepted: 1 February 2022
Published: 7 February 2022

Publisher's Note: MDPI stays neutral with regard to jurisdictional claims in published maps and institutional affiliations.

Copyright: © 2022 by the authors. Licensee MDPI, Basel, Switzerland. This article is an open access article distributed under the terms and conditions of the Creative Commons Attribution (CC BY) license (https://creativecommons.org/licenses/by/4.0/).

1. Introduction

An asphalt pavement can be recognized as a multiple-layer structure, which is typically composed of asphalt surface layers, road base layers and road sub-base layers. As a layered system, each layer carries and spreads loads from the above layer and passes the load to the next layer below [1–3]. Due to its high strength, good moisture stability and low cost, semi-rigid material is the most commonly used road base material for asphalt pavements in China. The semi-rigid base material is a type of hydraulic material which is produced by mixing predetermined ratios of water, aggregates (coarse and fine) and hydraulic materials, after a paving and compaction process to form a semi-rigid road base [4,5]. In general, the two commonly used semi-rigid road base materials are lime-fly ash stabilized aggregate (LFSG) and cement stabilized aggregate (CSG). Compared to flexible/asphalt base materials, LFSG and CSG can be more environmentally friendly as they consumed more fly ash and other waste materials [6,7].

In a semi-rigid pavement structure system, semi-rigid road base plays an important role for structural bearing capacity and has to sustain repeated traffic loading. One of main failure modes in the semi-rigid base is then the bending fatigue damage. Fatigue of semi-rigid pavement is a type of distress associated with the weakening behavior of semi-rigid base materials caused by repeatedly applied loads at certain stress levels [8]. The fatigue distress of semi-rigid base material is usually initiated in the form of microcracks and propagated to macrocracks due to repeated shear and tensile stresses in the semi-rigid base layer [9]. Propagation of cracks in the base layer is related to the fractures of both adhesive and cohesive modes in mastic films [10]. The fatigue life of the road base layer is strongly affected by the mechanical properties of semi-rigid base materials [11].

Mechanical properties of semi-rigid base materials are important for the pavement durability and many researchers have focused on this field. Sheng et al. investigated the modulus attenuation of semi-rigid base material and found that the mechanical behavior of pavement is stable when the modulus of semi-rigid road base is in a proper range [12]. Sheng et al. also studied the influence of elastic modulus on the pore pressure of the semi-rigid pavement and established correlations between semi-rigid base modulus and pore pressure [13]. Wang et al. investigated the dynamic compressive resilient modulus of semi-rigid base materials and demonstrated that the cement content and curing time are the main factors [14]. The research of Yao et al. reached the same conclusion [15–17]. Zhou et al. investigated the fatigue resistance of semi-rigid base material and established a fatigue performance model including aggregate gradation, cement dosage, water content as well as air void [18]. Sha et al., evaluated the fatigue performance of semi-rigid base material by using MTS and analyzed the influence of structure type and flexural strength on the fatigue life of semi-rigid base materials [19].

In summary, a significant amount of research has been conducted to understand the mechanical performance of semi-rigid road base materials. However, the dynamic modulus and fatigue resistance of semi-rigid road base materials still need to researched because of the discreteness and variability.

The objective of this paper is to further understand the dynamic performance of different semi-rigid base materials. A comparative study of two commonly used semi-rigid base materials (LFSG and CSG) was conducted by characterizing the dynamic compressive resilience modulus and four-point bending fatigue life. The dynamic compressive resilience modulus of semi-rigid base materials was evaluated by using uniaxial dynamic compression test through a Universal Testing Machine (UTM). The four-point bending test was performed by using a Universal Testing Machine (UTM) to measure the fatigue resistance of semi-rigid base materials. Finally, a general modulus degradation model of these two semi-rigid road base materials was developed based on fatigue test results and mechanical principle of continuous damage.

2. Materials and Experimental

2.1. Materials

2.1.1. Cement

Ordinary Portland cement of 425 type supplied by Lvzhou cement factory (Jinan, China) was used to prepare cement stabilized aggregate (CSG). The property of this cement was evaluated in accordance with the Chinese Standard JTG E30-2005 [20], and the results are listed in Table 1.

2.1.2. Lime

A type of calcareous hydrated lime supplied by Jinan lime factory (Jinan, China) was used to prepare lime-fly ash stabilized aggregate (LFSG). The physical and chemistry properties of this lime were characterized according to the Chinese Standard JTG E51-2009 and the results are listed in Table 2.

2.1.3. Fly Ash

The fly ash used for LSFG preparation was supplied by Laiwu coal-fired power plants in Shandong Province (Jinan, China). The properties including loss on ignition, particle size and chemical components were tested in accordance with the Chinese Standard JTG E51-2009 [21], and the results are presented in Table 3.

Table 1. Physical properties of cement used.

Inspection Items			Test Results	Technical Requirement
Fineness (%)			Qualified	≤10
Water quantity for standard consistence (%)			28	Actual measurement
Stability (Standardized Approach)			Qualified	≤5 mm
Setting Time (min)	Initial setting		210	≥45 min
	Final setting		280	≤12 h
Cement mortar strength (MPa)	Compressive strength	3 d	18.9	≥16.0
		28 d	43.1	≥42.5
	Bending Strength	3 d	4.5	≥3.5
		28 d	8.23	≥6.5

Table 2. Physical and chemistry properties of lime.

Inspection Items	Test Results	Test Method
CaO content (%)	52.6	T 0811-1994
MgO content (%)	3.79	T 0812-1994
CaO + MgO content (%)	56.4	T 0813-1994d

Table 3. Physical chemistry properties of fly ash.

Inspection Items		Test Results	Technical Requirement
Loss on ignition (%)		3.32	≤20
Passing rate (%)	0.3 mm	99.20	≥90
	0.075 mm	72.30	≥70
Chemical component (%)	SiO_2	56.62	$\sum > 70$
	Fe_2O_3	7.05	
	Al_2O_3	27.40	

2.1.4. Aggregate

The aggregate used for CSG and LSFG preparation was obtained from a limestone quarry named Changqing in Shandong Province (Xinli stone factory, Jinan, China). The mineral composition of this type of limestone was evaluated by using an X-ray Diffraction (XRD) test (D2 PHASER, Bruker, Billerica, MA, USA). The dominant composition is calcite with the content of 99.15 wt.%, and another two low-content components are quartz (0.81 wt.%) and iron-oxide (0.04%). The physical properties of fine and coarse aggregate were characterized according to the Chinese standard JTG E20-2011 [22], and the results are listed in Table 4.

Table 4. Physical properties of aggregates.

Inspection Items	Test Result		Technical Requirement
	Fine Aggregate	Coarse Aggregate	
Specific gravity (g/cm^3)	2.73	2.73	≥2.6
Water absorption (%)	0.75	0.38	≤2.0
LA Abrasion loss (%)	-	17.60	≤28
Crushing value (%)	-	14.30	≤26

2.2. Mix Design

2.2.1. Aggregate Gradation

Dense gradations with a maximum particle size of 31.5 mm were chosen in this research to prepare the LFSG and CSG, according to the Chinese Standard JTG D50-

2017 [23]. The aggregate size distribution of LFSG and CSG is presented in Tables 5 and 6, respectively.

Table 5. Aggregate gradation passing rate of LFSG.

Sieve Pit (mm)	31.5	26.5	19	9.5	4.75	2.36	1.18	0.6	0.075
Composed gradation	100.0	97.2	68.1	33.5	15.6	10.2	8.1	6.1	2.2
Lower limit	100	95	48	24	11	6	2	0	0
Upper limit	100	100	68	34	21	16	12	6	3

Table 6. Aggregate gradation passing rate of CSG.

Sieve Pit (mm)	31.5	19	9.5	4.75	2.36	0.6	0.075
Composed gradation	100.0	81.1	50.5	26.4	17.6	9.0	3.1
Lower limit	100	68	38	22	16	8	0
Upper limit	100	86	58	32	28	15	3

2.2.2. Cement Dosage and Optimum Moisture Content

A compaction test was conducted to decide the cement dosage as well as the optimum moisture content of LFSG and CSG in accordance with the Chinese standard JTG E51-2009, and their determined values are presented in Table 7. The significantly higher unconfined compressive strength of CSG showed its higher pavement structural capacity than that of LFSG.

Table 7. Mixture design of LFSG and CSG.

Stabilized Aggregate	Mixture Ratio (by Weight)	Moisture Content (%)	Maximum Dry Density (g/m^3)	7d Compressive Strength (MPa)
LFSG	lime: fly ash: aggregate = 6: 12: 82	6.20	2.13	0.95
CSG	cement: aggregate = 4.5: 100	5.30	2.33	6.70

2.3. Test Methods

2.3.1. Dynamic Modulus Test

The dynamic modulus of LFSG and CSG was evaluated by using a Universal Testing Machine with the maximum force of 10 kN (UTM-100, IPC global Co., Ltd., Milan, Italy). The dynamic modulus of the stabilized material is defined as the ratio of the axial stress to the recoverable axial strain. Before testing, cylinder specimens with dimensions of φ 150 mm × 150 mm were first prepared by using a static pressing method according to the predetermined moisture dosage and maximum dry density. Then, these specimens were put in a curing chamber with a curing temperature of (20 ± 2) °C and humidity of >95% (T 0845-2009). The dynamic modulus of the stabilized material specimens was measured at a temperature of 20 °C and loading frequency from 0.1 to 25 Hz. During testing, the UTM loading head generated a sinusoidal loading wave and the strain level was controlled at 25 μm. The stress was recorded by the UTM stress sensor (UTM-100, IPC, Milan, Italy) and the strain was measured by the Linear Variable Differential Transformer (LVDT, UTM-LVDT, IPC, Milan, Italy) attached on the specimen surface. The dynamic modulus of semi-rigid materials at different loading cycle and different frequency was calculated automatically by the UTM [14].

2.3.2. Four-Point Bending Fatigue Test

Fatigue cracking is a major type of distress observed in pavements and the fatigue resistance of semi-rigid road base material plays an important role in the service duration of asphalt pavements. The four-point bending is a commonly used test to assess the fatigue life of asphalt mixtures. According to the JTG E51-2009 standard, the fatigue resistance of CSG and LSFG was carried out after 90 d age and 180 d age, respectively. Before testing, beam specimens with dimensions of 380 mm × 63.5 mm × 50 mm need to be prepared. Big semi-rigid road base material beams with dimensions of 150 mm × 150 mm × 550 mm were first prepared and then separated by cutting machine to obtain required dimensions. With the view of neglecting the dispersion of test results, two beam specimens from the same layer were performed for flexural-tensile strength and fatigue evaluation, respectively, as shown in Figure 1.

Figure 1. Testing beam specimens from a large beam specimen for the strength and fatigue test.

The flexural-bending strength of CSG and LSFG was evaluated by using a three-point pressure at a constant loading rate of 50 mm/min and the detailed procedure was described in the JTG E51-2009 standard. Four replicate tests were performed on each stabilized material. Based on this test, the ultimate failure load was recorded and the flexural-bending strength was calculated according to Equation (1). This parameter provides a standard index for the determination of loading level in the fatigue test.

$$R_s = \frac{PL}{b^2h} \quad (1)$$

where, R_s is the flexural-bending strength (MPa), P is the ultimate failure load (N), L is the span of the beam (mm), b is the width of the beam (mm), and h is the height of the beam (mm).

The four-point bending test was conducted using a Universal Testing Machine (UTM) at a controlled stress mode to evaluate the fatigue resistance of semi-rigid materials. The target stress–strength ratios of LSFG were 0.6, 0.65, 0.7 and 0.75, while the target stress ratios of CSG were 0.55, 0.6, 0.65 and 0.7. Six replicate tests were performed at each stress ratio. Before testing, samples need to be conditioned at the target temperature (20 °C) for at least 4 h. During testing, a sinusoidal loading pattern was induced to the specimen at a loading cycle frequency of 10 Hz. The failure life was defined when the beam specimen was brittle break. Physical parameters such as the initial flexural modulus (MPa), initial flexural strain (μm), critical failure flexural modulus (MPa), and flexural fatigue life (times) were calculated automatically by the UTM.

2.3.3. Full-Scale Accelerated Loading Test

In order to investigate the fatigue deterioration and verify the fatigue model of semi-rigid road base materials in the full-scale pavement, the accelerated loading test was performed [24]. The full-scale accelerated loading test was conducted on an experimental road in Jinan, China. The accelerated loading system was employed to induce fatigue damage for the experimental CSG and LFSG materials. The dimensions of the accelerated loading system are 26.24 × 3.48 × 7.92 m. A 20.2 t axle load with dual-tire wheels and 1.15 MPa tire pressure was applied to the experimental pavement at an average speed of 23 km/h, as shown in Figure 2. In each day, around 50,000 loading passes were applied.

Figure 2. Accelerated loading system.

The full-scale pavement had three different pavement structures with CSG and LFSG, as shown in Figure 3. The thicknesses of the pavement layers in Structure 1 and 2 included 40 mm of asphalt surface layer and 290 mm of CSG or LFSG base over a 260 mm of subgrade treated with lime soil. The thicknesses of the pavement layers in Structure 3 included 40 mm of asphalt surface layer and 180 mm of LFSG base over a 260 mm of subgrade treated with lime soil.

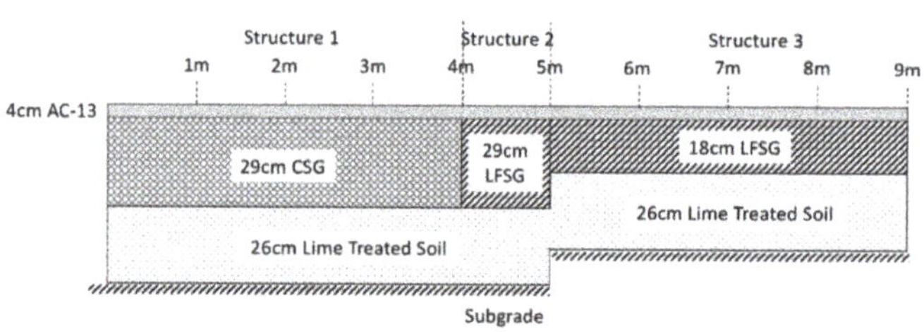

(a)

(b)

Figure 3. Full-scale pavement structure design (**a**) and FWD test in the full-scale pavement (**b**).

The pavement deflection was detected after every 70,000 loading cycles by Dynatest 8000, Falling Weight Deflectometer (FWD, Dynatest 8000, Ballerup, Denmark). As shown in Figure 3, three different pavement structures were prepared and their deflection on the wheel pass was detected. The diameter of FWD-bearing plates is 30 cm. FWD had nine deflection sensors and three levels of loading with 566, 707 and 848 kPa, respectively. The modulus of CSG and LFSG bases was back calculated by the pavement deflection using SIDMOD software (version 1.0, Zhengzhou University, Zhengzhou, China).

3. Results and Discussion

3.1. Dynamic Modulus of LFSG and CSG

The dynamic modulus of LFSG and CSG at different loading frequencies was measured and the results are presented in Figure 4. The coefficient of variation (the ratio of standard deviation to the average value) of the dynamic modulus of LFSG and CSG is shown in Figure 5.

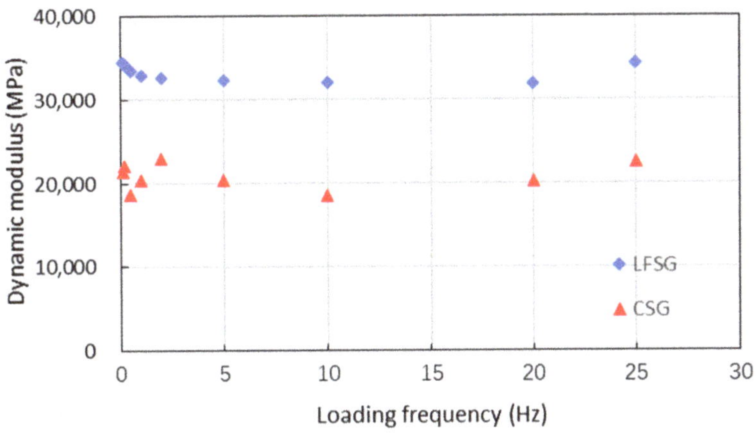

Figure 4. Dynamic modulus of CSG and LSFG.

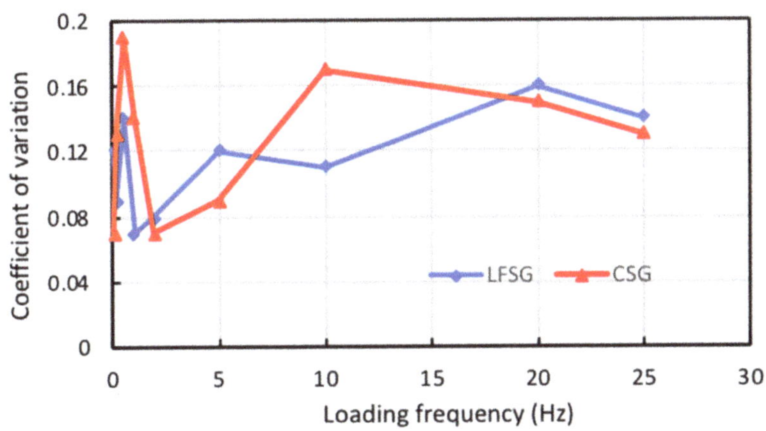

Figure 5. Coefficient of variation of dynamic modulus for CSG and LSFG.

As shown in Figure 4, the dynamic modulus of LFSG is in the range of 32,000–35,000 MPa, which is obviously higher than that of CSG with the values between 18,000 and 22,000 MPa. This indicates the LFSG road base course can generate less vertical strain than that of CSG under the same loading level. This can supply two benefits to the pavement structure. Firstly, the traffic load can be transferred to the roadbed area more evenly and reduce the pressure in roadbed area. Secondly, less vertical strain of base course can reduce the strain level of upper asphalt surface, which in turn improves the fatigue life of the asphalt layer.

With respect to the same semi-rigid material, the influence of loading frequency on the dynamic modulus seems not obvious. After considering the coefficient of variation in Figure 5, the data fluctuation of dynamic modulus can be neglected. It indicates that the dynamic modulus of semi-rigid material remains unchanged at different frequencies.

3.2. Flexural-bending Strength of LFSG and CSG

The flexural-bending strength of stabilized materials is an indicator of resistance to bending and tension in the pavement layer. Figure 6 presents the flexural-bending strengths of CSG and LSFG. It can be seen that the flexural-bending strength of LFSG is 3.12 MPa, which is higher than that of CSG, at 2.59 MPa. The higher flexural-bending strength of LFSG shows its better resistance to high traffic loading. When used for the base course of pavement, the LFSG can supply a higher bearing capacity than that of CSG, which in turn has better resistance to reflection crack.

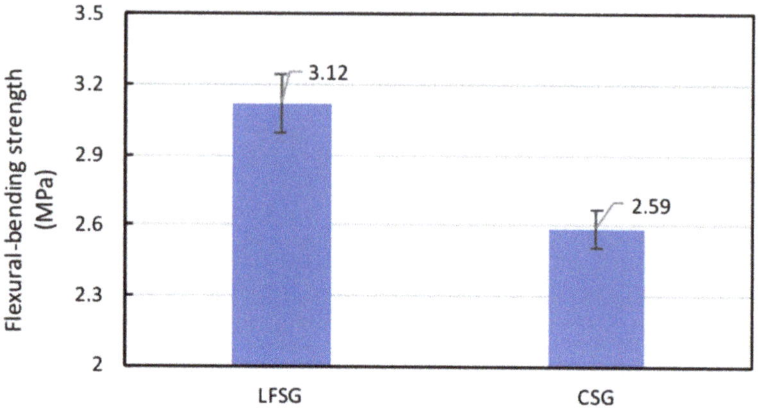

Figure 6. Flexural-bending strength of CSG and LSFG.

As expected, based on previous studies [25], two types of CSG with the same cement contents had different flexural strength. The CSG in this study was 119% higher than the previous study. The result showed the flexural strength of CSG was influenced by different aggregate gradation.

3.3. Fatigue Behavior of CSG and LSFG

In this research, the flexural stiffness at the 100th cycle of repeated loading was defined as the initial flexural modulus. This parameter reflects the capability of the beam to resist the flexural deformation. Figure 7 presents initial flexural modulus of CSG and LFSG. It can be found that the initial flexural modulus of CSG was higher than that of LFSG at any stress–strength level, which indicates better resistance of CSG to the flexural deformation. With respect to the same stabilized material, the initial flexural modulus first increased with the increase of stress–strength ratio and followed by a rapid decline once the peak value reached. This is because high stress level may cause potential damage of stabilized material, which in turn decrease the flexural stiffness. Based on the curves presented in Figure 7, the peak initial flexural modulus value of CSG occurs at the stress–strength ratio of around 0.65, while the LFSG reached to the peak initial flexural modulus at the stress–strength ratio of around 0.675. This means that the LFSG is more flexible than CSG. In combined with the analysis in Figure 6, the LFSG has both better ability of deformation and toughness.

In this research, the fatigue life of the stabilized material beam was defined as the occurrence of brittle fracture. Figure 8 shows the fatigue life of CSG and LFSG and the average fatigue life of these two materials was presented in Figure 9. As shown in Figure 8, the fatigue life of these two stabilized materials decreased with the increase of stress–strength ratio. The stress level significantly influenced the fatigue life of these two materials. It was found that a higher traffic load will accelerate the fatigue deterioration of the stabilized materials in the road base course. With respect to the same stress–strength ratio, some fatigue life values of these two materials nearly overlapped. With respect to the average fatigue life in Figure 9, the influence of stress level on the fatigue life is obvious. For instance, when the stress ratio increased from 0.6 to 0.65, it can result in around a five-times decline of the fatigue life. As for these two materials, the LFSG obtained slightly higher fatigue life than that of CSG under the same stress ratio. It is suggested that the LFSG is expected to be more fatigue-resistant than CSG. This phenomenon is well corresponded to the initial flexural modulus in Figure 7. It is demonstrated that stabilized materials with more flexibility have a positive effect on its fatigue resistance.

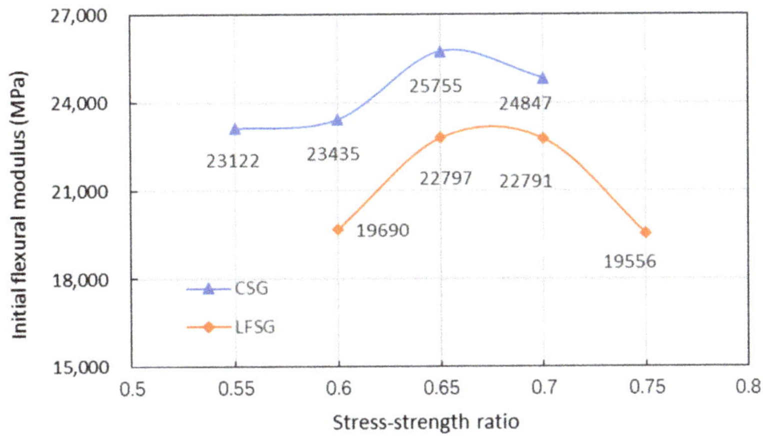

Figure 7. Initial flexural modulus of CSG and LFSG with different stress–strength ratio.

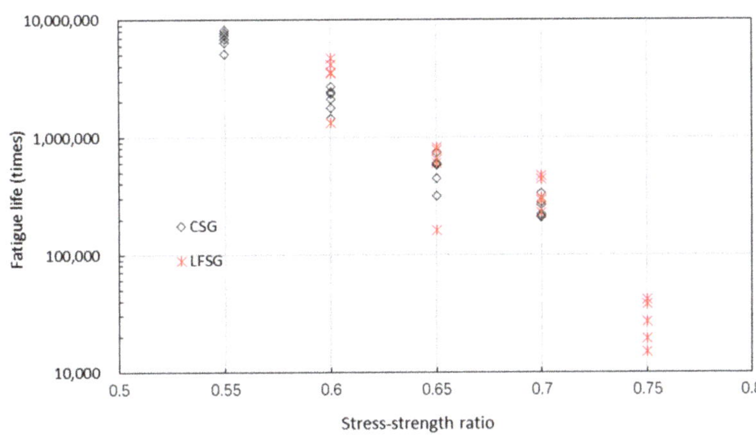

Figure 8. Fatigue life of CSG and LFSG in different stress–strength ratios.

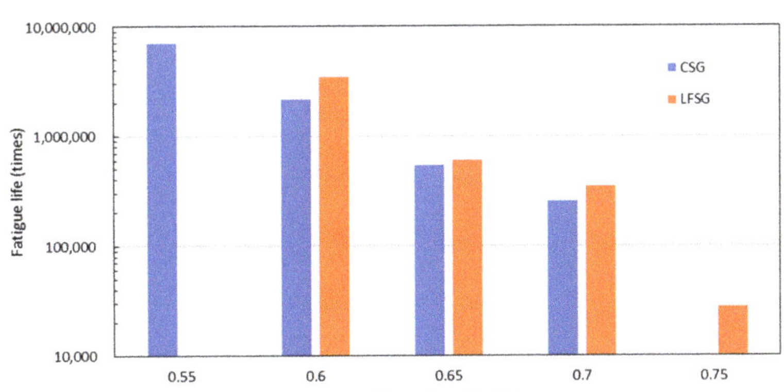

Figure 9. Average fatigue life of CSG and LFSG in different stress–strength ratios.

3.4. Analysis on Fatigue Damage of Semi-rigid Road Base Materials

3.4.1. Introduction of Fatigue Damage Model

The fatigue deterioration behavior of semi-rigid road base materials is usually represented by an empirical method, such as an *S-N* curve, Basquin equation and Goodman method [25,26]. The *S-N* curve is based on the calculation of accumulated fatigue damage and indicated the relationship between fatigue stress levels and number of loading cycles at failure. The parameter S represents the stress level which is indicated by the stress–strength ratio. N represents the fatigue life which is indicated by the number of loading cycles at fatigue failure. *S-N* curves can be represented by Equations (2)–(4), respectively.

$$S = a - b\log N \tag{2}$$

$$\log S = a - b\log N \tag{3}$$

$$(S - S_0)^a N = b \tag{4}$$

where, S is stress–strength ratio, N is the number of loading cycles at fatigue failure, a and b are calculated by linear regression, S_0 is fatigue limit stress.

3.4.2. Fatigue Damage Model Construction

The four-point bending fatigue test results of LFSG and CSG are regressed by Equation (5), which is based on Equation (2).

$$\log N = a + b(\sigma/S) \tag{5}$$

where, N denotes number of loading cycles at fatigue failure, σ denotes flexural-tensile stress (MPa), S denotes flexural-tensile strength (MPa), a and b denote regression coefficients.

Fitting curves of logarithm fatigue life with stress–strength ratio of LFSG and CSG are shown in Figures 10 and 11, respectively. According to the two fitting curves, fatigue damage models of Equations (6) and (7) are established. And the results of significance test are shown in Tables 8 and 9. The results show that the effects of curve regression are significant, and the values of a and b are reliable.

$$\log N = 13.76 - 12.23(\sigma/S) \tag{6}$$

$$\log N = 12.26 - 9.96(\sigma/S) \tag{7}$$

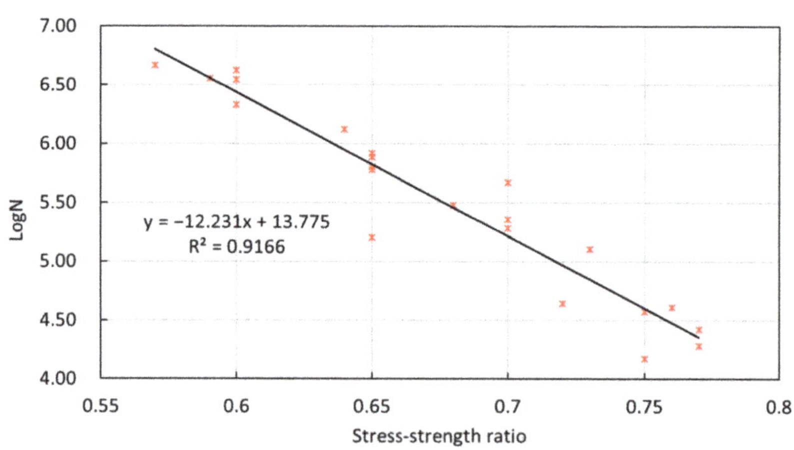

Figure 10. Fitting curves of LFSG fatigue test in different stress–strength ratios.

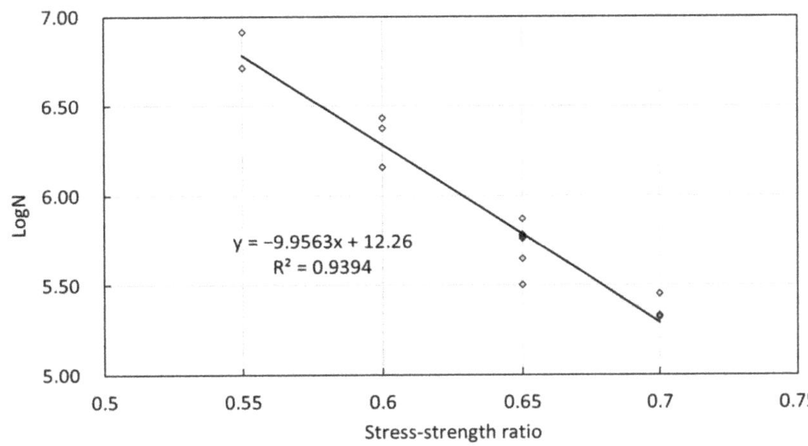

Figure 11. Fitting curves of CSG fatigue test in different stress–strength ratios.

Table 8. Mixture design of LFSG and CSG.

	Quadratic Sum	Degree of Freedom	Standard Deviation	Statistics F	Confidence Interval $F\alpha$ (1, 20) $\alpha = 0.01$
Regression	12.40	1			
Residue	1.12	20	0.24	219.75	8.10
Sum	13.53	21			Significant

Table 9. Significance test of CSG fatigue model.

	Quadratic Sum	Degree of Freedom	Standard Deviation	Statistics F	Confidence Interval $F\alpha$ (1, 12) $\alpha = 0.01$
Regression	3.19	1			
Residue	0.21	12	0.13	185.97	9.33
Sum	3.39	13			Significant

The fatigue test results of LFSG and CSG are also regressed by Equation (8) which is based on Equation (3).

$$\log(\sigma/S) = a - b\log N \tag{8}$$

where, N denotes number of loading cycles at fatigue failure, σ denotes flexural-tensile stress (MPa), S denotes flexural-tensile strength (MPa), a and b denote regression coefficients.

Fitting curves of logarithm fatigue life with logarithm stress–strength ratio of LFSG and CSG are shown in Figures 12 and 13, respectively. According to the two fitting curves, fatigue damage models of Equations (9) and (10) are constructed. And the results of significance test are shown in Tables 10 and 11. The results show that the effects of curve regression are significant, and the values of a and b are reliable.

$$\log N = 2.26 - 18.89(\sigma/S) \tag{9}$$

$$\log N = 3.09 - 14.33(\sigma/S) \tag{10}$$

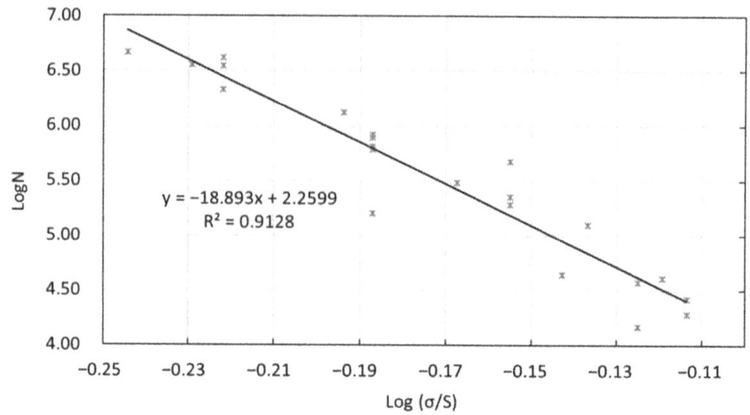

Figure 12. Fitting curves of LFSG fatigue test.

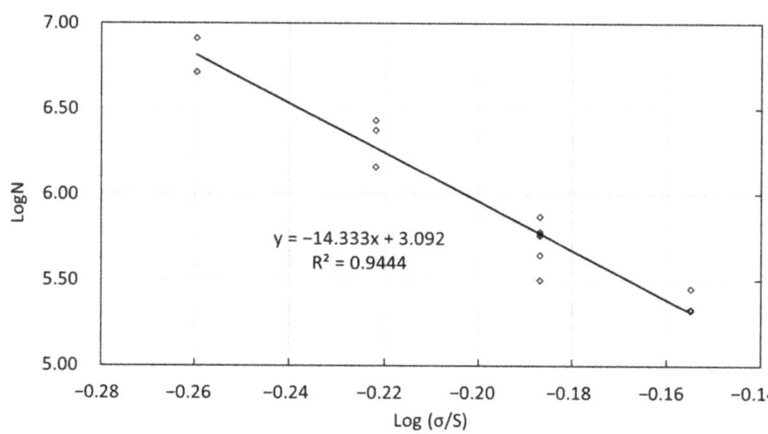

Figure 13. Fitting curves of CSG fatigue test.

Table 10. Significance test of CSG fatigue model.

	Quadratic Sum	Degree of Freedom	Standard Deviation	Statistics F	Confidence Interval $F\alpha$ (1, 20) $\alpha = 0.01$
Regression	19.16	1			
Residue	5.63	20	0.53	68.09	8.10
Sum	13.53	21			Significant

Table 11. Significance test of CSG fatigue model.

	Quadratic Sum	Degree of Freedom	Standard Deviation	Statistics F	Confidence Interval $F\alpha$ (1, 12) $\alpha = 0.01$
Regression	3.20	1			
Residue	0.19	12	0.13	203.64	9.33
Sum	3.39	13			Significant

3.5. Analysis on Attenuation of Bending Modulus of Semi-Rigid Road Base Materials

As for the beam specimen in the four-point bending test, the bending moment in the one third of the middle region is the same. So, the deterioration mechanism of the material modulus in this region is also the same. During the four-point bending testing, parameters such as the load and the displacement on each loading cycle were recorded by sensors, and the bending elastic modulus of beam specimen at each loading cycle was calculated. Based on the four-point bending fatigue test, the bending elastic modulus of every loading cycle under each stress–strength ratio can be obtained. It attempts to establish the relation models of the bending elastic modulus attenuation, loading cycles and stress ratio [27,28].

3.5.1. Introduction of Modulus Degradation Model

Incremental-recursive mechanics of continuous damage was applied in this research to establish a modulus degradation model related to the stress ratio. As a type of algorithm, incremental-recursive has been utilized to perform pavement structure analysis. For instance, mechanical-empirical Pavement Design Guide (MEPDG) used the incremental-recursive to analyze the fatigue and permanent deformation of asphalt pavement [29]. The characteristic of this method is to separate the whole process into several stages. The deterioration at this stage was defined as an increment to iterate with the summation of previous stages [30]. As a non-linear fatigue damage curve in Figure 14, the E_i is the initial modulus of the material, N_p is the accumulated loading cycles, dE is the decrement of modulus. The incremental-recursive model is establishing the relationship between the modulus degradation rate (dE/E_i) and loading cycles (N) (Equation (11)):

$$\frac{dE}{E_i} = F(N, S) \tag{11}$$

where, S is the mechanical response under loading, this parameter can be selected as required (stress, strain, or stress ratio).

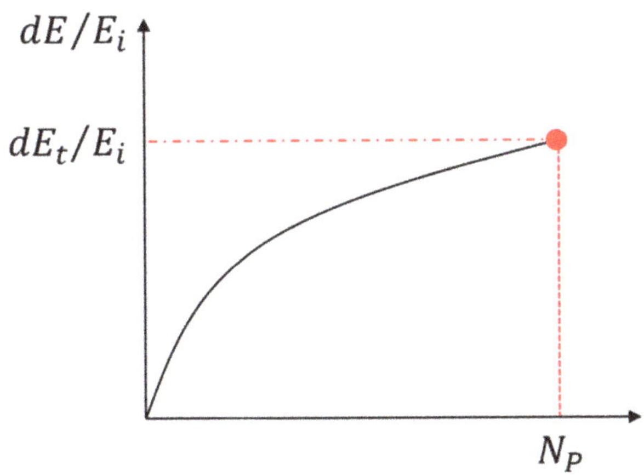

Figure 14. Model of non-linear fatigue damage.

According to the theory of continuum damage mechanics, cracks result from the growth and accumulation of micro-cracks in material microstructures [31,32].

In the uniaxial tensile test, the occurrence of fracture will cause the decrease of effective cross-section, and the stress must be transferred to the remaining intact cross-section. So, the damage can be defined as the relative value of the effective cross-sectional reduction, as shown in Equation (12):

$$\omega = \frac{A_0 - A}{A_0} = \frac{dA}{A_0} \tag{12}$$

where, ω is the damage, A_0 is the initial cross-section, A is the retained cross-section.

The damage can also be explained as the attenuation of material modulus, as shown in Equation (13):

$$\omega = \frac{E_0 - E}{E_0} = \frac{dE}{E_0} \tag{13}$$

where, ω is the damage, E_0 is the initial material modulus, E is the modulus after material damage.

The development of material damage is a function of the actual stress, as shown in Equation (14):

$$\omega = \left(\frac{N}{10^6}\right)^{\alpha} \times \left(\frac{\sigma}{S}\right)^{\beta} \times (1-\omega)^{\gamma} \tag{14}$$

where, ω is the damage, σ/S is the stress ratio, N is the loading cycles, and α, β, γ are constant.

The modulus attenuation model related to the mechanical index of fatigue equation can be established by using Equation (14).

3.5.2. Modulus Attenuation Model Construction

Fatigue damage of semi-rigid base material accumulates continuously under the repeated loading. When the loading cycle accumulates to a certain degree, the material will eventually lead to fatigue damage. As the accumulation of fatigue damage in the semi-rigid base material is a gradual process, it is impossible to observe the damage process directly. The deterioration of bending elastic modulus provides a feasible approach to evaluate the damage degree of semi-rigid base material. Therefore, the change of elastic modulus is very important to evaluate the damage state of semi-rigid materials and structures.

During the model establishment, the development of bending elastic modulus under different stress ratios and different loading cycles need to be considered. The attenuation of bending elastic modulus had three stages included rapid decline stage, slow decline stage and destruction stage, respectively. The rapid decline and slow decline stages were prediction regions. The destruction stage was the discarded stage, as shown in Figure 15. In order to ensure the fitting accuracy, the bending elastic modulus data in the destruction stage were neglected. So, the incremental-recursive model only predicts the degradation of bending elastic modulus from the initial modulus to the decay of the modulus before failure.

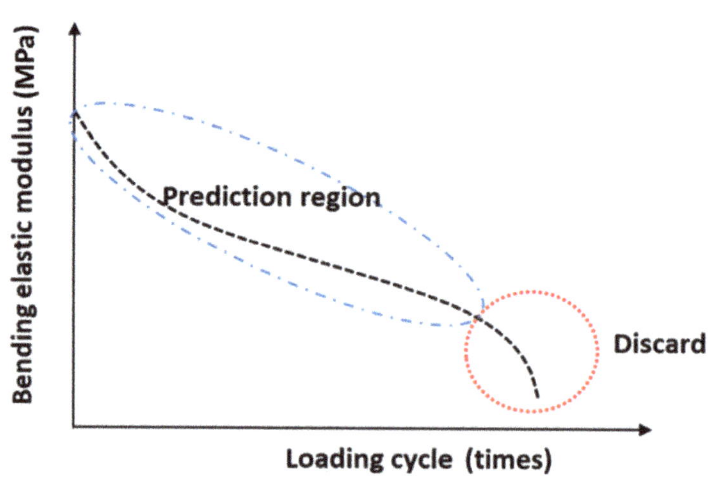

Figure 15. Development of bending elastic modulus in two regions.

According to the data of bending elastic modulus and the incremental-recursive model, constants in Equation (5) were obtained by using Levenberg–Marquardt fitting algorithm method. The fitting equations of LFSG and CSG are presented in Equations (15) and (16), respectively. The relatively high correlation coefficients R^2 (0.9674 and 0.8488) of these two equations indicate this model can achieve reliable fitting effect.

$$\omega = \left(\frac{N}{10^6}\right)^{0.33} \times \left(\frac{\sigma}{S}\right)^{3.73} \times (1-\omega)^{-1.58} \quad (15)$$

$$\omega = \left(\frac{N}{10^6}\right)^{0.31} \times \left(\frac{\sigma}{S}\right)^{2.78} \times (1-\omega)^{-0.78} \quad (16)$$

This modulus degradation model can not only be used to predict the fatigue deterioration of semi-rigid materials under different stress levels, but is also able to calculate the current bending elastic modulus based on the initial modulus value [33]. So, the fatigue life of LFSG and CSG under different loading levels can be predicted, which in turn supplies a reliable model and indexes for the fatigue damage analysis of semi-rigid base course of pavement structure.

As expected based on previous studies [25], two types of CSG with same cement contents had different modulus degradation models. The results showed the stress–strength ratio should be considered.

3.5.3. Modulus Attenuation Model Verification

The modulus of three pavement structures with CSG and LFSG is back calculated after different loading cycles. Meanwhile, the modulus of CSG and LFSG under the same loading cycles is calculated by the two modulus attenuation models. The modulus of back calculation and degradation model in three pavement structures are shown in Figures 16–18. In these figures, the deflection back calculate modulus was back calculated by the surface deflection basin data in the three pavement structures. The degradation modulus was calculated by the Equations (15) and (16). The deflection back calculates the modulus based on the actual measurement deflection basins. However, the degradation modulus is based on the fitting models.

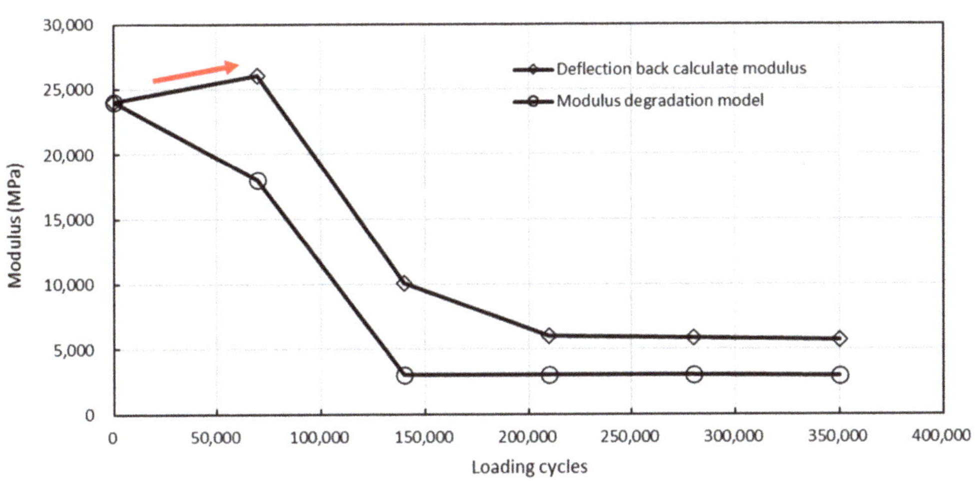

Figure 16. The modulus of deflection back calculation and degradation model in Structure 1.

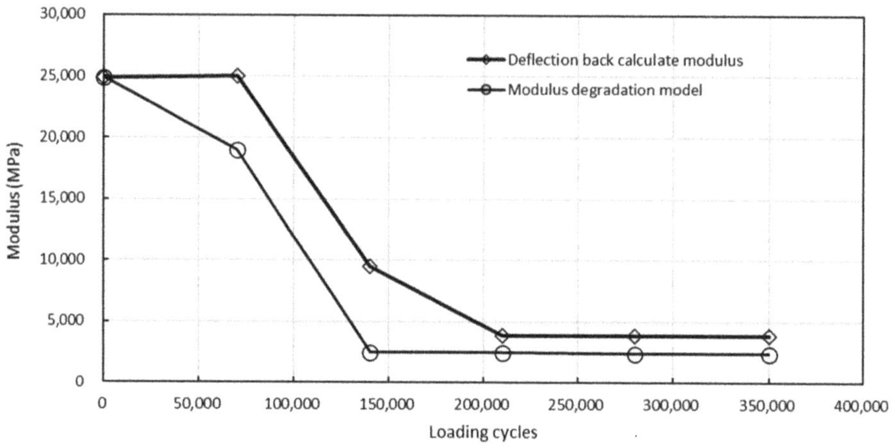

Figure 17. The modulus of deflection back calculation and degradation model in Structure 2.

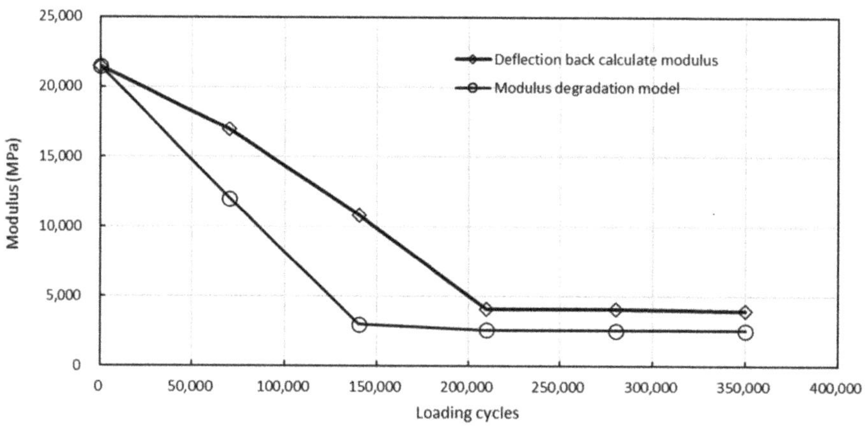

Figure 18. The modulus of deflection back calculation and degradation model in Structure 3.

According to the back-calculation modulus of Structure 1 to 3, the modulus degradation of semi-rigid base can be classified as three stages, which are the fatigue cracking generation stage, cracking development stage and cracking stable stage, in the cyclic loading. The time of fatigue cracking generation stage is equal to the fatigue life of the four-point bending fatigue test. The four-point bending fatigue test is influenced by specimen dimensions and loading modes. As microcracks appear on the bottom of specimen, the cracks will spread across the specimen rapidly and lead to the specimen failure. However, the full-scale accelerate loading test is different. Cracks through the semi-rigid base require a long time as microcracks appear and this is the cracking development stage. When the cracks go through the semi-rigid base completely, the modulus of the semi-rigid base will be stable. It is the cracking stable stage.

In cracking stable stage, the modulus of deflection back-calculation is decreased to 18.3%, 15.7% and 18.8% of the initial modulus in Structure 1 to Structure 3, respectively. The average modulus of deflection back-calculation reduced to 17.6% of its initial modulus. The modulus calculated by modulus degradation models is less than the modulus of deflection back-calculation. However, in the full-scale accelerated test, the results of the modulus prediction model still have good correlation with the results of FWD.

4. Conclusions

This research assessed the mechanical performance of semi-rigid road base materials (LFSG and CSG) under the dynamic load. The dynamic compression test, the four-point bending test and full-scale accelerate loading test were performed to evaluate the dynamic compressive resilience modulus and fatigue resistance of semi-rigid base materials, respectively. The following conclusions were given based on the results in this study:

1. From the dynamic compression test, the LFSG had a 59% higher dynamic modulus than that of the CSG. The load frequency had less influence on the dynamic modulus of these two semi-rigid road base materials;
2. The LFSG obtained a 19.7% higher flexural-bending strength than that of the CSG, which indicated its higher bearing capacity as a road base course material. The LFSG had the potential to resist flexural tensile failure as a road base course;
3. The fatigue life of LFSG and CSG decreased with the increase of stress ratio. It is indicated that a higher traffic loading will accelerate the fatigue deterioration of the semi-rigid base course. Slightly higher fatigue life of LFSG indicated its better fatigue resistance than that of the CSG;
4. A fatigue damage model was established by *S-N* curve. This model can be used to predict the fatigue life of CSG and LFSG materials under different stress–strength ratios; and
5. A modulus degradation model was developed based on the incremental-recursive mechanics. This model can not only be used to predict the fatigue deterioration of semi-rigid materials under different stress levels, but is also able to calculate the current bending elastic modulus based on the initial modulus value.

The results of this study will help to evaluate the performance decay and predict the residual life of road accurately in the future. The results will also guide the preventive maintenance of the highway. However, these results could only predict the fatigue life of semi-rigid road base asphalt pavement. In the future works, the investigation of fatigue life with different types of asphalt mixtures was recommended.

Author Contributions: Conceptualization, S.Y. and J.W.; methodology, X.Z.; software, X.W.; validation, T.C., Q.S. and L.C.; formal analysis, Y.Z.; investigation, X.Z.; resources, W.W.; data curation, T.C.; writing—original draft preparation, X.Z.; writing—review and editing, W.W.; visualization, X.W.; supervision, J.W.; project administration, S.Y.; funding acquisition, L.C. All authors have read and agreed to the published version of the manuscript.

Funding: This research was funded by the National Key R&D Program of China, grant number 2018YFB1600103 and project ZR2020QE271 was supported by Shandong Provincial Natural Science and Foundation and Key Research and Development Program of Shandong Province, grant number 2019GSF109020.

Institutional Review Board Statement: Not applicable.

Informed Consent Statement: Not applicable.

Data Availability Statement: Data sharing is not applicable to this article.

Conflicts of Interest: The authors declare no conflict of interest.

References

1. Zhang, J.Z.; Li, P.Z.; Liang, M.; Jiang, H.G.; Yao, Z.Y.; Zhang, X.M.; Yu, S. Utilization of red mud as an alternative mineral filler in asphalt mastics to replace natural limestone powder. *Constr. Build. Mater.* **2020**, *237*, 117821. [CrossRef]
2. Zhang, J.Z.; Sun, C.J.; Li, P.Z.; Liang, M.; Jiang, H.G.; Yao, Z. Experimental study on rheological properties and moisture susceptibility of asphalt mastic containing red mud waste as a filler substitute. *Constr. Build. Mater.* **2019**, *211*, 159–166. [CrossRef]
3. Zhang, J.Z.; Sun, H.; Jiang, H.G.; Xu, X.B.; Liang, M.; Hou, Y.; Yao, Z. Experimental assessment of reclaimed bitumen and RAP asphalt mixtures incorporating a developed rejuvenator. *Constr. Build. Mater.* **2019**, *215*, 660–669. [CrossRef]
4. Li, S.L.; Wu, G.M.; Wu, H.S. Study on acoustic emission characteristics of a semirigid base of dense skeleton type during complete uniaxial compression tests. *Adv. Mater. Sci. Eng.* **2016**, *6*, 1–8. [CrossRef]

5. Xuan, D.X.; Houben, L.J.M.; Molenaar, A.A.A.; Shui, Z.H. Mechanical properties of cement-treated aggregate material-A review. *Mater. Design* **2012**, *33*, 496–502. [CrossRef]
6. Huang, B.; Dong, Q.; He, W. Laboratory evaluation of reclaimed asphalt pavement used as unbound base material. In Proceedings of the 87th Annual Meeting of Transportation Research Board 2008, National Research Council, Washington, DC, USA, 13–17 January 2008.
7. Xuan, D.X.; Molenaar, A.A.A.; Houben, L.J.M. Shrinkage cracking of cement treated demolition waste as a road base. *Mater. Struct.* **2016**, *49*, 631–640. [CrossRef]
8. Fallon, E.; McNally, C.; Gibney, A. Evaluation of fatigue resistance in asphalt surface layers containing reclaimed asphalt. *Constr. Build. Mater.* **2016**, *128*, 77–87. [CrossRef]
9. Moghaddam, T.B.; Karim, M.R.; Abdelaziz, M. A review on fatigue and rutting performance of asphalt mixes. *Sci. Res. Essays* **2011**, *6*, 670–682. [CrossRef]
10. Underwood, B.S.; Guido, Z.; Gudipudi, P.; Feinberg, Y. Increased costs to US pavement infrastructure from future temperature rise. *Nat. Clim. Change* **2017**, *7*, 704–710. [CrossRef]
11. Micaelo, R.; Pereira, A.; Quaresma, L.; Cidade, M.T. Fatigue resistance of asphalt binders: Assessment of the analysis methods in strain-controlled tests. *Constr. Build. Mater.* **2015**, *98*, 703–712. [CrossRef]
12. Sheng, Y.P.; Li, H.B.; Zhao, H.S.; Chang, M.F. Effect on mechanical behavior of asphalt pavement structure based on semi-rigid base modulus attenuation. *J. Hebei University of Tech.* **2016**, *45*, 101–107.
13. Sheng, Y.P.; Chen, S.F.; Wang, D.; Wang, L.B. Pore water pressure characteristics of semi-rigid base for cement concrete pavement. *J. Transp. Eng.* **2012**, *12*, 6–12.
14. Wang, Y.Q.; Tan, Y.Q.; Guo, M.; Liu, Z.Y.; Wang, X.L. Study on the dynamic compressive resilient modulus and frost resistance of semi-rigid base materials. *Road Mater. Pavement* **2017**, *18*, 259–269. [CrossRef]
15. Yao, K.; Chen, Q.; Xiao, H.; Liu, Y.; Lee, F.H. Small-strain shear modulus of cement-treated marine clay. *J. Mater. Civil Eng.* **2020**, *32*, 04020114. [CrossRef]
16. Yao, K.; Li, N.; Chen, D.H.; Liu, Y. Generalized hyperbolic formula capturing curing period effect on strength and stiffness of cemented clay. *Constr. Build. Mater.* **2019**, *199*, 63–71. [CrossRef]
17. Yao, K.; Chen, Q.; Ho, J.; Xiao, H.; Lee, F.H. Strain-dependent shear stiffness of cement-treated marine clay. *J. Mater. Civil Eng.* **2018**, *30*, 04018255. [CrossRef]
18. Zhou, H.; Sha, A.M. Analysis on the influence of material composition on semi-rigid base fatigue property. *J. Wuhan University of Tech.* **2012**, *34*, 41–45. [CrossRef]
19. Sha, A.M.; Jia, K.; Li, X.G. Fatigue performances of semi-rigid base course materials. *J. Transp. Eng.* **2009**, *9*, 29–33. [CrossRef]
20. *Standard, No. JTG E30-2005*; Test Methods of Cement and Concrete for Highway Engineering. Renmin Communication Press: Beijing, China, 2005.
21. *Standard, No. JTG E51-2009*; Test Methods of Materials Stabilized with Inorganic Binders for Highway Engineering. Renmin Communication Press: Beijing, China, 2009.
22. *Standard No. JTG E20-2011*; Standard Test Methods of Bitumen and Bituminous Mixtures for Highway Engineering. Renmin Communication Press: Beijing, China, 2011.
23. *Standard, No. JTG D50-2017*; Specifications for Design of Highway Asphalt Pavement. Renmin Communication Press: Beijing, China.
24. Ozer, H. Prediction of pavement fatigue cracking at an accelerated testing section using asphalt mixture performance tests. *Int. J. Pavement Eng.* **2018**, *9*, 264–278. [CrossRef]
25. Lv, S.T.; Xia, C.D.; Liu, H.F.; You, L.Y.; Qu, F.T.; Zhong, W.L.; Yang, Y.; Washko, S. Strength and fatigue performance for cement-treated aggregate base materials. *Int. J. Pavement Eng.* **2019**, *22*, 690–699. [CrossRef]
26. Ma, Y. The bending fatigue performance of cement-stabilized aggregate reinforced with polypropylene filament fiber. *Constr. Build. Mater.* **2015**, *83*, 230–236. [CrossRef]
27. Zhang, J.P.; Cu, S.C.; Cai, J.; Pei, J.Z.; Jia, Y.S. Life-cycle reliability evaluation of semi-rigid materials based on modulus degradation model. *KSCE J. Civ. Eng.* **2018**, *22*, 2043–2054. [CrossRef]
28. Xue, J.; Jiang, Y. Analysis on the fatigue properties of vertical vibration compacted lime-fly ash-stabilized macadam. *Constr. Build. Mater.* **2017**, *155*, 531–541. [CrossRef]
29. ARA, Inc. *ERES Consultants Division, Guide for Mechanistic-Empirical design of new and rehabilitated pavement structure*; NCHRP 1-37A Final Report; ARA, Inc.: Villa Park, IL, USA, 2004.
30. ARA, Inc. *ERES Division. Guide for Mechanistic-Empirical Design of New and Rehabilitated Pavement Structures 2003*; ARA, Inc.: Villa Park, IL, USA, 2003.
31. Kachanov, L.M. *Mechanics of Elastic Stability*; Martinus Nijhoff Publishers: Leiden, The Netherlands, 1986.
32. Li, S.L.; Wu, G.M.; Wu, H.S. Acoustic emission characteristics of semi-rigid bases with three moisture conditions during bending tests. *Road Mater. Pavement* **2019**, *20*, 187–198. [CrossRef]
33. Naser, M.Z.; Alavi, A.H. Error Metrics and Performance Fitness Indicators for Artificial Intelligence and Machine Learning in Engineering and Sciences. *Archit. Struc. Constr.* **2021**, 1–19. [CrossRef]

coatings

MDPI

Article

Study on Performance Optimization of Composite Natural Asphalt Modified Gussasphalt Mix

Huadong Sun [1], Peng Jiang [1,2,*], Yongling Ding [1,*], Laixue Pang [1], Yinbin Liu [3], Yafei Wang [4] and Xuhe Shang [4]

1 School of Civil Engineering, Shandong Jiaotong University, Jinan 250357, China; 204113@sdjtu.edu.cn (H.S.); 202120657@mail.sdu.edu.cn (L.P.)
2 School of Civil Engineering, Shandong University, Jinan 250100, China
3 Jinan Zhangqiu District Highway Development Center, Jinan 250203, China; 201914559@mail.sdu.edu.cn
4 Shandong Lianhai Construction Technology Co., Ltd., Jinan 250000, China; 18011035@stu.sdjtu.edu.cn (Y.W.); 201720392@mail.sdu.edu.cn (X.S.)
* Correspondence: jiangp2020@163.com (P.J.); 204130@sdjtu.edu.cn (Y.D.); Tel.: +86-0531-8068-7901 (P.J.); +86-0531-8068-7901 (Y.D.)

Abstract: In order to systematically study and develop a type of gussasphalt (GA) mix with superior performance, namely GA-10; the effect of Qingchuan Rock Asphalt (QRA) and Trinidad Lake asphalt (TLA) on the GA-10 mix was assessed based on the study of composite natural asphalt modified gussasphalt (CNAMGA) binder. Various analytical tests were used to evaluate the engineering properties, thermal stability and microstructure of CNAMGA mix. The results indicate that the stability of QRA modified binder and TLA modified binder in the normal temperature range and the high temperature range have been improved, and the temperature susceptibility is reduced. The optimal asphalt–aggregate ratio of the GA mix is determined to be 9.7%, which has good high-temperature stability, low-temperature crack resistance and construction workability. The QRA mix has better high-temperature stability than the TLA mix, whereas the low-temperature cracking resistance of the TLA mix is better than that of the QRA mix. The two kinds of GA-10 mix have similar construction workability. The fact that the abundant fine aggregates wrapped in binder fill the coarse aggregates surface contributes to the better adhesion of the GA asphalt concrete. The distribution of aggregate and binder is relatively uniform with fewer pores, and the overall proportion of the binder is greater than that of aggregate.

Keywords: gussasphalt; natural asphalt; thermal stability; engineering properties; microstructure

Citation: Sun, H.; Jiang, P.; Ding, Y.; Pang, L.; Liu, Y.; Wang, Y.; Shang, X. Study on Performance Optimization of Composite Natural Asphalt Modified Gussasphalt Mix. *Coatings* **2022**, *12*, 78. https://doi.org/10.3390/coatings12010078

Academic Editor: Valeria Vignali

Received: 4 December 2021
Accepted: 4 January 2022
Published: 11 January 2022

Publisher's Note: MDPI stays neutral with regard to jurisdictional claims in published maps and institutional affiliations.

Copyright: © 2022 by the authors. Licensee MDPI, Basel, Switzerland. This article is an open access article distributed under the terms and conditions of the Creative Commons Attribution (CC BY) license (https://creativecommons.org/licenses/by/4.0/).

1. Introduction

As a type of concrete classification based on construction technology [1], GA concrete presents a flowing state at the appropriate construction temperature, which can autonomously reach the required density and flatness without compaction [2]. Therefore, it is more suitable for bridge deck paving and other construction occasions in which heavy compaction machinery cannot be used. The overlays of an orthotropic deck, which is one of the most effective structure forms for long span bridges, has good durability, fatigue resistance and water resistance. The disposal of which, if not well regulated, could pose a huge concern [3–5]. The enclosed space structure of steel box girder bridges exposed to sunlight could lead the overlays' temperature to reach 70 °C and even higher, so it is extremely important that the overlays be armed with good thermal stability [6]. GA concrete not only meets these performance requirements, but can also effectively resist the stress caused by the partial deflection of steel plate. Simultaneously, it also possesses excellent adhesion and compliance towards the steel bridge deck. Thus, it is suitable for deck paving in the application of long-span and medium-span steel bridges [7,8]. GA was accepted as the preferred overlays for the Hong Kong-Zhuhai-Macao Bridge, which currently keeps the record of the world's longest cross-sea bridge [9].

Such excellent performance is attributed to the mix proportion design of GA concrete [10,11]. The mix proportion design of GA binder is characterized by the high content of mineral powder, the high content of asphalt binder, the high mixing temperature and the low content of coarse aggregate. Chen et al. [12], Wang and Li [13] investigated the effects of the content and type of binder, mineral filler, aggregate gradation and fabrication technics etc. on gussasphalt performance with fundamental approaches, and thereby obtained the recommended optimum formula, which suggests the TLA content of 25–30%. Different temperatures, asphalt types and filler-asphalt ratios on the high-temperature properties with laboratory testing, Xin et al. [6] found that NES-1 rock asphalt has more significant influence than TLA on improving the high-temperature performance, and Jin et al. [14] thought the generalized shear modulus can be regarded as a stiffness index to evaluate the high temperature stability. Qian et al. [15], Zhang et al. [16] and Wu et al. [17] focused on the influence of different strain levels of loading on the fatigue performance of GA mix. Results show that polymer modified agent can effectively improve the fatigue strength of GA mix, and the loading times can reach millions under 700 $\mu\varepsilon$. Modifying Burgers mechanical model, Xie et al. [18] and Luo et al. [19] derived the viscoelastic material parameters with laboratory and theoretical equations. They found that high temperature reduces the viscoelastic deformation recovery ability, increases the viscoelastic residual strain and accelerates the generation of permanent deformation. Based on the physical examinations, Zou et al. [9], Ke et al. [8] and Shan et al. [20] studied the thermo-oxidative aging properties of GA with Fourier transform infrared spectroscopy, the potential applications of GA concrete on bridge deck were evaluated. Given the practical applications on steel bridge deck overlays, generally, the vehicle speed on the bridge deck is faster than that on the common city road, thus, the requirements for rutting resistance of pavement are lower. However, steel deck, as we all know, is strongly susceptible to temperature variation, so it is important to know the stability of asphalt concrete under drastic temperature changes. Temperature susceptibility can be employed to quantify the thermal performance of asphalt concrete [6,21]. However, Zhao et al. [22] claimed that the GA applied in Jiangyin Bridge had a weaker high-temperature stability and less modulus [23], according to the analysis of long-term maintenance data. What can be seen is the urgent need to optimize the gussasphalt concrete to tolerate a more varied application environment.

Note that none of the above studies involves the microscopic study of GA concrete. Materials science research shows that the acquisition or improvement of each specific property is realized through the change of its microstructure [24]. The identification of the chemical components associated with the formation of microstructure can reveal asphalt's rheology [25–27]. The physical properties such as stiffness, elasticity and plasticity, adhesion, surface energy and healing mainly depend on the microstructure [28]. At present, most of the asphalt mix design methods adopt macroscopic parameters as the design index [13,23,27]. However, a mass of engineering practices has proved that current design method cannot control the material performance well. In some cases, the test indicators are the same but the performance varies greatly, implying that some macroscopic design indicators do not conform to the material damage mode [29–32].

The asphalt binder replaces the coarse aggregate acting as the skeleton structure, and the strength of GA concrete mainly depends on the binder's properties [33,34]. In addition, air void proportion is an important characteristic that affects the properties of GA concrete, for example, its strength and durability [35]. Caro et al. [36] and Ma et al. [37] revealed the effect of porosity on the fatigue life of asphalt concrete with discrete element and finite element models. In fact, errors in measuring air void proportion with traditional methods are inevitable, even some measures, such as coring, are destructive for pavement. Hoegh et al. [38] conducted a nondestructive assessment of air-void variability using an air-coupled step-frequency array system. Artificial neural networks and multiple linear regression were employed by Zavrtanik et al. [39] to establish the relationship between different parameters and air void content. Without a doubt, understanding the properties of GA concrete at the microscale with modern advanced techniques is beneficial to improving the knowledge of their macro-performance [40]. It was highly evaluated by Mahmud et al. [41] that X-ray Computerized Tomography (CT) provides an all-directional understanding of the distribution and properties of air voids, indicating strongly that CT might have great potential applications in the field.

Using different techniques to study the microstructure of asphalt is conducive to establishing the relationship between the microstructure and the physical properties of asphalt. After foundational laboratory tests, differential scanning calorimetry (DSC), scanning electron microscope (SEM) and X-ray CT, the thermal properties and microstructure and the engineering properties of the composite natural asphalt modified gussasphalt (CNAMGA) mix were revealed. The optimal collocation of CNAMGA mix was determined and verified. What this study indicates can serve as a basis for further research as regards GA.

2. Experimental Materials

2.1. Base Asphalt

The 70 # and 90 # Kunlun-Asphalt (PetroChina Fuel Oil Co., Ltd, Panjin, China). are taken as base asphalt, relevant performance tests are conducted strictly according to the Chinese standard of E20-2011, Standard Test Methods of Bitumen and Bituminous Mixtures for Highway Engineering [42], the results are shown in Tables 1 and 2.

Table 1. Technical indexes of 70 # base asphalt.

Test Index		Unit	Specified Value	Test Results	Test Method
Penetration	15 °C, 100 g, 5 s	0.1 mm	-	32	T0604
	25 °C, 100 g, 5 s		60–80	62	
	30 °C, 100 g, 5 s		-	99	
Softening point (R&B)		°C	≥46	47.2	T0606
Ductility (15 °C, 5 cm/min)		cm	≥100	116	T0605
Density (25 °C)		g/cm^3	Actual measurement	0.988	T0603
Thin film oven test (TFOT) (163 °C, 300 min)	Mass loss	%	−0.8–0.8	0.01	T0610
	Penetration ratio (25 °C)	%	≥61	70	T0604
	Ductility (10 °C, 5 cm/min)	cm	≥6	6.7	T0605

Table 2. Technical indexes of 90 # base asphalt.

Test Index		Unit	Specified Value	Test Results	Test Method
Penetration	15 °C, 100 g, 5 s	0.1 mm	-	40	T0604
	25 °C, 100 g, 5 s		80–100	92	
	30 °C, 100 g, 5 s		-	151	
Softening point (R&B)		°C	≥45	50.5	T0606
Ductility (15 °C, 5 cm/min)		cm	≥100	108	T0605
Density (25 °C)		g/cm^3	Actual measurement	1.109	T0603
TFOT (163 °C, 300 min)	Mass loss	%	−0.8–0.8	0.04	T0610
	Penetration ratio (25 °C)	%	≥57	67	T0604
	Ductility (10 °C, 5 cm/min)	cm	≥8	16.7	T0605

2.2. Natural Asphalt

In recent years, natural asphalt has been extensively studied and widely used in asphalt pavement, especially the Trinidad Lake Asphalt (TLA) [6,12,43]; the composition of natural asphalt is similar to that of petroleum asphalt, but the content of hard components (resin and asphaltene) is higher in natural asphalt. In China, lake asphalt mainly depends on import from foreign countries. To explore more substitutes with high yield, stable properties and low cost, Chinese Qingchuan Rock Asphalt (QRA) was employed to the fabrication of GA in this study.

2.3. Aggregate and Mineral Filler

Three kinds of basalt aggregate (aggregate I: 0–3 mm, aggregate II: 3–5 mm, aggregate III: 5–10 mm) and limestone mineral filler are adopted for this study, relevant performance tests for aggregate and mineral filler are conducted according to the Chinese standard of E20-2011 [42], the results are provided in Tables 3 and 4.

Table 3. Density test of materials.

Material	Trial Value (g·m^{-3})		Specified Value
	Apparent Specific Gravity	Bulk Specific Gravity	
Aggregate I	2.802	-	≥2.5
Aggregate II	2.743	2.694	
Aggregate III	2.794	2.703	
Mineral filler	2.686	-	

Table 4. Passing rate of materials.

Material	Aperture Size (mm)								
	13.2	9.5	4.75	2.36	1.18	0.6	0.3	0.15	0.075
Aggregate I	100.0	95.1	1.4	0.4	0.1	0.0	0.0	0.0	0.0
Aggregate II	100.0	100.0	93.8	1.2	0.3	0.1	0.1	0.0	0.0
Aggregate III	100.0	100.0	100.0	84.1	55.8	41.2	29.9	20.9	14.2
Mineral filler	100.0	100.0	100.0	100.0	100.0	100.0	100.0	93.4	84.7

2.4. Modifying Agent

To enhance the performance and workability of the construction of the GA mix, modifying agents were necessary in the preparation of the GA mix. The linear structure SBS with a molecular weight of 220–260 kD, block styrene content of 40.4 wt%, Shore A hardness of 82; TR with a molecular weight of 650–2600 kD, specific gravity of 0.97, softening point of 110 °C, glass transition temperature of 84 °C; FEO with a density (20 °C) of 0.994 g/cm^3, flashing point of 251 °C, aniline point of 30 °C, kinematic viscosity (100 °C) of 35.36, aromatic hydrocarbon content of 82%, produced by Chong Qing Wu Sheng Building Materials Co., Ltd. (Chongqing, China) and the long-chain aliphatic alkanes SAW with a melting point of 115 °C, flashing point of 290 °C were employed in this study to prepare the modifier. With the consultation of relative references and basic test in the early stage of the study, the modifiers ratio was determined by SBS:SAW:TR:FEO = 2:2:0.5:0.5.

3. Test Methods

3.1. Differential Scanning Calorimetry (DSC)

DSC is a technique for measuring the power and temperature difference between the sample and the reference object under programmed temperature control. The enthalpy change size (the area of the heat absorption peak) and the location of the heat absorption peak (the temperature range of the heat absorption peak) can be realized by the DSC curve. With high accuracy and reproducibility, DSC provides an effective means to quantify the thermal analysis of the asphalt binder [44,45]. The range of measured temperature was between −40 and 140 °C. In a nitrogen atmosphere (flow rate of 20 mL/min), the heating rate was set to 10 °C. The heat flow value of different specimens was recorded at an increment of 20 °C.

3.2. Engineering Properties Examination

The evaluated specimen (70.7 mm × 70.7 mm × 70.7 mm) for the indentation test (Figure 1a) was maintained at a specified temperature (60 °C is used as a design index for the torrid zone in the summer in China, 50 and 40 °C are adopted for the hot zone and warm zone in the summer in China, respectively) in a water bath for at least one hour. The load of 52.5 kg was applied at 0.05 m^2 acting area on the specimen, and the depth of the indenter was recorded to the nearest 0.01 mm after 30 and 60 min, respectively. The depth of the indenter recorded at 30 min is defined as the indentation (technical requirements 1–4 mm), and the depth increment of the indenter between 30 and 60 min is defined as the indentation increment (technical requirements ≤0.4 mm).

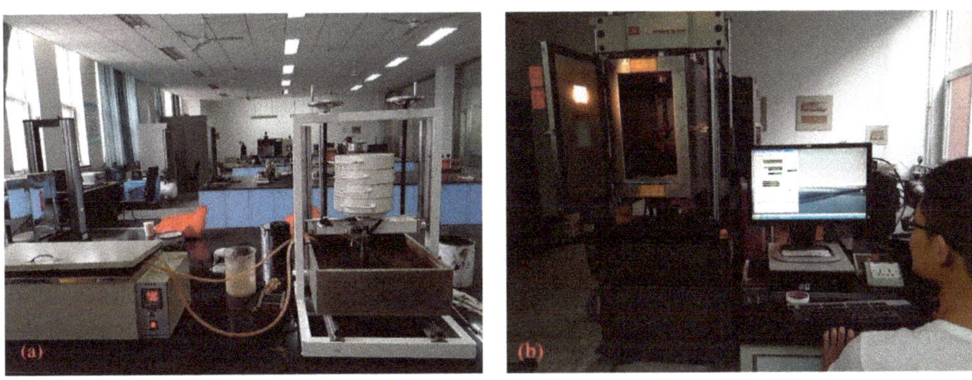

Figure 1. Performance test, (**a**) indentation test; (**b**) Low-temperature bending test.

The Lueer fluidity test instrument consists of a container, a support frame and a 995 g dombra-shaped drop hammer made of brass. An indicator of 50 mm was marked on the upper part of the bar of the drop hammer. The time required for the mix at 240 °C to pass from the container through the 50 mm indicator is called the Lueer fluidity [13,46], it is advisable to keep the fluidity within 3–20 s.

The low-temperature bending test (Figure 1b) was conducted according to the Technical Guide on Design and Construction of Bridge Deck Paving of Highway Steel Box Girder. At the specified temperature of −10 °C, the specimen (300 mm × 100 mm × 50 mm) was loaded at the loading velocity of 50 mm/min through the MTS810 testing machine. The data acquisition system automatically collected electrical signals from sensors and displacement meters. The ultimate tensile strain should be not less than 8×10^3 μ.

The wheel tracking test was conducted at 60 °C in a dry environment. A device like the Hamburg Wheel-Tracking device, a solid-steel wheel traveling at a speed of 42 cycles/min, was applied on a specimen (300 mm × 300 mm × 70 mm). The minimum requirement of the dynamic stability was 350 cycle/mm for a GA mix. Without an interlocking skeleton structure of coarse aggregates, the GA mix may show a weakness of rutting. Therefore, the dynamic stability requirement for a GA mix is lower than that of other common mixes, which is to be around 1500 cycle/mm.

The preparation of specimens for different tests is listed in Table 5.

Table 5. The preparation of specimens.

Test Item	Asphalt–Aggregate Ratio	Engineering Properties
Indentation test	25QRA	3QRA, 3TLA
Lueer fluidity test	5QRA	3QRA, 3TLA
Low-temperature bending test	5QRA	3QRA, 3TLA
Wheel tracking test	-	3QRA, 3TLA

3.3. *Scanning Electron Microscopy (SEM)*

SEM is the most widely used method for sample characterization, which is convenient for observing and analyzing the surface morphology and sample structure. Realizing the observation and analysis of samples from macro to micro dimensions, the magnification times of SEM can be continuously adjustable in the range from several times to several hundred thousand times. Additionally, SEM can be equipped with different detectors and can observe the dynamic process in real time. FEI Quanta 200 SEM (Hillsborocity, OR, USA) was employed for this study, a specimen taken from mix was subjected to the metal spraying process.

3.4. *X-ray Computerized Tomography (CT)*

Slice images obtained through X-ray CT can truly reflect the heterogeneous and multi-level structure characteristics of the bituminous mixture [47,48]. Taking the evaluated bituminous mixture specimen in this study as an example, after X-rays emitted by the Xradia 510 Versa microscope (Oberkochen, Germany) traverses the specimen, the recorded CT numbers are reconstructed and transformed into a series of visualized grayscale images (Figure 2a) by three-dimensional image processing techniques [49,50]. Denser materials (Figure 2(bi)) in asphalt mixture absorb more X-rays; therefore, they show greater intensity and appear to be brighter, and air voids (Figure 2(bii)) appear to be darker. The density of asphalt mastic (Figure 2(biii)) is between aggregate and air voids, it presents as gray and black in the image. The perspective of each phase of the asphalt mixture specimen can be quickly and easily presented and separated for further analysis after computer image processing.

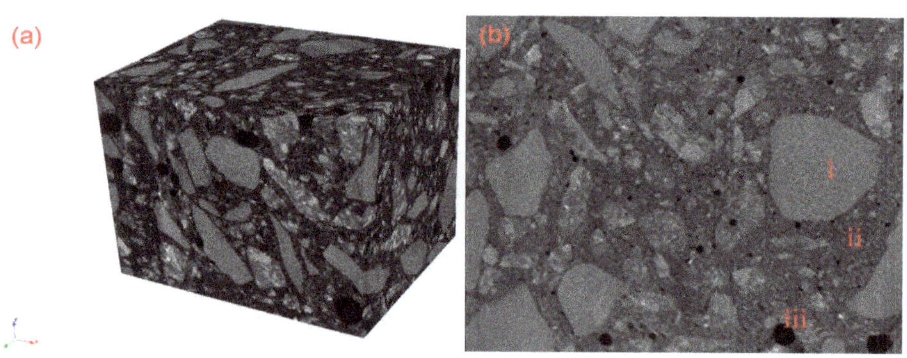

Figure 2. Grayscale treatment of CT, (**a**) 3D reconstruction; (**b**) CT slice image i: Aggregate, ii: Air void, iii: Asphalt mastic.

4. CNAMGA Binder

4.1. *Fabrication and Examination*

Two kinds of natural asphalt (QRA and TLA) and modifier (SBS:SAW:TR:FEO = 2:2:0.5:0.5) were used for specimen fabrication. Published work [12] indicated that a limited increase in the softening point was found at a TLA content of less than 20% when TLA only acted as a disperse additive, and reference [51] showed that the QRA content of 8% had no considerably adverse influence on low-temperature performance. In view of the addition of a modifier, 20% TLA for 70 # base asphalt and 12.5% QRA for 90 # base asphalt were determined. Standby base asphalt (70 # and 90 #) were heated to reach a flowing state in the constant temperature oven at 145 °C, then a proper amount of base asphalt was heated up to 165 °C in the magnetic stirring electric heating mantle. SBS was added bit by bit until all the SBS was added, while the binder was being stirred for 40–50 min, according to the content of natural asphalt (20% TLA for 70 # base asphalt and 12.5% QRA for 90 # base asphalt), a corresponding amount of asphalt was weighted. SAW, TR and QRA were added a small amount at a time until the required amount had been added, while the binder was being stirred for 1 h. Then, keeping the mixing temperature under 180 °C, the asphalt binder was stirred with homomixer at a speed of 3500 rpm for 40–50 min. Finally, specimens were prepared and subjected to a physical performance test.

The asphalt penetration test (25 °C, 100 g, 5 s), ductility test (10 °C, 5 cm/min) softening point test (R&B), RTFOT (163 °C, evaporation for 5 h) were conducted respectively according to the T0604-2011, T0605-2011 and T0606-2011 and T0608-1993 of E20-2011. The dynamic shear rheometer (DSR, Malvern, UK) test and bending beam rheometer (BBR) test were conducted in accordance with the procedures outlined in AASHTO T 315 [52] and AASHTO T 313 [53], respectively. On the basis of these previous works, the results are shown in Table 6. It can be seen from test results that all four kinds of collocation meet the technical requirements.

Table 6. CNAMGA binder properties.

Items	**70 # Modified Asphalt**		**90 # Modified Asphalt**		**Specified Value**
	0% TLA	**20% TLA**	**0% QRA**	**12.5% QRA**	
Penetration (0.1 mm)	51.3	27.6	44.5	38	10–40
Softening point (°C)	78.2	82.2	72.3	87.5	≥72
Ductility (mm)	20.5	14.2	19.7	12.5	≥10
Rutting factor (kPa)	1.39 (at 82 °C)	1.84 (at 88 °C)	1.46 (at 82 °C)	1.99 (at 88 °C)	≥1.0
Creep stiffness (MPa)	249 (at −18 °C)	288 (at −18 °C)	236 (at −18 °C)	285 (at −18 °C)	≤300
Creep rate m-value	0.348	0.306	0.356	0.317	-

4.2. Temperature Susceptibility Examination

The rheological properties and thermal stability are of significance for a GA binder; therefore, DSC was employed to explore the aggregation state condition of two types of natural asphalt in different modified asphalt and to characterize the asphalt binders. The DSC curve is presented and the determination of the baseline and peak area is shown in Figure 3.

An exothermic and endothermic effect is detected during the cooling and heating scan that represents the wax fractions in asphalt crystallization and melting, respectively. It can be seen from Figure 3a, in the temperature range of −40–−36 °C, that there is no obvious change between the curve of 90 # modified asphalt and QRA modified binder. With the increase in temperature, binder changes from a glass state to a viscoelastic state. The endothermic peak area of the QRA modified binder increased by 39.4% compared with that of the 90 # modified binder within the temperature range of −35–−20 °C. In the ranges of −20–40, 30–65 and 90–140 °C, compared with the 90 # modified binder, the endothermic peak area of the QRA modified binder decreased by 39.4%, 53.6% and 38.9%, respectively. The above changes in the endothermic peak area indicate that the change in enthalpy at 30–65 °C is smaller than that of other temperature ranges, which results in a smaller endothermic peak area of the QRA modified binder. It follows that most wax fractions in the QRA modified binder remains constant, the stability of the QRA modified binder in the normal temperature range and the high temperature range has been improved, and the temperature susceptibility is reduced.

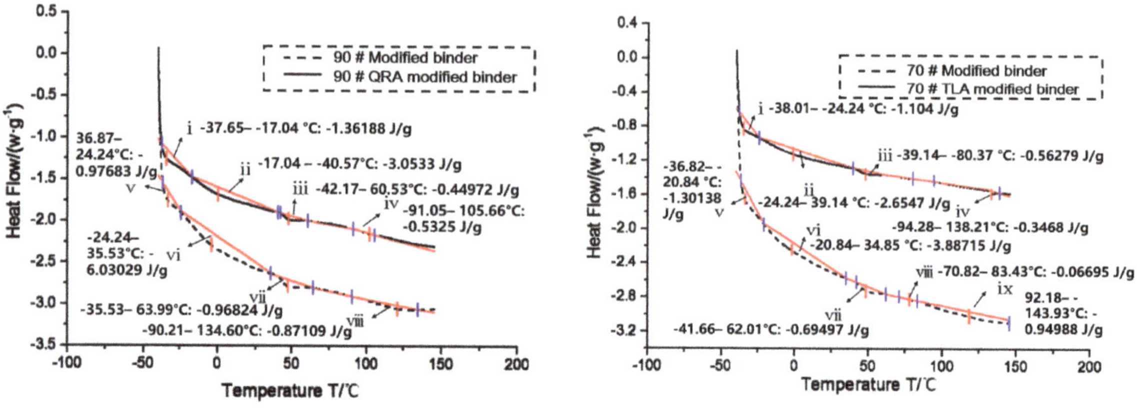

(a) (b)

Figure 3. The DSC curve of (**a**) 90 # modified asphalt and 90 # QRA asphalt; (**b**) 70# modified asphalt and 70# TLA asphalt.

In the temperature range of −40–−38 °C, there is no obvious change between the curve of the 90 # modified binder and the TLA modified binder. The fact that no endothermic peak occurs in this temperature range indicates that the glass transition temperature of asphalt is not changed by the incorporation of TLA. In the ranges of −21–35, −40–83 and 92–144 °C, compared with the 90 # modified binder, the endothermic peak area of the TLA modified binder decreased by 31.7%, 29.6% and 63.5%, respectively. The above changes in the endothermic peak area indicate that the endothermic peak area is smaller at 92–144 °C, where most wax fractions in the TLA modified binder remain constant. The changes indicate that the stability of the TLA modified binder in the high temperature range has clearly been improved, and that the temperature susceptibility is effectively reduced.

The above results of the experiment show that the modified asphalt forms a more stable aggregation structure due to the addition of TLA and QRA; therefore, the two binders are employed to prepare the GA-10 mix.

5. GA-10 Mix

5.1. Proportions of Composite Natural Asphalt Modified Gussasphalt Mix

GA contained a high content of asphalt binder and mineral powder, therefore, the mix design of the GA mix is different from that of other asphalt mix, different countries also have different design characteristics according to their own engineering practice experience and their own climatic conditions [10]. Basalt aggregate and limestone mineral filler have three particle sizes (5–10 mm, 3–5 mm, 0–3 mm), the proportions of TLA mix are aggregate III:aggregate II:aggregate I:mineral filler = 23%:11%:39%:27%, the proportions of the QRA mix are aggregate III:aggregate II:aggregate I:mineral filler = 27%:16%:30%:27%, the two kinds of GA-10 mix gradation is shown in Figure 4.

The GA mix with continuous gradation contains more fine aggregate, free asphalt binder also accounts for a large part of it, therefore, the internal friction angle of the mix is smaller; its shear strength mainly depends on the cohesive force of binder. It can be seen that the content of asphalt binder is an important factor affecting the performance of the GA mix. In this study, the asphalt–aggregate ratio sweeps from 9.1% to 10.3% at an increment of 0.3%. Because the type of binder has little influence on the pavement performance of the mix, based on the five kinds of asphalt–aggregate ratio, the QRA mix is taken as an example to be subjected to the performance test.

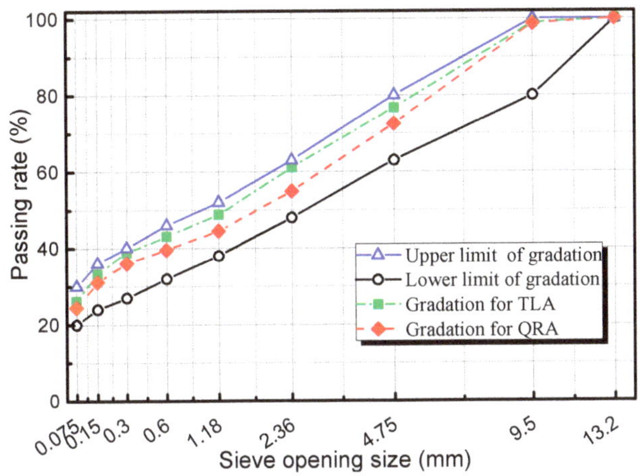

Figure 4. GA-10 gradation curve.

5.2. Asphalt–Aggregate Ratio

The indentation test was used to evaluate the mix strength and high-temperature deformation resistance. According to the different asphalt–aggregate ratios, the corresponding asphalt mix was prepared at the mixing temperature of 240 °C, and then the indentation test was carried out. Figure 5 depicts the results of the indentation test.

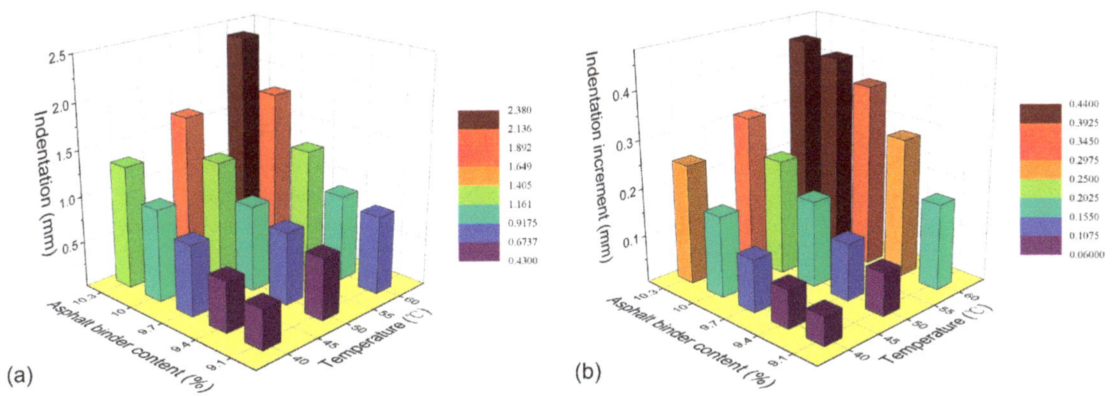

Figure 5. Influence of asphalt content on (**a**) indentation and (**b**) the indentation increment.

As can be seen from Figure 5, with the climbing asphalt content, the indentation and the indentation increment of the GA mix gradually increase at the same temperature, with the ascending test temperature; the indentation and indentation increment of the GA mix gradually rise at the same asphalt content. When the temperature goes up from 40 to 60 °C, the indentation of the five asphalt–aggregate ratios from 9.1 to 10.3% increase by 97.7%, 70.9%, 70.1%, 84% and 78.9%, respectively, the indentation increment of the five asphalt–aggregate ratios from 9.1 to 10.3% soar by 200%, 262.5%, 245.5%, 147.1% and 76%, respectively, this shows that the indentation with the asphalt–aggregate ratio of 9.4% or 9.7% is less affected by temperature, and that the indentation increment with the asphalt–aggregate ratio of 10.0% or 10.3% has weaker temperature susceptibility. According to the Technical Guide on Design and Construction of Bridge Deck Paving of Highway Steel Box Girder, the indentation and indentation increment with the asphalt–aggregate ratio of 9.7% can simultaneously meet the requirements for the torrid zone in the summer in China; when the asphalt–aggregate ratio is 10% or 10.3%, the indentation and indentation increment can simultaneously meet the requirements for the hot or warm zone in the summer in China.

Lueer fluidity is an important index to characterize the construction workability of the GA mix. Low-temperature crack resistance refers to ability of the asphalt mixture to resist cracking diseases at low temperatures, the low-temperature bending test and Lueer fluidity test results were performed and the results are given in Figure 6.

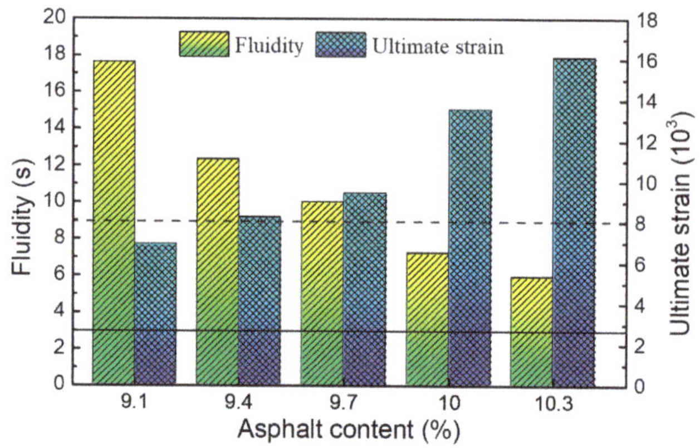

Figure 6. Influence of asphalt content on fluidity and ultimate strain. Note: fluidity value above the solid line, ultimate strain value above the dash line meets the requirements.

Figure 6 shows that the fluidity of the GA-10 mix declines with the ascending asphalt content, which indicates that the mix becomes easier to pour with the increase of the asphalt–aggregate ratio. Note that when the asphalt–aggregate ratio rises from 9.1% to 10.3%, the fluidity reduces from 17.6 to 5.9 s, with a decrease of 66.5%, but the fluidity still meets the requirements. The ultimate strain of the GA-10 mix rises with the increase of the asphalt content, which indicates that the low-temperature crack resistance of the mix has been improved with the climbing asphalt–aggregate ratio. The ultimate strain of the GA-10 mix rises from 6890 to 16,073 μ, with an increase of 133.3%, however, it cannot meet the requirements when the asphalt–aggregate ratio is less than 9.4%. Taking all the above factors into consideration, the optimal asphalt–aggregate ratio of the GA-10 mix is determined to be 9.7%, it meets the design requirements and has good high-temperature stability, low-temperature crack resistance and construction workability.

5.3. Verification Examinations

The wheel tracking test is a direct test to evaluate the performance of pavement under dynamic load, it is also used to evaluate the high-temperature stability of the GA mix. The road performance of the two GA-10 mixes was further compared and verified. The mix specimens with the optimal asphalt–aggregate ratio were tested. The test results are shown in Table 7.

Table 7. Performance of two kinds of GA-10.

Kind of Mix	Indentation Increment at 60 °C (mm)	Indentation Increment at 60 °C (mm)	Ultimate Strain (μ)	Fluidity (s)	Dynamic Stability (Cycle/mm)
QRA	1.84	0.38	9430	10.0	1309
TLA	2.50	0.34	15032	8.9	1245

As can be seen from Table 6, the indentation and indentation increment of the QRA mix at 60 °C test temperature are less than that of the TLA mix, the dynamic stability of the

QRA mix at 60 °C test temperature is larger than that of the TLA mix, which indicates that the QRA mix has better high-temperature stability than the TLA mix, which confirms the previous conclusion of the DSC. However, the ultimate strain of the TLA mix is 1.59 times as that of the QRA mix; thus, the low-temperature cracking resistance of the TLA mix is better than that of the QRA mix. The two kinds of GA-10 mix do not differ much in fluidity, which implies that they two have similar construction workability.

6. Microstructure Characterization

6.1. SEM Detection

The successful application of modern detection technologies has made the examination of the asphalt mix even better [54,55]; with the help of SEM, the morphology of the mix can be revealed distinctly [56–58]. The two kinds of GA-10 mix were subjected to the SEM detection; the effect of the modifier and natural asphalt on the asphalt mix microstructure was characterized and studied. The images are given in Figure 7.

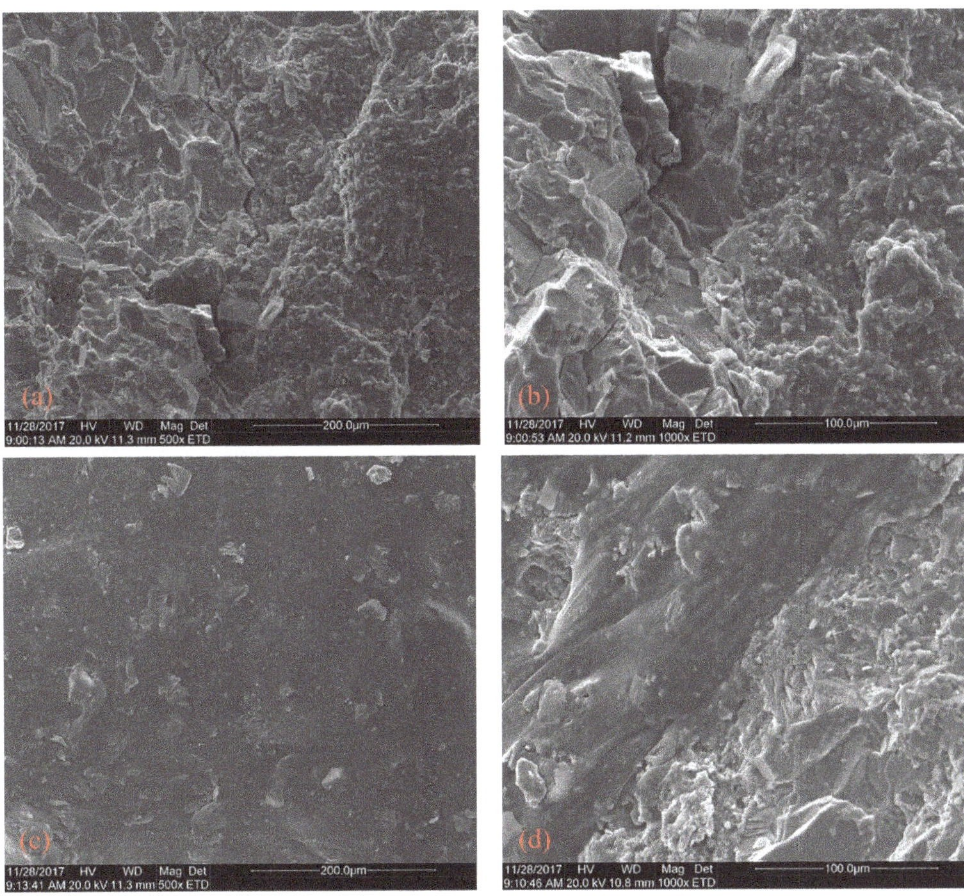

Figure 7. The SEM images of the (**a**) QRA mix (200 μm); (**b**) QRA mix (100 μm); (**c**) TLA mix (200 μm) and (**d**) TLA mix (100 μm).

Figure 7 demonstrates the accuracy of the size and appearance of the different components the two kinds of GA-10 mix; it shows that the modifier distributes and interconnects well with the asphalt in the mix. Figure 7a depicts that the fine aggregates wrapped by the binder contribute to the better adhesion of the GA asphalt concrete interior. Figure 7b further displays the stable interface between the coarse aggregates and binder; the fissures and holes on the aggregate surface are effectively filled and glued. Figure 7c,d also depict the phenomenon of the evenly distributed coarse aggregate densely packed by binder, the rough surface of coarse aggregate can be filled by abundant fine particles in binder, which is also conducive to the combination of the binder and aggregate. The whole structure network is established by the modifier and binder, which may, to a certain extent, be responsible for the physical properties, such as elastic-plastic, stiffness and surface healing.

6.2. CT Detection

The two kinds of GA-10 were subjected to CT detection. According to the distribution characteristics of gray value in the grey histogram, the objects distributed on the darker gray level form a wave crest, and the background distributed on the brighter gray level forms another crest. The gray value at the valley bottom between the two crests is used as the threshold value. Based on the self-compiled program of MATLAB, the threshold division range of aggregate, binder and air void could be determined according to the gray histogram. These components were then subjected to recognition and segmentation; the processing results of four arbitrary cross-sections are shown in Figure 8.

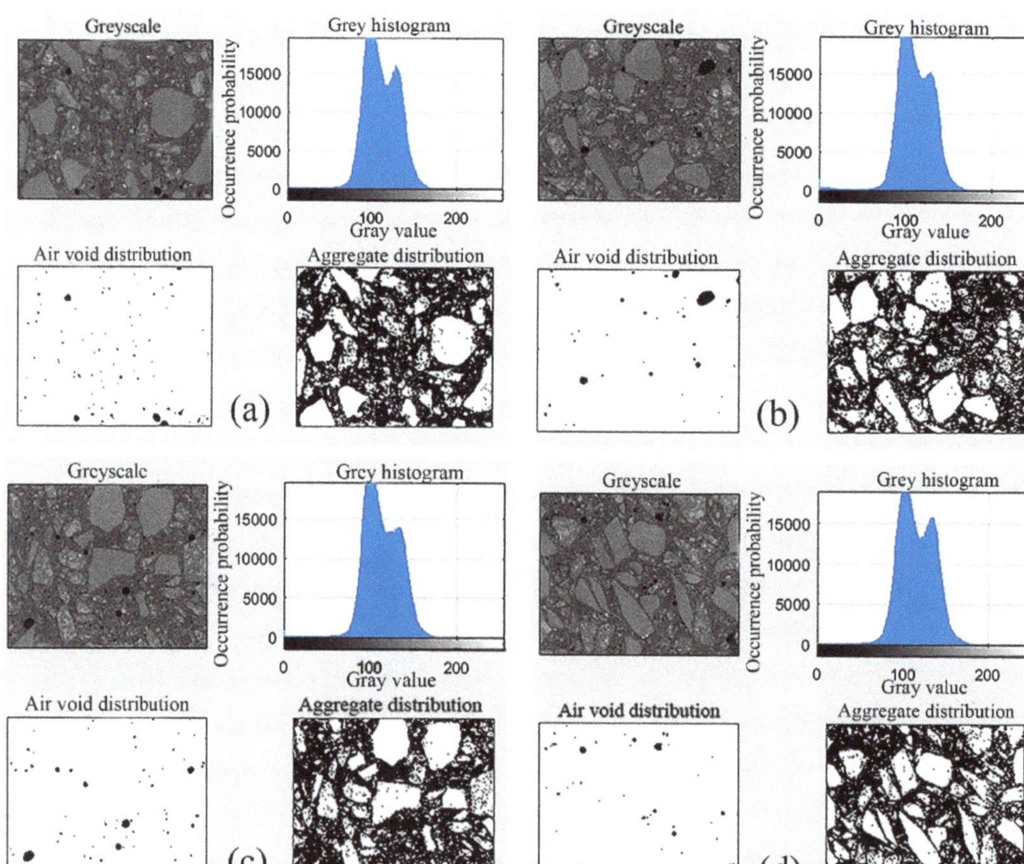

Figure 8. Images after greyscale treatment at the position of (**a**) one third of the TLA mix specimen length, (**b**) two thirds of the TLA mix specimen length, (**c**) one third of the QRA mix specimen length and (**d**) two thirds of the QRA mix specimen length.

As can be seen from Figure 8, the air void proportion is less; the threshold division range of air void is determined reasonably. In the aggregate distribution diagram, the aggregates are evenly distributed; the proportion of coarse and fine aggregates is appropriate, but there are some dots on the aggregates. The reason may be that the threshold is not determined properly, or there is asphalt binder adhering in the inner pore of aggregate. Ten pieces of CT cross-sections were taken from each of the two kinds of GA-10 mix at different positions. The proportion of air voids, aggregates and binder were calculated with repeatability tests; the results are given in Figure 9, on the left part of dash line in Figure 9 is the QRA mix, the right one is the TLA mix.

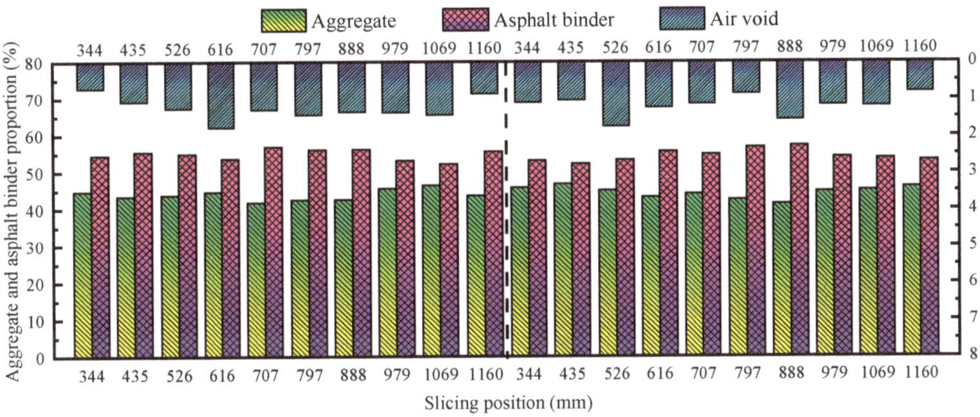

Figure 9. The proportion of air voids, aggregates and binder.

According to the data of twenty slicing images, the internal air void distribution of the two GA-10 mix specimens is relatively uniform with no obvious upward or downward trend, and the overall air void proportion fluctuates between 0.8% and 1.8%. The average air void proportion of the QRA mix is 1.27%, the average air void proportion of the TLA mix is 1.19%. The air void proportion at the upper and lower ends of specimen is relatively small and that of the middle part is larger. With the help of CT technology, it can be accurately seen that the air void proportion of the whole specimen is higher than the specified value by 1%. The aggregate proportion fluctuates between 41% and 47%, the average aggregate proportion of the QRA mix is 43.27%, the average aggregate proportion of the TLA mix is 44.44%. The asphalt binder proportion fluctuates between 52% and 57%, the average asphalt binder proportion of the QRA mix is 54.83%, the average asphalt binder proportion of the TLA mix is 54.37%. The maximum fluctuation range of the aggregate and binder proportion of the two GA-10 mix is not more than 6%, thus, it can be concluded that the distribution of aggregate and binder is relatively uniform with fewer pores and that the overall proportion of binder is greater than that of aggregate.

7. Conclusions

This study evaluated the thermal properties and microstructure, and engineering properties of the CNAMGA mix. The optimal collocation of the CNAMGA mix was determined and verified. The following conclusions can be drawn based on the aforementioned results and discussion.

(1) The asphalt forms a more stable aggregation structure due to the addition of natural asphalt. The stability of the QRA modified binder in the normal temperature range and the high temperature range has been clearly improved, the TLA modified binder in the high temperature range has been clearly improved, and the temperature susceptibility of both is reduced.

(2) The optimal asphalt–aggregate ratio of the GA mix is determined to be 9.7%, the QRA mix has better high-temperature stability than the TLA mix, which confirms the previous conclusion of the DSC. However, the low-temperature cracking resistance of the TLA mix is better than that of the QRA mix. The two kinds of GA-10 mix have similar construction workability.

(3) That fact that the abundant fine aggregates wrapped in binder fill the coarse aggregates surface contributes to the better adhesion of the GA asphalt concrete. The distribution of aggregate and binder is relatively uniform with fewer pores, and the overall proportion of binder is greater than that of aggregate.

Author Contributions: Conceptualization, H.S. and P.J.; methodology, Y.D.; software, P.J.; writing—original draft preparation, P.J.; writing—review and editing, H.S. and Y.L.; visualization, L.P. and Y.W.; supervision, H.S. and Y.L.; project administration, L.P. and X.S.; funding acquisition, H.S. All authors have read and agreed to the published version of the manuscript.

Funding: This research was funded by the China Postdoctoral Science Foundation (Grant No. 2019M662442), the National Natural Science Foundation of China (No. 51803109), and the Young Talent of Lifting engineering for Science and Technology in Shandong (SDAST2021qt01).

Institutional Review Board Statement: Not applicable.

Informed Consent Statement: Not applicable.

Data Availability Statement: The data used to support the findings of this study are available from the corresponding author upon request.

Acknowledgments: The authors wish to express their sincere gratitude for the support of the Shandong Jiaotong University and Shandong University.

Conflicts of Interest: The authors declare no conflict of interest.

References

1. Wu, W.; He, Z.; Hao, Z.; Zhang, H. Experimental study on fatigue performance of gussasphalt mixture. *J. Wuhan Univ. Technol. Mater. Sci. Ed.* **2014**, *29*, 745–750. [CrossRef]
2. Wang, M.; Zhang, H.; Zhu, M.; Hao, Z.; Xue, X. Research on structure and properties of embedded Gussasphalt. In Proceedings of the GeoHunan International Conference, Hunan, China, 9–11 June 2011.
3. Qiu, Y.; An, S.; Rahman, A.; Ai, C. Evaluation and optimization of bridge deck waterproof bonding system using multi-objective grey target decision method. *Road Mater. Pavement Des.* **2019**, *21*, 1844–1858. [CrossRef]
4. Jia, X.; Huang, B.; Chen, S.; Shi, D. Comparative investigation into field performance of steel bridge deck asphalt overlay systems. *KSCE J. Civ. Eng.* **2016**, *20*, 2755–2764. [CrossRef]
5. Liu, Y.; Qian, Z.; Hu, J.; Jin, L. Temperature behavior and stability analysis of orthotropic steel bridge deck during gussasphalt pavement paving. *J. Bridg. Eng.* **2018**, *23*, 04017117. [CrossRef]
6. Xin, C.; Lu, Q.; Ai, C.; Rahman, A.; Qiu, Y. Optimization of hard modified asphalt formula for gussasphalt based on uniform experimental design. *Constr. Build. Mater.* **2017**, *136*, 556–564. [CrossRef]
7. Luo, S.; Qian, Z.; Yang, X.; Lu, Q. Laboratory evaluation of double-layered pavement structures for long-span steel bridge decks. *J. Mater. Civ. Eng.* **2018**, *30*, 04018111. [CrossRef]
8. Ke, Z.; Mingzhi, S.; Yunjie, Q. Study on the Rheological Properties of Gussasphalt after Superheat Aging. In *IOP Conference Series: Earth and Environmental Science*; IOP Publishing: Bristol, UK, 2019.
9. Zou, G.; Xu, X.; Li, J.; Yu, H.; Wang, C.; Sun, J. The effects of bituminous binder on the performance of gussasphalt concrete for bridge deck pavement. *Materials* **2020**, *13*, 364. [CrossRef]
10. Zhang, R.; Huang, X.; Zhao, Y. Design of gussasphalt gradation. *Dongnan Daxue Xuebao (Ziran Kexue Ban)/J. Southeast Univ. (Nat. Sci. Ed.)* **2007**, *47*, 661–665.
11. Luo, S.; Qian, Z.; Yang, X.; Wang, H. Design of gussasphalt mixtures based on performance of gussasphalt binders, mastics and mixtures. *Constr. Build. Mater.* **2017**, *156*, 131–141. [CrossRef]
12. Chen, J.S.; Liao, M.C.; Huang, C.C.; Wang, C.H. Fundamental characterization of engineering properties of gussasphalt mixtures. *J. Mater. Civ. Eng.* **2011**, *23*, 1719–1726. [CrossRef]
13. Wang, H.; Li, G. Study of factors influencing gussasphalt mixture performance. *Constr. Build. Mater.* **2015**, *101*, 193–200. [CrossRef]
14. Jin, L.; Qian, Z.; Zheng, Y. High temperature performance and evaluation index of gussasphalt mortar based on DMA method. *Dongnan Daxue Xuebao (Ziran Kexue Ban)/J. Southeast Univ. (Nat. Sci. Ed.)* **2014**, *44*, 1062–1067. [CrossRef]
15. Qian, J.; Wang, Q.; Wu, W.; Zhang, H. Fatigue performance of gussasphalt concrete made from modified AH-70# asphalt. *Mater. Des.* **2013**, *52*, 686–692. [CrossRef]

16. Zhang, H.; Qian, J.S.; Wu, W.J.; Zhang, F.; Hao, Z.H. Fatigue life analysis of Gussasphalt concrete based on energy method. *Tumu Jianzhu yu Huanjing Gongcheng/J. Civ. Archit. Environ. Eng.* **2010**, *32*, 135–140.
17. Wu, W.J.; He, Z.Y.; Hao, Z.H.; Wang, M. Differences in fatigue behavior among the Gussasphalt, SMA and AC. In *Advanced Materials Research*; Trans Tech Publications, Ltd.: Bach, Swizerland, 2013.
18. Xie, F.; Zhang, D.; Zhou, A.; Ji, B.; Chen, L. On the viscoelastic parameters of gussasphalt mixture based on modified burgers model: Deviation and experimental verification. *Adv. Mater. Sci. Eng.* **2017**, *2017*, 4324765. [CrossRef]
19. Luo, R.; Shi, C.; Fan, X.; Feng, G. Research on permanent deformation of gussasphalt mixture under dynamic penetration. *Huazhong Keji Daxue Xuebao (Ziran Kexue Ban)/J. Huazhong Univ. Sci. Technol. (Nat. Sci. Ed.)* **2018**, *46*, 13–16. [CrossRef]
20. Shan, H.; Li, Q.; Mi, L.; Ye, Q.; Jia, X. Failure mechanism of shoving on steel bridge deck with gussasphalt overlay. *Tumu Jianzhu yu Huanjing Gongcheng/J. Civ. Archit. Environ. Eng.* **2016**, *38*, 104–109. [CrossRef]
21. Pouget, S.; Sauzéat, C.; Di Benedetto, H.; Olard, F. Numerical simulation of the five-point bending test designed to study bituminous wearing courses on orthotropic steel bridge. *Mater. Struct.* **2010**, *43*, 319–330. [CrossRef]
22. Zhao, Y.; Ni, F.; Zhou, L.; Jiang, J. Performance evaluation of long-span suspension bridge pavement based on long-term maintenance data. *J. Mater. Civ. Eng.* **2020**, *32*, 04019363. [CrossRef]
23. Yao, B.; Cheng, G.; Wang, X.; Cheng, C. Characterization of the stiffness of asphalt surfacing materials on orthotropic steel bridge decks using dynamic modulus test and flexural beam test. *Constr. Build. Mater.* **2013**, *44*, 200–206. [CrossRef]
24. Mazumder, M.; Ahmed, R.; Wajahat Ali, A.; Lee, S.J. SEM and ESEM techniques used for analysis of asphalt binder and mixture: A state of the art review. *Constr. Build. Mater.* **2018**, *186*, 313–329. [CrossRef]
25. Das, P.K.; Baaj, H.; Tighe, S.; Kringos, N. Atomic force microscopy to investigate asphalt binders: A state-of-the-art review. *Road Mater. Pavement Des.* **2016**, *17*, 693–718. [CrossRef]
26. Yu, X.; Burnham, N.A.; Tao, M. Surface microstructure of bitumen characterized by atomic force microscopy. *Adv. Colloid Interface Sci.* **2015**, *218*, 17–33. [CrossRef] [PubMed]
27. Singh, D.; Sawant, D. Understanding effects of RAP on rheological performance and chemical composition of SBS modified binder using series of laboratory tests. *Int. J. Pavement Res. Technol.* **2016**, *9*, 178–189. [CrossRef]
28. Kim, H.H.; Mazumder, M.; Torres, A.; Lee, S.-J.; Lee, M.-S. Characterization of CRM binders with wax additives using an atomic force microscopy (afm) and an optical microscopy. *Adv. Civ. Eng. Mater.* **2017**, *6*, 504–525. [CrossRef]
29. Yao, H.; You, Z. Effectiveness of micro-and nanomaterials in asphalt mixtures through dynamic modulus and rutting tests. *J. Nanomater.* **2016**, *2016*, 2645250. [CrossRef]
30. Du, S. Interaction mechanism of cement and asphalt emulsion in asphalt emulsion mixtures. *Mater. Struct.* **2014**, *47*, 1149–1159. [CrossRef]
31. Kim, H.H.; Lee, M.S.; Lee, S.J. Identification of the microstructural components of crumb rubber modified asphalt binder (CRMA) and the feasibility of using environmental scanning electron microscopy (ESEM) coupled with energy dispersive X-ray spectroscopy (EDX). *Int. J. Highw. Eng.* **2016**, *18*, 41–50. [CrossRef]
32. Cavalli, M.C.; Griffa, M.; Bressi, S.; Partl, M.N.; Tebaldi, G.; Poulikakos, L.D. Multiscale imaging and characterization of the effect of mixing temperature on asphalt concrete containing recycled components. *J. Microsc.* **2016**, *264*, 22–33. [CrossRef] [PubMed]
33. Widyatmoko, I.; Elliott, R.C.; Read, J.M. Development of heavy-duty mastic asphalt bridge surfacing, incorporating trinidad lake asphalt and polymer modified binders. *Road Mater. Pavement Des.* **2005**, *6*, 469–483. [CrossRef]
34. Artamendi, I.; Allen, B.; Allpress, C.; Phillips, P.; Wingrove, C. Resurfacing of the queen elizabeth ii bridge: Binder selection and evaluation. In *Bearing Capacity of Roads, Railways and Airfields, Proceedings of the 10th International Conference on the Bearing Capacity of Roads, Railways and Airfields, BCRRA 2017, Athens, Greece, 28–30 June 2017*; CSC Press: Boca Raton, FL, USA, 2017.
35. Nambiar, E.K.K.; Ramamurthy, K. Air-void characterisation of foam concrete. *Cem. Concr. Res.* **2007**, *37*, 221–230. [CrossRef]
36. Caro, S.; Diaz, A.; Rojas, D.; Nuñez, H. A micromechanical model to evaluate the impact of air void content and connectivity in the oxidation of asphalt mixtures. *Constr. Build. Mater.* **2014**, *61*, 181–190. [CrossRef]
37. Ma, T.; Zhang, Y.; Zhang, D.; Yan, J.; Ye, Q. Influences by air voids on fatigue life of asphalt mixture based on discrete element method. *Constr. Build. Mater.* **2016**, *126*, 785–799. [CrossRef]
38. Hoegh, K.; Khazanovich, L.; Dai, S.; Yu, T. Evaluating asphalt concrete air void variation via GPR antenna array data. *Case Stud. Nondestruct. Test. Eval.* **2015**, *3*, 27–33. [CrossRef]
39. Zavrtanik, N.; Prosen, J.; Tušar, M.; Turk, G. The use of artificial neural networks for modeling air void content in aggregate mixture. *Autom. Constr.* **2016**, *63*, 155–161. [CrossRef]
40. Yang, J.; Zhu, X.; Yuan, Y.; Li, L. Effects of aging on micromechanical properties of asphalt binder using AFM. *J. Mater. Civ. Eng.* **2020**, *32*, 04020081. [CrossRef]
41. Mahmud, M.Z.H.; Hassan, N.A.; Hainin, M.R.; Ismail, C.R. Microstructural investigation on air void properties of porous asphalt using virtual cut section. *Constr. Build. Mater.* **2017**, *155*, 485–494. [CrossRef]
42. Code of China. *JTG E20-2011, Standard Test Methods of Bitumen and Bituminous Mixtures for Highway Engineering*; China Communications Press: Beijing, China, 2011.
43. Luo, S.; Lu, Q.; Qian, Z.; Wang, H.; Huang, Y. Laboratory investigation and numerical simulation of the rutting performance of double-layer surfacing structure for steel bridge decks. *Constr. Build. Mater.* **2017**, *144*, 178–187. [CrossRef]
44. Das, P.K.; Kringos, N.; Birgisson, B. Microscale investigation of thin film surface ageing of bitumen. *J. Microsc.* **2014**, *254*, 95–107. [CrossRef] [PubMed]

45. Das, P.K.; Kringos, N.; Wallqvist, V.; Birgisson, B. Micromechanical investigation of phase separation in bitumen by combining atomic force microscopy with differential scanning calorimetry results. *Road Mater. Pavement Des.* **2013**, *14*, 25–37. [CrossRef]
46. Guilian, Z.; Xiaoning, Z.; Wu, C. Evaluation of steel bridge deck ma mixture properties during construction. *J. Mar. Sci. Technol.* **2015**, *23*, 293–301. [CrossRef]
47. Evans, C.L.; Wightman, E.M.; Yuan, X. Quantifying mineral grain size distributions for process modelling using X-ray micro-tomography. *Miner. Eng.* **2015**, *82*, 78–83. [CrossRef]
48. Cnudde, V.; Boone, M.N. High-resolution X-ray computed tomography in geosciences: A review of the current technology and applications. *Earth-Sci. Rev.* **2013**, *123*, 1–17. [CrossRef]
49. Jin, C.; You, Z.; Zhang, W.; Liu, K. Microstructural modeling method for asphalt specimens supporting 3D adaptive and automatic mesh generation. *J. Comput. Civ. Eng.* **2016**, *30*, 04015013. [CrossRef]
50. Chen, L.; Wang, Y. Improved image unevenness reduction and thresholding methods for effective asphalt X-ray CT image segmentation. *J. Comput. Civ. Eng.* **2017**, *31*, 04017002. [CrossRef]
51. Lu, Z.; He, Z.; Yu, Q.; Gang, H. Influence of rock asphalt modified asphalt on asphalt rheological property. *Chongqing Jiaotong Univ. (Nat. Sci. Version)* **2009**, *28*, 543–547.
52. American Association of State Highway and Transportation Officials. *AASHTO T 315, Standard Method of Test for Determining the Rheological Propertiesof Asphalt Binder Using a Dynamic ShearRheometer (DSR)*; American Association of State Highway and Transportation Officials: Washington, DC, USA, 2012.
53. American Association of State Highway and Transportation Officials. *AASHTO T 313, Standard Method of Test for Determining the Flexural Creep Stiffnessof Asphalt Binder Using the Bending BeamRheometer (BBR)*; American Association of State Highway and Transportation Officials: Washington, DC, USA, 2016.
54. Al-Rawashdeh, A.S.; Sargand, S. Performance assessment of a warm asphalt binder in the presence of water by using surface free energy concepts and nanoscale techniques. *J. Mater. Civ. Eng.* **2014**, *26*, 803–811. [CrossRef]
55. Nazzal, M.D.; Abu-Qtaish, L.; Kaya, S.; Powers, D. Using atomic force microscopy to evaluate the nanostructure and nanomechanics of warm mix asphalt. *J. Mater. Civ. Eng.* **2015**, *27*, 04015005. [CrossRef]
56. Fang, X.; Chen, Z.; Jiao, Z.; Xiao, G.; Shao, L. Study on microstructure, rheology and thermal stability of cement epoxy asphalt mortar multiphase materials. *Mag. Concr. Res.* **2013**, *65*, 475–485. [CrossRef]
57. Qin, Q.; Farrar, M.J.; Pauli, A.T.; Adams, J.J. Morphology, thermal analysis and rheology of Sasobit modified warm mix asphalt binders. *Fuel* **2014**, *115*, 416–425. [CrossRef]
58. Menapace, I.; Masad, E.; Bhasin, A.; Little, D. Microstructural properties of warm mix asphalt before and after laboratory-simulated long-term ageing. *Road Mater. Pavement Des.* **2015**, *16*, 2–20. [CrossRef]

coatings

MDPI

Article

Sustainability Evaluation of Cold In-Place Recycling and Hot Mix Asphalt Pavements: A Case of Qassim, Saudi Arabia

Fawaz Alharbi *, Fahad Alshubrumi, Meshal Almoshaogeh, Husnain Haider, Ahmed Elragi and Sherif Elkholy

Department of Civil Engineering, College of Engineering, Qassim University, Buraydah 51452, Saudi Arabia; 381125126@qu.edu.sa (F.A.); m.moshaogeh@qec.edu.sa (M.A.); husnain@qec.edu.sa (H.H.); afelragi@qec.edu.sa (A.E.); selkholy@qec.edu.sa (S.E.)

* Correspondence: f.a@qec.edu.sa

Abstract: The construction of conventional hot mix asphalt (HMA) pavements results in a number of economic and environmental issues, such as the cost of new overlays and associated impacts on natural resources. Although the cold recycling with an emulsified asphalt-recycling agent holds certain benefits over the HMA, its implementation on different road types, ranging from farm-to-market roads to expressways, is yet contentious due to the need for sophisticated equipment and trained workforce. The present research developed a methodology to evaluate all the three dimensions of sustainability, including economic (construction cost), environmental (natural resource depletion), and social (need for advanced equipment and skilled labor) of various scenarios of RAP and conventional asphalt pavements. The present study evaluated an equivalent thickness of the Cold In-place Recycling (CIR) pavement, which behaves similar to HMA pavement under the influence of different traffic loads. Fifty CIR and HMA scenarios for different traffic volumes and pavement layers thicknesses were developed. Finally, the sustainability of all the scenarios was evaluated for traffic designation in Saudi Arabia using fuzzy-based multicriteria analysis. Ranking of scenarios found CIR as a more sustainable overlay option for the feeders, collectors, main urban streets, expressways, and heavily trafficked highways in industrial areas where ESALs (Equivalent Single Axle Loads) range between 2,000,000 and >31,000,000. Considering the limited availability of advanced equipment and skilled labor for CIR pavements, HMA was found be a more sustainable option for farm-to-market roads with the "very light" traffic class. The methodology will help the pavement managers in decision making regarding the selection of sustainable pavement technologies for different road types in Saudi Arabia and the rest of the world.

Keywords: reclaimed asphalt pavement; sustainability evaluation; Hot-Mix Asphalt Pavement; cold in-place recycling pavement; sustainable pavements; fuzzy VIKOR

Citation: Alharbi, F.; Alshubrumi, F.; Almoshaogeh, M.; Haider, H.; Elragi, A.; Elkholy, S. Sustainability Evaluation of Cold In-Place Recycling and Hot Mix Asphalt Pavements: A Case of Qassim, Saudi Arabia. *Coatings* **2022**, *12*, 50. https://doi.org/10.3390/coatings12010050

Academic Editors: Leilei Chen and Giorgos Skordaris

Received: 17 November 2021
Accepted: 28 December 2021
Published: 1 January 2022

Publisher's Note: MDPI stays neutral with regard to jurisdictional claims in published maps and institutional affiliations.

Copyright: © 2022 by the authors. Licensee MDPI, Basel, Switzerland. This article is an open access article distributed under the terms and conditions of the Creative Commons Attribution (CC BY) license (https://creativecommons.org/licenses/by/4.0/).

1. Introduction

Increasing urban development trends have resulted in the construction of dense road infrastructure. Conventionally used Hot-Mix Asphalt Pavement (HMA) caused various environmental impacts from its material production to operations and disposal. This has prompted the government agencies and designers to search for appropriate ways to reduce the detrimental environmental impacts of road construction and maintenance, such as consumption of natural resources and production of emissions that are harmful to the health of the workers and residents. An increasing focus on the use of reclaimed asphalt pavement (RAP) in the construction and maintenance of roads around the world has been noticed in the recent past. RAP holds several environmental and economic benefits gained from recycling in comparison to other recycling technologies [1].

RAP materials are produced from road maintenance activities by milling or crushing the existing HMA. Reusing RAP as a surface course [2] or base and sub-base layers [3] is the primary advantage, while it can also be mixed with other materials to improve the performance, such as using crumb rubber [4] and Portland cement concrete [5]. Cold in-place recycling (CIR) technology is an alternate to make use of RAP. Saudi Arabia started exploring the possibility of using this technology on a large scale after launching the 2030 Vision that aims to establish a sustainable economy, preserve natural resources, and reduce harmful emissions. Besides the economic and environmental benefits of recycled pavements, the need for specialized equipment and skilled labor is among some of the anticipated barriers to their implementation. Evaluation of the recycled pavements, encompassing all the three dimensions of sustainability, has yet to be evaluated in Saudi Arabia.

CIR technology follows a continuous process of cold milling of the pavement surface and remixing with asphalt emulsion or other modifiers to improve the qualities of the reprocessed material, followed by screeding and compaction of the reprocessed materials [6]. Chemical additives are often used to improve the efficiency of the CIR process. The process starts with a milling machine pulverizing to deteriorate the top 50–100-mm layer of HMA. To achieve the necessary gradation, the milled material is crushed and screened on site. Subsequently, the milled grains are mixed with binding agents, such as emulsion, cement, lime, or fly ash. The mixture is reapplied to the roadway, which is subsequently graded to the final elevation [6]. CIR holds several advantages over HMA in terms of reducing aggregate usage, material transportation, and energy consumption. Furthermore, using CIR is environmentally sustainable as it reduces carbon dioxide (CO_2) emissions by 9% over the lifecycle as opposed to conventional mixtures; the CO_2 emission reduction is 54% when just considering the recycling process [7]. According to Schwartz (2016), CIR technology reduces CO_2 emissions by 80% as compared to conventional HMA applications and saves 60% of bitumen content [8]. Presently, the RAP is being used for low-traffic roads in Saudi Arabia, where a simple surface treatment is all that is required. Examples of such work include restoring a damaged pavement, excessive cracking, extreme rutting, and an unstable base or subgrade [9–11]. The quality of the old milling materials influences the strength of the cold recycled asphalt mixture [12]. The sustainability of CIR for the construction of major highways in Saudi Arabia has not been investigated to date.

To investigate how the RAP temperature affects the strength of the CIR mixture components, Kim and Lee (2011) prepared samples of foamed CIR at various temperatures and performed indirect tensile strength [13]. They determined that the temperature of RAP materials has a substantial effect on the wet indirect tensile strength of CIR foam mixtures, and the optimal foam quality varies with temperature after checking the samples. RAP sources and properties, as well as residual binders, influence CIR mechanical properties and efficiency. In comparison to traditional HMA blends, it is clear that the literature indicates that the CIR technology has quality engineering properties and field efficiency to be effectively used in constructing low-volume traffic roads.

Construction and maintenance of roads require an extensive amount of material and energy that significantly impact both the physical environment and natural resources. The Federal Highway Administration (FHWA) considers that a sustainable pavement should meet the basic human needs, use available resources effectively, and conserve adjacent environment [14]. Sustainability of pavements can be achieved during the entire life cycle, including all the processes initiating from material production, pavement design, construction, operation, and maintenance/rehabilitation, to the end of pavement life [15].

CIR has a lot of potential for repairing, strengthening, and recycling asphalt pavements. However, estimating the design thickness of the pavement layers from CIR compared to HMA for different traffic loads is not yet clear in the literature. The main objective of the present research was to develop a sustainability evaluation methodology for CIR and HMA pavements. The thickness of an equivalent CIR pavement that behaves similarly to the conventional HMA was estimated for different traffic loads. Scenario analysis of both the CIR and HMA paving structures for varying traffic classes and thicknesses for the base and sub-base layers was performed. The study also aimed to find the difference in thickness of the paving layers for each method and use this data to determine the difference in the cost and environmental and social impacts of CIR compared to HMA. Finally, sustainability of all the pavement scenarios was evaluated using fuzzy-based multicriteria analysis.

2. Materials and Methods

2.1. Study Area

The province of Qassim lies in the heart of Saudi Arabia (see Figure 1). Qassim holds a special geographical significance in the country as it joins the capital city Riyadh to Madinah that is famous for its religious importance. Due to the extensive agricultural activities in Qassim, both the rural and urban areas are interconnected through a blend of various road types ranging from farm-to-market roads to four-lane urban highways and intercity expressways. The capital of Qassim is Buraydah that is located at 26°19′16″ N and 43°57′32″ E. Frequent traffic movements due to ever-increasing agricultural and industrial activities demand regular rehabilitation of HMA pavements. In the absence of asphalt recycling practices, all the replaced asphalt is presently being disposed of in the open dumps located at various sites. Figure 1 shows one of the asphalt disposal sites located near a high-value residential neighborhood in Buraydah. Figure 1b illustrates the processes involved in a conventional CIR train. The current situation clearly demands for an investigative effort to come up with a sustainable solution for this useful yet out-of-the-place resource. For evaluating the sustainability of CIR and HMA pavements for the study area, cost data were obtained from the Department of Transportation in Qassim.

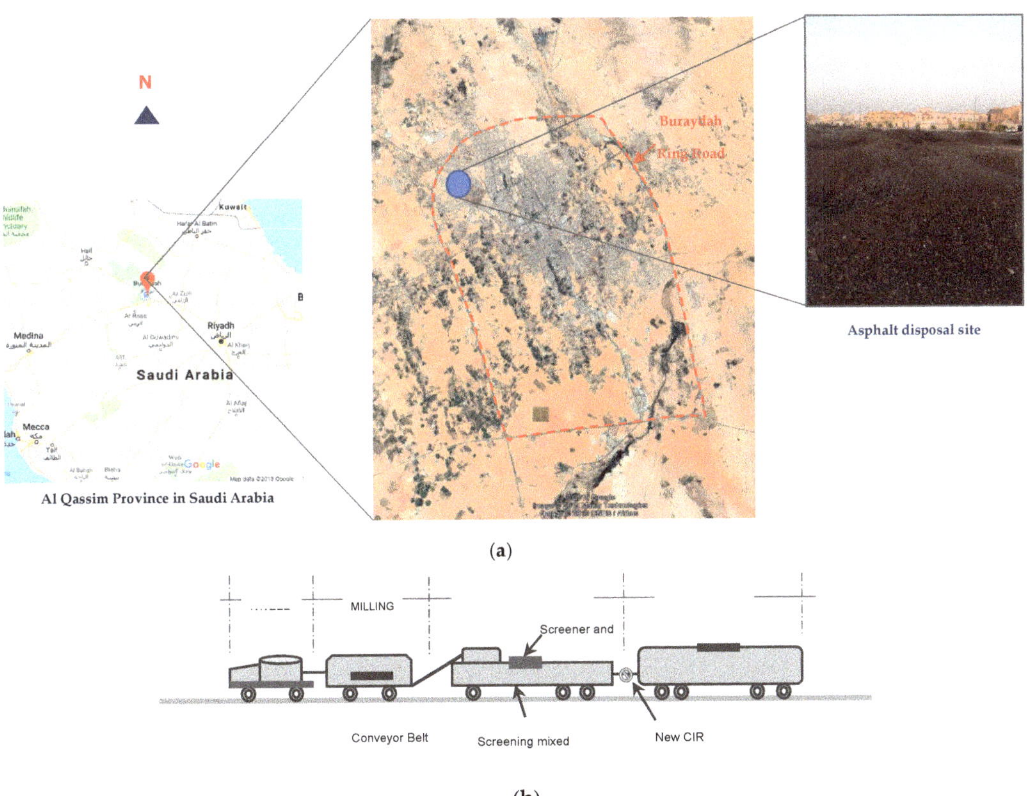

Figure 1. Study area and Cold in-place recycling (CIR) train, (**a**) Study area showing the open dumping at the asphalt disposal site in Buraydah, Qassim, (**b**) typical CIR train [16].

2.2. Pavement Design

The relative capacity of different road pavements to serve certain traffic over time determines their efficiency. According to the definition, serviceability is the ability of a specific section of the pavement to serve high-speed, high-volume, and mixed traffic in its existing condition [17]. The design procedure for flexible pavement, as recommended by the American Association of State Highway and Transportation Officials (AASHTO), requires several inputs to determine the appropriate thickness of the pavement (see Figure 2). Examples of the inputs used for the design of flexible pavement include traffic volume, performance period, resilient modulus of the soil on which the road will be constructed, and specifications of the materials used in paving for the surface, base, and sub-base layers. Evaluating the relative impact of each input can optimize the pavement design.

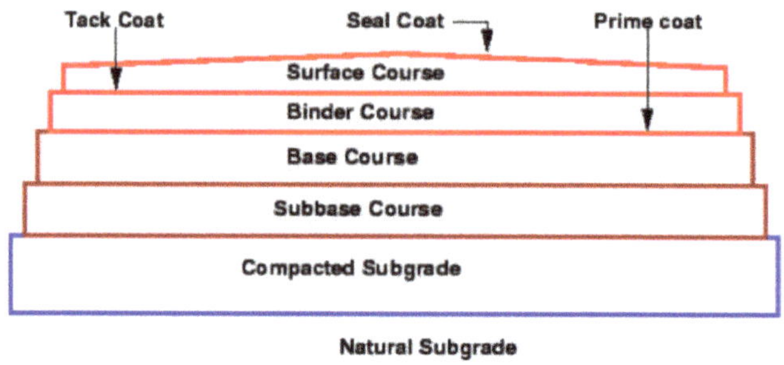

Figure 2. A typical road cross section that shows its various layers.

A prepared sub-grade layer, which is the roadbed soil or borrow material compacted to a defined density, is the foundation for a flexible pavement. On top of the prepared roadbed, a sub-base course is constructed. The base course is built on top of the sub-base course or directly on the roadbed soil if no sub-base is used. The commonly used aggregates are crushed stone, crushed gravel, and sand. The surface layer sitting on the top of the base course is normally made up of an asphalt concrete binder mixture.

Since a flexible pavement is a layered structure, it was designed step by step using a process called Layered Design Analysis [18]. The AASHTO versatile design procedure relies on a design equation formed after a series of road tests and several nomographs. However, some of the design equations' input parameters are either difficult to obtain or to choose definitively. The following subsections discuss the assumptions and correlations that were considered in the current study.

The equation incorporates the Structural Number (SN), which is an abstract number expressing the structural strength of a pavement layer system required for a given combination of soil support [19].

The AASHTO design equation for flexible pavement is presented as follows:

$$\log_{10}(W_{18}) = Z_R \times S_0 + 9.36\log_{10}(SN+1) - 0.2 + \frac{\log_{10}\left(\frac{\Delta PSI}{4.2-1.5}\right)}{0.4 + \frac{1094}{(SN+1)^{5.19}}} + 2.3\log_{10} M_R - 8.07 \quad (1)$$

where SN is the structural number, W_{18} is the accumulated 80-kN Equivalent Single Axle Loads (ESALs) over the life of the project, Z_R is the standard normal deviation, M_R is resilient modulus (PSI), S_0 denotes the standard deviation, and ΔPSI is the change in performance serviceability index.

In the present research, Equation (1) was used to design an equivalent CIR pavement that behaves similarly to the conventional HMA for different traffic loads. This equation was successfully applied by the Ministry of Transport for pavement design in Saudi Arabia with some local adjustments [20].

2.3. *Performance Evaluation Variables*

2.3.1. Performance Period

The performance period refers to how long an initial pavement construction will last until it needs to be repaired or how long it will last between repairs. It is the amount of time it takes for a new, rebuilt, or rehabilitated road structure to deteriorate from its original status to its terminal serviceability. In this research, a 10-year performance period was used since it is suitable for both HMA and CIR [6,21,22].

2.3.2. Traffic Loading

The design procedures are based on an overall estimated equivalent single-axle load (ESAL) of 80 kN [19]. The total ESALs over the analysis period is all that is needed if a pavement is built for the analysis period without any resurfacing or rehabilitation [18]. The Ministry of Transportation in the Kingdom of Saudi Arabia recommends using the information in Table 1 if the contract documents do not specify the ESALS applied to the project to determine the traffic classes needed for the design criteria [20].

Table 1. ESALs and traffic designation [20].

Traffic Class	ESALs Range	Road Grades	ESALs Used in the Present Study
Very Light	Less than 300,000	Agricultural roads with light traffic, local and city streets without trucks	250,000
Light	300,000 to 3 million	Agriculture, feeder, and collector roads	2,000,000
Medium	3 million to 10 million	Main roads and city streets	7,000,000
Heavy	10 million to 30 million	Highways and expressway	20,000,000
Very Heavy	More than 30 million	Heavily trafficked highways and industrial areas	31,000,000

2.3.3. Reliability

Generally, reliability is a way of integrating some level of certainty into the design process to ensure that the different design alternatives can last the lifetime period of the road. The reliability principle necessitates the selection of a standard deviation that is indicative of local conditions in order to be applied. The Federal Department of Transportation (FDOT) design guide suggests that a standard deviation of 0.45 be used for flexible pavements and reliability of 90% [23].

2.3.4. Serviceability

To compute the change in serviceability, PSI, used in Equation (1) and initial and terminal serviceability indexes must be created. According to the FDOT design guide, the typical initial value of PSI is 4.2 and the terminal value is 2.5.

2.3.5. Layer Coefficients

A material's relative potential to behave as a structural component of the pavement is measured by the layer coefficient of a unit thickness of material. Layer coefficients may also be calculated using road test or a correlation with the material's resilient modulus. The values of the layer coefficients are calculated using the AASHTO charts. Only granular base and sub-base materials are included in the guide equations relating the resilient modulus and the corresponding values of layer coefficients.

The equations used for each type of layer are as follows:

$$a_2 = 0.249 \log E_2 - 0.977 \quad (2)$$

$$a_3 = 0.227 \log E_3 - 0.839 \quad (3)$$

where E_2 and E_3 are resilient modulus (PSI) of unbound base layer materials and unbound sub-base layer materials, respectively. Column 4 of Table 2 presents the assumed and calculated values of the layer coefficient for various layers used in the present research.

Table 2. Resilient modulus and layer coefficients used in the present study.

No.	Pavement Layer	Resilient Modulus (MR) [a]	Layer Coefficient
1.	HMA	3500 MPa (507,632 PSI)	assumed 0.44 [b] for asphalt concrete corresponds to E = 3500 MPa
2.	CIR	320 MPa (46,412 PSI)	assumed 0.30 [c] corresponds to E = 320 MPa
3.	Granular base	200 MPa (29,000 PSI)	calculated 0.13 from Equation (2) corresponds to E_2 = 200 MPa
4.	Granular sub-base	100 MPa (14,500 PSI)	calculated 0.11 from Equation (3) corresponds to E_3 = 100 MPa

[a] [6,18,22–24]. [b] [18,23]. [c] [6,22,24].

2.4. Layer Thickness

Once the Required Structural Number (SN) has been determined, the AASHTO method uses a step-by-step method of analyzing layer thicknesses. Using acceptable layer coefficients, the structural number for each layer is converted into the corresponding thicknesses. The following formula is used to transform the structural numbers to layer thicknesses.

$$SN = (a_1 \times D_1) + (a_2 \times D_2) + (a_3 \times D_3) \quad (4)$$

where SN is the total calculated strength of the pavement layers and has units of inches; $a_1, a_2,$ and a_3 are the layer coefficients that represent the strength of the materials used in surface, base, and sub-base layers; and $D_1, D_2,$ and D_3 denote the actual thicknesses (in inches) of surface, base, and sub-base courses.

2.5. Development of Sustaianbility Index

2.5.1. Fuzzy Analytic Hierarchy Process

Fuzzy Analytic Hierarchy Process (Fuzzy-AHP) estimated the weights of the following sustainability criteria:

C1: Construction cost: The costs per m^3 of different pavement layers were obtained from the Department of Transportation in Buraydah, Qassim, Saudi Arabia. The cost including both material and labor costs were as follows 280 SR/m^3 for HMA, 64 SR/m^3 for CIR, 45 SR/m^3 for base layer, and 25 SR/m^3 for sub-base layer.

C2: Resource conservation: This criterion corresponds to the layers' thickness, obtained from Equation (4), as higher thicknesses required a large amount of gravel and other materials acquired from natural resources.

C3: Ease in construction: CIR pavement construction process is done with the help of a CIR train that involves a service of processes (e.g., milling, recycling, and emulsion), as shown in Figure 1b. Unlike HMA, the CIR train needs to be operated by trained personnel. Hence, this criterion entails the need for advanced equipment and skilled operators.

Fuzzy-AHP adopts a pairwise comparison using linguistic terms, such as equal importance and moderate importance. These terms are translated into triangular fuzzy numbers (TFN) to approximate the qualitative judgments of four decision makers from academia and field (engineers and managers working in construction and municipalities). Haider et al. (2020) used the α-cut approach-based Fuzzy-AHP to estimate the weights of water quality parameters for ranking of naturally contaminated groundwater wells in the Qassim Province of Saudi Arabia [25]. The same approach was adopted here and the detailed steps can be seen in [25,26] and Appendix A. Table A1 presents the nine-point rating scale used in pairwise comparison. The consistency of the pairwise matrix scored by each decision maker was checked through the consistency index (CI) and consistency ration (CR). The value of CR has to be less than '1' to ensure the consistency of the pairwise matrix scored by each decision maker.

2.5.2. Fuzzy VIKOR

Serafim Opricovic (1998) first developed the VIKOR ranking method to deal with conflicting criteria [27]. In the present study, the Fuzzy VIKOR, an extension of this method, was used to rank various scenarios of HMA and CIR with varying pavement thicknesses. Fuzzy VIKOR was adopted to accommodate the uncertainties in estimation of cost and layers' thicknesses (C1 and C2) and the vagueness in defining the ease in extension criteria (C3). The step-by-step procedure is outlined in the following [28].

Step 1: Fuzzy-AHP obtained the fuzzified weights for each sustainability criterion as $\widetilde{W_i} = (w_{li}, w_{lm}, w_{lu})$, where $w_{li}, w_{lm},$ and w_{lu} are the lower, medium, and upper limits of the criteria weights in the form of TFNs.

Step 2: As all the criteria are cost criteria and none of them is a benefit criterion, the positive triangular ideal solution ($\widetilde{f}_i^*$) for each criterion corresponds to the lowest possible values, while the negative triangular ideal solution ($\widetilde{f}_i^o$) corresponds to the highest values and is defined as:

$$\widetilde{f}_i^* = MAX_j\widetilde{f}_{ij}\ \ \widetilde{f}_i^o = MIN_j\widetilde{f}_{ij}\ \text{ for } i\ \in I^c \tag{5}$$

where I^c represents the set of WQPs as the cost criteria.

The $\widetilde{f}_i^*$ and $\widetilde{f}_i^o$ values for each sustainability criteria based on the data were obtained from Equation (5). The criteria with scores less than the lowest possible values were considered as the positive ideal solution.

Step 3: Using Equation (6), estimate the normalized fuzzy difference ($\widetilde{d}_{ij}$, $j = 1, \ldots, J$, $i = 1, \ldots, n$):

$$\widetilde{d}_{ij} = \left(\widetilde{f}_i^* \ominus \widetilde{f}_{ij}\right)/(r_i^* - l_i^o)\ \text{ for } i\ \in I^c \tag{6}$$

Step 4: Using Equation (7), determine the fuzzy weighted sum $\widetilde{S}$ and fuzzy operators MAX $\widetilde{R}$:

$$\widetilde{S}_j = \sum_{i=1}^{n} \bigoplus\left(\widetilde{w}_i \bigotimes \check{d}_{ij}\right) \text{ and } \widetilde{R}j = MAX_i\left(\widetilde{w}_i \bigotimes \check{d}_{ij}\right) \tag{7}$$

Step 5: To determine the final sustainability rank for each scenario, compute $\widetilde{Q}_j$ as:

$$\widetilde{Q}_j = v\left(\widetilde{S}_j \ominus \widetilde{S}^*\right)/(S^{or} - S^{*l}) \bigoplus (1-v)\left(\widetilde{R}j \ominus \widetilde{R}^*\right)/\left(R^{or} - R^{*l}\right) \tag{8}$$

where $\widetilde{S}^*$ is the MIN $\widetilde{S}_j$, S^{or} is MAX S_j^r, $\widetilde{R}^*$ is the MIN $\widetilde{R}_j$, SR^{or} is MAXj R_j^r, and v is the weight of the strategy.

Step 6: For the results obtained through Equation (9), perform the defuzzification using the following equation:

$$P\left(\widetilde{M}\right) = M = \frac{l + 4m + u}{6} \tag{9}$$

Step 7: As the high $\widetilde{Q}_j$ corresponds to the pavement scenario with low sustainability, final sustainability ranks (SR) were established using the following relationship for each scenario:

$$SR_i = 1 - \widetilde{Q}_j \tag{10}$$

3. Results

3.1. Development of Pavement Scenarios

The present research was an attempt to improve the existing practice of open dumping of the replaced asphalt (generated from rehabilitation activities) and the allied social and environmental consequences in Saudi Arabia. Although the CIR has a potential to mitigate these impacts and enhance the overall sustainability of the pavement construction process, the technology needs to be evaluated on equal technical grounds with the HMA construction. A thicker CIR surface layer is required to behave analogously to HMA for a given traffic class. Five traffic classes and five different combinations of base and sub-base thicknesses generated 25 scenarios for each type of pavement. Table 3 presents all the 50 scenarios evaluated for CIR and HMA. The first step of this research estimated the equivalent thicknesses of CIR for each traffic class and different thicknesses of base and sub-base layers. The last column of Table 1 shows the values of ESALs used in the current study.

Table 3 presents five different base and sub-base layers' thicknesses evaluated in this research. The inputs, including ESALs in Table 1 and layers' thicknesses in Table 3, were applied to Equation (1). Using 90% reliability with 0.45 standard deviation, 4.2 initial serviceability index, 2.5 terminal serviceability index, and 62 MPa (8992 psi) sub-grade resilient modulus, the resulted layers' thicknesses in the structural framework of HMA and CIR pavement are shown in Table 3 for the different scenarios. Table 3 also shows that, for most of the scenarios, design thickness of CIR-based pavement is higher than that of the HMA-based pavement at various traffic loads. This was expected because of the quality of the used materials, new versus recycled. Nevertheless, HMA needs new material for surface layers, while CIR uses the recycled material and does not need additional natural resources.

Table 3 illustrates the comparison between traditional HMA and recycled paving CIR according to the thickness of the paving layers. The table describes that, at lower ESALs (mostly up to 7,000,000), the thickness difference between the two techniques ranges from 0.0 cm for the lowest EASL to about 9.0 cm for medium ESAL. For higher ESALs, the difference significantly increases up to 12 cm. Moreover, it appears that the ratio between the thickness of CIR and HMA layers decreases with the increasing thickness of base and sub-base layers. For instance, the difference was 4 cm for very light and 12 cm for very heavy traffic classes with a 15-cm base layer without any sub-base while the difference reduced to 1 cm and 10 cm for the same traffic classes with a 10-cm sub-base and 20-cm base layer. For the equal thickness (20 cm) for both the base and sub-base, the difference reduced to 0 cm for very light and to 9 cm for very heavy traffic, which further reduced to 0 am and 8 cm with a 20-cm sub-base and 25-cm base layer. This may be attributed to the obvious effect of the foundation soil layers on the pavement design. Therefore, the design engineer would be able to develop other alternatives of road design to the client to choose among them based on the needed road class or grade and the available budget for the project.

3.2. Sustainability Evaluation of Pavement Scenarios

Based on the thicknesses of pavement layers, 10 scenarios were developed for each traffic class that came up with a total of 50 scenarios. Five scenarios each for HMA and CIR make a total count of 50 scenarios (see Table 4). The last three columns of Table 3 present the scores of the three sustainability criteria. Possible uncertainities in criteria weight estimation, construction cost estimation (C1), assumptions in calculating pavement thicknesses (C2), and subjective assessment of ease in construction (C3) were accomodated by the Fuzzy-AHP and Fuzzy-VIKOR methods. Consistency ratios for the four pairwise matrices were found to be less than '1' as per the Equation (A4) and Table A2 of the Fuzzy-AHP methodology described in Appendix A. The fuzzified weights of the crtieria were found to be C1: 0.44, 0.558, and 0.615; C2: 0.239, 0.269, and 0.283; and C3: 0.171, 0.172, and 0.178.

Table 3. Pavement Structure of HMA and CIR Scenarios.

Scenario	Traffic Class	ESALs	Thickness of Paving Layers (cm)			C1: Construction Cost (SR/m^3)	C2: Natural Resource Depletion [1]	C3: Need of Adv. Equipment and Skilled Operators [2]
			Sub-Base	Base	Surface	–	–	–
A1.1: 0SB-15B-16HMA	Very light	250,000	0	15	9	31.95	24	L
A1.2: 0SB-15B-29CIR	Very light	250,000	0	15	13	15.07	15	H
A2.1: 0SB-15B-16HMA	Light	2,000,000	0	15	16	51.55	31	L
A2.2: 0SB-15B-29CIR	Light	2,000,000	0	15	23	21.47	15	H
A3.1: 0SB-15B-16HMA	Medium	7,000,000	0	15	20	62.75	35	L
A3.2: 0SB-15B-29CIR	Medium	7,000,000	0	15	29	25.31	15	H
A4.1: 0SB-15B-16HMA	Heavy	20,000,000	0	15	24	73.95	39	L
A4.2: 0SB-15B-29CIR	Heavy	20,000,000	0	15	35	29.15	15	H
A5.1: 0SB-15B-16HMA	Very heavy	31,000,000	0	15	26	79.55	41	L
A5.2: 0SB-15B-29CIR	Very heavy	31,000,000	0	15	38	31.07	15	H
A1.3: 0SB-20B-19HMA	Very light	250,000	0	20	8	31.4	28	L
A1.4: 0SB-20B-27CIR	Very light	250,000	0	20	11	16.04	20	H
A2.3: 0SB-20B-19HMA	Light	2,000,000	0	20	14	48.2	34	L
A2.4: 0SB-20B-27CIR	Light	2,000,000	0	20	21	22.44	20	H
A3.3: 0SB-20B-19HMA	Medium	7,000,000	0	20	19	62.2	39	L
A3.4: 0SB-20B-27CIR	Medium	7,000,000	0	20	27	26.28	20	H
A4.3: 0SB-20B-19HMA	Heavy	20,000,000	0	20	22	70.6	42	L
A4.4: 0SB-20B-27CIR	Heavy	20,000,000	0	20	33	30.12	20	H
A5.3: 0SB-20B-19HMA	Very heavy	31,000,000	0	20	25	79	45	L
A5.4: 0SB-20B-27CIR	Very heavy	31,000,000	0	20	35	31.4	20	H
A1.5: 10SB-20B-17HMA	Very light	250,000	10	20	7	31.1	37	L
A1.6: 10SB-20B-24CIR	Very light	250,000	10	20	8	16.62	30	H
A1.5: 10SB-20B-17HMA	Light	2,000,000	10	20	12	45.1	42	L
A1.6: 10SB-20B-24CIR	Light	2,000,000	10	20	17	22.38	30	H
A3.5: 10SB-20B-17HMA	Medium	7,000,000	10	20	17	59.1	47	L
A3.6: 10SB-20B-24CIR	Medium	7,000,000	10	20	24	26.86	30	H
A4.5: 10SB-20B-17HMA	Heavy	20,000,000	10	20	21	70.3	51	L
A4.6: 10SB-20B-24CIR	Heavy	20,000,000	10	20	30	30.7	30	H
A5.5: 10SB-20B-17HMA	Very heavy	31,000,000	10	20	22	73.1	52	L
A5.6: 10SB-20B-24CIR	Very heavy	31,000,000	10	20	32	31.98	30	H

Table 3. *Cont.*

Scenario	Traffic Class	ESALs	Thickness of Paving Layers (cm)			C1: Construction Cost (SR/m^3)	C2: Natural Resource Depletion [1]	C3: Need of Adv. Equipment and Skilled Operators [2]
			Sub-Base	Base	Surface	–	–	–
A1.7: 20SB-20B-14HMA	Very light	250,000	20	20	7	33.6	47	L
A1.8: 20SB-20B-21CIR	Very light	250,000	20	20	7	18.48	40	H
A2.7: 20SB-20B-14HMA	Light	2,000,000	20	20	10	42	50	L
A2.8: 20SB-20B-21CIR	Light	2,000,000	20	20	14	22.96	40	H
A3.7: 20SB-20B-14HMA	Medium	7,000,000	20	20	14	53.2	54	L
A3.8: 20SB-20B-21CIR	Medium	7,000,000	20	20	21	27.44	40	H
A4.7: 20SB-20B-14HMA	Heavy	20,000,000	20	20	18	64.4	58	L
A4.8: 20SB-20B-21CIR	Heavy	20,000,000	20	20	27	31.28	40	H
A4.9: 20SB-25B-13HMA	Very heavy	31,000,000	20	20	20	70	60	L
A4.10: 20SB-25B-19CIR	Very heavy	31,000,000	20	20	29	32.56	40	H
A1.9: 20SB-25B-13HMA	Very light	250,000	20	25	7	35.85	52	L
A1.10: 20SB-25B-19CIR	Very light	250,000	20	25	7	20.73	45	H
A2.9: 20SB-25B-13HMA	Light	2,000,000	20	25	8	38.65	53	L
A2.10: 20SB-25B-19CIR	Light	2,000,000	20	25	12	23.93	45	H
A3.9: 20SB-25B-13HMA	Medium	7,000,000	20	25	13	52.65	58	L
A3.10: 20SB-25B-19CIR	Medium	7,000,000	20	25	19	28.41	45	H
A5.7: 20SB-20B-14HMA	Heavy	20,000,000	20	25	17	63.85	62	L
A5.8: 20SB-20B-21CIR	Heavy	20,000,000	20	25	25	32.25	45	H
A5.9: 20SB-25B-13HMA	Very heavy	31,000,000	20	25	19	69.45	64	L
A5.10: 20SB-25B-19CIR	Very heavy	31,000,000	20	25	27	33.53	45	H

[1] Total thickness of all layers for each scenario. [2] Subjective criteria.

Table 4. Scenarios for all traffic classes.

Very Light [1]	Rank	Heavy	Rank
A1.1: 0SB-15B-16HMA	2	A4.1: 0SB-15B-16HMA	10
A1.2: 0SB-15B-29CIR	1	A4.2: 0SB-15B-29CIR	1
A1.3: 0SB-20B-19HMA	3	A4.3: 0SB-20B-19HMA	8
A1.4: 0SB-20B-27CIR	7	A4.4: 0SB-20B-27CIR	4
A1.5: 10SB-20B-17HMA	4	A4.5: 10SB-20B-17HMA	6
A1.6: 10SB-20B-24CIR	5	A4.6: 10SB-20B-24CIR	2
A1.7: 20SB-20B-14HMA	6	A4.7: 20SB-20B-14HMA	5
A1.8: 20SB-20B-21CIR	10	A4.8: 20SB-20B-21CIR	7
A1.9: 20SB-25B-13HMA	9	A4.9: 20SB-25B-13HMA	9
A1.10: 20SB-25B-19CIR	8	A4.10: 20SB-25B-19CIR	3
Light	–	**Very Heavy**	–
A2.1: 0SB-15B-16HMA	9	A5.1: 0SB-15B-16HMA	9
A2.2: 0SB-15B-29CIR	1	A5.2: 0SB-15B-29CIR	1
A2.3: 0SB-20B-19HMA	8	A5.3: 0SB-20B-19HMA	10
A2.4: 0SB-20B-27CIR	6	A5.4: 0SB-20B-27CIR	4
A2.5: 10SB-20B-17HMA	3	A5.5: 10SB-20B-17HMA	5
A2.6: 10SB-20B-24CIR	2	A5.6: 10SB-20B-24CIR	2
A2.7: 20SB-20B-14HMA	4	A5.7: 20SB-20B-14HMA	6
A2.8: 20SB-20B-21CIR	10	A5.8: 20SB-20B-21CIR	7
A2.9: 20SB-25B-13HMA	7	A5.9: 20SB-25B-13HMA	8
A2.10: 20SB-25B-19CIR	5	A5.10: 20SB-25B-19CIR	3
Medium	–	–	–
A3.1: 0SB-15B-16HMA	9	–	–
A3.2: 0SB-15B-29CIR	1	–	–
A3.3: 0SB-20B-19HMA	10	–	–
A3.4: 0SB-20B-27CIR	4	–	–
A3.5: 10SB-20B-17HMA	6	–	–
A3.6: 10SB-20B-24CIR	2	–	–
A3.7: 20SB-20B-14HMA	5	–	–
A3.8: 20SB-20B-21CIR	7	–	–
A3.9: 20SB-25B-13HMA	8	–	–
A3.10: 20SB-25B-19CIR	3	–	–

[1] SB: Sub-base, B: Base, HMA: Hot Mix Asphalt, and CIR: Cold In-place Recycling.

In the subsequent step, the fuzzy-VIKOR aggregated the criteria scores to estimate the sustainability ranks for all the traffic classes' scenarios. Keeping in view the space limitations, Table 5 presents results for the 'medium' traffic class as an example. Table 5 presents the fuzzified scores of all the criteria for scenario numbers 1.3, 2.3, 3.3, 4.3, and 5.3. The last two columns of Table 5 provide the positive triangular ideal solution ($\widetilde{f}_i^*$) and the negative triangular ideal solution ($\widetilde{f}_i^o$) for each criterion. As all are the cost criteria, $\widetilde{f}_i^*$ corresponds to the 'best-case scenario' and $\widetilde{f}_i^o$ to the 'worst-case senario'.

Table **??** presents the fuzzy results for Sj, Rj, and Qj, using the steps descibed in Equations (6)–(8). Equation (9) defuzzified the final scores and the results are given in Table **??**. The defuzzified Q_j scores are essentially the aggregated performance ranks for each pavement scenario. Finally, Equation (10) established the sustainability ranks (also mentioned in Table 4), and Figure 3 summarizes these results for all the scenarios evaluated in the present study.

Table 5. Example of scoring matrix for nine scenarios (No. 10 not included due to space limitations) for medium traffic class.

Criteria	A3.1: 0SB-15B-16HMA			A3.2: 0SB-15B-29CIR			A3.3: 0SB-20B-19HMA			A3.4: 0SB-20B-27CIR			A3.5: 10SB-20B-17HMA			A3.6: 10SB-20B-24CIR			A3.7: 20SB-20B-14HMA			A3.8: 20SB-20B-21CIR			A3.9: 20SB-25B-13HMA		
	l	m	r	l	m	r	l	m	r	l	m	r	l	m	r	l	m	r	l	m	r	l	m	r	l	m	r
f1 = CC	57.75	62.75	67.75	20.31	25.31	30.31	57.2	62.2	67.2	21.28	26.28	31.28	54.1	59.1	64.1	21.86	26.86	31.86	48.2	53.2	58.2	22.44	27.44	32.44	47.65	52.65	57.65
f2 = RC	30	35	40	10	15	20	34	39	44	15	20	25	42	47	52	25	30	35	49	54	59	35	40	45	53	58	63
f3 = EC	1	2	3	8	9	10	1	2	3	8	9	10	1	2	3	8	9	10	1	2	3	8	9	10	1	2	3
f1+	10	15	20	10	15	20	10	15	20	10	15	20	10	15	20	10	15	20	10	15	20	10	15	20	10	15	20
f1−	75	80	85	75	80	85	75	80	85	75	80	85	75	80	85	75	80	85	75	80	85	75	80	85	75	80	85
ri ° − li *	75	75	75	75	75	75	75	75	75	75	75	75	75	75	75	75	75	75	75	75	75	75	75	75	75	75	75
dij=	0.503	0.637	0.770	0.004	0.137	0.271	0.496	0.629	0.763	0.017	0.150	0.284	0.455	0.588	0.721	0.025	0.158	0.291	0.376	0.509	0.643	0.033	0.166	0.299	0.369	0.502	0.635
f2+	10	15	20	10	15	20	10	15	20	10	15	20	10	15	20	10	15	20	10	15	20	10	15	20	10	15	20
f2−	60	65	70	60	65	70	60	65	70	60	65	70	60	65	70	60	65	70	60	65	70	60	65	70	60	65	70
ri * − li °	60	60	60	60	60	60	60	60	60	60	60	60	60	60	60	60	60	60	60	60	60	60	60	60	60	60	60
dij =	0.167	0.333	0.500	−0.17	0.000	0.167	0.233	0.400	0.567	−0.083	0.083	0.250	0.367	0.533	0.700	0.083	0.250	0.417	0.483	0.650	0.817	0.250	0.417	0.583	0.550	0.717	0.883
f3+	1	2	3	1	2	3	1	2	3	1	2	3	1	2	3	1	2	3	1	2	3	1	2	3	1	2	3
f3−	8	9	10	8	9	10	8	9	10	8	9	10	8	9	10	8	9	10	8	9	10	8	9	10	8	9	10
ri ° − li *	9	9	9	9	9	9	9	9	9	9	9	9	9	9	9	9	9	9	9	9	9	9	9	9	9	9	9
dij=	−0.22	0.000	0.222	0.556	0.778	1.000	−0.22	0.000	0.222	0.556	0.778	1.000	−0.22	0.000	0.222	0.556	0.778	1.000	−0.22	0.000	0.222	0.556	0.778	1.000	−0.22	0.000	0.222

Table 6. Ranking of pavement scenarios (No. 10 not included due to space limitations) using fuzzy-VIKOR.

–	A3.1: 0SB-15B-16HMA			A3.2: 0SB-15B-29CIR			A3.3: 0SB-20B-19HMA			A3.4: 0SB-20B-27CIR			A3.5: 10SB-20B-17HMA			A3.6: 10SB-20B-24CIR			A3.7: 20SB-20B-14HMA			A3.8: 20SB-20B-21CIR			A3.9: 20SB-25B-13HMA		
–	l	m	r	l	m	r	l	m	r	l	m	r	l	m	r	l	m	r	l	m	r	l	m	r	l	m	r
Sj	0.223	0.445	0.655	0.057	0.211	0.391	0.236	0.459	0.669	0.082	0.240	0.423	0.250	0.472	0.681	0.126	0.290	0.475	0.243	0.459	0.666	0.169	0.339	0.527	0.256	0.473	0.680
Sj Crisp	–	0.44	–	–	0.22	–	–	0.46	–	–	0.25	–	–	0.47	–	–	0.29	–	–	0.46	–	–	0.34	–	–	0.47	–
Rj	0.503	0.637	0.770	0.004	0.137	0.271	0.496	0.629	0.763	0.556	0.778	1.000	−0.22	0.000	0.222	0.025	0.158	0.291	−0.22	0.000	0.222	0.556	0.778	1.000	0.369	0.502	0.635
Rj Crisp	–	0.64	–	–	0.14	–	–	0.63	–	–	0.78	–	–	0.00	–	–	0.16	–	–	0.00	–	–	0.78	–	–	0.50	–
Qj	−0.10	0.424	0.909	−0.417	0.037	0.492	−0.09	0.437	0.922	−0.24	0.244	0.724	−0.27	0.279	0.788	−0.34	0.127	0.586	−0.28	0.266	0.771	−0.15	0.349	0.835	−0.11	0.417	0.899
Qj Crisp	–	0.414	–	–	0.037	–	–	0.426	–	–	0.243	–	–	0.268	–	–	0.126	–	–	0.256	–	–	0.347	–	–	0.407	–

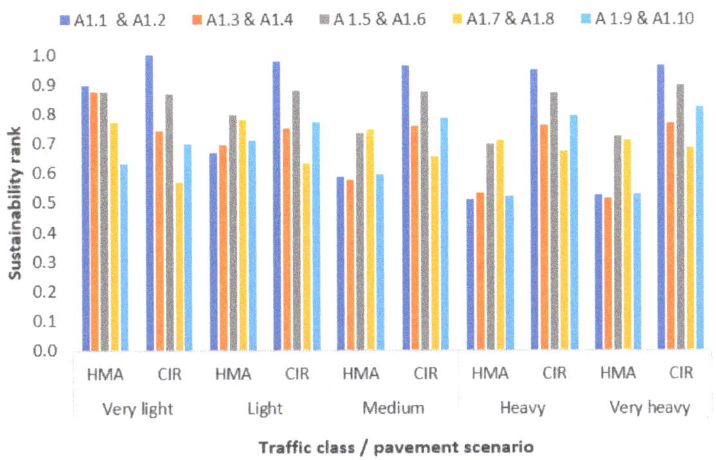

Figure 3. Satiability ranks for all scenarios defined in Table 3. The legend is showing the scenarios' numbers for the very light traffic class only. The same color scheme is applicable to all traffic classes (also see Table 4).

4. Discussion

The results presented in Figure 3 show an overall supersedence of CIR over the HMA pavement for all traffic classes due to its low cost and minimal use of natural resources. Interestingly, in the case of 'very light' traffic class, HMA with a 0-cm sub-base, 15-cm base, and 16-cm wearing coarse (A1.1: 0SB-15B-16HMA) came out to be the second-best scenario based on the overall sustainability rank. The sustainability rank score of the top-ranked CIR scenario with the same thicknesses of sub-base and base and almost two times thicker wearing course (A1.2: 0SB-15B-29CIR) was almost 10% higher (1.0) than the HMA scenario (0.894). The results could be different if a higher relative weight is given to C3: need for advanced equipment and skilled labour. In the same traffic class, the third and fourth ranks were also obtained by the HMA. These results suggest the use of HMA for the farm-to-market roads with very-light traffic (<300,000 ESALs), where the availability of both the advanced equipment and skilled labors could be a primary constraint to using CIR pavements.

In the case of the 'light' traffic class, the CIR scenarios superseded the HMA scenarios with the top two sustainability ranks. In the comparative evaluation of the two pavement types for a given thickness of sub-base and base, CIR scenarios also obtained higher ranks. For instance, A2.5:10SB-20B-17HMA ranked at third place while the CIR scenario (A2.6:10SB-20B-24CIR) obtained second rank. Similarly, A10:20SB-25B-19CIR outperformed A2.9:20SB-25B-113HMA by getting the fifth rank in comparison to the seventh rank of its counterpart. CIR attained the first four ranks in the overall classification for the medium traffic class (7,000,000 ESALs), except for an equal thickness of sub-base and base (20 cm) where HMA got fifth rank and CIR the seventh. Figure 3 illustrates an analogous behavior for 'heavy' and 'very heavy' traffic classes where all the CIR scenarios were visibly surpassing their comparable HMA scenarios.

The proposed evaluation of CIR and HMA clearly revealed that CIR technology is comparable to HMA for all traffic classes with some additional thickness of the surface layer. This approach will not only minimize the environmental impacts on non-renewable natural resources (aggregates) but also mitigate the visual nuisance due to open dumping of the asphalt mix generated from road maintenance. The proposed methodology will help the pavement engineers and managers for applying CIR as a sustainable construction and maintenance technology in Saudi Arabia and elsewhere.

5. Conclusions

The present research investigated the design thickness of the recycled-based pavement (CIR) equivalent to the conventional HMA pavement under the influence of different traffic loading classes. Five design scnearios with varying thicknesses of sub-base, base, and wearing course for each traffic class were investigated that generated a total of 50 scenarios, 25 each for CIR and HMA pavement types. All the scenarios were evalauted for their economic (construction cost), enviornmental (natural resource depletion), and social (need of advanced equipment and skill labour) sustainbility.

A thicker CIR surface layer behaves identically to that of HMA. Moreover, the ratio between the CIR and HMA layers reduces as the thickness of base and sub-base layers increases. The difference of 4 cm for very light (250,000 ESALs) and 12 cm for very heavy (31,000,000 ESALs) traffic classes was obtained for a 15-cm base layer in the absence of a sub-base. The difference reduced to 1 cm for very light and 10 cm for very heavy traffic classes for a 10-cm sub-base and 20-cm base layer. Interestingly, the same thickness (7 cm) was found for both the CIR and HMA for very light traffic with equal thickness (20 cm) for both the base and sub-base, while 9 cm was found for very heavy traffic.

The results of the sustainability evaluation menifest a clear dominance of CIR over the HMA for all classes except 'very light' traffic. This means that CIR is a more sustainable pavement overlay for feeders, collectors, main urban streets and highways, expressways, and heavily trafficked highways in industrial areas where ESALs range from 2,000,000 to more than 31,000,000. In the particular case of the 'very light' traffic class, HMA was found be a more sustainable option, keeping in view the limited availability of advanced equipment and skilled labour for construction of farm-to-market-roads. A recycled pavement mixture could present an economic design alternative to the conventional mixtures and should be considered when designing and rehabilitating roads. This sustainable alternative is not only economically attractive but also it conserves the environment with less use of natural materials. The study will help the design engineers and infrastructure asset managers for planning pavement construction and rehabilitation programs in Saudi Arabia and elsewhere. Future studies can investigate deatiled lifecycle costing and lifecycle assessment of CIR pavements.

Author Contributions: F.A. (Fawaz Alharbi), funding, supervision, conceptualization, methodology, writing—original draft preparation; F.A. (Fahad Alshubrumi), data collection, analysis, validation, writing—original draft preparation, M.A., conceptualization, supervision, writing—review and editing; H.H., supervision, methodology, writing—original draft preparation; A.E., conceptualization, methodology, supervision; S.E., conceptualization, supervision, writing—review and editing. All authors have read and agreed to the published version of the manuscript.

Funding: This research was funded by the Deanship of Scientific Research, Qassim University with Grant No. 9909-qec-2019-1-1-Q.

Institutional Review Board Statement: Not applicable.

Informed Consent Statement: Not applicable.

Data Availability Statement: All the shareable data are given in the main text.

Acknowledgments: The authors gratefully acknowledge Qassim University, represented by the Deanship of Scientific Research, on the financial support for this research under the number 9909-qec-2019-1-1-Q during the academic year 1440 AH/2019 AD.

Conflicts of Interest: The authors declare no conflict of interest.

Appendix A

The procedure of α-cut-based Fuzzy-AHP is as follows [27].

Step 1: Develop the pairwise comparison matrix.

K denotes the number of decision makers (DMs) who completed the pairwise comparison matrix. Using the nine-point rating scale given in Table A1, the fuzzy reciprocal judgment matrix $\widetilde{A}^k$ was developed:

$$\widetilde{A}^k = \left[\widetilde{a}_{ij}\right]^k \tag{A1}$$

where i and j represent the criteria and number of the criteria in the matrix, respectively, and $j = 1, 2, \ldots, n$.

The complete fuzzy reciprocal matrix $\widetilde{R}^k$ is defined as:

$$\widetilde{R}^k = \left[\widetilde{r}_{ij}\right]^k \tag{A2}$$

where $\widetilde{r}_{ij}$ is the relative importance difference between the criteria i and j. It is represented by the triangular fuzzy numbers (TFN), illustrated in Figure A1, as

$$\widetilde{r}_{ij} = \left(l_{ij}, m_{ij}, u_{ij}\right). \text{ Here } \widetilde{r}_{11} = (1,1,1),\ \forall i = j \text{ and } \widetilde{r}_{ij} = \frac{1}{\widetilde{r}^k_{ij}},\ \forall i = j = 1, 2, \ldots, n.$$

Step 2: Perform consistency check.

$\widetilde{R}^k = \left[\widetilde{r}_{ij}\right]$ represents the fuzzy positive reciprocal matrix where $\widetilde{r}_{ij} = \left(\alpha_{ij}, \beta_{ij}, \gamma_{ij}\right)$ and its consistency is checked for each DM in the α-cut approach using the following equation:

$$CI = \frac{\lambda_{max} - n}{n - 1} \tag{A3}$$

where λ_{max} is the dimension of the matrix and is the maximum eigenvalue.

Equation (A4) calculates the consistency ratio (CR):

$$CR = \frac{CI}{RI} \tag{A4}$$

where RI represents the random index in Table A2 and is found for the number of sustainability criteria (i.e., n).

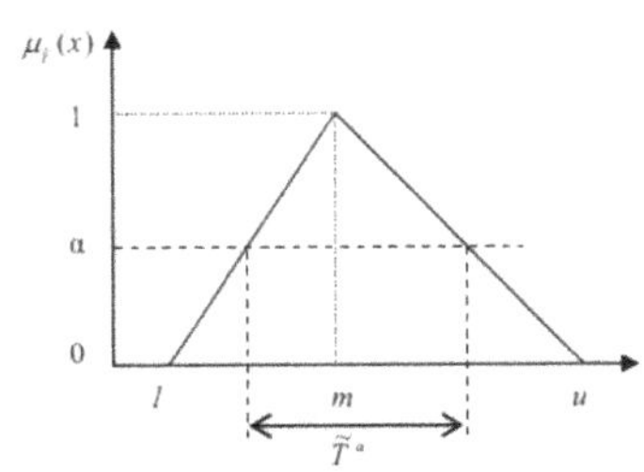

Figure A1. The α-cut of a triangular fuzzy number $\widetilde{T}$ Source: [26].

Table A1. Fuzzy scales and triangular fuzzy numbers (TFN) used for linguistic variables.

Linguistic Term	Fuzzy Number	TFN (l, m, u)	Linguistic Term	Fuzzy Number	TFN (l, m, u)
Extreme unimportance	$\widetilde{9}^{-1}$	1/9, 1/9, 1/9	Intermediate value between $\widetilde{1}$ and $\widetilde{3}$	$\widetilde{2}$	1, 2, 3
Intermediate values between $\widetilde{7}^{-1}$ and $\widetilde{9}^{-1}$	$\widetilde{8}^{-1}$	1/9, 1/8, 1/7	Moderate importance	$\widetilde{3}$	2, 3, 4
Very unimportance	$\widetilde{7}^{-1}$	1/8, 1/7, 1/6	Intermediate value between $\widetilde{3}$ and $\widetilde{5}$	$\widetilde{4}$	3, 4, 5
Intermediate value between $\widetilde{5}^{-1}$ and $\widetilde{7}^{-1}$	$\widetilde{6}^{-1}$	1/7, 1/6, 1/5	Essential importance	$\widetilde{5}$	4, 5, 6
Essential unimportance	$\widetilde{5}^{-1}$	1/6, 1/5, $\frac{1}{4}$	Intermediate value between $\widetilde{5}$ and $\widetilde{7}$	$\widetilde{6}$	5, 6, 7
Intermediate value between $\widetilde{3}^{-1}$ and $\widetilde{5}^{-1}$	$\widetilde{4}^{-1}$	1/5, 1/4, 1/3	Very vital importance	$\widetilde{7}$	6, 7, 8
Moderate unimportance	$\widetilde{3}^{-1}$	1/4, 1/3, $\frac{1}{2}$	Intermediate value between $\widetilde{7}$ and $\widetilde{9}$	$\widetilde{8}$	7, 8, 9
Intermediate and $\widetilde{3}^{-1}$	$\widetilde{2}^{-1}$	1/3, 1/2, 1	Extreme importance	$\widetilde{9}$	9, 9, 9
Equally importance	$\widetilde{1}$	1, 1, 1	–	–	–

Table A2. Randomly generated values of consistency index (*RI*).

n	1	2	3	4	5	6	8	8	9	10
RI	0	0	0.58	0.90	1.12	1.24	1.32	1.41	1.45	1.49

Step 3: Estimate the fuzzy weights.
Equation (A5) finds the positive matrix '*k*':

$$\widetilde{T}^{\alpha} = [(m-1)\alpha + l,\ u-(u-m)\alpha], 0 \leq \alpha \leq 1 \tag{A5}$$

$\widetilde{R}^k_m = [\widetilde{r}_{ij}]^k_m$ can be calculated by setting $\alpha = 1$, while the lower and upper bounds $\widetilde{R}^k_l = [\widetilde{r}_{ij}]^k_l$ and $\widetilde{R}^k_u = [\widetilde{r}_{ij}]^k_u$ can be found by setting $\alpha = 0$.

Next, estimate the criteria weights using Equations (A1) and (A6) for all the DMs:

$$w_i = \frac{\left(\prod_{j=1}^{n} a_{ij}\right)^{1/n}}{\sum_{j=1}^{n}\left(\prod_{j=1}^{n} a_{ij}\right)^{1/n}} \tag{A6}$$

where w_i represents the criteria weight and the weight vector $W = (w_i)$, $i = 1, 2, \ldots, n$.

By applying Equation (A6) to l, m, and u bounds, the weight vertices were calculated as $W^k_l = (w_i)^k_l$, $W^k_m = (w_i)^k_m$, *and* $W^k_u = (w_i)^k_u$.

The smallest possible constant S^k_l and the largest possible constant S^k_{lu} for minimizing the fuzziness of the weights were estimated using Equations (A7a) and (A7b):

$$S^k_l = min\left\{\left(\frac{w^k_{im}}{w^k_{il}}\middle| 1 \leq i \leq n\right)\right\} \tag{A7a}$$

$$S^k_u = max\left\{\left(\frac{w^k_{im}}{w^k_{iu}}\middle| 1 \leq i \leq n\right)\right\} \tag{A7b}$$

The following equations estimated the lower and upper bounds of the weight vector:

$$w^{*k}_{il} = S^k_l w^k_{il}, i = 1, 2, \ldots, n \tag{A8a}$$

$$w^{*k}_{iu} = S^k_u w^k_{iu}, i = 1, 2, \ldots, n \tag{A8b}$$

Finally, the fuzzy weight matrix was developed for each DM as:

$$\widetilde{W}^k_i = \left(w^{*k}_{il}, w^{*k}_{im}, w^{*k}_{iu}\right),\ i = 1, 2, \ldots, n \tag{A9}$$

Step 4: Combine the judgment of all the DMs.

This step aggregates the fuzzy weights' matrices obtained from Equation (A9) using Equation (A10):

$$\widetilde{\widetilde{W}}_i = \frac{1}{K}\left(\widetilde{W}^1_i \oplus \widetilde{W}^2_i \oplus \ldots \oplus \widetilde{W}^k_i\right) \tag{A10}$$

where $\widetilde{\widetilde{W}}_i$ is the combined fuzzified weight of the criterion *i* estimated by gathering the *K* number of DMs' judgments. The combined fuzzy weights from Equation (A10) were used in Fuzzy-VIKOR for sustainbility ranking of pavement scenarios.

References

1. Guo, M.; Liu, H.; Jiao, Y.; Mo, L.; Tan, Y.; Wang, D.; Liang, M. Effect of WMA-RAP technology on pavement performance of asphalt mixture: A state-of-the-art review. *J. Clean. Prod.* **2020**, *266*, 121704. [CrossRef]
2. Xudong, H.; Yihua, N.; Yu, F.; Zheng, Q. Pavement performance of asphalt surface course containing reclaimed asphalt pavement (RAP). *J. Test. Eval.* **2012**, *7*, 1162–1168.

3. Arshad, M. Laboratory investigations on the mechanical properties of cement treated RAP-natural aggregate blends used in base/subbase layers of pavements. *Constr. Build. Mater.* **2020**, *254*, 119234. [CrossRef]
4. Saberi, F.; Fakhri, M.; Azami, A. Evaluation of warm mix asphalt mixtures containing reclaimed asphalt pavement and crumb rubber. *J. Clean. Prod.* **2017**, *165*, 1125–1132. [CrossRef]
5. Al-Qraimi, S.; Hago, A.; Hassan, H. Recycling of reclaimed asphalt pavement in portland cement concrete. *J. Eng. Res.* **2009**, *6*, 37–45.
6. Asphalt Recycling & Reclaiming Association. *Basic Asphalt Recycling Manual*; Federal Highway Adminstration: Washington, DC, USA, 2015.
7. Giani, M.; Dotelli, G.; Brandini, N.; Zampori, L. Comparative life cycle assessment of asphalt pavements using reclaimed asphalt, warm mix technology and cold in-place recycling. *Resour. Conserv. Recycl.* **2015**, *104 Pt A*, 224–238. [CrossRef]
8. Schwartz, C. Structural characteristics and environmental benefits of cold-recycled asphalt paving materials. In Proceedings of the National Pavement Preservation Conference, Nashville, TN, USA, 10–11 November 2016.
9. Bhavsar, J. Comparing Cold In-Place Recycling (cir) and Cold In-Place Recycling with Expanded Asphalt Mixture (CIREAM). Master's Thesis, University of Waterloo, Waterloo, ON, Canada, 2015.
10. Kim, J.; Lee, H.; Jahren, C.; Heitzman, M.; Chen, D. Long-Term field performance of cold in-place recycled roads in iowa. *J. Perform. Constr. Facil.* **2010**, *24*, 265–274. [CrossRef]
11. Kim, Y.; Lee, H.; Heitzman, M. Dynamic modulus and repeated load tests of cold in-place recycling mixtures using foamed asphalt. *J. Mater. Civ. Eng.* **2009**, *21*, 279–285. [CrossRef]
12. Wang, H.; Zhao, Y.; Huang, X.; Pi, Y. Strength mechanism and influence factors for cold recycled asphalt mixture. *Adv. Mater. Sci. Eng.* **2015**, *2015*.
13. Kim, Y.; Lee, H. Influence of reclaimed asphalt pavement temperature on mix design process of cold in-place recycling using foamed asphalt. *J. Mater. Civ. Eng.* **2011**, *23*, 961–968. [CrossRef]
14. Ozer, H.; Al-Qadi, I.; Harvey, J. *Strategies for Improving the Sustainability of Asphalt Pavements*; FHWA-HIF-16-012; US Department of Transportation: Washington, DC, USA, 2016.
15. Al-Jumaili, M.A.H. Sustainability of asphalt paving materials containing different waste materials. In *IOP Conference Series: Materials Science and Engineering, Proceedings of the International Conference on Robotics and Mechantronics (ICRoM 2017), Hong Kong, China, 12–14 December 2017*; IOP Publishing: Bristol, UK, 2018.
16. Department of Public Roads; Los Angelos County. Cold In-Place Recycling (CIR). Available online: https://dpw.lacounty.gov/gmed/lacroads/treatmentcoldinplace.aspx (accessed on 13 December 2021).
17. Highway Research Board. *The AASHO Road Test. Report 5—Pavement Research*; Highway Research Board, National Academy of Sciences, National Research Councel: Washington, DC, USA, 1962.
18. Huang, Y. *Pavement Analysis and Design*; Prentice-Hall, Inc.: Englewood Cliffs, NJ, USA, 1993.
19. AASHTO. *AASHTO Guide for Design of Pavement Strucutres*; AASHTO: Washington, DC, USA, 1993.
20. Ministry of Transport and Logitics Services in KSA. *Hot Mix Asphalt Design System Using SuperPave System*; MOT: Riyadh, Saudi Arabia, 2006.
21. Zhou, Q.; Okte, E.; Sen, S.; Ozer, H.; Al-Qadi, I.; Roesler, J.; Chatti, K. *Development of a Life-Cycle Assessment Tool for Pavement Preservation and Maintenance on Flexible and Rigid Pavement*; Center for Highway Pavement Preservation: East Lansing, MI, USA, 2019.
22. Cross, S. *Cold In-Place Recycling (CIR) Study*; Federal Highway Administration: Lakewood, CO, USA, 2012.
23. FDOT. *Flexible Pavement Design Manual*; Florida Department of Transportation: Tallahassee, FL, USA, 2020.
24. Díaz-Sánchez, M.A.; Timm, D.H. Structural coefficients of cold central-plant recycled asphalt mixtures. *J. Transp. Eng.* **2017**, *143 Pt A*, 6. [CrossRef]
25. Haider, H.; Ghumman, A.R.; Al-Salamah, I. Assessment framework for natural groundwater contamination in arid regions: Development of indices and wells ranking system using fuzzy VIKOR method. *Water* **2020**, *12*, 423. [CrossRef]
26. Wang, C.-H.; Chou, M.-Y.; Pang, C.-T. Applying fuzzy analytic hierarchy process for evaluating service quality of online auction. *Int. J. Comput. Inf. Eng.* **2012**, *6*, 586–593.
27. Opricovic, S. Multicriteria Optimization of Civil Engineering Systems. Ph.D. Thesis, Faculty of Civil Engineering of the University of Belgrade, Belgrade, Serbia, 2 January 1998; pp. 5–21.
28. Opricovic, S. Fuzzy VIKOR with an application to water resources planning. *Expert Syst. Appl.* **2011**, *38*, 12983–12990. [CrossRef]

coatings MDPI

Article

Development of a Mechanistic Method to Obtain Load Position Strain in Instrumented Pavement

Guiling Hu [1,*], Wenyang Han [2], Jincheng Wei [2], Deqing Wang [3], Xiaomeng Zhang [2], Wenjun Hu [1] and Wei Chen [1]

1 School of Transportation Engineering, Shandong Jianzhu University, Jinan 250101, China; huwenjun@sdjzu.edu.cn (W.H.); chenwei13572468@163.com (W.C.)
2 Shandong Transportation Institute, Jinan 250101, China; hanwenyang@sdjtky.cn (W.H.); weijincheng@sdjtky.cn (J.W.); zhangxiaomeng@sdjtky.cn (X.Z.)
3 Shandong Huitong Construction Group Co., Ltd., Jinan 250032, China; wdeqing1216@163.com
* Correspondence: huguilingtech@foxmail.com

Abstract: To study the in-situ response and performance of asphalt pavement, instrumenting pavement with a variety of sensors has become one of the most important tools in the field or accelerated load facilities. In the dynamic response collection process, engineers are more concerned with the load position strain of the pavement structure due to wheel wander. This paper proposes a method to obtain the load position and the strain at the load position when there is no lateral-axis positioning system based on multilayer elastic theory. The test section was paved in the field with installed strain sensors to verify and apply the proposed method. The verification results showed that both the calculated load position and load position strain matched the measured values with an absolute difference range of 5–60 mm, 0.5–2.5 με, respectively. The application results showed that the strain at the load position calculated by the proposed method had a good correlation with the temperature, as expected.

Keywords: instrumented pavement; wheel wander; strain gauge; offset; load position strain

Citation: Hu, G.; Han, W.; Wei, J.; Wang, D.; Zhang, X.; Hu, W.; Chen, W. Development of a Mechanistic Method to Obtain Load Position Strain in Instrumented Pavement. *Coatings* **2022**, *12*, 14. https://doi.org/10.3390/coatings12010014

Academic Editor: Valeria Vignali

Received: 1 December 2021
Accepted: 20 December 2021
Published: 23 December 2021

Publisher's Note: MDPI stays neutral with regard to jurisdictional claims in published maps and institutional affiliations.

Copyright: © 2021 by the authors. Licensee MDPI, Basel, Switzerland. This article is an open access article distributed under the terms and conditions of the Creative Commons Attribution (CC BY) license (https://creativecommons.org/licenses/by/4.0/).

1. Introduction

In the past two decades, measuring the dynamic strains of asphalt pavement structures has gradually become one of the most important means to evaluate the performance of a pavement and validate pavement distress models from a mechanistic viewpoint [1,2]. The instrumentation used at the Virginia Smart Road was to measure flexible pavement response to loading [3]. Twelve asphalt strain gauges were installed to measure longitudinal and transverse strains at the bottom of the Hot Mixture Asphalt (HMA) layer in an instrumented Test Section located in McClain County, Oklahoma [4]. Asefzadeh, A. and Hashemian, L. developed empirical statistical pavement temperature prediction models based on two years of field strains collected from an instrumented test road located in Edmonton, Alberta, Canada [5]. Hossain, N. and Singh, D. conducted a study to better understand the cause of pavement failure under actual traffic loading and environmental conditions in an instrumented test section on I-35 in McClain County [6].

It can be seen from the above, to measure dynamic strains in the field, it is necessary to install sensors in the pavement. However, when using these strains from the sensors to evaluate the pavement performance, one factor that must be taken into account is whether the wheel of the vehicle passes directly above the sensor. Engineers are most concerned with the strain at the load position, i.e., the maximum strain that could primarily cause pavement deterioration.

Timm and Priest carried out an investigation at the National Center for Asphalt Technology (NCAT) test track. The results showed that the offset has a strong relationship with the strain response, and the strain response was consistent with predictions from multilayer elastic theory [7]. Shafiee and Nassiri studied the effect of lateral wheel wander on the tensile and vertical strains at the bottom of a hot mix asphalt layer [8]. Chen and

Song estimated the effect of an autonomous truck's lateral distribution on the rutting depth and fatigue damage of flexible pavement by finite element analysis [9]. Noorvand and Karnati found that with the optimal wheel wander distribution in autonomous vehicles (AVs), the pavement service life can be potentially extended [10]. These studies highlighted that wheel wander has a significant influence on the pavement structure strain response and that obtaining the strain at the load position is important.

In practice, it is difficult to determine whether the wheels only pass above the position where the sensors are installed in instrumented pavement because of wheel wander, especially if a lateral-axis positioning system is not installed. This means that the strain measurements from the strain gauges are not always the desired values. Little research has been done on obtaining the strain at the load position. Current studies on this were mainly from the NCAT and the Minnesota Department of Transportation, both of which have test tracks. Initially, Willis and Timm defined the strain at the load location as the response that yielded the highest strain reading [11]. The 95th percentile of 450 readings was used to represent the "best-hit" response for the longitudinal strain in a subsequent study [12]. The Minnesota Department of Transportation (MN/DOT) developed a computer program for selecting peak dynamic sensor responses from pavement testing, but the load position strain was not mentioned in this report [13].

To address the issue described above, the objective of this study was to develop a mechanistic method for obtaining the strain at the load position without a lateral positioning system in instrumented pavement. One test section in the field with instrumentation was paved to carry out this research.

2. Proposal of the Mechanistic Method to Obtain Load Position Strain

2.1. Preliminaries

To measure the dynamic strain at the bottom of an asphalt layer, pavement is usually instrumented with strain gauges, as shown in Figure 1. At least two longitudinal gauges and two transverse strain gauges were installed at the bottom of asphalt layer in this study. The strain gauges were arranged in pairs on the wheel path to authenticate the data. To obtain more effective strain data, the lateral distance between the strain gauges should cover the wheel wander as much as possible. Many studies have reported that about 90% of the wheels pass within a lateral distance of 80 cm and that wheel wander obeys normal distribution [7,14,15]. Therefore, the recommended lateral distance between the strain gauges installed in pairs is at least 80 cm, so that as many wheels as possible pass between the two strain gauges.

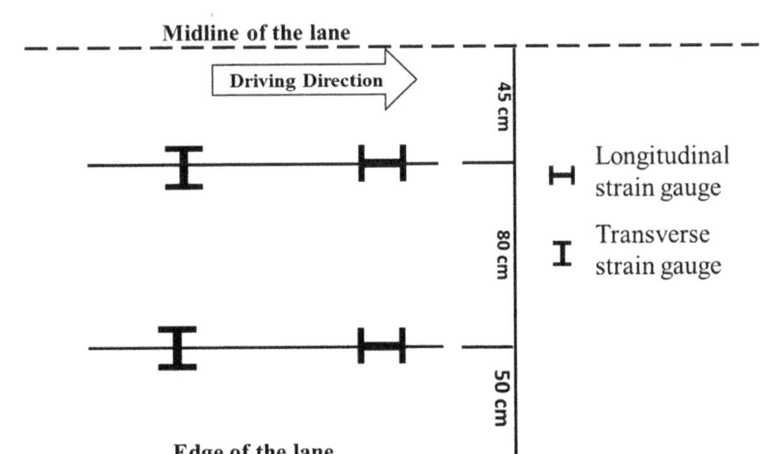

Figure 1. Strain gauge array.

2.2. Mechanistic Method to Obtain Load Position Strain

As pointed out earlier, the main factor of interest when designing a pavement structure and evaluating its fatigue performance is the load position strain at the bottom of the asphalt layer. Due to wheel wander, it is difficult to determine if the measured strains were exactly the strains at the load position. A mechanistic method was proposed to obtain the strain at load position by using the measured strains and the strain gauge arrangement shown in Figure 1. This mechanistic method correlated the theoretical strain with the measured strain. For flexible pavements, elastic multilayer theory is the most commonly used model to describe mechanical performances of pavement layers [16], so the theoretical strain was calculated based on the elastic multilayer theory in this paper.

The strain at the load position, $\varepsilon_{\mathrm{mlp}}$, was obtained as follows.

Step 1. Gathering measured data.

When a vehicle wheel passed by, voltage signals from strain gauge were collected by a high-speed data acquisition system. Once the raw data was collected, it went through the processing stage where it was recognized by wheel axle type and was filtered to reduce noise by using data processing software, such as origin 2021 (OriginLab, Northampton, MA, USA) or the proposed method in previous study [17]. Then, the biggest amplitude of the waveform was chosen as the strain caused by the passing wheel load. Theoretically, the data points should all be strains in pairs based on the gauge array, but due to the variability of the field test conditions, there may be data points with only a single strain measurement or no data. Therefore, a data trimming process was performed to eliminate these data points for the calculations described below.

Step 2. Development of relationship between offset and ratio of paired strain gauges.

When the wheel load was at position i (i = 1,2,3 … n) between the paired gauges, as shown in Figure 2, the theoretical paired strains of the two gauges, ε_{1i} and ε_{2i}, were obtained using the BISAR3.0 software (a software developed by SHELL Company, Hague, The Netherlands based on elastic multilayer theory). The input parameters for BISAR3.0 mainly included each layer's modulus, Poisson's ratio, and thickness. Meanwhile, the offset, D_i (from strain gauge 1 to the load position), was recorded.

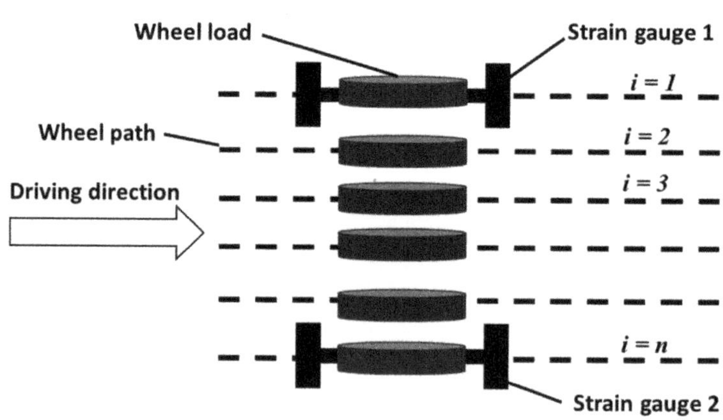

Figure 2. Different offsets of the wheel load.

The results of the strains and offsets were fit to a second-order polynomial:

$$D_i = a_1 R_i^2 + b_1 R_i + c_1 \quad (1)$$

$$R_i = \frac{\varepsilon_{1i}}{\varepsilon_{2i}} \quad (2)$$

where i is the load position, D_i is the wheel offset in the theoretical calculation, i.e., the distance from strain gauge 1 to the load position i, a_1, b_1, and c_1 are fitting coefficients, R_i is the ratio of the theoretical paired strains, and ε_{1i} and ε_{2i} are the theoretical paired strains of strain gauges 1 and 2 when the load is at position i, respectively.

Step 3. Field offset calculation.

The field offset was obtained by substituting the ratio of the paired strains measured in Step 1 into the regression equation in Step 2:

$$R_m = \frac{\varepsilon_{m1}}{\varepsilon_{m2}} \quad (3)$$

where R_m is the ratio of the measured paired strains, and ε_{m1} and ε_{m2} are the measured paired strains of strain gauges 1 and 2 in Step 1.

Step 4. Development of relationship between offset and ratio of strain at strain gauge 1 to load position strain.

The ratio of strain gauge 1 to the load position strain, R_{ilp}, can be obtained using the theoretical strain ε_{m1} obtained in Step 2 and ε_{lp}:

$$R_{ilp} = \frac{\varepsilon_{1i}}{\varepsilon_{lp}} \quad (4)$$

where R_{ilp} is the ratio of strain gauge 1 to the load position strain when the load is at position i, ε_{1i} is the theoretical strain of gauge 1 when the load is at position i, and ε_{lp} is the theoretical load position strain, calculated by BISAR3.0.

The results of R_{ilp} and the offsets were also fitted with a second-order polynomial:

$$R_{ilp} = a_2 D_i^2 + b_2 D_i + c_2 \quad (5)$$

where a_2, b_2, and c_2 are fitting coefficients.

Step 5. Load position strain calculation.

The ratio of the strain measured at gauge 1 to the load position strain, R_{mlp}, can be calculated by inserting the field offset from Step 3 into the regression function from Step 4. The load position strain can then be calculated as follows:

$$\varepsilon_{mlp} = \frac{\varepsilon_{m1}}{R_{mlp}} \quad (6)$$

where R_{mlp} is the measured ratio of the strain at gauge 1 to the load position strain, and ε_{mlp} is the load position strain in the field.

3. Verification of Proposed Method in the Field

3.1. Test Plan

A test section located in Qinglin Road, Shandong Province, China was established for evaluating the performance of the full-depth asphalt pavement structure. Verification of the proposed method was part of the objective. As shown in Figure 3, the whole thickness of the section was 68 cm, with a 48 cm thick asphalt layer. Four longitudinal asphalt gauges were installed at the bottom of the asphalt layer, as shown in Figure 4. Compared with Figure 1, two more gauges were added to increase the amount of data and verify the method.

During the test, a vehicle with a standard axle weight was driven on a fine sand-covered road surface to record the offset. The vehicle was driven six times for each offset position. Given that 90% of the wheels passed within the paired strain gauges (as mentioned in Section 2.1), the horizontal position of SG1 was defined as the initial position, i.e., the position at which the wheel offset was 0. The speed of the vehicle was 80 km/h. Meanwhile, the dynamic strains of the strain gauges were collected for each offset. Tables 1 and 2 present the results of the manually recorded offsets and the dynamic strains.

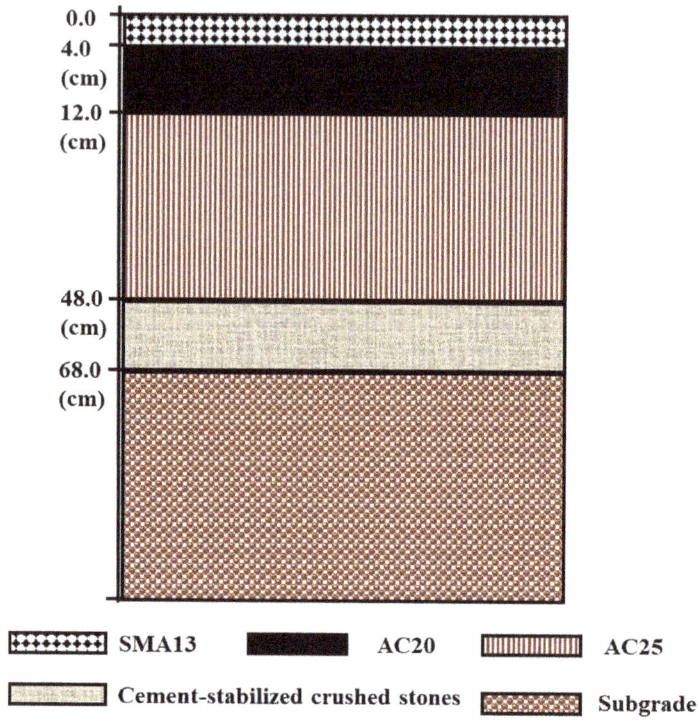

Figure 3. Structure of test section.

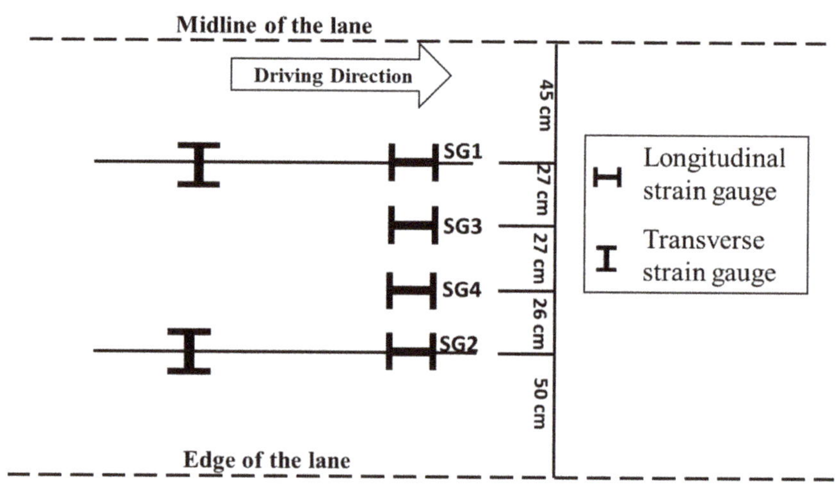

Figure 4. Strain gauge array of the test section.

Table 1. Dynamic strains with manually recorded offsets for offset verification.

		Dynamic Strain, με											
Test Number		1		2		3		4		5		6	
Strain Gauge		SG1	SG2	SG1	SG2	SG1	SG2	SG1	SG2	SG1	SG2	SG1	SG2
Offset, mm	0	34.09	21.69	34.81	21.29	34.08	21.69	32.94	21.64	33.38	20.85	34.99	21.64
	100	34.20	23.52	33.91	24.12	34.36	23.26	32.74	23.02	34.52	24.56	35.21	24.26
	200	34.06	25.83	33.92	26.47	33.66	26.83	32.52	26.45	33.95	25.73	32.48	25.48
	270	32.07	28.57	33.59	28.38	34.01	27.77	32.60	28.34	34.01	27.89	33.45	28.86
	300	33.27	29.42	32.86	30.50	32.75	29.08	33.11	28.00	32.45	29.96	34.04	28.46
	400	32.19	31.31	31.64	32.56	31.41	30.08	32.54	30.81	30.41	29.85	32.27	30.02
	540	29.56	33.95	29.53	33.10	28.22	33.59	28.63	33.47	27.73	33.91	27.72	32.18
	600	26.43	33.33	27.01	34.43	25.86	34.45	26.59	34.53	26.37	34.76	26.67	32.61
	700	23.77	34.16	22.97	34.61	22.58	34.14	23.24	35.10	25.06	34.18	23.45	32.86
	800	20.62	34.29	21.32	33.49	20.92	33.59	21.61	34.09	19.14	33.19	20.31	34.68

Table 2. Dynamic strains with manually recorded offsets for load position strain verification.

Offset, mm	Strain Gauge	Test Number					
		1	2	3	4	5	6
		Load Position Strain, με					
0	SG1	34.09	34.81	34.08	32.94	33.38	34.99
27	SG3	33.31	33.10	32.63	35.28	33.03	33.77
54	SG4	34.14	35.22	33.67	33.31	34.13	35.15
80	SG2	34.29	33.49	33.59	34.09	33.19	34.68

3.2. Offset Verification

By following Steps 1 and 3 of the proposed method, the relationship between the offset and ratio of the paired strain gauges was established. First, the theoretical paired strains (SG1 and SG2) with different offsets were calculated by BISAR 3.0. The structural input parameters of the test section are shown in Table 3. The results of the strains and offsets were fit to a second-order polynomial, and the resulting regression equation is as follows:

$$D_i = 279.22R_i^2 - 1348.1R_i + 1479.4 \quad (7)$$

Table 3. Input parameters of the test section for BISAR 3.0.

Pavement Structure	Materials	Thickness, cm	Modulus, MPa	Poisson's Ratio
Surface course	SMA-13	4	8700	0.25
Upper binder course	AC-20	8	9000	0.25
Lower binder course	AC-25	36	9800	0.25
Base course	Cement stabilized crush stones	20	12,000	0.25
Subgrade	Compacted subgrade	-	60	0.4

The measured paired strains in the field then were inserted into Equation (7) to obtain the field offset.

Figure 5 shows that the values calculated using the proposed method compared favorably to the measured values. The data points were evenly distributed on both sides of the parity line and were close to it. To further qualify the accuracy of the proposed method, the absolute difference and percent difference (%difference) between the measured and calculated offsets were investigated. The percent difference was calculated as follows:

$$\%\text{difference} = \left(\frac{\text{offset}_{\text{measured}} - \text{offset}_{\text{calculated}}}{\text{offset}_{\text{measured}}}\right) \times 100 \quad (8)$$

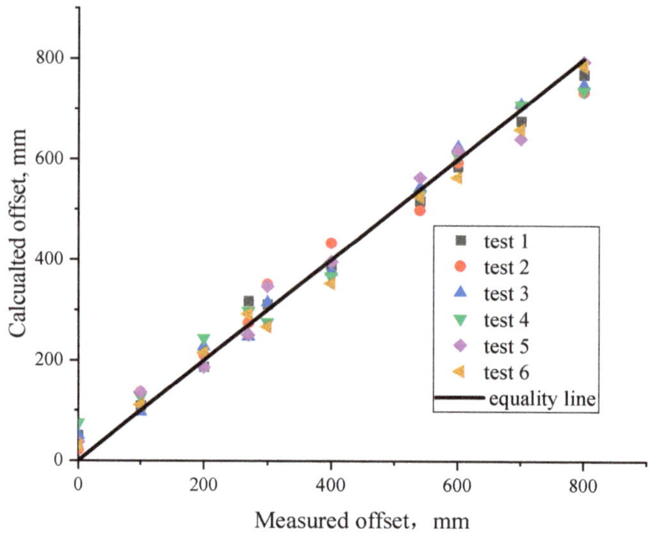

Figure 5. Calculated offset vs. measured offset.

As shown in Figures 6 and 7, the range of the absolute difference was 5–60 mm, which should be viewed as fairly small. Most of the %difference values were in the range of 2–17%. The %difference decreased as the offset increased, because the offset is in the denominator in Equation (8). A one-way analysis of variance was performed on the measured and calculated offsets, and results showed that there was no significant difference between the two in Table 4.

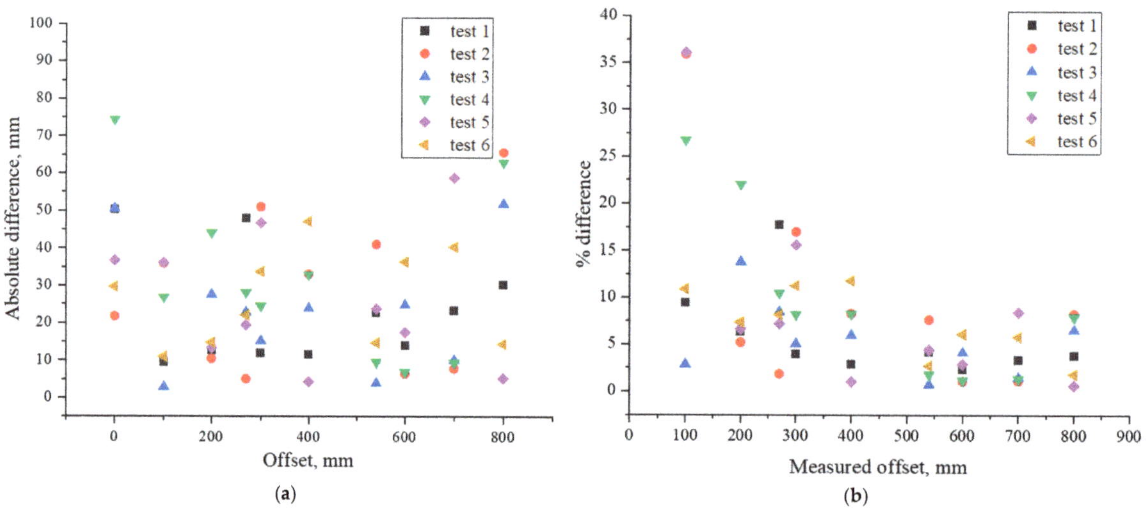

Figure 6. Difference between calculated and measured offsets: (**a**) absolute difference and (**b**) percent difference (%difference).

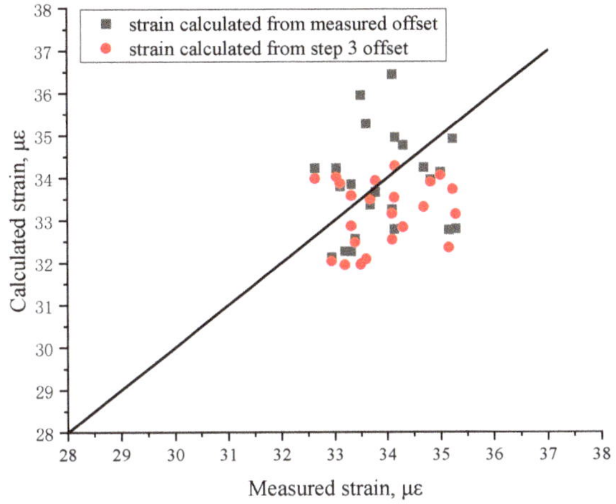

Figure 7. Calculated strain vs. measured strain.

Table 4. Results of one-way analysis of variance between calculated and measured offsets.

	Degrees of Freedom	Sum of Squares	Mean Square	F Value	Prob > F
Model	1	79.29785	79.29785	0.00134	0.97084
Error	118	6,970,853.425	59,075.03	-	-
Sum	119	6,970,932.723	-	-	-

3.3. Load Position Strain Verification

Following Steps 4 and 5, the offset and ratio of the theoretical strain at gauge 1 to the load position strain were fit to a second-order polynomial, and the regression formula is as follows:

$$R_{ilp} = -8 \times 10^{-7} D_i^2 + 10^{-4} D_i + 1.025 \quad (9)$$

The measured and calculated offsets from Step 3 when the vehicle passed directly above strain gauges SG1, SG2, SG3, and SG4 were inserted into the above equation to obtain the ratio of the strain at gauge 1 to the load position strain. In the last step, the load position strain was obtained by Equation (6).

Figure 7 shows that the calculated strains from the Step 3 offsets and the measured offsets were close to the parity line. The absolute and %differences between the measured and calculated strains were also calculated, and the results are shown in Figure 8. The ranges of the differences between the measured strains and those calculated from the offset were 0.5–2.5 με for the absolute difference and 0.5–16% for the %difference. From the perspective of the absolute difference and %difference, the strains calculated from the measured offset were slightly more accurate than those from the Step 3 offset, but a one-way analysis of variance showed there was no significant difference between the three strain acquisition methods (Table 5). It should be noted that the strain at the load position cannot be calculated by the measured offset in practice, because the pavement surface is not covered by fine sand.

Table 5. Results of one-way analysis of variance between calculated and measured offsets.

	Degrees of Freedom	Sum of Squares	Mean Square	F Value	Prob > F
Model	2	4.22862	2.11431	2.53545	0.08659
Error	69	57.53906	0.8339	-	-
Sum	71	61.76768	-	-	-

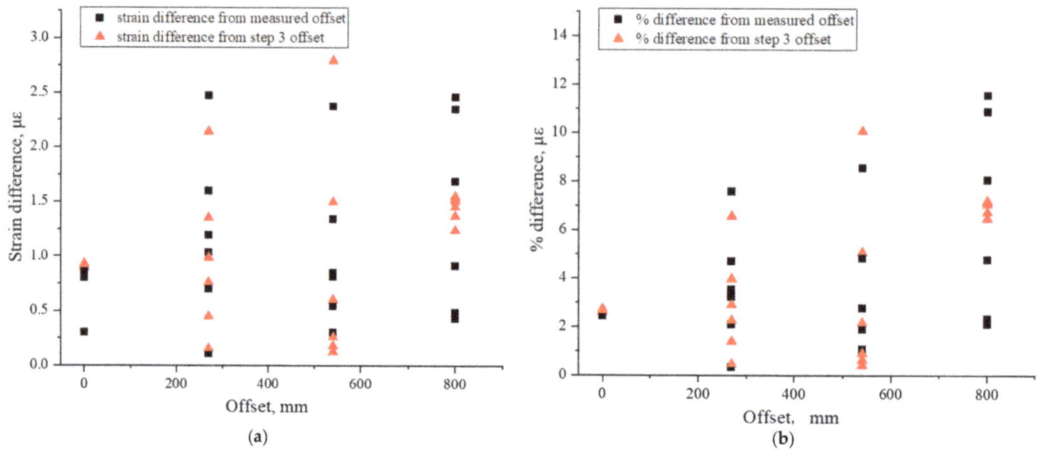

Figure 8. Difference between calculated and measured strains: (**a**) absolute difference and (**b**) %difference.

4. Application of the Proposed Method

After the test section was opened to traffic, the strain at the bottom of the asphalt layers under the standard load was collected at 2 pm each day from April to July. As mentioned earlier, these strains were not always the load position strains due to wheel wander. The proposed method was used to obtain the load position strain.

Figure 9 illustrates the strains directly measured from the strain gauges and the strains at the load position obtained from the proposed method. The stains from the strain gauges did not show a tendency to gradually increase with temperature. Most of the strains captured by SG1 were greater than those captured by SG2, which means most of the wheels were closer to SG1 as they passed between the two gauges. As expected, the load position strains calculated from the proposed method showed a good correlation with temperature, which showed that this method maintained a good consistency in the verification test.

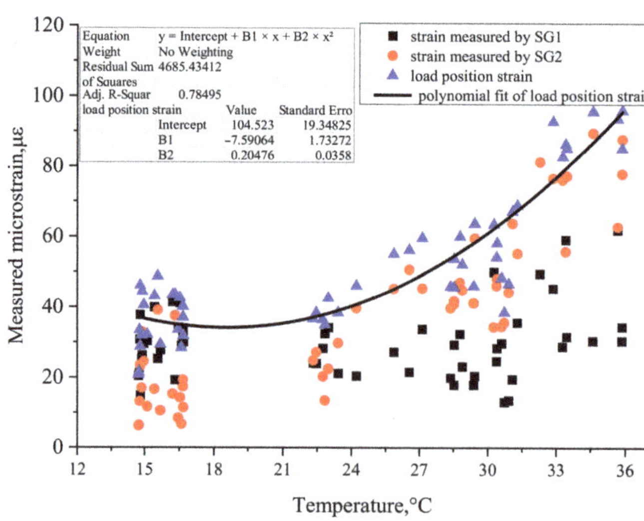

Figure 9. Load position strain calculated from proposed method with different temperatures from April to July.

5. Discussion

The primary objective of this research was to propose a method to obtain load position strain. Using the sensors array, the relationship between the two theoretical strains from sensors and wheel offset was established. The theoretical strains were calculated from BISAR 3.0 based on the elastic multilayer theory. In fact, other mechanical response calculation software like WESLEA also can be used to calculate the response [18,19]. Then the theoretical strain was replaced by the measured strain in the field to obtain the offset and load position strain in field.

The proposed method was verified by a test section with the installation of sensors. The results of verification show that there is no significant difference between calculated and measured strains, and this can be viewed the proposed method is feasible. Then the method was applied to the load position strain acquisition for four months in practical and the result shows that strain has a good correlation with temperature. This is consistent with the most previous studies [3,7,20,21].

At last, the applicability of the proposed method in other pavement structures needs to be further explored, because only one test structure was used for the verification. One of the possible reasons is that the second-order polynomial model may be not suitable for the establishment of the relationship between strain ratio and offset in other pavement structure.

6. Conclusions

This paper proposed a mechanistic method that can help obtain load position strain at the bottom of asphalt layer in instrumented pavement when the lateral axis position system is not installed. Based on the proposed method and the verification process, the following conclusions were drawn:

1. The proposed method was based on the elastic multilayer theory and correlated the theoretical strain with the measured strain by a second-order polynomial model. It allows the calculation of the load position stain without installation of lateral position system in instrumented pavement.
2. In the proposed method, it was recommended that at least two gauges should be installed into the pavement with the lateral space 80 cm at least to cover the wheel wander range.
3. The result of the verification shows that the ranges of the differences between the measured strains and those calculated from the offset were 0.5–2.5 $\mu\varepsilon$ for the absolute difference and 0.5–16% for the %difference. The one-way analysis of variance showed there is no significant difference between the measured and calculated from the proposed method.
4. The result of the application shows that load position strain calculated from proposed method has a good correlation with temperature which means this method maintains a good consistency with the results of verification.
5. Due to the limited validation and application data, the applicability of this method in other pavement structures needs to be further verified.

Author Contributions: G.H., J.W. and W.H. (Wenyang Han) proposed the method. G.H. and J.W. designed the experiments. W.H. (Wenyang Han) D.W. and W.C. performed the experiments. G.H., X.Z. and D.W. analyzed the data. W.H. (Wenjun Hu), G.H. and W.H. contributed model establishment. G.H. and W.H. (Wenyang Han) wrote the paper. G.H. and J.W. revised the paper. All authors have read and agreed to the published version of the manuscript.

Funding: This research was funded by the Shandong Provincial Natural Science and Foundation, grant number ZR2020QE272, the Shandong Province Higher Educational Science and Technology Program, grant number J18KA216, the Shandong Jianzhu University Doctoral research foundation, grant number X18073Z and the Science and technology innovation project of Shandong Provincial Department of industry and information technology, grant number 202160101777.

Institutional Review Board Statement: Not applicable.

Informed Consent Statement: Not applicable.

Data Availability Statement: Not applicable.

Conflicts of Interest: The authors declare no conflict of interest.

References

1. Timm, D.H.; Priest, A.L.; McEwen, T.V. *Design and Instrumentation of the Structural Pavement Experiment at the NCAT Test Track. Report No. 04-01*; National Center for Asphalt Technology, Auburn University: Auburn, AL, USA, 2004.
2. Selvaraj, S.I. Review on the Use of Instrumented Pavement Test Data in Validating Flexible Pavement Mechanistic Load Response Models. In Proceedings of the 8th International Conference on Traffic and Transportation Studies, Changsha, China, 1–3 August 2012.
3. Al-Qadi, I.L.; Loulizi, A.; Elseifi, M.; Lahouar, S. The Virginia Smart Road: The impact of pavement instrumentation on understanding pavement performance. *J. Assoc. Asph. Paving Technol.* **2004**, *73*, 427–465.
4. Nur, H.E.I.; Singh, D.; Musharraf, Z.P.E. Dynamic Modulus-based Field Rut Prediction Model from an Instrumented Pavement Section. *Procedia-Soc. Behav. Sci.* **2013**, *104*, 129–138.
5. Asefzadeh, A.; Hashemian, L.; Bayat, A. Fatigue Life Evaluation of Pavement Embankments made with Tire Derived Aggregates. *Can. J. Civ. Eng.* **2018**, *45*, 795–802. [CrossRef]
6. Solanki, P.; Zaman, M.; Muraleetharan, K.K. *Field Performance Monitoring and Modeling of Instrumented Pavement on I-35 in McClain County—Construction and Instrumentation Report*; School of Civil Engineering and Environmental Science, University of Oklahoma: Norman, OK, USA, 2008.
7. Timm, D.H.; Priest, A.L. *Wheel Wander at the NCAT Test Track (No. 05–02)*; National Center for Asphalt Technology: Auburn, AL, USA, 2005.
8. Shafiee, M.H.; Nassiri, S.; Eng, P.; Bayat, A. Field investigation of the effect of operational speed and lateral wheel wander on flexible pavement mechanistic responses. In Proceedings of the Transportation 2014: Past, Present, Future—2014 Conference and Exhibition of the Transportation Association of Canada, Montreal, QC, Canada, 28 September–October 2014.
9. Chen, F.; Song, M.; Ma, X.; Zhu, X. Assess the impacts of different auftonomous trucks' lateral control modes on asphalt pavement performance. *Transp. Res. Part C. Emerg. Technol.* **2019**, *103*, 17–29. [CrossRef]
10. Noorvand, H.; Karnati, G.; Underwood, B.S. Autonomous vehicles: Assessment of the implications of truck positioning on flexible pavement performance and design. *Transp. Res. Rec.* **2017**, *2640*, 21–28. [CrossRef]
11. Willis, J.R.; Timm, D.H. Repeatability of Asphalt Strain Measurements Under Full-Scale Dynamic Loading. *J. Transp. Res. Board* **2008**, *2087*, 40–48. [CrossRef]
12. Timm, D.H.; Robbins, M.M.; Willis, J.R.; Tran, N.; Taylor, A.J. *Field and Laboratory Study of High-Polymer Mixtures at the NCAT Test Track Interim Report*; National Center for Asphalt Technology: Auburn, AL, USA, 2012.
13. Burnham, T.R.; Tewfik, A.; Srirangarajan, S. *Development of a Computer Program for Selecting Peak Dynamic Sensor Responses from Pavement Testing*; Minnesota Department of Transportation: Saint Paul, MN, USA, 2007.
14. Benekohal, R.F.; Hall, K.T.; Miller, H.W. Effect of lane widening on lateral distribution of truck wheels. *Transp. Res. Rec.* **1990**, *1286*, 57–66.
15. Buiter, R.; Cortenraad, W.M.; Van Eck, A.C.; Van Rij, H. Effects of Transverse Distribution of Heavy Vehicles on Thickness Design of Full-Depth Asphalt Pavements. *Transp. Res. Rec.* **1989**, *1227*, 66–74.
16. Bruno, S.; Del Serrone, G.; Di Mascio, P.; Loprencipe, G.; Ricci, E.; Moretti, L. Technical Proposal for Monitoring Thermal and Mechanical Stresses of a Runway Pavement. *Sensors* **2021**, *21*, 6797. [CrossRef] [PubMed]
17. Han, H.; Han, W.; Ma, S.; Hu, G. Enhanced Processing of Low Signal-to-Noise-Ratio Dynamic Signals from Pavement Testing *Measurement* **2021**, *182*, 109697. [CrossRef]
18. Mazumder, M.; Kim, H.; Lee, S.J. Perpetual Pavement: Future Pavement Network. *J. Adv. Constr. Mater.* **2015**, *19*, 1.
19. Van Cauwelaert, F.J.; Lequeux, D. *Computer Programs for the Determination of stresses and Displacements in four Layered Structures*, Water Experiment Station, U.S. Army Corps of Engineers: Vicksburg, MS, USA, 1986.
20. Yang, Y.; Wei, J.; Lin, W.; Timm, D.; Huber, G. Binzhou Perpetual Pavement Test Road Dynamic Response of Pavement Under Very Heavy Loads. *Int. J. Road Mater. Pavement Des.* **2009**, *10*, 151–166. [CrossRef]
21. Yang, Y.S.; Wang, L.; Wei, J.C.; Ma, S.J. Typical Pavement Structure Dynamic Response Data Collection and Analysis under Heavy Vehicle Loading. *J. Highw. Transp. Res. Dev.* **2010**, *27*, 10–16.

coatings

MDPI

Article

Comparative Study on Complex Modulus and Dynamic Modulus of High-Modulus Asphalt Mixture

Licheng Guo [1], Qinsheng Xu [2], Guodong Zeng [3], Wenjuan Wu [2,*], Min Zhou [3], Xiangpeng Yan [2], Xiaomeng Zhang [2] and Jincheng Wei [2]

1 Foshan Highway & Bridge Construction Co., Ltd., Foshan 528313, China; hao_li_20007@163.com
2 Shandong Transportation Institute, Jinan 250102, China; 15966652536@163.com (Q.X.); yanxiangpeng@sdjtky.cn (X.Y.); zhangxiaomeng@sdjtky.cn (X.Z.); weijincheng@sdjtky.cn (J.W.)
3 Foshan Transportation Science and Technology Co., Ltd., Foshan 528315, China; wutongshuxi_a@163.com (G.Z.); yuan370370@163.com (M.Z.)
* Correspondence: wuwenjuan@sdjtky.cn

Abstract: In the French high-modulus asphalt mixture design system, the complex modulus of the mixture under the conditions of 15 °C and 10 Hz is taken as the design index. However, in China, the dynamic modulus under the conditions of 15 °C, 10 Hz, 20 °C, 10 Hz and 45 °C, 10 Hz was taken as the stiffness modulus index of high-modulus asphalt mixture. The difference in modulus values between the two systems caused the pavement structure layer to be thicker and the construction cost to be higher in China. In order to find out the appropriate modulus value of high-modulus asphalt mixture suitable for China's modulus parameter conditions to better carry out the reasonable design and evaluation of high-modulus asphalt mixture in China, the modulus of four types of high-modulus asphalt mixtures under the two systems through the two-point bending complex modulus test of the CRT-2PT trapezoidal beam and the SPT uniaxial compression dynamic modulus test were analyzed in this paper. Under the premise of meeting the stiffness modulus index of the French high-modulus asphalt mixture, the relationship conversion models between the dynamic modulus and complex modulus of high-modulus asphalt mixture under different temperatures were established. According to the conversion models, the design evaluation value range of dynamic modulus suitable for China's condition was recommended. It is recommended that the dynamic modulus of China's high-modulus asphalt mixture at 15 °C and 10 Hz is not less than 16,000 MPa, the dynamic modulus at 20 °C and 10 Hz is not less than 14,000 MPa, and the dynamic modulus at 45 °C and 10 Hz is not less than 2500 MPa. Five kinds of high-modulus asphalt mixtures used in actual road engineering were tested to verify the reliability of the recommended dynamic modulus values based on the modulus conversion model, and the results are consistent with the recommended value range of the model.

Keywords: low-mark high-modulus mixture; complex modulus of trapezoidal beam; uniaxial compression dynamic modulus; conversion model

Citation: Guo, L.; Xu, Q.; Zeng, G.; Wu, W.; Zhou, M.; Yan, X.; Zhang, X.; Wei, J. Comparative Study on Complex Modulus and Dynamic Modulus of High-Modulus Asphalt Mixture. *Coatings* **2021**, *11*, 1502. https://doi.org/10.3390/coatings11121502

Academic Editor: Valeria Vignali

Received: 13 October 2021
Accepted: 25 November 2021
Published: 6 December 2021

Publisher's Note: MDPI stays neutral with regard to jurisdictional claims in published maps and institutional affiliations.

Copyright: © 2021 by the authors. Licensee MDPI, Basel, Switzerland. This article is an open access article distributed under the terms and conditions of the Creative Commons Attribution (CC BY) license (https://creativecommons.org/licenses/by/4.0/).

1. Introduction

High-modulus mixture refers to an asphalt mixture with a complex modulus ≥14,000 MPa under the conditions of 15 °C and 10 Hz [1]. It was first proposed by French road workers and has been used in France for more than 20 years [2–7]. High-modulus mixture has become a research hotspot due to its excellent high temperature stability, strong resistance to water damage and good fatigue resistance, which has great advantages in reducing the thickness of pavement and saving resources [8,9]. Jong Lee [10] studied the use of high-modulus mixtures for long-life asphalt pavements. His work showed that the stiffness modulus of high-modulus mixtures is 50% higher than that of conventional asphalt mixtures, and that other road performances have been improved accordingly. The tensile strain of the high-modulus asphalt mixture structure layer is less than the allowable value of the long-life asphalt pavement. Arnold et al. [11]

developed the anti-rutting performance of high-modulus mixture EME (Enrobés à Module Elevé Class) for the rutting disease of road intersections. The test results show that the rutting resistance of the high-modulus mixture is 100 times higher than that of other mixtures. At the same time, this high-modulus mixture has a higher asphalt content, and the improvement of modulus and fatigue performance can reduce the thickness of the pavement by one third. Amjad et al. [12] studied the various aspects of hard asphalt high-modulus asphalt mixtures, which showed good fatigue and permanent deformation characteristics. Compared with conventional asphalt mixtures, its anti-fatigue damage ability is increased by 9.3%, and the modulus of resilience at 60 °C increased by 63%. Based on the Miner fatigue cumulative damage criterion, Jianping Xu [13] estimated the fatigue life of high-modulus pavement, and believed that the fatigue life of high-modulus pavement was 23.9% longer than that of conventional asphalt pavement.

In recent years, high-modulus asphalt mixtures have become more and more widely used in China. According to the LPC (Laboratoire Central des Ponts et Chaussées) Bituminous Mixtures Design Guide, the judgment of high-modulus mixture and the selection of modulus index should refer to the complex modulus of the trapezoidal beam under 15 °C and 10 Hz conditions [14]. However, In China, engineers and researchers believe that the dynamic modulus can better characterize the performance of asphalt mixtures [15–17]. They use the dynamic modulus of the mixture as the main reference parameter. In the engineering application of high-modulus mixtures, the selection of the stiffness modulus index of the mixture is more the dynamic modulus under the conditions of 15 °C, 10 Hz, 20 °C, 10 Hz and 45 °C, 10 Hz [18–21]. The CRT-2PT (Cooper Technology Two Point Trapezoidal Bending Test) complex modulus test and the SPT (Simple Performance Tester) dynamic modulus test are two different stiffness modulus tests; the SPT uniaxial compression dynamic modulus test is compression mode, while CRT-2PT trapezoidal beam complex modulus test is bending-tension mode. Different loading methods will cause big differences in test results. In China, if the high-modulus mixture is designed based on the complex modulus index value, the design of the pavement structure layer thickness will be too large, due to the small modulus value, and the construction cost will be greatly increased. Moreover, the complex modulus test equipment for trapezoidal beams is expensive, which is difficult to popularize in China. Some scholars have carried out a comparative study of dynamic modulus tests and trapezoidal beam complex modulus tests. Huang You et al. [22–24] selected four mixtures for the dynamic modulus test and the trapezoidal beam complex modulus test. The two test methods were evaluated from the test principle, operation process and test data. It is recommended to use the SPT dynamic modulus test for the compressive upper layer asphalt mixture, and the trapezoidal beam complex modulus test for the middle and lower layer mixture, which is mainly subjected to bending and tension. Although a large number of experimental studies have been conducted on the modulus of the mixture under the two test systems, there is no clear answer as to whether the modulus index of the French high-modulus asphalt mixture can be reached, and the range of the dynamic modulus technical index of high-modulus mixture suitable for Chinese characteristics has not been proposed.

Therefore, the modulus under the two evaluation systems through the two-point bending complex modulus test of CRT-2PT trapezoidal beams and the SPT uniaxial compression dynamic modulus test was analyzed in this paper. Under the premise of satisfying the complex modulus $\geq$14,000 Mpa under the conditions of 15 °C and 10 Hz, the conversion relationship model between the complex modulus of the trapezoidal beam and the uniaxial compression dynamic modulus was established. Based on the conversion model, a range of values suitable for the design and evaluation of the dynamic modulus of high-modulus mixtures in China was recommended and verified, which provides a reference for the design and evaluation of high-modulus mixtures in China using dynamic modulus parameters. It can effectively promote the wide application of high-modulus mixtures in engineering construction in China.

2. Materials and Experimental

2.1. Materials

The crushed limestone and its ground ore powder were used as aggregates and filler, the technical indicators of which meet the requirements of standard JTG-F40-2004 [21], respectively, as shown in Table 1.

Table 1. Properties of coarse and fine aggregate.

	Technical Index		Test Results	Index Requirements
Coarse aggregate	Apparent relative density	15–20 mm 10–15 mm 5–10 mm 3–5 mm	2.762 2.732 2.746 2.758	≤2.50
	Water absorption rate (%)	15–20 mm 10–15 mm 5–10 mm 3–5 mm	0.45 0.32 0.45 0.63	≤3.0
	Crushing value (%)	-	19.2	≤28
	Needle flake content (%)	15–20 mm 10–15 mm 5–10 mm	3.5 9.3 10.7	≤18
	Soft stone content (%)		0.55	≤5
	Washing method <0.075 mm particle content (%)	15–20 mm 10–15 mm 5–10 mm 3–5 mm	0.3 0.6 0.4 0.9	≤1.0
	Adhesion to asphalt, grade		5	≥4
Fine aggregate	Apparent relative density Sand equivalent (%) Angularity of fine aggregate (s) Methylene blue (g/Kg)		2.644 67 36 7	≥2.5 ≥60 ≥30 ≤25

Two types of low-grade asphalt were used in the French high-modulus mixture. One is the hard asphalt with the label of 10/20, namely France 15# hard asphalt, the other is China 15# hard asphalt, which has similar performance to France 15#. The technical indicators of the two types of hard bitumen meet the requirements of standard JTG-F40-2004, and the test results are shown in Tables 2 and 3.

Table 2. Asphalt conventional technical indicators.

Pilot Projects		China 15#	France 15#
Penetration (100 g, 5 s, 25 °C)/0.1 mm		15.2	16.8
Softening Point (5 °C)/°C		64.25	66.6
Ductility (5 cm/min,15 °C)		Brittle	Brittle
Solubility/%		99.6	99.78
density (15 °C)/g/cm3		1.029	1.033
Flash point/°C		280	344
dynamic viscosity/Pa·s		6076	7219
RTFOT	Quality change/%	−0.31	0.051
	Residual penetration ratio/%	78	71.6

Table 3. Maximum deformation and elastic recovery rate.

Asphalt Material	9.75Psi		20Psi	
	Maximum Creep Deflection (mm)	Elastic Recovery Rate (%)	Maximum Creep Deflection (mm)	Elastic Recovery Rate (%)
China 15#	0.0039	82.1	0.0128	61.7
France 15#	0.0046	79.5	0.0104	65.4

2.2. Mix Design

EME-14 and AC-20, two types of typical high-modulus mixture gradations for France and China, suitable for the middle and lower layers, were selected in this study. According to the aggregate screening results and the actual high-modulus road engineering application situation, the design composite gradation was shown in Table 4.

Table 4. Synthetic gradation design of high-modulus mixture.

Particle Diameter (mm)	Mass Percentage of Particles Smaller Than a Certain Particle Diameter (%)				
	EME-14	HMAC-20	Design Upper Limit	Design Lower Limit	Design Median
26.5	100	100	100	100	100
19	100	93.8	100	90	95
16	100	88.4	-	-	-
13.2	99.4	81.3	-	-	-
9.5	90.2	69.2	82	66	74
4.75	48.8	43.6	64	41	52.5
2.36	33.1	29.6	43	28	35.5
1.18	26.4	24	-	-	-
0.6	16.6	16.3	-	-	-
0.3	11.2	12.1	-	-	-
0.15	8.5	7.9	-	-	-
0.075	6.5	6	8	6	7

The asphalt binder dosage is determined by the abundance coefficient K. The abundance coefficient K is a ratio of the conventional thickness of the asphalt film attached to the surface of the aggregate with an asphalt binder [25]. K has nothing to do with the density of gravel; according to the LPC Bituminous Mixtures Design Guide, the amount of asphalt is calculated and estimated by the abundance coefficient K, which satisfies $K \geq 3.4$. The asphalt binder dosage of high-modulus mixture was calculated according to Equations (1)–(3), shown in Table 5.

Table 5. Abundance coefficient calculation table.

Mixture Type	G (%)	S (%)	s (%)	f (%)	ρG	Oil-Stone Ratio/%	Abundance Coefficient
EME-14	37	50.4	4.2	7.7	2.730	5.7	3.46
HMAC-20	58.6	33.2	3.4	4.8	2.730	5.3	3.44

The relationship between the abundance coefficient K and the oil-stone ratio:

$$TL_{ext} = K \times \alpha \sqrt[5]{\sum} \quad (1)$$

$$100\sum = 0.25G + 2.3S + 12s + 150f \quad (2)$$

$$\alpha = \frac{2.65}{\rho G} \quad (3)$$

where:

TL_{ext} is the percentage of admixture (whetstone ratio), %
K is specific surface area, m^2/kg

G is the ratio of aggregates with a particle size greater than 6.3 mm, %
S is the ratio of aggregates with a particle size of 0.25 mm~0.63 mm, %
s is aggregate ratio with a particle size of 0.063 mm~0.25 mm, %
F is the ratio of aggregates with a particle size of less than 0.063 mm, %
α is correlation coefficient related to aggregate density
ρG is density of aggregate, g/cm^3

2.3. *Test Methods*

2.3.1. High-Modulus Mixture Performance Index Verification Test

The performance of four kinds of high-modulus mixtures, EME-14 + France 15#, EME-14 + China 15#, HMAC-20 + France 15#, HMAC-20 + China 15#, under level 1 to level 4, was verified according to the LPC Bituminous Mixtures Design Guide to evaluate whether it meets the performance index requirements of high-modulus mixtures. The test methods were shown in Table 6. Under the premise of meeting the performance index of high-modulus mixture, using the SPT and CRT-2PT performance testing machines, the mixtures were subjected to a uniaxial compression dynamic modulus test and a two-point bending complex modulus test at different temperatures and frequencies. The modulus of the mixture under the two systems was analyzed according to the test schemes as shown in Table 7.

Table 6. High-modulus mixture performance index verification test plan.

Test Level	Test Item	Test Methods
Level 1	Rotational compaction void ratio (%)	EN 12697-31 [26]
	Water stability: Durex test (Compressive strength ratio, %)	EN 12697-12 (Method B) [27]
Level 2	High temperature stability: French wheel rutting test (30,000 times, 60 °C)	EN 12697-22 [28]
Level 3	Stiffness modulus: complex modulus (MPa, 15 °C, 10 Hz/0.02 s)	EN 12697-26 [29]
Level 4	Fatigue life: two-point bending fatigue of trapezoidal beam (10 °C, 25 Hz, 106 times, 130 με)	EN 12697-24 [30]

Table 7. Test schemes of mixture stiffness modulus under different evaluation systems.

Test Parameters	Test Plan	
	Complex Modulus of Trapezoidal Beam	Uniaxial Compression Dynamic Modulus
Test piece size	Height: 250 ± 1 mm, thickness: 25 mm ± 1 mm Top bottom: 25 ± 1 mm, bottom: 56 mm/70 mm ± 1 mm	Diameter: 100 mm, height: 150 mm standard specimen
Strain level	30 με	75–125 με
Test temperature	5, 15, 20, 30, 45 °C	5, 15, 20, 30, 45 °C
Test frequency	1, 5, 10, 15, 20, 25 Hz	0.1, 0.2, 0.5, 1, 2, 5, 10, 20, 25 Hz
Number of parallel tests	6	

2.3.2. Complex Modulus Test

According to the European Design Code [25,29], the CRT-2PT two-point trapezoidal beam bending test machine shown in Figure 1 was used to conduct the complex modulus test, which is realized by the two-point bending loading method. CRT-2PT adopts the strain control mode with sine wave waveform. The 30με strain level, 5 test temperatures of 5, 15, 20, 30, 45 °C and 6 loading frequencies of 1, 5, 10, 15, 20, 25 Hz were selected in the test. At the beginning of the test, the trapezoidal beam specimen was vertically installed and fixed in the temperature control box, and the top of the specimen was subjected to a constant sine wave load and reciprocated in the horizontal direction. At the same time, shear stress and bending moment were generated inside the specimen. The force state of the specimen was shown in Figure 2, close to the force state of the actual pavement.

During the test, the displacement and force changes of the specimen during the horizontal reciprocating movement were recorded by the displacement sensor and the stress sensor, which simulates the bending and tensile deformation of the bottom of the asphalt layer on the road surface to obtain the complex modulus of the mixture.

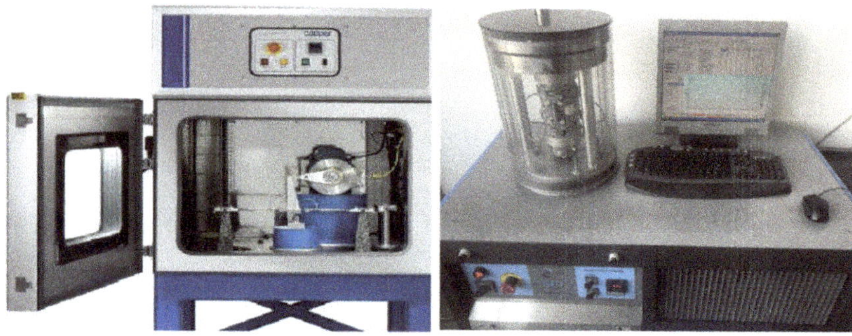

(a) (b)

Figure 1. Schematic diagram of 2PT (**a**) and SPT (**b**) loading equipment.

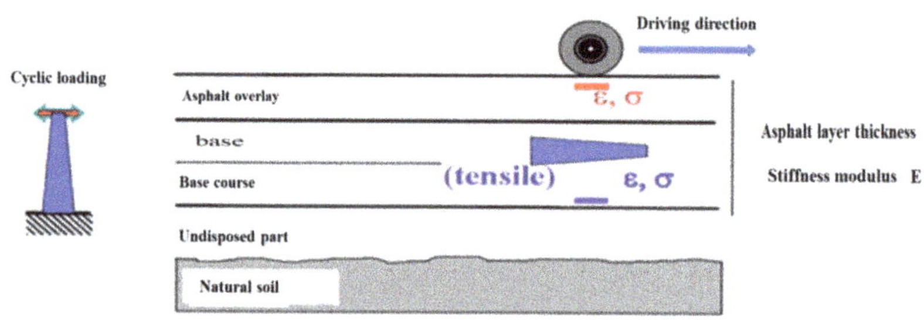

Figure 2. The principle of loading the complex modulus of trapezoidal beam in two-point bending.

During the test, the real and imaginary parts of the complex modulus of the asphalt mixture were calculated according to Equations (4) and (5).

$$E_1 = \gamma \times \frac{F}{z} \times \cos(\phi) \times \frac{\mu}{10^3} \times \omega^2 \tag{4}$$

$$E_2 = \gamma \times \frac{F}{z} \times \sin(\phi) \tag{5}$$

where: γ is shape factor; F is load force; z is displacement of the top of the specimen; μ is quality factor; ω is angular velocity; φ is the phase difference of the top displacement of the specimen/°.

γ and μ is calculated according to Equations (6) and (7).

$$\gamma = \frac{12L^2}{b(h_1 - h_2)} \times \left[(2 - \frac{h_2}{2h_1}) \times \frac{h_2}{h_1} - \frac{3}{2} - Ln\frac{h_2}{h_1} \right] \tag{6}$$

$$\mu = 0.135M + m \tag{7}$$

The meaning of h_1, h_2 and b is shown in Figure 3.

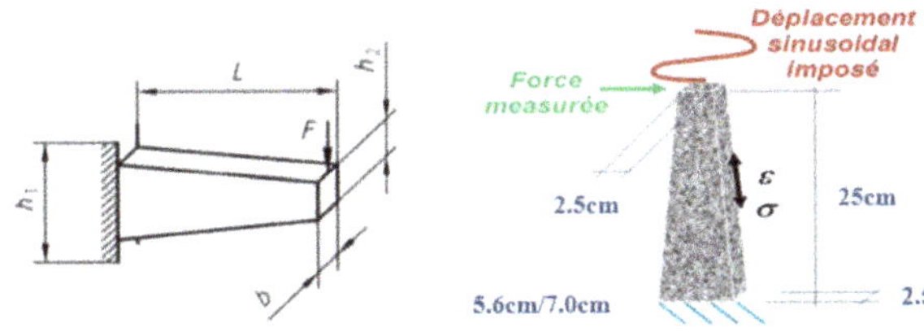

Figure 3. Dimensions of trapezoidal beam specimen.

The modulus and phase angle of the mixture's modulus were calculated according to Equations (8) and (9).

$$|E^*| = \sqrt{E_1^2 + E_2^2} \tag{8}$$

$$\phi = \arctan\frac{E_2}{E_1} \tag{9}$$

2.3.3. Dynamic Modulus Test

The SPT testing machine produced by IPC Global in Australia, as shown in Figure 1, was used to conduct the uniaxial compression dynamic modulus test. It applied axial and sinusoidal periodic loads to a standard cylindrical specimen with a height of 150 mm and a diameter of 100 mm, at different temperature and loading frequency, as shown in Figure 4. SPT adopted the strain control mode with a half-sine wave loading waveform. The SPT testing machine cannot accurately control the strain level, so the strain control value range 75–125 με recommended by the NCHRP9-29 was used in the test. A total of 5 temperatures of 10, 15, 20, 30, 45 °C and 9 loading frequencies of 0.1, 0.2, 0.5, 1, 2, 5, 10, 20, 25 Hz were selected. The sample was tested after keeping it at each temperature for 5 h. The sinusoidal periodic load can be calculated by Equations (10) and (11). By measuring the load and strain acting on the specimen during the last 5 loading cycles, the axial stress and strain amplitude were obtained, and then the axial recoverable deformation of the specimen and the dynamic modulus and phase angle were calculated according to Equations (12)–(14).

$$\sigma(t) = \sigma_0 \sin \omega t \tag{10}$$

$$\omega = 2\pi f \tag{11}$$

where: ω is loading angular frequency; t is time; f is loading frequency.

$$\sigma_0 = \frac{\overline{P}}{A} = 4\frac{\overline{P}}{\pi D^2} \tag{12}$$

where: σ_0 is stress amplitude; $\overline{P}$ is the average load amplitude; A is loading area; D is specimen diameter.

$$\varepsilon_0 = \frac{\Delta\bar{t}}{t_0} \tag{13}$$

where: ε_0 is strain amplitude; $\Delta\bar{t}$ is the average value of the axial deformation amplitude; t_0 is the measurement distance.

$$E_d = \frac{\sigma_0}{\varepsilon_0} \tag{14}$$

where: E_d is dynamic modulus.

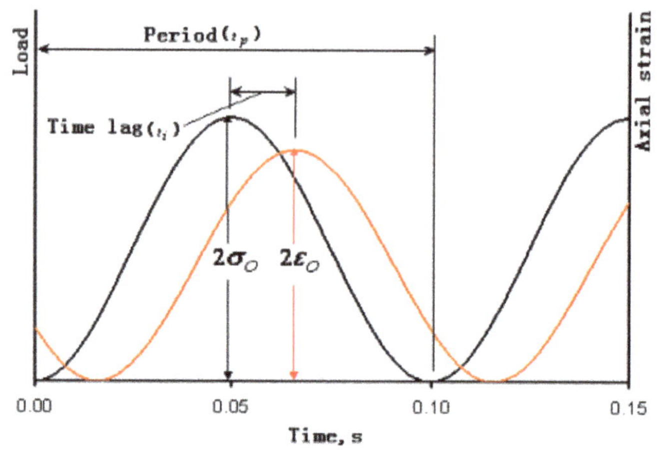

Figure 4. Loading mode of SPT uniaxial compression dynamic modulus test.

3. Test Results and Analysis

3.1. Performance Verification of High-Modulus Mixture

The performance of four types of high-modulus mixtures, EME-14 + France 15#, EME-14 + China 15#, HMAC-20 + France 15#, HMAC-20 + China 15#, under level 1 to level 4, is displayed in Table 8. The designed mixtures all meet the performance index requirements of high-modulus mixtures.

Table 8. Performance test results of level 1 to level 4.

Test Level	Test Item	Test Results				Technical Index Requirements
		China 15#		France 15#		
		EME-14	HMAC-20	EME-14	HMAC-20	
Level 1	Rotational compaction void ratio (%)	3.35	3.82	3.27	3.69	<6%
	Durex test (ITSR, %)	100.23	101.25	101.22	101.92	≥70%
Level 2	French wheel track test (30,000 times, 60 °C, deformation rate)	4.32	5.13	4.17	4.92	≤7.5%
Level 3	Complex modulus (15 °C, 10 Hz, MPa,)	16,219	14,027	16,139	14,333	≥14,000 MPa
Level 4	Two-point bending fatigue of trapezoidal beam (10 °C, 25 Hz, 130 με, times)	1,220,320	1,082,310	1,340,350	1,108,630	$\geq 10^6$

3.2. Comparative Analysis of Modulus

The CTR-2PT trapezoidal beam two-point bending complex modulus test and the SPT uniaxial compression dynamic modulus test were carried out on four kinds of high-modulus mixtures. The modulus and phase angle results of the high-modulus mixture under the two test modes at the main evaluation temperature are shown in Tables 9 and 10. The modulus decreased with the increase of temperature, and increased with the increase of loading frequency. Under the same temperature and loading frequency, the dynamic modulus was higher than the complex modulus. Taking EME-14 + China 15# as an example, the dynamic modulus is 17.6% higher than the complex modulus at 15 °C and 25 Hz loading frequency.

Table 9. Dynamic modulus results of different types of low-mark high-modulus mixtures.

Types	Void Ratio (%)	Loading Frequency (Hz)	Dynamic Modulus (MPa)			Phase Angle (°)		
			15 °C	20 °C	45 °C	15 °C	20 °C	45 °C
EME-14 + China 15#	3.76	25	22,232	18,591	5741	10.41	11.96	28.94
		20	21,631	17,998	5358	10.74	12.39	28.76
		10	19,932	16,340	4279	11.84	13.83	30.28
		5	18,258	14,707	3367	13.17	15.41	31.54
		2	15,973	12,549	2398	15.06	17.73	32.56
		1	14,240	11,009	1838	16.61	19.52	32.33
		0.5	12,570	9496	1402	18.32	21.40	31.71
		0.2	10,442	7694	985	20.66	23.99	30.6
		0.1	8934	6493	762	22.5	25.82	29.07
EME-14 + France 15#	3.82	25	20,456	18,836.5	5888.5	9.33	11.47	29.27
		20	20,117	18,355.5	5450.5	9.78	11.71	29.26
		10	18,771	16,816	4255	11.22	12.965	29.98
		5	17,273	15,267	3141.5	12.54	14.425	30.45
		2	15,254	13,230	2121.5	14.39	16.455	30.97
		1	13,669	11,631.5	1730.5	15.82	18.125	30.47
		0.5	12,116	10,194.5	1315	17.32	19.83	30.04
		0.2	10,163	8385.5	935.1	19.49	22.21	29.22
		0.1	8744	7168.5	734.1	21.12	23.94	27.24
HMAC-20 + China 15#	3.93	25	19,958	17,915	5955	10.99	11.96	28.94
		20	19,065	17,246	5139	11.24	12.18	28.76
		10	17,885	15,747	4261	12.49	13.44	30.28
		5	16,254	14,280	3285	13.86	14.99	31.54
		2	14,071	12,389	2197	15.93	16.98	32.56
		1	12,466	10,797	1597	17.58	18.67	32.33
		0.5	10,944	9445	1244	19.26	20.4	31.71
		0.2	9015	7756	862.1	21.56	22.85	30.6
		0.1	7722	6615	655.2	23.23	24.65	29.07
HMAC-20 + France 15#	3.76	25	19,675	17,374	6122	11.29	12.67	28.88
		20	18,865	16,940	5612	12.21	13.07	29.05
		10	17,227	15,382	4349	13.17	14.42	30.93
		5	15,643	13,835	3478	14.31	15.91	31.72
		2	13,552	11,906	2046	15.98	17.91	32.76
		1	12,052	10,478	1804	17.26	19.49	32.59
		0.5	10,620	9112	1386	18.57	21.16	31.16
		0.2	8829	7426	1008	20.38	23.34	30.29
		0.1	7604	6306	813	22.68	24.78	29.82

3.3. *Master Curve Analysis*

The master curve can fully and intuitively reflect the mechanical properties of the asphalt mixture [31–33]. When estimating the permanent deformation and fatigue life of an asphalt pavement, the modulus and phase angle of the asphalt mixture under different temperature and loading frequency conditions are usually considered. According to the data in Tables 9 and 10, the modulus master curve and phase angle master curve of four high-modulus mixtures at standard temperature, which is 15 °C for the trapezoidal beam and 20 °C for the SPT, were established by using the principle of time-temperature equivalence. The main curves of the modulus and phase angle for the trapezoidal beam complex modulus test and the SPT dynamic modulus test were shown in Figures 5 and 6, respectively.

The overall change trend of the modulus master curve obtained by the two test methods is consistent, but the frequency domain of the SPT dynamic modulus master curve is wider, which is caused by the test loading frequency range of the two. The minimum loading frequency of the SPT test was 0.01 Hz, while that of the trapezoidal beam test was 5 Hz. In contrast, the main modulus curve obtained by the trapezoidal beam does not extend to the low frequency region.

Table 10. Complex modulus results of different types of low-mark high-modulus mixtures at 15 °C.

Types	Void Ratio (%)	Loading Frequency (Hz)	Complex Modulus (MPa)	Phase Angle (°)
EME-14 + China 15#	4.12	25	18,300	9.189264
		20	17,684	9.547917
		15	16,848	10.24761
		10	16,219	10.40017
		5	15,001	11.44451
		1	12,085	11.96734
EME-14 + France 15#	3.93	25	17,226	8.582561
		20	17,064	9.387128
		15	16,550	9.942839
		10	16,139	10.0863
		5	15,010	10.78818
		1	13,124	11.56282
HMAC-20 + China 15#	4.07	25	15,956	9.218034
		20	14,842	10.34659
		15	14,329	11.08765
		10	14,027	10.79887
		5	12,929	11.81575
		1	10,284	12.015642
HMAC-20 + France 15#	4.15	25	16,228	9.977964
		20	15,725	9.873416
		15	14,835	10.23612
		10	14,333	9.987529
		5	13,255	10.85202
		1	10,990	11.36823

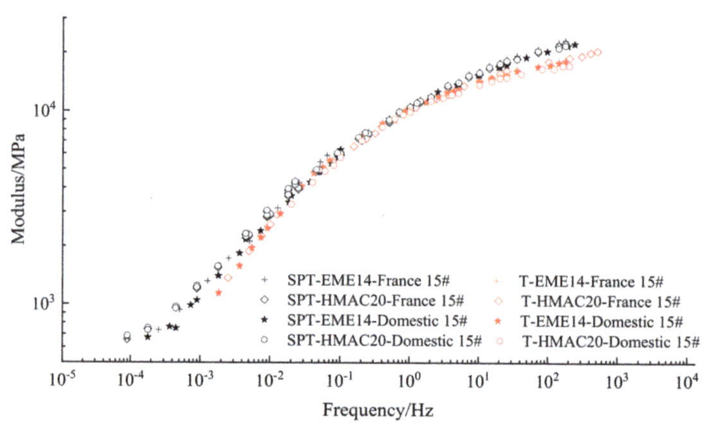

Figure 5. Modulus master curves of high-modulus mixtures with different test method.

The SPT modulus master curve established based on the respective standard temperature translation is higher than the CRT-2PT complex modulus master curve, which shows that the dynamic modulus obtained by the SPT test is greater than the complex modulus obtained by the trapezoidal beam test. It is mainly due to the fact that the compressed mode was used in the SPT test, while the bending-tension mode was used in the CRT-2PT trapezoidal beam complex modulus test.

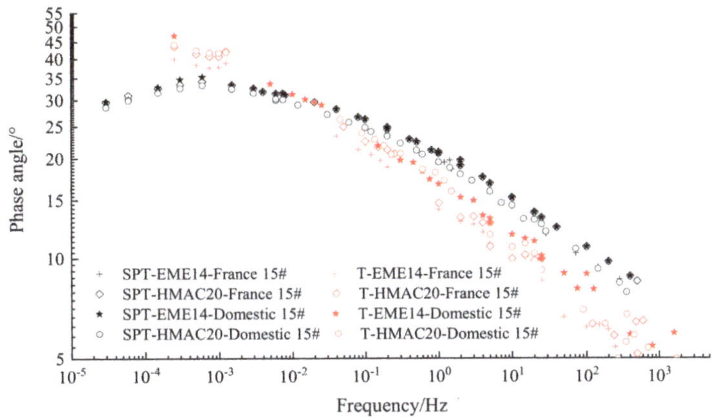

Figure 6. Phase angle master curves of high-modulus mixtures with different test method.

The specimen was continuously compacted under the load in the SPT test, and became increasingly denser without damage, causing the obtained modulus value to be too large. However, the specimen was in the state of bending-tension, and the load effect is unfavorable to the specimen, causing the measured modulus value to be too small [34–37].

Compared with the main curve of the modulus, the main curve of the phase angle of the mixture under the two loading modes is also different; however, the overall change trend is the same. The phase angle of the mixture decreases with the increase of frequency. The phase angle data is more discrete, and the main curve fitting result is not ideal. This was caused by the test equipment, the limitation of the loading frequency range of the two-point trapezoidal beam bending test machine, the molding of the test piece, and the test operation.

3.4. Conversion Relation Model between Complex Modulus and Dynamic Modulus

In the French high-modulus mixture design and evaluation system, the stiffness modulus index refers to the complex modulus of the trapezoidal beam under the conditions of 15 °C and 10 Hz. However, in the design of the Chinese high-modulus mixture, the dynamic modulus of uniaxial compression under the conditions of 15 °C and 10 Hz, 20 °C and 10 Hz, 45 °C and 10 Hz is often used for design and evaluation. Combined with the China high-modulus mixture modulus parameter conditions, under the premise that the performance index of the mixture meets the complex modulus index requirement of the French high-modulus mixture, the modulus conversion model between the complex modulus of the CRT-2PT trapezoidal beam and the dynamic modulus of the SPT uniaxial compression was established, as shown in Figures 7–10. When the complex modulus of the mixture trapezoidal beam is 14,000 MPa, the dynamic modulus value under different temperature conditions was calculated according to the conversion model. According to the calculated results, the value range of the dynamic modulus under the condition modulus parameter in China was recommended. The results are shown in Table 11.

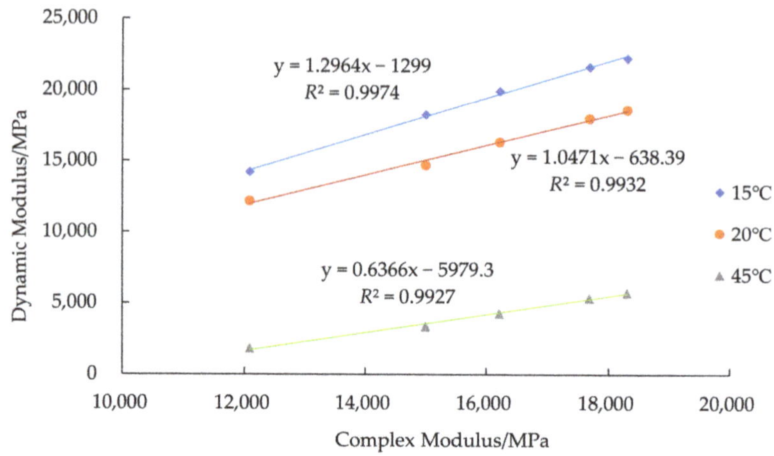

Figure 7. EME-14-China 15#.

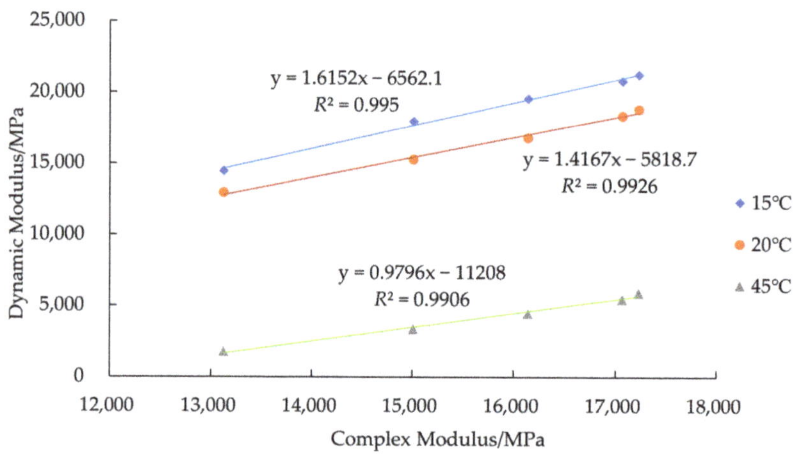

Figure 8. EME-14- France 15#.

It can be seen from Figures 5–8 that the complex modulus of different types of high-modulus mixtures was positively correlated with the dynamic modulus, and the correlation coefficients of the fitted linear equations were all above 0.99, showing a good correlation between the modulus obtained by these two test methods.

Different types of high-modulus mixtures have different dynamic modulus evaluation values according to the established conversion model.

Taking the error of the fitting formula or other factors into account, it is recommended when using the dynamic modulus parameter to evaluate high-modulus mixtures that the dynamic modulus at 15 °C and 10 Hz is not less than 16,000 MPa, the dynamic modulus at 20 °C and 10 Hz is not less than 14,000 MPa and the dynamic modulus at 45 °C and 10 Hz is not less than 2500 MPa, as shown in Table 11.

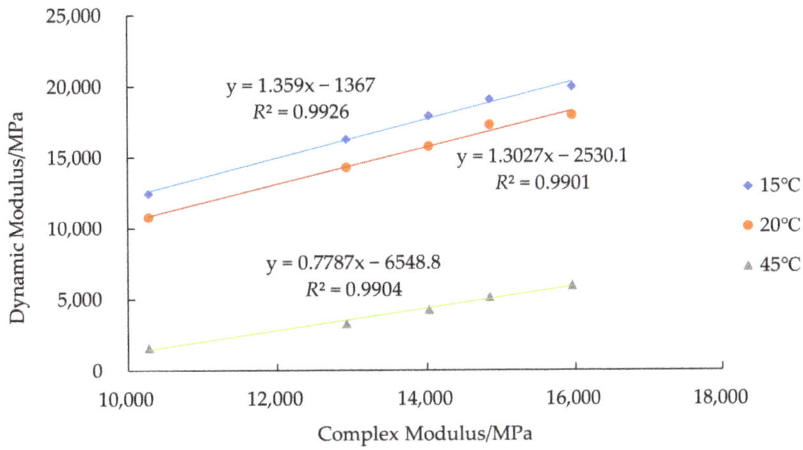

Figure 9. HMAC-20-China 15#.

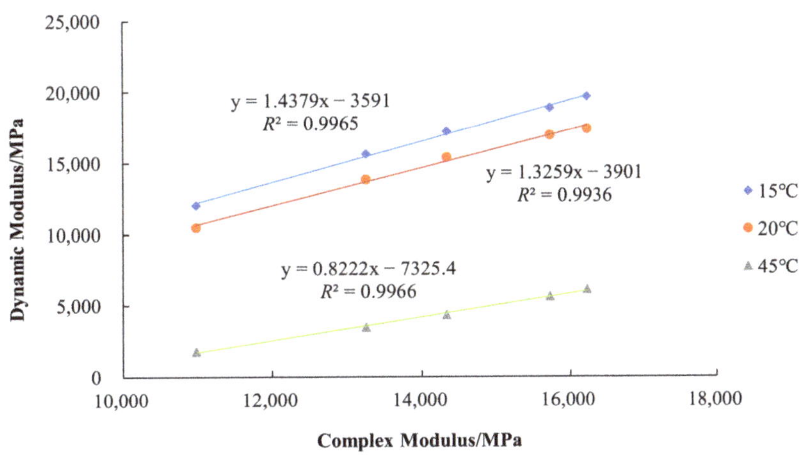

Figure 10. HMAC-20-France 15#.

3.5. Determination of Dynamic Modulus Evaluation Range of High-modulus Mixture

In order to verify the reliability of the recommended dynamic modulus evaluation value based on the modulus conversion model, 5 kinds of high-modulus mixtures used in actual road engineering, AC-20-SBS+PR MODULE, AC-20-SBS+EME high modulus modifier, AC-20-70#+PR MODULE, AC-20-70#+EME high modulus modifier and AC -20+high-modulus were carried out using the CRT-2PT trapezoidal beam two-point bending complex modulus and the SPT uniaxial compression dynamic modulus test. The relationship between the complex modulus at 15 °C,10 Hz and the corresponding uniaxial compression dynamic modulus at 15 °C, 10 Hz, 20 °C, 10 Hz and 45 °C, 10 Hz were shown in Figures 11–13.

Table 11. High-modulus mixture uniaxial compression dynamic modulus evaluation index under different temperature conditions.

Mixture Type	Test Conditions	Conversion Model	Calculated	Recommended Value
EME-14 + China 15#	15 °C, 10 Hz	y = 1.2964x − 1299	16,850	≥16,000
EME-14 + France 15#		y = 1.6152x − 6562.1	16,050	
HMAC-20 + China 15#		y = 1.359x − 1367	17,659	
HMAC-20 + France 15#		y = 1.4379x − 3591	16,539	
EME-14 + China 15#	20 °C, 10 Hz	y = 1.0471x − 638.39	14,021	≥14,000
EME-14 + France 15#		y = 1.4167x − 5818.7	14,015	
HMAC-20 + China 15#		y = 1.3027x − 2530.1	15,707	
HMAC-20 + France 15#		y = 1.3259x − 3901	14,661	
EME-14 + China 15#	45 °C, 10 Hz	y = 0.6366x − 5979.3	2933	≥2500
EME-14 + France 15#		y = 0.9796x − 11208	2506	
HMAC-20 + China 15#		y = 0.7787x − 6548.8	4353	
HMAC-20 + France 15#		y = 0.8222x − 7325.4	4185	

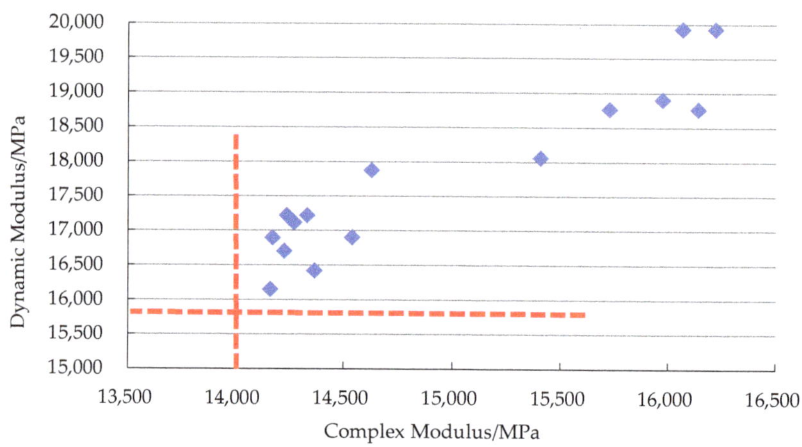

Figure 11. Corresponding relationship of modulus at 15 °C, 10 Hz.

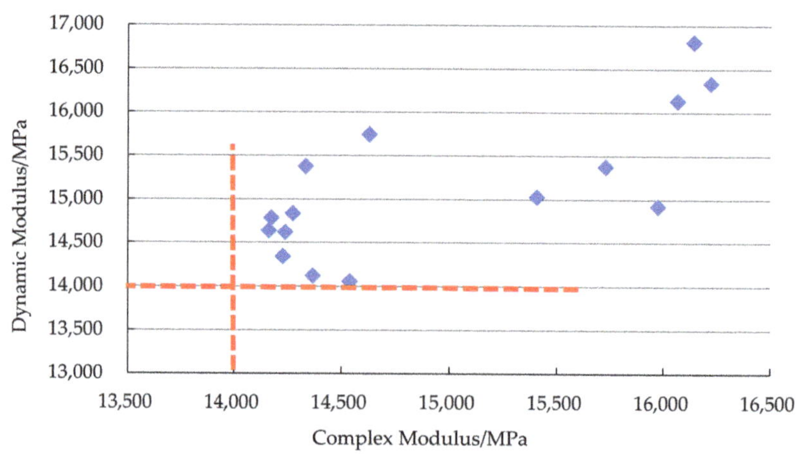

Figure 12. Corresponding relationship of modulus at 20 °C, 10 Hz.

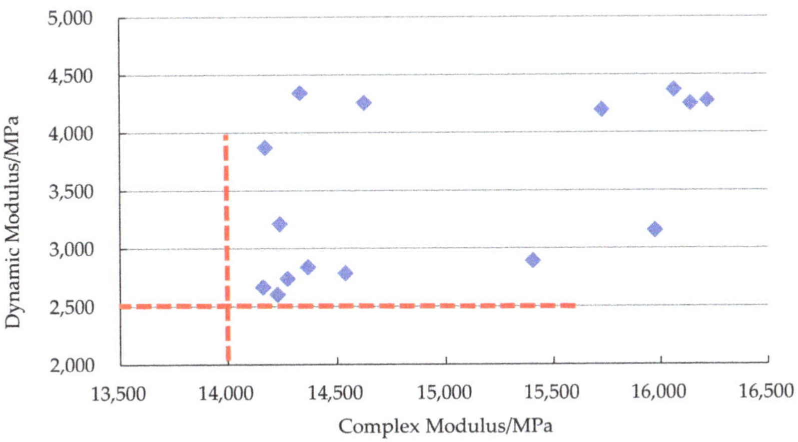

Figure 13. Corresponding relationship of modulus at 45 °C, 10 Hz.

It can be seen that, under the premise that the complex modulus of the trapezoidal beam is ≥14,000 Mpa, different types of high-modulus mixtures have a corresponding dynamic modulus ≥16,000 MPa at 15 °C,10 Hz, ≥14,000 MPa at 20 °C,10 Hz and ≥2500 Mpa. at 45 °C,10 Hz. The test results were consistent with the recommended dynamic modulus evaluation values calculated by the conversion model.

4. Conclusions

1. The performance of four types of high-modulus mixtures, EME-14 + France 15#, EME-14 + China 15#, HMAC-20 + France 15#, and HMAC-20 + China 15#, under level 1 to level 4, was verified; all met the performance index requirements of high-modulus mixtures.
2. Different loading control modes have a greater impact on the modulus and phase angle of the mixture. The frequency domain threshold of the master curve obtained by the SPT test is wider, the modulus value is larger, and the phase angle data dispersion is smaller than the CRT-2PT trapezoidal beam complex modulus test.
3. The relationship conversion model between dynamic modulus and complex modulus was established with good correlation. Based on the established correlation model, the dynamic modulus evaluation range of high-modulus mixtures under different evaluation temperature conditions was recommended.
4. Combined with the application of high-modulus mixtures in actual road engineering, the reliability of the recommended dynamic modulus range based on the calculation of the conversion model has been properly verified. The test results are all within the range of the dynamic modulus evaluation value given by the model.

Author Contributions: L.G., Q.X., W.W. and X.Z. conceived and designed the experiments. Q.X, X.Y. and J.W. performed the experiments. L.G., G.Z., W.W. and M.Z. analyzed the data. L.G., Q.X. and M.Z. contributed reagents/materials/analysis tools. Q.X. and W.W. wrote the paper. Q.X., W.W. and X.Z. revised the paper. All authors have read and agreed to the published version of the manuscript.

Funding: This research was funded by the National Key R & D Program of China, grant No.2018YFB1600103, the National Natural Science Foundation of China, grant No. 42107213, Shandong Provincial Natural Science and Foundation, grant No. ZR2020QE271 and Shandong Provincial Key Research and Development Program, grant No. 2019GSF109020.

Institutional Review Board Statement: Not applicable.

Informed Consent Statement: Not applicable.

Data Availability Statement: Data sharing is not applicable to this article.

Conflicts of Interest: The authors declare no conflict of interest.

References

1. *NF P98-141: 1999 Asphalt Mixture-Wearing Course and Base Course: High Modulus Asphalt Mixture-Definition-Classification-Characteristics-Production and Construction*; French Standardization Association: Paris, France, 1999.
2. Xia, X.P.; Zhang, R.N.; Fu, H.W. Characteristics and new development of high modulus asphalt mixture. *Chin. Foreign Highw.* **2005**, *2*, 123–127.
3. Serfass, P.; Bense, P.; Pellevoisin, P. Properties and New Developments of High Modulus Asphalt Concrete. In Proceedings of the Lecture Series 8th International Conference on Asphalt Pavements, Seattle, WA, USA, 10–14 August 1997.
4. Corte, J.F. Development and uses of hard grade asphalt and of high modulus asphalt mixes in France. *Transp. Res. Circ.* **2001**, *503*, 12–31.
5. Zheng, J. Research on Durable High Modulus Asphalt Pavement Structure. Master's Thesis, Chongqing Jiaotong University, Chongqing, China, 2013.
6. Olidid, C.; Hein, D. Guide for Mechanistic-Empirical Design of New and Rehabilitated Pavement Structures. In Proceedings of the 2004 Annual Conference and Exhibition of the Transportation Association of Canada-Transportation Innovation-Accelerating the Pace, Québec City, QC, Canada, 19–22 September 2004.
7. Zhou, F.; Scullion, T. Preliminary Field Validation of Simple Performance Tests for Permanent Deformation: Case Study. *Transp. Res. Rec.* **2003**, *1832*, 209–216. [CrossRef]
8. Zhou, Q.H.; Sha, A.M.; Yang, Q. Experimental study on mechanical properties of high modulus asphalt concrete. *J. Zhengzhou Univ.* **2008**, *3*, 128–131.
9. Witczak, M. *Simple Performance Tests: Summary of Recommended Methods and Database*; NCHRP-547; Transportation Research Board: Washington, DC, USA, 2005; pp. 2–7.
10. Lee, H.J.; Lee, J.H.; Park, H.M. Performance evaluation of high modulus asphalt nlixtllres for long life asphalt pavements. *Constr. Build. Mater.* **2007**, *21*, 1079–1087. [CrossRef]
11. Arnold, G.; Darcy, R.; Hall, S.; Mudgway, G. High Modulus Asphalt to Prevent Rutting at Intersections. In Proceedings of the 17th AAPA International Flexible Pavements Conference, Victoria, Australia, 13–16 August 2017; pp. 1–11.
12. Amjad, H.K.; Lateif, R.H. Evaluating the Performance of High Modulus Asphalt Concrete Mixture for Base Course in Iraq. *J. Eng.* **2017**, *23*, 14–33.
13. Xu, J.; Zhao, Y.; Liang, N. Life prediction of high modulus asphalt pavement based on fatigue cumulative damage. *J. Chang'an Univ. Nat. Sci. Ed.* **2017**, *38*, 26–33.
14. France RST "Asphalt Mixture Design" Working Group. *LPC Bituminous Mixtures Design Guide*; Laboratoire Central des Ponts et Chaussées: Paris, France, 2010.
15. Wang, X.D.; Sha, A.M.; Xu, Z.H. *Dynamics Characteristics and Dynamic Parameters of Asphalt Pavement Materials*; China Communications Press: Beijing, China, 2002; pp. 102–106.
16. Zou, H.Z. Experimental research on dynamic modulus of asphalt mixture. Master's Thesis, Chang'an University, Chang'an, China, 2013.
17. You, Y.J. Research on the dynamic modulus of asphalt mixture. Master's Thesis, Shandong Jianzhu University, Jinan, China, 2017
18. *GB/T 36143-2018 High Modulus Anti-Fatigue Asphalt Mixture for Road*; People's Communications Press: Beijing, China, 2018.
19. *DB 13/T 2823-2018 Technical Guidelines for Highway High Modulus Asphalt Pavement Construction*; Hebei Provincial Bureau of Quality and Technical Supervision: Hebei, China, 2018.
20. *DB 21/T1754-2009 Construction Specification for High Modulus Asphalt Mixtures*; Renmin Communication Press: Shenyang, China, 2009.
21. *JTG F40-2004 Technical Specifications for Construction of Highway Asphalt Pavements*; People's Communications Press: Beijing, China, 2005.
22. Huang, Y. Experimental Research on Dynamic Modulus of Asphalt Mixture Trapezoidal Beam. Master's Thesis, Changsha University of Science and Technology, Changsha, China, 2015.
23. Huang, Y.; Liu, Z.H.; Wang, X.D.; Li, S. Comparative study of dynamic modulus trapezoidal beam and SPT test of asphalt mixture. *J. Cent. South Univ.* **2017**, *48*, 3092–3099.
24. Huang, Y.; Wang, X.D.; Liu, Z.H.; Zhou, X.Y.; Zhang, L. Experimental study on dynamic modulus trapezoidal beam of SAC13 asphalt mixture. *Highw. Transp. Sci. Technol.* **2017**, *34*, 7–14.
25. Corté, J.-F.; Serfass, J.-P. The French approach to asphalt mixtures design: A performance-related system of specifications. *Assoc. Asph. Paving Technol.* **2000**, *69*, 794–834.
26. *CSN EN 12697-31 Bituminous Mixtures-Test Methods for Hot Mix Asphalt-Part 31: Specimen Preparation by Gyratory Compactor*; The Commonwealth Standards Network: London, UK, 2007.

27. *CSN EN 12697-12 Bituminous Mixtures-Test Methods for Hot Mix Asphalt Part 12: Determination of the Water Sensitivity of Bituminous Specimens*; The Commonwealth Standards Network: London, UK, 2008.
28. *CSN EN 12697-22 Bituminous Mixtures-Test Methods for Hot Mix Asphalt-Part 22: Wheel Tracking*; The Commonwealth Standards Network: London, UK, 2020.
29. *CSN EN 12697-26 Bituminous Mixtures-Test Methods for Hot Mix Asphalt-Part 26: Stiffness*; The Commonwealth Standards Network: London, UK, 2003.
30. *CSN EN 12697-24 Bituminous Mixtures-Test Methods for Hot Mix Asphalt Part 24: Resistance to Fatigue*; The Commonwealth Standards Network: London, UK, 2012.
31. Zhao, Y.Q.; Wu, J.; Wen, J. Determination and analysis of dynamic modulus of asphalt mixture and its master curve. *Highway* **2006**, *8*, 163–166.
32. Ma, X.; Ni, F.J.; Chen, Y.S. Dynamic modulus test of asphalt mixture and prediction model. *Chin. J. Highw. Transp.* **2008**, *21*, 35–39.
33. Luo, S.; Qian, Z.D.; Harvey, J. Research on dynamic modulus for epoxy asphalt mixtures and its master curve. *Chin. J. Highw. Transp.* **2010**, *23*, 16–20.
34. Ma, L.; Zhang, X.N. Comparison of HMA dynamic moduli between indirect tension and uniaxial compression test modes. *J. Highw. Transp. Res. Dev.* **2009**, *26*, 11–17.
35. Li, Q.; Li, G.F.; Wang, H.C. Effect of loading modes on dynamic moduli of asphalt mixtures. *J. Build. Mater.* **2014**, *17*, 816–822.
36. Liu, Y. *Research on Dynamic Response and Fracture Performance Based of Asphalt Mixture on Semi-Circular Bending Test*; Harbin Institute of Technology, School of Transportation Science and Engineering: Harbin, China, 2009; pp. 3–7.
37. Qian, Z.D.; Yang, Y.N.; Chen, T.J. Dynamic response of asphalt pavement under moving loads with low and variable speed. *J. Cent. South Univ.* **2015**, *46*, 1140–1146.

coatings

MDPI

Article

Mechanism of Polyurethane Binder Curing Reaction and Evaluation of Polyurethane Mixture Properties

Min Sun [1,*], Yufeng Bi [2,*], Wei Zhuang [2], Sai Chen [2], Pinhui Zhao [1,3,*], Dezheng Pang [4] and Wensheng Zhang [4]

[1] School of Transportation Engineering, Shandong Jianzhu University, Jinan 250101, China
[2] Shandong Provincial Communications Planning and Design Institute Co., Ltd., Jinan 250031, China; zhuangweijtky@163.com (W.Z.); chensaiyile@hotmail.com (S.C.)
[3] Shandong Runxingcheng Road Construction Materials Research and Development Center, Yantai 264000, China
[4] Wanhua Chemical Co., Ltd., Yantai 265505, China; dzpang@whchem.com (D.P.); wszhangb@whchem.com (W.Z.)
* Correspondence: sunmin20@sdjzu.edu.cn or 15253170143@163.com (M.S.); biyf@163.com or 18866130036@163.com (Y.B.); zhaopinhui08@163.com or 13671@sdjzu.edu.cn (P.Z.)

Abstract: This study focuses on analyzing the curing reaction mechanism of polyurethane (PU) binders and comprehensively evaluating the PU mixture's properties. The former was investigated by conducting a Fourier transform infrared spectroscopy (FTIR) test on PU binders with different curing times. The volume change characteristics, construction operation time, and strength formation law were clarified through the splitting tensile test of PU mixtures under different environmental conditions. The optimal PU mixture stacking time and curing time under different environmental conditions were determined. The properties of the PU mixture and asphalt mixture were evaluated and compared through a rutting test, low-temperature bending test, freeze–thaw splitting test, and four-point bending fatigue test. The results show that the physical and chemical curing of the PU binder occurred within the first 24 h of curing, and the reaction speed gradually accelerated to form a polyurea structure 24 h later. It is recommended to stack the PU mixture for 4 h before compaction and to cure it for 2 days before opening under the conditions of 50% humidity and 15–40 °C surrounding temperature. The PU mixture shows better temperature stability and fatigue resistance than the asphalt mixture, and the splitting tensile strength of the PU mixture before and after the freeze–thaw splitting test is also higher. It is clear that the PU mixture is a green road building material with good performance.

Keywords: PU binder; curing reaction mechanism; PU mixture; properties; stacking time; curing time

Citation: Sun, M.; Bi, Y.; Zhuang, W.; Chen, S.; Zhao, P.; Pang, D.; Zhang, W. Mechanism of Polyurethane Binder Curing Reaction and Evaluation of Polyurethane Mixture Properties. *Coatings* **2021**, *11*, 1454. https://doi.org/10.3390/coatings11121454

Academic Editor: Fengwei (David) Xie

Received: 10 November 2021
Accepted: 24 November 2021
Published: 26 November 2021

Publisher's Note: MDPI stays neutral with regard to jurisdictional claims in published maps and institutional affiliations.

Copyright: © 2021 by the authors. Licensee MDPI, Basel, Switzerland. This article is an open access article distributed under the terms and conditions of the Creative Commons Attribution (CC BY) license (https://creativecommons.org/licenses/by/4.0/).

1. Introduction

Asphalt pavement is widely used owing to its low cost, in addition to its characteristics of low noise and good driving quality; nearly 90% of the roads worldwide comprise asphalt mixtures [1]. In recent years, severe loads and extreme temperatures have occurred frequently; therefore, special roads—such as long and wide longitudinal slope highways, port area highways, port special roads, heavy-duty bridge deck pavement, and airport runways—have higher requirements for pavement materials [2]. However, asphalt is a temperature-sensitive material, which makes asphalt mixtures prone to rutting, cracks, water damage, and other forms of damage that can affect their service life [2]. Furthermore, hot-mix asphalt mixtures consume significant amounts of fuel to heat mineral aggregates and asphalt during the production process [3]. A large amount of carbon dioxide, sulfur dioxide, PM2.5 dust, and smoke are also emitted into the atmosphere during the construction process, which imposes a significant burden on the surrounding environment [1,3]. Therefore, new energy-saving and emission-reducing green pavement construction materials are required for the aforementioned special types of road [4]. To mitigate these problems, scholars have proposed the concept of polyurethane (PU) mixture pavement, where high molecular weight polymer PU is used to replace asphalt as the binder for pavements [5]. Since the end of the 1970s, researchers have begun to use one- or two-component PU to replace asphalt as the binder for functional pavement [6–23], pavement structural layers [24–26], bridge deck pavement [27], etc.

The PU mixture can be prepared, transported, and compacted at room temperature, save energy and emission, and has good temperature stability and durability, which can greatly reduce the frequency and cost of road maintenance. The active end –NCO group of the PU binder reacts with the water adsorbed on the substrate surface and the active hydrogen, such as –OH species, on the surface to form urea bonds and promote cross-linking and curing, gradually increasing the strength [28]. During the formation of the PU mixture, the environmental conditions affect the curing reaction of the PU binder [29]. When the ambient temperature is high, the functional group reaction is vigorous, and the opening time of the PU binder is short, which affects the construction operable time of the mixture [30]. When the ambient humidity is high, the curing reaction rate is accelerated, and carbon dioxide (CO_2) is released during the reaction which leads to PU mixture volume expansion [30]. After the formation of the PU mixture, water immersion can easily cause the plasticization and hydrolysis of PU, and the temperature will further accelerate its oxidation, plasticization, and hydrolysis [31]. Wang et al. [5,17,18] determined the strength and modulus of a PU macadam mixture based on the cube unconfined compression test and rectangular beam bending test, investigating the strength characteristics and failure mechanisms of porous PU mixtures (PPMs) based on the load stress test. Wang et al. designed porous elastic road surfaces (PERSs) based on PU binders, which not only showed good noise-reduction performance, but also high low-temperature tensile performance, polishing resistance, and rutting resistance [7,8]. Cong et al. [4] analyzed the impact of water immersion damage on the mixture and found that, in comparison to traditional porous asphalt mixtures, an open-graded mixture designed with PU instead of asphalt can obtain a higher effective void ratio (greater than 20%). The mechanical, functional, and mesoscopic properties of a high-performance PU bounded pervious mixture (PUPM) were further clarified by Li et al., who found that PUPM exhibits excellent ant-stripping performance, and has a larger equivalent radius, effective void fraction, and tortuosity [12]. Hong et al. [32] studied the evolution law of the performance of the PU binder under ultraviolet (UV) aging by using a UV aging chamber to simulate the natural aging process; the tensile properties and viscoelasticity were taken as the evaluation indexes, and the PU dense-graded concrete prepared using the vacuum-assisted resin transfer molding technology was evaluated [33]. However, current research mainly focuses on the mixture composition and performance evaluation of PU mixtures [34]. Research on the main reaction types and key group transformations of different PU binder curing reaction stages has not been conducted in-depth. The strength formation mechanism of polyurethane mixture is not clear, which affects the determination of composition design index of polyurethane mixture, thus the mixture with the best performance cannot be obtained. Furthermore, the influence law of environmental factors on the strength formation law and volume change characteristics of PU mixtures needs to be further investigated. As the basis for guiding the connection time of mixing, paving, compaction, and other processes in the construction of polyurethane mixture, guide the construction of polyurethane mixture pavement. The temperature stability, water stability, and fatigue resistance of PU mixtures require more accurate quantitative characterization to comprehensively evaluate the specific applicable working conditions.

Therefore, this study aims to explore the curing reaction mechanism, strength formation law, and performance of PU mixtures. The curing reaction mechanism of PU binders was detected using Fourier transform infrared spectroscopy (FTIR). The strength formation law was clarified using the splitting tensile test of PU mixtures under different environmental conditions, and the curing and stacking time under these conditions was proposed. The properties of the PU mixture were evaluated through the rutting test, low-temperature bending test, freeze–thaw splitting test, and four-point bending fatigue test, etc. The PU binder curing mechanism and PU mixture strength formation mechanism provide basis for the composition design and construction process arrangement of PU mixture. The performance evaluation results of PU mixture help to determine the construction conditions of PU mixture and recommend PU pavement structure. This study's results provide a new idea for the research and development of high-performance road materials, and is conducive to the development of green construction technology in road engineering.

2. Materials and Methodology

2.1. Materials

2.1.1. Raw Material

The materials used herein mainly included PU, 70# matrix asphalt, styrene butadiene styrene (SBS) modified asphalt, and aggregates. The mineral aggregates were 0–3 mm, 3–5 mm, 5–10 mm, and 10–15 mm basalt aggregates and limestone mineral powder. The SBS modified asphalt and 70# matrix asphalt were produced by Shandong Huarui Road Materials Co., Ltd. (Zibo, China). The PU binder was a linear organic polymer embedded in soft and hard segments, produced by Wanhua Chemical Co., Ltd. (Yantai, China). The soft segment of PU was composed of polyol, and the hard segment was composed of the diphenylmethanc-4,4′-diisocyanate (MDI), carbamate, and urea groups. The specific technical indexes of PU binders are given in Table 1.

Table 1. Technical indexes of PU binder.

Technical Indicators		Unit	Technical Requirement
Surface drying time		min	40 ± 10
Tensile strength		MPa	≥15.0
Fracture elongation		%	≥100
Molecular weight		/	13,000–17,000
UV aging	Tensile strength	MPa	≥9.0
	Fracture elongation	%	≥40

2.1.2. Mixture Composition and Production

The composition design results for the PU mixture (SPU-13), SBS modified asphalt mixture with a skeleton dense structure (SMA-13), and suspended dense structure matrix asphalt mixture (AC-13) are summarized in Table 2.

Table 2. Mineral aggregate gradation and binder content of mixtures.

Sieve size (mm)	Cumulative Passing Percentage of Each Sieve (%)		
	SMA-13	AC-13	SPU-13
16	100	100	100
13.2	81.8	99.4	95.9
9.5	61.2	88.1	67.8
4.75	24.2	74.1	31
2.36	19.9	41.9	21.4
1.18	16.8	33.5	16.7
0.6	14.5	24.5	11.5
0.3	12.4	17.9	8.6
0.15	10.9	11.7	6.1
0.075	10.1	7.7	2.4
Binder Content (%)	5.9	4.8	4.4

The preparation of AC-13 and SMA-13 mixtures was carried out in accordance with the methods specified in "Standard Test Methods of Bitumen and Bituminous Mixtures for Highway Engineering" (JTG E20-2011) [35]. The SPU-13 mixture was prepared using an asphalt mixture mixer (BH-20, Cangzhou Huayun Experimental Instrument Co., Ltd.

Cangzhou, China); however, the aggregates and binder were not heated before mixing. First, the specified proportion of aggregates were mixed in the mixing pot for 20–30 s. Next, the corresponding proportion of PU binder was added and mixed for 30–50 s. Finally, the corresponding proportion of mineral powder was added and mixed for 30–50 s, and the smoothness of the mixture was checked.

2.2. Methodology

2.2.1. Fourier Transform Infrared Spectroscopy Test for PU Binder

The EQUINOX-55 FTIR spectrometer (Bruker, Karlsruhe, Germany) was used to characterize the original, curing, and cured PU binder, and to identify chemical groups according to ASTM D5477. The PU binder was dried in a 120 °C vacuum oven (VO-60T, Shanghai Dengsheng Instrument Manufacturing Co., Ltd., Shanghai, China) for 1 h. The PU binder was coated on KBr sheet to prepare FTIR specimens. The specimens were cured at 20 °C and 50% ambient humidity for 0 h, 2 h, 4 h, 24 h, 72 h, and 168 h, following which FTIR tests were conducted. The scanning range was 4000–400 cm^{-1} and the resolution was 4 cm^{-1}, with 32 scanning iterations.

2.2.2. Basic Performance Test Method for the Mixtures

The basic performance test of the mixtures was conducted in accordance with the methods specified in "Standard Test Methods of Bitumen and Bituminous Mixtures for Highway Engineering" (JTG E20-2011).

(1) Marshall specimen height test

The PU mixtures were stacked under different environmental conditions and were then compacted 50 times on each side to form a Marshall test piece. Then, the height of the test piece after curing for 3 h, 6 h, 12 h, 24 h, 36 h, 48 h, 72 h, and 168 h was measured. The stacking conditions and serial numbers are shown in Table 3.

Table 3. Stacking conditions and serial numbers of PU mixtures.

Temperature (°C)	Humidity (%)	Stacking Time (h)	Serial Number
20	0.5	0	SPU-1
20	0.5	4	SPU-2
20	0.5	8	SPU-3
20	0.95	4	SPU-4
40	0.5	4	SPU-5

(2) Splitting tensile test

The splitting tensile tests of the PU mixtures were carried out according to the provisions of T 0716-2011. The loading rate was 50 mm/min. There were three specimens in each group, and the average value was taken as the splitting strength.

(3) Rutting test

The rutting tests of the three mixtures were carried out according to the methods of T0719-2000. Each group had three parallel specimens with a size of 15 cm × 15 cm × 5 cm, with a test temperature of 60 °C, pressure conditions of 0.7 MPa, and a traveling speed of 42 cycles/min.

(4) Low-temperature bending test

To characterize the low-temperature crack resistance of the three mixtures, the bending tests were carried out in accordance with regulations of T0715-2011. Each group had three parallel specimens with a size of 250 mm × 30 mm × 35 mm [35], a loading speed of 50 mm/min, and a test temperature of −10 °C.

(5) Freeze–thaw splitting test

To characterize the water stability of the three mixtures, freeze–thaw splitting tests were carried out according to the methods in T0729-2000, and each group had three parallel specimens [35].

(6) Four-point bending fatigue test

Under the action of a standard load, the maximum tensile strain at the bottom of the pavement was between 300–600 με. As an excellent pavement building material, PU mixtures have excellent fatigue resistance and self-healing properties; however, they are expensive [7,8]. Their use is suitable for special roads such as long and long longitudinal slope pavement, heavy load pavement, and long-life pavement [12–15]. The maximum tensile strain at the bottom of the surface layer of such sections is often greater than that for ordinary pavement [33]. At present, the SMA-13 mixture is mainly used for these special sections. Therefore, four-point bending fatigue tests on SMA-13 and SPU-13 mixtures with 600 με, 800 με, and 1000 με strain level were conducted. The fatigue tests were carried out according to the provisions of T0739-2011, the test temperature was 15 °C ± 0.5 °C and the loading frequency was 10 Hz ± 0.1 Hz [35]. Three parallel tests were carried out at the same strain level, and the size of the test piece was 380 mm × 65 mm × 50 mm.

3. Results and Discussion

3.1. Curing Reaction Mechanism of the PU Binder

3.1.1. Fourier Transform Infrared Spectroscopy Analysis of the Initial PU Binder

The infrared spectrum of the initial PU binder sample is shown in Figure 1. The infrared spectrum bands and the types of functional groups represented are summarized in Table 4. The C=O stretching vibration from the –NHCOO functional group appeared at about 1728.9 cm^{-1}, and there was an obvious characteristic band of the –N=C=O functional group at 2267.9 cm^{-1}, indicating that there was an unreacted –N=C=O functional group in the molecular structure of the initial PU sample [34]; further, there was a free ethoxy characteristic band at 1103.1 cm^{-1}, indicating that the PU binder was polyether PU, so the PU binder has good hydrolysis stability, good flexibility and elongation, and good low temperature resistance.

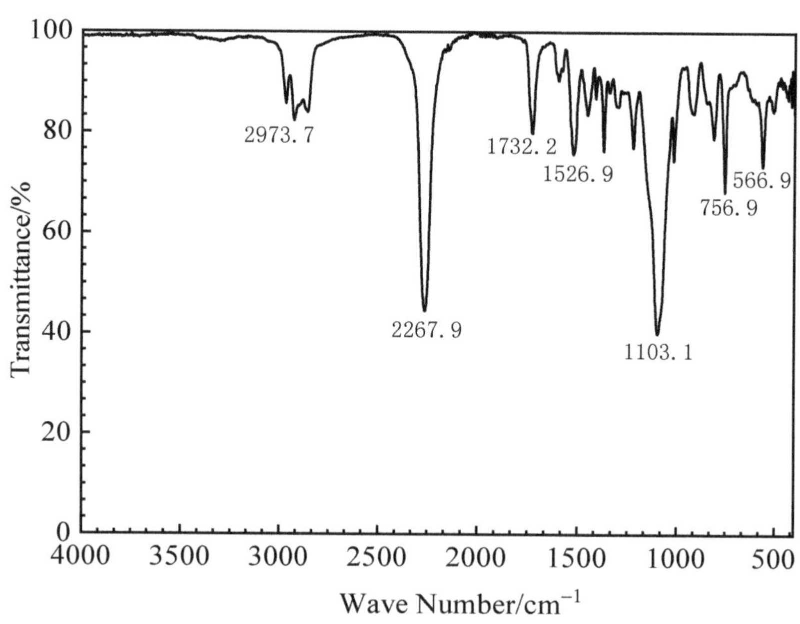

Figure 1. Infrared spectrum of the original PU binder.

Table 4. Functional groups represented by the infrared spectrum bands of the PU binder [34].

Wave Bands (cm^{-1})	Functional Group Category
2973.7, 2927.9, 2857.7	Telescopic vibration from $-CH_2$
2267.9	$-N=C=O$ antisymmetric stretching vibration band
1732.2	C=O telescopic vibration from the –NHCOO functional group
1599.0	Bending vibration band from –NHCOO
1526.9	Bending vibration band from –NH
1454.5	Shear vibration from $-CH_2$
1413.1	–COO vibration band
1372.7	Symmetrical bending vibration from $-CH_3$
1103.1	Free ether oxygen characteristic band
1017.7	–C–C absorption band

3.1.2. Reaction Characteristics of PU Binder during Curing

The infrared spectrum test results of the original PU binder and that cured for 2 h, 4 h, 24 h, 72 h, and 168 h are shown in Figure 2.

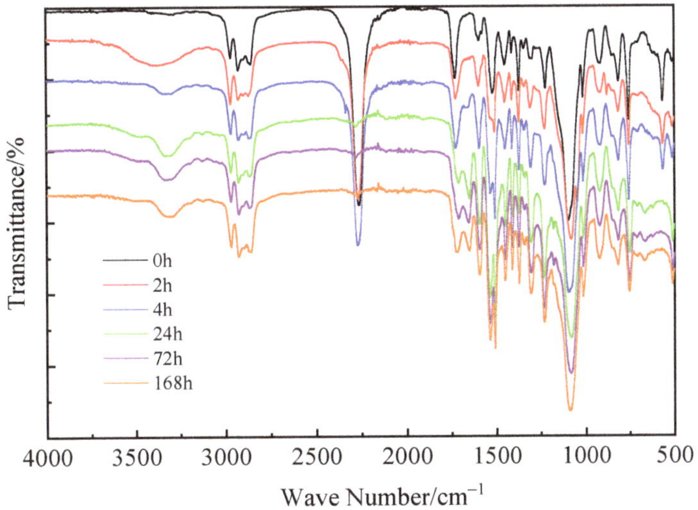

Figure 2. Infrared spectrum of PU binder at different curing times.

When comparing the infrared spectra of the original and cured for 2 h, 4 h, and 24 h PU binders, it is clear that the –NH stretching vibration band appeared and strengthened near 3328.1 cm^{-1} and increased with curing reaction time. The –NCO characteristic band near 2274.6 cm^{-1} did not weaken significantly within the first 4 h of curing time; however, the band weakened significantly after 24 h of curing. The bending vibration band of NHCOO near 1599 cm^{-1} exhibited a gradually increasing trend, indicating that the physical and chemical curing of the PU binder shown in Figure 3a occurred in the first curing 24 h. It is shown that Carbon dioxide (CO_2) was produced in the process of the chemical curing reaction, which could have caused the volume expansion of the PU mixture with a dense structure.

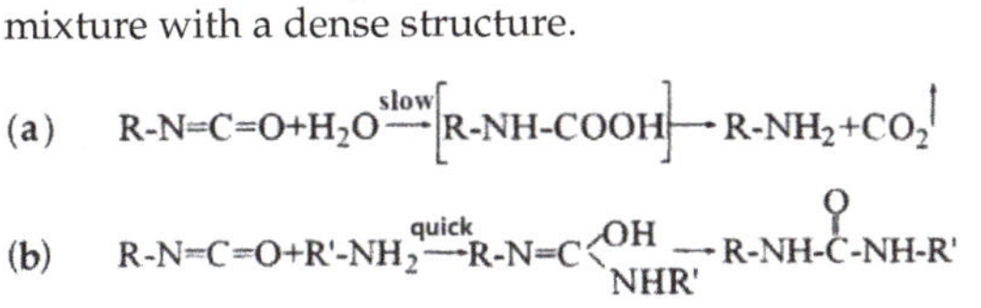

Figure 3. (**a**) Preliminary curing of PU binder. (**b**) Subsequent curing of PU binder.

The C=O stretching vibration band of urea appeared at 1654.1 cm^{-1} in the 24 h curing infrared spectrum; however, there was no such characteristic band in the 2 h and 4 h curing infrared spectrum, indicating that the reaction shown in Figure 3b began after 24 h of curing, and then gradually accelerated to form a polyurea structure. The polyurea bond structure had high strength and cohesion energy, resulting in a sharp increase in the bonding strength of the binder. It implied that the strength of polyurethane mixture increased greatly after 24 h of curing reaction.

3.1.3. Main Group Changes of the PU Binder during Curing

The vibrational absorption band of the NCO group was strong and sharp and therefore cannot be affected by the absorption of other chemical functional groups, the NCO group was taken as the research object to study the curing process of the PU binder [28,29]. The FTIR curves of the NCO group with different curing times are shown in Figure 4. It can be seen that the characteristic band of the NCO group weakened gradually with time, and the curing reaction rate over the first 72 h was significantly higher than that in the later stages. In the first 72 h, the NCO group band changed significantly, and the reaction rate then slowed down. It may be because the linear polymer chain increased with the progress of the reaction in the first 72 h, and then the reaction speed slowed down because some unreacted NCO groups were wrapped in the adhesive system, blocking the contact between NCO groups and water.

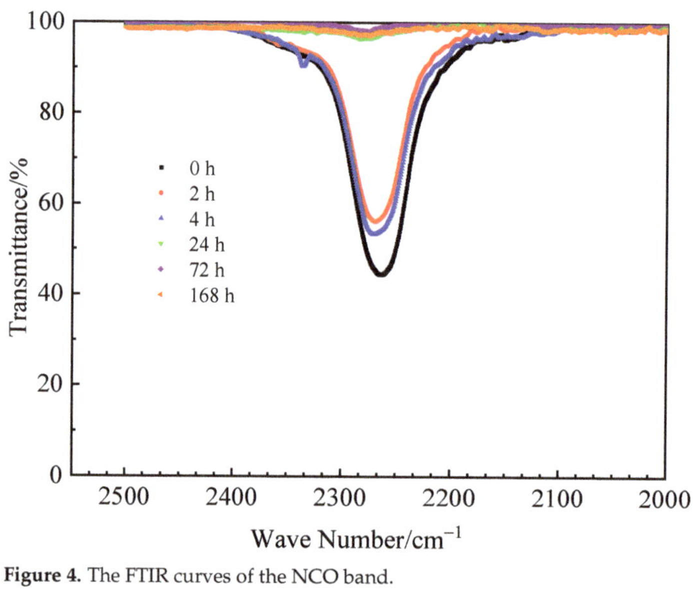

Figure 4. The FTIR curves of the NCO band.

The FTIR curves of the NH band are shown in Figure 5. It can be seen that the characteristic band of NH is not stable in the first 24 h of the curing reaction, while the NH band was practically stable after 72 h of curing. The characteristic band of the infrared spectrum moved from 3332.4 cm^{-1} to 3295.8 cm^{-1}, indicating that the internal hydrogen bond of PU was still gradually enhanced in the later stages of chemical curing. Both the migration of NCO and the penetration of water need time. Therefore, in the later stages of curing, the bond strength of the binder increased slowly; this process usually lasted for a long time. The reaction created more hydrogen bonds between polar functional groups in the PU molecular structure and increased its linear polymer chains.

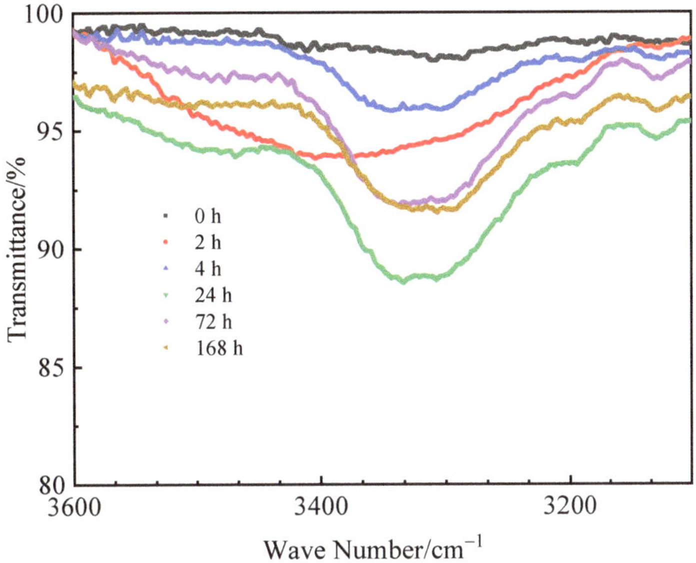

Figure 5. The FTIR curves of the NH band.

3.2. *Volume Characteristics and Strength Formation Law of the PU Mixture*

3.2.1. Volume Characteristics of the PU Mixture during Curing

The height of specimens during curing was also measured. The height differences of the specimens were obtained by subtracting the initial height of the specimens before curing and are shown in Figure 6.

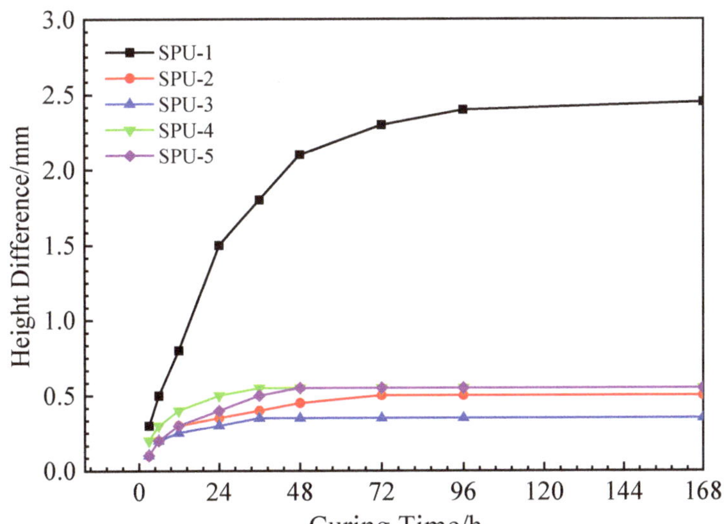

Figure 6. Height difference of specimens during curing.

According to the SPU-1 curve, the Marshall specimen that was directly formed after the preparation of the PU mixture suffered from the issue of volume expansion, and the height differences of the specimen were basically stable and maintained at about 2.4 mm after 96 h of curing. However, the test curves of SPU-2, SPU-3, SPU-4, and SPU-5 showed that the volume of the specimens stabilized after 72 h of curing; the height differences of

SPU-2, SPU-4, and SPU-5 specimens stabilized at about 0.5 mm; and the height differences of SPU-3 specimens stabilized at about 0.3 mm. This indicated that the proper stacking of the prepared PU mixture can effectively control the volume expansion of the PU mixture. This is because, after the preparation of the PU mixture, the PU binder reacts according to Figure 3a within the first 4 h, releasing CO_2, and resulting in volume expansion. After stacking for 4 h or 8 h, compaction molding discharges CO_2 in the mixture, thus inhibiting the volume expansion of the mixture.

Within the first 24 h of curing, the height difference growth rate of the SPU-4 specimen was greater than that of SPU-2 and SPU-5. After 72 h of curing, the height difference of SPU-4 was 0.05 mm greater than that of SPU-2 and SPU-5, indicating that a high ambient humidity accelerates the curing reaction rate of the PU mixture and affects the subsequent reaction behavior, resulting in different reaction endpoints. Within the first 24 h of curing, the growth rate of SPU-4 height difference was greater than that of SPU-2; however, the height difference of SPU-2 and SPU-4 was the same after 72 h of curing, indicating that the ambient temperature accelerated PU mixture curing reaction rate, but did not change the subsequent reaction behavior; the reaction endpoints of the two mixtures were the same. Therefore, it was suggested that the PU mixture was properly stacked after preparation to control the volume expansion of the PU mixture, and the stacking time is consistent with the transportation, paving, and other construction procedures.

3.2.2. Permissible Stacking Time of the PU Mixture

The PU binder gradually solidifies in the stacking process, it is necessary to study the influence of stacking time and ambient temperature on the permissible stacking time of the PU mixture. The prepared PU mixture was stacked at 50% humidity and 15 °C, 20 °C, 40 °C, 60 °C, and 80 °C ambient temperatures; then, the Marshall specimens were formed. The Marshall specimens were cured at 15 °C and 50% humidity for 4 days, and the splitting tensile strength was then tested. The results are shown in Figure 7.

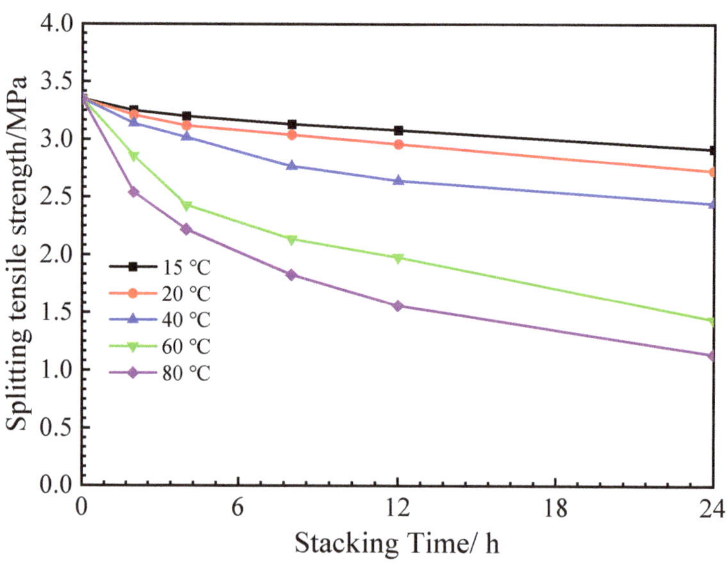

Figure 7. Splitting tensile strength of the PU mixture at different stacking times.

According to Figure 7, the higher the natural stacking time, the lower the splitting tensile strength of the cured PU mixture. This is because, during the stacking process, the curing reaction of the PU binder occurred, the subsequent compaction destroyed the formed strength to a certain extent, resulting in a reduction in the splitting tensile strength. The higher the ambient temperature, the more obviously the splitting tensile strength was affected by the stacking time. This is because, when the temperature was higher, the chemical condensation reaction of the PU mixture was more intense, leading to a higher strength during stacking. Taking an 80% reduction of the splitting strength as the standard for the construction operability of the PU mixture, the permissible stacking time of the PU mixture at the temperatures of 15 °C, 20 °C and 40 °C was 4 h, and the permissible stacking times of the PU mixture at the construction temperatures of 60 °C and 80 °C were 3 h and 2 h, respectively. Based on the volume expansion law, the permissible stacking time, and the plant mixing construction technology of the PU mixture under different environmental conditions, it was recommended to stack the PU mixture for 4 h before compaction, in an environment of 50% humidity and 15–40 °C temperature.

3.2.3. Strength Formation Law of the PU Mixture

The PU mixture was prepared at 20 °C and a humidity of 50%. The Marshall specimens were formed when the PU mixture was prepared and stacked for 4 h, which were then cured at 50% and 95% of air humidity. The splitting tensile strength of the specimens is shown in Figure 8.

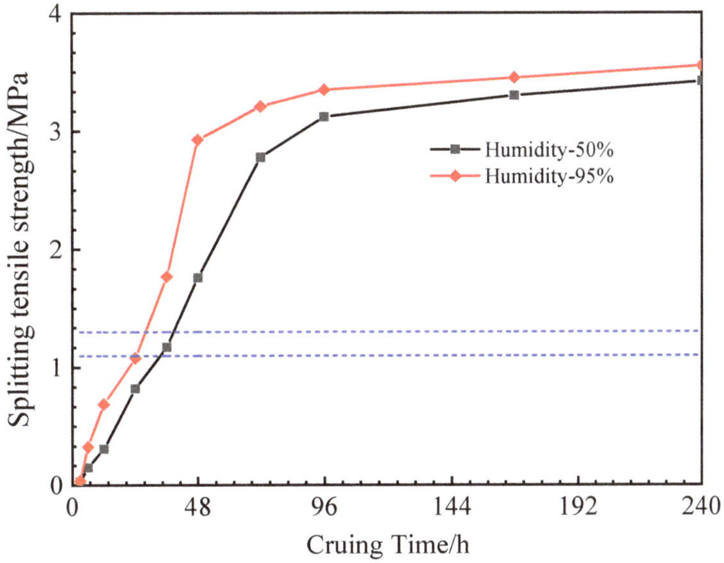

Figure 8. Splitting strength of the PU mixture with different curing humidities.

It can be seen in Figure 8 that the splitting tensile strength of the PU mixture increased rapidly in the first 48 h and 72 h of curing in environments of 95% and 50% air humidity. After 96 h, the curing of the PU mixture was basically complete. The final splitting tensile strength of the PU mixture at the two humidities stabilized at about 3.5 MPa. The results showed that water was one of the key factors affecting the curing of the PU binder. The higher the air humidity, the higher the curing rate. Previous studies have shown that the splitting tensile strength of the AC-13 and SMA-13 mixture stabilized at about 1.1 MPa and 1.3 MPa, respectively [31,32]. According to the curves, the splitting tensile strength of the SPU-13 mixture with 48 h curing could meet this requirement. Therefore, it was recommended that the PU mixture be opened to traffic after curing for 48 h.

3.3. Basic Performance of the PU Mixture

3.3.1. Temperature Stability of the PU Mixture

The results of the 60 °C rutting test and low-temperature bending test of the three mixtures are shown in Figure 9. The dynamic stability of SPU-13 was 40.9 times and

10.7 times that of AC-13 and SMA-13, respectively, and the low-temperature bending strain of SPU-13 was 2.4 times and 1.8 times that of AC-13 and SMA-13, respectively, indicating that the PU mixture had good high- and low-temperature stability. This is because the PU binder reacted with the water in the air and the hydroxyl groups on the surface of aggregates to form a bonded network structure. In this process, the internal hydrogen bond of PU was gradually strengthened, gradually strengthening the PU mixture. Therefore, the PU mixture was less affected by high and low temperatures.

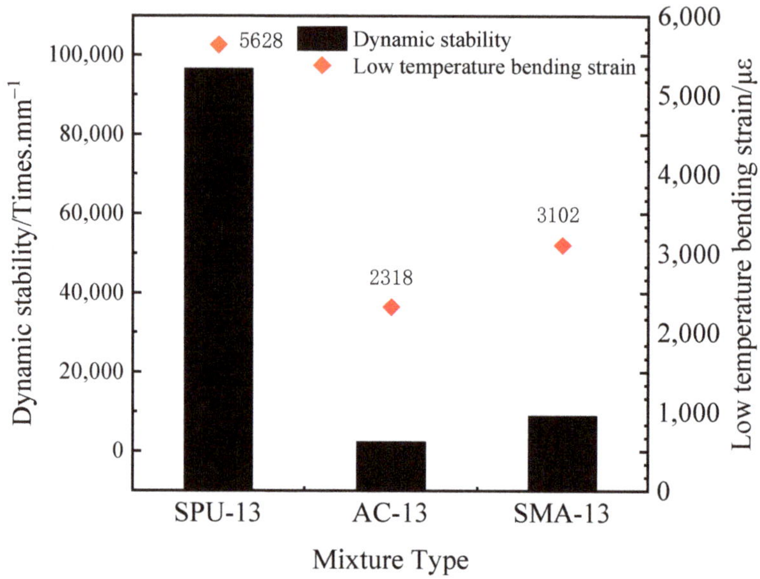

Figure 9. High- and low-temperature test results for the three mixtures.

3.3.2. Water Stability of the PU Mixture

The freeze–thaw splitting test results of different types of mixtures are shown in Figure 10. The freeze–thaw splitting strength ratio of the SPU-13 mixture was lower than that of the SMA-13 and AC-13 mixtures, which did not meet the provisions of the "Technical code for construction of highway asphalt pavement" (JTG F40-2004) [36]. However, the splitting tensile strength of the SPU-13 mixture before and after the freeze–thaw action was higher than that of the SMA-13 mixture. This indicated that, although the freeze–thaw cycle caused great damage to the PU mixture, the strength of the PU mixture after the freeze–thaw cycle could still meet the requirements of pavement materials. The reason for this was that, as a viscoelastic–plastic heterogeneous material, the weak points in the molecular chain (C–O bond, unsaturated C=C double bond, etc.) of the PU mixture could have led to chemical changes, such as bond breaking after freeze–thaw action, which affected the internal stress transmission and coordinated strain development of the PU mixture.

3.3.3. Fatigue Resistance of the PU Mixture

The bending stiffness modulus of the SPU-13 and SMA-13 mixtures during the fatigue test are shown in Figures 11 and 12, respectively. Taking the cyclic loading times corresponding to the attenuation of the bending stiffness modulus of the initial modulus, the fatigue life results of the two mixtures under different strain levels are shown in Figure 13.

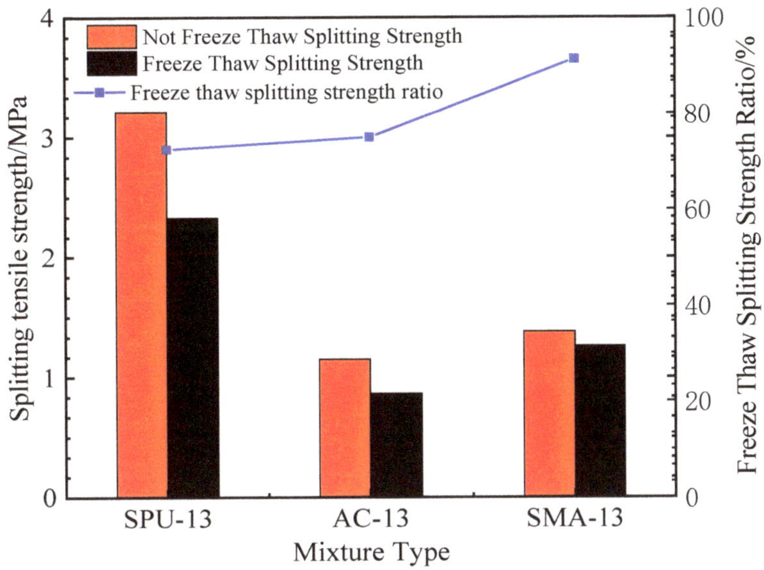

Figure 10. Freeze-thaw splitting test results for different mixtures.

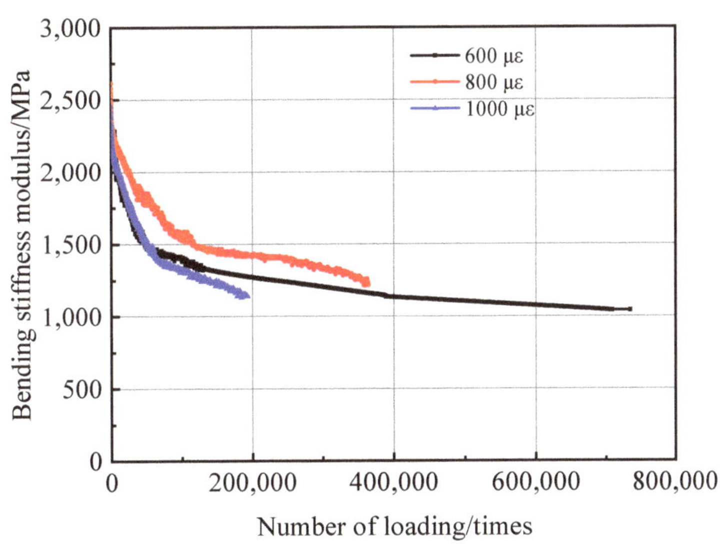

Figure 11. Bending stiffness modulus of the SPU-13 mixture.

Figures 11–13 show that the bending stiffness modulus of the two mixtures decreased sharply at the initial stages, and then increased slowly. The stiffness modulus curves gradually exhibited a stable trend until the bending stiffness modulus decreased to 50% of the initial modulus, and there was no inflection point in the stiffness modulus curves, implying that the fatigue failure of the two mixtures did not occur until the end of the test. Therefore, it was conservative to use the 50% reduction of the initial flexural tensile stiffness modulus of the mixture as the termination condition of the fatigue test and the determination standard for fatigue life.

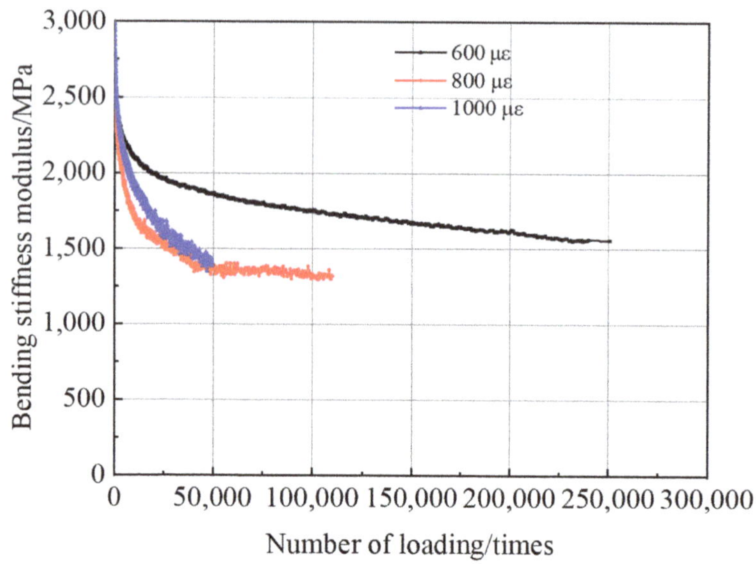

Figure 12. Bending stiffness modulus of the SMA-13 mixture.

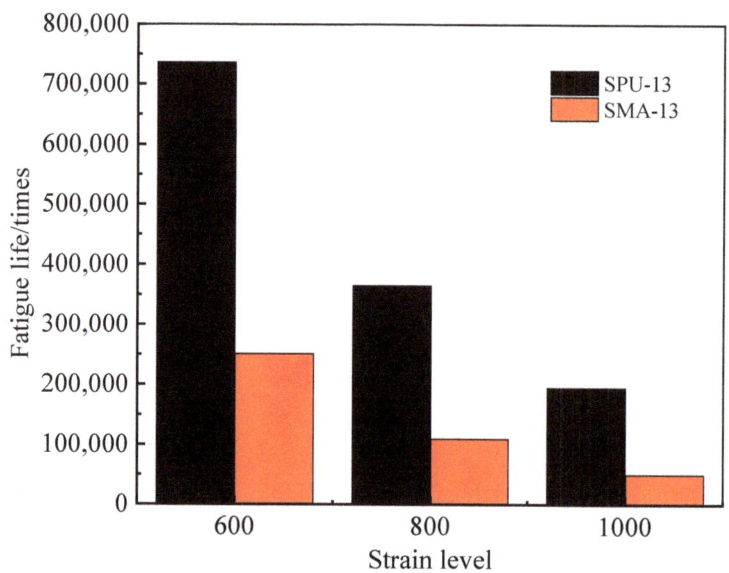

Figure 13. Fatigue life test results of the SPU-13 and SMA-13 mixtures.

For the same mixture, the higher the strain level, the faster the flexural tensile stiffness modulus decreases and the earlier the fatigue test terminates, indicating that the ability of the mixture to resist repeated loads at high strain levels was weak and the fatigue life was low. At the same strain level, the durations of the initial sharp decline stages of the two mixtures' flexural stiffness modulus curves were basically the same; however, the duration of the slow fatigue development stage of the SPU-13 mixture was longer than that of the SMA-13 mixture, and the flexural tensile stiffness modulus of the SPU-13 mixture decreased very slowly at this stage. This shows that the speed of fatigue damage of the SPU-13 mixture specimen under alternating loads was relatively slow, and the instances of the initiation and development of microcracks in the specimen were fewer. The fatigue lives of the SPU-13 mixture under strain levels of 600 με, 800 με, and 1000 με were 2.93 times, 3.31 times, and 3.89 times that of the SMA-13 mixture, respectively, indicating that the PU mixture had a strong ability to resist repeated loading, which meets the needs of special pavement such as long longitudinal slopes, heavy load road, etc.

4. Conclusions

This study contributed to the comprehensive understanding of the curing reaction mechanism of PU binder, strength formation law and the properties of PU mixtures. The following conclusions can be drawn:

1. The internal hydrogen bond of PU was formed in the first 24 h, and then the reaction slowed down. The linear polymer chain increased with the progress of the reaction in the first 72 h, and then the reaction speed slowed down because some unreacted NCO groups were wrapped in the adhesive system, blocking the contact between-NCO groups and water.
2. It was recommended to stack the PU mixture for 4 h before compaction under conditions of 50% humidity and 15–40 °C temperature. Furthermore, it is recommended that the PU mixture be cured for 2 days before being opened to traffic.
3. The PU mixture has excellent high and low temperature stability and fatigue stability. Though the freeze–thaw stability is not good, the splitting strength of the PU mixture before after freeze–thaw can meet the requirements of pavement materials. Thus, the PU mixture can be applied in pavement materials to prolong the service life of pavement and reduce energy consumption and emission.
4. The freeze-thaw damage mechanism and its enhancement mechanism are still unclear. More meso and micro tests should be carried out to improve the freeze-thaw stability of PU mixture in the future so to apply it to more complex working conditions.

Author Contributions: Conceptualization, Y.B.; methodology, S.C. and D.P.; validation, S.C. and P.Z.; formal analysis, M.S. and D.P.; investigation, Y.B.; resources, W.Z. (Wensheng Zhang) and P.Z.; data curation, W.Z. (Wei Zhuang); writing—original draft preparation, M.S. and W.Z. (Wei Zhuang); writing—review and editing, Y.B.; supervision, Y.B.; project administration, Y.B. and W.Z. (Wensheng Zhang). All authors have read and agreed to the published version of the manuscript.

Funding: This work was financially supported by the National Natural Science Foundation of China (No. 51608511), Key Scientific Research Projects in the Transportation Industry of the Ministry of Transport (2019MS2028), Shandong Expressway Group Project (HSB 2021-72).

Institutional Review Board Statement: Not applicable.

Informed Consent Statement: Not applicable.

Data Availability Statement: This original copy does not include distributed Figures, Tables, and Charts before, thus all Figures, Tables, and Charts of this original copy are unique.

Acknowledgments: We would like to recognize numerous co-workers, students, and research facility associates for giving specialized assistance on instrument examination.

Conflicts of Interest: The authors declare no conflict of interest.

References

1. Xiao, F.; Yao, S.; Wang, J.; Wei, J.; Amirkhanian, S. Physical and chemical properties of plasma treated crumb rubbers and high temperature characteristics of their rubberised asphalt binders. *Road Mater. Pavement Des.* **2020**, *21*, 587–606. [CrossRef]
2. Jiang, W.; Xiao, J.; Yuan, D.; Lu, H.; Xu, S.; Huang, Y. Design and experiment of thermoelectric asphalt pavements with power-generation and temperature-reduction functions. *Energy Build.* **2018**, *169*, 39–47. [CrossRef]
3. Guo, T.; Fu, H.; Wang, C.; Chen, H.; Chen, Q.; Wang, Q.; Chen, Y.; Li, Z.; Chen, A. Road Performance and Emission Reduction Effect of Graphene/Tourmaline-Composite-Modified Asphalt. *Sustainability* **2021**, *13*, 8932. [CrossRef]
4. Cong, L.; Guo, G.; Yu, M.; Yang, F.; Tan, L. The energy consumption and emission of polyurethane pavement construction based on life cycle assessment. *J. Clean. Prod.* **2020**, *256*, 120395. [CrossRef]
5. Wang, H.M.; Li, R.K.; Wang, X.; Ling, T.Q.; Zhou, G. Strength and road performance for porous polyurethane mixture. *China J. Highw. Transp.* **2014**, *27*, 24–31. (In Chinese)

6. Chen, J.; Yin, X.; Wang, H.; Ding, Y. Evaluation of durability and functional performance of porous polyurethane mixture in porous pavement. *J. Clean. Prod.* **2018**, *188*, 12–19. [CrossRef]
7. Chen, J.; Ma, X.; Wang, H.; Xie, P.; Huang, W. Experimental study on anti-icing and deicing performance of polyurethane concrete as road surface layer. *Constr. Build. Mater.* **2018**, *161*, 598–605. [CrossRef]
8. Chen, J.; Yao, C.; Wang, H.; Huang, W.; Ma, X.; Qian, J. Interface Shear Performance between Porous Polyurethane Mixture and Asphalt Sublayer. *Appl. Sci.* **2018**, *8*, 623. [CrossRef]
9. Wang, D.; Schacht, A.; Leng, Z.; Leng, C.; Kollmann, J.; Oeser, M. Effects of material composition on mechanical and acoustic performance of poroelastic road surface (PERS). *Constr. Build. Mater.* **2017**, *135*, 352–360. [CrossRef]
10. Wang, D.; Liu, P.; Leng, Z.; Leng, C.; Lu, G.; Buch, M.; Oeser, M. Suitability of PoroElastic Road Surface (PERS) for urban roads in cold regions: Mechanical and functional performance assessment. *J. Clean. Prod.* **2017**, *165*, 1340–1350. [CrossRef]
11. Lu, G.; Renken, L.; Li, T.; Wang, D.; Li, H.; Oeser, M. Experimental study on the polyurethane-bound pervious mixtures in the application of permeable pavements. *Constr. Build. Mater.* **2019**, *202*, 838–850. [CrossRef]
12. Li, T.; Lu, G.; Wang, D.; Hong, B.; Tan, Y.Q.; Oeser, M. Key properties of high-performance polyurethane bounded pervious mixture. *China J. Highw. Transp.* **2019**, *32*, 158–169. (In Chinese)
13. Lu, G.; Liu, P.; Wang, Y.; Faßbender, S.; Wang, D.; Oeser, M. Development of a sustainable pervious pavement material using recycled ceramic aggregate and bio-based polyurethane binder. *J. Clean. Prod.* **2019**, *220*, 1052–1060. [CrossRef]
14. Törzs, T.; Lu, G.; Monteiro, A.; Wang, D.; Grabe, J.; Oeser, M. Hydraulic properties of polyurethane-bound permeable pavement materials considering unsaturated flow. *Constr. Build. Mater.* **2019**, *212*, 422–430. [CrossRef]
15. Lu, G.; Liu, P.; Törzs, T.; Wang, D.; Oeser, M.; Grabe, J. Numerical analysis for the influence of saturation on the base course of permeable pavement with a novel polyurethane binder. *Constr. Build. Mater.* **2020**, *240*, 117930. [CrossRef]
16. Li, R.; Wang, H.; Zhou, G. Experimental study on the strength and influencing factors of multi-pore polyurethane gravel mixture. *Chin. Foreign Road* **2015**, *35*, 244–247. (In Chinese)
17. Wang, H.; Hu, Q.; Li, R. Study on hydrothermal stability of permeable polyurethane mixture. *Highw. Eng.* **2014**, *39*, 246–250. (In Chinese)
18. Li, R.; Wang, X.; Wang, H.; Zhou, G. Experimental study on glue dosage of permeable polyurethane pavement. *Highw. Eng.* **2015**, *40*, 105–108. (In Chinese)
19. Cong, L.; Wang, T.; Tan, L.; Yuan, J.; Shi, J. Laboratory evaluation on performance of porous polyurethane mixtures and OGFC. *Constr. Build. Mater.* **2018**, *169*, 436–442. [CrossRef]
20. Guo, G.; Cong, L.; Yang, F.; Tan, L. Application Progress of Polyurethane Material in Pavement Engineering. *J. Highw. Transp. Res. Dev.* **2020**, *37*, 1–9. (In Chinese)
21. Cong, L.; Yang, F.; Guo, G.; Ren, M.; Shi, J.; Tan, L. The use of polyurethane for asphalt pavement engineering applications: A state-of-the-art review. *Constr. Build. Mater.* **2019**, *225*, 1012–1025. [CrossRef]
22. Xiao, Q.; Liu, M.; Wang, Y.; Li, W.; Wang, Q. Mechanism analysis of snow and ice removal of new pavement material stabilized by polyurethane and rubber particles. *Chin. Foreign Highw.* **2015**, *35*, 258–261. (In Chinese)
23. Zhang, D.; Ning, B.; Wu, J.; Chen, F. The Performance of Permeable Thin Layer Anti-Sliding Material. *J. Chongqing Jiaotong Univ. (Nat. Sci.)* **2019**, *38*, 61–67. (In Chinese) [CrossRef]
24. Sun, M. Preparation of Polyurethane Binder and Composition Design and Performance Evaluation of its Mixture. Ph.D. Thesis, Chang'an University, Xian, China, 2019. (In Chinese)
25. Min, S.; Bi, Y.; Zheng, M.; Chen, S.; Li, J. Evaluation of a Cold-Mixed High-Performance Polyurethane Mixture. *Adv. Mater. Sci. Eng.* **2019**, *2019*, 1507971. [CrossRef]
26. Sun, M.; Bi, Y.; Zheng, M.; Wang, J.; Wang, L. Performance of Polyurethane Mixtures with Skeleton-Interlocking Structure. *J. Mater. Civ. Eng.* **2020**, *32*, 04019358. [CrossRef]
27. Xu, S.; Zhang, Y.; Guo, Y.; Ma, C.; Gao, D.; Peng, G. Determination of Polyurethane Concrete Compaction Timing Based on Penetration Resistance Test System. *China J. Highw. Transp.* **2021**, *34*, 226–235. (In Chinese)
28. Kong, X.; Liu, G.; Curtis, J.M. Characterization of canola oil based polyurethane wood adhesives. *Int. J. Adhes. Adhes.* **2011**, *31*, 559–564. [CrossRef]
29. Wang, Q.; Chen, S.; Wang, T.; Zhang, X. Damping, thermal, and mechanical properties of polyurethane based on poly (tetramethylene glycol)/epoxy interpenetrating polymer networks: Effects of composition and isocyanate index. *Appl. Phys. A* **2011**, *104*, 375–382. [CrossRef]
30. Xia, L.; Cao, D.; Zhang, H.; Guo, Y. Study on the classical and rheological properties of castor oil-polyurethane pre polymer (CPU) modified asphalt. *Constr. Build. Mater.* **2016**, *112*, 949–955. [CrossRef]
31. Shen, D.; Shi, S.; Xu, T.; Huang, X.; Liao, G.; Chen, J. Development of shape memory polyurethane based sealant for concrete pavement. *Constr. Build. Mater.* **2018**, *174*, 474–483. [CrossRef]
32. Hong, B.; Lu, G.; Gao, J.; Wang, C.; Wang, D. Anti-ultraviolet Aging Performance of Polyurethane Binders Used in Roads. *China J. Highw. Transp.* **2020**, *33*, 240–253. (In Chinese)
33. Hong, B.; Lu, G.; Gao, J.; Wang, D. Evaluation of polyurethane dense graded concrete prepared using the vacuum assisted resin transfer molding technology. *Constr. Build. Mater.* **2021**, *269*, 121340. [CrossRef]
34. Hong, B. Study on the Resistance to Water, Alkali and Salt Solutions of Pultruded Polyurethane-Based CFRP Plates. Ph.D. Thesis, Harbin Institute of Technology, Harbin, China, 2018. (In Chinese).

35. JTG E20-2011. *Standard Test Methods of Bitumen and Bituminous Mixtures for Highway Engineering*; Ministry of Transport of the People's Republic of China: Beijing, China, 2011.
36. JTG F40-2004. *Technical Specifications for Construction of Highway Asphalt Pavements*; Ministry of Transport of the People's Republic of China: Beijing, China, 2004.

coatings

Article

Study on the Optimum Steel Slag Content of SMA-13 Asphalt Mixes Based on Road Performance

Wei Chen [1], Jincheng Wei [1,2,*], Xizhong Xu [2], Xiaomeng Zhang [2], Wenyang Han [2], Xiangpeng Yan [2], Guiling Hu [1] and Zizhao Lu [3]

1 School of Transportation Engineering, Shandong Jianzhu University, Jinan 250101, China; chenwei13572468@163.com (W.C.); huguilingtech@Foxmail.com (G.H.)
2 Science and Technology Innovation Center, Shandong Transportation Institute, Jinan 250102, China; xxz137152@163.com (X.X.); zhangxiaomeng@sdjtky.cn (X.Z.); hanwenyang@sdjtky.cn (W.H.); yanxiangpeng336@163.com (X.Y.)
3 Shandong Transportation Planning and Design Research Institute Co., Ltd., Jinan 250031, China; sdjtlzz@163.com
* Correspondence: c2523860697@163.com

Abstract: To reduce the use of aggregates such as limestone and basalt, this paper used steel slag to replace some of the limestone aggregates in the production of SMA-13 asphalt mixes. The optimum content of steel slag in the SMA-13 asphalt mixes was investigated, and the performance of these mixes was evaluated. Five SMA-13 asphalt mixes with varying steel slag content (0%, 25%, 50%, 75%, and 100%) were designed and prepared experimentally. The high-temperature stability, low-temperature crack resistance, water stability, dynamic modulus, shear resistance, and volumetric stability of the mixes were investigated using the wheel tracking, Hamburg wheel tracking, three-point bending, freeze–thaw splitting, dynamic modulus, uniaxial penetration, and asphalt mix expansion tests. The results showed that compared to normal SMA-13 asphalt mixes, the high-temperature stability, water stability, and shear resistance of the SMA-13 asphalt mixes increased and then decreased as the steel slag content increased. All three performance indicators peaked at 75% steel slag content, and the dynamic stability, freeze–thaw splitting ratio, and uniaxial penetration strength increased by 90.48%, 7.39%, and 88.08%, respectively; however, the maximum bending tensile strain, which represents the low-temperature crack resistance of the asphalt mix, decreased by 5.98%. The dynamic modulus of the SMA-13 asphalt mixes increased with increasing steel slag content, but the volume expansion at a 75% steel slag content was 0.446% higher than at a 0% steel slag content. Based on the experimental results, the optimum content of steel slag for SMA-13 asphalt mixes was determined to be 75%.

Keywords: different steel slag content; SMA-13 asphalt mixture; pavement performance; optimal steel slag content

Citation: Chen, W.; Wei, J.; Xu, X.; Zhang, X.; Han, W.; Yan, X.; Hu, G.; Lu, Z. Study on the Optimum Steel Slag Content of SMA-13 Asphalt Mixes Based on Road Performance. *Coatings* **2021**, *11*, 1436. https://doi.org/10.3390/coatings11121436

Academic Editors: Leilei Chen, Ming Liang, Dongyu Niu, Shisong Ren, Ruxin Jing and Valeria Vignali

Received: 26 October 2021
Accepted: 21 November 2021
Published: 23 November 2021

Publisher's Note: MDPI stays neutral with regard to jurisdictional claims in published maps and institutional affiliations.

Copyright: © 2021 by the authors. Licensee MDPI, Basel, Switzerland. This article is an open access article distributed under the terms and conditions of the Creative Commons Attribution (CC BY) license (https://creativecommons.org/licenses/by/4.0/).

1. Introduction

With the rapid development of economic globalization, the level of each industry has also increased, and the consumption of steel materials in economic construction is particularly prominent. According to statistics, world crude steel production reached 1808.6 million tons in 2018, with China accounting for more than 50%. Steel slag is a by-product of the steelmaking process, accounting for 13% of the scrap produced in steelmaking [1]. The utilization rate of steel slag in the United States, Europe, and other developed countries is as high as 70–80% [2]. However, the utilization rate of steel slag resources in China is only 29.5%, of which 50% is used for recycling in road works, sintering, iron-making, and steel-making plants. The remaining steel slag resources are disposed of in large-scale open piles or directly in landfills, which not only occupy land resources but also cause severe pollution to the environment [3]. This is contrary to the sustainable development goal that many countries are striving to achieve, so the question of how to effectively use steel slag remains [4,5].

Steel slag has better properties, such as roughness and crush resistance, than limestone and basalt [6,7]. Steel slag has abundant metal oxides with rough surfaces due to their aggregation, and the rugged texture of these oxides effectively protects the pyroxene crystals located between them from abrasion and provides a rough and undulating microscopic morphology to the slag surface, causing greater surface friction between steel slag [6].

In recent years, the incorporation of steel slag aggregates into asphalt mixtures has greatly attracted road researchers' interest. The microstructure and surface characteristics of steel slag were analyzed by X-ray diffraction and scanning electron microscopy. The results show that steel slag aggregate has many tiny pores on its surface, which can efficiently adsorb asphalt binder material [7], and the resulting asphalt pavement has higher stability and skid resistance [8–11]. It has been shown that the use of steel slag as coarse aggregate in asphalt concrete pavements can improve the performance of the pavement, especially in terms of durability and stability [1,12–14]. Liapis et al. [9,15] compared paved steel slag asphalt mix sections with conventional limestone asphalt mix pavements and found that the surface macrostructure, as well as the skid resistance of the tested sections, were better. Arbani et al. conducted Marshall tests, dynamic creep tests, and indirect tensile tests on asphalt mixtures incorporated with steel slag to evaluate its mechanical properties and deformation resistance. The results showed that the incorporation of steel slag can effectively improve the Marshall stability and fatigue resistance of asphalt mixes and delay the time of permanent deformation of the mixes. The shortcoming is that the low-temperature crack resistance of steel slag asphalt mixes is lagging [16]. Some researchers have considered grinding steel slag into fine particles and adding it to the hot-mix asphalt mixture in the form of fine aggregates. Test results showed that the steel slag asphalt mixture enhanced the rutting resistance, but the amount of asphalt was significantly increased because of the use of fine steel slag aggregates [5,17]. Kavussi et al. [18] demonstrated the better fatigue resistance of steel slag asphalt mixes by doing four-point bending fatigue tests on steel slag mixes. Chen et al. [19] prepared asphalt mixes by blending gneiss and steel slag aggregates, and they studied the water stability of asphalt mixes by high-temperature damage and low-temperature damage modes. The results indicated that the incorporation of steel slag in asphalt mixes can prevent water damage. Masoudi et al. [20] found that adding steel slag to warm asphalt mixes for aging tests can slow down their short-term and long-term aging. Behnood et al. [21] added steel slag to Stone Mastic Asphalt (SMA) mixes and concluded that the elastic properties of steel slag asphalt mixes were better through performance tests such as Marshall stability and rebound modulus. Phan et al. [22] used infrared camera and microwave heating techniques to analyze the self-healing properties of steel slag asphalt mixes. The results showed that the best ductility and crack self-healing properties were achieved when 30% of average coarse aggregate was replaced by steel slag. However, there are many irregular protrusions and tiny pores on the surface of steel slag aggregates, which cause asphalt mixes to be difficult to be compacted and have a high void ratio [23], and the pores can absorb large amounts of asphalt, resulting in increased asphalt usage [5]. Another problem is that the chemical composition of the steel slag aggregate contains substances such as free f-CaO and f-MgO that react with CO_2 in the air, mainly to form $CaxMg_1\text{-}xCO_3$. If these components are hydrated, they can cause pavement cracking [24]. Therefore, steel slag aggregates should be placed in an open environment or immersed in water for at least six months to allow them to fully react before they are recycled and then made into asphalt mixes [25] for better results. Research has shown that a short-term rutting cycle number of 2520 cycles can simulate heavy traffic volumes, predict long-term rutting, and also reflect the performance of asphalt mixtures; however, the A/C index, complex stability index, and shear index were proven to be reliable indicators to verify the performance of the recycled asphalt mix [26]. Researchers evaluated the effect on rutting resistance and fatigue cracking using the locking point concept of aggregates through performance tests of asphalt mixtures, and the concept of interlocking points during compaction of the mix was verified [27].

In summary, mixing steel slag aggregates into asphalt mixes can improve their skid resistance, stability, water damage resistance, self-healing, and rutting resistance. However, there are few studies on the optimal amount of steel slag incorporation for SMA-13 asphalt mixes. In this paper, SMA-13 asphalt mixes prepared with different amounts of steel slag aggregates instead of common limestone coarse aggregates were used and analyzed for road performance and volume expansion. It was concluded that an SMA-13 asphalt mixture with a certain amount of steel slag doping had the best road performance, which provides the theoretical basis for later application and development in practical engineering.

The objectives of this study are: to evaluate the mechanical properties of five types of asphalt mixes with steel slag by means of wheel tracking, Hamburg wheel tracking (HWT), three-point bending, freeze–thaw splitting, dynamic modulus, and uniaxial penetration tests. The purpose is to determine the optimum amount of steel slag for SMA-13 asphalt mixes to provide a theoretical basis for better application to road projects.

2. Materials and Methods

2.1. Materials

The asphalt binder used in this paper was SBS-modified asphalt produced by JingBo Petrochemical Company (Binzhou, China). According to the Chinese standard JTG E20-2011 [28], the conventional properties of the asphalt were tested, and the specific technical indices are shown in Table 1, which all meet the requirements of certain specifications.

Table 1. Technical index of SBS modified asphalt.

Items	Test Values	Specification [29]
Penetration (25 °C, 0.1 mm)	70.1	60–80
Softening point (°C)	64.5	≥55
Flash point (°C)	272	≥230
Ductility (5 °C, cm)	46.3	≥30

The steel slag was made of hot-sealing steel slag aggregate produced by the Rizhao Iron and Steel Plant in Rizhao, China. To prevent the volume expansion caused by the reaction of steel slag with water, it was placed in a natural environment while exposed to rain and air for eight months, which basically eliminated volume instability. The limestone used was high-quality limestone from Jinan, China. Based on the Chinese standard JTG E42-2005 [30], each functional index of coarse aggregate and fine aggregate was tested, and the specific indices are shown in Tables 2–5, which met the specification requirements If steel slag fine aggregate is used for an asphalt mixture, it may cause a larger volume change and an additional increase in the amount of asphalt used [31]. Therefore, in this study, fine limestone aggregate was used as the fine aggregate.

Table 2. Properties of coarse aggregates.

Items	Unit	Steel Slag Test Results	Limestone Test Results	Specification [29]
Apparent relative density	g/cm^3	3.543	2.726	≥2.6
Water absorption	%	1.930	0.556	≤2.0
Crush value	%	9.60	19.8	≤26
Abrasion value	%	11.1	22.3	≤28
Soundness	%	2.7	6.0	≤12

Table 3. Properties of fine aggregates.

Items	Unit	Test Values	Specification [29]
Apparent relative density	g/cm^3	2.725	≥2.5
Sand equivalent	%	73	≥60
Soundness	%	15	≥12
Angularity	s	46	≥30

Table 4. Properties of fillers.

Items	Unit	Test Values	Specification [29]
Apparent relative density	t/m^3	2.700	≥2.50
Water content	%	0.1	≤1
Appearance	-	No agglomerates	No agglomerates
Hydrophilic coefficient	-	0.49	<1
Plasticity index	%	2.2	<1

Table 5. Steel slag SMA-13 gradation composition design.

	Steel Slag Content 0%	Steel Slag Content 25%	Steel Slag Content 50%	Steel Slag Content 75%	Steel Slag Content 100%
Steel slag 10–15 mm/%	0	11	22	32	40
Limestone 10–15 mm/%	38	28	18	8	0
Steel slag 5–10 mm/%	0	11	22	31	40
Limestone 5–10 mm/%	39	28	17	9	0
Limestone 0–3 mm/%	13	12	12	11	11
mineral powder/%	10	10	9	9	9
optimum asphalt content/%	5.14%	5.25%	5.34%	5.50%	5.61%

The mineral powder used in this study was made from ground limestone, acting as a filler in the asphalt mix with the aim of reducing the voids in the asphalt concrete. The mineral powder and asphalt together form an asphalt mastic, which improved the strength and stability of the asphalt concrete; the specific indicators are shown in Table 4.

To determine the optimal steel slag incorporation for the SMA-13 asphalt mixes, a total of five mix gradations were designed with different steel slag incorporation levels [32]: 0%, 25%, 50%, 75%, and 100% steel slag content. The coarse aggregate part of the SMA-13 asphalt mixture was formed by combining the steel slag and limestone with particle sizes of 5–10 mm and 10–15 mm at certain ratios. The fine aggregate part was limestone with a particle size of 0–3 mm. The volumetric method [33,34] was used to replace the limestone coarse aggregates with steel slag coarse aggregates at different dosing levels while preventing excessive density differences between the two, resulting in deviations between the actual synthetic gradation curve and the target gradation curve. The Marshall compaction test was used to determine the best amount of asphalt. Road performance verification was performed, and the specific grade composition is shown in Table 5.

2.2. Experimental Methods

Performance verification of the SMA-13 asphalt mixes was conducted using the wheel tracking, HWT, three-point bending, freeze–thaw splitting, dynamic modulus, uniaxial penetration, and asphalt mix expansion tests, as shown in Figure 1.

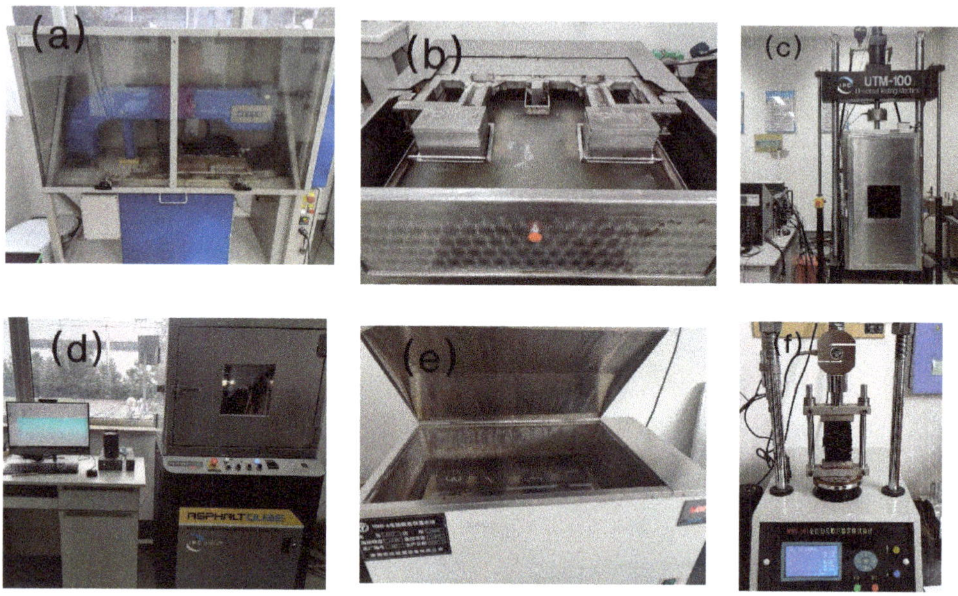

Figure 1. Experimental images of asphalt mixture tests used in this research: (**a**) wheel tracking test, (**b**) Hamburg wheel tracking test, (**c**) three-point bending test, (**d**) dynamic modulus test, (**e**) asphalt mix expansion test, and (**f**) freeze–thaw splitting test.

2.2.1. Wheel Tracking Test

The incorporation of steel slag into SMA-13 asphalt mixes usually improves their rutting resistance due to the strength of the slag itself [35]. According to the Chinese standard JTG E20-2011 (T0719) [28], the rutting resistance of asphalt mixture is evaluated by the wheel tracking test, and its dynamic stability can better reflect the ability of asphalt pavement to resist rutting formation under high-temperature conditions in summer. The test used standard asphalt mixture specimens with a length of 300 mm, width of 300 mm, and height of 50 mm. Before the test began, the mixture specimen was placed in a thermostat at 60 ± 1 °C for 6 h to ensure that the internal temperature was stable at 60 °C. Then, the specimen was placed at the test temperature of 60 °C, the contact pressure between a solid tire made of rubber and the specimen was 0.7 MPa, and the round-trip crimping speed was 42 ± 1 time/min. A linear variable differential transformer (LVDT) collected the rut depth change and calculated the dynamic stability based on the change of time and rut depth, as shown in Equation (1):

$$DS = \frac{(t_2 - t_1) \times N}{d_2 - d_1}, \tag{1}$$

where DS is the dynamic stability (cycles/mm); N is the loading peed (42 cycles/min); d_1 is the amount of deformation corresponding to time t_1 (mm); and d_2 is the amount of deformation corresponding to time t_2 (mm).

2.2.2. Hamburg Wheel Tracking (HWT) Test

The HWT test is considered in the literature [36] as a reproducible and reliable method for characterizing the premature failure of asphalt mixtures. According to the standard JTG E20-2011 [28], the HWT test can evaluate the rutting resistance and water stability of asphalt mixtures. Cylindrical specimens of asphalt mixture with a height of 65 mm and a diameter of 150 mm formed by a rotary compactor were first cut into Hamburg standard specimens using a cutting machine [37]. Then, we placed a mold where the test piece was installed into 50 ± 1 °C water. A steel wheel was made to perform roller compaction on the specimen at a rate of 52 ± 2 times per minute. When the steel wheel crushed the specimen 20,000 times or the rutting depth measured by the LVDT reached 12.5 mm, the rutting meter automatically stopped and saved the data. This test was mainly used to determine the early damage of the asphalt mixture by measuring the rutting depth and loading times.

2.2.3. Freeze–Thaw Splitting Test

The freeze–thaw splitting test is based on the standard JTG E20-2011 [28]. It evaluates the resistance of asphalt mixtures to water damage by the splitting tensile strength ratio of asphalt mixture specimens before and after freezing and thawing [38]. A standard specimen undergoes Marshall compaction 50 times on each side, and the specimen size is 100 mm in diameter and 63.5 ± 2.5 mm in height. The specimens were divided into two groups; one group was placed in a water bath at 25 °C for 2 h ± 10 min before the test, while another group was first placed in a vacuum. The degree of saturation was at 70–80% before placing the test specimen into a plastic bag filled with 10 mL of water. Then, the plastic bag was placed in a thermostat at –18 °C for at least 16 h, chilled, and then placed in a water bath at 60 °C for 24 h. Then, the specimens were moved to a water bath at 25 °C for 2 h, and the splitting strength of the Marshall specimens was measured according to a loading rate of 50 mm/min.

2.2.4. Dynamic Modulus Test

The dynamic modulus test applies an offset sine wave axial compressive stress to the specimen according to a specific temperature and loading frequency to measure the recoverable strain of the specimen, which is an important method to simulate the actual road performance of asphalt mixtures. According to the Chinese standard JTG E20-2011 (T0738) [28], cylindrical specimens of 150 mm in diameter and 170 mm in height formed by a rotary compaction apparatus were cut into standard specimens of 100 mm in diameter and 150 mm in height by coring. The specimens were held at the test temperature for at least 4 h before the test. The AST material testing machine was used for testing. The dynamic modulus can be calculated by Equation (2):

$$|E^*| = \frac{\sigma_0}{\varepsilon_0}, \tag{2}$$

where $|E^*|$ is the dynamic modulus (MPa), σ_0 is the applied maximum stress (MPa), and ε_0 is the measured peak strain.

2.2.5. Three-Point Bending Test

The three-point bending test was used to evaluate the performance of asphalt mixtures against bending and tensile damage at low temperatures. According to the Chinese specification JTG E20-2011 (T0715) [28], the asphalt mixture specimen (300 mm × 300 mm × 50 mm) was cut into small, prismatic beam specimens with a length of 250 ± 2 mm, width of 30 ± 2 mm, and height of 35 ± 2 mm. Before the test, the specimens were placed in insulation at –10 ± 0.5 °C for at least 45 min to ensure that its internal temperature was uniform. Then, the specimen was placed on a base with a span diameter of 200 mm, and a load was applied to the middle of the specimen at a loading speed of 50 mm/min. The mid-span deflection was measured using the LVDT. To reduce the occurrence of errors, four replicate tests were used for each asphalt mixture, and the flexural tensile strength, maximum flexural strain, and bending stiffness modulus were calculated using Equations (3)–(5). The low-temperature durability of the steel slag SMA-13 asphalt mixes was determined by measuring the crack resistance:

$$R_B = \frac{3LP_B}{2bh^2}, \tag{3}$$

$$\varepsilon_B = \frac{6hd}{L^2}, \tag{4}$$

$$S_B = \frac{R_B}{\varepsilon_B}, \tag{5}$$

where R_B is the flexural tensile strength of the specimen at the time of damage (MPa), ε_B is the maximum bending strain (με), S_B is the modulus of bending stiffness of the specimen at the time of damage (MPa), b is the width of the span section of the specimen (mm), h is the height of the beam (mm), L is the span of the testing fixture, 200 mm, P_B is the maximum load when the specimen is damaged, and N is the mid-span deflection of the specimen when breaking the ring (mm).

2.2.6. Uniaxial Penetration Test

Referring to the Chinese specification JTG D50-2017 [39], a standard cylindrical specimen with a diameter of 150 mm and a height of 100 mm was placed in a constant-temperature chamber at 60 ± 0.5 °C and insulated for 5–6 h. A UTM-100 tester was used to maintain a loading rate of 1 mm/min to apply a 42 mm diameter loading indenter to the specimen before stopping the test when the stress value dropped to 90% of the extreme value point. To reduce the occurrence of errors, four replicate tests were used for each asphalt mixture. The penetration strength of the specimen is calculated using Equations (6) and (7):

$$R_l = f_l \sigma_P, \tag{6}$$

$$\sigma_P = \frac{P}{A}, \tag{7}$$

where R_l is the penetration strength (MPa), σ_P is the penetration stress (MPa), P is the ultimate load when the specimen is damaged, N, A is the cross-sectional area of the indenter (mm^2), and f_l is the penetration stress coefficient. When the diameter of the test piece was 150 mm, f_l = 0.35.

2.2.7. Asphalt Mix Expansion Test

According to the Chinese specification JTG E42-2005 [30], when steel slag is used as a material for road asphalt layers, its activity and swelling must be tested to see if it meets the use standards. At least three standard Marshall specimens were made for each steel slag content. The diameters and heights of the specimens were measured with vernier calipers at three and four places, respectively, to calculate the initial volume V1. Then, the specimens were placed into a constant-temperature water bath at 60 ± 1 °C for 72 h before removal and cooling to room temperature. Then, the appearance of cracks or bulging phenomenon were observed, and the new volume of the specimen V2 was measured according to the same method as before. Equation (8) was used to calculate the expansion of the steel slag asphalt mixture:

$$C = \frac{v_2 - v_1}{v_1} \times 100, \tag{8}$$

where C is the expansion amount of the steel slag asphalt mixture (%), v_1 is the volume of the specimen before the water bath (cm^3), and v_2 is the volume of the specimen after the water bath (cm^3).

3. Results

3.1. Study on High-Temperature Stability

The high-temperature rutting resistance of the SMA-13 asphalt mixes with different steel slag contents was analyzed by the dynamic stability and rutting depth obtained from the wheel tracking and HWT tests with the incorporation of steel slag. Figure 2 shows the variations in dynamic stability and rutting of the asphalt mixes at high-temperature conditions with the increase in steel slag content. The test results showed that the dynamic stability of the mixes increased with the increase in steel slag content, and all samples had a stability that was greater than the specification value of 3000 times/mm [29]. The dynamic stability reached a peak of 8000 times/mm when the slag content was 75% before declining. However, the SMA-13 asphalt mixes with 25%, 50%, 75%, and 100% steel slag were all at least 25% more dynamically stable than those without steel slag. The rutting depth of the SMA-13 asphalt mixture with 75% steel slag was 1.19 mm, which was smaller than the other samples. Figure 3 shows that the rutting depth of the HWT test conducted at 50 °C in a water bath decreased and then increased with the increase in steel slag content, reaching a minimum value of 2.25 mm when the steel slag content reached 75%. The addition of steel slag can improve the high-temperature rutting resistance of SMA-13 asphalt mixes because steel slag is harder and more angular than limestone, and it has a stronger ability to resist pressure. The compacted coarse aggregate can form a tightly embedded locking structure, and steel slag is alkaline, has more internal pores, and bonds more efficiently with asphalt, thus improving the high-temperature stability of the asphalt mixture. However, as the steel slag content increased to 100%, the amount of asphalt required increased, and the asphalt inside the pores of the slag reached saturation, which is more likely to lead to flooding oil and rutting under the same compaction conditions. As the amount of steel slag content increases, the porosity and compaction may be more difficult to control due to the unique angularity of the slag leading to a more difficult mix structure to be compacted, resulting in reduced high-temperature stability.

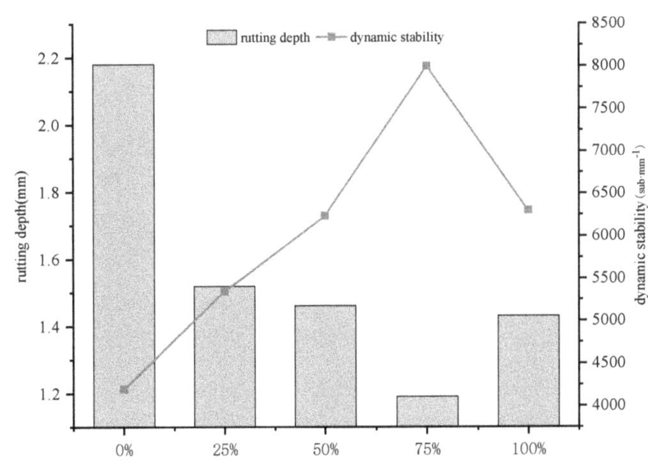

Figure 2. Wheel tracking test results of asphalt mix with different steel slag contents.

3.2. Study on Low-Temperature Crack Resistance

The maximum flexural tensile strength and maximum flexural strain of SMA-13 asphalt mixes were analyzed according to the Chinese standard JTG D50-2017 [39] to evaluate their low-temperature crack resistance. Asphalt mixes with good low-temperature crack resistance can be characterized by maximum bending and tensile strains. According to Table 6, the maximum bending and tensile strains of the samples were better than the specifications [29]. However, the maximum bending and tensile strains of the SMA-13 asphalt mixture with 0% steel slag content were the largest. These values started to decrease with the increase in steel slag content, and they decreased by 5.98% when the steel slag content was 75%. Mainly because of the steel slag in the open storage process, its surface will accumulate many tiny dust particles adsorbed in the pores of the steel slag, reducing the adhesion with asphalt. With the increase in steel slag, the amount of asphalt increases, but the above phenomenon will be more obvious. Therefore, with the increase in steel slag content, the low-temperature crack resistance of the SMA-13 asphalt mixture will decrease. However, when comparing a steel slag content of 75% and 0%, the low-temperature crack resistance did not decrease much.

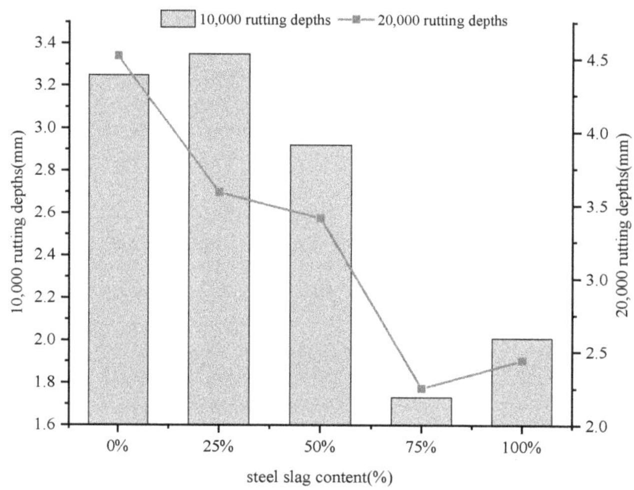

Figure 3. Hamburg wheel tracking test results of asphalt mix with varying steel slag content.

Table 6. Results of low-temperature bending tests on SMA-13 asphalt mixes with varying steel slag content.

Steel Slag Content (%)	Flexural Tensile Strength (MPa)	Maximum Bending Strength (με)	Bending Stiffness Modulus (MPa)
0	16.72	3091.74	4402.20
25	15.04	2979.04	4563.85
50	14.29	2925.42	4883.31
75	13.27	2906.60	5048.87
100	12.55	2852.06	5408.72

3.3. *Study on Water Stability*

The test conditions of the freeze–thaw splitting test were more stringent than the general water immersion test. The purpose was to test the resistance of the asphalt mixes to water damage while varying the steel slag content. From Figure 4, it can be seen that the freeze–thaw splitting strength ratio of the asphalt mix gradually increased with the incorporation of steel slag, but the increase was not significant. The freeze–thaw splitting strength ratio of the asphalt mixture reached a maximum of 91.5% when the steel slag content reached 75%, which was 7.39% higher than that of 0% slag, followed by a 3.5% decrease in the freeze–thaw splitting strength ratio when the steel slag content was 100%. This indicated that the incorporation of steel slag can improve the water stability of SMA-13 asphalt mixes. The main reason for this is that there are many tiny pores inside the steel slag, which is also alkaline, and the asphalt combined more densely, Thus, when combined with asphalt, the steel slag can increase the adhesion force and improve the water stability of the mixture. However, there are harmful impurities in the steel slag such as CaO and MgO that have not fully reacted, and a reaction with water will generate $Ca(OH)_2$ and $Mg(OH)_2$, which affects the water stability of the steel slag SMA-13 asphalt mixture.

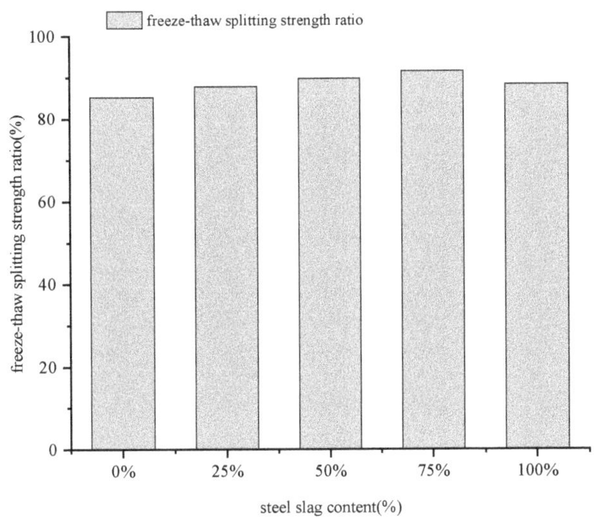

Figure 4. Freeze–thaw splitting strength ratio of SMA-13 under varying steel slag content.

3.4. Study on Dynamic Modulus

The main reason for using the uniaxial compression dynamic modulus as an essential performance parameter for asphalt mixture materials is that it is closer to the actual material response characteristics of pavement and to the assumption of an elastic layered system. Figure 5 represents the increase in the dynamic modulus of the asphalt mixture with the increase in loading frequency at a constant test temperature. As the material properties of asphalt mixture are viscoelastic, there is a certain delay in its deformation under the external load stress. Part of the transient energy release is not sufficient, and the accumulation of energy with the increase in loading frequency gradually increases, which also leads to a gradual increase in the dynamic modulus $|E^*|$. The dynamic modulus decreases with increasing temperature when the loading frequency is the same, as was seen in the SMA-13 asphalt mixtures. This showed that the increase in temperature affected the dynamic modulus of steel slag SMA-13 asphalt mixes because the adhesion between the internal pores of steel slag and asphalt decreased when the temperature increased, and the incorporation of steel slag will be accompanied by a higher asphalt dosage. The above phenomenon will be more obvious. According to the Chinese specification JTG D50-2017 [39], the dynamic compression modulus of the designed SMA-13 asphalt mixture measured at 20 °C and 10 Hz needs to fall in the range of 7500–12,000 MPa. Therefore, as can be seen from Figure 6, a steel slag content of 0% and 25% did not meet the specification requirements. However, the dynamic modulus measured at 50%, 75%, and 100% steel slag content met the specification requirements. Therefore, incorporating an appropriate amount of steel slag into SMA-13 asphalt mixes can improve the dynamic modulus values. In summary, it can be seen that the dynamic modulus when the steel slag content was 50%, 75%, and 100% met the specification requirements, where the dynamic modulus was larger at 75% and 100%. However, the dynamic modulus decreased more rapidly with the increase in temperature when the steel slag content was 100%. Thus, the recommended amount of steel slag content for SMA-13 asphalt mixes is 75%.

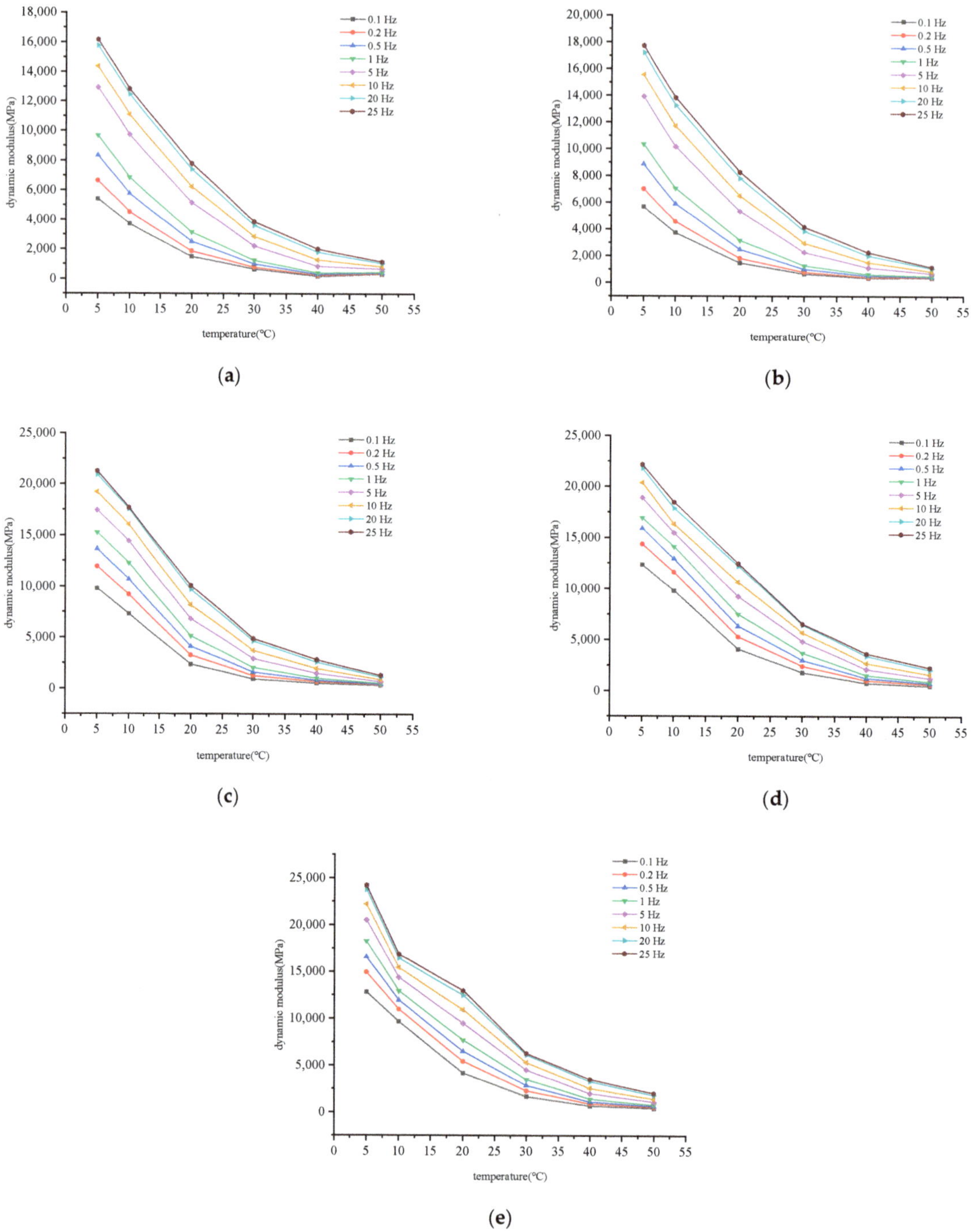

Figure 5. Dynamic modulus test results of asphalt mixes with varying steel slag content: (**a**) 0%; (**b**) 25%; (**c**) 50%; (**d**) 75%; and (**e**) 100% steel slag content.

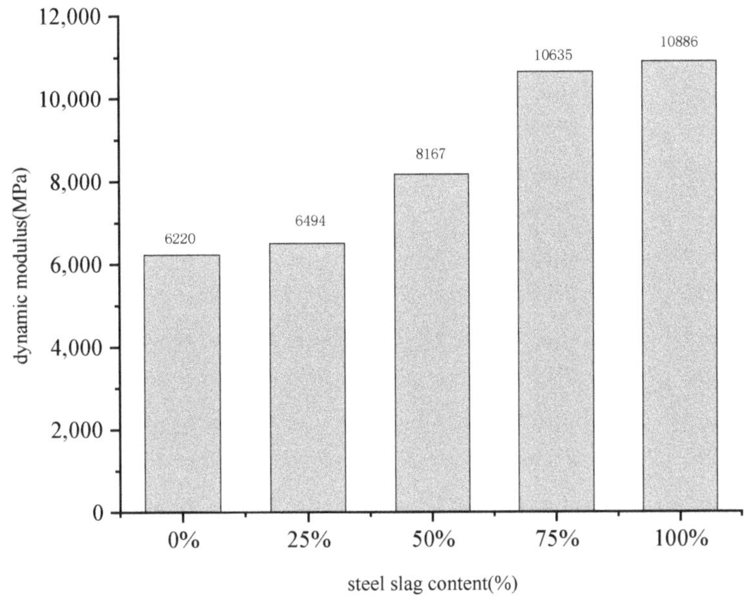

Figure 6. Experimental results on the relationship between dynamic modulus and steel slag content at 20 °C and 10 Hz.

3.5. Study on Shear Performance

Based on the penetration strength calculated from the uniaxial penetration test, the change in shear resistance of the SMA-13 asphalt mixture was analyzed for different amounts of steel slag. The test results are shown in Figure 7. With the increase in steel slag content, the penetration strength of the SMA-13 asphalt mixture first increased and then gradually decreased. The penetration strength of the asphalt mixture reached its peak when the steel slag was mixed at 75%. Steel slag is an alkaline material, and asphalt easily reacts with alkaline materials to form chemical bonds. As a result, the contribution to the shear strength of the mixture is greater, and as the proportion of coarse aggregate in SMA-13 asphalt mixes is increasingly replaced by steel slag, the cohesion increases. Steel slag has irregular angularity, so the inlaid friction force and sliding friction force between steel slag and steel slag is larger than that of limestone aggregates. In addition, the strong locking structure formed by the asphalt mixture after grinding has a greater gain on the internal friction angle; however, when the amount of steel slag exceeds a certain percentage, it will cause the asphalt mixture to be less easily compacted. Furthermore, the sliding friction force between the aggregates will be relatively increased, resulting in a decrease in the internal friction angle. Comprehensive analysis of the shear resistance of the mixture is better when the amount of steel slag is 75%.

3.6. Study on Volume Stability

The contents of f-CaO and f-MgO present in steel slag will increase with the increase in steel slag content, and they will swell to different degrees when mixed with water, which will lead to the appearance of cracks and other diseases in the asphalt mixture [5] and reduce the performance of the road. The asphalt mixture needs to be tested for expansion to check whether its volume expansion rate meets the specification requirements. According to the requirements of Chinese specification JTG E42-2005 [30], the volume change before and after the test is not to be more than 1.5%. The test results are shown in Table 7. The volumetric expansion of the SMA-13 asphalt mixture increased gradually with the increase in steel slag content. The main reason is that with the increase in steel slag admixture, the content of f-CaO and f-MgO inside the slag also increased and reacted with water to produce $Ca(OH)_2$ and $Mg(OH)_2$ [38], which led to volume expansion and affected the volumetric stability of the asphalt mixture. The volume expansion rate of the SMA-13 asphalt mixtures with varying steel slag content were all less than 0.9%, which met the specification requirements. No cracks on the surface or other undesirable diseases appeared. It may be that the treatment with the steel slag reduced the content of f-CaO and f-MgO more effectively, and asphalt wrapped around the surface of the steel slag, reducing the chance of contact with moisture, thus reducing the occurrence of swelling. The volumetric stability of the steel slag SMA-13 asphalt mixture is inversely proportional to the amount of steel slag contained in the mixture.

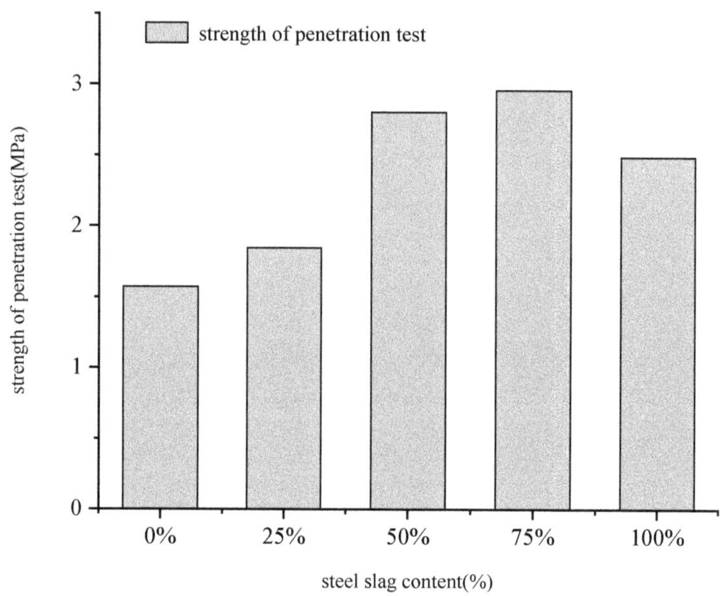

Figure 7. Test results of penetration strength of SMA-13 asphalt mix with different steel slag contents.

Table 7. Results of swelling tests on SMA-13 asphalt mixes with different steel slag contents.

	Volume before Immersion (cm^3)	Volume after Immersion (cm^3)	Volume Expansion Ratio (%)	Surface State of Specimen after Immersion
Steel slag content 0%	511.069	512.352	0.251	No cracks on the surface, no obvious drumming
Steel slag content 25%	512.123	513.931	0.353	
Steel slag content 50%	508.177	510.347	0.427	
Steel slag content 75%	508.052	511.593	0.697	
Steel slag content 100%	523.717	528.200	0.856	

4. Conclusions

In this study, the limestone coarse aggregate in SMA-13 asphalt mixes was replaced with steel slag. Five SMA-13 asphalt mixes prepared with varying steel slag content (0%, 25%, 50%, 75%, and 100%) were tested by the wheel tracking, Hamburg wheel tracking, three-point bending, freeze–thaw splitting, dynamic modulus, uniaxial penetration, and asphalt mix expansion tests to investigate the effect of increasing steel slag content on performance. Based on the results of the various road performance tests, the following conclusions can be drawn.

(1) Compared to the steel slag content of 0%, the dynamic stability of SMA-13 asphalt mixes with a steel slag content of 25%, 50%, 75%, and 100% increased by 27.33%, 48.5%, 90.48%, and 50.00%, respectively. When the steel slag content was 75%, the rutting depths for both the wheel tracking test and the Hamburg wheel tracking test reached minimum values of 1.19 mm and 2.25 mm, respectively. Demonstrating that the incorporation of steel slag can improve the high-temperature stability of SMA-13 asphalt mixes, the best improvement was achieved when the steel slag content was 75%. However, an increased steel slag content can lead to a reduction in the low-temperature crack resistance of the asphalt mix.

(2) The water stability of the asphalt mixes tended to increase and then decrease with the incorporation of steel slag. The freeze–thaw splitting ratio was 91.5%, and the water stability was optimal when the steel slag content was 75%. The presence of hazardous substances in steel slag can lead to poor volumetric stability of the asphalt mix, which decreases with increasing amounts of steel slag.

(3) The dynamic modulus of the asphalt mix increased as the steel slag content increased. At a temperature of 20 °C and a loading frequency of 10 Hz, the dynamic modulus of the SMA-13 asphalt mixes with 75% steel slag content increased by 70.98% compared to those with 0% steel slag content. The shear resistance of the SMA-13 asphalt mixes was strongly influenced by the embedded effect between the steel slag aggregates. As the slag content increased, the shear resistance tended to increase and then decrease. When the slag content was 75%, the penetration strength of the mix reached a maximum of 2.957 Mpa with an optimal shear resistance.

By analyzing the road performance of SMA-13 asphalt mixes with varying steel slag content, the optimum content was finally determined to be 75%.

Author Contributions: W.C.: Writing—original draft; X.Z., W.C.: Data curation; X.X., W.H., Z.L. and G.H.: Methodology; W.H., X.X., Z.L. and G.H.: Project administration; X.Y., J.W. and X.Z.: investigation; X.Y. and J.W.: Supervision. All authors have read and agreed to the published version of the manuscript.

Funding: This research was funded by National Key R&D Program of China, grant number (2018YFB1600100), Shandong Natural Science Foundation Committee (ZR2020QE271, ZR2020KE024) and Shandong Province Key R&D Program (2019GSF109020, 2019GGX101042).

Institutional Review Board Statement: Not applicable.

Informed Consent Statement: Not applicable.

Data Availability Statement: Data sharing is not applicable to this article.

Acknowledgments: The authors would like to thank the Shandong Transportation Institute for their support.

Conflicts of Interest: The authors declare no conflict of interest.

References

1. Ahmedzade, P.; Sengoz, B. Evaluation of steel slag coarse aggregate in hot mix asphalt concrete. *J. Hazard. Mater.* **2009**, *165*, 300–305. [CrossRef]
2. Yi, H.; Xu, G.P.; Cheng, H.G.; Wang, J.S.; Wan, Y.F.; Chen, H. An overview of utilization of steel slag. In Proceedings of the 7th International Conference on Waste Management and Technology (ICWMT), Beijing, China, 5–7 September 2012; pp. 791–801.
3. Guo, J.L.; Bao, Y.P.; Wang, M. Steel slag in China: Treatment, recycling, and management. *Waste Manag.* **2018**, *78*, 318–330. [CrossRef]
4. Nejad, F.M.; Azarhoosh, A.R.; Hamedi, G.H. The effects of using recycled concrete on fatigue behavior of hot mix asphalt. *J. Civ. Eng. Manag.* **2013**, *19*, S61–S68. [CrossRef]
5. Kavussi, A.; Qazizadeh, M.J. Fatigue characterization of asphalt mixes containing electric arc furnace (EAF) steel slag subjected to long term aging. *Constr. Build. Mater.* **2014**, *72*, 158–166. [CrossRef]
6. Woodward, W.; Phillips, E.; Sinhal, W.; Woodward, D. The effect of aggregate type and size on the performance of thin surfacing materials. *Saferroads Org* **2008**, *32*, 65–71.
7. Wu, S.; Xue, Y.; Ye, Q.; Chen, Y. Utilization of steel slag as aggregates for stone mastic asphalt (SMA) mixtures. *Build. Environ.* **2007**, *42*, 2580–2585. [CrossRef]

8. Mladenovič, A.; Turk, J.; Kovač, J.; Mauko, A.; Cotič, Z. Environmental evaluation of two scenarios for the selection ofmaterials for asphalt wearing courses. *J. Clean. Prod.* **2015**, *87*, 683–691. [CrossRef]
9. Liapis, I.; Likoydis, S. Use of Electric Arc Furnace Slag in Thin Skid–Resistant Surfacing. *Procedia- Soc. Behav. Sci.* **2012**, *48*, 907–918. [CrossRef]
10. Mladenovic, A.; Sofilic, T.; Sofilic, U. Defining of EAF steel slag application possibilities in asphalt mixture production/Elk plieno slako panaudojimo gaminant asfalto misini galimybes. *J. Environ. Eng. Landsc. Manag.* **2011**, *19*, 148–157.
11. Kehagia, F. Skid resistance performance of asphalt wearing courses with electric arc furnace slag aggregates. *Waste Manag. Res. J. Int. Solid Wastes Public Clean. Assoc. Iswa* **2009**, *27*, 288–294. [CrossRef] [PubMed]
12. Zhu, J.; Wu, S.; Zhong, J.; Wang, D. Investigation of asphalt mixture containing demolition waste obtained from earthquake-damaged buildings. *Constr. Build. Mater.* **2012**, *29*, 466–475. [CrossRef]
13. Setién, J.; Hernández, D.; González, J.J. Characterization of ladle furnace basic slag for use as a construction material. *Constr. Build. Mater.* **2009**, *23*, 1788–1794. [CrossRef]
14. Xie, J.; Wu, S.; Lin, J.; Cai, J.; Chen, Z.; Wei, W. Recycling of basic oxygen furnace slag in asphalt mixture: Material characterization & moisture damage investigation. *Constr. Build. Mater.* **2012**, *36*, 467–474.
15. Wen, H.; Wu, S.; Bhusal, S. Performance Evaluation of Asphalt Mixes Containing Steel Slag Aggregate as a Measure to Resist Studded Tire Wear. *J. Mater. Civ. Eng.* **2016**, *28*, 04015191. [CrossRef]
16. Arabani, M.; Azarhoosh, A.R. The effect of recycled concrete aggregate and steel slag on the dynamic properties of asphalt mixtures. *Constr. Build. Mater.* **2012**, *35*, 1–7. [CrossRef]
17. Ameri, M.; Hesami, S.; Goli, H. Laboratory evaluation of warm mix asphalt mixtures containing electric arc furnace (EAF) steel slag. *Constr. Build. Mater.* **2013**, *49*, 611–617. [CrossRef]
18. Kavussi, A.; Jalili, Q.M.; Hassani, A. Fatigue behavior analysis of asphalt mixes containing electric arc furnace (EAF) steel slag. *J. Rehabil. Civ. Eng.* **2015**, *3*, 74–86.
19. Chen, Z.; Jiao, Y.; Wu, S.; Tu, F. Moisture-induced damage resistance of asphalt mixture entirely composed of gneiss and steel slag. *Constr. Build. Mater.* **2018**, *177*, 332–341. [CrossRef]
20. Masoudi, S.; Abtahi, S.M.; Goli, A. Evaluation of electric arc furnace steel slag coarse aggregate in warm mix asphalt subjected to long-term aging. *Constr. Build. Mater.* **2017**, *135*, 260–266. [CrossRef]
21. Behnood, A.; Ameri, M. Experimental investigation of stone matrix asphalt mixtures containing steel slag. *Sci. Iran.* **2012**, *19*, 1214–1219. [CrossRef]
22. Phan, T.M.; Park, D.W.; Le, T.H.M. Crack healing performance of hot mix asphalt containing steel slag by microwaves heating. *Constr. Build. Mater.* **2018**, *180*, 503–511. [CrossRef]
23. Pasetto, M.; Baldo, N. Performance comparative analysis of stone mastic asphalts with electric arc furnace steel slag: A laboratory evaluation. *Mater. Struct.* **2012**, *45*, 411–424. [CrossRef]
24. Chen, Z.; Wu, S.; Wen, J.; Zhao, M.; Yi, M.; Wan, J. Utilization of gneiss coarse aggregate and steel slag fine aggregate in asphalt mixture. *Constr. Build. Mater.* **2015**, *93*, 911–918. [CrossRef]
25. Alnadish, A.; Aman, Y. A study on the economic using of steel slag aggregate in asphalt mixtures reinforced by aramid fiber. *ARPN J. Eng. Appl. Sci.* **2018**, *13*, 276–292.
26. Javilla, B.; Fang, H.; Mo, L.; Shu, B.; Wu, S. Test evaluation of rutting performance indicators of asphalt mixtures. *Constr. Build. Mater.* **2017**, *155*, 1215–1223. [CrossRef]
27. Polaczyk, P.; Ma, Y.; Xiao, R.; Hu, W.; Jiang, X.; Huang, B. Characterization of aggregate interlocking in hot mix asphalt by mechanistic performance tests. *Road Mater. Pavement Des.* **2021**, *22*, S498–S513. [CrossRef]
28. JTG E20-2011. *Standard Test Methods of Bitumen and Bituminous Mixtures for Highway Engineering*; Ministry of Transport, PR China: Beijing, China, 2011.
29. JTG F40-2004. *Technical Specifications for Construction of Highway Asphalt Pavenments*; Ministry of Transport, PR China: Beijing, China, 2004.
30. JTG E42-2005. *Standard Test Methods of Aggregate for Highway Engineering*; Ministry of Transport, PR China: Beijing, China, 2005.
31. Xue, Y.; Wu, S.; Hou, H.; Jin, Z. Experimental investigation of basic oxygen furnace slag used as aggregate in asphalt mixture. *J. Hazard. Mater.* **2006**, *138*, 261–268. [CrossRef]
32. Asi, I.M.; Qasrawi, H.Y.; Shalabi, F.I. Use of steel slag aggregate in asphalt concrete mixes. *Can. J. Civ. Eng.* **2007**, *34*, 902–911 [CrossRef]
33. Gao, J.; Sha, A.; Wang, Z.; Tong, Z.; Liu, Z. Utilization of steel slag as aggregate in asphalt mixtures for microwave deicing. *J Clean. Prod.* **2017**, *152*, 429–442. [CrossRef]
34. Tataranni, P.; Sangiorgi, C. Synthetic Aggregates for the Production of Innovative Low Impact Porous Layers for Urban Pavements *Infrastructures* **2019**, *4*, 48. [CrossRef]
35. Pasetto, M.; Baldo, N. Mix design and performance analysis of asphalt concretes with electric arc furnace slag. *Constr. Build Mater.* **2011**, *25*, 3458–3468. [CrossRef]
36. Chaturabong, P.; Bahia, H.U. Mechanisms of asphalt mixture rutting in the dry Hamburg Wheel Tracking test and the potential to be alternative test in measuring rutting resistance. *Constr. Build. Mater.* **2017**, *146*, 175–182. [CrossRef]
37. The Hamburg rutting test—Effects of HMA sample sitting time and test temperature variation. *Constr. Build. Mater.* **2016**, *108*, 22–28. [CrossRef]

38. Wang, W.; Cheng, Y.; Ma, G.; Tan, G.; Sun, X. Further Investigation on Damage Model of Eco-Friendly Basalt Fiber Modified Asphalt Mixture under Freeze-Thaw Cycles. *Appl. Sci.* **2018**, *9*, 60. [CrossRef]
39. JTG D50-2017. *Specifications for Design of Highway Asphalt Pavenment*; Ministry of Transport, PR China: Beijing, China, 2017.

coatings

MDPI

Article

Homogeneity Assessment of Asphalt Concrete Base in Terms of a Three-Dimensional Air-Launched Ground Penetrating Radar

Xiaomeng Zhang [1], Wenyang Han [1], Luchuan Chen [2], Zhengchao Zhang [1], Zhichao Xue [3], Jincheng Wei [1], Xiangpeng Yan [1] and Guiling Hu [4,*]

1 Shandong Transportation Institute, Jinan 250101, China; zhangxiaomeng@sdjtky.cn (X.Z.); hanwenyang@sdjtky.cn (W.H.); zhangzhengchao@sdjtky.cn (Z.Z.); weijincheng@sdjtky.cn (J.W.); yanxiangpeng@sdjtky.cn (X.Y.)
2 Shandong Hi-Speed Group Co., Ltd., Jinan 250101, China; chenluchuan01@163.com
3 Shandong Hi-Speed Infrastructure Construction Co., Ltd., Jinan 250101, China; 36186216@163.com
4 School of Transportation Engineering, Shandong Jianzhu University, Jinan 250101, China
* Correspondence: huguilingtech@foxmail.com

Abstract: Obtaining the required homogeneity, including uniform thickness and density, is very crucial for controlling the quality of flexible asphalt layers. Although non-destructive testing (NDT) methods are time-saving and less labor-intensive, they only provide indirect measurement data under testing area conditions and strongly depend on the explanations by prediction models. In this study, in terms of the three-dimensional air-launched Ground Penetrating Radar (GPR) technique, the dielectric constant of asphalt concrete base with dry conditions in pavements was detected and calculated by different methods (the Coring Method, Reflection Amplitudes Method and Common Mid-Point Method). According to the calculated dielectric constant, the thickness and density of asphalt concrete base were further calculated and assessed. Comparing with the Coring Method, the Common Mid-Point Method was recommended to calculate dielectric constants in order to obtain reliable thickness of asphalt pavement base. Among the Birefringence, Boettcher, Linearity indicator, and Rayleigh models, the Rayleigh model was suggested to predict the density, and the predicted density exhibited a good correlation coefficient with the measured one. Furthermore, by choosing these proper calculation methods, an assessment was successfully conducted to evaluate homogeneity of a constructed field pavement in practice.

Keywords: ground penetrating radar; asphalt concrete base; homogeneity assessment; dielectric constant; model

Citation: Zhang, X.; Han, W.; Chen, L.; Zhang, Z.; Xue, Z.; Wei, J.; Yan, X.; Hu, G. Homogeneity Assessment of Asphalt Concrete Base in Terms of a Three-Dimensional Air-Launched Ground Penetrating Radar. *Coatings* **2021**, *11*, 1398. https://doi.org/10.3390/coatings11111398

Academic Editors: Leilei Chen, Ming Liang, Dongyu Niu, Shisong Ren, Ruxin Jing and Valeria Vignali

Received: 29 September 2021
Accepted: 15 November 2021
Published: 17 November 2021

Publisher's Note: MDPI stays neutral with regard to jurisdictional claims in published maps and institutional affiliations.

Copyright: © 2021 by the authors. Licensee MDPI, Basel, Switzerland. This article is an open access article distributed under the terms and conditions of the Creative Commons Attribution (CC BY) license (https://creativecommons.org/licenses/by/4.0/).

1. Introduction

Asphalt pavement layers are widely used in modern road networks. In order to achieve good quality control and quality assurance, destructive and non-destructive testing (NDT) methods have been implemented and used for assessing homogeneity of asphalt pavement layers after construction [1]. The traditional method is to obtain core samples from the road and then carry out controlled tests in a laboratory, which is widely accepted by transport departments [2]. However, this traditional method is time-consuming, labor-intensive, and destructive to the asphalt pavement. Moreover, cores can be drilled only at a very small percentage of asphalt pavements, which only provides limited information to present whole testing area conditions [3–5].

The NDT methods have thus been promoted in the last decades and could effectively avoid destruction of asphalt pavement, which is timesaving and less labor-intensive. Among NDT methods, ground penetrating radar (GPR) is one of the most popular surface geophysical and NDT methods, which can avoid traffic disrupting [6,7]. The applications of GPR also included early cracking identification of cement concrete pavements and moisture analysis in the masonry arch bridge [8,9]. GPR transmits electromagnetic (EM) waves signals into pavement layers and detects the reflection waves from the surface and interface of asphalt pavement. The frequency of GPR signal is typically in the range of 10 to 5000 MHz [10,11]. The EM waves of GPR are formed by coupled electric and magnetic fields propagating into asphalt pavement materials. Changes in dielectric constants of the materials scatter can be reflected by the EM waves. The GPR receiver detects these scattered and reflected signals [1,12–14]. These reflected waves are processed through digital signal processing to calculate physical parameters and generate two or three dimensional (2D/3D) images of asphalt pavements. In the 1980s, the Federal Highway Administration (FHWA) developed one of the first vehicle mounted GPR system for pavement detection [15]. During the 1990s, as the development of processing software with high-resolution, time-efficient and reliability, GPR technology for road detection was developed to have faster systems operating at higher frequencies [16–19].

Previous studies thus showed that GPR can be used to assess the thickness and density of asphalt concrete layers [20–25]. The information from the GPR should be explained and then used to calculate and predict dielectric constant, thickness, and density by using prediction models. As an important parameter detected from GPR, previous studies found that the dielectric constants were influenced by pavement conditions (i.e., the overlapping structure and the presence of surface moisture). Moreover, the dielectric constants calculated by using different methods (i.e., the Coring Method, Reflection Amplitudes Method and Common Mid-Point Method) influence the accuracy of the predicted thickness [26,27]. Furthermore, when using different density prediction models to estimate density, such as the Birefringence model, Boettcher model, Linearity indicator model and Rayleigh model [28–30], it might vary considerably. In literature, less information is given to present the estimation variation of such different prediction models and to choose an optimum existing model for the explanation of the EM waves from GPR.

The objective of this study was to choose usable and accurate methods and models to evaluate the thickness and density homogeneity of an asphalt concrete base in terms of GPR for quality control and quality assurance. The actual dielectric constants of asphalt concrete base were the fundamental parameter for calculate thickness and density by GPR. The coring method is able to obtain the dielectric constants accurately, but inconveniently. In order to find efficient methods, three different calculation methods for the dielectric constant by the GPR data were considered and four density prediction models (Birefringence, Boettcher, Linearity indicator, and Rayleigh) were chosen to estimate the density for comparison. The GPR was used to test an asphalt concrete base (a reconstructed highway located in Jinan, Shandong, China with 50 m in length as a field test) to choose the best dielectric constant method and density model.

2. Principle of Three-Dimensional GPR

The typical GPR consists of antennas with a transmitter and a receiver, a computer as the control system, and the power supply. In Figure 1, the antennas transmit EM waves penetrating through pavement structure layers. When the EM waves touch the interface of two-layer materials with different dielectric constants, part of the EM waves will reflect back from the interface of the two layers to the receiver antennas. The other EM waves are transmitted across the two-layer interface to the next layer. The GPR software stores the EM waves back and arrival time. The GPR also stores the amplitude of the reflected EM waves. The dielectric constant represents the ability of the material to absorb and release EM waves. The difference of dielectric constants in two adjacent layers affects the amplitude of the EM waves. As debonding occurs in the asphalt concrete layer, the amplitude of the reflected EM waves increases because of different dielectric constants with air, water, and asphalt concrete. The dielectric constant of the asphalt pavement contains information of asphalt concrete integrity, including water contents, chemical composition, and volume fraction of asphalt mixtures. For dry asphalt mixtures, the dielectric constants are between 2 and 4; for the wet asphalt mixtures, the dielectric constant are between 5 and 12.

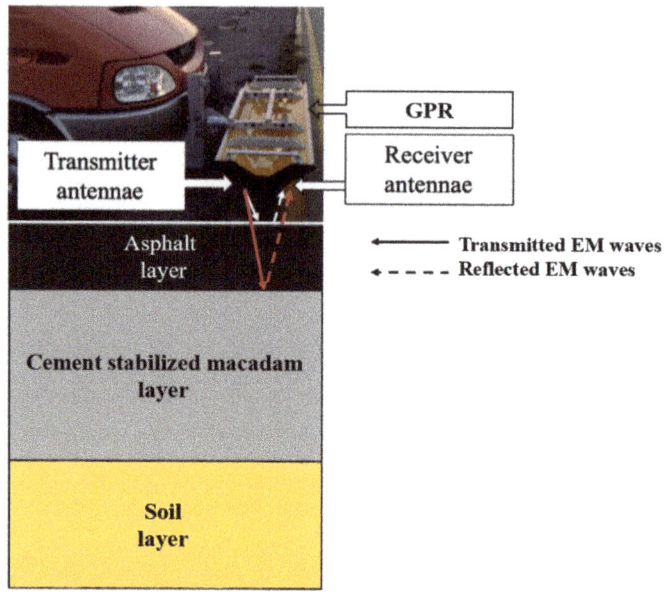

Figure 1. GPR system.

The 2D GPR operates a scan of only one single line with a transmitter and a receiver. It is unable to obtain information from different depths of the pavements with a single frequency. The 3D three-dimensional GPR (3D GPR) provides step-frequency EM waves and can solve the contradiction between frequency and the depth. Moreover, the 3D GPR has the ability to operate a scan of multiple lines simultaneously with the antenna.

2.1. 3D GPR Antenna

The antenna arrays of 3D GPR are suitable for different applications and operational requirements. The 3D-GPR antenna is air-coupled and has bow-tie monopole pairs. The air-coupled antenna installs up to the ground with 50 cm and can be detected with high-speed.

The antenna array consists of 11 pairs of transmitting and receiving antennas. The 21 parallel scan lines in one detection is conducted by the antenna array continuously. Since there is only one transmitting antenna and one receiving antenna for 2D GPR—the range of detection results by 2D GPR is a line, while 3D GPR detection results are represented by a surface range.

2.2. Step-Frequency

Step-frequency is a radar waveform consisting of a series of linear increasing frequency sine waves. The data of the phase and amplitude at each frequency are measured by GPR Then, the data uses Fourier transform to establish a time domain result. Therefore, data in the frequency domain is switched to the data in the time domain through computer processing. The stepping frequency waveform has the best source characteristics of a uniform spectrum. Step-frequency data can be stored as both frequency domain data and time domain data after the transform. However, frequency data allow for a wider range of processing possibilities in the frequency domain.

3. Test Methods

3.1. GPR Data Acquisition and Processing

Detection data of asphalt pavement was conducted using a GeoScope three-dimensional GPR, 3D-RADAR Co., Ltd., Trondheim, Norway, equipped with a frequency 200 MHz to

3000 MHz air-launched bowknot monopole antennae matrix (DX1821, 3D-RADAR Co., Ltd.) that was mounted to the front of the vehicle with an effective detective depth of approximately 1.2 m. The GeoScope GPR has step-frequency technology, which could change the frequency of electromagnetic waves in the work. The antennae matrix includes 21 pairs of parallel launching and receiving antennae in order to acquire three-dimension image and data of road structure layers. The width of the antennae matrix is 1.8 m. A global positioning system (GPS) antenna was installed on the roof of the vehicle, and a mechanical distance measuring instrument (DMI) unit was installed on the vehicle left rear wheel. The GPR detection took place in dry weather (i.e., the surface of asphalt is dry) and temperature was 25 °C. The average speed of the vehicle was 70 km/h. The GPR parameters are shown in Table 1.

Table 1. GPR parameters.

Parameters	Minimum Frequency	Maximum Frequency	Frequency Step	Dwell Time	Sampling Time	Sampling Distance
Value	60 MHz	2980 MHz	20 MHz	3.0 μs	25 ns	20.4 mm

The GPR data were processed using a software 3dr Examiner (Version 2.83, 2013, 3D-RADAR Co., Ltd., Trondheim, Norway), which has a filter tool for noise and interference reduction. The reflections of the GPR signal from various layer interfaces within the pavement material were visually identified.

3.2. *Three Calculation Methods of Dielectric Constants*

For homogeneity assessment of thickness and density of the tested road, the key parameter is the dielectric constant of the asphalt concrete layer. According to the processed GPR detection data, three different calculation methods for dielectric constant of asphalt concrete layer were considered within this study, including the Coring Method, the Reflection Amplitudes Method, and the Common Mid-Point Method [26,29,30].

3.2.1. The Coring Method

The Coring method that is destructive can be used to calculate the dielectric constant of the asphalt concrete layer at the selected cored points and this can be considered as reference. For application of the Coring Method, the thickness of the asphalt concrete layer was measured from the cores. The electromagnetic wave reflection time was recorded from the GPR data detected.

The five different points in the tested area were cored and the thickness of the samples were measured. The dielectric constants of asphalt concrete layers were calculated using Equation (1) [1]:

$$\varepsilon_{AC} = \frac{t^2 c^2}{4H^2} \quad (1)$$

where, ε_{AC} denotes the dielectric constant, t (s) denotes the electromagnetic wave reflection time, c (m/s) denotes the velocity of light in vacuum, and H (m) denotes the thickness of asphalt concrete layer, respectively.

The standard deviation and coefficient of the variation of the test results were calculated by Equations (2) and (3):

$$Standard\ Deviation = \sqrt{\frac{\sum_{i=1}^{N}(\varepsilon_{ACi} - \mu)^2}{N}} \quad (2)$$

$$Coefficient\ of\ Variation = \frac{Standard\ Deviation}{\mu} \quad (3)$$

where, ε_{ACi} denotes the i time dielectric constant calculated by the mothed, μ denotes the average dielectric constant value, and N denotes the test number.

3.2.2. The Reflection Amplitudes Method

The Reflection Amplitudes Method relies on the comparison of the amplitude from the surface of asphalt pavement (A_1) with the amplitude from a metal plate reflection on the pavement surface (A_m); this is because the metal plate offers 100% reflection to the GPR signals. The amplitude of the reflection from the metal plate was obtained from the data collected before the GPR detect using the same configuration applied during the test. The amplitude of the surface reflection and the time through the asphalt concrete layer were obtained from the 3dr Examiner processing software. As shown in Figure 2, the amplitudes showcased as red lines were the boundary of different material layers. The curves on the right represent the actual radar signal and representation at a point on the radar image on the left. Real (red line) represents the waveform of the real part of the radar signal, while magnitude (green line) represents the amount of energy. This procedure was able to eliminate noise and detect significant signals. This signal recorded the corresponding time and depth, using reflections recorded from the metal plate as reference. With both reflection amplitudes, the dielectric constant was calculated with Equation (4) [1].

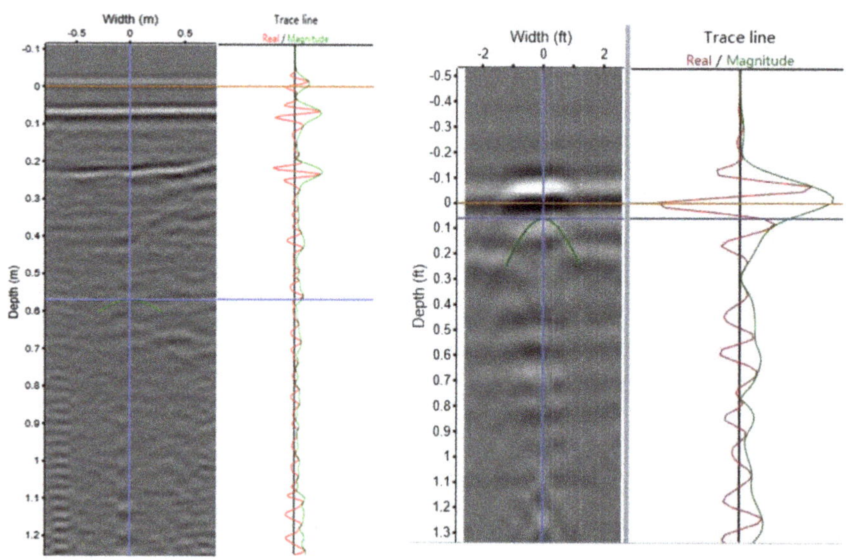

(a) **(b)**

Figure 2. The amplitude of the GPR test. (**a**) Reflection on the surface; (**b**) Reflection on a metal plate.

The dielectric constant of the asphalt concrete layer can be calculated by the Reflection Amplitudes Method using Equation (4):

$$\varepsilon_{AC} = \left(\frac{1 + \frac{A_1}{A_m}}{1 - \frac{A_1}{A_m}} \right)^2 \tag{4}$$

where, A_1 and A_m denote the reflection EM wave amplitudes of the asphalt pavement surface and metal plate placed on the surface of the pavement, respectively.

3.2.3. Common-Mid-Point (CMP) Method of 3D GPR

The CMP means several pairs of antenna excitation points are symmetrically distributed at the center of the test area, called the M point. The R point is called the common midpoint of these excitation points which is at the interface just below the M point. The velocity of the EM waves propagating through the layers is calculated by the distance between the transmitting and receiving antenna. In other words, the CMP method calculates the velocity of the EM waves propagating through the layers in terms of the distance between the transmitting and receiving antenna, and then calculates the dielectric constant of the EM waves.

With the principle of CMP, 2D radar with single channel cannot use the CMP method because it only has one antenna with the function of transmit and receive at the same time. 3D-Radar has an antenna array based on different distance and independent multiple frequency transmit receive antenna, which can measure simultaneously. Each receiving antenna receives electromagnetic wave from the transmitting antenna, which is reflected in each layer interface. The CMP detection technique using the 3D-Radar is shown in Figure 3. T and R represent the transmitting antenna and receiving antenna, respectively. With the CMP method, the dielectric constant of each pavement layer can be measured.

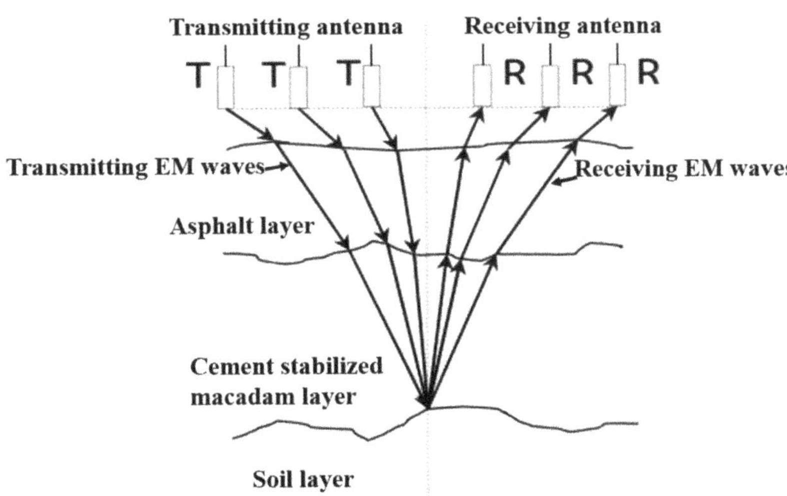

Figure 3. Schematic diagram of CMP.

The dielectric constant of the asphalt concrete layer was calculated by the Common Mid-Point Method by Equation (5) [1]:

$$\varepsilon_{AC} = \frac{c^2}{x^2}\left({t_2}^2 - {t_1}^2\right) \tag{5}$$

where, x (m) denotes the distance between two antennas, t_1 and t_2 denote the time of EM wave transits in asphalt concrete layer by receiving and transmitting one-part type antennas and separating type antennas, respectively.

3.3. Calculation Methods of Thickness

The thickness of the asphalt concrete layer depends on the dielectric constant of the asphalt mixture and can be calculated by an Equation that was changed from Equation (6) [1]:

$$H = \frac{c}{2\sqrt{\varepsilon_{AC}}}\Delta t \tag{6}$$

where, ε_{AC} can be obtained from the Reflection Amplitudes Method and the Common Mid-Point Method, Δt (s) denotes the transmitted time of EM waves in the asphalt concrete layer.

3.4. Calculation Methods of Density

For the density prediction of asphalt concrete layer, the four prediction models chosen from literature—the Birefringence, Boettcher, Linearity indicator, and Rayleigh models—

are presented in Equations (7)–(10), respectively, which can calculate the bulk density of the asphalt concrete layer using the dielectric constant [20–24].

$$G_{mb} = \frac{\sqrt{\varepsilon_{AC}} - 1}{\frac{P_b}{G_b}\sqrt{\varepsilon_b} + \frac{(1-P_b)}{G_{se}}\sqrt{\varepsilon_s} - \frac{1}{G_{mm}}} \quad (7)$$

$$G_{mb} = \frac{\frac{\varepsilon_{AC}-\varepsilon_b}{3\varepsilon_{AC}} - \frac{1-\varepsilon_b}{1+2\varepsilon_{AC}}}{\left(\frac{\varepsilon_s-\varepsilon_b}{\varepsilon_s+2\varepsilon_{AC}}\right)\frac{(1-P_b)}{G_{se}} - \left(\frac{1-\varepsilon_b}{1+2\varepsilon_{AC}}\right)\left(\frac{1}{G_{mm}}\right)} \quad (8)$$

$$G_{mb} = \frac{\varepsilon_{AC} - 1}{\frac{P_b}{G_b}\varepsilon_b + \frac{(1-P_b)}{G_{se}}\varepsilon_s - \frac{1}{G_{mm}}} \quad (9)$$

$$G_{mb} = \frac{\frac{\varepsilon_{AC}-\varepsilon_b}{\varepsilon_{AC}+2\varepsilon_b} - \frac{\varepsilon_a-\varepsilon_b}{\varepsilon_a+2\varepsilon_b}}{\left(\frac{\varepsilon_s-\varepsilon_b}{\varepsilon_s+2\varepsilon_b}\right)\frac{(1-P_b)}{G_{se}} - \left(\frac{\varepsilon_a-\varepsilon_b}{\varepsilon_a+2\varepsilon_b}\right)\left(\frac{1}{G_{mm}}\right)} \quad (10)$$

where, ε_a is the dielectric constant of air, ε_b is the dielectric constant of the bituminous binder, ε_s is the dielectric constant of aggregate, G_{mb} (g/cm^3) is the bulk density of asphalt mixture, G_{se} (g/cm^3) is the specific gravity of aggregate, G_{mm} (g/cm^3) is the maximum theoretical density of asphalt mixture, P_b is the bituminous binder content. According to the results of laboratory test, ε_b = 3.5, ε_s = 7, P_b = 0.06, G_{se} = 2.66 g/cm^3, G_{mm} = 2.565 g/cm^3, respectively.

3.5. Calibration Method of Actual Density

The non-nuclear density gauge, PQI380, TransTech, Greenville, SC, USA, was used to detect the density of asphalt concrete base in the field test in order to confirm the appropriate calculation method from the four density calculation models. The detected data of the non-nuclear density gauge was first calibrated by the coring samples. A clean and dry area with 1 × 1 m^2 was selected for the calibration. Three circle areas were tested for density by the non-nuclear density gauge. Each circle area had five test points and cored a sample in the middle of the circle (Figure 4). Three coring samples in total were tested to obtain the average density in a laboratory. The results of density calibration are shown in Table 2. It was found that the density tested by the coring samples was comparable to the results of the non-nuclear density gauge. The non-nuclear density detection results may be adopted to verify the estimated values by four density prediction models.

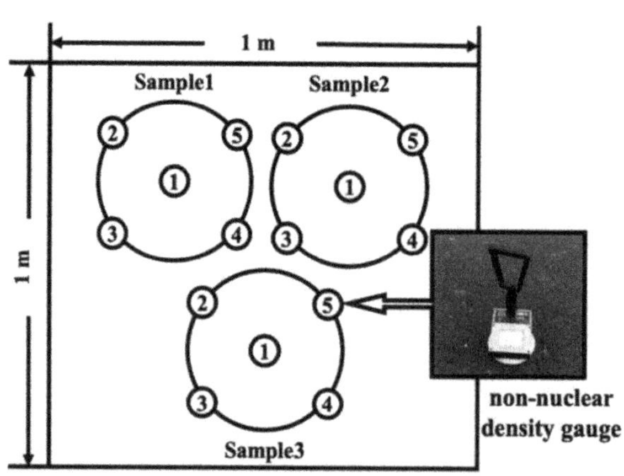

Figure 4. Schematic of density calibration in the field test.

Table 2. Density calibration results of the non-nuclear density gauge.

Test Content	Density (g/cm³)		
	Sample 1	Sample 2	Sample 3
Test Point 1	2.34	2.39	2.34
Test Point 2	2.37	2.41	2.44
Test Point 3	2.29	2.50	2.43
Test Point 4	2.37	2.51	2.54
Test Point 5	2.50	2.44	2.42
Mean of Test Points	2.37	2.45	2.44
Coring Sample	2.35	2.36	2.35

3.6. Field Test in a Newly Constructed Pavement

The field test conducted a continuous real-time GPR test on a reconstruction and extension project on the G20 highway, which was located in Jinan, Shandong Province, China, as shown in Figure 5. The asphalt concrete base of this reconstruction and extension project had already finished. The width and length of the asphalt pavement in this test were 10.5 m and 50 m, respectively. The design thickness of the asphalt concrete base was 9 cm. The asphalt mixture used in the asphalt concrete base was asphalt-treated base (ATB 25) with 70 bitumen.

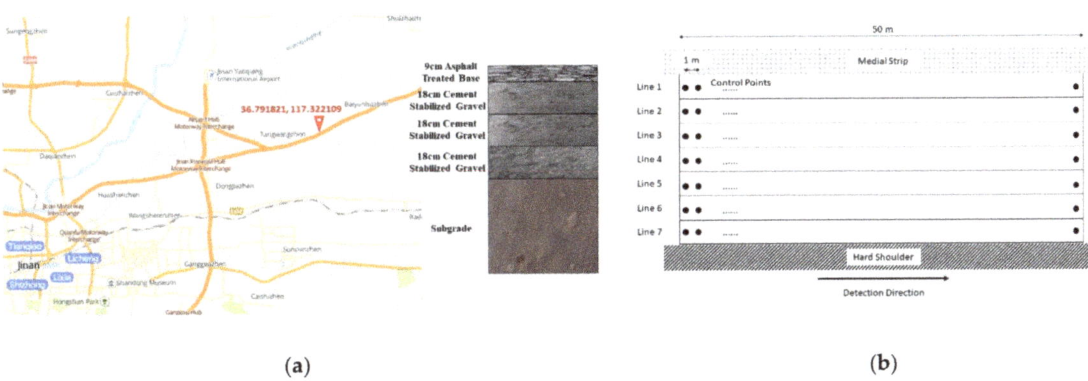

Figure 5. Location and pavement structure of the field test. (**a**) Location and structure; (**b**) Test area.

In order to detect the thickness of the test area, seven parallel detection lines (Line 1 to Line 7) with a width of 1.5 m were established. Line 1 was close to the medial strip and Line 7 was located in the hard shoulder. Fifty control points of GPR detection with space between one meter per line were acquired, as shown in Figure 4. The 350 thickness values of the asphalt concrete base were calculated by Equations (4) and (5). Three coring samples of the test area were cored in order to verify the prediction thickness values.

4. Results and Discussion

4.1. Comparison of the Dielectric Constants Estimated by Different Methods

The results of the dielectric constants of the asphalt concrete layer by the Coring Method, Reflection Amplitudes Method, and The Common Mid-Point Method are shown in Table 3. For the Coring method, the dielectric constant calculated ranged from 4.55 to 5.67 with an average value of 4.94. The coefficient of variation was 8%.

Table 3. Dielectric constants of asphalt concrete calculated by three calculation methods.

Sample	Dielectric Constant			Error	
	Coring Method	Reflection Amplitudes Method	Common Mid-Point Method	(εRAM − εCore)/εCore	(εCMP − εCore)/εCore
1	4.94	2.56	5.06	−48%	2%
2	5.67	2.66	5.48	−53%	−3%
3	4.75	3.45	4.84	−27%	2%
4	4.55	6.73	4.85	48%	7%
5	4.81	4.40	4.83	−9%	0%
Average Value	4.94	3.96	5.01	−20%	1%
Standard Deviation	0.38	1.53	0.25	-	-
Coefficient of Variation	0.08	0.39	0.05	-	-

Based on the Reflection Amplitudes Method, the dielectric constant ranged from 2.56 to 6.73 with an average value of 3.96. And the coefficient of variation was 39%, considerably compared to the Coring Method. In general, the dielectric constant calculated by the Reflection Amplitudes Method are lower than the coring method [2]. The highest error between the dielectric constant, determined from the Coring method and the Reflection Amplitudes Method for each sample, occurred at sample 2 with −53%.

The dielectric constant calculated using the Common Mid-Point Method, ranged from 4.83 to 5.48 with an average value of 5.01. The coefficient of variation was 5%. Compared to the dielectric constant results from the Coring Method, the highest error occurred at sample 4 with 7%.

As listed in Table 2, compared to the three calculation methods, the Reflection Amplitudes Method had the highest coefficient of variation when analyzing all five points. The Coring Method and the Common Mid-Point Method had similar average values and coefficient of variation for calculating dielectric constants of the asphalt concrete layer. The Coring Method was the most accurate way because it had a real parameter from the cores. However, the Coring Method was also inefficient with coring in the test area.

According to the results of calculation, the Common Mid-Point Method showed reasonable agreement with cores and acquired dielectric constants of the asphalt concrete layer in all the areas. In this study, the Common Mid-Point Method was then recommended.

4.2. Assessment of Thickness Homogeneity

In terms of the dielectric constants calculated by the Common Mid-Point Method, the thickness distribution of the asphalt concrete base in the tested area is shown in Figure 6. In total, there were 350 tested points, where 7 intervals with 1.5 m along the width (line 1, line 2, line 3, line 4, line 5, line 6 and line 7) and 50 intervals with 1 m along the 50 m long pavement were chosen equally, respectively. It was found that the thickness of 88 tested points of thickness was less than 9 cm. Its proportion of total test area was 25%. The thickness correction factor and the mean value of the corrected thickness are shown in Tables 4 and 5.

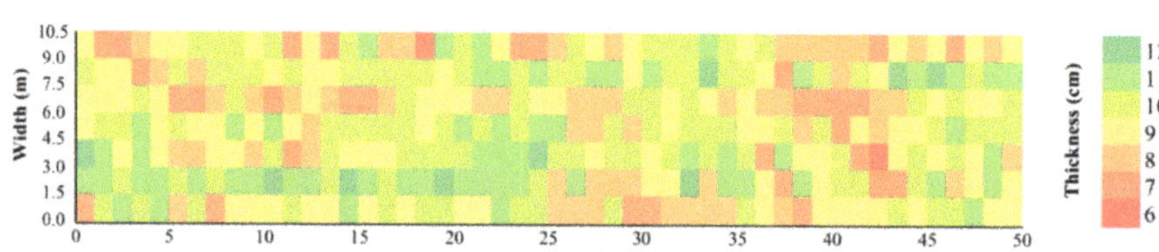

Figure 6. Thickness distribution of the asphalt concrete base in the tested area.

Table 4. Thickness correction factor.

Test Contents	Sample 1	Sample 2	Sample 3
Coring Thickness	11.2 cm	9.2 cm	10.6 cm
Calculated Thickness	12.0 cm	10.0 cm	12.0 cm
Coring/Calculated	0.933	0.920	0.883
Correction Factor		Mean = 0.912	

Table 5. Mean value of corrected thickness.

Test Contents	Thickness (cm)						
	Line 1	Line 2	Line 3	Line 4	Line 5	Line 6	Line 7
Mean	8.92	9.96	8.7	9.67	9.59	10.09	9.21
Standard Deviation	1.15	1.02	0.93	0.87	1.18	1.27	1.03
Coefficient of Variation	0.13	0.10	0.11	0.09	0.12	0.13	0.11
All Area Mean				9.45			
All Area Standard Deviation				1.17			
All Area Coefficient of Variation				0.12			

In Table 5, the maximum mean value of thickness was Line 6 and the minimum one was Line 3. The mean value, the standard deviation, and the coefficient of variation of thickness in all the test areas were 9.45, 1.17, and 0.12 cm, respectively. The results showed that the thinner areas were located in Line 1 and Line 3, which was close to the edge of the pavement and the middle of the asphalt paver.

In order to assess the thickness homogeneity of the asphalt concrete base more qualitatively, the test area (Line 1 to Line 7) was divided into 70 units, and each unit was 1.5 m in width and 5 m in length, shown in Figure 5b. The mean thickness values of every area were calculated and are shown in Figure 7. The red line was the design thickness of the asphalt concrete base. The results of all lines showed thickness variation, and the thickness of Line 3, Line 1, and Line 7 were lower than that of the designed thickness value. The Line 5 thickness had the highest variation. The thickness of the asphalt concrete base showed that the thinner area was close to the edge of the pavement and the middle section of the asphalt paver.

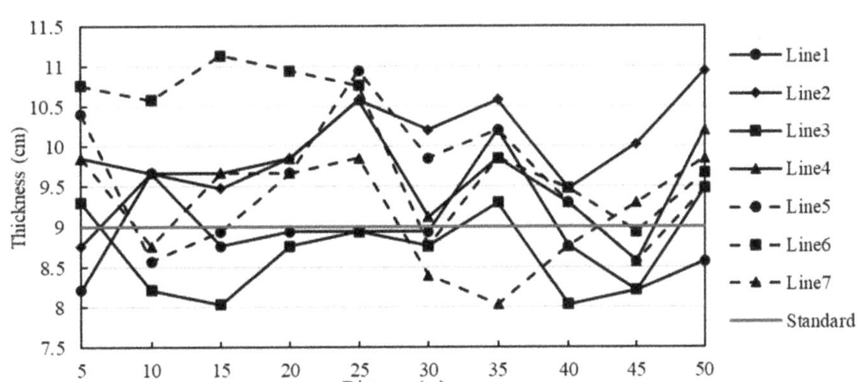

Figure 7. Mean values of thickness in 70 evaluated areas.

Figure 8 is the thickness coefficient of variation of the test area. The mean coefficient of variation in Line 4 was 0.073 and showed the best thickness homogeneity.

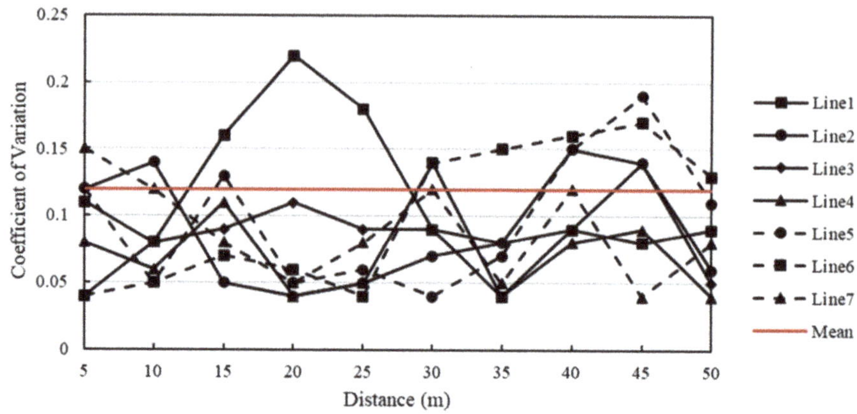

Figure 8. Thickness coefficient of variation in 70 evaluated areas.

4.3. *Assessment of Density Homogeneity*

In terms of the four density prediction models, the predicted bulk densities of asphalt concrete base in the field test are shown in Table 6. In order to not affect the construction period, Line 4 was selected and used to predict the density. The predicted density compared with the measurement density by non-nuclear density gauge is shown in Figure 9. The correlation coefficient of prediction density with measurement density was calculated and is shown in Table 7. The predicted density of the four models had a similar correlation coefficient with the measured density. The density by the Birefringence model had an obvious volatility and the difference value between maximum density and minimum density was 0.71. The results of the Linearity Indicator model deviated highly with the measured density than the other models. The Boettcher model and the Rayleigh model had similar results, but the density predicted by the Rayleigh model closed to the measured density.

Table 6. Density of the asphalt concrete layer in Line 4 calculated by the four prediction models.

Location	Dielectric Constant	Birefringence Model	Boettcher Model	Linearity Indicator Model	Rayleigh Model
		Bulk Density (g/cm^3)			
5 m	4.54	1.95	1.88	1.55	1.91
10 m	4.90	2.23	2.00	1.71	2.04
15 m	5.07	2.36	2.06	1.79	2.10
20 m	5.08	2.37	2.06	1.79	2.10
25 m	5.46	2.66	2.19	1.96	2.23
30 m	5.28	2.52	2.13	1.88	2.17
35 m	5.20	2.46	2.11	1.84	2.14
40 m	4.86	2.20	1.99	1.69	2.02
45 m	5.13	2.41	2.08	1.81	2.12
50 m	4.89	2.22	2.00	1.71	2.03

Table 7. Correlation coefficient of the predicted density with the measured density.

Prediction Model	Birefringence Model	Boettcher Model	Linearity Indicator Model	Rayleigh Model
Correlation Coefficient	−0.712	−0.714	−0.713	−0.719

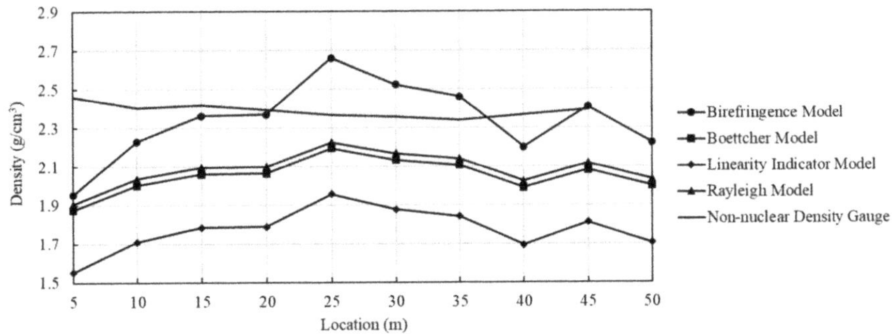

Figure 9. Comparison of the predicted density by the four prediction models and the measured density by the non-nuclear density gauge.

The Rayleigh model presented the highest correlation coefficient when compared with the results from the four prediction models. This model had the potential way to calculate and corrected the density by GPR data. The Rayleigh model was then used to predict the density and the correction factor. The correction factor was the ratio of density obtained by the non-nuclear density gauge and the Rayleigh model (G Non-nuclear/G Rayleigh). Therefore, the continuous prediction of density and the correction factor compared to the non-nuclear density gauge results in the test field are shown in Table 8. The predicted density results had an average correction factor of 1.15, compared to the measured density from the non-nuclear density gauge. With this correction factor, the GPR data and Rayleigh model was able to calculate the accurate density of asphalt pavement without coring or non-nuclear density gauge. In the application of practical road projects, this method could save a lot of testing time and improve the efficiency of quality control.

Table 8. Density correction factor of the Rayleigh model.

Location	Rayleigh Model	Non-Nuclear Density Gauge	Correction Factor (G Non-Nuclear/G Rayleigh)	Mean
5 m	1.91	2.46	1.29	1.15
10 m	2.04	2.41	1.18	
15 m	2.10	2.42	1.15	
20 m	2.10	2.39	1.14	
25 m	2.22	2.37	1.06	
30 m	2.17	2.36	1.09	
35 m	2.14	2.34	1.09	
40 m	2.02	2.37	1.17	
45 m	2.12	2.39	1.13	
50 m	2.03	2.39	1.17	

The Rayleigh model was then used to predict the density of the parallel test lines in the tested area. The density distributions of the asphalt concrete base in the seven tested lines are presented in Figure 10. Line 1, Line 6, and Line 7 had relatively low density in the asphalt concrete base. These results indicate that the edge of asphalt concrete base showed segregation of the asphalt mixture. Insufficient compaction of the asphalt mixture was one of the reasons for low-density distribution. According to the results of density detection by the GPR and Rayleigh model, the construction technology of paving and compaction could be improved.

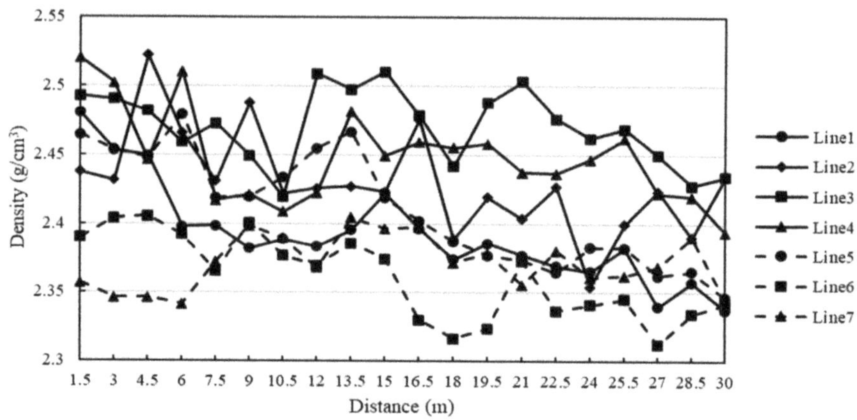

Figure 10. Density distribution of the tested area.

5. Conclusions

Although the air-launched 3D GPR is nondestructive, time-saving, and less labor-intensive to assess density and thickness, it can only provide indirect measurement data under testing area conditions and strongly depends on the explanations by prediction models. This study focused on the adaptability and accuracy of different prediction models to calculate the dielectric constant, thickness, and density of an asphalt concrete base. By choosing proper calculation methods, the thickness and density of this layer were predicted, and homogeneity was assessed. The main findings were given below:

(1) The Common Mid-Point Method, using the air-launched bowknot monopole antennae matrix, had a satisfactory performance in determination of the dielectric constant of the asphalt concrete base and presented an error of 1%, when compared with the calculated dielectric constant from cores. The results indicated potential of the Common Mid-Point Method to be adopted for providing reliable estimation results on the dielectric constant of an asphalt concrete layer and obtaining accurate calculated thickness.
(2) In terms of GPR and the Common Mid-Point Method, thickness homogeneity assessment on the tested filed found that the thickness of an asphalt concrete base showed the thin area to be close to the edge of the pavement and the middle section of the asphalt concrete base.
(3) Among the Birefringence, Boettcher, Linearity indicator, and Rayleigh models, the Rayleigh model was suggested to predict density, and the predicted density exhibited a good correlation coefficient with the measured density. The results of the Rayleigh model presented an average correction factor of 1.15, when compared to the density from the non-nuclear density gauge.

However, the Common Mid-Point Method was only able to detect a line area and the detection dielectric constants were a representative value of small areas. The variation of dielectric constants in the small area should be considered. In this study, only asphalt pavement layers were researched. The applications of GPR also included cement concrete pavements.

Author Contributions: X.Z., W.H., J.W. and L.C. conceived and designed the experiments. X.Z., Z.Z., X.Y. and Z.X. performed the experiments. J.W., Z.X. and X.Y. analyzed the data. W.H., L.C. and G.H. contributed reagents/materials/analysis tools. X.Z. and W.H. wrote the paper. G.H. and Z.Z. revised the paper. All authors have read and agreed to the published version of the manuscript.

Funding: This research was funded by the National Key R&D Program of China, grant number 2018YFB1600103, project ZR2020QE271 was supported by Shandong Provincial Natural Science and Foundation and Shandong province key research and development program, grant number 2019GSF109020, No. 2019GGX101042.

Institutional Review Board Statement: Not applicable.

Informed Consent Statement: Not applicable.

Data Availability Statement: Data sharing is not applicable to this article.

Conflicts of Interest: The authors declare no conflict of interest.

References

1. Wang, S.Q.; Zhao, S.; Al-Qadi, I. Continuous real-time monitoring of flexible pavement layer density and thickness using ground penetrating radar. *NDT E Int.* **2018**, *100*, 48–54. [CrossRef]
2. Zhao, S.; Al-Qadi, I. Development of regularization methods on simulated ground-penetrating radar signals to predict thin asphalt overlay thickness. *Signal Process.* **2017**, *132*, 261–271. [CrossRef]
3. Leng, Z.; Al-Qadi, I.; Shangguan, P.C.; Son, S. Field application of Ground-Penetrating Radar for measurement of asphalt mixture density case study of Illinois Route 72 overlay. *Transp. Res. Rec. J. Transp. Res. Board* **2012**, *2304*, 133–141. [CrossRef]
4. Al-Qadi, I.; Lahouar, S. Measuring layer thicknesses with GPR-theory to practice. *Constr. Build. Mater.* **2005**, *19*, 763–772. [CrossRef]
5. Loizos, A.; Plati, C. Accuracy of pavement thicknesses estimation using different ground penetrating radar analysis approaches. *NDT E Int.* **2007**, *40*, 147–157. [CrossRef]
6. Aleksey, K.; Aleksandra, V.; Evgeniy, V.T.; Anderson, N.L.; Sneed, L.H. Utilization of air-launched ground penetrating radar (GPR) for pavement condition assessment. *Constr. Build. Mater.* **2017**, *141*, 130–139.
7. Wallace, W.L.; Xavier, D.; Peter, A. A review of Ground Penetrating Radar application in civil engineering: A 30-year journey from Locating and Testing to Imaging and Diagnosis. *NDT E Int.* **2018**, *96*, 58–78.
8. Mezgeen, R.; Vega, P.; Mercedes, S.; Jorge, C.; Pais, F.; Caio, S. An experimental and numerical approach to combine Ground Penetrating Radar and computational modelling for the identification of early cracking in cement concrete pavements. *NDT E Int.* **2020**, *115*, 102293.
9. Mercedes, S.; Susana, L.; Belen, R.; Henrique, L. Non-destructive testing for the analysis of moisture in the masonry arch bridge of Lubians (Spain). *Struct. Control Health Monit.* **2013**, *20*, 1366–1376.
10. Saarenketo, T.; Scullion, T. Road evaluation with ground penetrating radar. *J. App. Geophys.* **2000**, *43*, 119–138. [CrossRef]
11. Al-Qadi, I.; Leng, Z.; Lahouar, S.; Baek, J. In-place hot-mix asphalt density estimation using ground-penetrating radar. *Transp. Res. Rec. J. Transp. Res. Board* **2010**, *2152*, 19–27. [CrossRef]
12. Shangguan, P.C.; Al-Qadi, I.; Coenen, A.; Zhao, S. Algorithm development for the application of ground-penetrating radar on asphalt pavement compaction monitoring. *Int. J. Pavement Eng.* **2016**, *17*, 189–200. [CrossRef]
13. Zhao, S.; Shangguan, P.C.; Al-Qadi, I. Application of regularized deconvolution technique for predicting pavement thin layer thicknesses from ground penetrating radar data. *NDT E Int.* **2015**, *73*, 1–7. [CrossRef]
14. Zhao, S.; Al-Qadi, I. Development of an analytic approach utilizing the extended common midpoint method to estimate asphalt pavement thickness with 3-D ground-penetrating radar. *NDT E Int.* **2016**, *78*, 29–36. [CrossRef]
15. Plati, C.; Loizos, A. Using ground-penetrating radar for assessing the structural needs of asphalt pavement. *Nondestr. Test. Eval.* **2012**, *27*, 273–284. [CrossRef]
16. Maser, K.R.; Scullion, T. Automated pavement subsurface profiling using radar: Case studies of four experimental field sites. *Transp. Res. Rec. J. Transp. Res. Board* **1992**, *1344*, 148–154.
17. Poikajarvi, J.; Peisa, K.; Herronen, T.; Aursand, P.O.; Maijala, P.; Narbro, A. GPR in road investigations-equipment tests and quality assurance of new asphalt pavement. *Nondestr. Test. Eval.* **2012**, *27*, 293–303. [CrossRef]
18. Goel, A.; Das, A. Nondestructive testing of asphalt pavements for structural condition evaluation: A state of the art. *Nondestr. Test. Eval.* **2008**, *23*, 121–140. [CrossRef]
19. Spagnolini, U. Permittivity measurements of multilayered media with monostatic pulse radar. *IEEE Trans. Geosci. Remote* **1997**, *35*, 454–463. [CrossRef]
20. Zhao, S.; Al-Qadi, I. Algorithm development for real-time thin asphalt concrete overlay compaction monitoring using ground-penetrating radar. *NDT E Int.* **2019**, *104*, 114–123. [CrossRef]
21. Plati, C.; Georgouli, K.; Cliatt, B.; Loizos, A. Incorporation of GPR data into genetic algorithms for assessing recycled pavements. *Constr. Build. Mater.* **2017**, *154*, 1263–1271. [CrossRef]
22. Eskelinen, P.; Pellinen, T. Comparison of different radar technologies and frequencies for road pavement evaluation. *Constr. Build. Mater.* **2018**, *164*, 888–898. [CrossRef]
23. Tosti, F.; Ciampoli, L.B.; D'Amico, F.; Alani, A.M.; Benedetto, A. An experimental-based model for the assessment of the mechanical properties of road pavements using ground-penetrating radar. *Constr. Build. Mater.* **2018**, *165*, 966–974. [CrossRef]

24. Araujo, S.; Beaucamp, B.; Delbreilh, L.; Dargent, É.; Fauchard, C. Compactness/density assessment of newly-paved highway containing recycled asphalt pavement by means of non-nuclear method. *Constr. Build. Mater.* **2017**, *154*, 1151–1163. [CrossRef]
25. Yao, K.; Chen, Q.; Xiao, H.; Liu, Y.; Lee, F.H. Small-strain shear modulus of cement-treated marine clay. *J. Mater. Civ. Eng.* **2020**, *32*, 04020114. [CrossRef]
26. Marecos, V.; Fontul, S.; Solla, M.; de Lurdes Antunes, M. Evaluation of the feasibility of Common Mid-Point approach for air-coupled GPR applied to road pavement assessment. *Measurement* **2018**, *128*, 295–305. [CrossRef]
27. Hu, J.H.; Vennapusa, P.; White, D.J.; Beresnev, I. Pavement thickness and stabilized foundation layer assessment using ground-coupled GPR. *Nondestr. Test. Eval.* **2016**, *31*, 267–287. [CrossRef]
28. Marecos, V.; Fontul, S.; Antunes, M.; Solla, M. Evaluation of a highway pavement using non-destructive tests Falling Weight Deflectometer and Ground Penetrating Radar. *Constr. Build. Mater.* **2017**, *154*, 1164–1172. [CrossRef]
29. Dong, Z.H.; Ye, S.B.; Gao, Y.Z.; Fang, G.; Zhang, X.; Xue, Z.; Zhang, T. Rapid detection methods for asphalt pavement thicknesses and defects by a vehicle-mounted Ground Penetrating Radar (GPR) system. *Sensors* **2016**, *16*, 2067. [CrossRef]
30. Kassem, E.; Chowdhury, A.; Scullion, T.; Masad, E. Application of ground-penetrating radar in measuring the density of asphalt pavements and its relationship to mechanical properties. *Int. J. Pavement Eng.* **2016**, *17*, 503–516. [CrossRef]

coatings

Article

Evaluation of Properties and Mechanisms of Waste Plastic/Rubber-Modified Asphalt

Xiaorui Zhang [1,*], Chao Han [1], Frédéric Otto [2] and Fan Zhang [1]

1 School of Transportation, Southeast University, Nanjing 211189, China; hc1527@jsti.com (C.H.); 101004978@seu.edu.cn (F.Z.)
2 Chair and Institute of Highway Engineering, Mies-van-der-Rohe-Straße 1, 52074 Aachen, Germany; otto@isac.rwth-aachen.de
* Correspondence: zxr@seu.edu.cn; Tel.: +86-19852839976

Abstract: Waste plastic, such as polyethylene (PE), and waste rubber tires, are pollutants that adversely affect the environment. Thus, the ways these materials are used are important in realizing the goals of reduced CO_2 emissions and carbon neutrality. This paper investigates the fundamental properties, compatibility, and interaction mechanism of waste plastic/rubber-modified asphalt (WPRMA). Dynamic shear rheology, fluorescence microscopy, a differential scanning calorimeter, and molecular dynamic simulation software were used to evaluate the properties and mechanisms of WPRMA. The results show that the anti-rutting temperature of WPRMA with different waste plastic contents is higher than 60 °C and the optimal dosage of waste PE in WPRMA is 8%, which can enhance the high-temperature properties and compatibility of rubber-modified asphalt. The temperature can directly promote the melting and decomposition of the functional groups in WPRMA and thus must be strictly controlled during the mix production process. The interaction mechanism suggests that waste plastic can form networks and package the rubber particles in rubber-modified asphalt. The main force between waste plastic and rubber is Van der Waals force, which rarely occurs in chemical reactions.

Keywords: rubber-modified asphalt; waste plastic; polyethylene; molecular simulation; interaction mechanism

Citation: Zhang, X.; Han, C.; Otto, F.; Zhang, F. Evaluation of Properties and Mechanisms of Waste Plastic/Rubber-Modified Asphalt. *Coatings* **2021**, *11*, 1365. https://doi.org/10.3390/coatings11111365

Academic Editors: Valeria Vignali and Giorgos Skordaris

Received: 24 September 2021
Accepted: 3 November 2021
Published: 7 November 2021

Publisher's Note: MDPI stays neutral with regard to jurisdictional claims in published maps and institutional affiliations.

Copyright: © 2021 by the authors. Licensee MDPI, Basel, Switzerland. This article is an open access article distributed under the terms and conditions of the Creative Commons Attribution (CC BY) license (https://creativecommons.org/licenses/by/4.0/).

1. Introduction

Waste plastic has caused great concern because it pollutes the soil, sea, and atmosphere, especially when it is in the form of micro-nano plastic [1]. Micro-nano plastic is emitted into the atmosphere and thus directly and negatively affects human health, leading to diseases such as pneumoconiosis and cancer [2]. The total quantity of emitted waste plastic is 8 million tons each year around the world [3]. Therefore, positive ways to reuse waste plastic have become a major research focus. For example, researchers have found that waste plastic can act as a modifier to improve the properties of asphalt, which is an effective way to reuse waste plastic [4].

The various types of waste plastic and their compositions are complex in terms of their properties and structure [5]. Polyethylene (PE) is the most common waste plastic and is versatile [6]. PE can be categorized as high-density PE, medium-density PE, and low-density PE, but only medium-density PE and low-density PE can be used in modified asphalt [7]. The different structures of PE provide different modification effects for asphalt. For example, medium-density PE-modified asphalt offers the best high-temperature properties and low-density PE-modified asphalt offers the best low-temperature properties [8]; that is, the inclusion of PE or waste PE into the mix can control the high-temperature or low-temperature properties of asphalt. Therefore, if waste PE is used to modify rubber-modified asphalt, then the properties of the asphalt and the interaction between the waste plastic and rubber-modified asphalt need to be investigated.

In recent years, many research efforts have focused on rubber-modified asphalt. For example, Jitsangiam et al. [9] investigated the effects of rubber latex on rubber-modified asphalt based on macro and micro-observations and found that the performance of rubber-modified asphalt depends not only on the rubber latex content but also on the various internal interactions between the rubber latex and asphalt binder. Wang et al. [10] evaluated the properties of rubber asphalt that had been modified by bio-oil and found that swine manure bio-oil can reduce the viscosity of the binder and improve its workability. In other work, Wang et al. [11] studied the fundamental properties and mechanisms of rubber asphalt that had been modified by tetraethyl orthosilicate and showed that the physical and rheological properties of the tetraethyl orthosilicate and rubber-modified asphalt could be improved due to the formation of a network structure between the silicon -oxygen bonds (Si–O–Si) and rubber powder. Cao et al. [12] investigated the flue gas composition of waste rubber-modified asphalt and found that the main hazardous flue gases are sulfur-containing organics, benzene homologues, and hydrogen sulfide (H_2S). Ren et al. [13] examined the effects of the swelling-degradation degree of crumb rubber on crumb rubber styrene-butadiene-styrene (CR/SBS) -modified porous asphalt binder and mixture. Ren et al.'s results showed that CR/SBS-modified bitumen could be used as a high-viscosity binder in porous asphalt. Yin et al. [14] studied the mechanical properties and reaction mechanism of microwave-activated crumb rubber-modified asphalt and found that the microwave activation of the crumb rubber improved the anti-aging properties of the asphalt. Gui et al. [15] evaluated the rheological characteristics of warm-mix crumb rubber-modified asphalt and found that the combination of crumb rubber and warm-mix crumb rubber-modified asphalt exhibits a positive synergistic effect on the asphalt's rheological properties. The aforementioned studies suggest that adding rubber particles alone cannot achieve the excellent properties of rubber-modified asphalt, because rubber-modified asphalt has poor compatibility and low-temperature properties. Therefore, the key to the effective use of rubber-modified asphalt is to improve its low-temperature and compatibility properties.

Previous research has shown that low-density PE can improve the high-temperature rheological properties and low-temperature crack resistance of rubber-modified asphalt, as well as the compatibility between the rubber particles and asphalt binder [16]. Liang et al. [17] investigated the utilization of wax residue as a compatibilizer for rubber/recycled PE-modified asphalt and found that wax residue can enhance the storage stability of asphalt with a composite modifier and that the proper amount of wax residue improves the mixture's low-temperature behavior. Yan et al. [18] found that the inclusion of low-density waste PE can enhance the high-temperature properties of rubber-modified asphalt but diminishes its low-temperature properties. Gibreil and Feng [19] found that high-density PE can improve rubber-modified asphalt's resistance to moisture, while Alghrafy et al. [20] studied the rheological properties and aging performance of sulfur-extended asphalt modified with waste PE and found that the waste PE made the asphalt stiffer and thus better resistant to rutting/fatigue cracking. The aforementioned studies confirm that the inclusion of waste PE in the mix can improve the rheological properties and compatibility of rubber-modified asphalt. However, the precise effects of waste PE on the rheological properties and interaction mechanism of rubber-modified asphalt remain unclear.

In this study, the waste PE were added into rubber-modified asphalt to improve the properties of rubber-modified asphalt. The properties of waste plastic/rubber-modified asphalt (WPRMA) such as rheological property and storage stability are evaluated by using dynamic shear and softening point test. And also the interaction mechanism of WPRMA mechanisms are analysized by using fluorescence microscopy, a differential scanning calorimeter (DSC), and molecular dynamic simulation software.

2. Materials and Methods

The materials used in the study, as well as the laboratory tests conducted, are discussed in this section. This includes the raw materials, test methods, and preparation of WPRMA.

2.1. Raw Materials

The following raw materials were purchased for this study: waste PE from Taiyuan, Shanxi Province, China, rubber powder (80 mesh) from Changzhou Rongao Chemical New Material Co., Ltd., Changzhou, China. Solubilizer was bought from Sinopharm Chemical Co., Ltd., Shanghai, China, and the base asphalt SK-70 from Panjin Northern Asphalt Co., LTD, Panjin, China. The fundamental properties of the base asphalt are the softening point of 46 °C, penetration of 62 dmm, and ductility (15 °C) of 108 cm.

2.2. WPRMA Preparation

The WPRMA was prepared as follows. Firstly, the base asphalt SK-70 was heated to 135 °C in a stainless-steel asphalt binder extractor. Then, the rubber powder was added into the base asphalt slowly and stirred at 600 rpm for 1 h. Next, the mix was heated to 175 °C and put into the waste PE. The waste PE was then sheared for 1.5 h at a high-speed 5000 rpm shear for 1 h until the waste PE was completely dissolved in the asphalt binder. The waste rubber powder content was 20% and the dosages of the waste PE were 0%, 2%, 4%, 6%, 8%, and 10%, respectively.

2.3. Laboratory Experimentation and Testing

The fundamental properties of the WPRMA were tested as follows. The penetration of the WPRMA was measured at 25 °C according to ASTM D5/D5M-20. The softening point and ductility of the WPRMA were tested according to ASTM D36/D36M-20 and ASTM D113, respectively. The dynamic viscosity of the WPRMA was measured at 135 °C using a Brookfield viscometer in accordance with ASTM D4402. The rheological properties of the WPRMA were tested using an Anton Paar dynamic shear rheometer in accordance with ASTM D7175-15. The high-temperature rheological properties were tested using a 25-mm diameter parallel plate with a gap of 1 mm, and the low-temperature rheological properties were tested using an 8 mm diameter parallel plate with a gap of 2 mm. The tests were performed under controlled strain conditions. Temperature sweeps were applied with a fixed frequency of 10 rad/s, and the temperature ranged from 10 to 30 °C and 30 to 80 °C. The storage stability of the WPRMA was measured using a NETZSCH differential scanning calorimeter (DSC). The test temperature ranged from 50 to 240 °C and the heating rate was 10 °C/min. The softening point difference also was used to evaluate the storage stability of the WPRMA. The micro-morphology of the WPRMA was measured using an optical microscope (Shangguang, XSP-63, Shanghai, China) at 100× magnification.

2.4. Molecular Dynamic Modeling

For the molecular simulations of the WPRMA, PE ($C_{1600}H_{3208}$) was used as the waste plastic and butyl rubber (C_9H_{16}) represented the rubber powder. The asphalt was selected using the four components model which had been used in previous research [21]. Molecular models of WPRMA were built using layered modules. Figure 1 presents the molecular models of the WPRMA with 0%, 2%, 4%, 6%, 8%, and 10% waste PE.

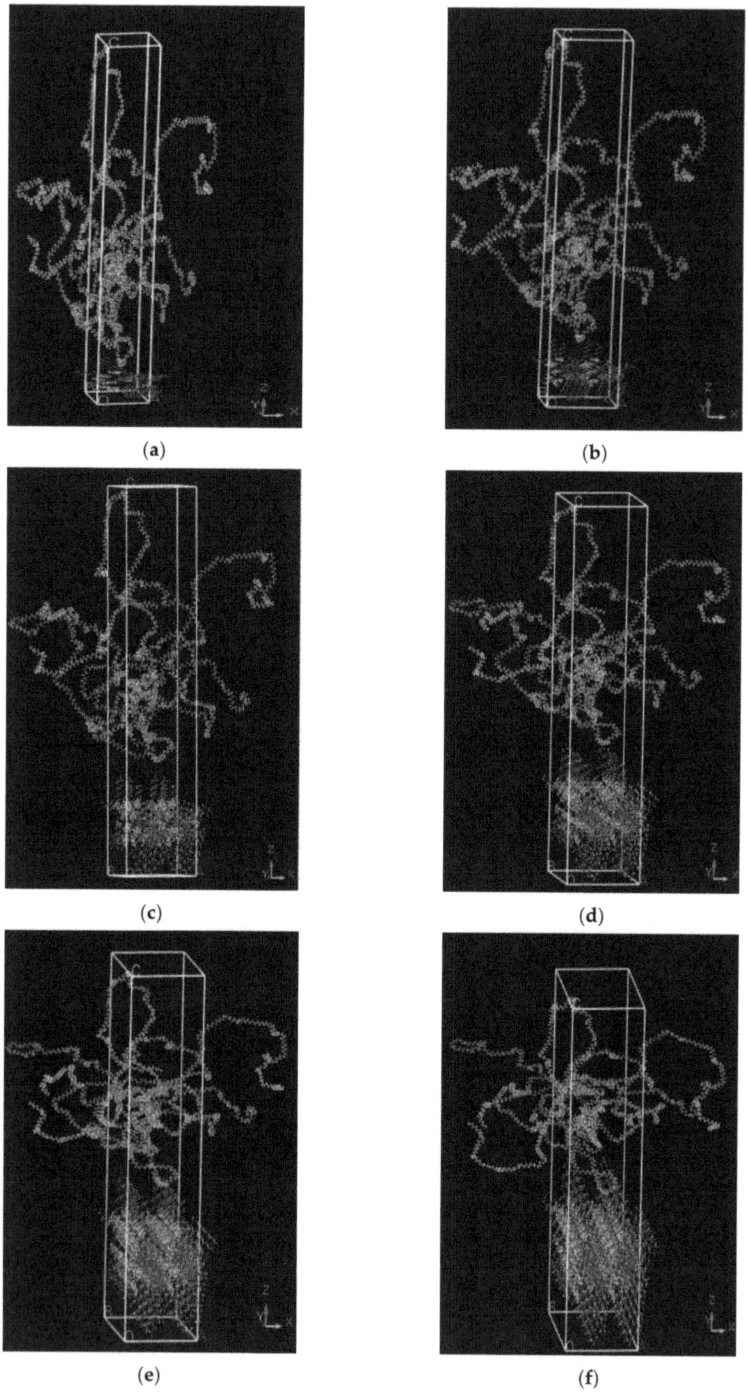

Figure 1. Molecular models of waste plastic/rubber-modified asphalt at various percentages of waste polyethylene: (a) 0% waste PE; (b) 2% waste PE; (c) 4% waste PE; (d) 6% waste PE; (e) 8% waste PE; (f) 10% waste PE.

During the molecular simulation process, the molecular models of the WPRMA were first treated using geometry optimization. Then, the cohesive energy density formula was used to calculate the cohesive energy density between the waste PE and rubber-modified asphalt and to analyze the compatibility of the WPRMA. Molecular dynamic simulations were performed to evaluate the diffusion coefficients, interaction force, and structural changes of the WPRMA using the Forcite module with NVT and NPT ensembles. The total simulation time was 100 ps and the time step was 1 fs.

3. Results and Discussion

3.1. Assessment of Reliability of Molecular Simulation Results

The energy and density results were used to assess the reliability of the molecular simulations. Figure 2a shows that the energy of the WPRMA decreases with an increase in the simulated time and that the energy fluctuation of the WPRMA trends to zero when the simulated time is five steps. The results show that the total time of 100 steps is acceptable because only the energy fluctuation trends to zero, which indicates that the system is stable and the results for this time interval are reasonable. Figure 2b shows that the density of the WPRMA decreases, then increases sharply, then decreases slightly, and finally trends to a constant (1.02 g/cm^3). Moreover, the constant density values are equal to the real density values of asphalt, which suggests that the molecular models can be used to simulate the properties of WPRMA.

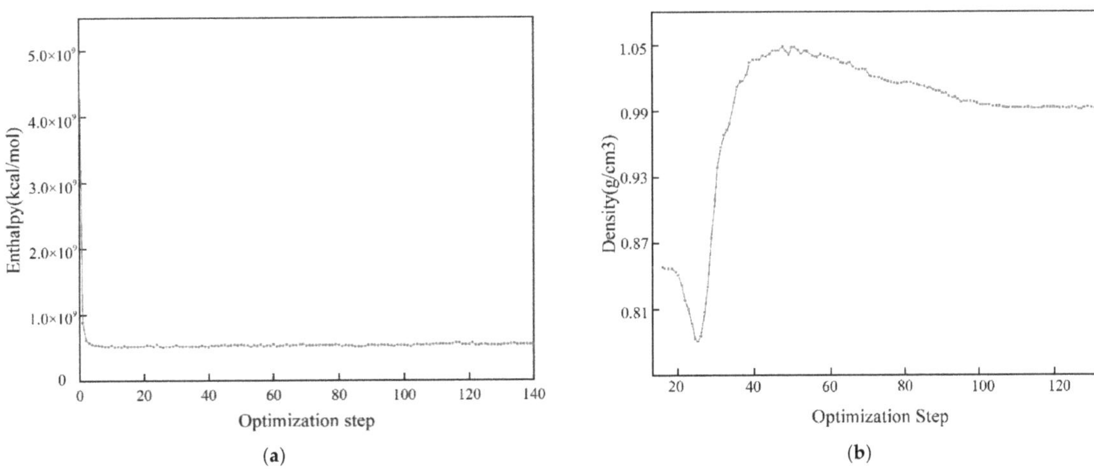

Figure 2. Reliability assessment of energy and density results obtained from molecular simulations: (**a**) energy curves, and, (**b**) density curves.

3.2. Assessment of WPRMA Properties via Molecular Simulations

Molecular simulations also can be used to evaluate the properties of WPRMA, i.e., the diffusion coefficient, viscosity, glass transition temperature, free volume, and cohesive energy density, as discussed in the following paragraphs.

The diffusion coefficient can be used to analyze the movement of molecules and evaluate the self-healing properties of the mix. For the molecular simulations, the diffusion coefficient can be calculated by Equation (1).

$$D = \frac{1}{6T}|r(t) - r(0)|^2 \tag{1}$$

where D is the diffusion coefficient, T is the total time of the molecular movement, $r(t)$ is the displacement at time t, and $r(0)$ is the displacement at time 0.

Figure 3 shows that the diffusion coefficient of WPRMA decreases with an increase in the dosage of the waste PE and that the rate of decrease of the diffusion coefficient slows when the waste PE dosage is above 8%. The reason for this outcome is that, owing to the increase in its dosage, the waste PE curls up and becomes entwined with the rubber-modified asphalt, which leads to the slow molecular movement of the waste PE. Thus, these results show that waste PE can inhibit the movement of WPRMA and diminish the self-healing properties of WPRMA.

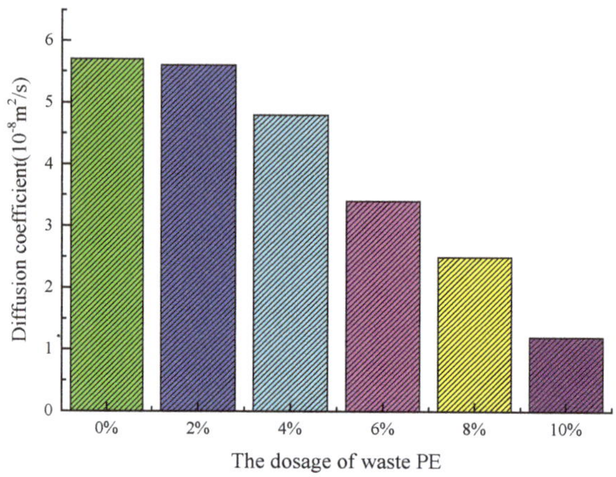

Figure 3. Diffusion coefficient of waste polyethylene in waste plastic/rubber-modified asphalt.

Viscosity is a key parameter that can be used to evaluate the viscoelastic properties of asphalt. For the molecular simulations, viscosity can be calculated by Equation (2).

$$\eta = \frac{\rho RT{R_g}^2}{6MD} \quad (2)$$

where η is viscosity, ρ is density, R is constant (8.314), T is the total time, ${R_g}^2$ is average radius of gyration, M is the molecular mass, and D is the diffusion coefficient.

Figure 4 shows that the simulated viscosity of WPRMA increases with an increase in the dosage of waste PE. In this research, it selected the optimal dosage of waste PE as 8% because, as shown in the figure, the viscosity of the WPRMA increases slowly when the dosage is above 8%. These results indicate that waste PE can increase the viscosity and enhance the adhesive properties of WPRMA. Previous research found that viscosity is related to the high-temperature properties of asphalt, whereby the greater the viscosity, the better the high-temperature properties of the asphalt [22]. Similarly, it can be concluded that waste PE can promote improvement in the high-temperature properties of WPRMA. Previous research also found that the viscosity of rubber-modified asphalt ranges from 300 to 400 cp. Compared with these earlier results, it was found that the simulated viscosity of rubber-modified asphalt is 320 cp, which indicates that the simulated results are reliable.

Our previous research found that density-temperature curves can be used to indicate the real glass transition temperature via molecular simulations [23]. Similarly, in this study, it can be obtained the glass transition temperature (T_g) from density-temperature curves using molecular simulations. Figure 5 shows that the glass transition temperature of WPRMA increases with an increase in the dosage of waste PE, while the glass transition temperature of WPRMA is lower than that of rubber-modified asphalt. This finding indicates that waste PE can increase the glass transition temperature of WPRMA and improve its low-temperature properties. Previous research also found that low-density PE could improve the low-temperature properties of asphalt [24]. This finding verifies the reliability of the simulation results for the glass transition temperature of WPRMA.

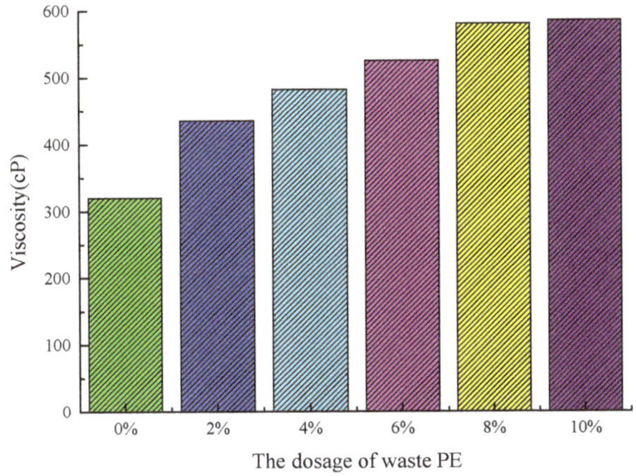

Figure 4. Viscosity of waste plastic/rubber-modified asphalt at various dosages of waste polyethylene.

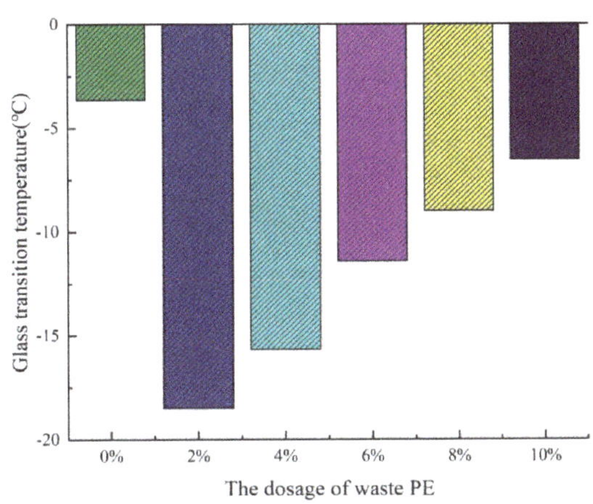

Figure 5. Glass transition temperatures of waste plastic/rubber-modified asphalt at various dosages of waste polyethylene.

Free volume can be used to evaluate the mobility properties of WPRMA. Figure 6 presents the free volume results of WPRMA with different dosages of waste PE. As shown, the Van der Waal forces with free volume percentages of 0%, 2%, 4%, 6%, 8%, and 10% WPRMA are 158,490.24 Å^3, 171,425.57 Å^3, 197,199.04 Å^3, 220,463.42 Å^3, 247,293.81 Å^3, and 270,809.26 Å^3, respectively. The free volume of the WPRMA is shown to increase with the increase in the waste PE dosage, which indicates that waste PE can increase the free volume of WPRMA and promote the mobility properties of WPRMA.

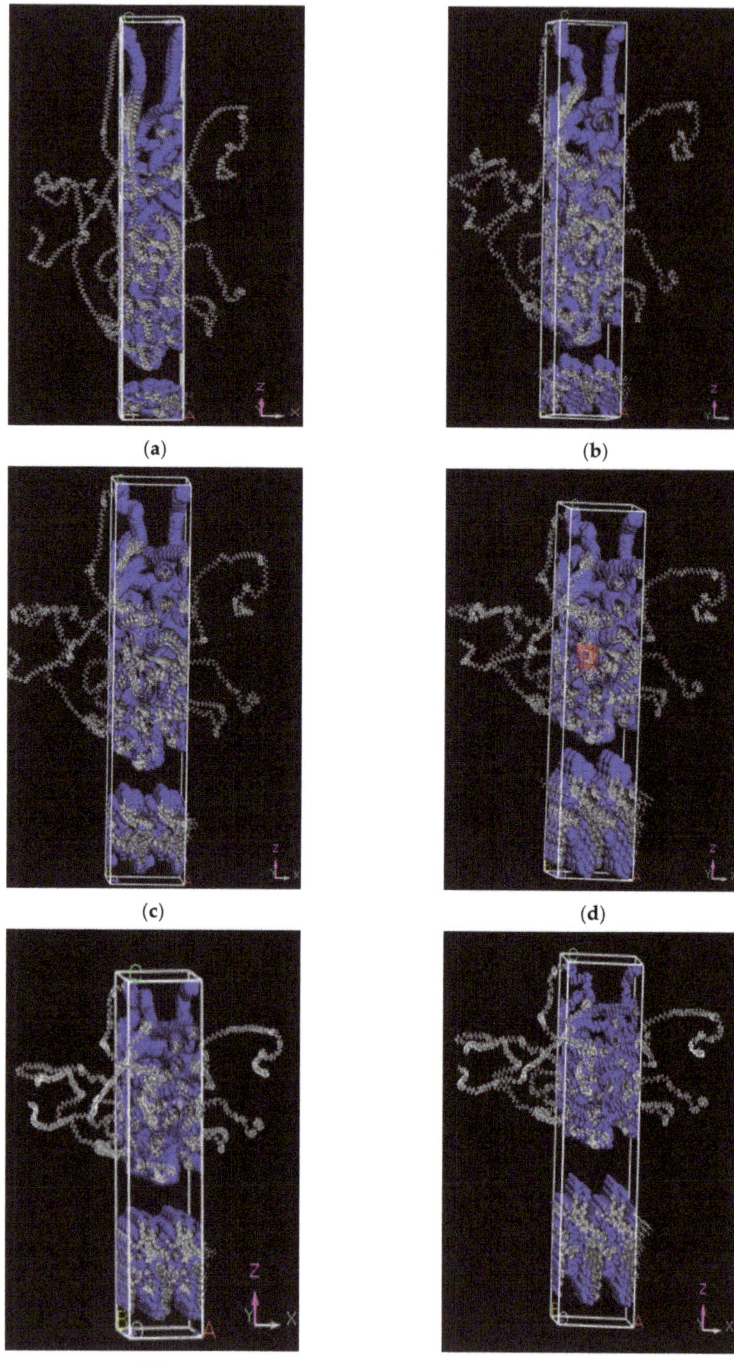

Figure 6. Free volume of waste plastic/rubber-modified asphalt with different percentages of waste polyethylene: (**a**) 0% waste PE; (**b**) 2% waste PE; (**c**) 4% waste PE; (**d**) 6% waste PE; (**e**) 8% waste PE; (**f**) 10% waste PE.

Cohesive energy density can be used to evaluate the interaction force of molecules and to calculate the solubility parameter of the molecules. The solubility parameter of WPRMA can be calculated by Equation (3).

$$\delta = k\sqrt{CED} \quad (3)$$

where δ is the solubility parameter, k is the constant number, and CED is the cohesive energy density.

Figure 7a shows that the cohesive energy density values for 0%, 2%, 4%, 6%, 8%, and 10% WPRMA are 2.09×10^7 J/m^3, 3.12×10^7 J/m^3, 3.43×10^7 J/m^3, 3.45×10^7 J/m^3, 3.48×10^7 J/m^3, and 4.84×10^6 J/m^3, respectively. The cohesive energy density of WPRMA first increases, then decreases with an increase in the waste PE dosage, which indicates that waste PE can promote an increase in the interaction force of WPRMA when the dosage is less than 8%. However, waste PE will diminish the interaction force of WPRMA when its dosage is 10%. This outcome indicates that the optimal dosage of waste PE is 8%, which provides the best cohesive energy density. Figure 7b shows that the solubility parameter of WPRMA follows a similar law to that of cohesive energy density. The solubility parameter of WPRMA first increases and then decreases with an increase in the waste PE. These results suggest that waste PE can promote the dissolution of WPRMA to a certain extent. If the waste PE dosage is above 8%, then the waste PE will inhibit the dissolution of the WPRMA.

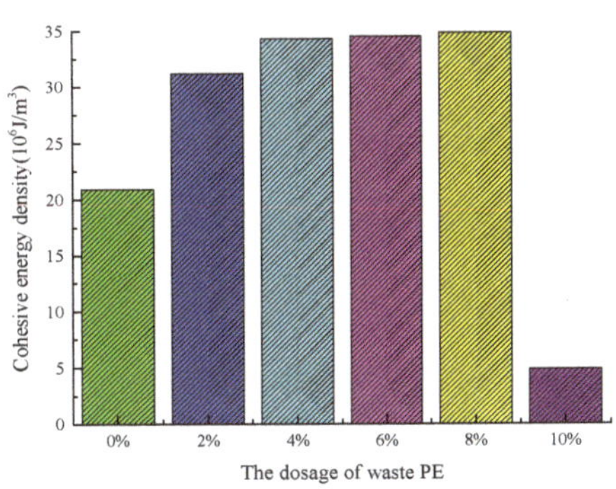

(a)

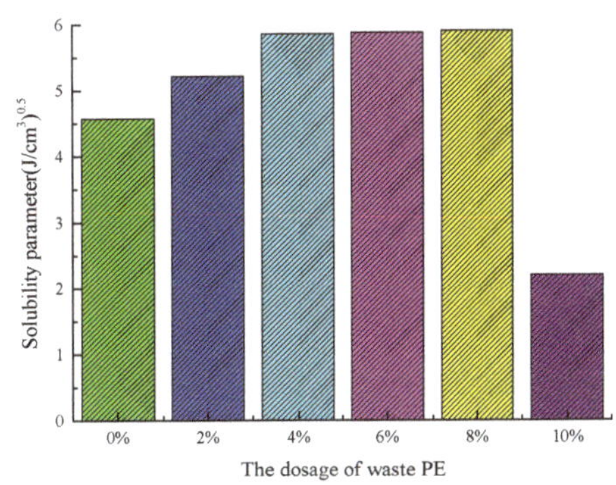

(b)

Figure 7. (**a**) Cohesive energy density and (**b**) solubility parameter of waste plastic/rubber-modified asphalt at various dosages of waste polyethylene.

3.3. *Assessment of WPRMA Rheological Properties and Storage Stability via Tests*

Figure 8a presents the high-temperature scan curves that show a decrease in the complex modulus value of the WPRMA with an increase in temperature and an increase in the complex modulus value of the WPRMA with an increase in the waste PE dosage. The reason for this outcome is that the fill in waste PE can enhance the complex modulus of WPRMA. These results show that waste PE can enhance the high-temperature properties of WPRMA, but high temperatures would diminish the viscoelastic properties of WPRMA. The anti-rutting temperatures at 0%, 2%, 4%, 6%, 8%, and 10% WPRMA are 62 °C, 64 °C, 65 °C, 67 °C, 68 °C, and 69 °C, respectively. These temperatures clearly are higher than 60 °C. Figure 8b presents the low-temperature scan curves that indicate that the complex modulus values of the WPRMA decrease with an increase in temperature and increase with an increase in waste PE. The reason for this outcome is that the PE in waste PE plays an important role by enhancing the viscoelastic properties of WPRMA. These results show that the PE in waste PE can enhance the low-temperature viscoelastic properties and crack resistance of WPRMA.

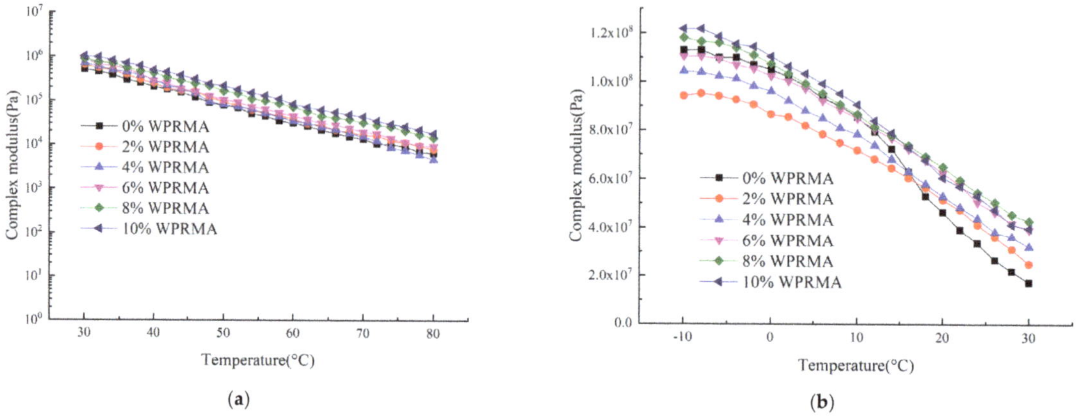

Figure 8. Temperature scan curves of waste plastic/rubber-modified asphalt at: (**a**) high temperatures and (**b**) low temperatures.

The softening point difference can be used to evaluate the storage stability of WPRMA. Figure 9 shows that the difference in softening point of the rubber-modified asphalt is greater than 2 °C, which indicates that the storage stability of the rubber-modified asphalt is poor. The softening point difference of WPRMA is less than that of rubber-modified asphalt, that is, lower than 2 °C, except for when the asphalt contains 10% waste PE. The softening point difference of WPRMA increases with an increase in the waste PE dosage. These results suggest that waste PE can improve the storage stability of WPRMA and promote the solubility of rubber powder in asphalt. However, once the waste PE dosage is above 8%, the inclusion of waste PE will diminish the storage stability of WPRMA.

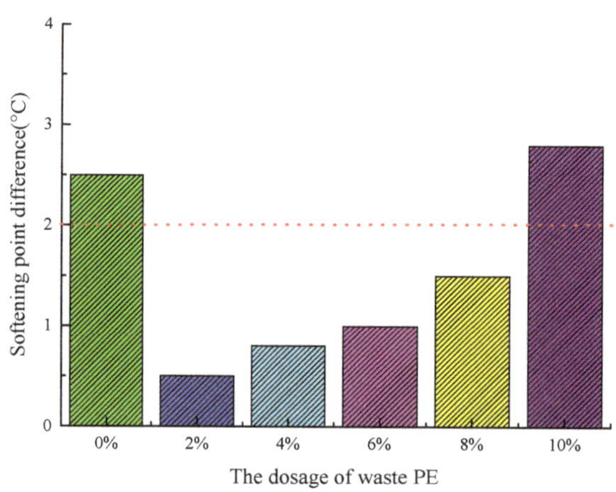

Figure 9. Softening point difference of waste plastic/rubber-modified asphalt at various dosages of waste polyethylene (PE).

3.4. Interaction Mechanism of WPRMA

Figure 10 presents the DSC curves of the WPRMA and the endothermic melting peaks when the temperature reaches about 90 °C. After the heating process, the melting curve is basically stable, which indicates that most of the functional groups of waste PE undergo a phase transformation at about 90 °C and that this kind of material absorbs heat easily and then melts. The endothermic melting peaks of the rubber-modified asphalt appear at 120 °C and the endothermic melting peaks of 6% WPRMA appear at 160 °C, which indicates that the waste PE degraded first, then the asphalt, and finally, the rubber powder. Therefore, we recommend a temperature range from 100 °C to 160 °C for WPRMA. WPRMA has different melting peaks, which suggests that the waste PE dosage can significantly affect the degradation of WPRMA and that a synergistic effect is present between waste PE and rubber powder.

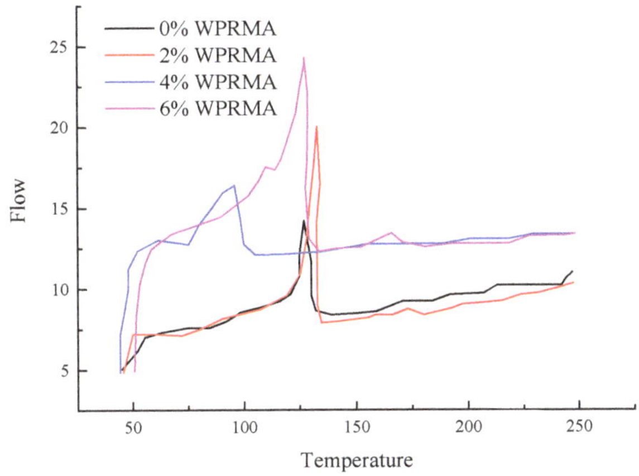

Figure 10. Differential scanning calorimeter curves of waste plastic/rubber-modified asphalt.

Figure 11 presents fluorescence microtopographic images of WPRMA that show the significant change in micro-morphology and increase in the movement phase with an increase in the waste PE. These results show that waste PE can improve the micro-morphology and flow of WPRMA. The micro-morphology of 2% or 4% WPRMA still shows an obvious granular shape, indicating that the waste PE dosage is too low and cannot completely melt the compound additive. The movement phase in 6% WPRMA indicates that waste PE and rubber powder can melt completely. Networks appear when the waste PE dosage is 8%, suggesting that 8% WPRMA offers the best compatibility, whereas the separation phase of 10% WPRMA indicates that 10% waste PE can diminish the compatibility of WPRMA.

The micro-mechanisms of WPRMA suggest the presence of (1) a granular shape when the waste PE dosage is low (2–4%), (2) a movement phase when the waste PE dosage is 6%, (3) networks when the waste PE dosage is 8%, and (4) a separation phase when the waste PE dosage is 10%. In short, this micro-morphology change law indicates a granular shape, movement phase, networks, and separation phase. Temperature is the factor that directly promotes the melting and decomposition of the functional groups in WPRMA, and thus, the temperature must be strictly controlled during the mixture production process.

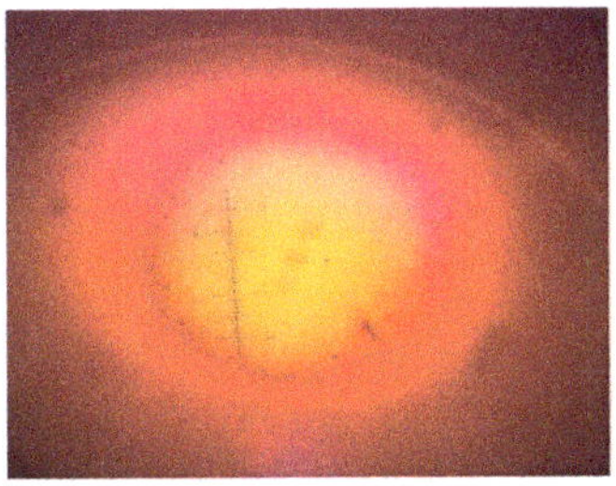

(**a**)

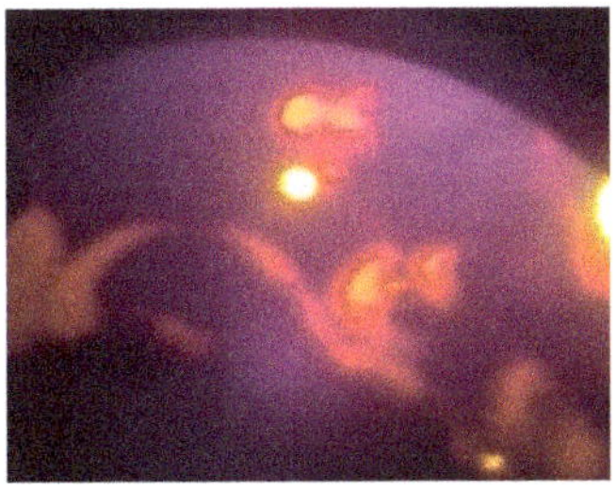

(**b**)

Figure 11. *Cont.*

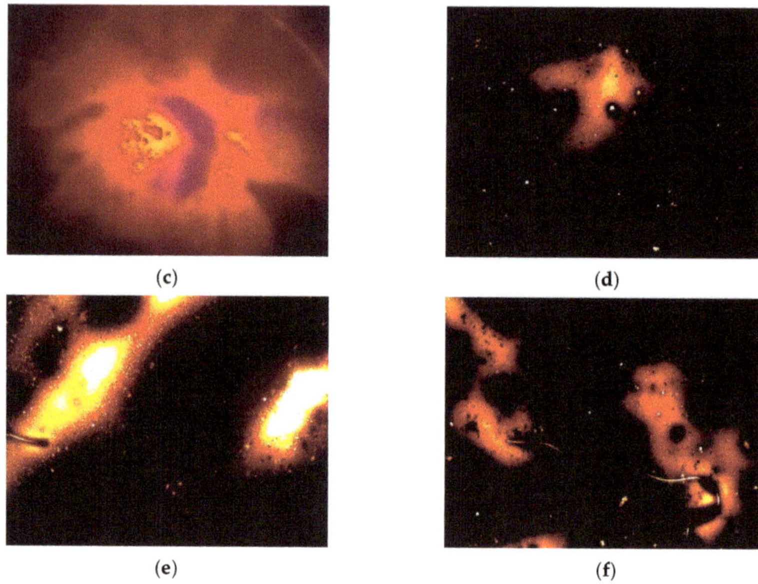

Figure 11. The micro-morphology of waste plastic/rubber-modified asphalt at various dosages of waste polyethylene: (a) 0% waste PE; (b) 2% waste PE; (c) 4% waste PE; (d) 6% waste PE; (e) 8% waste PE; (f) 10% waste PE.

4. Conclusions

In this study, the properties and mechanisms of WPRMA were evaluated using dynamic shear rheology, fluorescence microscopy, a DSC, and molecular dynamic simulation software. The following conclusions can be drawn from the results.

- The inclusion of waste PE in the mix can inhibit the movement of WPRMA and diminish its self-healing properties. However, the inclusion of waste PE can increase the viscosity and enhance the adhesive properties of WPRMA. Furthermore, waste PE can enhance the high-temperature properties of WPRMA, but high temperatures can diminish the viscoelastic properties of WPRMA due to the presence of fill in waste PE.
- The inclusion of waste PE in the mix can increase the free volume of WPRMA and promote its mobility properties. During the degradation process of WPRMA, first the waste PE degrades, then the asphalt, and finally the rubber powder. The dosage of the waste PE can significantly affect the degradation of WPRMA because waste PE and rubber powder are synergetic.
- The micro-morphology of WPRMA in terms of change regularly indicates a granular shape, movement phase, networks, and a separation phase. Temperature is the factor that directly promotes the melting and decomposition of the functional groups in WPRMA. Therefore, temperature must be strictly controlled during the mixture production process.

Author Contributions: Conceptualization, X.Z.; methodology, X.Z.; software, C.H.; validation, F.Z. investigation, C.H.; data curation, C.H.; writing—original draft preparation, X.Z.; writing—review and editing, F.Z.; visualization, F.Z.; supervision F.O.; funding acquisition, C.H. All authors have read and agreed to the published version of the manuscript.

Funding: This work was supported by the Special Foundation of Achievements Transformation Guide of the Department of Science and Technology of Shanxi Province (No. 201804D131034).

Institutional Review Board Statement: Not applicable.

Informed Consent Statement: Not applicable.

Data Availability Statement: The data presented in this study are available on request from the corresponding author.

Conflicts of Interest: The authors declare no conflict of interest.

References

1. Wang, J.; Emmerich, L.; Wu, J.; Vana, P.; Zhang, K. Hydroplastic polymers as eco-friendly hydrosetting plastics. *Nat. Sustain.* **2021**, *7*, 877–883. [CrossRef]
2. Haritz, S.; Andrew, P. Plastics recycling with a difference. *Science* **2018**, *360*, 380–381.
3. Yuan, L.; Buzoglu, K.; Tang, C. Alternative plastics. *Nat. Sustain.* **2021**, *4*, 2121. [CrossRef]
4. Tsang, Y.; Kumar, V.; Smadar, P.; Yang, Y.; Leed, J.; Ok, Y.S.; Song, H.; Kim, K.-H.; Kwonf, E.E.; Jeon, Y.J. Production of bioplastic through food waste valorization. *Environ. Int.* **2019**, *119*, 625–644. [CrossRef]
5. Lam, S.; Mahari, W.; Ok, Y.; Penga, W.; Chong, C.T.; Ma, N.L.; Chase, H.A.; Liew, Z.; Yusup, S.; Kwon, E.E.; et al. Microwave vacuum pyrolysis of waste plastic and used cooking oil for simultaneous waste reduction and sustainable energy conversion: Recovery of cleaner liquid fuel and techno-economic analysis. *Renew. Sustain. Energy Rev.* **2019**, *115*, 109359. [CrossRef]
6. Wang, C.; Xian, Z.; Jin, X.; Liang, S.; Chen, Z.; Pan, B.; Wu, B.; Ok, Y.S.; Gu, C. Photo-Aging of polyvinyl chloride microplastic in the presence of natural organic acids. *Water Res.* **2020**, *183*, 116082. [CrossRef]
7. Du, Z.; Jiang, C.; Yuan, J.; Xiao, F.; Wang, J. Low temperature performance characteristics of polyethylene modified asphalts—A review. *Constr. Build. Mater.* **2020**, *264*, 120704. [CrossRef]
8. Coralie, J.; Haritz, S. Dynamic polymer network points the way to truly recyclable plastics. *Nature* **2019**, *568*, 467–468.
9. Jitsangiam, P.; Nusit, K.; Phenrat, T.; Kumlai, S.; Pra-Ai, S. An examination of natural rubber modified asphalt: Effects of rubber latex contents based on macro- and micro-observation analyses. *Constr. Build. Mater.* **2021**, *289*, 123158. [CrossRef]
10. Wang, H.; Jing, Y.; Zhang, J.; Cao, Y.; Lyu, L. Preparation and performance evaluation of swine manure bio-oil modified rubber asphalt binder. *Constr. Build. Mater.* **2021**, *294*, 123584. [CrossRef]
11. Wang, Z.; Xu, X.; Wang, X.; Jinyang, H.; Guo, H.; Yang, B. Performance of modified asphalt of rubber powder through tetraethyl orthosilicate (TEOS). *Constr. Build. Mater.* **2021**, *267*, 121032. [CrossRef]
12. Cao, L.; Yang, C.; Li, A.; Wang, P.; Zhang, Y.; Dong, Z. Flue gas composition of waste rubber modified asphalt (WRMA) and effect of deodorants on hazardous constituents and WRMA. *J. Hazard. Mater.* **2021**, *403*, 123814. [CrossRef]
13. Ren, S.; Liu, X.; Xu, J.; Lin, P. Investigating the role of swelling-degradation degree of crumb rubber on CR/SBS modified porous asphalt binder and mixture. *Constr. Build. Mater.* **2021**, *300*, 124048. [CrossRef]
14. Yin, L.; Yang, X.; Shen, A.; Wu, H.; Lyu, Z.; Li, B. Mechanical properties and reaction mechanism of microwave-activated crumb rubber-modified asphalt before and after thermal aging. *Constr. Build. Mater.* **2020**, *267*, 120773. [CrossRef]
15. Gui, W.; Liang, L.; Wang, L.; Gao, X.; Zhang, F. Performance evaluation of warm-mixed crumb rubber modified asphalt based on rheological characteristics. *Constr. Build. Mater.* **2021**, *285*, 122881. [CrossRef]
16. Duarte, G.M.; Faxina, A.L. High-temperature rheological properties of asphalt binders modified with recycled low-density polyethylene and crumb rubber. *Constr. Build. Mater.* **2021**, *298*, 123852. [CrossRef]
17. Liang, M.; Sun, C.; Yao, Z.; Jiang, H.; Zhang, J.; Ren, S. Utilization of wax residue as compatibilizer for asphalt with ground tire rubber/recycled polyethylene blends. *Constr. Build. Mater.* **2020**, *230*, 116966. [CrossRef]
18. Yan, K.; Xu, H.; You, L. Rheological properties of asphalts modified by waste tire rubber and reclaimed low density polyethylene. *Constr. Build. Mater.* **2015**, *83*, 143–149. [CrossRef]
19. Gibreil, H.; Cheng, P. Effects of high-density polyethylene and crumb rubber powder as modifiers on properties of hot mix asphalt. *Constr. Build. Mater.* **2017**, *142*, 101–108. [CrossRef]
20. Alghrafy, Y.; Alla, E.; El-Badawy, S. Rheological properties and aging performance of sulfur extended asphalt modified with recycled polyethylene waste. *Constr. Build. Mater.* **2020**, *273*, 121771. [CrossRef]
21. Zhou, X.; Moghaddam, T.B.; Chen, M.; Wu, S.; Adhikari, S.; Wang, F.; Fan, Z. Nano-scale analysis of moisture diffusion in asphalt-aggregate interface using molecular simulations. *Constr. Build. Mater.* **2021**, *285*, 122962. [CrossRef]
22. Yang, C.; Zhang, J.; Yang, F.; Cheng, M.; Wang, Y.; Amirkhanian, S.; Wu, S.; Wei, M.; Xie, J. Multi-Scale performance evaluation and correlation analysis of blended asphalt and recycled asphalt mixtures incorporating high RAP content. *J. Clean. Prod.* **2021**, *317*, 128278. [CrossRef]
23. Zhou, X.; Zhang, X.; Xu, S.; Wu, S.; Liu, Q.; Fan, Z. Evaluation of thermo-mechanical properties of graphene/carbon-nanotubes modified asphalt with molecular simulation. *Mol. Simul.* **2017**, *43*, 312–319. [CrossRef]
24. Liang, M.; Xin, X.; Fan, W.; Zhang, J.; Jiang, H.; Yao, Z. Comparison of rheological properties and compatibility of asphalt modified with various polyethylene. *Int. J. Pavement Eng.* **2021**, *22*, 11–20. [CrossRef]

coatings

MDPI

Article

Research on the Mechanism of Surfactant Warm Mix Asphalt Additive-Based on Molecular Dynamics Simulation

Pinhui Zhao [1,2,*,†], Mingliang Dong [1,†], Yansheng Yang [3], Jingtao Shi [2,4], Junjie Wang [1], Wenxin Wu [1], Xingchi Zhao [1], Xu Zhou [1] and Chenlong Wang [1]

1 School of Transportation Engineering, Shandong Jianzhu University, Jinan 250101, China; dongmingliang10@163.com (M.D.); wangyongjian111@163.com (J.W.); 2165690@163.com (W.W.); zxc19819366968@163.com (X.Z.); zhouxu201026@163.com (X.Z.); 17866618231@163.com (C.W.)
2 State Key Laboratory of Heavy Oil Processing, College of Chemical Engineering, China University of Petroleum (East China), Qingdao 266580, China; sjt0822@163.com
3 R&D Center of Beijing Municipal Road and Bridge Building Materials Group Co., Ltd., Beijing 102600, China; yys_upc@163.com
4 Petrochina Fuel Oil Co., Ltd., Research Institute, Beijing 100195, China
* Correspondence: zhaopinhui08@163.com
† Authors contributes equally to the paper.

Abstract: Warm mix asphalt (WMA) technology can bring certain environmental and technical benefits through reducing the temperature of production, paving, and compaction of mixture asphalt. Recent studies have shown that some WMA additives are able to reduce the temperature by increasing the lubricating properties of asphalt binder-based on the tribological theory, this paper studied the mechanism of adsorbing and lubricating film of base asphalt and WMA on the surface of stone by molecular dynamics (MD) simulation method, and the effect of surfactant WMA additive on the lubrication performance of the shear friction system of "stone–asphalt–stone". The model of base asphalt lubricating film, including saturates, aromatics, resin and asphaltene, as well as the model of warm mix asphalt lubricating film containing imidazoline-type surfactant WMA (IMDL WMA) additive molecule, were established. The shear friction system of "stone–asphalt–stone" of base asphalt and warm mix asphalt was built on the basis of an asphalt lubrication film model and representative calcite model. The results show that the addition of IMDL WMA additive can effectively improve the lubricity of asphalt, reduce the shear stress of asphalt lubricating film, and increase the stability of asphalt film. The temperature in the WMA lubricating film rises, while the adsorption energy on the stone surface decreases with the increase of shear rate, indicating that the higher the shear rate is, the more unfavorable it is for the WMA lubricating film to wrap on the stone surface. In addition, the shear stress of the WMA lubricating film decreased with increasing temperature, while the shear stress of the base asphalt lubricating film increased first and then decreased, demonstrating that the compactability of the asphalt mixture did not improve linearly with the increase of temperature.

Keywords: warm mix asphalt; additive; molecular dynamics simulation; lubrication; mechanism

Citation: Zhao, P.; Dong, M.; Yang, Y.; Shi, J.; Wang, J.; Wu, W.; Zhao, X.; Zhou, X.; Wang, C. Research on the Mechanism of Surfactant Warm Mix Asphalt Additive-Based on Molecular Dynamics Simulation. *Coatings* **2021**, *11*, 1303. https://doi.org/10.3390/coatings11111303

Academic Editor: Diego Martinez-Martinez

Received: 3 October 2021
Accepted: 22 October 2021
Published: 27 October 2021

Publisher's Note: MDPI stays neutral with regard to jurisdictional claims in published maps and institutional affiliations.

Copyright: © 2021 by the authors. Licensee MDPI, Basel, Switzerland. This article is an open access article distributed under the terms and conditions of the Creative Commons Attribution (CC BY) license (https://creativecommons.org/licenses/by/4.0/).

1. Introduction

Warm mix asphalt (WMA) refers to a form of asphalt mixture with mixing and compaction temperatures 20–55 °C lower than typical hot mix asphalt (HMA) [1–3]. The goal of WMA is to produce mixtures with substantially the same road performance and construction workability as HMA under significantly reduced production temperatures [4–8]. Due to the decrease of production temperature, WMA technology boasts many environmental benefits such as reducing fuel and energy consumption, cutting down harmful gas emissions and preventing construction personnel from toxic asphalt smoke [4,5,9–12], which has been developed rapidly in the past decades. The decrease of temperature also has many technological benefits such as delaying aging of asphalt cement, extending transport distance, improving compaction effect during cold weather, reducing pavement cooling time for faster opening to traffic, and using higher blends of recycled asphalt pavement material (RAP) in the mixture [13–15]. It also brings potential economic benefits for the reduction of energy consumption, and less wear of asphalt plant equipment and extension of service life of asphalt pavement.

At present, WMA technology can be divided into three categories [16,17]: foaming asphalt, wax (or organic) additives and chemical additives. However, thus far, the mechanism by which WMA technology makes the mixture easy to mix and compacted at low temperatures is not fully understood [18]. It was found that the reduction in binder viscosity caused by WMA additives was insufficient to explain the decrease in the production and compaction temperature of the final mixture in the field [2,19,20]. Conversely, in some cases, the addition of WMA additives strengthens the viscosity of the binder [18]. In addition, in some studies on the workability of WMA and HMA mixtures, it has been found that the compactability of the mixtures does not linearly improve with the increase of temperature, but becomes worse above a certain temperature [21,22], which makes it clear that viscosity reduction is not the only mechanism responsible for the performance of WMA at low temperatures. Recent studies have shown that improving the lubricity of asphalt binder and reducing the dynamic friction between stones may be a further mechanism of WMA [1,21–23]. The frictional behavior or relative motion of the binder as a lubricating film between two stones may play an important role in the mixing and compaction of WMA. Therefore, it is important to study the lubricating effect of WMA additives to further reveal the workability of warm asphalt mixes at low temperatures.

Dynamic friction is a physical phenomenon caused by the contact and relative movement of two or more objects, which leads to energy consumption and wear. Lubricants are widely used in many fields to reduce dynamic friction, and the performance of lubricants is usually characterized by appropriate tribological tests [21]. Tribology is the study of the interaction between surfaces in relative motion. The friction of a material placed between two solid surfaces in relative motion is usually described by a Stribeck curve [24]. As shown in Figure 1, Stribeck curve can be divided into four lubricative regions [21,24,25]:

- In the mixed state (B), the lubricating oil forms fluid dynamic pressure as the velocity increases, which reduces direct contact between the two surfaces, thereby easing friction.
- In the elastohydrodynamic state (C), friction is minimized when the surfaces are no longer in contact.
- In the hydrodynamic region (D), friction depends primarily on the viscous resistance of the lubricant at high sliding speeds. Specifically, the friction of the whole system rises again as the internal friction of the lubricant increases.

Tribology has a wide range of engineering applications in materials science, but it is an emerging discipline in asphalt technology [1,21–23]. Ingrsia et al. characterized the properties of asphalt binders with chemical WMA and wax WMA additives-based on the concept of tribology. They found that asphalt binders containing WMA additives had lower friction [1]. Gallego et al. introduced the lubrication test method for determining the influence of surfactant additives on the lubricity of asphalt binder. The study shows that the use of surfactant additives in asphalt binders can reduce the friction coefficient [23]. Canestrari et al. reviewed the latest progress of tribological tests used in the study of asphalt binder, as well as beneficial suggestions for improving these test methods [21]. Bairgi et al. studied the effect of foaming on the friction and rheological properties of foam asphalt. The results show that the foaming process changes the frictional resistance of foamed asphalt binder, especially in the state of elastohydrodynamic and fluid lubrication, the foaming process reduces the friction coefficient. Compared with rheological properties, the friction properties of foam WMA are more related to the improvement of working performance [22].

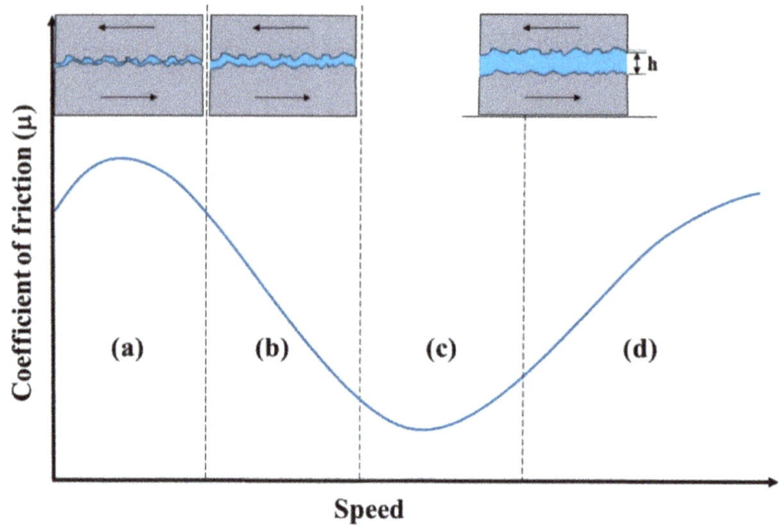

Figure 1. Stribeck curve: coefficient of friction as a function of speed (**a**) the boundary regime); (**b**) the mixed regime; (**c**) the elasto-hydrodynamic regime; (**d**) the hydrodynamic regime).

Furthermore, asphalt acts as a lubricant during asphalt mixing and compaction in the field of road materials, reducing friction between the stones, and its lubrication properties have a significant impact on the energy required. In accordance with recent studies, some WMA additives can potentially reduce asphalt mixture production and compaction temperatures (thus reducing energy consumption) by improving the lubricating effect of the binder [1,23]. Therefore, the tribological test has been introduced to characterize the lubrication properties of asphalt binder recently. The friction test of asphalt binders is mainly carried out by a tribological device mounted on a dynamic shear rheometer (DSR) with a specific geometry. However, studies have shown that attention should be paid to the geometric shape of the friction device, the selection of the matrix, the wear condition, and the determination of test procedures and conditions when studying the performance of lubricants through friction experiments [21]. Meanwhile, Puchalski found that the viscosity value of asphalt is low at high temperatures (such as those used for mixing and compacting asphalt mixture), and the chemical WMA additive can form a boundary lubricating film to protect aggregate particles from direct contact with solids, thus reducing the friction between particles [26].

At the same time, previous studies have shown that the surfactant WMA additive has little influence on the viscosity of asphalt binder within the recommended dosage range [19]. It can be reasonably speculated that the lubrication form of WMA additive between stones belongs to boundary lubrication with low temperature and high load, that is, adsorption lubrication. It is mainly completed by the physical or chemical adsorption film formed by the WMA additive adsorbed on the surface of the stone. In addition macroscopic lubrication problems can still be studied by friction experiment, which requires a precise experimental condition [21]. While in view of the boundary lubrication of asphalt binder at high temperature, the traditional macroscopic lubrication concept theory and related experimental methods are difficult to reveal the lubrication mechanism accurately and deeply, as the adsorption lubrication film of binder will have a special size effect different from the macroscopic lubrication film. With the development of computer technology, molecular dynamics (MD) simulation has become a powerful and feasible method for material design and performance prediction, as well as a hot spot in the fields of materials, solutions, surfaces, biochemistry and drugs. MD simulation is an important method to study the interface behavior and micro-lubrication because of its mature theory high computational efficiency and incomparable advantages in experiments. Luo et al. studied the adsorption lubrication mechanism and failure mechanism of low-sulfur diesel components on iron surface by means of molecular simulation, and investigated the influence mechanism of anti-wear additives on lubrication performance [27]. Chen et al. studied the adsorption lubrication performance of six esters/N-tetradecane on iron surface by molecular simulation method. The results reflected that the lubrication performance of low-sulfur diesel oil could be improved through increasing ester volume fraction and changing ester components [28]. Chen et al. carried out a quantitative study of the friction coefficient and mechanism of MoS_2 nanoparticles on the basis of molecular dynamics (MD) simulation [29]. Under the conditions of hydrodynamic and boundary lubrication, Godlevskii et al. considered the possibility of determining the characteristics of the lubricating layer by computational MD, and provided an example of solving the adsorption, supramolecular self-organization and rheological parameters of the model lubricant in the lubricating layer [30]. However, there is no research report on the application of molecular simulation in the asphalt–stone lubrication friction.

The purpose of this paper is to study the interface lubrication characteristics between asphalt and stone surface by molecular dynamics (MD) simulation, as well as build lubricating films containing asphalt and WMA (asphalt + WMA additive) in accordance with four fractions including saturates, aromatics, resin and asphaltene, as well as imidazoline type surfactant WMA additive. Then the shear lubrication simulation model of "stone–asphalt–stone" is constructed-based on asphalt lubrication film and typical calcite mineral-matrix. The molecular model is verified by literature reports or laboratory results, and the action mechanism of warm mix asphalt is studied by investigating shear rate, shear stress, WMA additive distribution and adsorption energy.

Asphalt is a complex chemical mixture composed of a variety of hydrocarbons and heteroatoms such as sulfur, oxygen and nitrogen [31–33]. Therefore, the specific chemical composition of asphalt cannot be accurately identified and quantified by the currently available analytical and experimental instruments [34]. Nevertheless, modern separation technology can divide asphalt into different components-based on the similarity of the components in polarity and molecular characteristics. According to the four-component classification method proposed by the American Society for Testing and Materials (ASTM) D4124-09, asphalt can be divided into saturates, aromatics, resin and asphaltene, commonly known as four SARA components [31,34]. This paper studied the lubrication and friction behavior between asphalt and aggregate consistent with the four-component asphalt model proposed by Guo et al. [35].

Although surfactant WMA additives have been used for more than a decade, the composition and structure of WMA additives have been rarely reported. Taking palmitic acid and organic polyamines as raw materials, Zhao et al. synthesized a new imidazoline type surfactant WMA (IDML WMA) additive. The research results showed that IMDL WMA additive had good warm mixing effect [19]. Therefore, IMDL WMA additive molecular model was adopted as WMA additive in the molecular simulation model of WMA lubricating film in this paper.

The molecular models of natural rocks have different structural characteristics due to different rock-forming minerals and diagenetic conditions. In road engineering, stone is usually divided into acid stone and alkaline stone. In China, alkaline stone is often used to prepare asphalt mixture in practical projects because of the better adhesion between alkaline stone and asphalt. The main mineral composition of alkaline stone is calcite (chemical formula is $CaCO_3$). The mineral surface model in this paper is a calcite crystal retrieved from the Cambridge Structural Database (CSD). After the mineral is selected, the mineral crystals need to be dissociated to expose the natural surface containing in the original crystal form and the broken surface caused by the machining process. The stone used in asphalt mixture is mainly processed natural stone. When the external force is applied, the stone minerals break along the surface with weak interlayer bonding strength. Studies have indicated that the exposed surface is closely related to layer spacing, bonding properties, surface electrical properties and surface free energy [34,36]. With the increase of layer spacing, the mineral surface is more easily exposed in nature as the binding strength between layers and surface free energy gradually decreases. As part of the trigonal system, calcite's surface structure and growth mechanism have been extensively studied, including low energy electron diffraction (LEED), X-ray photoelectron spectroscopy (XPS), scanning electron microscopy (SEM), X-ray diffraction (XRD) and atomic force microscopy (AFM) [34]. Studies show that the calcite surface [1] is the most exposed dissociation surface for the layer spacing is the largest in the exposed surface, the bonding strength between layers is the weakest. It was also confirmed by relevant molecular simulations and surface energy calculations on the calcite surface [37]. Therefore, this paper takes calcite crystal (104) as the friction surface to study the lubrication friction of asphalt.

2. Calculation Method of MD Simulation

2.1. Calculation Parameters

In this paper, two "stone–asphalt–stone" molecular lubrication friction models are constructed. The difference between the two models is that the asphalt lubrication film of the intermediate layer is the base asphalt model and the WMA (base asphalt + WMA additive) model, respectively. The four-component model (including saturates, aromatics, resin and asphaltene) proposed by Guo et al. was adopted for the asphalt model [35] (Figure 2a), and IMDL WMA additive proposed by Zhao et al. was used for WMA additive [19] (Figure 2b). The stone cell, the friction surface, the size of friction surface system and the thickness of the model is calcite, calcite (104), 2.4287 nm × 2.4750 nm × 1.6738 nm and 6 layers respectively (Figure 2c). All calculations are carried out in the forcite module of the Material Studio 8.0 software package. The selected force field is Compass II derived from ab initio method of quantum mechanics. In the geometric optimization of a single molecule, except that Atom-based algorithm is used to calculate the interaction of Van der Waals force and Coulomb force, Ewald algorithm and Atom-based algorithm is adopted to calculate the Coulomb force and Van der Waals force respectively. The geometric optimization of the asphalt four-component molecule and the WMA additive molecule are conducted by Geometry Optimization of Forcite, and the optimized molecular configuration is shown in Figure 2. Andersen and Berendsen methods were used to control the temperature and pressure during the MD simulation, and the count step was set to 1.0 fs.

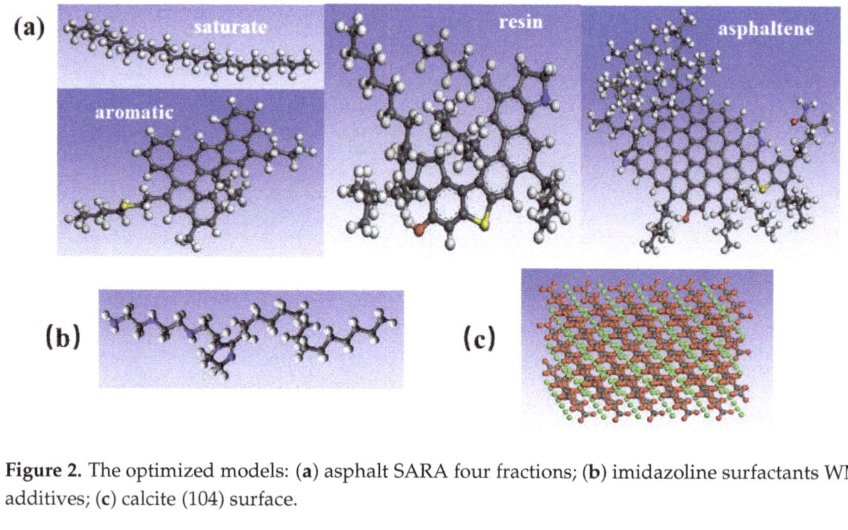

Figure 2. The optimized models: (**a**) asphalt SARA four fractions; (**b**) imidazoline surfactants WMA additives; (**c**) calcite (104) surface.

2.2. Details of MD Simulation

2.2.1. Construction of Asphalt Lubricating Film Cell

Before constructing the shear model, it is indispensable to establish the asphalt lubrication film model and determine the density of the asphalt lubrication film. Two asphalt lubricating films were used in shear simulation calculation to compare the difference of lubricating properties between WMA and base asphalt, and analyze the influence of WMA additive on the lubricating properties: One is the base asphalt lubricating film without WMA additive, the ratio of the number of molecules is saturates: aromatics: resin: asphaltene = 5:7:3:1; the other is WMA lubrication film containing IMDL WMA additive, the ratio of the number of molecules is saturates: aromatics: resin: asphaltene: WMA additive = 5:7:3:1:3. The Amorphous cells of asphalt lubrication film were constructed with Amorphous Cell modules and the green cells were geometrically optimized. Then the cells were pre-balanced at 200 ps by isobaric–isothermal ensemble (NPT) at set temperature, and the dynamic pre-equilibrium was carried out at 300 ps after annealing. In order to check the equilibrium state of the above molecular model, Figure 3 shows the change of energy (including potential energy, kinetic energy, non-bond energy and total energy) of WMA lubricating film system with simulation time during NPT simulation. During the NPT simulation of 298 K, all types of energy were stable when the simulation time exceeded 20 ps. The density of asphalt cell at the corresponding temperature is 1.018 g/cm^3, which is nearly consistent with the asphalt density data measured in the laboratory (1.02 $\pm$ 0.02 g/cm^3), indicating that the calculation parameters of molecular simulation are reasonable and effective.

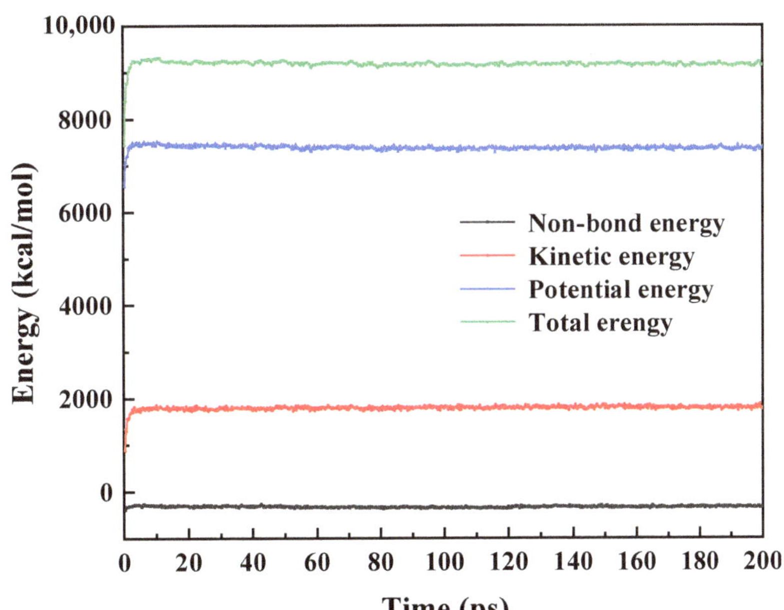

Figure 3. Energy of warm mix asphalt lubricating film model during NPT equilibrium at simulation temperature of 298 K.

2.2.2. Construction of Calcite Cell

Calcite is selected to establish the mineral matrix model. The chemical formula of this mineral is $CaCO_3$. In the chemical composition, CaO and CO_2 accounts for 56.03% and 43.97% respectively. As a typical alkaline stone, it is the most popular stone type in China. The calcite model was imported from the Material Studio 8.0 software package and cut with the surface module. The cutting surface was calcite (104), and the size of calcite was expanded to keep consistent with the size of asphalt cell.

2.2.3. Construction of Shear Simulation Model of "Stone–Asphalt–Stone"

After optimizing the asphalt cell, the amorphous cell of asphalt and WMA is constructed by Amorphous cell module as the nano friction layer of stone. The shear friction system model of "stone–asphalt–stone" is constructed by the build layer tool. The model is divided into three layers, in which the middle layer is asphalt lubrication mold, the upper and lower layers are the surface layer of calcite crystal cells, and a 2.0 nm vacuum layer is set on the upper calcite surface to avoid the interaction between friction surfaces due to periodic structure. The "stone–asphalt–stone" shear friction system model is shown in Figure 4.

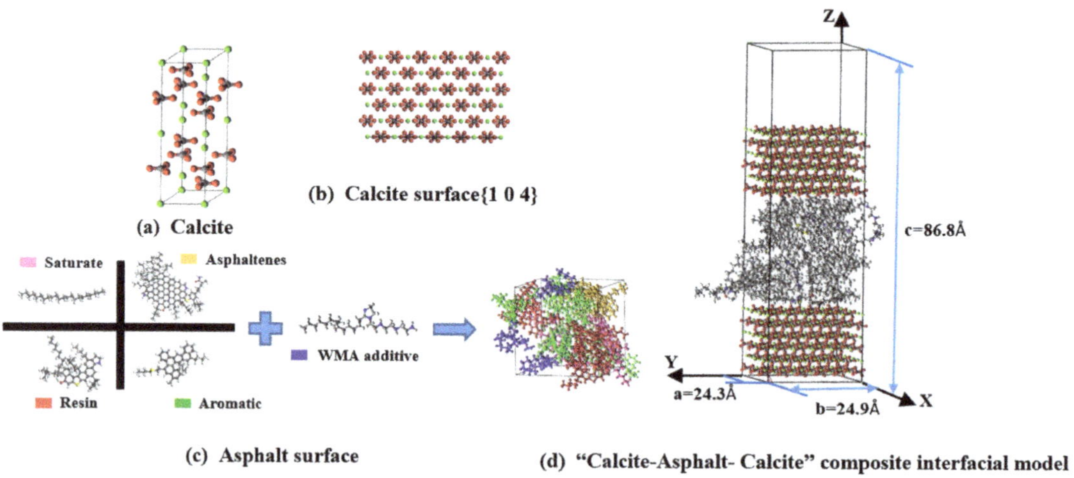

Figure 4. The structure model of the shear flow of asphalt confined between calcite (104) surfaces (**a**) calcite; (**b**) calcite (104) surface; (**c**) asphalt surface; (**d**) "calcite–asphalt–calcite" composite interfacial model.

2.2.4. Simulation Process of Shear Molecular

After the construction of the molecular simulation model of shear friction, the canonical ensemble (NVT) was adopted to calculate the molecular dynamics of the shear friction system at 100 ps at a set temperature, such that the shear friction system can fully relax Then, the Confined Shear method of Forcite module was used to simulate the lubrication of asphalt film between stone surfaces. In the shear simulation process, the atoms on the surface of the stone cell can only move in a simple harmonic manner at their equilibrium position except moving along the x-axis as a whole, and the shape and temperature of the whole stone cell surface remain basically unchanged. Temperature scan and shear rate scan were carried out in the simulation process. The simulated shear temperature range was 100 to 170 with 10 °C temperature interval. The shear rates were 1, 2, 3, 5, 10, 20, 40, 60, 80 and 100 m/s respectively, and the corresponding shear simulation time was set as 2000, 1000, 666, 400, 200, 100, 50, 33, 25 and 20 ps to ensure the consistency of the relative displacement between the upper and lower stone surfaces after the completion of the simulation. Taking the shear rate of the WMA lubrication film at 20 m/s as an example, the situation of the "stone–asphalt–stone" shear friction model at different times is shown in Figure 5.

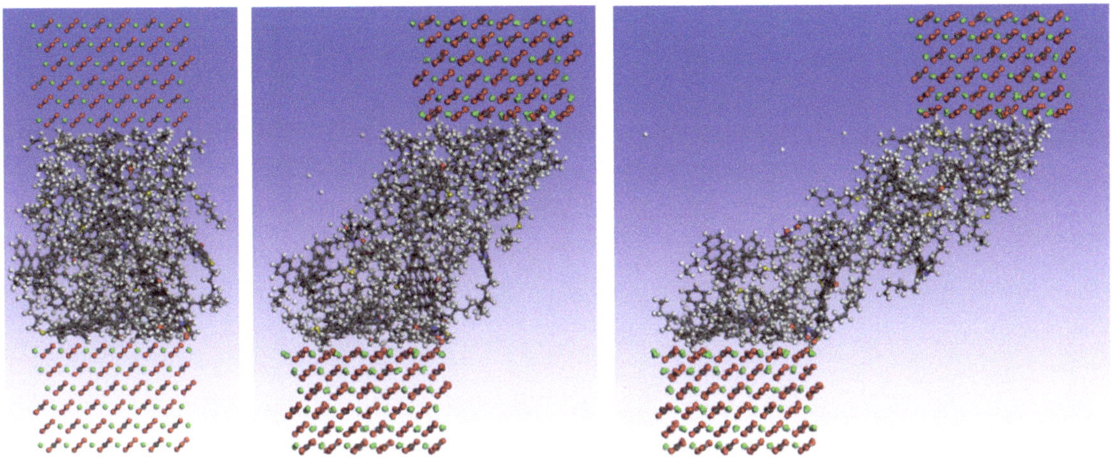

Figure 5. The configuration of friction system at different times under shear rate of 20 m/s ((**a**) 0 ps; (**b**) 50 ps; (**c**) 100 ps).

3. Results and Discussion

3.1. Effect of Shear Rate

The factors influencing the compaction process of asphalt mixture mainly include: compaction work, system temperature, asphalt dosage, etc. Meanwhile, the tribological experiments of asphalt were introduced to further explore the mechanism of WMA additive in recent years. In the tribological experiments of asphalt, the shear rate, temperature and the amount of asphalt were the main factors affecting the experimental results. Therefore, the design idea of asphalt tribological test was used for reference in the shear molecular simulation study of asphalt lubricating film, and the lubrication of asphalt lubricating film on stone in the "stone–asphalt–stone" shear friction system was investigated under different shear rates.

After the completion of the confined shear molecular dynamics simulation of the friction system, the simulation parameters such as the shear stress of the asphalt lubricating film, the temperature distribution and the rate distribution inside the lubricating film can be obtained, and the response process of the asphalt lubricating film under the shear friction on the stone surface and the mechanism of WMA additive can be analyzed based on the above-mentioned parameters.

3.1.1. Variation Rule of Shear Stress

Shear stress is the resultant force that the molecules of the contact surface at the interface of the asphalt lubrication film are subjected to by the internal molecules along the shear direction (refers to the x-axis in this paper) in unit area when the asphalt lubrication film shear motion along the stone surface. It is divided into upper and lower surface shear stress. Shear stress is an important index to evaluate the lubrication effect of asphalt lubrication film on friction between stones. Figure 6 shows the change of the shear stress on the upper and lower surfaces of the WMA lubrication film along the x-axis with time, where the shear rate is 20 m/s and the shear temperature is 120 °C.

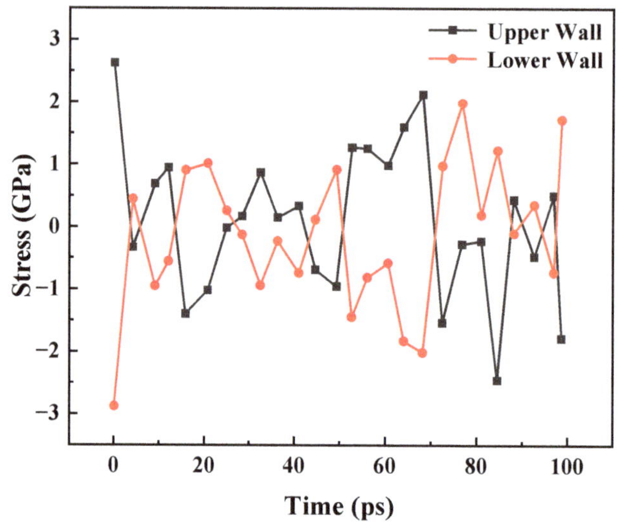

Figure 6. Changes of stress of warm mix asphalt lubricating film with simulation time at shear rate of 20 m/s.

As can be seen from Figure 6, the lubricating film bears a heavy load with over 2 GPa shear stress when the thickness of asphalt lubricating film reaches nanometer level. The strength of the shear stress on the upper and lower surfaces of the asphalt lubrication film is basically equal while the direction is opposite, that is, the shear stress on the two surfaces is basically balanced. However, the shear stress on the upper and lower surfaces of the WMA lubrication film fluctuates with the time, and there is no obvious rule. It is found that the shear stress tends to a fixed value after noise reduction. Therefore, it is significant to take the average value when investigating the changes of shear stress at different shear rates. Base asphalt lubricating film at 170 °C and WMA lubricating film at 120 °C were selected for analysis in accordance with the mixing temperature of above-mentioned asphalt. The average shear stress on the upper and lower surfaces of the two asphalt lubricating films along the x-axis with the change of shear rate is shown in Figure 7.

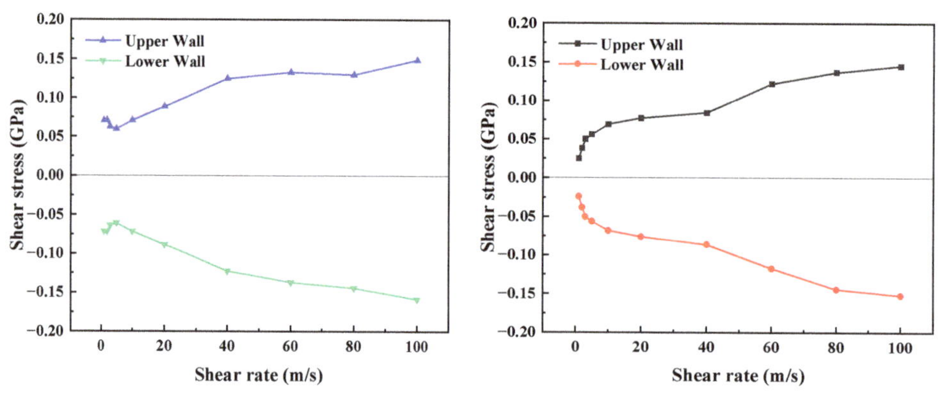

Figure 7. Effect of shear rate on the stress of the (**a**) base asphalt lubricating film at simulation temperature of 443 K; (**b**) warm mix asphalt lubricating film at simulation temperature of 393 K.

As shown in Figure 7, the shear stresses on the upper and lower surfaces of the base asphalt lubricating film and the WMA lubricating film are basically equal in magnitude and opposite in direction at all shear rates, that is, the shear stresses shall maintain balance. In addition, the shear stress of lubricant film of base asphalt and WMA shows different variation trends with the increase of shear rate. In the friction system of base asphalt lubricating film (Figure 7a), the shear stress decreases and then increases with the increase of shear rate. The adsorption of asphalt lubricating film on the stone surface needs to go through a process because of large molecules of asphalt lubricating film. At the beginning of shear, base asphalt molecules adsorb on the stone surface and form a lubricating film with the increase of shear speed. The friction and shear stress reduced as the contact between stone surfaces gradually decreases. When the shear rate reaches 5 m/s, the surface of the stone is completely covered by the asphalt lubrication film and the lubrication film is in the most stable state with minimum friction force and shear stress. When the shear rate is greater than 5 m/s in the working area, the shear stress on the lubrication film and the friction on the stone surface begin to rise with the continuous increase of the shear rate, resulting in lubrication failure for the asphalt lubrication film may not be functioned. However, compared with the friction system of base asphalt, the friction system of WMA lubrication film (Figure 7b) shows different variation trends, that is, the shear stress rises with the increase of shear rate. The growth rate of shear stress is inconsistent over different shear rate ranges. When the shear rate is less than 10 m/s, the shear stress of WMA lubricating film is greatly affected by the shear rate with fast increase trend, while in the range of 10–40 m/s, the shear stress of WMA lubrication film is less affected by the shear rate and changes slowly. However, when the shear rate is higher than 40 m/s, the shear stress rises linearly with the increase of the shear rate. This reflects that the lubrication film of WMA is subjected to a certain shear friction stress during shearing, which rises rapidly with the increase of shear rate. When the shear rate is 10–40 m/s, the lubrication film of WMA and the shear stress are relatively stable under the action of WMA additives. When the shear rate is higher than 40 m/s, the friction stress of the warm mixed asphalt lubricating film on the stone surface begins to increase, while the lubrication effect of the warm mixed asphalt lubricating film on the stone is weakened, resulting the effect of WMA additives fail to work.

In the entire range of shear rate examined, the shear stress of the base asphalt lubricating film is always greater than that of the WMA lubricating film at the same shear rate, even though the former is at a relatively high temperature (the higher the temperature, the smaller the viscosity and viscous resistance). This indicates that the addition of WMA additive reduces the friction stress of the lubricating film of the binder, thus revealing the mechanism of warm mixing action of WMA additive at the molecular level. At the same time, the addition of WMA additive creates a stable working area for the lubricating film of the binder, in which the shear stress hardly rises with the increase of shear rate, corresponding to the working area where WMA additive plays a warm mixing role.

3.1.2. Influence of Shear Rate on Temperature Distribution in Asphalt Lubricating Film

In the process of lubrication and friction between asphalt binder lubricating film and stone surface, the lubricating film will be subjected to the shear action of stone surface and the interaction between molecules in the binder, which will inevitably generate heat and lead to the increase of asphalt lubricating film temperature. The temperature of each position of the lubricating film is different as the temperature transfer will be affected by the heat transfer rate of asphalt lubricating film. In this paper, the influence of shear rate on the temperature distribution of warm mixed asphalt lubricating film along the film thickness direction (z-axis) was investigated, as shown in Figure 8.

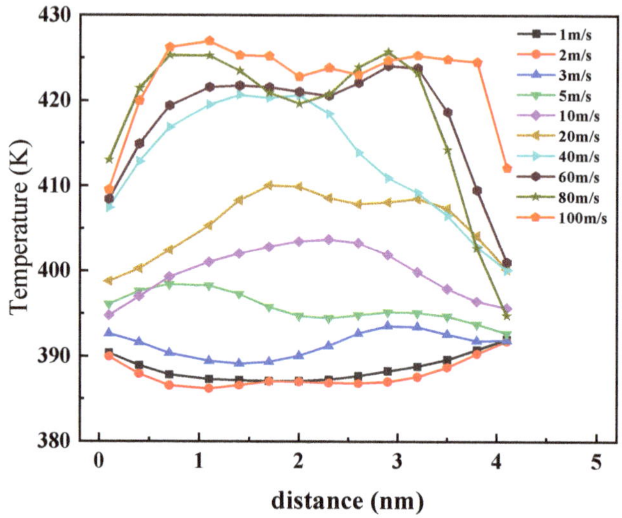

Figure 8. Temperature distribution curves of warm mix asphalt lubricating film under different shear rates.

As can be seen from Figure 8, the temperature distribution in asphalt lubricating film during the shear simulation presents a complex variation rule with the increase of shear rate. When the shear rate is less than 5 m/s, the internal temperature of the lubricating film hardly changes without obvious regularity because of the small amount of friction force and the heat generated in the friction process. When the shear rate is between 10 m/s and 40 m/s, the temperature of the lubricating film increases as enough heat generated by the friction process. However, the heat generated by friction cannot be transmitted out in time because the thermal conductivity of asphalt lubricating film is poor. Therefore, the temperature of the lubricating film tends to gradually rise from the adsorption interface to the interior, and the temperature reaches the maximum at the central position of the film. According to the study in Section 3.1.1, the stable working area of WMA lubricating film is 10–40 m/s, where the lubricating film plays a stable lubrication role. At the same time, the fluidity of the whole lubricating film is enhanced due to the increase of temperature, which contributes to the mutual movement of stones in the process of mixing and compaction. When the shear rate is higher than 40 m/s, a large amount of heat is generated in the friction process, which leads to a sharp rise in the temperature of the lubricating film. The temperature of the lubricating film in the whole thickness range reaches equilibrium quickly, and the temperature of the lubricating film near the center is close. It is also found that in the whole shear rate range, the temperature in the asphalt lubrication film rises with the increase of the shear rate.

In the process of mixing and compaction of asphalt mixture, the asphalt film can act as a lubricant between the stones when the asphalt has fluidity through heating. It is found in the simulation that process the shear stress on the lubricating film rises with the increase of shear rate, that is, the temperature in the asphalt lubricating film rises as the friction force on the stone surface and the heat generated by friction increases. The viscosity of asphalt lubrication film fell sharply when the temperature inside the asphalt lubrication film is too high. At this point, the stone between asphalt lubrication film is in a state of dilution, the lubrication effect on stones is weakened or invalid as the internal pressure is reduced, and the stones are easier to rub, which is not conducive to asphalt mixing and compaction. This is the fundamental reason that the compactibility of the mixture failures to improve linearly with the increase of temperature, while it becomes worse above a certain temperature as mentioned in related studies [21,22]. It is further shown that reducing the viscosity of asphalt is not the only mechanism for WMA workability at low temperature.

3.2. Effect of System Temperature on Shear Stress

The boundary lubrication is directly related to the equilibrium of adsorption and desorption of the lubricating film on the adsorption surface. Langmuir adsorption theory believes that temperature is the main factor affecting adsorption, and the higher the temperature is, the higher the adsorption rate of lubricating film molecules on the adsorption surface will be. Meanwhile, higher temperature will break the dynamic equilibrium process of adsorption and desorption as adsorption is an exothermic process, resulting in the increase of adsorption capacity. Therefore, the effect of system temperature (373, 383, 393, 403, 413, 423, 433, 443 K) on the shear friction of WMA lubricating film and base asphalt lubricating film was further studied at the shear rate of 20 m/s in this paper. The average shear stress of the upper and lower surfaces of WMA lubrication film and base asphalt lubrication film along the *x*-axis at different temperatures is shown in Figure 9.

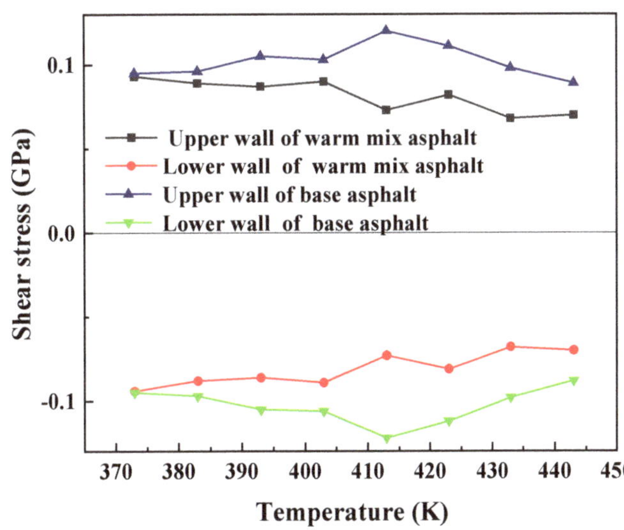

Figure 9. Effects of temperature on the stress of the film.

According to the results in Figure 9, the shear stress on the upper and lower surfaces of the WMA and the base asphalt lubricating film is equal within the temperature range of 373–443 K. The shear stress of the WMA film is less than that of the base asphalt film at the same temperature, and the difference between above-mentioned stress is widening with the increase of temperature, which further indicates that the addition of WMA additive reduces the shear stress of the lubricating film. In addition, the shear stress of lubricating film of WMA and base asphalt shows different trends with temperature change. With the increase of temperature, the shear stress of the lubricating film of WMA shows a decreasing trend, while the shear stress of the lubricating film of base asphalt shows a trend of first increasing and then decreasing with maximum shear stress at 413 K. This shows that the compactibility of the mixture does not improve linearly with the increase of temperature, but becomes worse above a certain temperature, which is consistent with the previous research results [21,22]. The change of shear stress is not apparent in the entire temperature range, indicating that in a certain temperature range, the system temperature is not the main factor affecting the lubrication effect of WMA film on stone.

3.3. Effects of WMA Additives

Many existing studies have shown that adding surfactant WMA additives to asphalt can improve the lubricity of asphalt lubricating film and form boundary lubricating film to reduce the friction between particles through protecting aggregate particles from the direct contact of solids [26]. However, the reason why surfactant WMA additives can promote the lubricity of asphalt has not been concluded, and there is a lack of theoretical research in this field.

3.3.1. Effect of WMA Additive on Shear Stress of Friction System

In this paper, the effect of imidazoline type surfactant WMA (IMDL WMA) additive on asphalt lubricity was investigated by molecular dynamics simulation, and the warm mixing mechanism of surfactant WMA additive was studied theoretically. Figure 10 shows the shear stress on the upper and lower surfaces of WMA and base asphalt lubricating film at different shear rates at 120 °C.

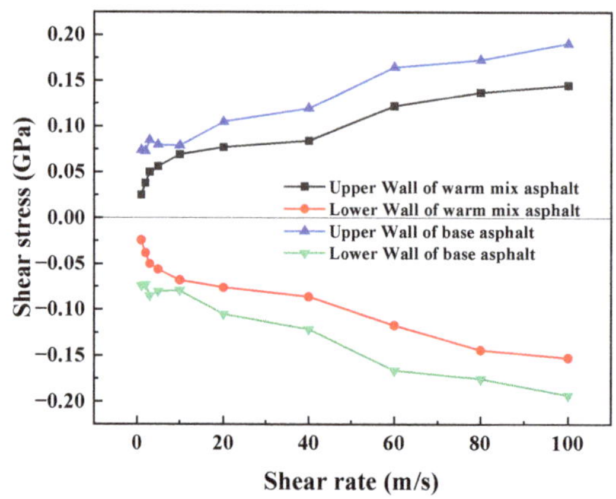

Figure 10. Effects of the IMDL WMA additive on the stress of the film.

As shown in Figure 10, due to the addition of IMDL WMA additive, the shear stress of WMA lubricating film is significantly lower than that of base asphalt lubricating film at the same shear rate. Therefore, the addition of IMDL WMA additive can improve the lubricity of asphalt binder, reduce the friction between asphalt and stone, and promote the mixture to be better mixing and compaction, which theoretically explains the mechanism of surfactant WMA additive. It can be seen that the shear force of the WMA lubricating film changes little when the shear rate is in the range of 10–40 m/s, which is basically not affected by the shear rate. Therefore, this area is the stable working area of the WMA lubricating film. The stable working area of the base asphalt lubrication film is in the shear rate range of 5–10 m/s, which is less than that of the WMA. Therefore, it can be concluded that the stability of the boundary lubrication film of asphalt binder improved by the addition of IMDL WMA additive, such that it can provide stable lubrication under a larger and wider load range, it is easier to meet the mixing and compaction conditions of asphalt mixture in actual production.

3.3.2. Density Distribution of IMDL WMA Additive Molecules between Frictional Surfaces

According to boundary lubrication theory, the strength of the adsorbed lubricating film is far better than that of the flow since the boundary adsorption film loses its mobility by adsorbing on the surface of the lubricating substrate, it can be kept on the friction surface more stably. Therefore, the reason why the adsorption film can maintain stable and reliable lubrication under high shear is inseparable from its adsorption on the substrate surface.

The IMDL WMA additive enjoys high polarity due to its imidazoline group, which belongs to the basic group. Compared with asphalt, it is easier to form a lubrication film on the stone surface to effectively prevent the direct contact between the friction surface. The density distribution of IMDL WMA additive molecules in the WMA lubricating film along the z-axis between the friction surfaces of the stone is shown in Figure 11.

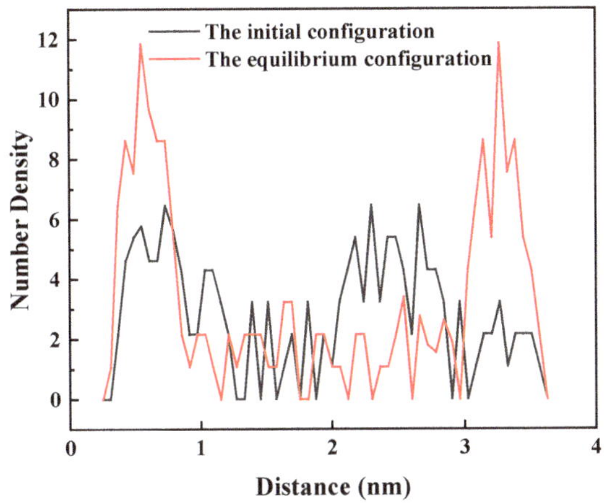

Figure 11. Number density profile of WMA additive molecules in the warm mix asphalt lubricating film along the z-axis.

As can be seen from Figure 11, compared with the initial configuration, IMDL WMA additive molecules moves toward the friction interface after the friction system is fully relaxed and reaches equilibrium. Accordingly, WMA additive molecules preferentially adsorb the stone surface to form a lubricating film, which separates the stone from the asphalt friction surface and provides an interface with low shear resistance. This phenomenon further explains the reason why the lubricating effect of WMA lubricating film is better than that of base asphalt lubricating film at the same temperature and shear rate from a micro level.

Figure 12 shows the effect of adding IMDL WMA additive on the number of molecular density distribution of asphalt components. As shown in Figure 12, compared with the base asphalt, the number density distribution of asphalt molecules on the friction surface in the WMA decreased significantly after the addition of IMDL WMA additive. This is because the IMDL WMA additive would preferentially move to the friction surface and replace the asphalt molecules on the friction surface. Therefore, the distribution of WMA additive molecules in asphalt lubricating film was indirectly verified.

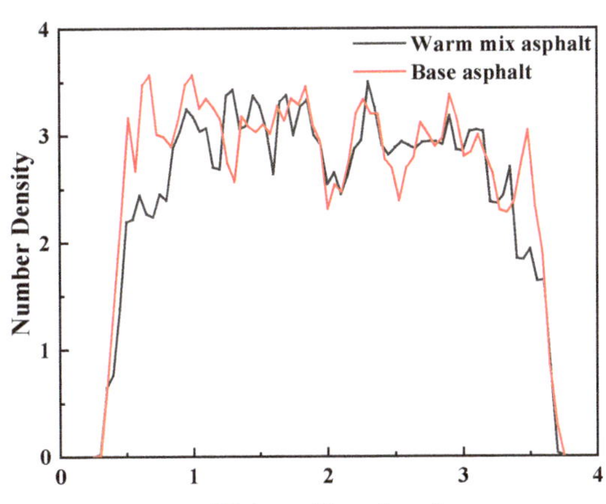

Figure 12. Number density profile of asphalt component molecules in the warm mix asphalt lubricating film along the z-axis.

3.3.3. Adsorption Energy of Asphalt Lubrication Film on Friction Surface

The adsorption property of asphalt lubrication film on stone surface has an important impact on its lubrication effect. These properties unique to asphalt lubrication film under nanometer lubrication condition are determined by the interaction of asphalt lubrication film and stone friction surface in physical and chemical nature, which is mainly reflected by adsorption energy.

Adsorption energy is the energy required to divide the "stone–asphalt–stone" shear friction system into two independent systems. The bonding strength of the asphalt and stone surface is quantified by calculating the adsorption energy of the asphalt and stone surface, to reflect the lubrication effect of the asphalt lubrication film between the friction surfaces. The bond energy is constant in COMPASS force field. The change of total energy of asphalt and stone system and the "stone asphalt–stone" shear friction system are caused by the non-bond energy (van der Waals force and static electricity), that is to say, the adsorption energy of "stone–asphalt–stone" shear friction system can be determined by the non-bond energy. The adsorption energy of the "stone–asphalt–stone" shear friction system is calculated by Equation (1).

$$E_{binding} = (E_{asphalt} + E_{aggregate}) - E_{total} \quad (1)$$

where $E_{binding}$ is the adsorption energy between asphalt and stone, kcal/mol; $E_{asphalt}$ is the energy of asphalt lubrication film without stone, kcal/mol; $E_{aggregate}$ is the energy of stone surface without asphalt lubricating film, kcal/mol; E_{total} is the total energy of the "stone–asphalt–stone" friction system in the simulated output results, kcal/mol.

Adsorption energy is the result of energy dissipation in the process of thermodynamic equilibrium of bitumen–stone interface. According to Equation (1), the greater the difference between $E_{asphalt} + E_{aggregate}$ and E_{total}, the greater the adsorption energy of asphalt lubrication film on the stone surface. The adsorption energy of the WMA lubrication film on the stone surface in the shear simulation process calculated in accordance with Equation (1) is shown in Figure 13.

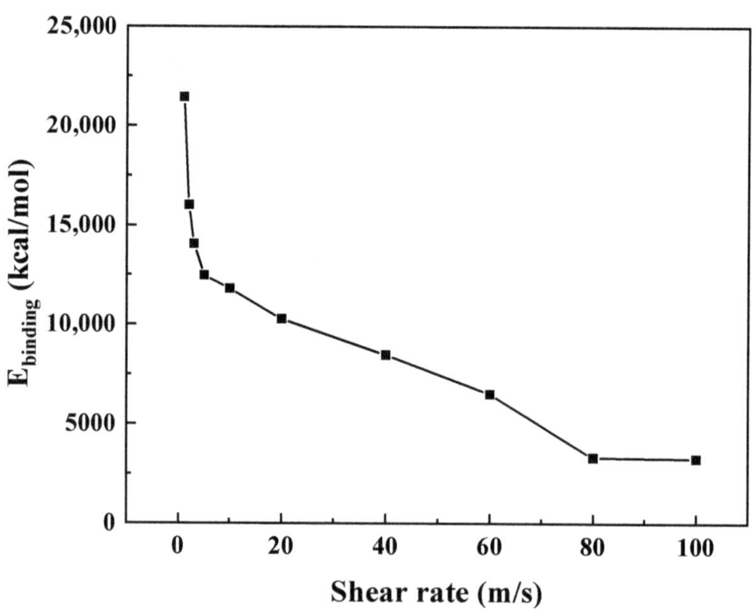

Figure 13. Binding energies ($E_{binding}$) of warm mix asphalt lubricating film between calcite (104) surfaces.

As shown in Figure 13, when the shear rate is lower than 5 m/s, the adsorption energy of the warm mixed asphalt lubrication film on the stone surface decreases rapidly. When the shear rate is higher than 5 m/s, the reduction rate of adsorption energy slows down, while when the shear rate reaches 80 m/s, the adsorption energy gradually tends to be stable. In general, the adsorption energy of the WMA lubricating film on the stone surface decreases with the increase of the shear rate, which indicates that the higher the shear rate is, the more unfavorable it is for the WMA lubricating film to be wrapped on the stone surface, or even to rupture, such that the lubrication effect of the WMA lubricating film on the friction surface is weakened.

4. Conclusions

This paper studied the mechanism of adsorption and lubrication between the base asphalt and WMA lubrication film on the stone surface, and the effects of shear rate, system temperature and WMA additive on the adsorption and lubrication performance-based on the tribological theory and molecular dynamics simulation method. The results show that:

(1) In the friction system of warm mix asphalt and base asphalt, the shear stress changes differently with the increase of shear rate. The WMA lubricating film remains stable when the shear rate is in the range of 10–40 m/s, which means the shear stress of the warm mixed asphalt lubricating film is less affected by the shear rate and changes slowly. However, when the shear rate is higher than 40 m/s, the shear stress rises linearly with the increase of the shear rate. In the friction system of base asphalt lubricating film, the shear stress decreases first and then increases with the increment of shear rate. When the shear rate reaches 5 m/s, the lubricating film is in the most stable state with minimum friction force and shear stress.

(2) The increment of shear rate leads to the increase of temperature in the lubricating film of WMA. When the shear rate is less than 5 m/s, the internal temperature of the lubricating film hardly changes without obvious regularity because of the small amount of friction force and the heat generated in the friction process. When the shear rate is in the range of 10–40 m/s, the temperature of the lubricating film increases gradually from the adsorption interface to the interior and reaches the maximum at the center of the film. When the shear rate is higher than 40 m/s, the temperature of the lubricating film rises sharply as a large amount of heat is generated during the friction process.

(3) With the increase of temperature, the shear stress of the lubricating film of WMA shows a decreasing trend, while the shear stress of the lubricating film of base asphalt shows a trend of first increasing and then decreasing with maximum shear stress at 413 K, which indicates that the compactability of the mixture does not improve linearly with the increase of temperature. In addition, system temperature exerts little impact on shear stress, which is not the main factor affecting the adsorption lubrication of WMA.

(4) The addition of IMDL WMA additive can effectively improve the lubricity of asphalt and stability of asphalt film by reducing the shear stress of asphalt lubricating film. After the friction system is balanced, the IMDL WMA additive molecules are mainly distributed at the friction interface and preferentially adsorbed on the stone surface to form a lubricating film. Besides, the adsorption energy of the WMA lubricating film on the stone surface decreases with the increase of the shear rate, which indicates that the higher the shear rate is, the more unfavorable it is for the WMA lubricating film to be wrapped on the stone surface, or even to rupture, such that the lubrication effect of the WMA lubricating film on the friction surface is weakened.

The above research results are of great significance in guiding the selection and development of WMA additives, as well as the determination of mixing and compaction temperatures of mixture, showing a broad application prospect. In the future research the authors will further explore the effects of different structures of WMA additives and different compositions of stones on the lubricating friction performance in combination with experiments.

Author Contributions: Conceptualization, Y.Y.; methodology, P.Z. and J.S.; software, M.D. and W.W.; validation, M.D., J.W., W.W., X.Z. (Xingchi Zhao), X.Z. (Xu Zhou) and C.W.; data curation, M.D., J.W. and W.W.; writing—original draft preparation, P.Z. and M.D.; writing—review and editing, P.Z.; supervision, P.Z.; project administration, P.Z.; funding acquisition, P.Z. All authors have read and agreed to the published version of the manuscript.

Funding: This study was monetarily bolstered by the Natural Science Fund project in Shandong province: ZR202102190324, Transportation Science and Technology Plan of Shandong Prov-ince(2020B10), Project of science and technology support for youth entrepreneurship in Colleges and universities of Shandong Province (2019KJG004).

Institutional Review Board Statement: Not applicable.

Informed Consent Statement: Not applicable.

Data Availability Statement: The information used to bolster the discoveries of this research are from prior detailed researches cited before. This original copy does not include distributed Figures, Tables, and Charts before, thus all Figures, Tables, and Charts of this original copy are unique.

Acknowledgments: We would like to recognize numerous co-workers, students, and research facility associates for giving specialized offer assistance on instrument examination. Mingliang Dong contributed equally to this work and should be considered co-first authors.

Conflicts of Interest: The authors have no conflict of interest.

Abbreviations

WMA	warm mix asphalt
MD	molecular dynamics
IMDL WMA	imidazoline type surfactant WMA
HMA	hot mix asphalt
RAP	recycled asphalt pavement
DSR	dynamic shear rheometer
ASTM	American Society for Testing and Materials
CSD	Cambridge Structural Database
LEED	low energy electron diffraction
XPS	X-ray photoelectron spectroscopy
SEM	scanning electron microscopy
XRD	X-ray diffraction
AFM	atomic force microscopy
NPT	isobaric-isothermal ensemble
NVT	canonical ensemble

References

1. Ingrassia, L.P.; Lu, X.H.; Canestrari, F.; Ferrotti, G. Tribological characterization of bituminous binders with Warm Mix Asphalt additives. *Constr. Build. Mater.* **2018**, *172*, 309–318. [CrossRef]
2. Shi, J.T.; Fan, W.Y.; Wang, T.H.; Zhao, P.H.; Che, F. Evaluation of the physical performance and working mechanism of asphalt containing a surfactant warm mix additive. *Adv. Mater. Sci. Eng.* **2020**, *2020*, 8860466. [CrossRef]
3. Shi, J.T.; Fan, W.Y.; Lin, Y.; Zhao, P.H.; Ouyang, J. Effects of surfactant warm-mix additives on the rheological properties of high-viscosity asphalt. *Adv. Mater. Sci. Eng.* **2020**, *2020*, 6506938. [CrossRef]
4. Rodrigues, Y.O.; Costa, D.B.; de Figueirêdo Lopes Lucena, L.C.; Lopes, M.C. Performance of warm mix asphalt containing Moringa oleifera Lam seeds oil: Rheological and mechanical properties. *Constr. Build. Mater.* **2017**, *154*, 137–143. [CrossRef]
5. Zelelew, H.; Paugh, C.; Corrigan, M.; Belagutti, S.; Ramakrishnareddy, J. Laboratory evaluation of the mechanical properties of plant-produced warm-mix asphalt mixtures. *Road Mater. Pavement Des.* **2013**, *14*, 49–70. [CrossRef]
6. Li, X.S.; Xie, Z.X.; Fan, W.Z.; Wang, L.; Shen, J.N. Selecting warm mix asphalt additives by the properties of warm mix asphalt mixtures—China experience. *Balt. J. Road Bridge Eng.* **2015**, *10*, 79–88. [CrossRef]
7. Liu, J.; Yan, K.; You, L.; Ge, D.; Wang, Z. Laboratory performance of warm mix asphalt binder containing polyphosphoric acid. *Constr. Build. Mater.* **2016**, *106*, 218–227. [CrossRef]

8. Wang, H.P.; Liu, X.; Apostolidis, P.; Scarpas, T. Review of warm mix rubberized asphalt concrete: Towards a sustainable paving technology. *J. Clean. Prod.* **2018**, *177*, 302–314. [CrossRef]
9. Zaumanis, M.; Olesen, E.; Haritonovs, V.; Brencis, G.; Smirnovs, J. Laboratory evaluation of organic and chemical warm mix asphalt technologies for SMA asphalt. *Balt. J. Road Bridge Eng.* **2012**, *7*, 191–197. [CrossRef]
10. Mohd Hasan, M.R.; You, Z.; Yang, X.; Heiden, P.A. Quantification of physicochemical properties, activation energy, and temperature susceptibility of foamed asphalt binders. *Constr. Build. Mater.* **2017**, *153*, 557–568. [CrossRef]
11. Jun Yoo, P.; Sik Eom, B.; Soo Park, K.; Hun Kim, D. Aggregate pre-coating approach using rubber- and silane-coupled thermoset polymer and emulsion for warm-mix asphalt mixtures. *Constr. Build. Mater.* **2017**, *152*, 708–714. [CrossRef]
12. Zhang, Y.; Leng, Z.; Zou, F.L.; Wang, L.; Chen, S.S.; Tsang, D.C.W. Synthesis of zeolite A using sewage sludge ash for application in warm mix asphalt. *J. Clean. Prod.* **2018**, *172*, 686–695. [CrossRef]
13. Tang, N.; Deng, Z.X.; Dai, J.G.; Yang, K.K.; Chen, C.Y.; Wang, Q. Geopolymer as an additive of warm mix asphalt: Preparation and properties. *J. Clean. Prod.* **2018**, *192*, 906–915. [CrossRef]
14. Behnood, A. A review of the warm mix asphalt (WMA) technologies: Effects on thermo-mechanical and rheological properties. *J. Clean. Prod.* **2020**, *259*, 120817. [CrossRef]
15. Guo, M.; Liu, H.Q.; Jiao, Y.B.; Mo, L.T.; Tan, Y.Q.; Wang, D.W.; Liang, M.C. Effect of WMA-RAP technology on pavement performance of asphalt mixture: A state-of-the-art review. *J. Clean. Prod.* **2020**, *266*, 121704. [CrossRef]
16. Rubio, M.C.; Martínez, G.; Baena, L.; Moreno, F. Warm mix asphalt: An overview. *J. Clean. Prod.* **2012**, *24*, 76–84. [CrossRef]
17. Zhang, S.W.; Wang, D.Y.; Guo, F.; Deng, Y.H.; Feng, F.M.; Wu, Q.Y.; Chen, Z.J.; Li, Y.B. Properties investigation of the SBS modified asphalt with a compound warm mix asphalt (WMA) fashion using the chemical additive and foaming procedure. *J. Clean. Prod.* **2021**, *319*, 128789. [CrossRef]
18. Hanz, A.J.; Faheem, A.; Mahmoud, E.; Bahia, H.U. Measuring effects of warm-mix additives: Use of newly developed asphalt binder lubricity test for the dynamic shear rheometer. *Transp. Res. Rec.* **2010**, *2180*, 85–92. [CrossRef]
19. Zhao, P.H.; Ren, R.B.; Zhou, H.; Ouyang, J.; Li, Z.G.; Sun, D.C. Preparation and properties of imidazoline surfactant as additive for warm mix asphalt. *Constr. Build. Mater.* **2021**, *273*, 121692. [CrossRef]
20. Hanz, A.J.; Bahia, H.U. Asphalt binder contribution to mixture workability and application of asphalt lubricity test to estimate compactability temperatures for warm-mix asphalt. *Transp. Res. Rec.* **2013**, *2371*, 87–95. [CrossRef]
21. Canestrari, F.; Ingrassia, L.P.; Ferrotti, G.; Lu, X.H. State of the art of tribological tests for bituminous binders. *Constr. Build. Mater.* **2017**, *157*, 718–728. [CrossRef]
22. Bairgi, B.K.; Mannan, U.A.; Tarefder, R.A. Influence of foaming on tribological and rheological characteristics of foamed asphalt. *Constr. Build. Mater.* **2019**, *205*, 186–195. [CrossRef]
23. Gallego, R.; Cidade, T.; Sanchez, R.; Valencia, C.; Franco, J.M. Tribological behaviour of novel chemically modified biopolymer-thickened lubricating greases investigated in a steel-steel rotating ball-on-three plates tribology cell. *Tribol. Int.* **2016**, *94*, 652–660. [CrossRef]
24. Mannan, U.A.; Tarefder, R.A. Tribological and rheological characterisation of asphalt binders at different temperatures. *Road Mater. Pavement Des.* **2018**, *19*, 445–452. [CrossRef]
25. Maru, M.M.; Trommer, R.M.; Almeida, F.A.; Silva, R.F.; Achete, C.A. Assessment of the lubricant behaviour of biodiesel fuels using Stribeck curves. *Fuel Process. Technol* **2013**, *116*, 130–134. [CrossRef]
26. Puchalski, S. *Investigation of Warm Mix Asphalt Additives Using the Science of Tribology to Explain Improvements in Mixture Compaction*; A dissertation submitted in partial fulfillment of the requirement for the degree of Master of Science; College of Engineering, University of Wisconsin-Madison: Madison, WI, USA, 2012.
27. Luo, H.; Fan, W.Y.; Li, Y.; Zhao, P.H.; Nan, G.Z. Effects of fatty acids on low-sulfur diesel lubricity: Experimental investigation, DFT calculation and MD simulation. *China Pet. Process. Petrochem. Technol.* **2013**, *15*, 74–81. [CrossRef]
28. Chen, T.T.; Mei, D.Q.; Ren, W.Y.; Wang, H.Q.; Yuan, Y.N.; Guo, R. Adsorption of esters contained fuel on the Fe surface: A molecular dynamics study. *Acta Pet. Sin. (Pet. Process. Sec.)* **2019**, *35*, 714–720. [CrossRef]
29. Chen, C.S.; Cian, H.J.; Yu, C.H.; Huang, C.W. Friction coefficient calculation and mechanism analysis for MoS_2 nanoparticle from molecular dynamics simulation. *Procedia Eng.* **2014**, *79*, 617–621. [CrossRef]
30. Godlevskii, V.A.; Fomichev, D.S.; Shilov, M.A.; Berezina, E.V.; Kuznetsov, S.A. Use of a computer molecular simulation to describe lubricating layer structure. *J. Frict. Wear.* **2009**, *30*, 7–11. [CrossRef]
31. Gao, Y.M.; Zhang, Y.Q.; Yang, Y.; Zhang, J.H.; Gu, F. Molecular dynamics investigation of interfacial adhesion between oxidised bitumen and mineral surfaces. *Appl. Surf. Sci.* **2019**, *479*, 449–462. [CrossRef]
32. Xu, G.X.; Wang, H. Study of cohesion and adhesion properties of asphalt concrete with molecular dynamics simulation. *Comp. Mater. Sci.* **2016**, *112*, 161–169. [CrossRef]
33. Huang, M.; Zhang, H.L.; Gao, Y.; Wang, L. Study of diffusion characteristics of asphalt–aggregate interface with molecular dynamics simulation. *Int. J. Pavement Eng.* **2021**, *22*, 319–330. [CrossRef]
34. Chu, L.; Luo, L.; Fwa, T.F. Effects of aggregate mineral surface anisotropy on asphalt-aggregate interfacial bonding using molecular dynamics (MD) simulation. *Constr. Build. Mater.* **2019**, *225*, 1–12. [CrossRef]
35. Guo, M.; Tan, Y.Q.; Bhasin, A.; Wei, J.M.; Hou, Y. Using molecular dynamics to investigate interfacial adhesion between asphalt binder and mineral aggregate. In Proceedings of the The 4th Chinese–European Workshop on Functional Pavement Design (CEW 2016), Guangzhou, China, 24 June 2016.

36. Cooper, T.G.; De Leeuw, N.H. A Computer modeling study of the competitive adsorption of water and organic surfactants at surfaces of the mineral scheelite. *Langmuir* **2004**, *20*, 3984. [CrossRef] [PubMed]
37. De Leeuw, N.H.; Parker, S.C. Surface structure and morphology of calcium carbonate polymorphs calcite, aragonite, and vaterite: An atomistic approach. *J. Phys. Chem. B* **1998**, *102*, 2914–2922. [CrossRef]

coatings

Review

Utilization of Steel Slag in Road Semi-Rigid Base: A Review

Haibin Li [1,2], Canyang Cui [1,2], Jun Cai [3], Mingming Zhang [1,2] and Yanping Sheng [4,*]

[1] School of Architecture and Civil Engineering, Xi'an University of Science and Technology, Xi'an 710054, China; lihaibin1212@126.com (H.L.); 20204228114@stu.xust.edu.cn (C.C.); 20204228117@stu.xust.edu.cn (M.Z.)
[2] Road Engineering Research Center, Xi'an University of Science and Technology, Xi'an 710054, China
[3] Qinghai Xihu Expressway Management Co., Ltd., Xining 810001, China; ycc0128@126.com
[4] School of Materials Science and Engineering, Chang'an University, Xi'an 710064, China
* Correspondence: syp@chd.edu.cn

Abstract: Steel slag (SS) is industrial waste, and there is a large amount of SS to be treated in China. Its disposal generates severe environmental pollution. One of the best ways to use SS is as a road base material. This paper reviews the possibility of using SS in semi-rigid base and evaluates the performance of SS base course. The interaction between three stabilizers (cement, lime–fly ash, and cement–fly ash) and SS is analyzed, and the influence of modifier content on the performance of base course is evaluated. The potential laws between SS, curing time, and unconfined compressive strength, as well as drying shrinkage and temperature shrinkage, are discussed and their effects on the performance of the base course are revealed. The finite element method, discrete element method, and molecular dynamics can be used to analyze the freeze-thaw, rutting resistance, and crack development of SS base. In addition, compared with traditional macadam base, the CO_2 emissions for the use of SS base are slightly more, one of the disadvantages of its use in production, transportation, and compaction. However, considering the overall mechanical, economic, and environmental benefits, it is recommended to use SS in semi-rigid base course. The future research scope of SS as base material is suggested.

Keywords: steel slag; semi-rigid base; performance evaluation; numerical simulation; environmental impacts

Citation: Li, H.; Cui, C.; Cai, J.; Zhang, M.; Sheng, Y. Utilization of Steel Slag in Road Semi-Rigid Base: A Review. *Coatings* **2022**, *12*, 994. https://doi.org/10.3390/coatings12070994

Academic Editors: Andrea Nobili and Valeria Vignali

Received: 15 June 2022
Accepted: 6 July 2022
Published: 15 July 2022

Publisher's Note: MDPI stays neutral with regard to jurisdictional claims in published maps and institutional affiliations.

Copyright: © 2022 by the authors. Licensee MDPI, Basel, Switzerland. This article is an open access article distributed under the terms and conditions of the Creative Commons Attribution (CC BY) license (https://creativecommons.org/licenses/by/4.0/).

1. Introduction

As an essential symbol of the national industrial development level, steel slag (SS) is one of the most residual solid wastes in the rapid development of its economy and society [1]. In developed countries and regions, SS is widely used in road engineering, concrete aggregate, cement production, fertilizer, and soil modifier [2–5]. The production of SS in China has been ranked first globally since the 21st century. In recent years, the steel output was close to 50% of the world's output [6]. The national steel output was 1.33 billion tons in 2020, and according to the traditional steelmaking method commonly used in China's iron and steel enterprises, the SS produced per ton of iron and steel is approximately 100–150 kg [7,8]. In addition, a total of 133–199 million tons of SS have been produced by 2020. However, less than 30% will be utilized in China. The Chinese Scrap Iron and Steel Application Association has put forward the goal that the total utilization rate of SS will reach 60% in 2025 [9].

Due to its large consumption of aggregate resources, there is a potential market to reuse the solid waste SS as subgrade material in road engineering. With certain technical advantages and market capacity, the road construction industry has gradually becoming the main battlefield of solid waste reuse [10]. However, compared with Japan, Europe, and the United States, the utilization rate of SS in China lags far behind those developed countries (regions). The percentage of SS used in road construction in Japan, Europe, and the United States is 32%, 43%, and 50%, respectively, while only 3% is used in China [11].

The utilization of SS has been studied in developed countries and regions earlier. According to the data from the Japan Nippon Slag Association in 2016, the utilization rate of SS in Japan was as high as 98.4% [11]. Of this, 30.9% of SS was used for soft soil foundation, road subgrade, and embankment [12]. In the United States, the majority of SS was used for road construction, while the remainder was used for marine filler, asphalt concrete aggregate, cement production, and the rest was used for soil pH adjustments and steel reproduction [3]. As the earliest industrialized region in the world, the data of the European SS Association showed that the utilization rate of SS reached 87.0% in 2012 [11]. SS was widely used in geotechnical engineering, road engineering, water conservancy engineering, and railway engineering in Germany early in 2005. In road engineering, Germany used SS for base course and surface course, and tried to mix it with other metallurgical slags as a bearing layer with high bearing capacity [13]. As another industrial power, the UK has widely used SS in the production of cement, highway engineering concrete, and coarse aggregate for road construction [14,15].

Researchers from other developing countries have also carried out research on the utilization of SS in road engineering. Behiry [16] of Egypt mixed SS with limestone aggregate, exploring the influence of SS content on improving the density and strength of base course, and analyzed the relationship between mixture and deflection, stress, strain, and other factors under different content. Kumar and Kumar [17] summarized and analyzed the utilization of SS in India, using SS in road base, and conducted continuous observation after construction, finding that the performance of SS road base was excellent. Karatağ, et al. [18] studied the feasibility of combining SS and macadam in Vietnam for road base, and their results showed that the mechanical properties and porosity were good.

SS is an excellent substitute for natural aggregate [18], which has the advantages of high strength, good wear resistance, durability, and strong stability. It can meet various indexes when used in road base and surface course, and performs better in mechanical properties than natural aggregate [19,20]. Compared to natural aggregate, SS has more apparent edges and corners, a rougher surface texture, and better wear resistance [21]. At the same time, the micropores on the surface of SS are conducive to combining with cement and improving its strength [22], showing its potential utilization in road base.

As the bearing layer of the road, the road base can be divided into rigid base, semi-rigid base, and flexible base according to the different mechanical characteristics. Since the 1970s, semi-rigid base has been rapidly promoted in China because of high bearing capacity, high stiffness, comprehensive material sources, and low cost. Recently, semi-rigid stabilized base asphalt pavement has become one of the typical structures of high-grade roads in China [23]. Various types of the semi-rigid bases have received extensive attention [24], such as cement stabilized base [25,26], lime stabilized base [27], lime–fly ash stabilized base [28,29], and cement–fly ash stabilized base [30]. However, the aggregate utilization of semi-rigid base is extensive, which has a significant impact on the ecological environment. With the continuous attention to global sustainable development and cleaner production [31], how to reduce environmental pollution and save resources has also becoming a critical research direction of road researchers. In recent years, researchers have increased the use of industrial by-products [32–35] in road base, such as SS, fly ash, coal gangue, etc. Alternatively, waste base materials are recycled [36,37].

However, little research has been done on the utilization of SS as semi-rigid road base material. The paper aims to review and highlight the possibility of using SS as semi-rigid base material, evaluating the performance of SS base and its impact on the environment. The flowchart of the review is shown in Figure 1.

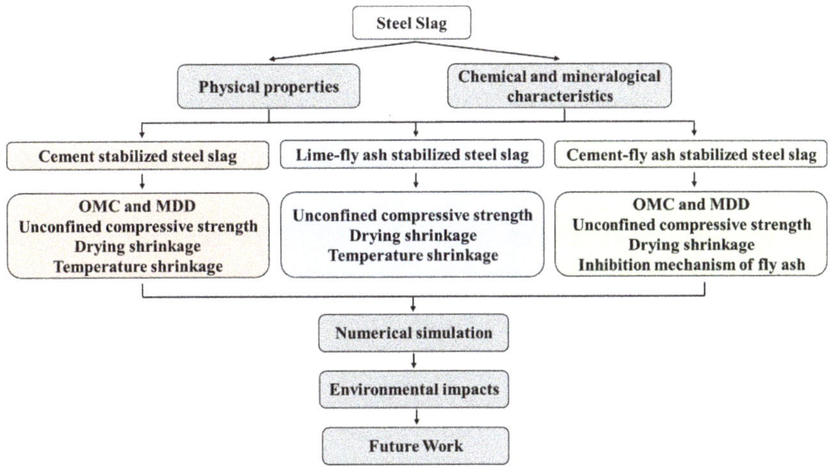

Figure 1. Flowchart of the review.

2. Materials

2.1. Physical Characteristics of SS

The physical and mechanical properties of SS from the present paper are shown in Table 1. In most of the relevant literature, the apparent specific gravity of SS is shown to be higher than 3.20 g/cm^3. The crushing value of the aggregate is used to measure the crushing resistance of the aggregate under increasing load. It is a vital index to measure the mechanical properties of the stone to evaluate its applicability in road engineering. The crushing value of SS ranges within 10%–21%, and all meet the requirements of less than 22% in the standard [24]. SS has a porous surface, with more open pores and higher water absorption than ordinary macadam. Elongated and flaky particles (EFP) will form large pores, greatly weakening the intercalation of aggregates and affecting the road performance of the mixture. According to Table 1, the EFP content of SS is very low, almost all particle sizes are less than 10%.

Table 1. Physical properties of SS from the researched literature.

Particle Size (mm)	ASG (g/cm^3)	CV (%)	LA (%)	WA (%)	EFP (%)	References
4.75–9.5	-	10.3	8.5	-	10.1	[38]
≥9.5	-				6.7	
0–4.75	3.37	23.5	-	1.33	0.0	[39]
4.75–16	3.48					
16–26.5	3.47					
0.06–50	3.34	-	20	0.85	-	[16]
0.075–31.5	3.31	16.7	21.5	0.43	5.7	[40]
<1	3.35	-	-	0.40	-	[41]
0.075–31.5	3.50	16.3	12.05	2.00	1.1	[42]
0–25	2.90–3.20	-	-	≤1.50	-	[43]
5–10	3.25	13.6	13.2	2.70	5.1 (4.75–9.5mm)	[44]
10–16	3.26			1.96	4.3 (>9.5mm)	
0.075–31.5	3.24–3.36	17.5	18.5	0.46	4.6	[45]
0.075–31.5	3.55	14.5	11.8	2.60	4.7	[46]
0–5	-	-	-	-	-	[47]
0–25	-	19.7	-	-	-	[48]
0.075–31.5	-	20.7	-	-	-	[49]
0–10	3.14	-	-	4.44	-	[50]

2.2. *Chemical and Mineralogical Characteristics of SS*

Generally, SS can be categorized into four types based on the production process: blast furnace slag, basic-oxygen furnace slag, electric-arc furnace slag, and ladle furnace slag [51]. However, in the selection of semi-rigid base materials, little information about the types of SS is mentioned in the previously referenced published paper. It means that SS used for semi-rigid base is not usually considered with regards to production process, but rather regarding its basic performance. The chemical composition of SS from the previous studies is shown in Table 2. SS has a higher density than limestone and granite because of its high iron content [52]. The relative proportion of Al_2O_3 and SiO_2 is high, and tricalcium silicate and active mineral calcium aluminate are formed in the later stage. When SS contains a greater proportion of CaO, then it has a greater alkalinity and has a greater activity [53]. CaO can react with SS and water, and the pH can reach 10–12, which has a certain anti-corrosion effect [40]. Free calcium oxide (f-CaO) is generally considered to be the main problem in terms of swelling in water [54,55]. The content of f-CaO in the researched literature is lower than the recommended limit [56].

Table 2. Chemical composition of SS from the researched literature.

CaO	Fe_2O_3+FeO	SiO_2	MgO	MnO	P_2O_5	Al_2O_3	SO_3	Na_2O+K_2O	f-CaO	References
51.23	16.78	8.59	3.93	2.26	1.52	1.29	0.30	-	2.47	[39]
-	89.00	1.00	-	-	-	-	-	-	-	[16]
66.58	3.20	21.86	1.39	-	-	5.46	0.32	0.91	-	[46]
52.52	0.33	23.72	4.50	0.97	-	2.26	0.76	-	-	[47]
40–50	7–10	10–15	10.82	-	3–5	-	-	-	2.10	[48]
21.08	12.84	14.21	13.49	-	-	5.83	-	-	-	[49]
24.98	25.45	17.08	10.58	8.91	-	5.40	0.25	0.25	-	[57]
45.90	11.20	16.30	6.50	-	4.10	9.70	-	-	-	[58]
42.72	25.02	15.77	4.79	4.79	0.86	6.07	-	0.08	0.71	[50]
19.59	40.74	22.69	0.23	0.13	0.60	12.75	-	-	<0.01	[59]
35.30	1.20	33.40	6.50	0.75	-	18.20	-	-	-	[34]

3. Performance Evaluation and of SS Mixtures

3.1. *Cement Stabilized Steel Slag (CSS)*

The volume expansion of SS can be reduced by natural weathering, aging, or chemical treatment [11,60,61]. Table 3 shows that part of SS is aged or pretreated before being used for base course to reduce volume expansion. The combination of SS and macadam cannot only be applied for the full utilization of the active components of SS and avoid overstrength, but also for good stability [49]. Through the mechanical test, it is found that adding SS can improve the mechanical properties of the mixture, which is better than the traditional semi-rigid base [13]. It is possible to improve the performance of the base course using SS, but consuming a great deal of SS and improving resource utilization can also be achieved by using SS as a partial or complete replacement for macadam. However, the research on CSS started late and there are few relevant engineering and technical standards.

3.1.1. OMC and MDD of CSS

There are many factors affecting OMC and MDD determined by the compaction test especially the influence of compaction work and the particle size gradation of SS. The compaction work is defined, and the transmission of compaction work and the movement effect among SS particles in different gradations are different, so the OMC and MDD under different SS gradations are different. MDD and OMC play an important role in the compaction, which many mechanical property tests are based on. Therefore, it is essential to study OMC and MDD of CSS.

Table 3. Components and properties of CSS from the researched literature.

Treatment	SS: Macadam Content (%)	Cement Content (%)	OMC (%)	MDD (g/cm^3)	References
-	69:31	5	4.60	-	[38]
-	30:70, 50:50, 70:30	3.5/4/4.5	4.72–5.14	2.39–2.56	[39]
-	100:0	3/4/5	6.72–8.52	2.72–2.75	[40]
Pretreat	100:0, 90:10, 70:30, 50:50, 0:100	4/5/6	4.90–5.85	2.31–2.38	[41]
Aged 1 year, Pretreat (fly ash)	75:25, 50:50, 25:75, 0:100	-	4.71–5.37	2.49–2.74	[42]
-	100:0, 60:40	5	4.70–5.60	2.65–2.66	[44]
-	100:0	3/4/5	6.73–8.51	2.71–2.75	[45]
Aged	85:15, 65:35, 55:45, 0:100	4/5/6	3.80–5.30	2.23–2.68	[46]
Pretreat (NaOH)	100:0	5/8/10/14	-	-	[47]
Aged 6 months	100:0	1/2/3/4/5	7.70–8.30	2.62–2.64	[48]
Aged	75:25, 50:50, 25:75	3	5.50–7.00	2.25–2.41	[49]
-	100:0, 75:25, 50:50, 25:75, 0:100	2/4/6	-	2.26–2.53 (7d) 2.28–2.60 (28d)	[57]
-	100:0, 75:25, 50:50, 25:75, 0:100	3	4.30–7.00	2.27–2.63	[58]

Figure 2 shows that the law of OMC and MDD of CSS with different proportions is significant. The higher the SS content, the greater OMC of the mixture and MDD. It can be explained that the water absorption of SS is smaller than that of crushed stone, which is also consistent with the physical characteristics of the SS in Section 2.1. The OMC and MDD of 0% SS mixture are significantly less than 100% SS.

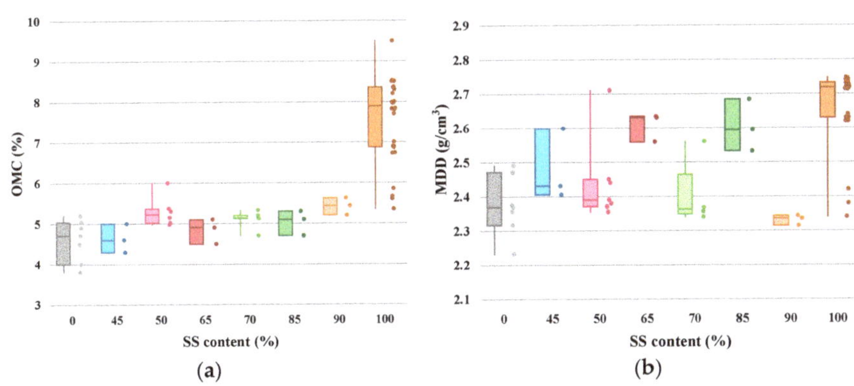

Figure 2. The relationship between (**a**) OMC, (**b**) MDD and different SS content (Adapted from [39–42,44–46,48,49,58]).

The difference of SS gradation will cause the change of its OMC and MDD, as shown in Figure 3a. The particle size ranges of SS for coarse, fine and intermediate are 0.075–31.5 mm [40,49]. According to the specification of China, the upper, middle, and lower limits of the gradation range are selected as three kinds of gradations [40]. The finer the SS gradation, the greater OMC and the smaller MDD, but the change range of MDD is small. It can be explained that the finer gradation contains more fine materials and has a more prominent specific surface area. The fine SS particles have high water absorption, so the water content of the mixture will become larger. The coarser gradation contains more coarse materials, and the skeleton structure formed between the coarse materials makes MDD larger than that of the finer gradation. However, the fine gradation is more likely to develop a dense structure due to the large amount of fine materials, so MDD is similar.

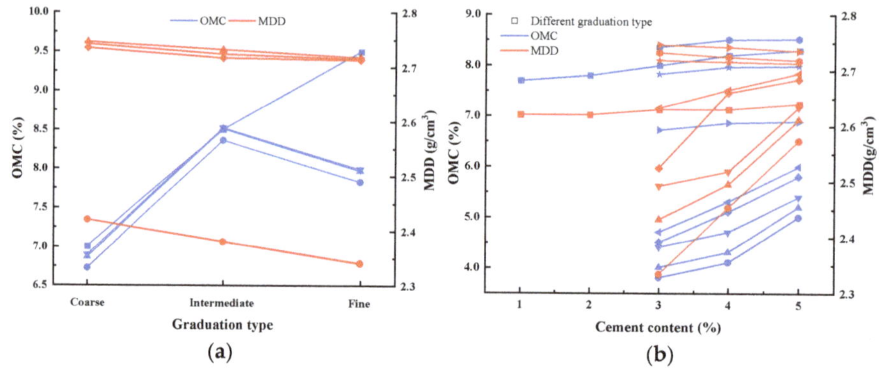

Figure 3. (**a**) The relationship between OMC, MDD and grading particle size of SS (Adapted from [40,49]) (**b**) OMC, MDD of different SS contents (Adapted from [40,48,62]).

Figure 3b shows that OMC increases with the addition of cement content. However, due to different materials and graduation types, the growth range of OMC is 0.4%–31.6%, when the cement content increases from 3% to 5%. The MDD shows a similar trend with the OMC. The increasing rate of most MDD ranges from 1.9% to 10.3%. MDD values dropped slightly in two sets, but fluctuated within 0.5%. Due to the difference of materials, it is considered negligible.

3.1.2. UCS of CSS

With the increase of SS, UCS increases continuously, as shown in Figure 4a. UCS of 7d meets the requirements of medium and light traffic of high-grade roads in China. When the content of cement exceeds 5% and SS exceeds 40%, it meets the requirements of heavy traffic (4 MPa) [63].

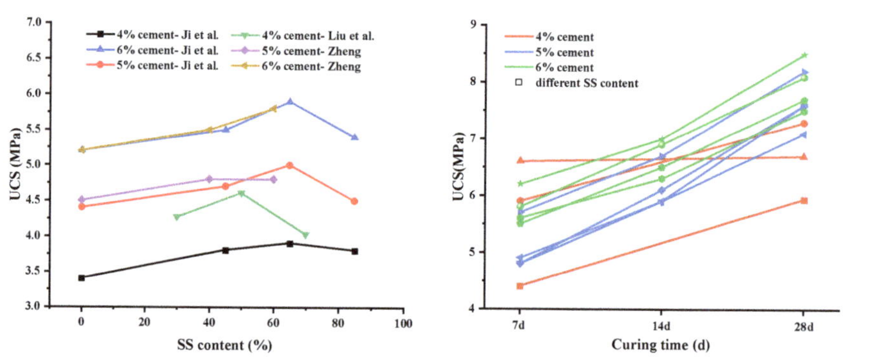

Figure 4. (**a**) UCS of different SS and cement contents (Adapted from [39,46,62]) (**b**) UCS of different curing time (Adapted from [39,46,62]).

Some studies show that UCS increases with the SS content increasing [49,62,64]. The addition of crushed stone directly reduces SS, and the reduction of active components in SS will inevitably have a negative impact on the compressive strength. When the crushed stone exceeds 75%, its compressive strength is almost the same as 100% crushed stone [49]. The reason is that a small amount of SS does not play a decisive role in the strength, and only when the amount of SS accounts for a large proportion will the 7d compressive strength be greatly improved.

Other researchers have found an optimal SS content between 40% and 80%, within which the UCS value can reach the maximum [39,42,46]. When the SS content increases from zero to the optimal content, the damage resistance is effectively improved. This is because the high strength and rough surface structure of SS cause the UCS to increase continuously. In addition, C_3S and C_2S cementitious components in SS form C-S-H gel to make the system stronger [65]. When the SS content is further increased, CSS will produce more harmful pores [66]. CSS has high porosity and water absorption, leading to a large internal gap of the sample, which is not conducive to the formation of strength.

When the cement content is low, the 7d UCS of CSS is low. It can be explained by the fact that there are many pores on the surface of SS, and some cement particles gather on the surface of SS so that this part of the cement cannot be hydrated quickly. In the case of low cement content, the adsorption of SS reduces the amount of reactive cement, which is not conducive to improving the compressive strength of the mixture.

With the same cement content and SS content, the UCS of CSS rises with the curing time increasing, as shown in Figure 4b. The appearance of SS can be seen clearly as shown in Figure 5a. It can be seen that there is a small amount of white floccules on the outer surface of SS, which is C-S-H gel. The number of products in the field of vision is not dense, indicating that SS has just begun to react with water. In the first 7 days of curing, the reaction inside CSS is relatively slow, and there are not many substances generated. Meanwhile, there are many pores on the surface of SS, and some cement particles gather on the surface of SS which cannot be hydrated quickly [62]. The adsorption of SS reduces the amount of reactive cement, which is not conducive to improving the compressive strength of the mixture. Therefore, the compressive strength of the mixture cannot be significantly improved at the initial stage. After 28 days, the white floccules have basically wrapped the SS, and the original appearance of the SS can hardly be seen, as shown in Figure 5b. The strength of CSS is mainly provided by skeleton strength, hydration, and pozzolanic reaction of SS. There are similar active ingredients in cement and SS, which will hydrate calcium silicate, hydrated calcium aluminate, hydrated calcium ferrite, and calcium hydroxide after reacting with water. The active silicon dioxide and aluminum oxide components in SS can react with calcium hydroxide to produce hydrated calcium silicate and hydrated calcium aluminate. As the curing time increases, a large number of substances are generated in CSS, which are closely connected to the SS and crushed stone particles to form a firm structure and improve the overall strength of the mixture.

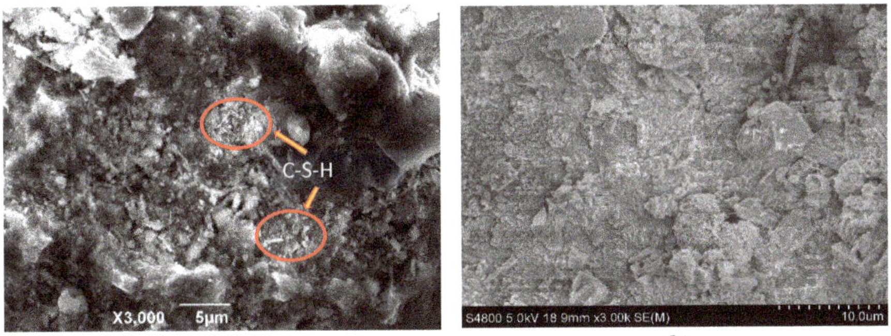

(a) (b)

Figure 5. (**a**) After 7d; (**b**) After 28d; SEM images of CSS (×3000 times) (Adapted from [62]).

With the increase of cement, the UCS of CSS also increases. The cement is continuously hydrated in the mixture, and the compressive strength of the mixture increases. There are few reaction products at first. The SS activity in the mixture is low, and the reaction speed is slow. However, with the increase of curing time, the activity of SS is slowly stimulated, the reaction is constantly accelerated, and the cementitious products are also continuously increased, increasing the compressive strength of the mixture [67,68]. Despite reducing cement hydration rates in the early stages, SS can boost cement hydration later. In addition, SS will also react with water to increase the compressive strength of the mixture. Therefore, at the same cement content, the more macadam replaced by SS, the greater the UCS of the mixture.

3.1.3. DS of CSS

The cement stabilized base course is easy to crack under long-term load or when the ambient temperature changes. After the cracking of the base course, the integrity of the base course can be destroyed with the cracking of the pavement surface. The relationship among water loss rate, DSC, and time is shown in Figure 6. The water loss rate and DSC of different mixing ratios show similar growth trends, and they are close to the level with the increase over time [39–41,62,64]. The water loss rate of the CSS mixture is greater than that of the CSM mixture, indicating that the water absorption performance of SS is better than that of macadam, and the hydration effect of CSS is more evident than that of CSM [40]. Therefore, when SS is used as base material, special attention should be paid to field construction in the first two weeks to avoid adverse impact of water loss on the performance of base [39,58]. The high-water loss rate in the early stage occurs when the specimen is saturated, the cement and SS in the mixture react with water to consume water, and a large amount of free water will evaporate from the specimen. The greater the internal reaction of the mixture in the early stage, the greater the water loss, making the water loss rate in the early stage more prominent. After that, the specimen will consume more water in the early stage, and the products generated by hydration reaction are wrapped on the surface of cement and SS particles. Therefore, the hydration reaction rate of cement and SS is slowed down and the water consumption is reduced. Thus, the water loss rate will gradually decrease in the later stage.

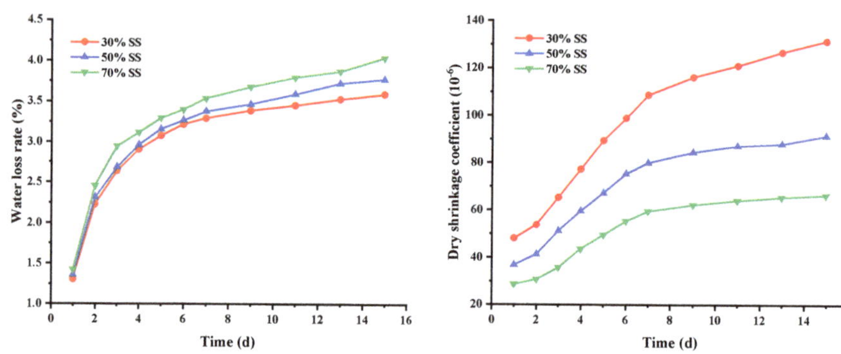

Figure 6. (**a**) The water loss rate of CSS with curing time (Adapted from [39]). (**b**) Dry shrinkage development of CSS with different SS contents (Adapted from [39]).

At the same cement content, the greater the amount of SS, the greater the water loss rate of the mixture and the smaller the DSC [39–41,62,64]. The DS of materials is caused by the shrinkage of pores caused by water evaporation, resulting in the change in overall macro volume. The micro expansion of active components such as f-CaO in SS in contact with water makes up for this defect, thus reducing the DS of the material [39,40,62,64,69]. Therefore, CSS as pavement base or subbase material can effectively slow down the formation of pavement reflection cracks. According to the research of Xu [70], DSC becomes 42% less with 50% SS added after 90 days of curing.

Under the same SS content, the more the cement content, the greater the water loss rate of the mixture, and the DSC increases to a certain extent, but it is not significant [62,70]. Based on 40% and 60% SS content, Zheng [62] increased the cement content by 1%, resulting in the DSC increasing by 7.1% and 9.7%, respectively. Similarly, the cement content was increased by 1%–2%, which caused a 10% increase in the DSC of the specimens with four SS contents [70].

3.1.4. TS of CSS

The average value of TSC corresponding to each temperature interval in the TS test results represents the overall level of shrinkage performance of SS mixtures related to temperature change, as shown in Figure 7. The development law of TSC is opposite to that of DSC. The increase of temperature causes the expansion of water, and the expansion pressure of water increases the particle spacing, resulting in the expansion of base materials [40]. With the same cement content, the TSC of CSS mixture increases with the increase of the content of SS. Due to the SS reaction with water, the hydration products and secondary minerals in CSS mixture will increase. Compared with macadam, it is more sensitive to the change in temperature [39,40,62]. The research results of Zheng [62] are shown as follows: under the content of 5% cement, the average TSC of CSM was $7.5 \times 10^{-6}/°C$. When the content of SS was 40% and 60%, the TSC of CSS increased by 14.7% and 25.3%, respectively. Similarly, under the content of 6% cement, the average TSC of CSM was $8.7 \times 10^{-6}/°C$. When the content of SS is 40% and 60%, the TSC of CSS increased by 19.5% and 29.9%, respectively. The cement content will affect the TS deformation of base material [40,62]. In the case of the same SS content, adding 1% cement would increase the TSC by 20.9% and 20.2%, respectively.

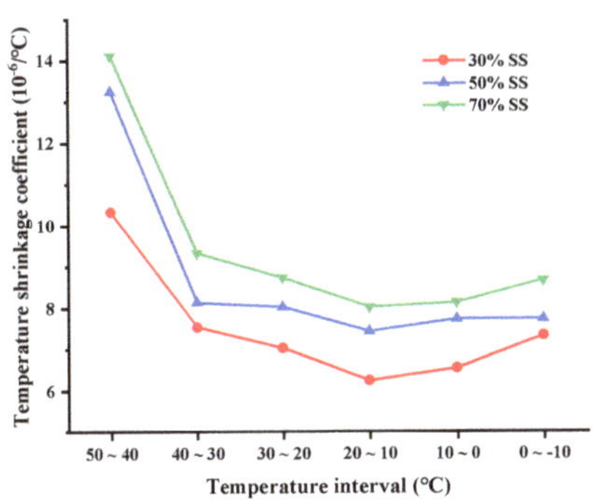

Figure 7. Temperature shrinkage of CSS with different SS contents (Adapted from [39]).

The cracking of CSM base is mainly caused by DS. It can be seen from the comparison between Figures 6b and 7 that the DS of CSS base is much greater than the TS. Although the TSC of the mixture increases with adding SS into CSM, the increased value is far less than the decrease of DSC. Therefore, due to the addition of SS into CSM, the DS of base course is greatly reduced, which indicates that the crack resistance of base course is enhanced. Furthermore, SS added to CSM improves the strength of the base course as well. In conclusion, on the premise of meeting the strength design requirements of the base course, the amount of cement in CSS mixture can be appropriately reduced, and the strength still meets the requirements. By reducing cement content, the economic cost of CSS will be reduced, and the DS and TS of the base will also be reduced so that crack resistance will be improved.

3.2. Lime–Fly Ash Stabilized Steel Slag (LFSS)

Fly ash is another type of industrial waste. The combination of fly ash and SS can not only obtain better environmental benefits but also inhibit the expansion of SS. The use of fly ash as a stabilizer for SS base has been widely studied, as in LFSS [59], lime–fly ash stabilized steel slag and crushed stone [71], fly ash–steel slag soil [72], and fly ash–steel slag–phosphogypsum [73]. Due to a relative scarcity of studies on other types, this paper mainly focuses on LFSS with 100% SS as aggregate. The traditional lime–fly ash soil base shows low early strength, and the addition of SS can effectively improve the early strength of the base for the characteristics of SS. Components and properties of LFSS from the researched literature are shown in Table 4.

Table 4. Components and properties of LFSS from the researched literature.

Lime: Fly Ash Content (%)	Binder: Aggregate Content (%)	OMC (%)	MDD (g/cm^3)	References
1:3	16:84	5.60–6.10	2.29–2.58	[58]
1:3	1:2.5, 1:3, 1:3.5, 1:4	8.10–10.60	2.07–2.28	[59]
6:19, 1:3, 4:11	25:75, 20:80, 15:85	8.7–14.0	1.69–2.00	[74]
1:3	25:75, 20:80, 15:85	-	-	[75]

3.2.1. UCS of LFSS

When the lime-to-fly ash ratio is 1:3, the UCS relationship of LFSS of various mix proportions is shown in Figure 8. It can be seen that the 7d UCS exceeds 4 MPa, which is much higher than the requirements for the extra heavy traffic on expressways in the specification (1.1 MPa) [63]. In LFSS mixture, SS is used as the skeleton to provide strength and stiffness for the mixture. Lime and fly ash as fillers must also maintain a certain proportion to ensure the long-term strength and durability of the LFSS mixture. The increase of the mixture's compressive strength mainly depends on the pozzolanic reaction of lime and fly ash [59]. With the increase of aggregate proportion, UCS values continue to improve. Pai, et al. [59] found that when the binder-to-aggregate ratio increased from 1:4 to 1:2.5, the UCS decreased by 20% (7d) and 19% (14d), respectively. Because the pozzolanic reaction is slow, the increase of curing time can significantly improve UCS value [59]. When the binder-to-aggregate ratio is 1:3, the UCS of 14d and 28d was 64% and 83% higher than that of 7d. When the binder-to-aggregate ratio is 1:4, the UCS value of 14d and 28d was 33% and 64% higher than that of 7d.

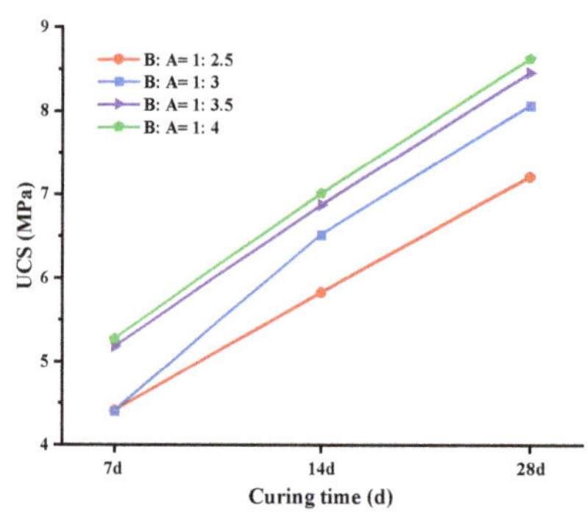

Figure 8. UCS of different curing time (lime-to-fly ash ratio = 1:3, B: A denotes binder-to-aggregate ratio) (Adapted from [59]).

As shown in Figure 9a, when the ratio of lime-to-fly ash is approximately 1:3, the 7d compressive strength of the binder is the largest. The addition of little lime cannot fully stimulate the activity of fly ash, resulting in a slow reaction of pozzolanic ash and low early strength. The influence of the lime-to-fly ash ratio on the compressive strength of the mixture can be seen from Figure 9b. The compressive strength of all proportions of the LFSS mixture is higher than 1.1 MPa, meeting the specification requirements [63]. Although the ratio of lime-to-fly ash can be from 1:2 to 1:5, due to quality and economic reasons, the ratio of 1:3 is more adopted, which is consistent with the data of Table 4. When the lime-to-fly ash ratio is between 5:15 and 4:16, the compressive strength result of LFSS mixture is better, which further explains the rationality of the lime-to-fly ash ratio.

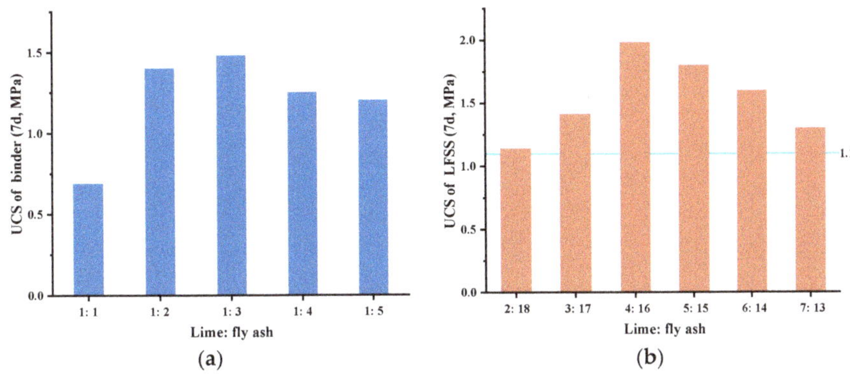

Figure 9. (**a**) UCS of binder; (**b**) UCS of LFSS; UCS of different binder of LFSS (binder-to-aggregate ratio= 1:4) (Adapted from [74]).

3.2.2. DS of LFSS

The water loss rate and DSC of LFSS are shown in Figure 10. Because all materials in LFSS have high water absorption, it shows significant water loss rate in the early stage. On the third day, the maximum water loss rate exceeded 5%. On average, the water loss rate in the first week exceeded 90% of the whole observation period, and the second week accounted for less than 10%. In comparison with CSS, LFSS exhibits less shrinkage deformation. The DS of LFSS is mainly affected by the water in SS and lime–fly ash. In general, the more the content of fine aggregate, lime and fly ash, the greater the DSC. In addition, the increase of dry density may also lead to the increase of DSC. Because the initial water content is high, mainly due to the rapid evaporation of free water, and the DS is less affected, and the DSC shows a downward trend.

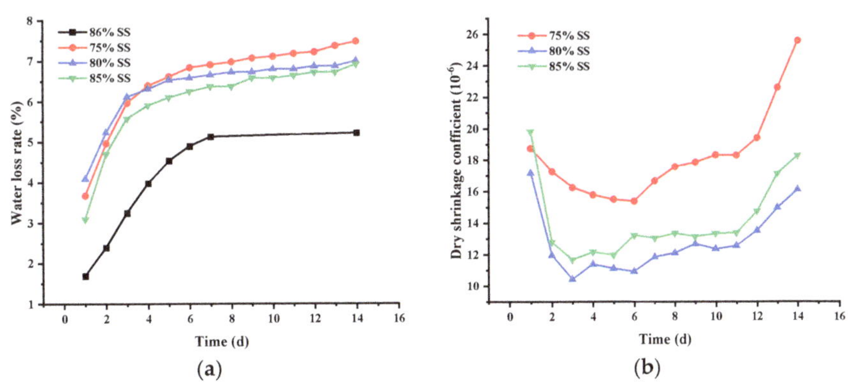

Figure 10. (**a**) The water loss rate of LFSS with curing time (lime-to-fly ash ratio = 1:3) (Adapted from [58,74]) (**b**) Dry shrinkage development of LFSS with different SS contents (Adapted from [74]).

3.2.3. TS of LFSS

The TS of lime–fly ash stabilized mixture is explained by the fact that each component of the material has thermal expansion and shrinkage when it meets heat and cold. There are many minerals in LFSS mixture, which can be mainly divided into aggregate and new cementitious minerals. The TSC varies between different materials. According to the existing study, the thermal expansion and shrinkage of new cement minerals is much greater than that of raw material minerals. Therefore, with the increase of SS in the mixture, the internal hydration products also increase, which improves the sensitivity of the mixture to temperature [76]. SS is more sensitive to temperature, which is consistent with the TS of CSS, as shown in Figure 11. The existing study results show that the TSC of LFSS shows different trends. However, it is noteworthy that there are inflection points around 0–10 °C. From high temperature interval to low temperature interval, TSC first showed a downward trend, and then increased [74]. The maximum values of different SS contents appeared at between −10 and −20 °C. The maximum values of TSC were 130.8%, 84.3%, and 93.6% higher than the minimum values, respectively. According to Li, et al. [58], TSC first increased with temperature decrease, then decreased. The difference between the maximum value and the minimum value of TSC is only 21.4%. It indicated that the effect of temperature in CFSS is not significant. In general, the change trend of TSC is complex and needs to be further studied.

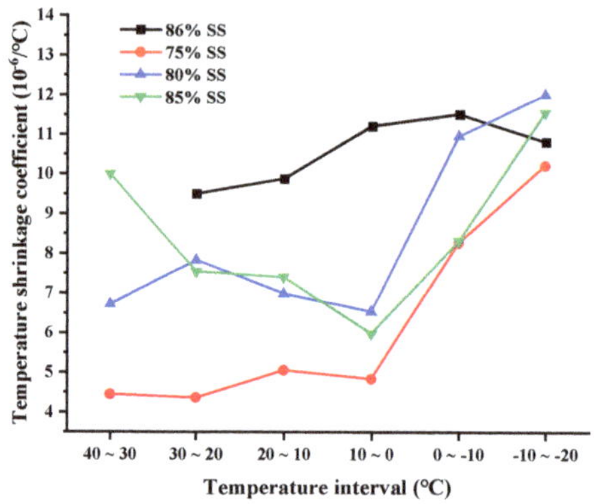

Figure 11. Temperature shrinkage of LFSS with different SS contents (Adapted from [58,74]).

3.3. Cement–Fly Ash Stabilized Steel Slag (CFSS)

In CFSS mixtures, the gradation and strength of the SS itself caused the SS skeleton to have a certain strength. In addition, among the SS particles, due to the active substances provided by cement hydration, physical and chemical reactions such as ion exchange reaction, $Ca(OH)_2$ crystallization reaction, and pozzolanic reaction are carried out in the water environment in the mixture. The main function of fly ash was to provide the active silica and alumina required for the pozzolanic reaction. The activity was stimulated by the action of cement and alkaline substances to promote the pozzolanic reaction. The reaction of cement fly ash in the mixture increased with age, and various reactions continued. The products generated by the pozzolanic reaction were aggregated, forming a chaotic spatial network connection between the SS particles, and the connection strength and stiffness were enhanced so that the CFSS had high strength. Components and properties of CFSS are shown in Table 5.

Table 5. Components and properties of CFSS from the researched literature.

Cement Content (%)	Fly Ash/SS Content (%)	OMC (%)	MDD (g/cm^3)	References
1/2/3/4	10/90, 20/80	7.90–10.10	2.31–2.63	[48]
4	0/100, 10/90, 20/80, 30/70, 40/60, 50/50, 60/40	7.00–15.80	1.64–2.30	[50]
2/4/6/8	0/100, 10/90, 20/80, 30/70, 40/60	-	-	[34]
4/6/8/10	36/60, 34/60, 32/60, 26/70, 24/70, 22/70, 20/70, 16/80, 14/80, 12/80, 10/80	15.30–23.20	1.53–1.84	[77]

3.3.1. OMC and MDD of CFSS

In Figure 12, the cement was calculated by external admixture. In CFSS mixtures OMC was significantly larger than that of CSS mixtures due to the presence of fly ash With the increase of cement content, both OMC and MDD showed an increasing trend

From 1%–4%, OMC increased by approximately 5%, and MDD increased by 3% and 4%, respectively. Compared with cement, the change of SS content had a more significant effect on OMC and MDD. The SS content changed from 80% to 90%, and the OMC decreased by 21%–24%. The OMC was changed since the WA of fly ash was much larger than that of SS. Conversely, the specific gravity of SS was slightly larger than that of fly ash, so the addition of more SS led to an increase in the density of the mixture. Therefore, the SS content changed from 80% to 90%, and the MDD increased by 9%–11%. At the same compaction energy, the mixture was compacted to a higher density, resulting in an increase in MDD and a decrease in OMC. Therefore, replacing fly ash with an equivalent amount of SS will inevitably increase the dry density of CFSS.

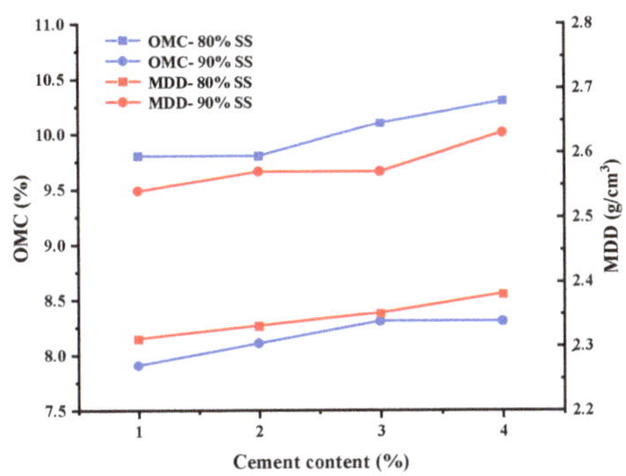

Figure 12. OMC, MDD with different SS and cement contents (Adapted from [48]).

3.3.2. UCS of CFSS

Figure 13a shows the UCS of CFSS at 4% cement content. Singh, et al. [34] and Yu [77] found the UCS development law of 0%–40% and 6%–80% SS content, respectively. The results showed that with the increase of SS, the UCS increases continuously. Zhou, et al. [50] found that with the increase of fly ash content, the UCS of CFSS first increased and then decreased. When the content of SS was 80%, the UCS reached the maximum value. In the mixtures, the cement provided the active substance, prompting the crystallization reaction of $Ca(OH)_2$, the pozzolanic reaction, and other reactions. Fly ash provided the active silica and alumina required for the pozzolanic reaction. The activity was stimulated under the action of cement and alkaline substances to promote the pozzolanic reaction.

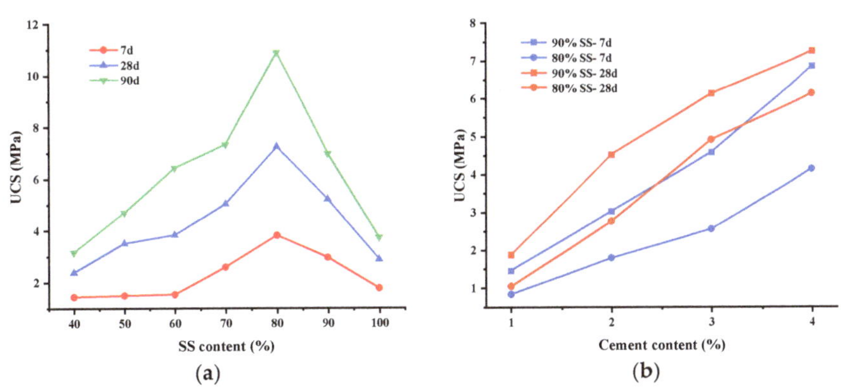

Figure 13. (**a**) The effect of fly ash addition on the UCS of CFSS (Adapted from [50]) (**b**) The effect of cement contents on the UCS of CFSS (Adapted from [48]).

Compared with no addition of fly ash (100% SS), the incorporation of 20% fly ash (80% SS) increased the 7d, 28d, and 90d UCS of CFSS by 126%, 157%, and 197%, respectively [50].

This can be explained by the fact that the surface of the SS particles without fly ash had a thin binder coating, resulting in a small contact area between the SS particles. In this case, it was easy to cause stress concentration and damage under pressure. With the addition of fly ash, the thickness of the binder coating gradually increased, making it increasingly dispersed and subject to compressive stress.

The compressive strength of CFSS increased with curing time. The reaction of cement fly ash in the mixture continues with the curing time, and the amount of the generated cementitious substances also increases. This made a strong connection between the SS particles, and the strength and stiffness of the association were gradually increased so that the compressive strength of the CFSS was also gradually increased.

With the increase of cement content, UCS of CFSS also increases as shown in Figure 13b. This can be explained by the fact that that the increase of curing time and binder content will increase the amount of C-S-H gel, thus improving the compressive strength of the mixture. In the case of the same cement content, the compressive strength was also significantly improved by adding SS. SS could therefore play a role in increasing bearing load and providing strength, which improved the UCS of CFSS.

3.3.3. DS of CFSS

Figure 14 reflects the relationship between moisture loss rate and curing time. The moisture loss rate of the mixture increased with time. In the first ten days, the water loss rate was faster, and its curve increased almost linearly. As the moisture content in the mixture was consistent with the surrounding environment, the average moisture loss hardly changed. In addition, there was also a certain relationship between the moisture loss rate and the OMC of the mixture. The greater the OMC of the mixture, the greater the water loss rate.

The DSC decreased with curing time. As shown in Figure 15a, the maximum values of DSC for different blend ratios all appeared in the early stage. This was the reason for the enormous DS strain of the mixture in the early stage. From approximately the eighth day, the DSC began to tend to the level and maintained at around 122×10^{-6}, 94×10^{-6}, and 101×10^{-6}, respectively [77] Compared to the first day, DSC decreased by 64%, 68%, and 51%, respectively. With the increase of SS content, the DSC decreased to a certain extent. For example, in the previous week of curing time, the DSC of 80% SS was more significant than that of 70% SS. The changes in DSC and cement content after the seventh day are shown in Figure 15b, when the content of SS was 70%. During the first week, the regularity of DSC development was not significant, but after a week, the mixture with lower cement content and more fly ash showed less DS. Among them, the DSC difference of 8% cement content and 10% mixture was not significant.

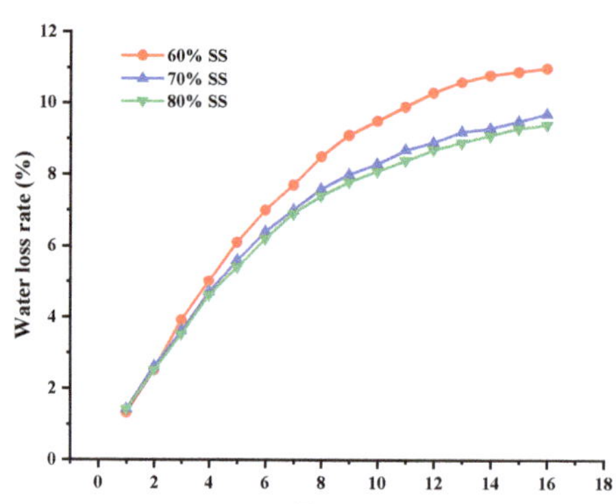

Figure 14. The water loss rate of LFSS with curing time (6% cement) (Adapted from [77]).

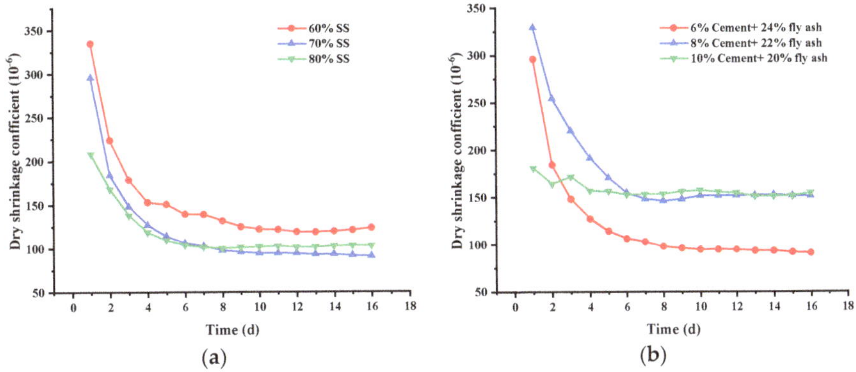

Figure 15. (**a**) different SS content; (**b**) different cement and fly ash contents; Dry shrinkage development of CFSS (Adapted from [77]).

3.3.4. Inhibition Mechanism of Fly Ash on CFSS

Figure 16 showed the SEM image results of the side of the cement–fly ash binder in contact with the SS. The results showed that many needle-like, fibrous, and sheet-like C-S-H phases were formed on the surface of the binder [50]. The hydration of f-CaO in SS did not form larger $Ca(OH)_2$ crystals [56]. The increase of fly ash content reduced the possibility of SS forming a skeleton and prevented the concentration of expansion stress between SS. The low elastic modulus of the cement–fly ash binder relieved the expansion stress formed by the $Ca(OH)_2$ expansion component. The binder reacted with the products produced by the expansion of the SS, preventing the further growth of $Ca(OH)_2$ [78,79].

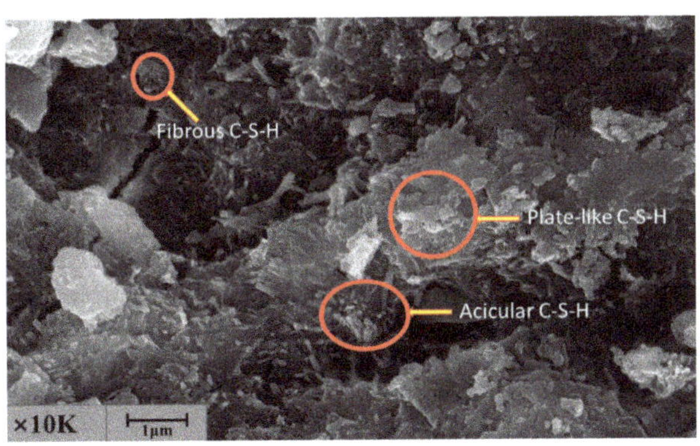

Figure 16. The SEM analysis of the binder on the side where the binder contacted and SS (Adapted from [50]).

4. Numerical Simulation

With the rapid development of computer hardware and related scientific theories, numerical simulation technology has been widely used in materials research, such as finite element method (FEM), discrete element method (DEM), and molecular dynamics (MD) [18,39,80–84]. The experimental cost of CSS is relatively large, the testing period is long, and there are many uncontrollable external factors. Therefore, numerical simulation technology has been adopted by some scholars in the study of CSS.

Wu [83] conducted a freeze–thaw cycle test on CSS specimens to consider the damage to the CSS base in a seasonally frozen area. A simulation model of the freeze–thaw cycle test suitable for CSS was constructed using ANSYS Workbench. Through the numerical comparison analysis of the compressive strength test and simulation undergoing different freezing and thawing cycles (0, 5, 10, 20 times), it was found that the accuracy exceeded 92%. Through the numerical simulation of the freeze–thaw cycle test of CSS specimens, the results showed that the stress and strain change periodically with temperature and are positively correlated. With the increase of freezing and thawing times, the specimens were subjected to stress and gradually diffused inward. The maximum stress on the inside and outside of the samples increased accordingly. By comparing and calculating the temperature, stress, and strain of each pavement structure with and without bed course, it was recommended to set a cushion to protect the CSS in low-temperature areas.

Karatağ, et al. [18] used Plaxis software to perform a plane-strain FEM analysis of the rutting characteristics of the SS base. A hardening soil model with small-strain stiffness was used to examine the deformation behavior of layered pavement whereas asphalt concrete modeled using the linear elastic model. After 10,110 triangular load cycles, the rutting rate of the asphalt pavement on the SS base was approximately 33% lower than that on the gravel base. It could be seen that SS was a better substitute for natural aggregate in terms of deformation properties.

MD is an emerging technology to describe interactions between substances [85–87]. MD simulation has been widely used in materials science, biology, medicine, physical chemistry, and other fields [88]. Liu, et al. [39] selected the condensed-phase optimized molecular potentials for atomistic simulation study for cement–water–SS interfacial simulation. The results showed that the interfacial binding energy of the SS mixture reached 2781.6 MJ/m^2, which was much larger than that of limestone (830.5 MJ/m^2). Since both limestone and SS are alkaline stones, Coulomb electrostatic interactions account for most of the interfacial binding energy. A total of 91.80% of the cement–limestone interfacial energy was composed of electrostatic energy, and in the cement–SS system, the electrostatic energy increases to 99.55%. It could be concluded that during the stirring process, due to the large positive charge contained in the SS, the electrostatic energy increased, resulting in enhanced adhesion between the cement and the SS. At the same time, the displacement velocity of the particles indicated that the diffusion rate of cement molecules in the SS mixture was faster than that of limestone. This showed that the cement wrapped the SS earlier and underwent hydration, which was beneficial to forming the early strength of the SS mixture.

Wang, et al. [84] argued that traditional finite element methods focused on the macroscopic behavior of materials, while ignoring the evolution of the microstructure. Therefore, based on 3D scanning technology and in combination with physical test results, a DEM of CSS with actual SS shape was constructed using Particle Flow Code 3D. During the CSS failure process, microcracks went through four stages: initiation, propagation, convergence, and penetration, as shown in Figure 17. The leading cause of CSS failure was the absorption and dissipation of energy within the structure. With the increase of SS content, the number of microcracks gradually increased and the width of the crack surface increased after the failure of the specimen. It was worth noting that the SS content, SS particle size, SS roughness, and interface tensile strength had a great influence on the CSS interface transition zone.

FEM, DEM, and MD have been adopted in different scales for SS base simulation research, but they are still in the exploratory stage. FEM and DEM are mainly used to simulate the cracks, rutting, freeze–thaw, anti-scour, and so on. MD is often selected to simulate the diffusion of binder in the mixture to analyze the strength formation mechanism. Numerical simulation results cannot be well connected with mechanical properties and durability, so they cannot be well applied.

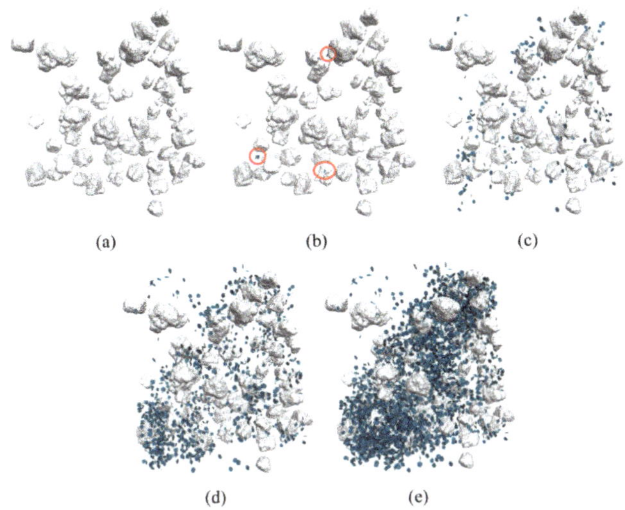

Figure 17. Crack evolution process of CSS (**a**) original model (**b**) initiation (**c**) expansion (**d**) convergence (**e**) penetration (Reprinted with permission from [84] Copyright Elsevier).

5. Environmental Impacts

Road engineering is one of the major consumers of natural resources [89,90] and one of the major producers of CO_2 emissions [91,92]. Life cycle assessment (LCA) is a set of standard procedures for evaluating environmental–economic efficiency in engineering with the aim of finding better alternatives [93]. Based on the LCA assessment, the utilization of SS has considerable environmental benefits [94]. Researchers have demonstrated that SS utilization can significantly reduce greenhouse gases and ecological impacts [94,95]. Kang, et al. [96] build a SS prediction model based on a convolutional neural network by relying on a big data acquisition and storage system to recommend the optimal utilization of SS.

Anastasiou, et al. [97] evaluated the pavement construction with fly ash as binder and SS as aggregate. Fly ash can significantly reduce the environmental impact of road construction, of which binder replacement rate is the most crucial factor. The environmental benefits of replacing 50% of cement with fly ash are far greater than the long-distance transportation of fly ash [98]. The CO_2 emissions of long-distance transport aggregates are similar, and they account for a small proportion of the total emissions from road construction. SS as aggregate can minimize the consumption of ecological resources, but it is not recommended to transport SS over long distances.

Liu, et al. [39] evaluated CO_2 emissions from material production and transportation for CSM and CSS at 50% SS. The results showed that due to the large power consumption in the cement production process, the emission of CO_2 accounts for the largest proportion. The transport emissions of SS and gravel mixture accounted for 28.9% and 19.3% of the total emissions, respectively. Throughout the process, CSS (188, 368 kg) produced 0.25% more CO_2 than CSM (187, 896 kg).

Gao, et al. [38] discussed the compaction process of SS base and andesite (a natural aggregate) base. The 98% compaction degree in the standard is used as the evaluation index [99]. Based on the comparison of the three compaction methods, the results show that strong vibration should be avoided or reduced and the static pressure exerted by the roller compactor should be increased during SS compaction. This approach mitigates the unfavorable "hard-to-hard" effect between adjacent slag particles, as shown in Figure 18. The "hard-to-hard" effect refers to the deformation of SS particles under strong vibration, followed by continuous elastic recovery and expansion. The final SS particle gap is more significant than the initial one, which leads to a decrease in the density of the mixture after compaction. Compared with the andesite base, the diesel consumption of the compactor at the SS base is higher, the energy consumption increases by 2.67 MJ/m^3, and the CO_2 equivalent emission increases by 0.20 kg/m^3.

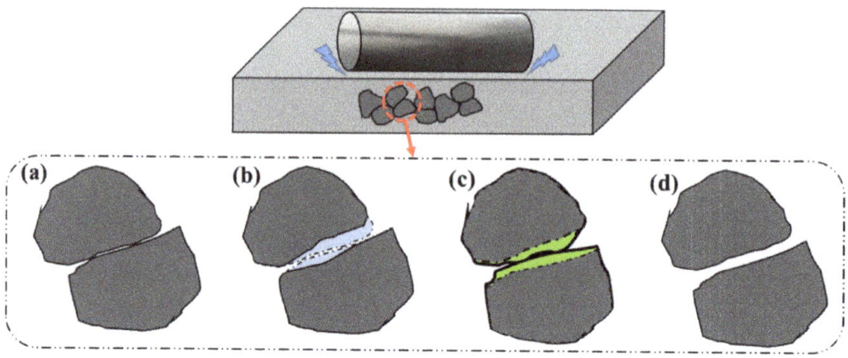

Figure 18. "Hard-to-hard" effect in SS compaction (**a**) the particles close to each other (**b**) deformation under external strong vibration (**c**) an elastic recovery expansion (**d**) the displacement of the particles results in a macroscopic decrease in density (Adapted from [38]).

At present, the environmental impact assessment of SS base mainly focuses on transportation and compaction. From the perspective of construction economic cost, it is recommended to use SS only in a short distance, and the environmental benefits generated by higher content of solid waste will be more significant. For the compaction process of road base, the CO_2 emission of SS base course is slightly more than that of traditional base.

6. Conclusions

This paper focuses on a critical on the utilization of SS as a road material in semi-rigid base course. The physical, mechanical, and durability properties indicate that SS is suitable for semi-rigid base. It is usually the basic performance of SS that is considered when used for semi-rigid base, rather than its production process. However, it is essential to pay attention to the high-water absorption and calcium oxide content of SS.

The utilization of SS as a road base material in the literature can be mainly classified into three aspects, consisting of CSS, LFSS, and CFSS. The SS content will significantly increase OMC and MDD in CSS. The gradation of CSS will affect OMC, but has little effect on MDD. The UCS rises with the SS content increasing. The water loss rate and DSC showed a similar growth trend as time increases. The increase of SS and cement content will increase the water loss rate, and DSC shows the opposite trend. After the addition of SS, the TSC of the mixture rises, but the increased value is far less than the decrease of DSC, which reduces the DS of the base course. In LFSS, when the content ratio of lime-to-fly ash is 1:3, the maximum strength is achieved, and this proportion is recommended. The ratio of binder-to-SS will also have a significant impact on UCS values, and around 20:80 is recommended. Due to the slow pozzolanic reaction, the increase of curing time can significantly improve the UCS value. In comparison with CSS, LFSS exhibits less DS deformation. The law of TSC is not clear, but there is an inflection point between 0–10 °C. In CFSS, the change of SS content has a more significant impact on OMC and MDD than that of cement content. The UCS increases with the increase of SS content, and there may be an optimal content. Cement, fly ash, and curing time have a positive effect on the UCS. As time increases, DSC decreases continuously, and large DS strains will be apparent during the early stage of the mixture. There is no obvious correlation between DSC and the content of cement and fly ash. FEM, DEM, and MD can be used to analyze the freeze–thaw, rutting resistance, and crack development of SS base. LCA is used to evaluate SS road base, which shows significant environmental significance. However, the CO_2 equivalent emission of SS base in the process of production, transportation, and construction is slightly higher than that of the traditional macadam base.

Based on this overview, the authors believe that further research is needed in certain directions so as to promote the application of SS as a road material in semi-rigid base:

(1) The SS base course is generally designed according to the traditional grading range recommended in the specification, and the particle size is within 0–31.5 mm. However, the particle size of SS produced in the iron and steel industry is too large, and secondary crushing is often required for the base course. After that, the SS mixture meets the requirements in various properties, and the gradation of SS itself is not fully utilized. If the grading design method based on the original grading of SS can be put forward, it will better promote the application of SS in the base course. In addition, the structural type is less considered, and it is not clear how the structural type affects the performance in the SS base.

(2) With the significant improvement of mechanical level and field construction technology, more projects adopt large thickness one-time molding to ensure the quality of road construction. However, in the laboratory test, heavy compaction and static pressure forming method are still widely used. It is worth noting that this method is matched with the static roller and cannot effectively guide the gradation design and field construction of a large thickness SS base course. In the laboratory test, it is necessary to select appropriate test methods of SS base to better simulate the field construction, so as to ensure the authenticity and reliability of the design results. At the same time, most of the existing studies focus on the mechanical properties, DS, and TS of the mixture, and there is little research on the anti-scour, fatigue performance, and frost resistance of the mixture. The applicability of SS base in rainy and cold areas is not clear.

(3) Moreover, it is time-consuming and expensive to test the performance of SS base course, and there are numerous uncontrollable factors. The numerical simulation of SS base course will be one of the important research directions in the future, but now the numerical simulation of SS base is still in the initial stage of development, and FEM, DEM, and MD have been adopted by researchers. How to combine numerical simulation with experimental characterization to explain the macro and micro mechanism and predict the performance of SS mixture remains to be explored. Meanwhile, based on LCA, the whole process of raw materials production, transportation, base course paving and use, and even abandonment needs to be evaluated. However, there is little research on the paving and using process of SS base. In addition to the compaction process, the high strength, high capacity and high-water absorption of SS will have different effects on the mixing and paving of the mixture. The impact of SS base on the environment has yet to be assessed.

Author Contributions: H.L.: conceptualization, methodology, validation, investigation, writing—original draft. C.C.: conceptualization, writing—review and editing. J.C.: supervision, project administration, resources, funding acquisition. M.Z.: methodology, writing—review and editing. Y.S.: writing—review and editing, supervision. All authors have read and agreed to the published version of the manuscript.

Funding: This research was funded by Key R & D and Transformation Plan of Qinghai Province (2021-SF-165), Key Research and Development Project of Shaanxi Province (2022SF-328), the Science and Technology Project of Henan Department of Transportation (2020J-2-3), Science and Technology Project of Shaanxi Department of Transportation (No.19-10K, No.19-28K).

Institutional Review Board Statement: Not applicable.

Informed Consent Statement: Not applicable.

Data Availability Statement: The information used to bolster the discoveries of this research are from prior detailed studies cited before.

Acknowledgments: We express our sincere gratitude to the experts, teachers, and students who have provided help for this article.

Conflicts of Interest: The authors declare no conflict of interest.

Abbreviations

ASG	Apparent specific gravity	LA	Los Angeles abrasion
CFSS	Cement–fly ash stabilized steel slag	LCA	Life cycle assessment
C-S-H	Calcium silicate hydrates	LFSS	Lime–fly ash stabilized steel slag
CSM	Cement stabilized macadam	MD	Molecular dynamics
CSS	Cement stabilized steel slag	MDD	Maximum dry density
CV	Crushing value	OMC	Optimum moisture content
DEM	Discrete element method	SS	Steel slag
DS	Drying shrinkage	TS	Temperature shrinkage
DSC	Drying shrinkage coefficient	TSC	Temperature shrinkage coefficient
EFP	Elongated and flaky particles	UCS	Unconfined compressive strength
FEM	Finite element method	WA	Water absorption

References

1. Li, B.; Tang, B.; Ma, Z.; Cheng, H.; Li, H. Physical and chemical properties of steel slag and utilization technology of steel slag at home and abroad. In Proceedings of the IOP Conference Series: Earth and Environmental Science, Moscow, Russia, 27 May–6 June 2019; p. 032012.
2. Song, Q.; Guo, M.; Wang, L.; Ling, T. Use of steel slag as sustainable construction materials: A review of accelerated carbonation treatment. *Resour. Conserv. Recycl.* **2021**, *173*, 105740. [CrossRef]
3. Gu, X.; Yu, B.; Dong, Q.; Deng, Y. Application of secondary steel slag in subgrade: Performance evaluation and enhancement. *J. Clean. Prod.* **2018**, *181*, 102–108. [CrossRef]
4. Gençel, O.; Karadag, O.; Oren, O.H.; Bilir, T. Steel slag and its applications in cement and concrete technology: A review. *Constr. Build. Mater.* **2021**, *283*, 122783. [CrossRef]
5. Hayashi, A.; Watanabe, T.; Kaneko, R.; Takano, A.; Takahashi, K.; Miyata, Y.; Matsuo, S.; Yamamoto, T.; Inoue, R.; Ariyama, T. Decrease of Sulfide in Enclosed Coastal Sea by Using Steelmaking Slag. *Tetsu-to-Hagane* **2012**, *98*, 207–214. [CrossRef]
6. Wang, Z.; Sohn, I. A review on reclamation and reutilization of ironmaking and steelmaking slags. *J. Sustain. Metall.* **2019**, *5*, 127–140. [CrossRef]
7. Gao, X.; Okubo, M.; Maruoka, N.; Shibata, H.; Ito, T.; Kitamura, S.-Y. Production and utilisation of iron and steelmaking slag in Japan and the application of steelmaking slag for the recovery of paddy fields damaged by Tsunami. *Miner. Process. Extr. Metall.* **2015**, *124*, 116–124. [CrossRef]
8. Semykina, A.; Shatokha, V.; Seetharaman, S. Innovative approach to recovery of iron from steelmaking slags. *Ironmak. Steelmak.* **2010**, *37*, 536–540. [CrossRef]
9. Wang, X.; Li, X.; Yan, X.; Tu, C.; Yu, Z. Environmental risks for application of iron and steel slags in soils in China: A review. *Pedosphere* **2021**, *31*, 28–42. [CrossRef]
10. Yu, H.; Ma, T.; Wang, D.; Wang, C.; Lu, S.; Zhu, X. Review on China's pavement engineering research·2020. *China J. Highw. Transp.* **2020**, *33*, 1–66.
11. Guo, J.; Bao, Y.; Wang, M. Steel Slag in China: Treatment, Recycling, and Management. *Waste Manag.* **2018**, *78*, 318–330. [CrossRef]
12. Yildirim, I.Z.; Prezzi, M. Chemical, Mineralogical, and Morphological Properties of Steel Slag. *Adv. Civ. Eng.* **2011**, *2011*, 463638. [CrossRef]
13. Motz, H.; Geiseler, J. Products of steel slags an opportunity to save natural resources. *Waste Manag.* **2000**, *21*, 285–293. [CrossRef]
14. Yi, H.; Xu, G.; Cheng, H.; Wang, J.; Wan, Y.; Chen, H. An overview of utilization of steel slag. *Procedia Environ. Sci.* **2012**, *16*, 791–801. [CrossRef]
15. Riley, A.L.; MacDonald, J.M.; Burke, I.T.; Renforth, P.; Jarvis, A.P.; Hudson-Edwards, K.A.; McKie, J.; Mayes, W.M. Legacy iron and steel wastes in the UK: Extent, resource potential, and management futures. *J. Geochem. Explor.* **2020**, *219*, 106630. [CrossRef]
16. Behiry, A.E.A.E.-M. Evaluation of steel slag and crushed limestone mixtures as subbase material in flexible pavement. *Ain Shams Eng. J.* **2013**, *4*, 43–53. [CrossRef]
17. Kumar, P.; Kumar, A. Steel industry waste utilisation in road sector of India. *J. Inst. Eng. (India) Civ. Eng. Div.* **2000**, *80*, 182–185.
18. Karataǧ, H.; Fırat, S.; Işik, N.S. Assessment of performance of steel slag used in road base by finite element analysis. In Proceedings of the 13th International Congress on Advances in Civil Engineering, Izmir, Turkey, 12–14 September 2018.
19. Pasetto, M.; Baldo, N. Mix design and performance analysis of asphalt concretes with electric arc furnace slag. *Constr. Build Mater.* **2011**, *25*, 3458–3468. [CrossRef]
20. Pasetto, M.; Baldo, N. Influence of the aggregate skeleton design method on the permanent deformation resistance of stone mastic asphalt. *Mater. Res. Innov.* **2014**, *18*, S3-96–S3-101. [CrossRef]
21. Bessa, I.S.; Castelo Branco, V.T.F.; Soares, J.B. Evaluation of polishing and degradation resistance of natural aggregates and steel slag using the aggregate image measurement system. *Road Mater. Pavement Des.* **2014**, *15*, 385–405. [CrossRef]
22. Wu, S.; Cui, P.; Xie, J.; Liu, Q.; Pang, L. Expansive inhibition method of steel slag aggregate and volume stability of mixture: A review. *China J. Highw. Transp.* **2021**, *34*, 166–179.
23. Sha, A. Material characteristics of semi-rigid base. *China J. Highw. Transp.* **2008**, *21*, 1.

24. *JTG/T F20-2015*; Technical Guidelines for Construction of Highway Roadbases. Ministry of Transport of the People's Republic of China: Beijing, China, 2015.
25. Deng, C.; Jiang, Y.; Tian, T.; Chen, Z. Resilient modulus and influencing factors of vertical vibration compacted cement-stabilized macadam. *Int. J. Pavement Eng.* **2021**, *22*, 1435–1445. [CrossRef]
26. Deng, C.; Jiang, Y.; Lin, H.; Ji, X. Mechanical-strength-growth law and predictive model for cement-stabilized macadam. *Constr. Build. Mater.* **2019**, *215*, 582–594. [CrossRef]
27. Mallela, J.; Quintus, H.V.; Smith, K. Consideration of lime-stabilized layers in mechanistic-empirical pavement design. *Natl. Lime Assoc.* **2004**, *200*, 1–40.
28. Xue, J.; Jiang, Y. Analysis on the fatigue properties of vertical vibration compacted lime–fly ash-stabilized macadam. *Constr. Build. Mater.* **2017**, *155*, 531–541. [CrossRef]
29. Deng, C.; Jiang, Y.; Yuan, K.; Tian, T.; Yi, Y. Mechanical properties of vertical vibration compacted lime–fly ash-stabilized macadam material. *Constr. Build. Mater.* **2020**, *251*, 119089. [CrossRef]
30. Lav, A.H.; Lav, M.A.; Goktepe, A.B. Analysis and design of a stabilized fly ash as pavement base material. *Fuel* **2006**, *85*, 2359–2370. [CrossRef]
31. Liu, Z.; Guan, D.; Moore, S.; Lee, H.; Su, J.; Zhang, Q. Climate policy: Steps to China's carbon peak. *Nature* **2015**, *522*, 279–281. [CrossRef]
32. Cetin, B.; Aydilek, A.H.; Guney, Y. Stabilization of recycled base materials with high carbon fly ash. *Resour. Conserv. Recycl.* **2010**, *54*, 878–892. [CrossRef]
33. Ferreira, C.; Ribeiro, A.; Ottosen, L. Possible applications for municipal solid waste fly ash. *J. Hazard. Mater.* **2003**, *96*, 201–216. [CrossRef]
34. Singh, S.; Tripathy, D.P.; Ranjith, P.G. Performance evaluation of cement stabilized fly ash–GBFS mixes as a highway construction material. *Waste Manag.* **2008**, *28*, 1331–1337. [CrossRef] [PubMed]
35. Mallick, R.B.; Hendrix, G., Jr. Use of foamed asphalt in recycling incinerator ash for construction of stabilized base course. *Resour. Conserv. Recycl.* **2004**, *42*, 239–248. [CrossRef]
36. Sun, Y.; Li, L. Strength assessment and mechanism analysis of cement stabilized reclaimed lime-fly ash macadam. *Constr. Build. Mater.* **2018**, *166*, 118–129. [CrossRef]
37. Li, Q.; Wang, Z.; Li, Y.; Shang, J. Cold recycling of lime-fly ash stabilized macadam mixtures as pavement bases and subbases. *Constr. Build. Mater.* **2018**, *169*, 306–314. [CrossRef]
38. Gao, B.; Yang, C.; Zou, Y.; Wang, F.; Zhou, X.; Barbieri, D.; Wu, S. Compaction Procedures and Associated Environmental Impacts Analysis for Application of Steel Slag in Road Base Layer. *Sustainability* **2021**, *13*, 4396. [CrossRef]
39. Liu, J.; Yu, B.; Wang, Q. Application of steel slag in cement treated aggregate base course. *J. Clean. Prod.* **2020**, *269*, 121733. [CrossRef]
40. Li, W.; Lang, L.; Lin, Z.; Wang, Z.; Zhang, F. Characteristics of dry shrinkage and temperature shrinkage of cement-stabilized steel slag. *Constr. Build. Mater.* **2017**, *134*, 540–548. [CrossRef]
41. Xiao, J.; Long, C.; He, J.; Chang, J.; Wu, C.; Liu, C.; Yan, W. Performance and micro characteristics of cement stabilized macadam with a large amount of Activated steel slag powder. *China J. Highw. Transp.* **2021**, *34*, 204.
42. Li, Q.; Li, B.; Li, X.; He, Z.; Zhang, P. Microstructure of pretreated steel slag and its influence on mechanical properties of cement stabilized mixture. *Constr. Build. Mater.* **2022**, *317*, 125799. [CrossRef]
43. Yoshida, N. Uniaxial compressive strength of hydraulic, graded iron and steel slag base-course material produced at different manufacturers and its increase with curing time. In Proceedings of the 15th Asian Regional Conference on Soil Mechanics and Geotechnical Engineering, ARC 2015, Fukuoka, Japan, 9–13 November 2015; pp. 1614–1618.
44. Li, C.; Xiang, X.; Liu, S.; Hua, Z.; Jiao, L. Experimental research on the basic performance of the semi-rigid base material with steel-slag. *Multipurp. Util. Miner. Resour.* **2015**, *3*, 73–77.
45. Li, W.; Lang, L.; Wang, Z.; Chen, J. Experimental research on cracking performance of semi-rigid steel slag base. *Constr. Technol.* **2017**, *46*, 47–52.
46. Ji, X.; Sheng, Y.; Lu, Z.; Xin, D.; Long, Y.; Chen, H. Properties of semi-rigid base material with steel slag. *J. Chang. Univ.* **2021**, *41*, 21–31.
47. Yang, Y.; Tan, B.; Li, Y.; Liu, Q. Preparation and properties of water stabilized base material of refined steel slag based on alkali excitation. *J. Guilin Univ. Technol.* **2021**, 1–8. Available online: http://kns.cnki.net/kcms/detail/45.1375.N.20211221.1348.002.html (accessed on 6 July 2022).
48. Lu, F.; Li, J. Application of Jinan Steel's converter slag in pavement base. *Highway* **2013**, *08*, 262–266.
49. Zeng, M.; Wu, S.; Hu, B.; Sun, Y.; Long, S. Experimental study on steel slag-crushed stone pavement base materials stabilized with portland cement. *Nat. Sci. J. Xiangtan Univ.* **2011**, *33*, 29–33.
50. Zhou, M.; Cheng, X.; Chen, X. Studies on the Volumetric Stability and Mechanical Properties of Cement-Fly-Ash-Stabilized Steel Slag. *Materials* **2021**, *14*, 495. [CrossRef]
51. O'Connor, J.; Nguyen, T.B.T.; Honeyands, T.; Monaghan, B.; O'Dea, D.; Rinklebe, J.; Vinu, A.; Hoang, S.A.; Singh, G.; Kirkham, M. Production, characterisation, utilisation, and beneficial soil application of steel slag: A review. *J. Hazard. Mater.* **2021**, *419*, 126478. [CrossRef]
52. Yildirim, I.; Prezzi, M. *Use of Steel Slag in Subgrade Applications*; Publication FWA: West Lafayette, Indiana, 2009.

53. Shi, C. Steel slag—Its production, processing, characteristics, and cementitious properties. *J. Mater. Civ. Eng.* **2004**, *16*, 230–236. [CrossRef]
54. Kandhal, P.S.; Hoffman, G.L. Evaluation of steel slag fine aggregate in hot-mix asphalt mixtures. *Transp. Res. Rec.* **1997**, *1583*, 28–36. [CrossRef]
55. Frías, M.; De Rojas, M.S.; Uría, A. Study of the instability of black slags from electric arc furnace steel industry. *Mater. Constr.* **2002**, *52*, 79–83. [CrossRef]
56. Wang, Q.; Wang, D.; Zhuang, S. The soundness of steel slag with different free CaO and MgO contents. *Constr. Build. Mater.* **2017**, *151*, 138–146. [CrossRef]
57. Barišić, I.; Dimter, S.; Rukavina, T. Elastic properties of cement-stabilised mixes with steel slag. *Int. J. Pavement Eng.* **2015**, *17*, 753–762. [CrossRef]
58. Li, F.; Chen, Y.; Gao, F.; Sun, Y.; Wu, C. Experiment and study on dry and temperature induced shrinkage properties of inorganic binder stabilized pavement base material with steel slag. *Highway* **2012**, *12*, 186–191.
59. Pai, R.R.; Bakare, M.D.; Patel, S.; Shahu, J.T. Structural Evaluation of Flexible Pavement Constructed with Steel Slag–Fly Ash–Lime Mix in the Base Layer. *J. Mater. Civ. Eng.* **2021**, *33*, 04021097. [CrossRef]
60. Jiang, Y.; Ling, T.; Shi, C.; Pan, S. Characteristics of steel slags and their use in cement and concrete—A review. *Resour. Conserv. Recycl.* **2018**, *136*, 187–197. [CrossRef]
61. Lun, Y.; Zhou, M.; Cai, X.; Xu, F. Methods for improving volume stability of steel slag as fine aggregate. *J. Wuhan Univ. Technol. Sci. Ed.* **2008**, *23*, 737–742. [CrossRef]
62. Zheng, W. *Application of Steel Slag in Cement Stabilized Crushed Stone Base*; Chang'an University: Xi'an, China, 2018.
63. *JTG D50-2017*; Specifications for Design of Highway Asphalt Pavement. Ministry of Transport of the People's Republic of China: Beijing, China, 2017.
64. Wu, S. *Experimental Study on Steel Slag-Crushed Stone Pavement Base Materials Stabilized with Portland Cement*; Hunan University: Changsha, China, 2011.
65. Patel, S.; Shahu, J. Resilient response and permanent strain of steel slag-fly ash-dolime mix. *J. Mater. Civ. Eng.* **2016**, *28*, 04016106. [CrossRef]
66. Wang, Q.; Yan, P. Early hydration characteristics and paste structure of complex binding material containing high-volume steel slag. *J. Chin. Ceram. Soc.* **2008**, *36*, 1406–1411.
67. Wang, Q.; Yan, P.; Han, S. The influence of steel slag on the hydration of cement during the hydration process of complex binder. *Sci. China Technol. Sci.* **2011**, *54*, 388–394. [CrossRef]
68. Zhuang, S.; Wang, Q. Inhibition mechanisms of steel slag on the early-age hydration of cement. *Cem. Concr. Res.* **2020**, *140*, 106283. [CrossRef]
69. Wang, G.; Wang, Y.; Gao, Z. Use of steel slag as a granular material: Volume expansion prediction and usability criteria. *J. Hazard. Mater.* **2010**, *184*, 555–560. [CrossRef] [PubMed]
70. Xu, B. *Application of Steel Slag in Cement-Stabilized Macadam Base*; Changsha University of Science and Technology: Changsha, China, 2017.
71. Zeng, M.; Ruan, W.; Meng, Y.; Lin, C. Design and performance of lime and fly-ash stabilized steel-slag and crushed-stone pavement base materials. *J. Hunan Univ.* **2012**, *39*, 1–6.
72. Rajakumaran, K. An experimental analysis on stabilization of expansive soil with steel slag and fly ash. *Int. J. Adv. Eng. Technol.* **2015**, *7*, 1745.
73. Shen, W.; Zhou, M.; Ma, W.; Hu, J.; Cai, Z. Investigation on the application of steel slag–fly ash–phosphogypsum solidified material as road base material. *J. Hazard. Mater.* **2009**, *164*, 99–104. [CrossRef] [PubMed]
74. Cao, B. *Application of Lime Fly Ash Steel Slag in Pavement Base*; Chang'an University: Xi'an, China, 2004.
75. Zhang, H.; Yu, M.; Li, X. Pavement Performance of Lime-Fly Ash Slag Mixture. *J. Chongqing Jiaotong Univ.* **2011**, *30*, 1344–1346.
76. Wu, R.; Zhang, L.; Han, Z.; Fan, J. Long-term water immersion and freeze-thaw cycles experiment of cement-stabilized macadam bases. *J. Huazhong Univ. Sci. Technol.* **2011**, *39*.
77. Yu, J. *Research on the Experiments about the Pavement Performance of Cement-Fly Ash Stabilized Steel Slag*; Nanjing Forestry University: Nanjing, China, 2010.
78. Özkan, Ö.; Sarıbıyık, M. Alkali silica reaction of BOF and BFS wastes combination in cement. *J. Civ. Eng. Manag.* **2013**, *19*, 113–120. [CrossRef]
79. Zhang, T.; Yu, Q.; Wei, J.; Li, J. Investigation on mechanical properties, durability and micro-structural development of steel slag blended cements. *J. Therm. Anal. Calorim.* **2012**, *110*, 633–639. [CrossRef]
80. Mechtcherine, V.; Gram, A.; Krenzer, K.; Schwabe, J.-H.; Shyshko, S.; Roussel, N. Simulation of fresh concrete flow using Discrete Element Method (DEM): Theory and applications. *Mater. Struct.* **2014**, *47*, 615–630. [CrossRef]
81. Wang, G.; Chen, X.; Dong, Q.; Yuan, J.; Hong, Q. Mechanical performance study of pervious concrete using steel slag aggregate through laboratory tests and numerical simulation. *J. Cleaner Prod.* **2020**, *262*, 121208. [CrossRef]
82. Wang, S.; Chen, G.; Zhang, L.; Yuan, J. Triaxial discrete element simulation of soil–rock mixture with different rock particle shapes under rigid and flexible loading modes. *Int. J. Geomech.* **2021**, *21*, 04021142. [CrossRef]
83. Wu, Y. *The Effect of Freeze-Thaw Cycles on the Performance of Cement Stabilized Steel Slag Base*; Hebei University of Engineering: Handan, China, 2020.

84. Wang, S.; Chen, G.; Zhang, L. Parameter inversion and microscopic damage research on discrete element model of cement-stabilized steel slag based on 3D scanning technology. *J. Hazard. Mater.* **2021**, *424*, 127402. [CrossRef]
85. Du, Z.; Zhu, X. Molecular Dynamics Simulation to Investigate the Adhesion and Diffusion of Asphalt Binder on Aggregate Surfaces. *Transp. Res. Rec. J. Transp. Res. Board* **2019**, *2673*, 500–512. [CrossRef]
86. Xu, G.; Wang, H. Molecular dynamics study of oxidative aging effect on asphalt binder properties. *Fuel* **2017**, *188*, 1–10. [CrossRef]
87. Xu, G.; Wang, H. Study of cohesion and adhesion properties of asphalt concrete with molecular dynamics simulation. *Comput. Mater. Sci* **2016**, *112*, 161–169. [CrossRef]
88. Huang, M.; Zhang, H.; Gao, Y.; Wang, L. Study of diffusion characteristics of asphalt–aggregate interface with molecular dynamics simulation. *Int. J. Pavement Eng.* **2021**, *22*, 319–330. [CrossRef]
89. Horvath, A. *Life-Cycle Environmental and Economic Assessment of Using Recycled Materials for Asphalt Pavements*; University of California Transportation Center: Berkeley, CA, USA, 2003.
90. Anastasiou, E.; Liapis, A.; Papagianni-Papadopoulou, I.; Memet, M. *Comparative Life Cycle Assessment of Concrete Road Pavements*; Aristotle University of Thessaloniki: Saronica, Greece, 2013.
91. Li, D.; Wang, Y.; Liu, Y.; Sun, S.; Gao, Y. Estimating life-cycle CO_2 emissions of urban road corridor construction: A case study in Xi'an, China. *J. Clean. Prod.* **2020**, *255*, 120033. [CrossRef]
92. Li, H.; Feng, Z.; Ahmed, A.T.; Yombah, M.; Cui, C.; Zhao, G.; Guo, P.; Sheng, Y. Repurposing waste oils into cleaner aged asphalt pavement materials: A critical review. *J. Clean. Prod.* **2022**, *334*, 130230. [CrossRef]
93. Pasetto, M.; Pasquini, E.; Giacomello, G.; Baliello, A. Life-Cycle Assessment of road pavements containing marginal materials: Comparative analysis based on a real case study. In *Pavement Life-Cycle Assessment*; CRC Press: Boca Raton, FL, USA, 2017; pp. 199–208.
94. Lee, K.-M.; Park, P.-J. Estimation of the environmental credit for the recycling of granulated blast furnace slag based on LCA. *Resour. Conserv. Recycl.* **2005**, *44*, 139–151. [CrossRef]
95. Saade, M.R.M.; da Silva, M.G.; Gomes, V. Appropriateness of environmental impact distribution methods to model blast furnace slag recycling in cement making. *Resour. Conserv. Recycl.* **2015**, *99*, 40–47. [CrossRef]
96. Kang, L.; Du, H.; Zhang, H.; Ma, W. Systematic research on the application of steel slag resources under the background of big data. *Complexity* **2018**, *2018*, 6703908. [CrossRef]
97. Anastasiou, E.; Liapis, A.; Papayianni, I. Comparative life cycle assessment of concrete road pavements using industrial by-products as alternative materials. *Resour. Conserv. Recycl.* **2015**, *101*, 1–8. [CrossRef]
98. O'Brien, K.R.; Ménaché, J.; O'Moore, L.M. Impact of fly ash content and fly ash transportation distance on embodied greenhouse gas emissions and water consumption in concrete. *Int. J. Life Cycle Assess.* **2009**, *14*, 621–629. [CrossRef]
99. *JTG E60-2008*; Code for Field Test of Highway Subgrade and Pavement. Ministry of Transport of the People's Republic of China: Beijing, China, 2008.

MDPI
St. Alban-Anlage 66
4052 Basel
Switzerland
www.mdpi.com

Coatings Editorial Office
E-mail: coatings@mdpi.com
www.mdpi.com/journal/coatings

Disclaimer/Publisher's Note: The statements, opinions and data contained in all publications are solely those of the individual author(s) and contributor(s) and not of MDPI and/or the editor(s). MDPI and/or the editor(s) disclaim responsibility for any injury to people or property resulting from any ideas, methods, instructions or products referred to in the content.

www.ingramcontent.com/pod-product-compliance
Lightning Source LLC
Chambersburg PA
CBHW061455220926
51850CB00004B/99
* 9 7 8 3 7 2 5 8 0 7 4 9 9 *